AF624958

Recent Trends in Carbohydrate Chemistry

Recent Trends in Carbohydrate Chemistry

Synthesis, Structure and Function of Carbohydrates

Volume 1

Edited by

Amélia Pilar Rauter

Bjørn E. Christensen

László Somsák

Paul Kosma

Roberto Adamo

ELSEVIER

Elsevier
Radarweg 29, PO Box 211, 1000 AE Amsterdam, Netherlands
The Boulevard, Langford Lane, Kidlington, Oxford OX5 1GB, United Kingdom
50 Hampshire Street, 5th Floor, Cambridge, MA 02139, United States

© 2020 Elsevier Inc. All rights reserved.

No part of this publication may be reproduced or transmitted in any form or by any means, electronic or mechanical, including photocopying, recording, or any information storage and retrieval system, without permission in writing from the publisher. Details on how to seek permission, further information about the Publisher's permissions policies and our arrangements with organizations such as the Copyright Clearance Center and the Copyright Licensing Agency, can be found at our website: www.elsevier.com/permissions.

This book and the individual contributions contained in it are protected under copyright by the Publisher (other than as may be noted herein).

Notices
Knowledge and best practice in this field are constantly changing. As new research and experience broaden our understanding, changes in research methods, professional practices, or medical treatment may become necessary.

Practitioners and researchers must always rely on their own experience and knowledge in evaluating and using any information, methods, compounds, or experiments described herein. In using such information or methods they should be mindful of their own safety and the safety of others, including parties for whom they have a professional responsibility.

To the fullest extent of the law, neither the Publisher nor the authors, contributors, or editors, assume any liability for any injury and/or damage to persons or property as a matter of products liability, negligence or otherwise, or from any use or operation of any methods, products, instructions, or ideas contained in the material herein.

Library of Congress Cataloging-in-Publication Data
A catalog record for this book is available from the Library of Congress

British Library Cataloguing-in-Publication Data
A catalogue record for this book is available from the British Library

ISBN: 978-0-12-817467-8

For information on all Elsevier publications visit our website at https://www.elsevier.com/books-and-journals

Publisher: Susan Dennis
Acquisitions Editor: Emily M McCloskey
Editorial Project Manager: Kelsey Connors
Production Project Manager: Omer Mukthar
Cover Designer: Victoria Pearson

Typeset by SPi Global, India

ELSEVIER Book Aid International
Working together to grow libraries in developing countries
www.elsevier.com • www.bookaid.org

Contents

Contributors

Mohammed Ahmar Univ Lyon, Institut de Chimie et Biochimie Moléculaires et Supramoléculaires, CNRS, Université Lyon 1, INSA Lyon, CPE Lyon, ICBMS, UMR 5246; Université Claude Bernard, Bâtiment Lederer, Villeurbanne Cedex, France

Éva Bokor Department of Organic Chemistry, University of Debrecen, Debrecen, Hungary

Anikó Borbás Department of Pharmaceutical Chemistry, University of Debrecen, Debrecen, Hungary

Bjørn E. Christensen Department of Biotechnology and Food Science, NTNU-Norwegian University of Science and Technology, Trondheim, Norway

Sébastien Comesse Laboratoire URCOM, EA 3221, INC3M-CNRS-FR 3038, Université Le Havre Normandie, Le Havre, France

Marianne Øksnes Dalheim Department of Biotechnology and Food Science, NTNU-Norwegian University of Science and Technology, Trondheim, Norway

Martina Delbianco Department of Biomolecular Systems, Max Planck Institute of Colloids and Interfaces, Potsdam, Germany

Supriya Dey Department of Organic Chemistry, Indian Institute of Science, Bangalore, India

Weigang Fan Univ Lyon, Institut de Chimie et Biochimie Moléculaires et Supramoléculaires, CNRS, Université Lyon 1, INSA Lyon, CPE Lyon, ICBMS, UMR 5246; Université Claude Bernard, Bâtiment Lederer, Villeurbanne Cedex, France

Andreia Fortuna Centro de Química e Bioquímica; Centro de Química Estrutural, Faculdade de Ciências, Universidade de Lisboa, Lisboa, Portugal

Ana M. Gómez Bioorganic Chemistry Department, Instituto Química Orgánica General (IQOG-CSIC), Madrid, Spain

Hubert Hettegger Department of Chemistry, Institute for Chemistry of Renewable Resources, University of Natural Resources and Life Sciences, Vienna (BOKU), Austria

Srinivas Hotha Department of Chemistry, Indian Institute of Science Education & Research, Pune, India

N. Jayaraman Department of Organic Chemistry, Indian Institute of Science, Bangalore, India

Viktor Kelemen Department of Pharmaceutical Chemistry, University of Debrecen, Debrecen, Hungary

Shinichi Kitamura Center for Research and Development of Bioresources, Organization for Research Promotion, Osaka Prefecture University, Osaka, Japan

Wolfgang Lindner Department of Analytical Chemistry, University of Vienna, Vienna, Austria

J. Cristóbal López Bioorganic Chemistry Department, Instituto Química Orgánica General (IQOG-CSIC), Madrid, Spain

Anup Kumar Misra Division of Molecular Medicine, Bose Institute, Kolkata, India

Florence Popowycz Univ Lyon, Institut de Chimie et Biochimie Moléculaires et Supramoléculaires, CNRS, Université Lyon 1, INSA Lyon, CPE Lyon, ICBMS, UMR 5246; Université Claude Bernard, Bâtiment Lederer, Villeurbanne Cedex, France

Yves Queneau Univ Lyon, Institut de Chimie et Biochimie Moléculaires et Supramoléculaires, CNRS, Université Lyon 1, INSA Lyon, CPE Lyon, ICBMS, UMR 5246; Université Claude Bernard, Bâtiment Lederer, Villeurbanne Cedex, France

Frédéric Renou Institut des Molécules et Matériaux du Mans, UMR CNRS 6283, Le Mans University, Le Mans, France

Thomas Rosenau Department of Chemistry, Institute for Chemistry of Renewable Resources, University of Natural Resources and Life Sciences, Vienna (BOKU), Austria

Gopal Ch Samanta Department of Organic Chemistry, Indian Institute of Science, Bangalore, India

Anshupriya Si Division of Molecular Medicine, Bose Institute, Kolkata, India

Shiho Suzuki Center for Research and Development of Bioresources, Organization for Research Promotion, Osaka Prefecture University, Osaka, Japan

Jia-Neng Tan Univ Lyon, Institut de Chimie et Biochimie Moléculaires et Supramoléculaires, CNRS, Université Lyon 1, INSA Lyon, CPE Lyon, ICBMS, UMR 5246; Université Claude Bernard, Bâtiment Lederer, Villeurbanne Cedex, France

Charlie Verrier Univ Lyon, Institut de Chimie et Biochimie Moléculaires et Supramoléculaires, CNRS, Université Lyon 1, INSA Lyon, CPE Lyon, ICBMS, UMR 5246; Université Claude Bernard, Bâtiment Lederer, Villeurbanne Cedex, France

Lianjie Wang Univ Lyon, Institut de Chimie et Biochimie Moléculaires et Supramoléculaires, CNRS, Université Lyon 1, INSA Lyon, CPE Lyon, ICBMS, UMR 5246; Université Claude Bernard, Bâtiment Lederer, Villeurbanne Cedex, France

Nuno M. Xavier Centro de Química e Bioquímica; Centro de Química Estrutural, Faculdade de Ciências, Universidade de Lisboa, Lisboa, Portugal

Yang Yu Department of Biomolecular Systems, Max Planck Institute of Colloids and Interfaces, Potsdam; Department of Chemistry and Biochemistry, Freie Universität Berlin, Berlin, Germany

Preface

Carbohydrate chemistry covers organic and organometallic chemistry addressing the properties and transformations of nature's ubiquitous, multifunctional, and stereochemically rich molecules, the carbohydrates. Since Emil Fischer, the first chemist to synthesize glucose in the 19th century, many pioneers launched this area of research, namely Melville L. Wolfrom, Roy Whistler, Jean Montreuil, Raymond Lemieux, and Derek Horton, just to mention a few of them. Carbohydrate chemistry became increasingly known, remarkable advances have been made, and unique findings have contributed to improve the quality of life in today's society.

This volume of "Recent Trends in Carbohydrate Chemistry" highlights the beauty of this chemistry, from new synthetic reactions and reaction mechanisms to novel biologically active derived molecules, generating a diversity of structures, often in a stereo- and/or regioselective manner as a result of carbohydrate structure and reactivity. Part One of this volume presents carbohydrate transformations in green and sustainable conditions, multicomponent reactions, *O*-, *S*-, *N*-, and *C*-glycosylations to link carbohydrates to a variety of aglycones via appealing synthetic methodologies toward structures with therapeutic potential. Part Two covers the synthesis and chemical modifications of polysaccharides, and contributions on amyloses, xanthan, and polysaccharide derivatization on silica.

We hope that this volume, covering attractive carbohydrate syntheses, will encourage, particularly the younger generation of carbohydrate chemists, to go on *playing around* with these sweet molecules of life, and contributing with their discoveries to unravel the secrets of Life!

Amélia Pilar Rauter
Bjørn E. Christensen
László Somsák
Paul Kosma
Roberto Adamo

Part One

Monosaccharide chemistry toward molecular diversity—Recent findings

Perspective on the transformation of carbohydrates under green and sustainable reaction conditions

1

Anshupriya Si, Anup Kumar Misra
Division of Molecular Medicine, Bose Institute, Kolkata, India

Chapter outline

1 Introduction

Functionalization of mono-, di-, and higher oligosaccharides using a variety of protecting groups is a fundamental requirement for the synthesis of oligosaccharides. Judicious selection of protecting groups plays important roles in the stereo- and regio-selective outcome of the glycosylations. Since the first report of the glycosylation by Fischer in 1893, several glycosylation techniques and functional group

Recent Trends in Carbohydrate Chemistry. https://doi.org/10.1016/B978-0-12-817467-8.00001-3
© 2020 Elsevier Inc. All rights reserved.

modifications have appeared in the literature. The reactivity and participation of the substrates in the glycosylation reaction is highly influenced by the protecting groups present in the molecules. However, it is quite difficult to predict the exact nature of a functionalized saccharide derivative in a chemical reaction due to the presence of multiple numbers of functional groups. In addition to the functionalization of carbohydrates for the synthesis of oligosaccharides, transformation of carbohydrates into a variety of compounds with biological and material science significance is another important area of research. Several molecules having pharmaceutical importance have been prepared from carbohydrates, for example, heterocycles,[1, 2] carbocycles,[3, 4] nucleosides,[5, 6] natural product like compounds,[7, 8] chalcogenides,[9, 10] and nitrogenous compounds[11, 12].

Traditional reaction conditions for the transformation of carbohydrates into different classes of molecules involve the use of volatile organic solvents and a variety of metallic reagents and catalysts. Similarly, functionalization of hydroxy and amino groups of carbohydrates such as multiple protections and deprotections[13, 14] and glycosylation for the preparation of oligosaccharides and glycoconjugates[15] also involve the use of hazardous reaction conditions. Most of the conventional reaction conditions for the aforementioned transformations do not meet the requirement of environmentally benign nature, are devoid of sustainability, and involve the use of hazardous chemicals/reagents, produce chemical wastes. Considering the concern for the environment, the current thrust in the organic chemistry is to develop sustainable processes using cost-effective, pollution-free synthetic strategies. In this context, the most attractive concept for sustainable chemical transformations is to adopt "Green chemistry" approach. In 1998, Paul Anastas and John Warner[16] proposed some pioneering principles for designing green and sustainable reaction conditions by reducing or eliminating the use or/and production of hazardous substances, which are presented:

1. "Prevention" such as designing of reaction methodologies without production of waste.
2. Highest level of atom economy for all components should be maintained in the final product.
3. Chemical synthetic methods should be designed to use and produce substances having low or no toxicity to the environment.
4. In order to make the synthetic method safer to the environment, reagents, catalyst, and solvents should be used in minimum quantity or preferably avoided.
5. Chemical reactions should be designed in such a way that they require minimum energy or can be carried out without requirement of any energy, for example, heating, electricity, and high pressure.
6. Renewable starting materials or feedstocks should be used in the chemical reactions.
7. Development of one-pot reaction conditions, iterative multistep reactions in one-pot, etc., reducing the number of steps for deprotections of functional groups.
8. Use of catalytic amount of catalysts/additives avoiding the use of their stoichiometric quantities.
9. Use of biodegradable chemicals in the reactions, which do not persist in the environment after their functional lifetime.
10. Controlling the reactions prior to the formation of hazardous by-products with the help of analytical techniques for real-time, in-process monitoring.
11. Selection of inherently safer chemicals/reagents in the synthetic strategies, causing less chance of chemical accidents.

Transformations of carbohydrates and their functionalizations related to the glycosylation reactions involve a wide range of chemical reactions. Therefore, it is pertinent to develop newer reaction conditions for the transformations/reactions of carbohydrates under environmentally benign sustainable reaction conditions avoiding organic solvents.

In the recent past, several reports on the development of sustainable reaction conditions using carbohydrates established that it is possible to build up sustainable strategies for the preparation of carbohydrate-derived molecules of interest in the laboratory as well as in industry adopting "green chemical approaches." A wide range of chemical reactions have been carried out using carbohydrate substrates since the inception of "Green chemistry"[16] such as use of natural carbohydrate polymers as renewable feedstocks; biocatalyzed chemical transformations; solvent-free reactions; development of multistep reactions maintaining atom economy; and use of water and biodegradable solvents, room temperature ionic liquids (RTILs), deep eutectic solvents (DESs), supercritical fluids (SCFs), etc., as reaction media. In recent decades, there has been several reviews regarding green solvents in carbohydrate chemistry.[17] In addition, a wide range of nonconventional energy sources have been applied to carry out reactions using carbohydrates, which include the application of microwave irradiation, ultrasonication, ball milling, photoinduced reactions, electrochemical reactions, reactions under high pressure, flow chemistry, and automation, to name a few. In this chapter, an attempt has been made to compile a large number of reports on the chemical transformations of carbohydrates having emphasis on Green and sustainable chemistry. The present perspective broadly focuses on two aspects of green chemical transformations of carbohydrates such as (a) use of green or biodegradable solvents and (b) application of modern technologies for conventional and solvent-free reactions.

2 Synthetic transformations of carbohydrates using nonhazardous environmentally benign solvents

Carrying out chemical reactions avoiding volatile organic solvents is an important principle of Green chemistry. However, there are many instances where solvents play an important role in the reactions, and their use is essential for satisfactory formation of products. Use of water or other biodegradable solvents replacing organic solvents without compromising the yield of the product is a challenge to the synthetic chemists. Following the Green chemistry perspective, several attempts have been made in the recent past to carry out organic reactions in water or other biodegradable solvents replacing organic solvents. Water can be considered as the best alternative reaction medium among the wide variety of environmentally benign solvents because of its compatibility with the Green chemistry principles. A diverse range of chemical transformations on carbohydrate substrates have been carried out in water. In addition to water, many reports appeared on the transformations of carbohydrates in other environmentally benign reaction medium such as RTILs),[18] DESs,[19–21] biomass-derived solvents (BDSs),[22] supercritical fluids (SCFs),[23] and polyethylene glycol[24]. Besides the chemical transformation of carbohydrates, several examples on the enzymatic transformations[25] of carbohydrates have also been reported in the recent past using green solvents. A set of the examples on various green solvent mediated transformations of carbohydrates is presented later.

3 Transformations of carbohydrates in water

Water can be the best suited solvent for carbohydrates because of its low cost abundance, high potential for solubilizing carbohydrates, and nontoxic nature to the environment.[26] However, there are some reactions such as glycosylations, esterification, and acetal formation where removal of moisture is an essential part of the reaction, and hence water is not considered as the best solvent for such reactions. In addition, many catalytic reactions are moisture sensitive and require inert anhydrous conditions and hence not suitable to carry out in water. Despite the limitations of water to be used as the solvent in the transformations of carbohydrates, a significant number of attempts have been made in the recent past with great success to carry out different reactions on carbohydrate substrates in water at atmospheric or high pressure. Selected examples of the transformations of carbohydrates in aqueous medium are presented later.

Lubineau and coworkers[27] developed an elegant approach for the preparation of β-glycosyl ketones from unprotected sugars in alkaline aqueous medium, which is a breakthrough on the green chemical approach for the carbohydrate transformations (Scheme 1). The reaction took place with high stereoselectivity and furnished exclusively β-glycosyl compounds. Earlier in 1986, Gonzalez and coworkers[28] reported the preparation of glycosylbarbituric acid derivatives in aqueous medium in excellent yields by the treatment of free sugars with barbituric acid derivatives (Scheme 2).

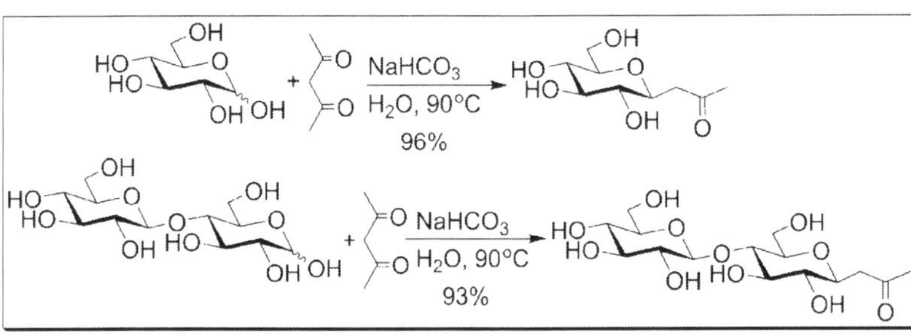

Scheme 1 Preparation of β-glycosyl ketones in aqueous media.

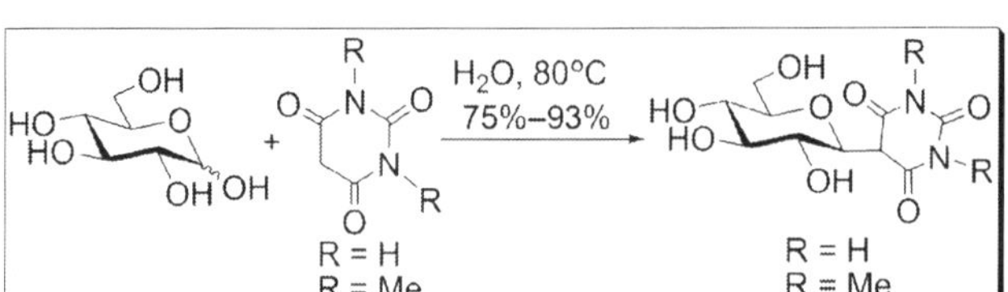

Scheme 2 Preparation of glycosylbarbituric acid derivatives in aqueous medium.

Bragnier and Schermann[29] applied similar reaction condition for the preparation of β-glycosyl ketones of 2-amino sugars. However, the yield of the reaction was moderate, and products were obtained as a mixture of *gluco-* and *manno-*isomers (Scheme 3).

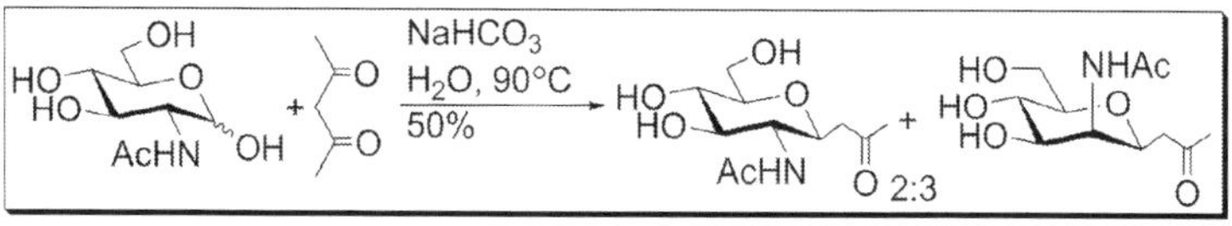

Scheme 3 Preparation of β-glycosyl ketones of D-glucosamine and D-mannosamine in aqueous media.

Besides the preparation of glycopyranosyl ketones, Wang et al.[30] reported the formation of glycofuranosyl ketones in high yields from pentose sugars under alkaline aqueous conditions. It was also observed that an exclusively single product was formed at elevated temperature and a mixture of unprecedented products was obtained at lower temperature (Scheme 4). It is noteworthy that glycopyranosyl ketones were formed at elevated temperature, whereas reactions at low temperature furnished glycofuranosyl ketones.

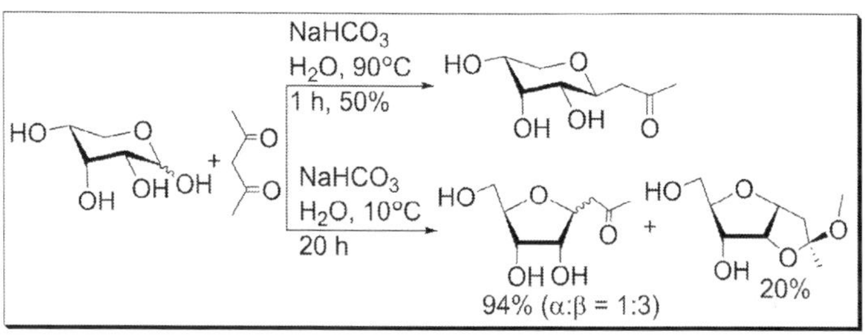

Scheme 4 Preparation of glycopyranosyl and furanosyl ketones in aqueous media.

Norsikian and coworkers[31] further transformed the glycosyl ketone to β-glycosylformaldehyde derivatives after some functional group modifications (Scheme 5).

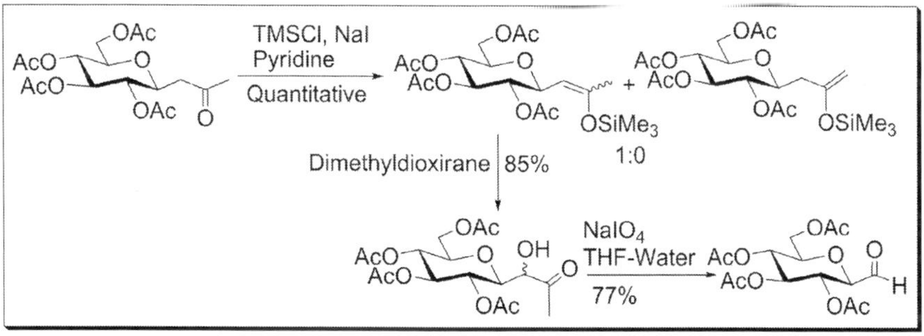

Scheme 5 Preparation of glycosylformaldehyde from glycosylpropan-2-one.

Tripathi and coworkers[32] treated β-glycosyl ketones with a variety of aldehydes in the presence of a base to provide phenyl substituted analogs of cinnamoylmethyl *C*-glycoside, which were further allowed to react with malononitrile in an alkaline aqueous medium to prepare the β anomer of diphenylmethyl *C*-glycopyranosides in very good yields (Scheme 6).

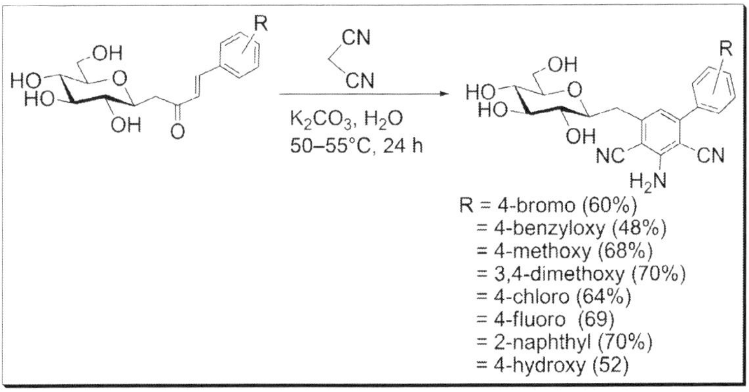

Scheme 6 Preparation of β anomer of the diphenylmethyl *C*-glycoside in aqueous medium.

Treatment of free sugars and 2-aminosugars with 1,3-diketones in acidic aqueous medium using $ZnCl_2$ or $Yb(OTf)_3$ led to the formation of low yields of polyhydroxylated furan, glycosylfuran, and pyrrole derivatives.[33] Taking clue from these studies, Misra and Agnihotri[34] prepared exclusively glycosylfuran derivatives by the treatment of unprotected sugars with 1,3-diketones or β-ketoesters in the presence of $CeCl_3{\cdot}7H_2O$ in aqueous medium (Scheme 7). Subsequently Yadav et al.[35] also reported the preparation of glycosylfuran derivatives using $InCl_3$ in aqueous conditions (Scheme 7).

C-glycosyl compounds have also been prepared in moderate yields directly by scandium cation exchanged montmorillonite catalyzed *C*-glycosylation of dimedone with free sugars in aqueous medium[36] (Scheme 8).

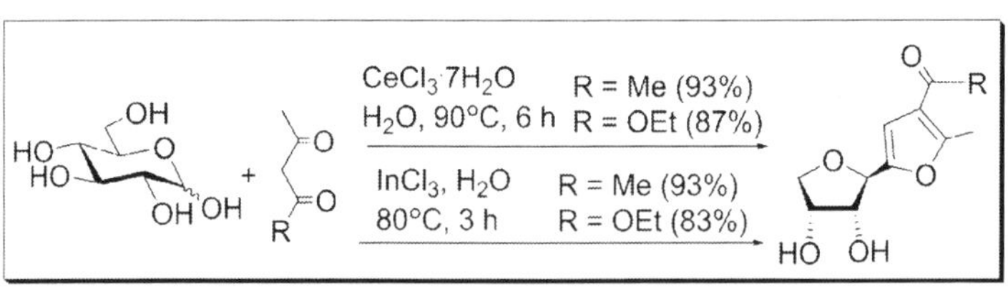

Scheme 7 Preparation of glycosylfuran derivatives in aqueous medium.

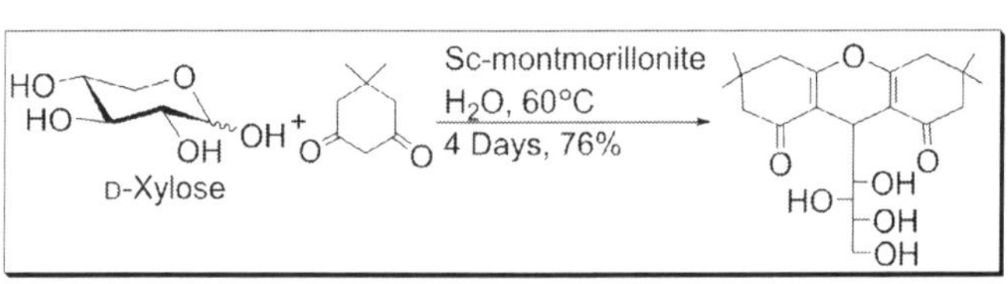

Scheme 8 Preparation of glycosyldimedone from free sugar by Sc-montmorillonite catalyzed condensation in water.

The β-glycosyl ketones prepared in aqueous medium have been used as precursors for the preparation of saturated and unsaturated fatty acid hydrazide derivatives of *C*-glycosyl glycolipid analogs in an aqueous medium[37] (Scheme 9).

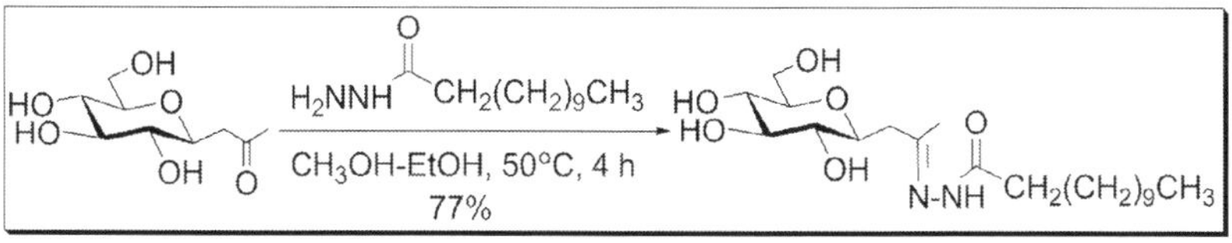

Scheme 9 Preparation of fatty acid hydrazones of *C*-glycosyl ketones in water.

Yadav and Rai[38] have developed an expeditious one-pot synthetic protocol for the preparation of benzoxazinone *C*-nucleosides via dehydrazinative β-glycosylation in alkaline aqueous media (Scheme 10).

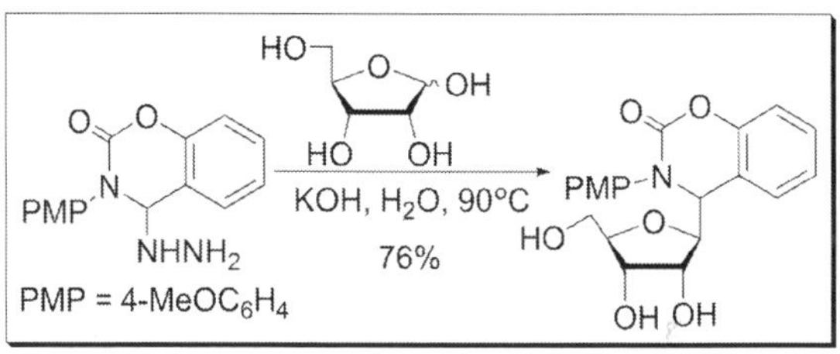

Scheme 10 Synthesis of benzoxazinone *C*-nucleosides in aqueous conditions.

Besides the preparation of *C*-glycosyl compounds, free sugars have been treated with homogeneous and heterogeneous catalysts in hot compressed water in the presence of an acid as well as a base. Treatment of D-glucose with H_2SO_4 in overheated compressed water (473 K) led to the formation of dehydrated product, 5-(hydroxymethyl)furfural (HMF), while isomerized product fructose was obtained by the treatment with sodium hydroxide in water at high pressure. Similar to the homogeneous conditions, treatment of D-glucose with a heterogeneous acidic catalyst, TiO_2, led to the formation of 5-(hydroxymethyl)furfural (HMF), whereas fructose was obtained by the treatment with a basic catalyst Zirconia (ZrO_2) (Scheme 11).[39]

Carbon-carbon bond formation of unprotected carbohydrates has been carried out applying Barbier reaction by the ultrasound-mediated treatment with allyl bromide and tin metal in ethanolic aqueous medium to obtain higher homologous sugars[40] (Scheme 12).

Serianni and coworkers[41] have successfully carried out the rearrangement of ketose and aldose sugars using a combination of paramolybdate anion exchange resin and formate anion exchange resin at an elevated temperature in water (Scheme 13).

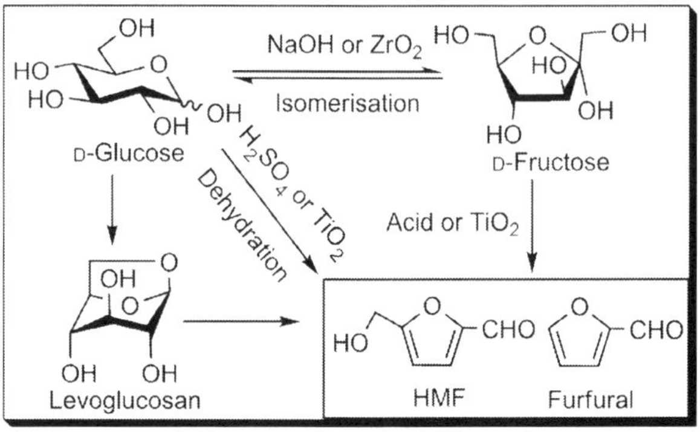

Scheme 11 Transformation of D-glucose in hot compressed water in the presence of homogeneous and heterogeneous acidic and basic catalysts.

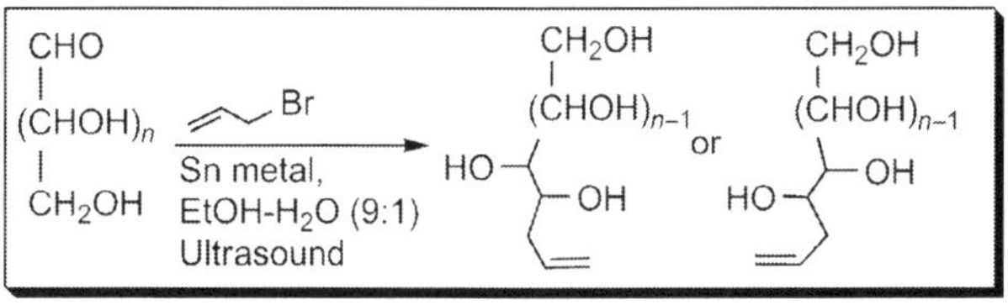

Scheme 12 Ultrasound mediated higher homologation of free sugar in aqueous reaction conditions.

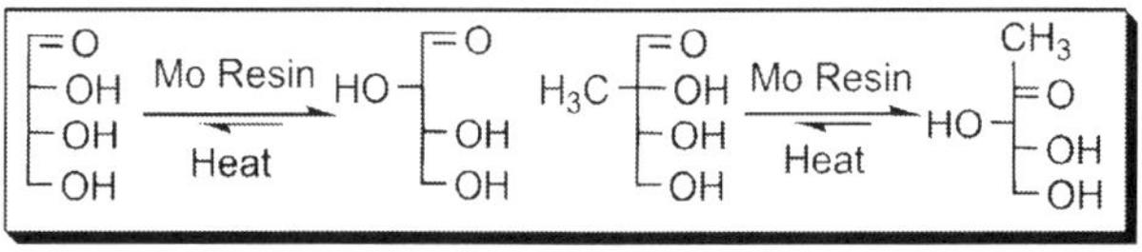

Scheme 13 Rearrangement of sugars in the presence of Mo resin in aqueous conditions.

Glycosylation is an important tool for the modification of the functional properties of bioactive molecules. Aoyama and Kobayashi[42] reported an elegant dehydrative glycosylation in water using a catalyst composed of a Brønsted acid and a surfactant. A diverse set of glycosides was synthesized by the dehydrative reaction of protected lactols with alcohols and nitrogen nucleophile (benzyl carbamate) with moderate stereoselectivity in the presence of dodecylbenzenesulfonic acid (DBSA) in water (Scheme 14).

Shoda and coworkers[43] reported high-yielding syntheses of β-glycosyl azides[43a]and aryl 1-thioglycosides[43b, c]directly from the unprotected sugars by the treatment with a combination of sodium azide or aromatic thiol and an ionic liquid, 2-chloro-1, 3-dimethylimidazolinium chloride (DMC). This transformation is an example of ionic liquid catalyzed reactions in aqueous medium (Scheme 15).

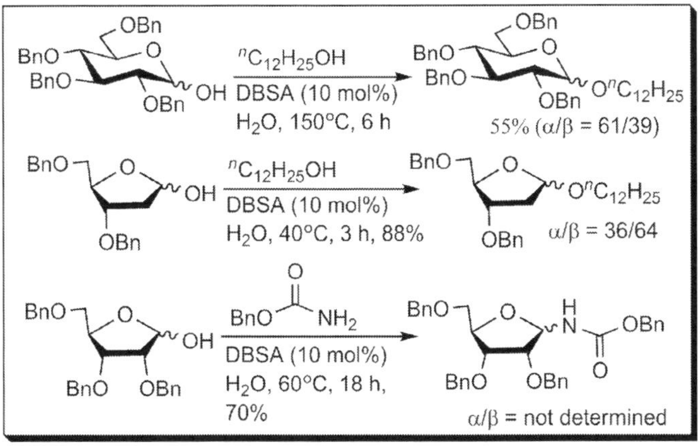

Scheme 14 Dehydrative glycosylation in aqueous medium.

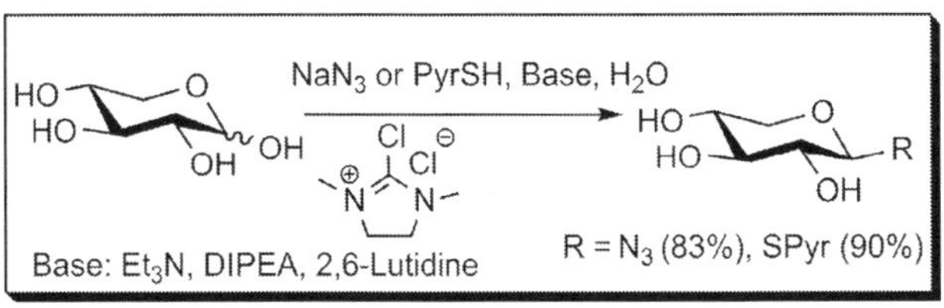

Scheme 15 Ionic liquid mediated preparation of glycosyl azides and thioglycosides in water.

Recently, Komarova et al.[44] demonstrated water-mediated reduction of carbohydrate azide derivatives using dithiothreitol in excellent yields. The reaction conditions are equally effective with mono- and disaccharide derivatives (Scheme 16). The reaction proceeded rapidly in the presence of water as a cosolvent, whereas the conversion was not complete in the absence of water.

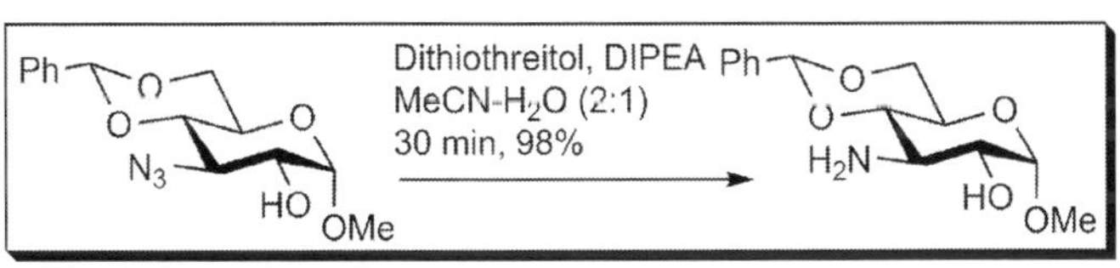

Scheme 16 Water-dependent reduction of an azido group in carbohydrate derivative.

Zhu and Schmidt[45] reported the synthesis of *S*-linked glycopeptide derivatives under phase transfer conditions. Since *S*-linked glycopeptides are important mimics of natural *O*-linked glycopeptides, high-yielding synthesis of such compounds is useful for the development of glycomimetics (Scheme 17).

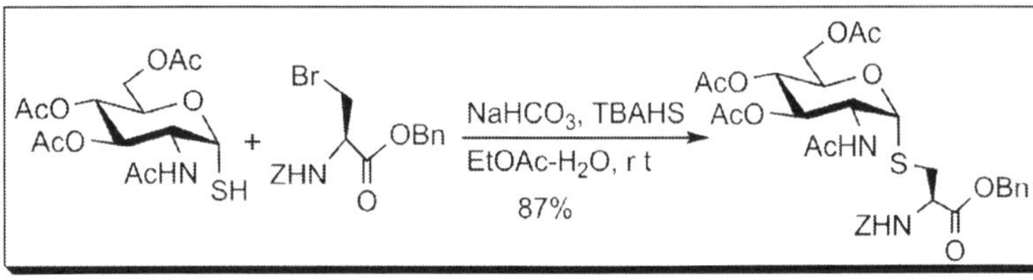

Scheme 17 Preparation of *S*-linked glycopeptides under phase transfer conditions.

In a different approach, Estrine et al.[46] demonstrated selective etherification of unprotected pentoses in water for the preparation of partially alkylated glycoside derivatives. The best yields were achieved by carrying out the reactions in the presence of a palladium catalyst and a crowded tertiary amine. The ratio of mono-, di-, tri-, and tetra-*O*-alkylated products was 4:57:37:2 in 45 min with 99% yield using $Me_2NC_{12}H_{25}$ as base (Scheme 18).

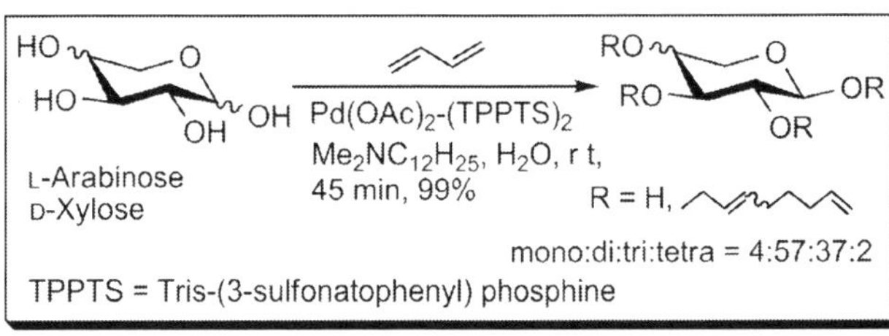

Scheme 18 Telomerization of butadiene with L-arabinose and D-xylose in water.

A variety of aromatic acids have been esterified by sugars using suitably protected sugar (D-glucose, D-galactose) chlorides in the presence of Cs_2CO_3, Aliquat 336 or TBAB and granular polytetrafluoroethylene (PTFE) under aqueous phase transfer conditions[47] (Scheme 19).

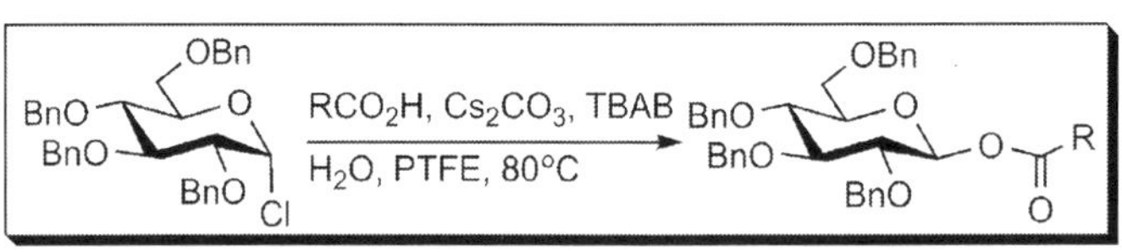

Scheme 19 Phase transfer catalytic esterification of carboxylic acids using a glycosyl chloride.

Wang and coworkers[48] reported the synthesis of azasugars via lanthanide salt-promoted aza-Diels-Alder (DA) reaction in aqueous medium. They have synthesized a series of chiral heterocyclic compounds by aza-DA reaction of sugar-derived chiral aldehydes and cyclopentadiene in the presence of benzylamine hydrochloride under aqueous conditions, which were transformed into azasugars as inhibitors of glycoprocessing enzymes (Scheme 20).

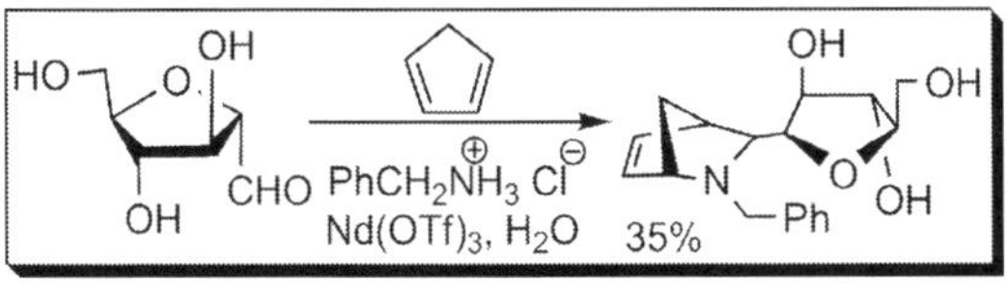

Scheme 20 Aza-Diels-Alder reaction on carbohydrate substrates in aqueous medium.

Ichikawa and coworkers[49] reported the preparation of urea-linked disaccharide derivatives in water via the formation of an oxazolidinone intermediate and its opening with an amino sugar. Unprotected glycosylated oxazolidinone intermediate derived from glycosyl azide furnished excellent yield of urea-linked pseudodisaccharides in the presence of an amine in aqueous medium (Scheme 21).

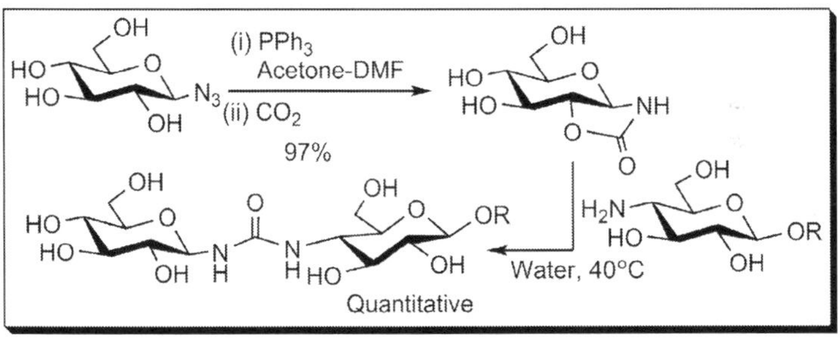

Scheme 21 Preparation of urea linked pseudodisaccharide derivatives in water.

Somsák and coworkers[50] reported the preparation of glycal or *C*-glycosyl derivatives by the treatment of thioglycosides or glycosyl sulfones with chromium(II) complexes in H_2O-DMF medium. It was proposed that the reaction proceeded via the formation of anomeric glycosyl radical (Scheme 22).

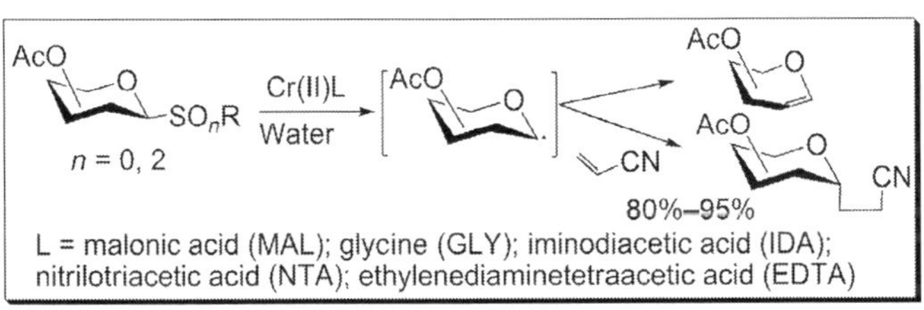

Scheme 22 Chromium(II)-mediated preparation of glycals and *C*-glycosyl derivatives from thioglycosides or glycosyl sulfones in aqueous medium.

Misra's group[51] reported a significant fast [1,3]-cycloaddition[52] between glycosyl azides and various alkynes under copper-mediated "click chemistry" conditions in aqueous medium to furnish glycosylated 1,2,3-triazole derivatives. They also reported the preparation of glycosylated 1,2,3-triazole derivatives directly from glycosyl bromides by a one-pot two-step reaction (Scheme 23).

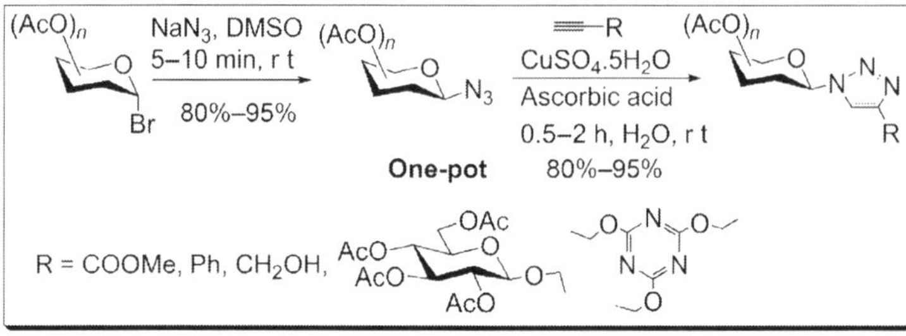

Scheme 23 Rapid one-pot preparation of 1-glycosyl-4-substituted 1,2,3-triazole conjugates from glycosyl bromides in aqueous medium.

Recently, Pelletier et al.[53] carried out glycosylation reactions between sucrosyl acceptors and glycosyl fluoride donors to yield trisaccharides. These reactions proceed at room temperature in an aqueous solvent mixture. Calcium salts and a tertiary amine promote the reaction with high site-selectivity for either the 3′-position or 1′-position of the fructofuranoside unit (Scheme 24).

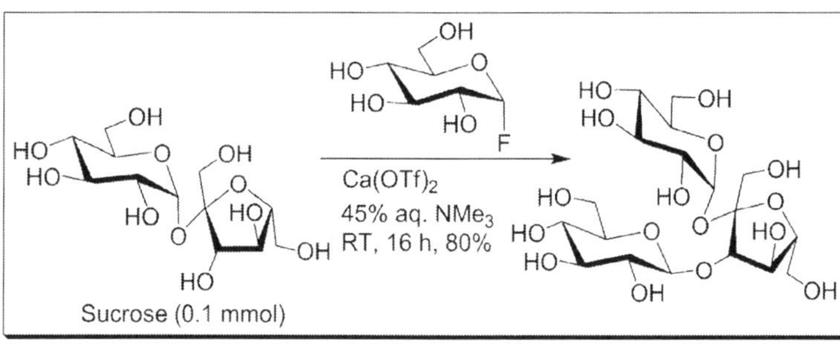

Scheme 24 Glycosylation reaction of sucrose with glycosyl fluoride in water.

Despite the possibility of hydrolysis in water, alkoxycarbonylation[54] of hydroxy groups of sucrose has been successfully carried out in aqueous reaction conditions (Scheme 25).

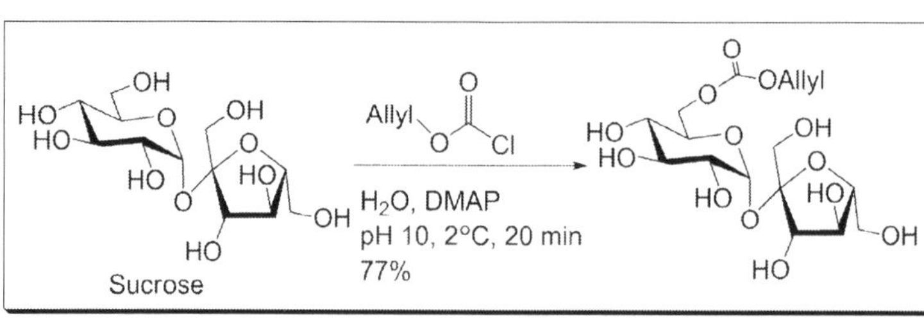

Scheme 25 Preparation of *O*-allyloxycarbonylated derivatives of sucrose in water.

In addition to the aforementioned examples, reports[55] are available dealing with the transformations of carbohydrates using aqueous reaction medium. It has been demonstrated that dehydrative reactions can also be successfully performed in aqueous medium by choosing appropriate reagents and conditions. Based on the success of the earlier experiments, it is expected that more sensitive reactions on carbohydrate substrates as well as stereoselective glycosylations will be possible in water by tuning the reaction conditions.

4 Transformations of carbohydrates in room temperature ionic liquids

Ionic liquids (ILs) are salts with low melting point, which exist as liquids consisting of poorly coordinating ion pairs, which can be tuned by altering the structures of anions and cations.[56] This class of compounds was known in the literature for a long time and introduced in the chemical reactions in the late 1990s. ILs are considered as promising alternatives to the organic solvents and sometimes termed as "neoteric solvents" and "designer solvents" due to their nonvolatile and nonflammable nature,[57] odorlessness, thermal stability, and recyclability.[58] Since the organic cation and inorganic anion present in the ILs are not closely packed, they can exist in liquid state at room temperature to elevated temperatures.[59] ILs can act as effective solvents for solubilizing polar as well as nonpolar compounds and do not involve in solvation or solvolysis.[60] After the pioneering report of Anastas and Warner in 1998 on the "Green chemistry" principles[16] in organic chemistry, ILs received attention as attractive alternatives to the conventional volatile, environmentally harmful organic solvents used in chemical transformations.[61] As a consequence, development of newer ILs and investigation on their properties and their applicability in the chemical reactions has increased exponentially over the last few years.

During the recent past, ILs have been used in carbohydrate chemistry as reaction solvents for dissolving high molecular weight polysaccharides and to perform their chemical and enzymatic transformations. They have also been used as solvents and catalysts in a wide variety of synthetic methodologies on carbohydrate substrates, which include protecting group manipulations, glycosylations, purification of reaction products, substitute of solid support, etc. Moreover, carbohydrate intermediates were used as sources for the preparation of ILs and chiral ILs.[62] Several reviews[63] have also appeared on the use of ILs in carbohydrate chemistry. Representative applications of ILs in carbohydrate chemistry have been compiled and presented below.

5 Use of ionic liquids as reaction solvents

The solubility of mono- and polysaccharides can be improved using ILs replacing conventional solvents due to their ionic nature. Besides solubilizing the carbohydrates,

ILs have been used as solvents for the preparation of carbohydrate based gels.[64] In addition, ILs can act as catalysts by accelerating the rate of a wide range of chemical transformations of carbohydrates such as acetylation, ortho-esterification, benzylidene acetal formation, and glycosylations, to name a few.

Forsyth and coworkers[65] reported a rapid and clean acetylation of unprotected mono-, di-, and trisaccharides using dicyanamide-based ionic liquid as solvent cum catalyst (Scheme 26). A series of reducing sugars, for example, D-glucose, D-galactose, D-mannose, L-rhamnose, L-fucose, lactose, maltose, and cellobiose were successfully acetylated under the reaction conditions.

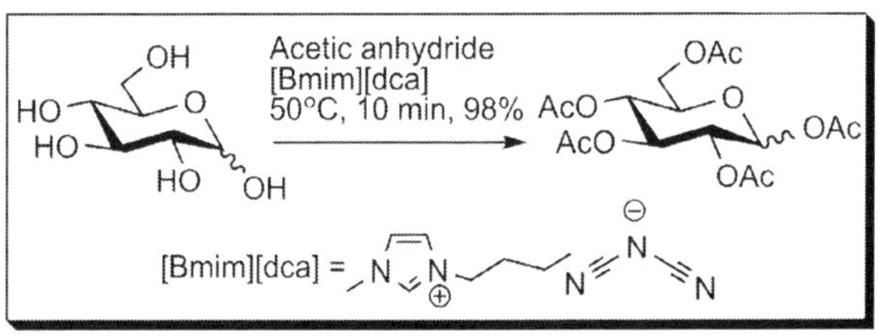

Scheme 26 Acetylation of unprotected sugar using dicyanamide-based ionic liquid.

Similarly, Linhardt et al.[66] demonstrated acetylation and benzoylation of simple sugars as well as sulfated sugars in high yields using 1-alkyl-3-methylimidazolium benzoate ([Emim][ba]) as task specific ionic liquid as solvent (Scheme 27).

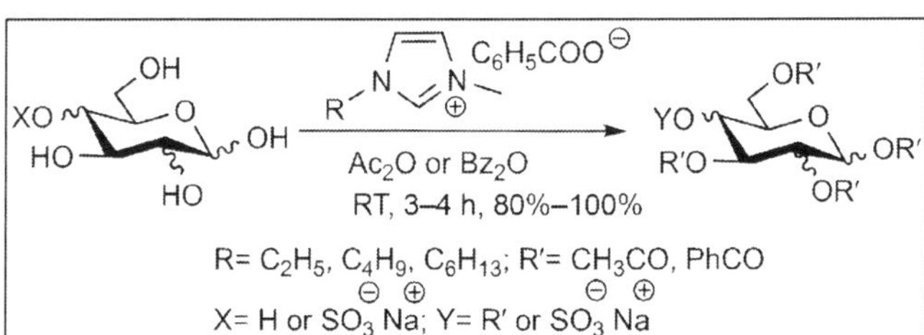

Scheme 27 Acetylation and benzoylation of simple sugars and sulfated sugars using ILs.

Abbott et al.[67] developed zinc-based ionic liquid for the acetylation of carbohydrates by mixing choline chloride and zinc chloride. A variety of reducing sugars have been acetylated using this IL. However, an anomeric mixture of acetylated sugar derivatives was obtained under the reaction conditions (Scheme 28).

Benzylidene acetal is an important protecting group frequently used for the simultaneous protection of two hydroxy groups in carbohydrate backbone. High-yielding preparation of benzylidene acetal containing carbohydrate derivatives from unprotected sugars using IL as solvent cum catalyst has been reported by Ragauskas's laboratory[68] (Scheme 29).

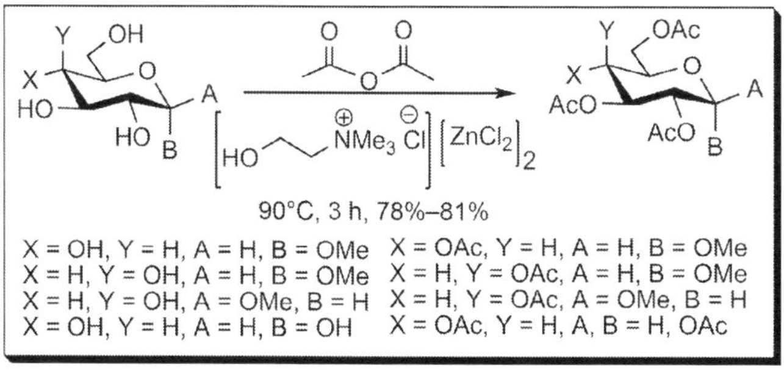

Scheme 28 Acetylation of monosaccharides using a zinc-based ionic liquid.

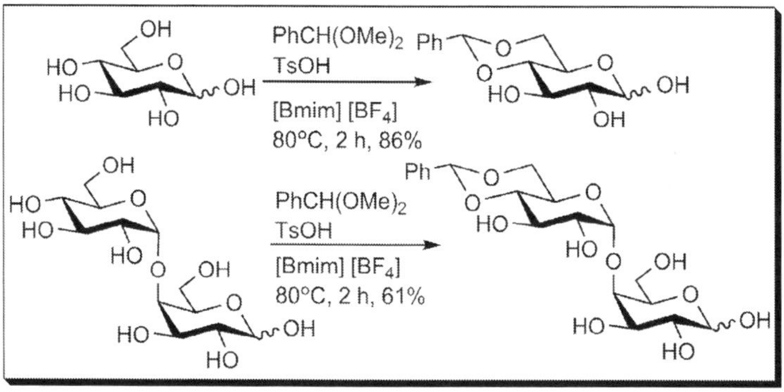

Scheme 29 Preparation of benzylidene acetals of unprotected sugars using ionic liquid.

Anas et al.[69] reported the application of IL as task-specific solvent for the preparation of acid susceptible orthoester derivatives of carbohydrates (Scheme 30). It is noteworthy that acid susceptible functional groups also remain unaffected under the reaction conditions using ILs.

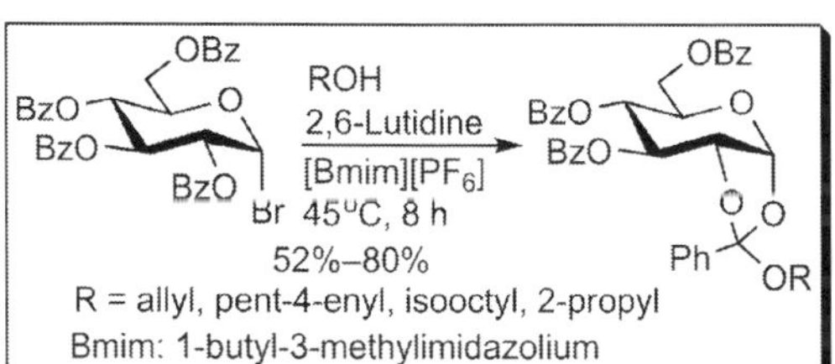

Scheme 30 Synthesis of 1,2-orthoester derivatives of carbohydrates using [Bmim][PF_6].

Thioglycosides[70] have been widely used as glycosyl donors for the preparation of complex oligosaccharides. Conventionally, they are prepared by the treatment of acetylated sugar derivatives with thiols in the presence of a Lewis acid in organic solvents such as CH_2Cl_2, $CHCl_3$, and CH_3CN. Misra et al.[71a] prepared thioglycosides in satisfactory yields by the reaction of glycosyl acetates with thiols in the presence of $BF_3{\cdot}OEt_2$ in [Bmim][BF_4] avoiding organic solvents. Later, an odorless preparation of thio- and selenoglycosides has also been reported by Misra and coworkers[71b] using IL as reaction solvent. Disulfides or diselenides were reduced to thiolates or selenolates in situ using a nonmetallic reducing agent (combination of Et_3SiH-$BF_3{\cdot}OEt_2$), which were reacted with glycosyl acetates to furnish thio- and selenoglycosides (Scheme 31).

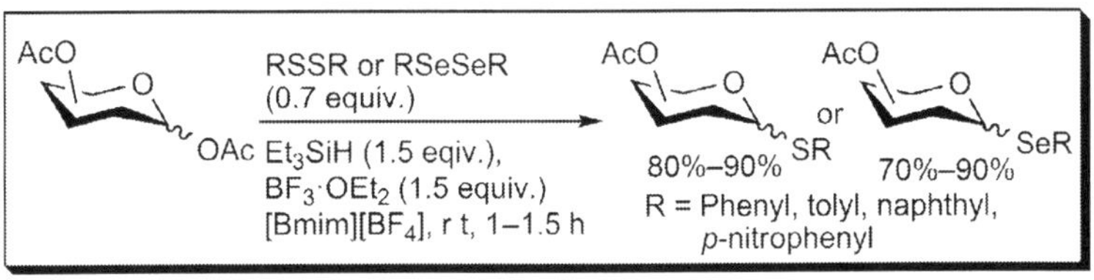

Scheme 31 One-pot preparation of thio-and selenoglycosides using [Bmim][BF_4].

In addition to the functional group transformation of carbohydrates, a number of reports appeared on the stereoselective glycosylation reactions using IL. Linhardt and coworkers[72] reported the synthesis of Fischer glycosylation products by the reaction of unprotected and unactivated reducing sugars with an alcohol in the presence of an acidic resin or protic acid in [Emim][ba]. A variety of disaccharide derivatives have also been prepared under the reaction conditions (Scheme 32).

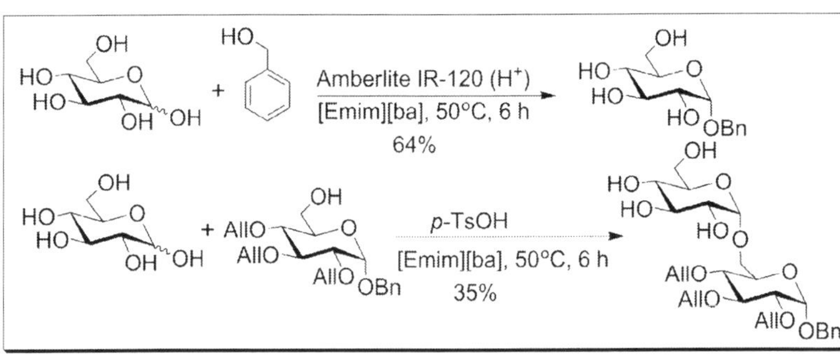

Scheme 32 Fischer glycosylation using acidic resin or protic acid in [Emim][ba].

Augé et al.[73] introduced IL-promoted atom economic glycosylation of unprotected sugars under Lewis acid catalysis. Treatment of free sugars with appropriate alcohols in the presence of a catalytic quantity of $Sc(OTf)_3$ (1 mol%) in [Bmim][OTf] furnished satisfactory yield of the glycosides as a mixture of α- and β-anomers. It was presumed that the oxycarbenium ion formed during the glycosylation was stabilized by the ionic liquid medium. The reaction proceeds to the thermodynamically favorable product and α-glycoside predominates at an elevated temperature (Scheme 33).

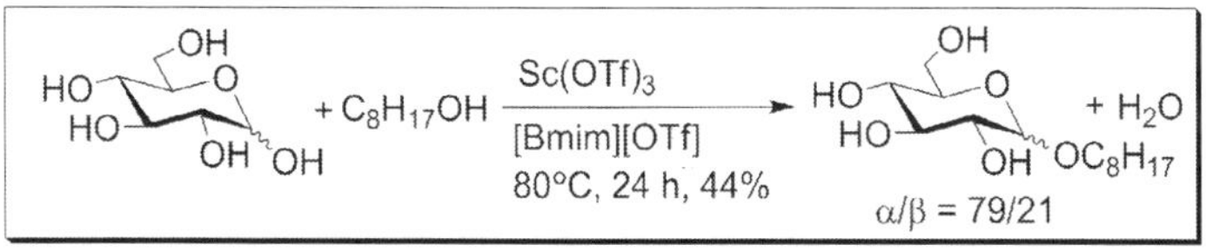

Scheme 33 Fischer glycosylation of free sugars catalyzed by $Sc(OTf)_3$ in [Bmim][OTf].

Yadav and colleagues[74] synthesized 2,3-unsaturated glycosides by the treatment of tri-*O*-acetylated glycal derivatives with a variety of aliphatic and aromatic alcohols in the presence of dysprosium triflate immobilized on [Bmim][PF_6] at room temperature (Scheme 34). This reaction have been successfully extended for the preparation of glycosylated amino acid derivatives.

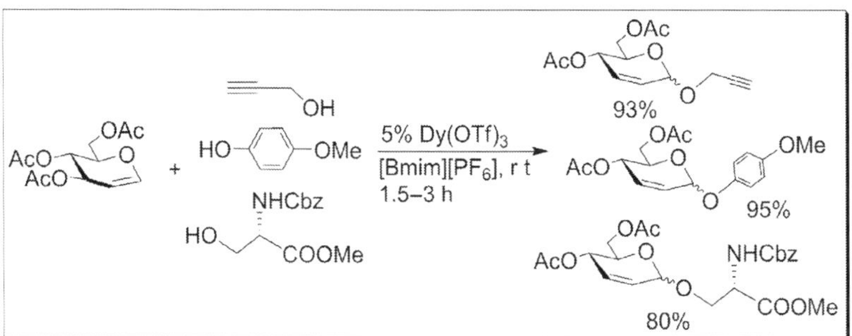

Scheme 34 Preparation of 2,3-unsaturated glycosides using $Dy(OTf)_3$ immobilized on [Bmim][PF_6].

Treatment of glycosyl trichloroacetimidates with alcohols in the presence of a catalytic Lewis acid (TMSOTf; 0.5 mol%) in a task-specific ionic liquid, [Bmim][PF_6] or [Emim][OTf], furnished excellent yields of glycosides with high stereoselectivity.[75, 76] The ILs used in the reactions were successfully recycled after isolating the products. Poletti and colleagues[77] demonstrated the influence of ILs on the stereochemical outcome of the glycosylations (Scheme 35).

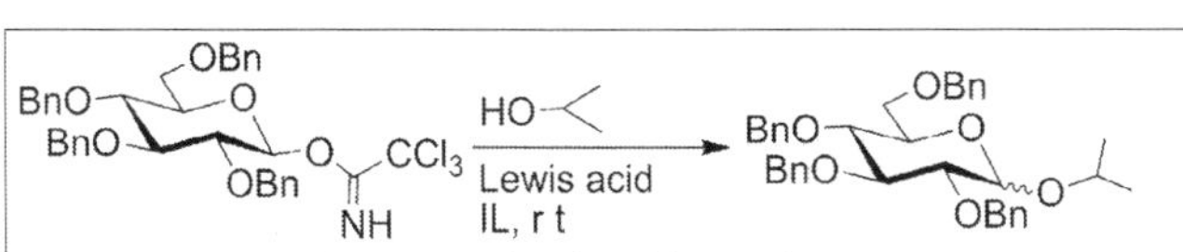

IL	Lewis acid	α:β	Time (h)	Yield (%)
[Bmim][PF_6]	TMSOTf	76:24	24	98
[Emim][OTf]	TMSOTf	45:55	20 min	Quant.
[Emim][OTf]	–	20:80	48	85

Scheme 35 Glycosylation with glycosyl trichloroacetimidate donor in ILs.

ILs have been successfully applied as the solvents for the preparation of oligosaccharides using glycosyl phosphite and fluoride derivatives as glycosyl donors activated by catalytic amount of protic acid. Toshima et al.[78,79] demonstrated that glycosylation with glycosyl phosphites in [1-hexyl-3-methylimidazolium][NTf_2] led to the formation β-glycosides, whereas the use of glycosyl fluorides led to the formation of α-products under similar reaction conditions (Scheme 36).

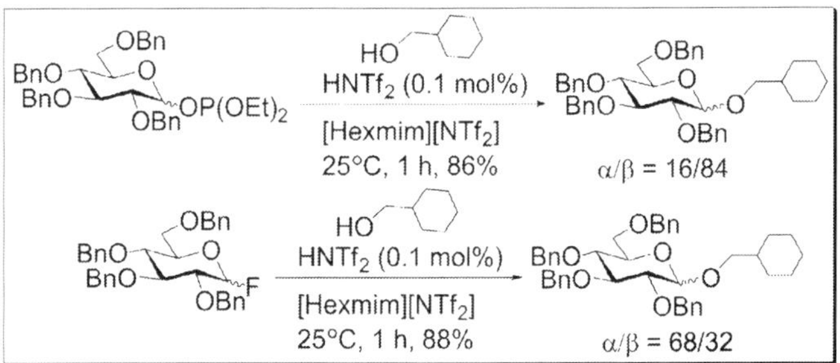

Scheme 36 Glycosylation using glycosyl phosphite and fluoride donors in IL.

Zhang et al.[80] reported stereoselective glycosylation of thioglycosides with a variety of alcohols using methyl triflate as the thiophilic activator in [Bmim][BF_4] (Scheme 37). A variety of thioglycosides were used for the glycosylation with a variety of acceptors.

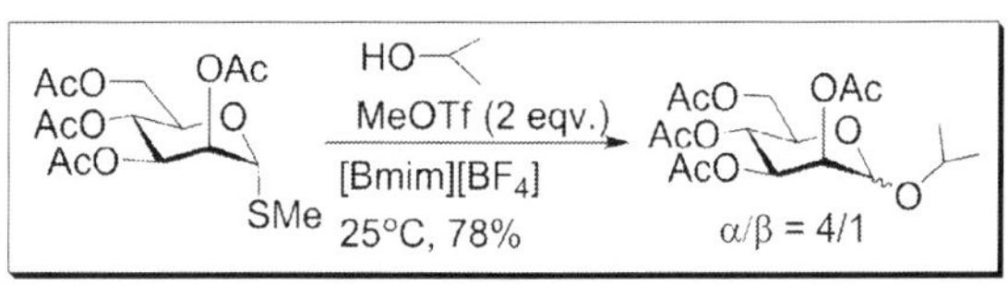

Scheme 37 Glycosylation of thioglycoside using MeOTf in [Bmim][BF_4].

Galan et al.[81] reported the use of a sulfonic acid-based surfactant ionic liquid as mild glycosylation promoter of thioglycosides in combination with *N*-iodosuccinimide (NIS) (Scheme 38). They have extended their methodology in the reactivity-based

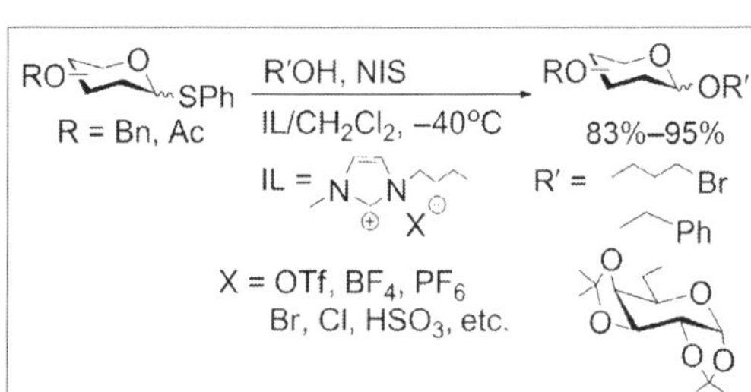

Scheme 38 Glycosylation of thioglycosides using NIS and surfactant ionic liquid.

one-pot glycosylation reactions for the preparation of linear and branched oligosaccharides using thioglycosides and trichloroacetimidate derivatives as glycosyl donors and [Bmim][OTf] as cosolvent and promoter[82] (Scheme 39).

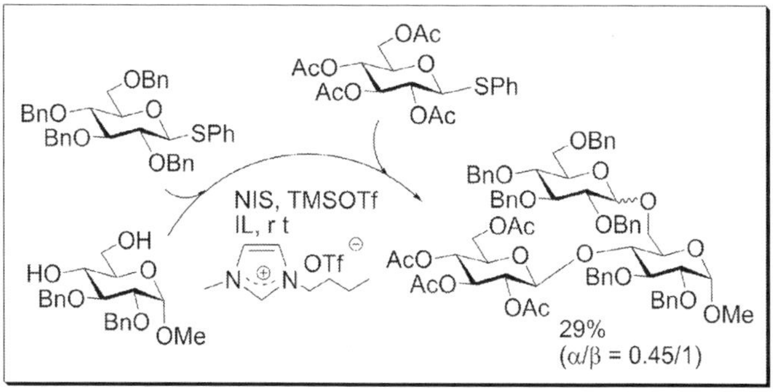

Scheme 39 Glycosylation of thioglycosides in one-pot using recyclable ionic liquid [Bmim][OTf].

In another aspect, Malhotra and coworkers[83] used ILs as a source of *N*-heterocyclic carbenes (NHCs).[84] They obtained *O*-aryl glycosides from acetobromosugars by the treatment with phenols in the presence of silver carbonate in [Bmim]½Cl at room temperature (Scheme 40). The *O*-aryl glycosides were obtained by the activation of glycosyl bromide with silver NHC complex formed in situ by the reaction of ILs with silver carbonate.

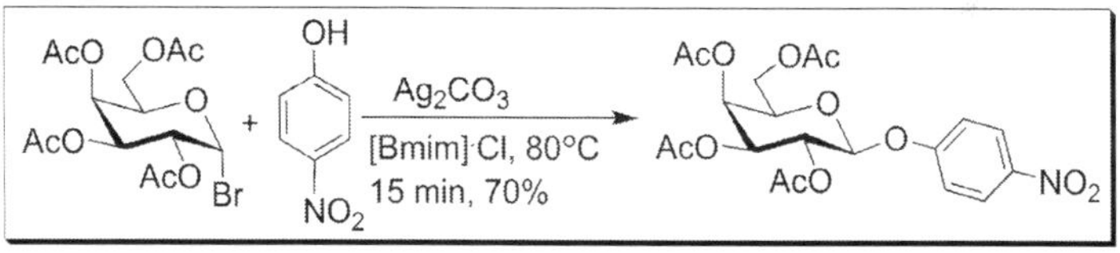

Scheme 40 Glycosylation of glycosyl bromide by in situ generated silver NHC complex.

Besides the application of ILs as solvents and catalysts, they have also been used as ionic support for the multistep synthesis of oligosaccharides. Chan's group and Huang's group independently reported the first oligosaccharide synthesis using ILs as soluble ionic tag. Chan et al.[85] used imidazolium cation as the ionic tag and used glycosyl sulfoxide as donor and thioglycoside as acceptor to provide an orthogonal glycosylation approach (Scheme 41). Huang's strategy[86] was to use glycosyl trichloroacetimidate as donor and thioglycoside as the acceptor which was tagged with the soluble imidazolium cation following the orthogonal glycosylation concept (Scheme 42).

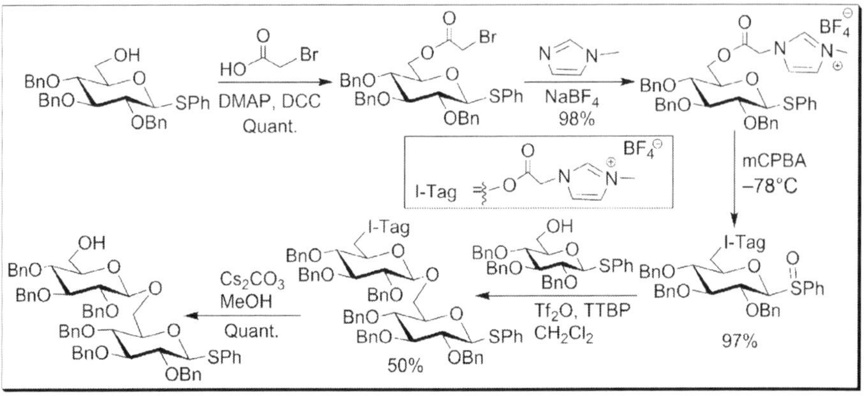

Scheme 41 An orthogonal glycosylation between thioglycoside and glycosyl sulfoxide using ionic tag by Chan's group.[85]

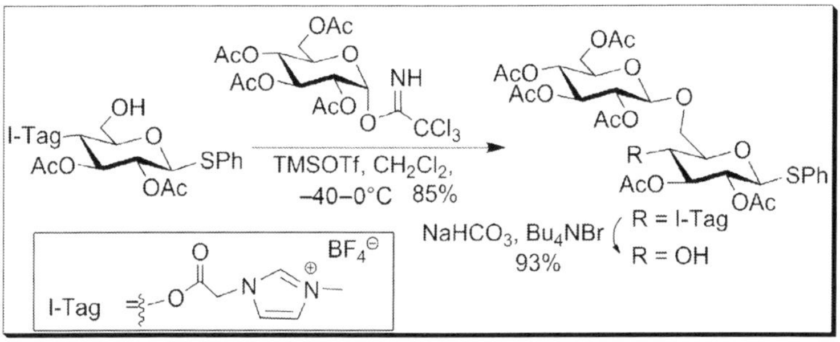

Scheme 42 An orthogonal glycosylation approach using ionic tag by Huang's group.[86]

Subsequently, Yerneni et al.[87, 88] and Pepin et al.[89] reported the synthesis of linear α-(1→6) linked oligomannan and α-(1→4) linked glucan, respectively, applying similar approaches involving soluble ionic liquid supported oligosaccharide synthesis (Schemes 43 and 44).

More recently, Galan and coworkers[90] developed an elegant approach for the ionic tag linked oligosaccharide synthesis. This "Ionic Catch and Release Oligosaccharide Synthesis" (ICROS) approach is equally suited for parallel and combinatorial synthesis of libraries of oligosaccharides. The linker is attached to the acceptor by anomeric glycosylation and after carbohydrate chain elongation the ionic tag is removed by conventional reaction conditions. Random synthesis of oligosaccharides was achieved using this strategy (Scheme 45).

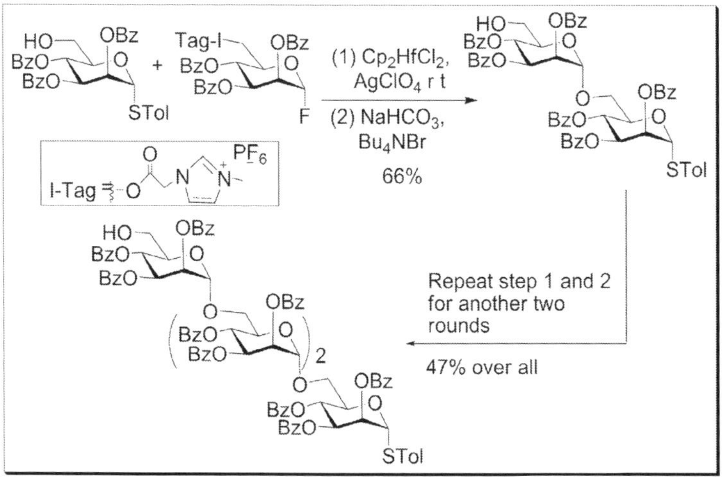

Scheme 43 Soluble IL-supported synthesis of linear tetramannoside.[87, 88]

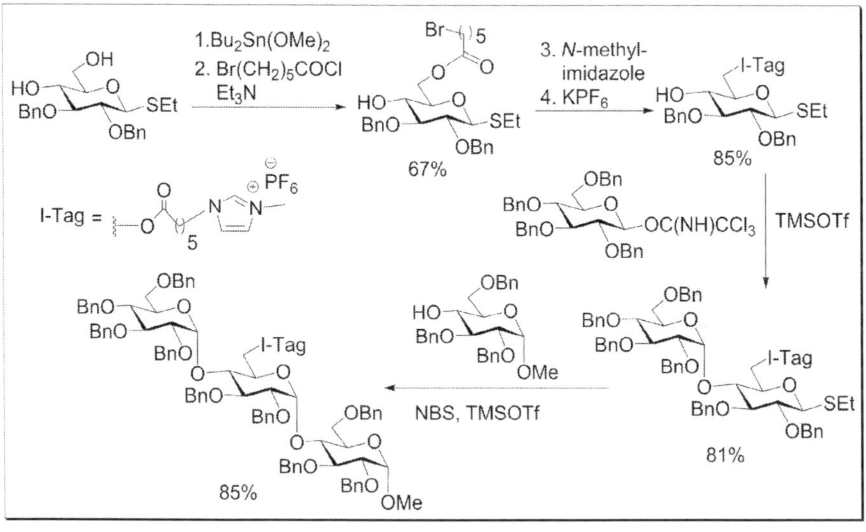

Scheme 44 Soluble IL-supported synthetic strategy for the preparation of homolinear α-(1→4)-glucosides.

ILs have also been successfully used in the preparation of *C*-glycosyl compounds. Toshima and coworkers[91] prepared *C*-aryl glycosides by the treatment of glycosyl fluorides and methyl glycosides with phenols in the presence of protic acids, HBF_4 and $HNTf_2$ in ILs such as $[C_6mim][BF_4]$ and $[C_6mim][NTf_2]$, respectively (Scheme 46).

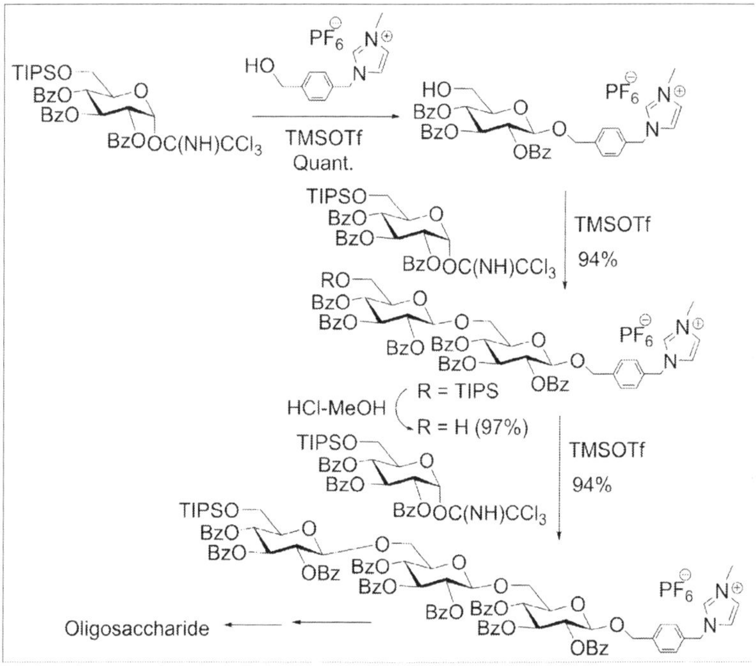

Scheme 45 An approach for an ionic catch and release oligosaccharide synthesis.

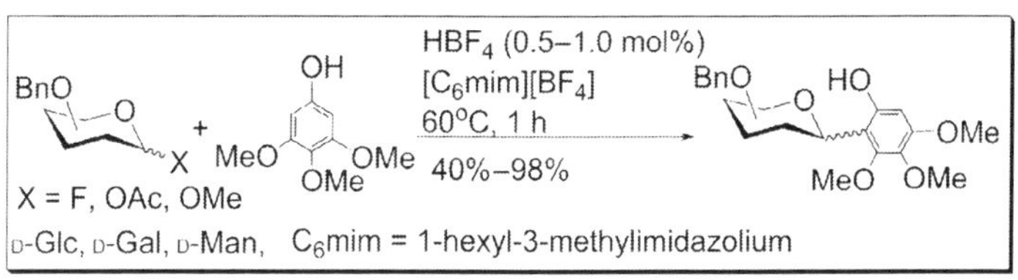

Scheme 46 Ionic liquid-mediated preparation of *C*-aryl glycoside.

6 Ionic liquid tags in enzymatic reactions

In a more recent report, the applicability and versatility of using I-Tags for the monitoring of enzymatic oligosaccharide synthesis as well as their usefulness in the purification has been demonstrated with the development of a new *N*-benzylsulfonyl-based ionic-liquid label (I-Tag).[92] The I-Tag linked monosaccharide moiety was prepared using a two-step procedure. Commercially available 4-(bromomethyl)benzene-1-sulfonyl chloride was conjugated with protected aminopropyl 2-acetamido-2-deoxyglucoside moiety under basic conditions and the halide ion was displaced with 1-methyl-1*H*-imidazole and KBF_4. The unmasking of the OH groups yielded I-Tag-labeled *N*-acetylglucosamine acceptor (GlcNAc), which was ready to be used in enzymatic reactions. The new I-Tag was compatible with the glycosyltransferases used in the study, thus opening the door for applications with other glycosyltransferases and other enzymes outside the area of oligosaccharide synthesis (Scheme 47).

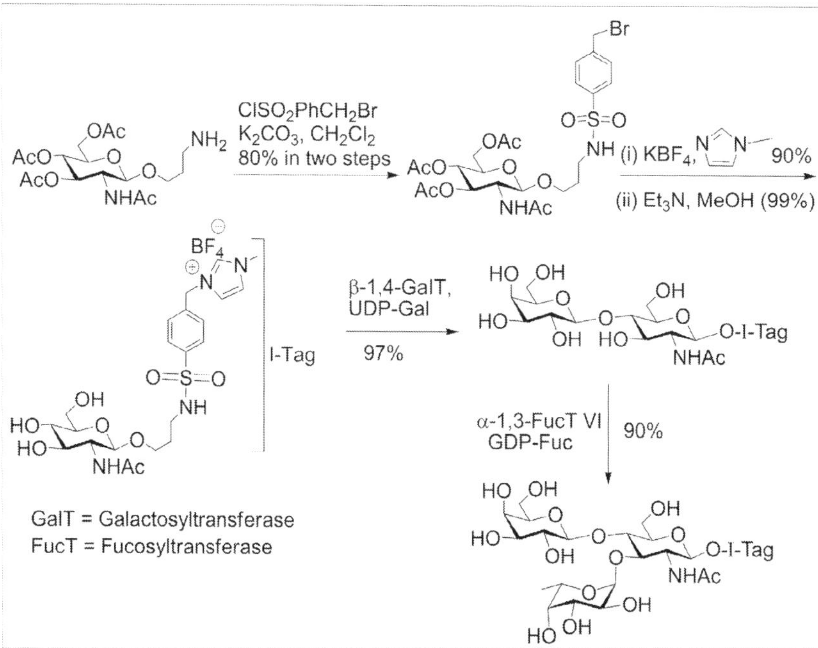

Scheme 47 Enzymatic synthesis of I-Tagged oligosaccharide.

7 Transformation of carbohydrates in supercritical fluids

In the quest of green and sustainable reaction conditions, supercritical fluids[93] (SCFs) have attracted special interest for their ability to act as solvents in a wide range of chemical reactions. Earlier, they have been used as solvents for the extraction of natural products, essential oils from the raw materials and their chromatographic purifications.[94] They can be considered as intermediates between liquids and gases since their properties (e.g., viscosity and density) can be changed by manipulating the pressure and temperature of the system. Because of their usefulness as green solvents, SCFs have been used in many organic reactions, which include reactions catalyzed by homogeneous organometallic complexes,[95] hydrogenation,[96]chemoselective methylation of amines and diols,[97]alkylation of aromatics,[98] epoxidation of alkenes,[99] and Beckmann and Pinacol rearrangements,[100] to name a few. SCFs have also found application in enzymatic reactions using immobilized enzymes. A variety of lipid-coated enzymes such as lipases, phospholipases, and glycosidases have been used in esterifications and transglycosylations in homogeneous aqueous medium (Scheme 48).[101]

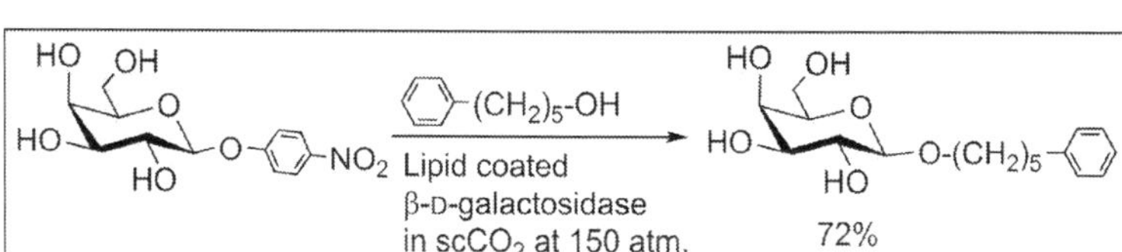

Scheme 48 Enzymatic *trans*-glycosylation in supercritical CO_2 ($scCO_2$).

Nishimura and coworkers reported stereoselective glycosylations of glycosyl trichloroacetimidate and fluoride derivatives with a series of simple alcohols and sugar acceptors in the presence of solid superacid in supercritical CO_2 ($scCO_2$).[102] The activation of per-*O*-acetyl galactosyl trichloroacetimidate with sulfated zirconia in $scCO_2$ furnished excellent yields of β-D-galactosides as the exclusive products. On the other hand, a mixture of α- and β-glycosides was obtained by the sulfated zirconia-catalyzed glycosylation of benzylated galactosyl fluoride, bromide, or trichloroacetimidate derivatives in $scCO_2$ (Scheme 49).

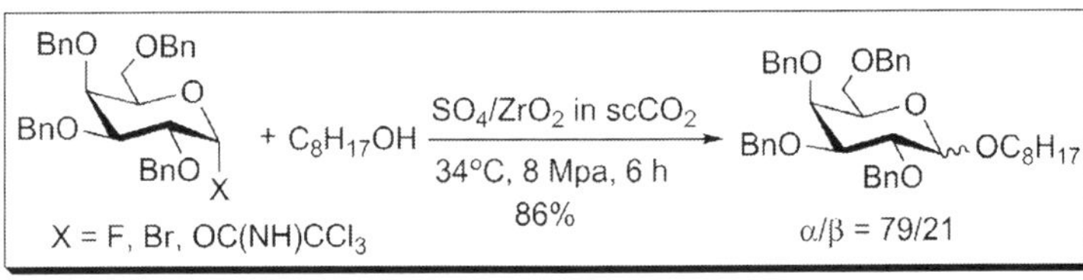

Scheme 49 Glycosylations promoted by sulfated zirconia in $scCO_2$.

Cardona et al.[103] demonstrated metal-free glycosylation reactions in $scCO_2$ avoiding the use of volatile organic solvents. Treatment of galactosyl halogenides with a variety of alcohols in the presence of MS 4Å in $scCO_2$ under 1500 psi furnished anomeric mixture of glycosides in excellent yields. In a separate experiment, galactosyl bromide resulted in the orthoester derivative in the presence of a base (2,6-lutidine) under similar reaction conditions (Scheme 50).

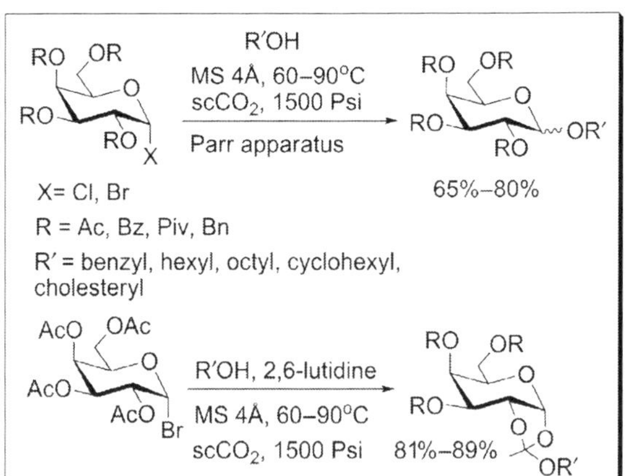

Scheme 50 Glycosylation and orthoester formation in high-pressure supercritical fluids.

8 Transformation of carbohydrates in deep eutectic solvents

Deep eutectic solvents (DESs) are liquids, which are prepared by combining two or three inexpensive and safe chemicals.[20] They are capable of self-assembling often through hydrogen bonding to make a eutectic mixture having lower melting point than the individual components.[104] Initially, DESs were prepared by mixing quaternary ammonium salts with metal salts in an appropriate ratio so that the resulting mixture became a liquid below 100°C. Later, several DESs were prepared by mixing quaternary ammonium salts [e.g., choline chloride (CC), which acts as hydrogen bond acceptor] with a variety of carboxylic acids, urea, amides, and alcohols (which act as hydrogen bond donor).[105]Although, DESs and ILs have several similarities in their nature and applications, the preparation of DESs is 100% atom economic, highly pure, and less expensive. In general, the preparation of DESs is involved with the handling of nontoxic and biodegradable chemicals.[106] In addition, several reducing monosaccharides have also been used as the hydrogen bond donor for the preparation of nontoxic, biodegradable DESs.[107]Although a wide range of organic reactions have been carried out in DESs, their applications in the transformation of carbohydrates are scarce. As a consequence, there is a large scope to work out newer methodologies for carbohydrate transformations/glycosylations in DESs.[108]

Xylose and its polymer xylan have been dehydrated to furnish furfural by treatment with metallic salts, for example, $FeCl_3$ and $CrCl_3$ in DES made with choline chloride (CC)-citric acid at 140°C (Scheme 51).[109]

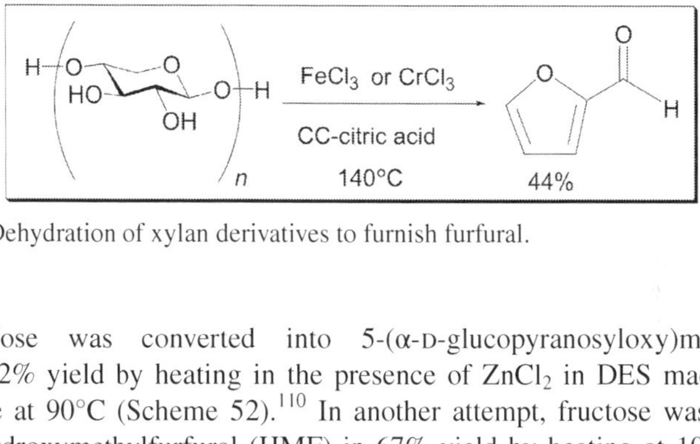

Scheme 51 Dehydration of xylan derivatives to furnish furfural.

Isomaltulose was converted into 5-(α-D-glucopyranosyloxy)methylfurfural (GMF) in 52% yield by heating in the presence of $ZnCl_2$ in DES made with CC-isomaltulose at 90°C (Scheme 52).[110] In another attempt, fructose was dehydrated to give 5-hydroxymethylfurfural (HMF) in 67% yield by heating at 100°C in DES prepared by mixing CC-*p*-TsOH (Scheme 52).[111]

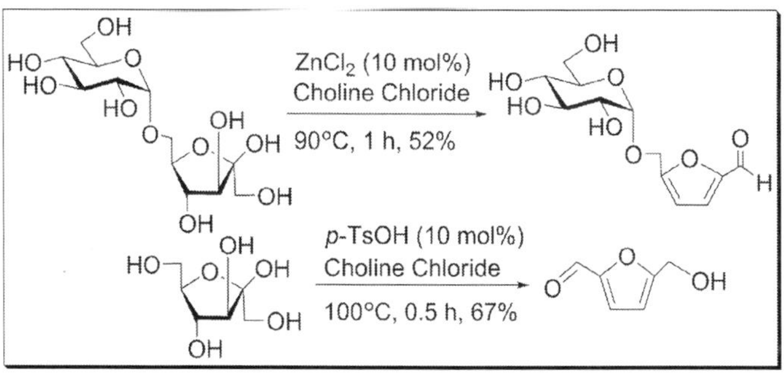

Scheme 52 Acid-catalyzed dehydration of isomaltulose and fructose in DES.

Synthesis of sugar annulated pyrrole derivatives was achieved under metal-free reaction conditions by heating a mixture of a free sugar (e.g, D-glucose, D-galactose, and D-mannose), aniline and 1,3-dicarbonyl compounds in CC-urea at 80°C (Scheme 53).[112]

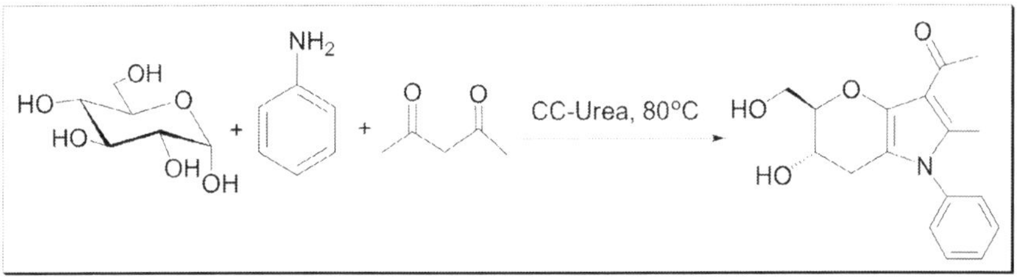

Scheme 53 Synthesis of sugar-annulated pyrroles in CC-urea.

DES made with CC-urea has been used as the medium for the Lipase (Novozyme 435) mediated selective acylation of free sugars at 70°C. Recently, Pohnlein and coworkers reported transesterification between glucose and vinyl hexanoate leading to the formation of a glucose-6-*O*-hexanoate using similar reaction conditions (Scheme 54).[113]

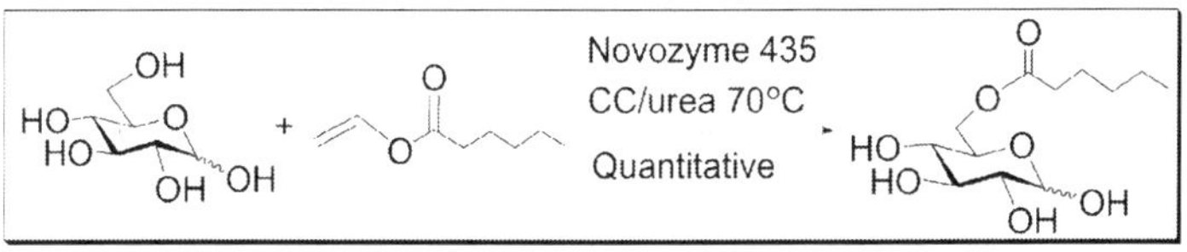

Scheme 54 Lipase-catalyzed transesterification between glucose and vinyl hexanoate in DES made with CC and urea.

DESs have also found applications in the functional group manipulations in carbohydrate substrates. Per-*O*-acetylated lactols have been prepared from free sugars in excellent yields by the treatment with acetic anhydride in DES consisting of choline chloride (CC) and $ZnCl_2$ (Scheme 55).[114]

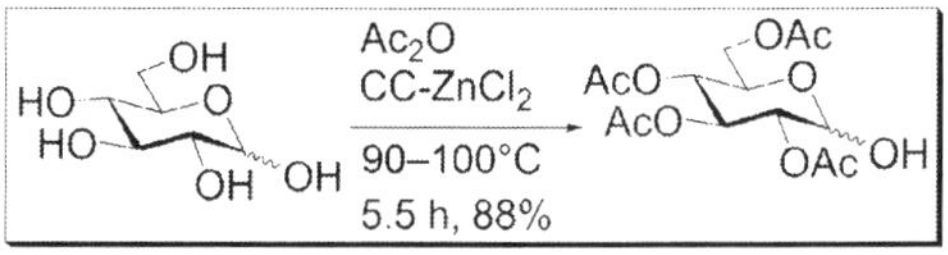

Scheme 55 Preparation of 2,3,4,6-tetra-*O*-acetyl-D-glucose in DES.

Ferrier rearrangement of glycal derivatives has been carried out in DES made up with CC and malonic acid to furnish 2,3-unsaturated dideoxyglycosides in excellent yields (Scheme 56).[115]

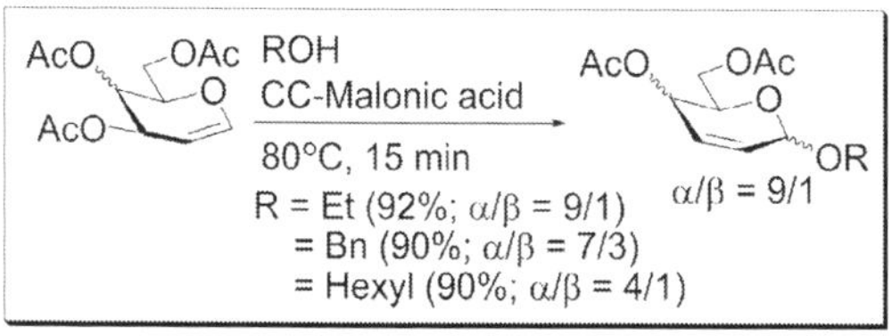

Scheme 56 Synthesis of 2,3-unsaturated glycosides from glycal using DES solvent.

It is noteworthy that in most of the reaction conditions reported so far, the DES has been recovered, reused, and recycled several times without any loss of the catalytic and solvent potential and hence could be considered as effective alternatives to the conventional volatile organic solvents.

9 Transformation of carbohydrates using fluorous solvents

The concept for the use of fluorous solvents in organic reactions and the use of a fluorous tag in solid and solution phase multistep organic syntheses have been introduced several years back by Horváth and Rábai.[116] Fluorous solvents and fluorinated functionalities have unique properties such as hydrophobicity, lipophobicity, inertness, nontoxic nature, and ease of phase separability, which allow them to be used in a wide range of chemical reactions and effective separation of products from the reaction mixtures.[117] Over the years, fluorous chemistry has been expanded into various directions of chemistry and biology such as development of fluorous ligands and catalysts; fluorous version of supramolecular, polymer, and materials chemistry; and bioorganic and medicinal chemistry.[118] The first "fluorous synthesis" was reported by Curran and coworkers[119] in which the target compounds were selectively soluble in the fluorous phase due to the presence of a fluorous tag attached to them. Subsequently, several fluorous reagents, protecting groups, and scavengers have been introduced based on the requirements in synthetic strategies. Fluorous compounds are termed as "heavy" and "light" fluorous compounds depending on the quantity of fluorine atoms present in the molecules.[120] The heavy fluorous compounds have poor solubility in organic solvents and require fluorinated solvent for their solubility such as benzotrifluoride. The light fluorous synthesis involves the use of a shorter fluorous component linked to the compounds to provide effective separation procedures. Fluorous technology has been used in homogeneous[121] as well as heterogeneous[122] chemical reactions, and a wide variety of compounds have been synthesized applying this technique. Fluorous phase synthesis has been successfully applied on carbohydrate substrates also. Although the "fluorous technology" has been introduced in carbohydrate chemistry by Curran and coworkers,[123] Seeberger's group applied fluorous tag synthesis technique in the solid phase oligosaccharide synthesis (SPOS) for the first time.[124] This approach has been successfully applied in the automated synthesis of linear and branched oligosaccharides as well as synthesis of oligosaccharides in microreactors. In the SPOS strategy, the unreacted immobilized acceptors were tagged with heptadecafluorodecyl diisopropyl silyl ether after each round of glycosylation. The desired products were isolated using fluorous-labeled silica gel chromatography after detachment from the resin. Despite the advantages for getting high levels of purity in the products, the need to use of excess fluorous label after each coupling step made the process quite expensive.

Zhang et al.[125] introduced a light-fluorous glycosyl donor and an orthogonal tagging strategy to synthesize oligosaccharides and glycoconjugates. In this synthetic strategy, the glycosyl donor is orthogonally protected with a C_8F_{17}-silyl tag, which was allowed to react with excess amount of glycosyl acceptor. Fluorous solid-phase extraction (FSPE) technique has been adopted to separate the glycosylated product and unreacted glycosyl acceptor (Scheme 57).

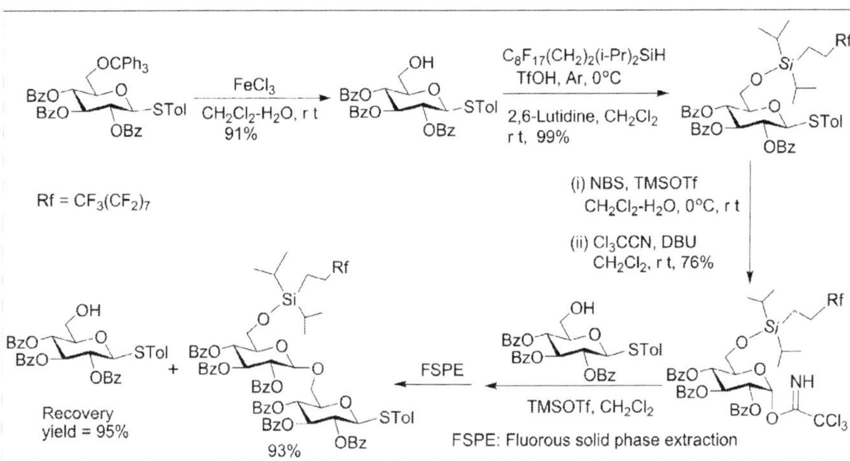

Scheme 57 Synthesis of fluorous tagged glycosyl donor and its use in glycosylation.

In 2010, Yang et al.[126] described an approach for the oligosaccharide synthesis combining the advantages of one-pot synthesis and fluorous phase separation. After achieving the oligosaccharides using one-pot iterative glycosylations, a fluorous tag has been introduced into the reaction mixture to selectively "catch" the desired oligosaccharide, which was rapidly separated from nonfluorous impurities by fluorous solid phase extraction (FSPE) technique. Subsequent "release" of the fluorous tag led to the rapid access to the pure oligosaccharides without the use of conventional column chromatography. Several linear and branched oligosaccharides (e.g., LewisX trisaccharide) have been synthesized applying this approach insignificantly less time (Scheme 58).

Fluorous technology has been used for the incorporation of reducing sugars into a microarray platform. Pohl's group[127] reported a fluorous tag-assisted rapid

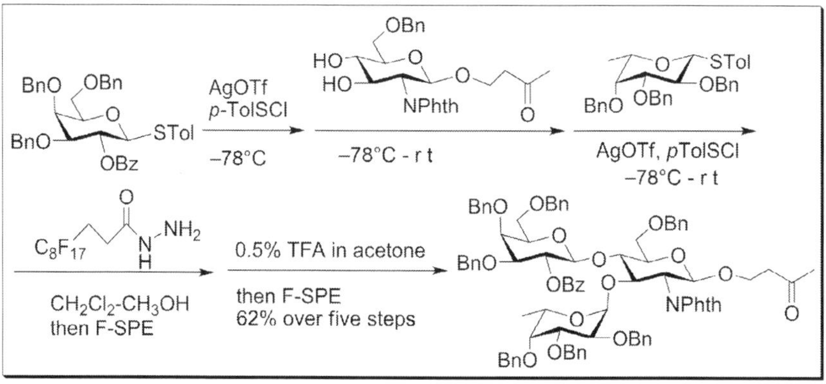

Scheme 58 Fluorous-assisted one-pot oligosaccharide (Lewis X trisaccharide) synthesis.

solution-phase synthesis of oligosaccharides in an automated fashion (Scheme 59). This approach has the added benefit of allowing direct incorporation of fluorous tagged sugars into a microarray platform for biological evaluation. A fluorous-tagged allyl linker[128, 129] was coupled to the glycosyl acceptor as glycoside, which was used as fluorous handle during oligosaccharide chain elongation and purification of the products and finally cleaved off from the product. Since the coupling reactions were performed in solution, they required only slight excess of the glycosyl donor and catalyst to furnish satisfactory yield of the products and their purification was done using fluorous solid-phase extractions (FSPEs).

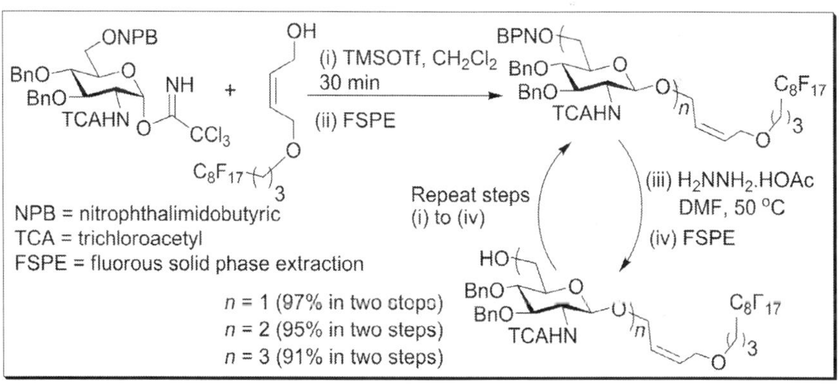

Scheme 59 Synthesis of D-glucosamine oligomers on a fluorous support using FSPE.

In addition, several reports[130] have appeared for the development of fluorous protecting groups for the oligosaccharide synthesis. Miura et al.[131] introduced a novel bisfluorous chain-linked propanoyl group (Bfp), which can be readily introduced to carbohydrate substrates and requires mild conditions for its removal. This protecting group can also be recycled in a new set of reaction after removal from the substrate. A tetrasaccharide has been synthesized using minimum chromatographic purifications using the substrates functionalized with Bfp group. Due to the presence of the Bfp group, the synthetic intermediates were purified by simple fluorous-organic solvent extraction without using chromatography (Scheme 60). A new type of fluorous protecting group (Froc) was also reported for peptide and oligosaccharide synthesis by Manzoni and Castelli.[132]

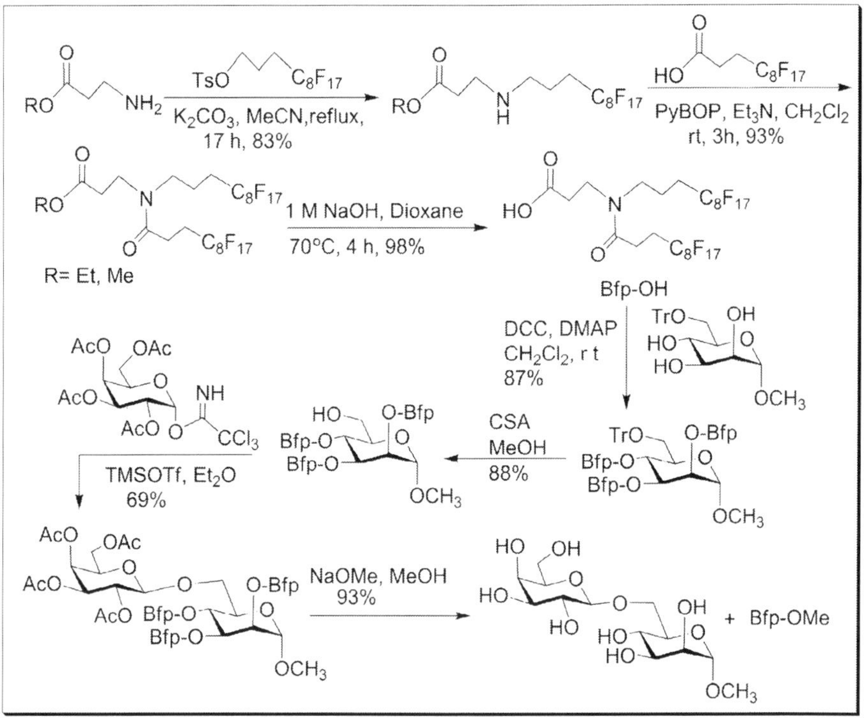

Scheme 60 Fluorous oligosaccharide synthesis using a novel fluorous protecting group.

Pohl and coworkers[133] reported the first synthesis of linear and branched mannose oligosaccharides using fluorous-tag assistance with fluorous reagents and FSPE protocols (Scheme 61). Later Boons and coworkers[134] reported fluorous-supported modular synthesis of heparan sulfate oligosaccharides by employing an anomeric aminopentyl linker protected by a benzyloxycarbonyl group modified by a perfluorodecyl tag, by which highly polar intermediates could be easily purified using fluorous solid phase extraction (FSPE).

Carrel and Seeberger[135]developed a new strategy in the solid phase oligosaccharide synthesis, which was termed as "cap-and-tag" synthesis. In this approach, the

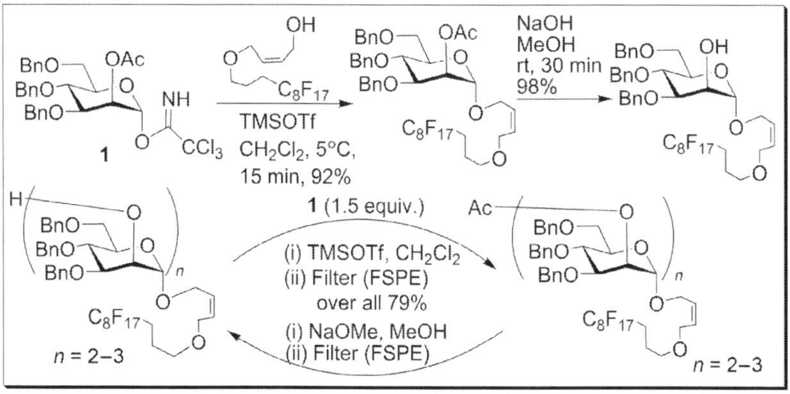

Scheme 61 Synthesis of a fluorous tagged mannose derivative and iterative synthesis of a linear oligomannoside.

by-products were removed by capping with acetyl group and the desired oligosaccharide was marked with a fluorous tag. Following the cleavage from the resin, the desired F-tagged oligosaccharide was separated from the acetyl-capped deletion sequences by fluorous solid phase extraction (FSPE). The methodology is illustrated by the solid phase synthesis of a model trisaccharide (Scheme 62).

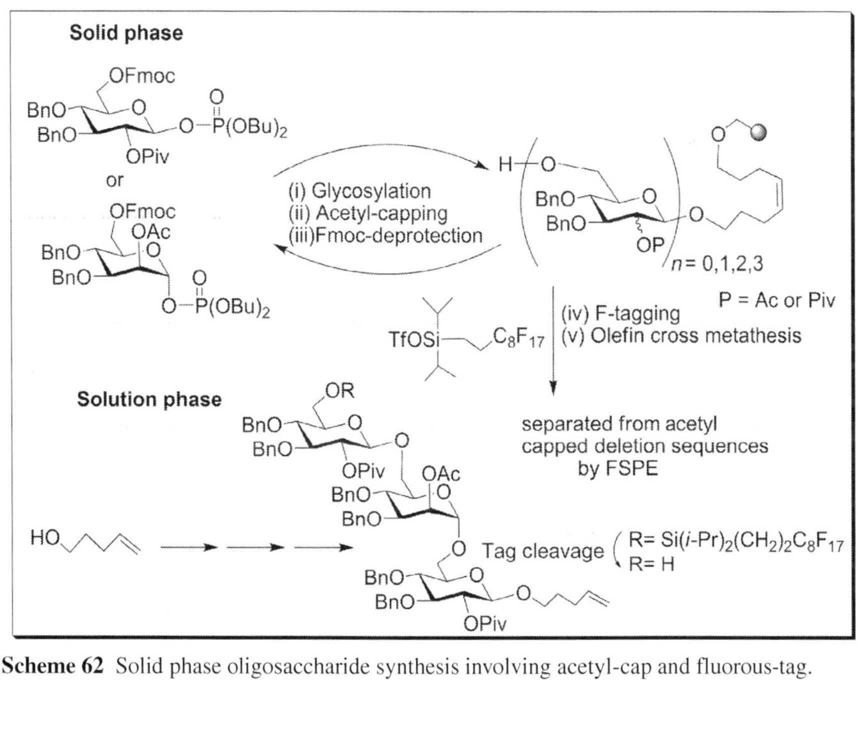

Scheme 62 Solid phase oligosaccharide synthesis involving acetyl-cap and fluorous-tag.

10 Transformation of carbohydrates using nonconventional energy sources

In parallel to the development of synthetic methodologies for the transformation of carbohydrates using environmentally benign solvents, several attempts were made to develop technology-based transformation of carbohydrates and oligosaccharide synthesis. The main emphasis was given to the application of nonconventional energy sources in the synthetic methodologies as well as development of solvent-free reaction conditions. In order to synthesize a wide range of carbohydrate-derived molecules, several technologies have been applied, which include (a) microwave heating, (b) ultrasonication, (c) photoinduced reactions, (d) ball milling, (e) electrochemistry, and (f) reactions under high pressure.

11 Oligosaccharide synthesis using microwave irradiation

Microwave irradiation has been applied to a variety of organic reactions including substitution, alkylation, condensation, cycloaddition, and introduction or removal of the protecting groups.[136–141] Microwave heating refers to the application of electromagnetic waves having the ranges of frequency from 300 MHz to 300 GHz and wave length of 0.01 to 1 m to generate heat in the material.[142] The basic principle behind the heating in a microwave oven is due to the interaction of charged particles of the reaction material with electromagnetic waves of particular frequency.[143] Carrying out organic reactions under microwave irradiation over the conventional heating has several advantages, such as operationally simple, clean, fast, efficient, and economical. Microwave irradiation accelerates the rate of reactions, which in turn reduce the reaction time and improve the yield of the product reducing the formation of by-products. In addition, application of this technique in organic reactions has been considered as a green chemical approach for the synthesis of a variety of compounds maintaining atom economy and less hazardous nature in terms of waste production.[144] Over the years, the application of microwave technology in chemical reactions has been modified from the use of domestic microwave oven to the programmable microwave cavity.[145] Although initially this technology has been used mainly in the preparation of heterocyclic compounds, a long list of reactions also can be found for its application in carbohydrate transformations. A few selected examples are presented in the text.

Simple alkyl and aryl glycosides were prepared using Fischer glycosylation by the treatment of free sugars with alcohols in the presence of an acid catalyst with or without solvent at high temperature. The reaction suffers from low yield, longer reaction time, and formation of by-products. Poulsen and coworkers[146] applied microwave heating for the speed-up preparation of Fischer glycosylation products of D-glucose, D-galactose, and D-mannose as well as *N*-acetyl-D-glucosamine and *N*-acetyl-D-galactosamine with a variety of alcohols (methanol, ethanol, Bn-OH, All-OH, etc.) (Scheme 63). The products were obtained in high yields without using any solvent in significantly less reaction time.

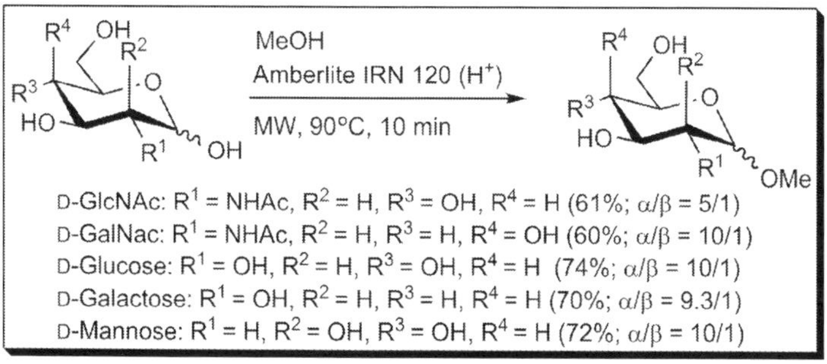

Scheme 63 Microwave irradiation accelerated Fischer glycosylation.

Lin and coworkers[147] demonstrated the microwave-assisted glycosylation of *exo*-glycals (as donors) to produce glycosides in high yields with shorter reaction time compared to the conventional methods without requirement of a catalyst (Scheme 64).

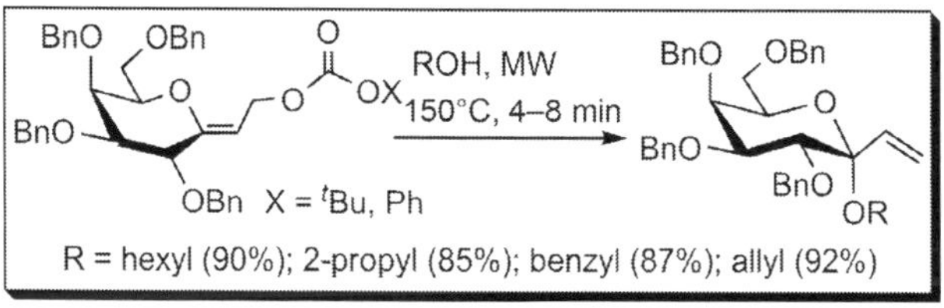

Scheme 64 Microwave-assisted glycosylation of *exo*-glycals.

Although methyl glycosides are not considered as efficient glycosyl donors, Yoshimura et al.[148] described an efficient microwave-assisted glycosylation of methyl glycopyranosides in the presence of $Yb(OTf)_3$ to furnish a variety of glycosides in high yields (Scheme 65).

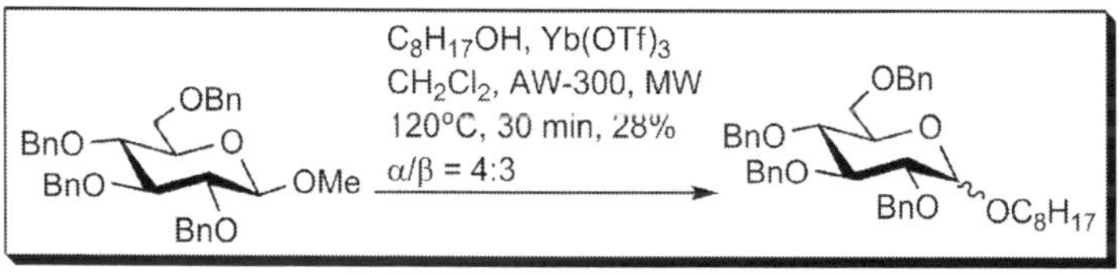

Scheme 65 Methyl glucoside as glycosyl donor under microwave irradiation.

Recently, efficient glycosylation and esterification of D-glucuronic acid and its 1,6-lactone under solvent-free microwave irradiation has been demonstrated by Wadouachi's group[149] (Scheme 66).

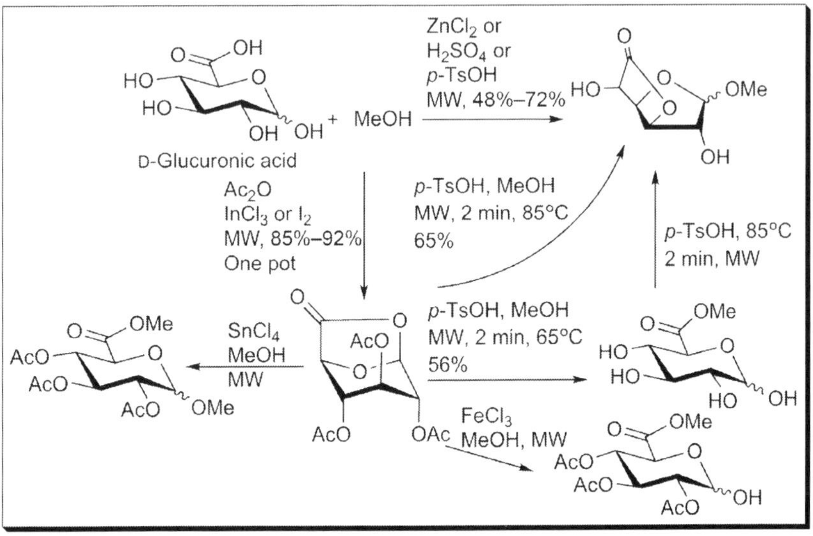

Scheme 66 Glycosidation and esterification of D-glucuronic acid under MW condition.

Nishimura's group[150] carried out efficient glycosylation applying programmable microwave irradiation at lower temperature. They were able to control the reaction pathways and reduced the formation of by-products (Scheme 67).

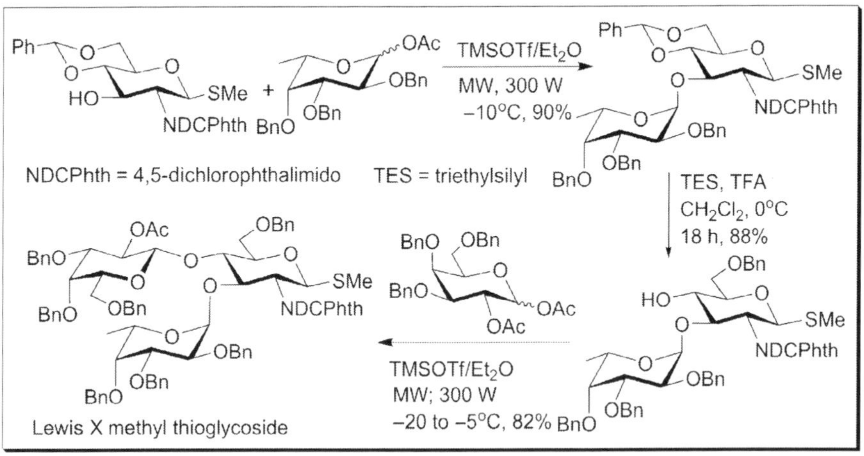

Scheme 67 Synthesis of a LewisX trisaccharide methyl thioglycoside under microwave irradiation at low temperatures.

Zhang et al.[151] have demonstrated glycal-based glycosylation method to produce 2,3-unsaturated-*O*-glycosides via Ferrier rearrangement of per-*O*-acetylated or per-*O*-benzylated glycal derivatives in the presence of $Fe_2(SO_4)_3{\cdot}xH_2O$ under microwave irradiation. The excellent α-selectivity, use of inexpensive and reusable catalyst, and environmentally benign reaction conditions make this an attractive method for the preparation of 2,3-unsaturated-*O*-glycosides from glycals (Scheme 68).

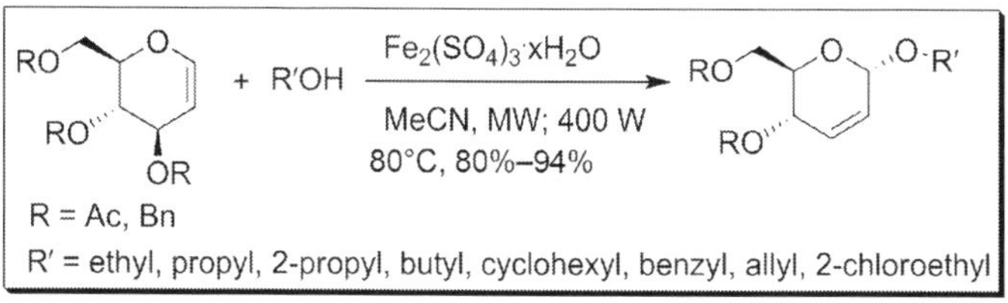

Scheme 68 Ferrier rearrangement of glucal derivatives catalyzed by $Fe_2(SO_4)_3{\cdot}xH_2O$ under microwave irradiation.

Recently, Carrasco et al.[152] reported a microwave-mediated chemoselective glycosylation of *N*-alkylaminoxy moiety with unprotected reducing sugars producing glycopeptoids. It was observed that microwave irradiation significantly increased the degree of glycosylation and shortened the reaction time (Scheme 69).

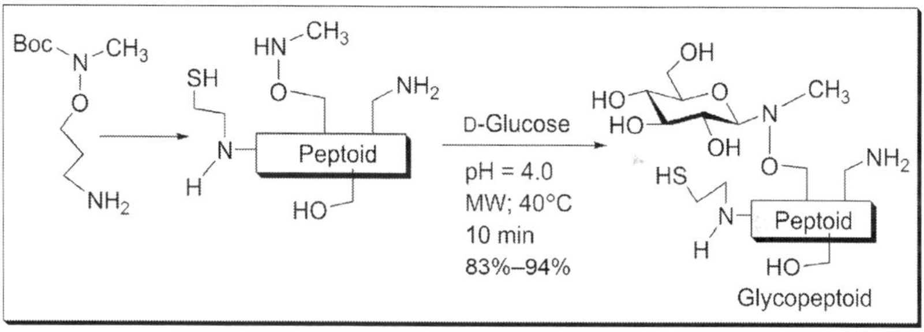

Scheme 69 Microwave mediated glycopeptoid synthesis.

Ko et al.[153]described the preparation of 1,6-anhydro sugars directly from free sugars using microwave-assisted synthesis, which allowed to get access to a variety of compounds such as thioglycosides, simple glycosides, and disaccharide derivatives after a series of transformations in one-pot. In the first step, per-*O*-trimethylsilylated monosaccharide intermediates were formed by the treatment of free sugars with hexamethyldisilazane (HMDS) in the presence of TMSOTf,[154] which were cyclized under microwave irradiation to give 1,6-anhydro sugar derivatives. Treatment of the 1,6-anhydro sugar derivatives with thiolating agents in the presence of ZnI_2 furnished thioglycosides, which were transformed into benzylidenated derivatives in one-pot reaction conditions.[155] In another approach, disaccharide derivatives were prepared directly from free sugars via the formation of 1,6-anhydro sugar derivatives followed by the reaction with trifluoroacetimidate derivatives (Scheme 70).[153]

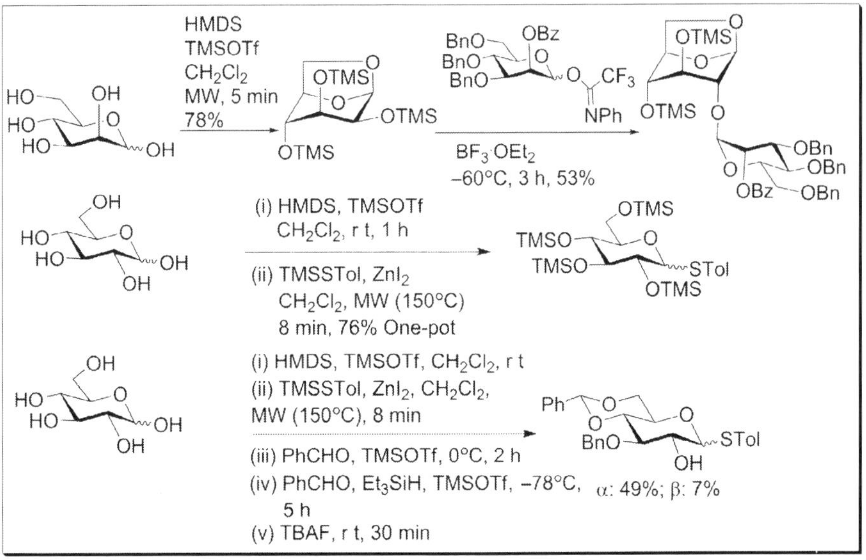

Scheme 70 Microwave-assisted one-pot functionalization of carbohydrates and glycosylation.

A variety of protecting group manipulations and glycosylation reactions were carried out in single reaction vessel under microwave irradiation.[156, 157] In short, application of microwave technology in protections-deprotections of functional groups of carbohydrate substrates as well as glycosylations allowed getting access to arrays of suitably protected intermediates and linear or branched complex oligosaccharides in high yields in shorter reaction time.

12 Transformation of carbohydrates using ball milling

Most often transformation of carbohydrates and synthesis of oligosaccharides involve reaction conditions using toxic and hazardous chemicals deviating the principles of "Green chemistry."[16] A large number of eco-friendly strategies for the preparation of a wide array of organic molecules have been developed during the last decade in the perspective of the environmentally benign chemical synthesis. Carrying out chemical reactions in "neat" to avoid the use of organic solvents is more eco-friendly in nature and fulfill the goals of Green chemistry. In this context, carrying out mechanochemical reactions[158–160] under solvent-free conditions with the help of a ball mill[161, 162] have become of increasing significance, and this technology has also been successfully employed in carbohydrate chemistry.

Recently, Kartha et al. described the synthesis of various *O*-glycosides from acetobromosugars and alcohols (ROH, where R = alkyl/substituted alkyl/alkenyl/alkynyl/glyceryl/cyclohexyl/steryl) under solvent-free conditions employing ball milling in the presence of metal carbonates as glycosylation promoter (Scheme 71).[163]

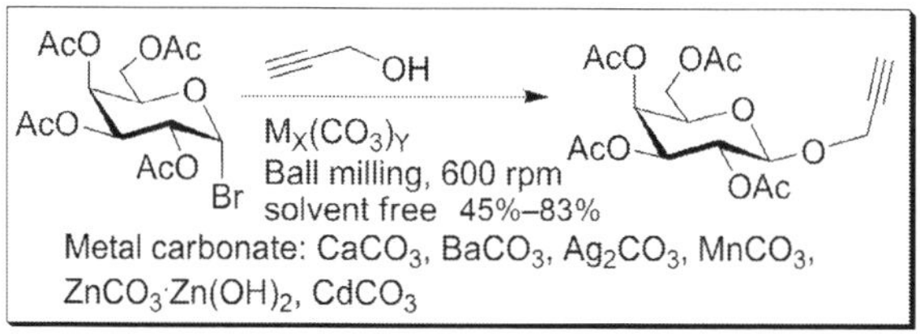

Scheme 71 Metal carbonate-promoted, solvent-free glycosylation using ball-milling technique.

Ball milling has been successfully applied for the preparation of glycosyl azides starting from glycosyl bromides or glycosyl chlorides and sodium azide under solvent-free conditions (Scheme 72).[164]

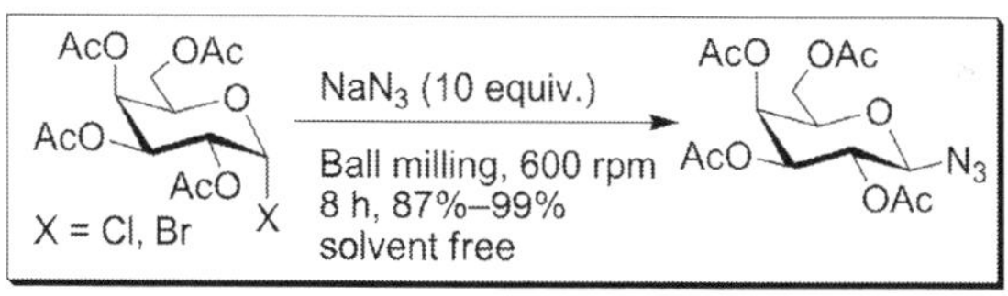

Scheme 72 Solvent-free preparation of glycosyl azides using ball-milling technique.

Efficient preparation of thioglycosides in high yield has also been achieved from glycosyl halides applying solvent-free ball milling mechanochemical conditions (Scheme 73).[165]

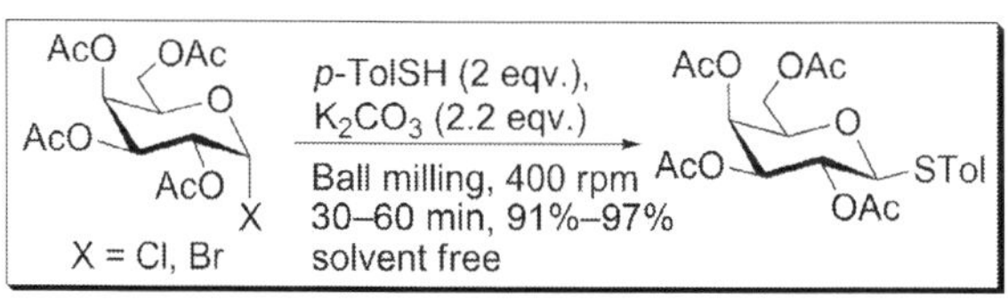

Scheme 73 Solvent-free, mechanochemical synthesis of aryl thioglycosides.

Kartha and co-workers[166] reported an efficient In(III) triflate-assisted removal of trityl group from carbohydrates, phenols, and alcohols under solvent-free mechanochemical conditions using ball milling. They have extended this method to a one-pot, two-step preparation of glycosides in excellent yield involving the removal of trityl group followed by reaction with an appropriate glycosyl donor under solvent-free conditions (Scheme 74). Furthermore, this group also reported highly efficient diastereoselective synthesis of thioglycosides by the reaction of glycosyl acetates with thiols in the presence of In(III) triflate by solvent-free grinding in a ball mill (Scheme 75).[167]

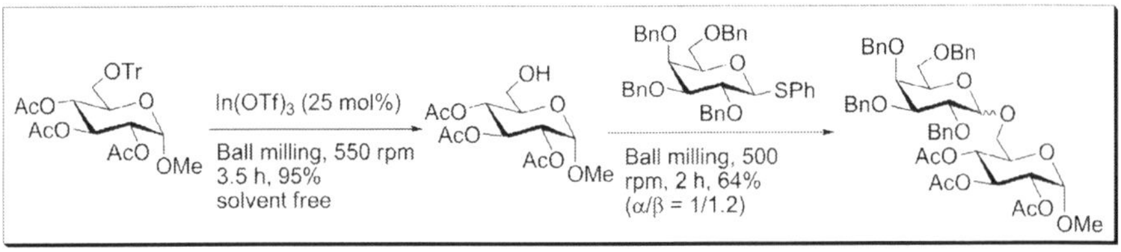

Scheme 74 One-pot detritylation and glycosylation using solvent-free ball-milling method.

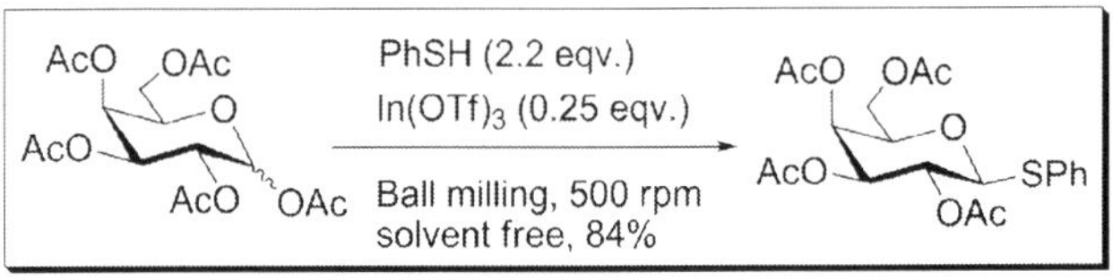

Scheme 75 Preparation of thioglycosides from sugar acetates using ball-milling conditions.

In another approach, Jerome and coworkers[168] reported the hydrolysis of cellulose followed by the production of butyl glycosides directly from cellulose in the presence of H_2SO_4 as catalyst under ball milling reaction setup (Scheme 76).

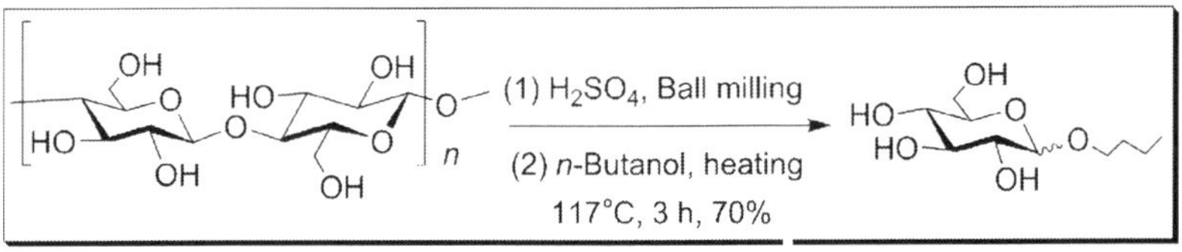

Scheme 76 Acid catalyzed ball milling of cellulose for the production of butyl glycosides.

Besides the examples discussed earlier, there are more reports[169] available on the application of ball milling technique in the transformation of carbohydrate derivatives and synthesis of oligosaccharides. Because of the advantages of this technique over the conventional approaches, ball-milling technique has become a useful alternative to carry out solvent-free chemical transformations of carbohydrates following Green chemistry principles.

13 Sonication-assisted transformations of carbohydrates

The application of ultrasound has emerged as one of the most useful alternative energy sources for the synthesis of carbohydrate-derived biologically and pharmaceutically potential compounds. Impressive advances have been made in the field of sonication-assisted organic reactions in the recent past.[170–172] This technique is quite effective in enhancing the reaction rate by decreasing the reaction time from days to minutes. Application of sonication in carbohydrate transformations has many advantages over the conventional approaches, which include superior yields, enhanced reactivity of the reactant, improved stereoselectivity, and shortened reaction times. Till date, several reports have appeared on the transformation of carbohydrates and synthesis of complex oligosaccharides with excellent stereoselectivity using ultrasound technology.[173, 174] In addition, sonication-mediated reactions have proven useful for a variety of chemo- and regioselective functional group transformations on carbohydrate substrates. Although a variety of sonication-mediated synthesis have appeared in the literature, some representative examples of functional group transformations of carbohydrates and synthesis of oligosaccharides under the influence of ultrasound are presented below.

14 Ultrasound-mediated functionalization of carbohydrates

Judicious functionalization of carbohydrates is one of the important criteria of the multistep synthesis of oligosaccharides. Since carbohydrates contain multiple hydroxyamine functionalities, several protecting groups have been developed for their functionalization, which include acyl groups, for example, acetyl, benzoyl, pivaloyl, chloroacetyl, and levulinyl as well as alkyl groups, for example, benzyl, allyl, and *p*-methoxybenzyl, together with some acetal functionalities such as benzylidene and isopropylidene.[175–179] In this context, Chang's laboratory[180] plays a leading role in applying sonochemical conditions in a wide variety of functional group transformations in carbohydrates. It was observed that acetylation of free sugars can be achieved in high yield in exceptionally short reaction time under sonication in comparison to several hours for conventional methods. In addition, acetylation of the hindered hydroxy groups present in di- and oligosaccharides such as trehalose and its derivatives can also be performed efficiently under sonication (Scheme 77).

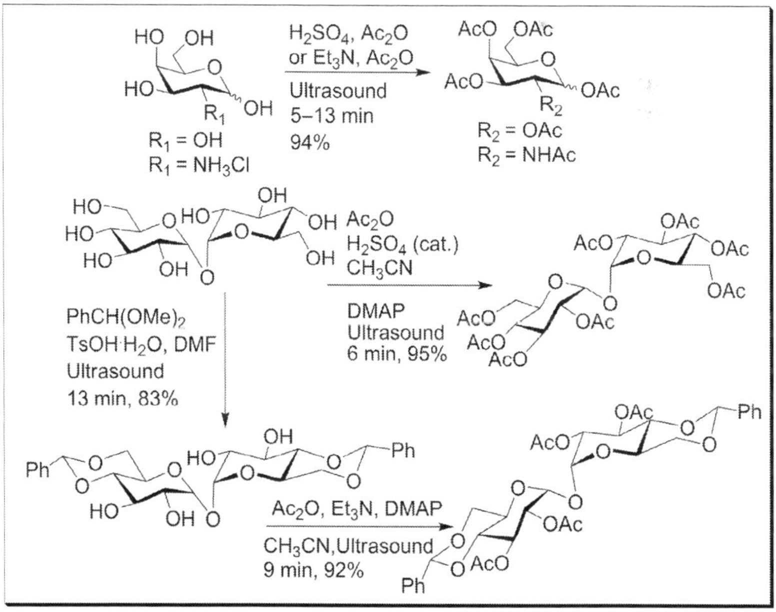

Scheme 77 Ultrasound mediated acetylation and benzylidenation of hydroxy groups of carbohydrates.

Sonochemical reaction has been successfully applied for the regio- and chemoselective removal of protecting groups in orthogonally functionalized carbohydrate derivatives. Sonochemical conditions are compatible with orthogonally protected substrates, which require mild conditions for the removal of the functional groups. In several instances, it was observed that removal of sterically hindered functional groups can be accomplished in a few minutes under sonication in comparison to the conventional reactions with longer reaction time (Scheme 78).[180, 181]

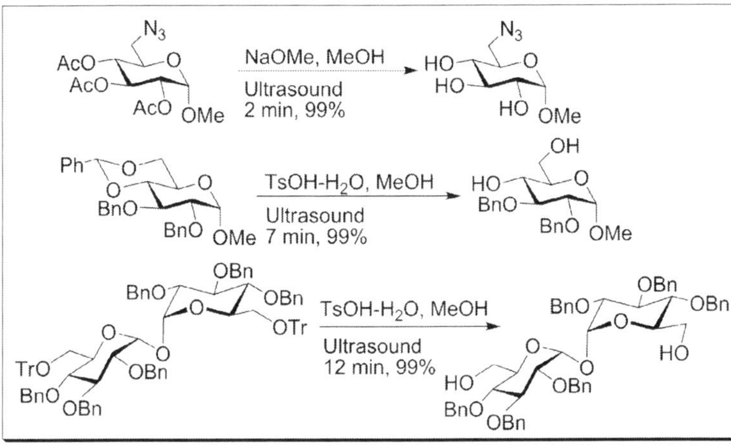

Scheme 78 Ultrasound mediated deprotection of functional groups in carbohydrates.[180]

Thioglycosides[70] have been extensively used as versatile glycosyl donors in glycosylations for the synthesis of a large number of linear and branched complex oligosaccharides. In general, they are prepared by the treatment of acetylated sugar derivatives with thiols in the presence of a Lewis acid (e.g., $BF_3 \cdot Et_2O$, TMSOTf, AgOTf, and TfOH). However, the conventional reactions require several hours or even a day. Chang and coworkers[180] achieved Lewis acid-mediated synthesis of thioglycosides from acetylated sugars in a few minutes under sonochemical conditions (Scheme 79). It is noteworthy that thioglycosides of 2-acetamido sugar derivatives can also be obtained in 2h instead of the conventional method with long reaction time.

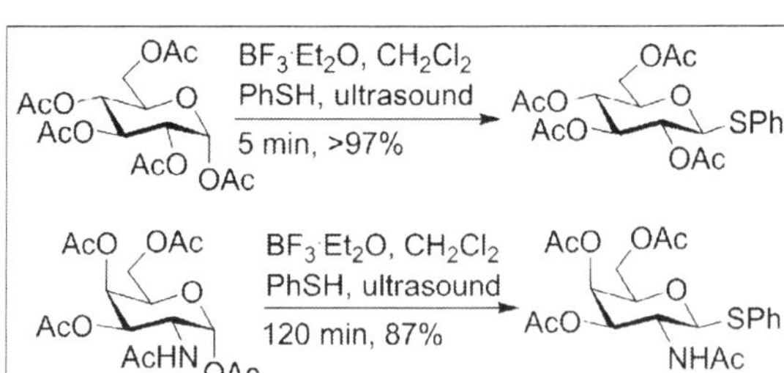

Scheme 79 Preparation of thioglycosides from sugar acetates under sonication.

Glycosyl azides can also be prepared from glycosyl bromides in significantly fast ultrasonic reaction conditions[180] without the requirement of any metallic salt as catalyst in comparison to the conventional reaction with long time at elevated temperature (Scheme 80).

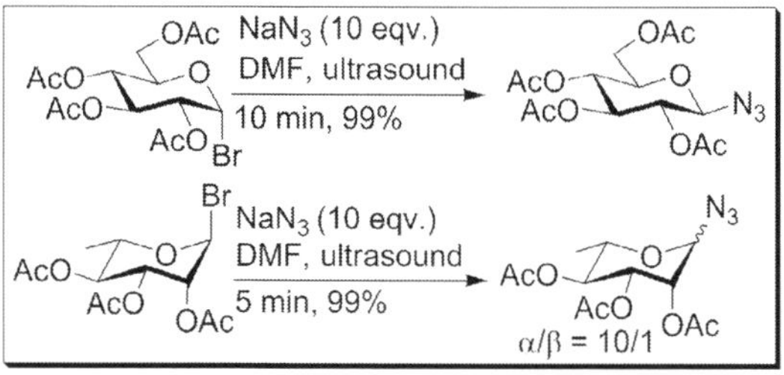

Scheme 80 Ultrasound-mediated preparation of glycosyl azides from glycosyl bromides.

Stereoselective glycosylation reactions play pivotal roles in the synthesis of complex oligosaccharides. Conventionally, a large variety of reaction conditions (such as use of different glycosyl donors and activators) have been developed for the preparation of complex oligosaccharides. In some cases, the glycosylation reactions require longer reaction time and proceed without stereoselectivity. Application of ultrasound in the stereoselective glycosylations was found encouraging as it reduced the reaction time significantly with satisfactory yield and stereochemical outcome in the products. A variety of glycosyl donors such as sugar acetates, thioglycosides, glycosyl bromides, and glycosyl trichloroacetimidates have been used in the sonochemical glycosylations in the presence of appropriate activators and the products were obtained within few minutes (Scheme 81).[181]

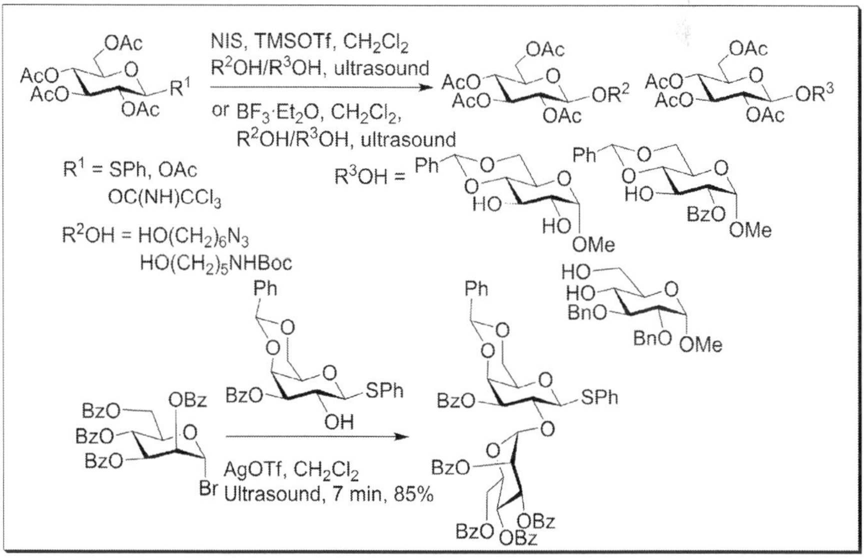

Scheme 81 Ultrasound-mediated stereoselective glycosylation reactions.

Significantly fast derivatization of carbohydrates has been efficiently performed in high yield by Srinivasan et al.[182] under sonochemical conditions using ionic liquid as a green solvent (Scheme 82).

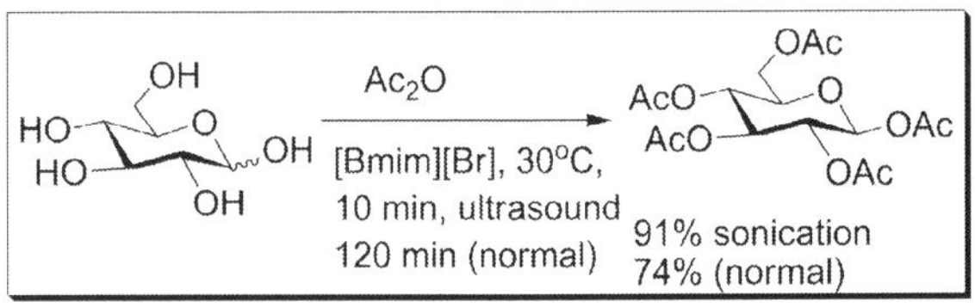

Scheme 82 Acetylation of glucose under sonication using ionic liquid as solvent.

Since chemical synthesis of oligosaccharides often requires selective protection-deprotection of carbohydrate intermediates, several selective functional group transformations have been achieved under sonochemical conditions. Selective removal of 4,4′-dimethoxytrityl ether (DMT) from a methyl glycoside using a combination of MeOH and CCl_4 has been successfully achieved in 89% yield under sonication without affecting the glycosidic bond, whereas normal reaction condition resulted in the degradation of glycosidic linkage (Scheme 83).[173]

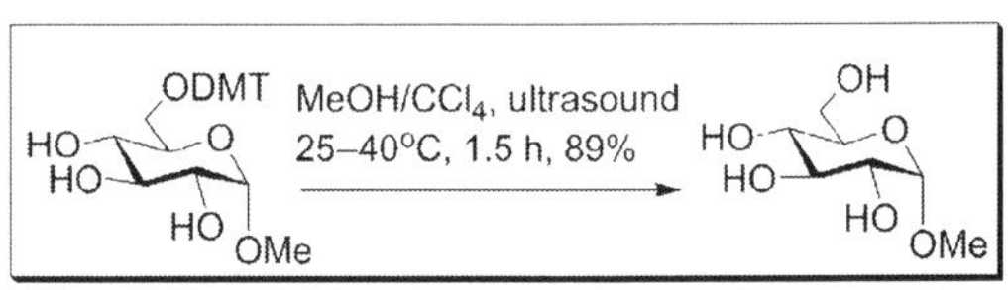

Scheme 83 Ultrasound-mediated selective removal of DMT group in carbohydrate.

Migration of acyl group from one vicinal hydroxy group to the other is often observed during the functionalization of carbohydrate intermediates.[183, 184] Recently, migration of acyl group in carbohydrate compounds was reported by Deng and group[180, 185] using silver oxide and TBAI under ultrasound-assisted reaction. Acetyl-migrated carbohydrate derivatives have been prepared in few minutes adopting sonochemical approach, in comparison to the conventional reaction conditions (Scheme 84).

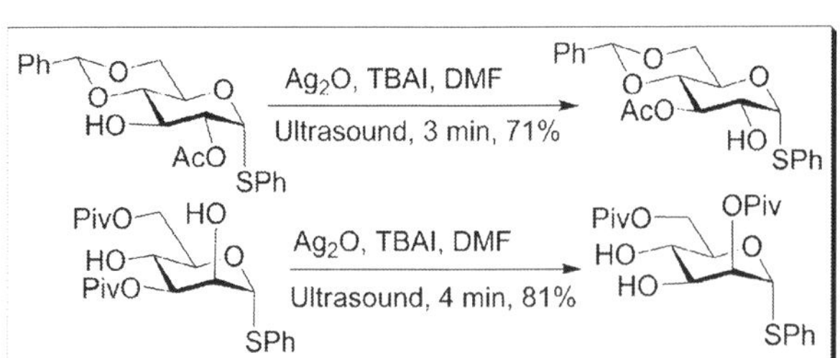

Scheme 84 Ultrasound-mediated migration of acyl groups.

Tatina et al.[186] developed 2,4,6-trichloro-1,3,5-triazine (TCT) catalyzed synthesis of arylidene acetals of a number of monosaccharides. Methyl α-D-glucopyranoside was reacted with aryl aldehyde dimethylacetal in the presence of TCT under sonication for 10 min to furnish 4,6-*O*-arylidene acetals in > 99% yields (Scheme 85), whereas the same reactions required 3 h in the absence of ultrasound. Further acetylation of arylidene dimethylacetal derivative using acetic anhydride catalyzed by TCT furnished acetylated product under sonication at 60°C for 10 min only. A wide variety of monosaccharide derivatives were functionalized using similar reaction conditions.

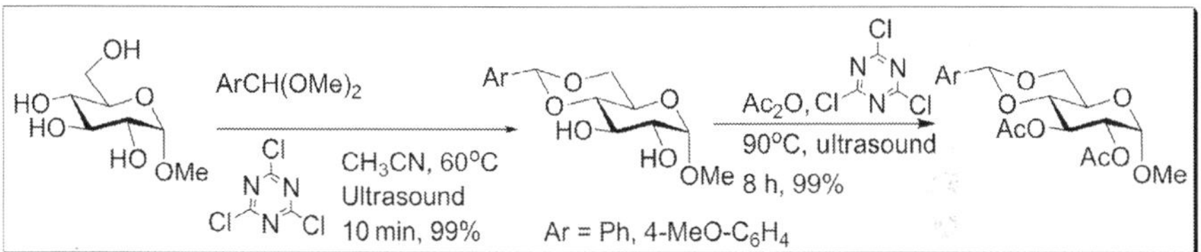

Scheme 85 Ultrasound-mediated synthesis of arylidene acetal derivatives and acetylation.

Zhao et al.[187] reported high yielding preparation of glycosyl orthoester derivatives from glycosyl bromides in significantly less time using sonochemical conditions (Scheme 86).

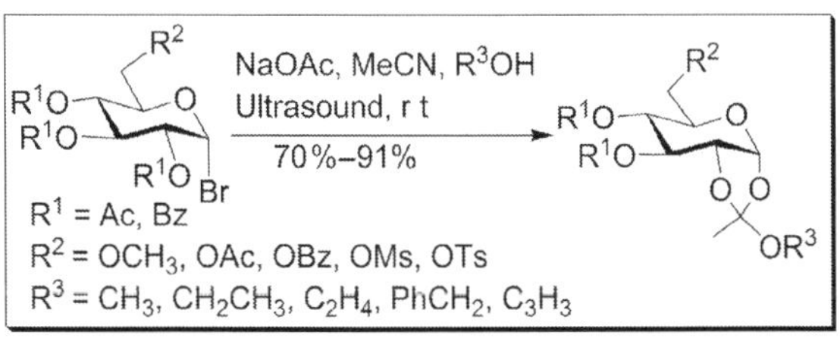

Scheme 86 Ultrasound-mediated preparation of glycosyl orthoesters from glycosyl bromides.

Sonochemical conditions have been successfully applied in the synthesis of complex oligosaccharides also. Recently, Galan et al.[188] reported ultrasound-mediated synthesis of a series of mucin-type oligosaccharide fragments containing α-linked azidopropyl spacers, which involve rate enhancement of several intermediate reaction steps (Scheme 87).

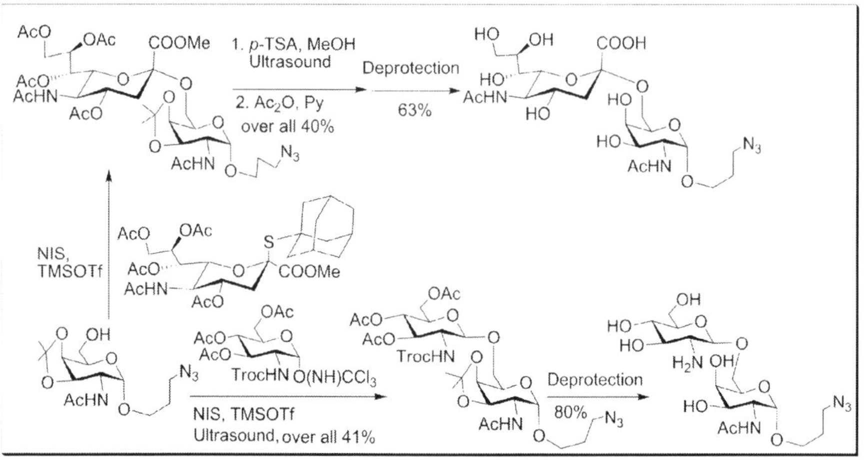

Scheme 87 Synthesis of mucin-type oligosaccharide fragments under sonication conditions.

Chang and coworkers[189] synthesized several medicinally important oligomannoside fragments with appropriate stereochemistry at the glycosyl linkages by sonochemical glycosylations in exceptionally short reaction time (Scheme 88).

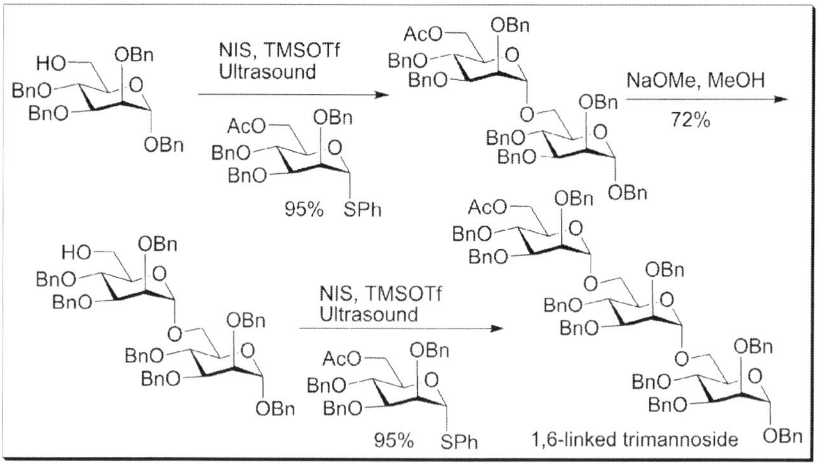

Scheme 88 Synthesis of an oligomannoside under sonication conditions.

Tony Mong and coworkers[190] reported the rate enhancement of NIS-TMSOTf-mediated activation of thioglycoside by the addition of DMF as additive under ultrasound irradiation. The influence of DMF on the reaction rate was observed in the absence of ultrasound as well. However, application of sonochemical conditions furnished much better yield of the products (Scheme 89).

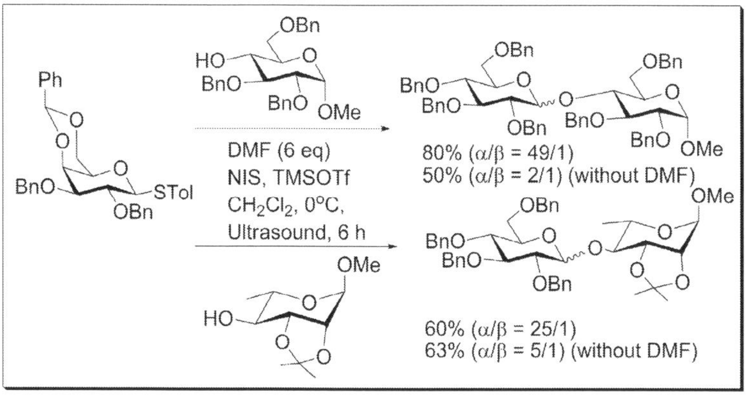

Scheme 89 Rate enhancement of the glycosylation by addition of DMF under sonication.

Shaikh et al.[191] reported Fischer glycosylation using ultrasonic technique accelerating the reaction time with good yield. Queneau and coworkers[192] developed a new technique for the oligomerization of D-glucose avoiding protection-deprotection steps. During the quest of protection group free preparation of glycosides by montmorillonite clay catalyzed Fischer glycosylation of free sugars, they ended up with the formation of self-glycosylated oligomer under ultrasound irradiation (Scheme 90).

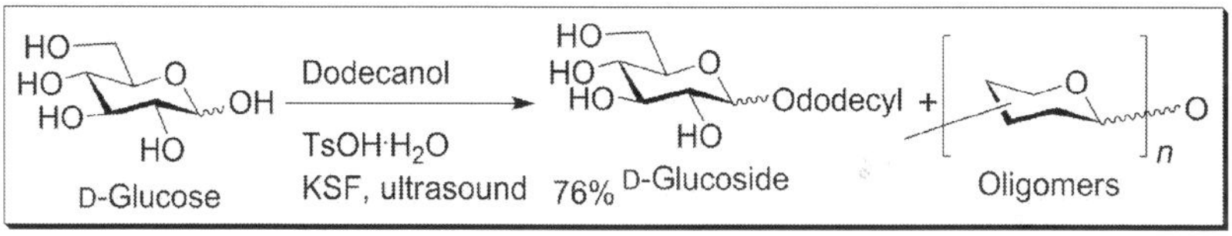

Scheme 90 Oligomerisation of D-glucose under sonication conditions.

Preparation of *C*-glycosyl linkage is one of the important areas of research in carbohydrate chemistry due to its potential application for the synthesis of bioactive compounds. Murphy and coworkers[193] observed significant rate enhancement in the TMSOTf catalyzed preparation of allyl and allenyl *C*-glycosyl compounds from glycosyl acetate or methyl *O*-glycoside under ultrasound irradiation. For example, the allylation of methyl 2,3,4,6-tetra-*O*-benzyl-α-D-glucopyranoside in the presence of TMSOTf and allyltrimethylsilane was completed within 15 min under ultrasound radiation, whereas the reaction was incomplete after 2 h using conventional heating (Scheme 91).

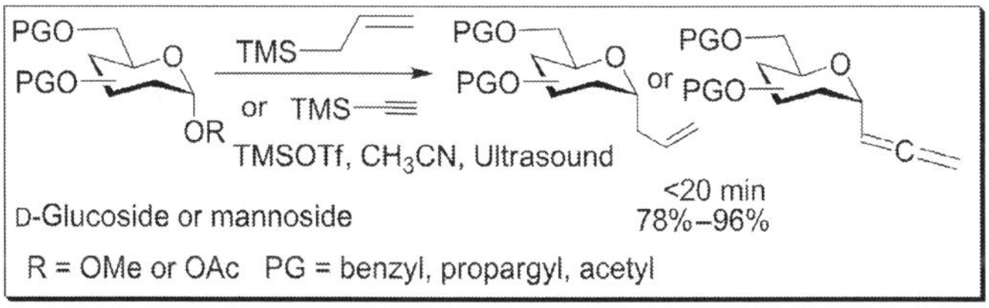

Scheme 91 Ultrasound-mediated synthesis of *C*-glycosyl derivatives.

The benefits of the sonochemical approach include better yields, shortening of reaction times, improved chemo-, regio-, and stereoselectivities for glycoside and oligosaccharide synthesis, and compatibility with different protection-deprotection reactions involved in multistep synthesis. In addition to the aforementioned examples ultrasound-induced energy, generated from acoustic cavitation has also been applied to click reactions (Scheme 92),[194, 195] enzyme-catalyzed esterification,[196] selective reduction of anomeric azides to amines[197]on carbohydrate substrates, as well as in the synthesis of natural product like molecules of medicinal interest (Scheme 93).[198] Indeed, ultrasound has emerged as a nonconventional technique with enormous scope in the synthetic carbohydrate chemistry.

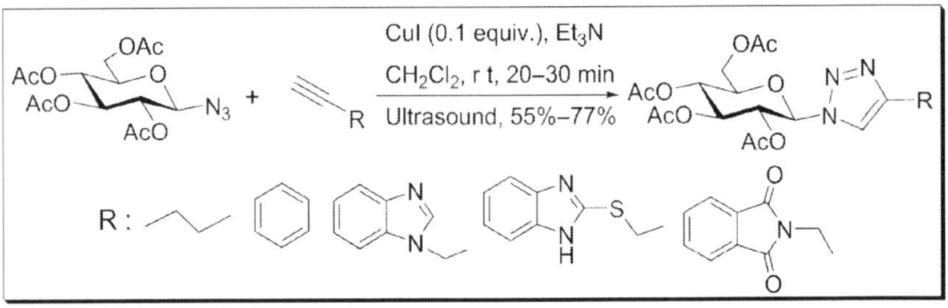

Scheme 92 Preparation of glycosylated triazole derivatives by the cycloaddition reaction of alkynes and azide derivatives under sonication.

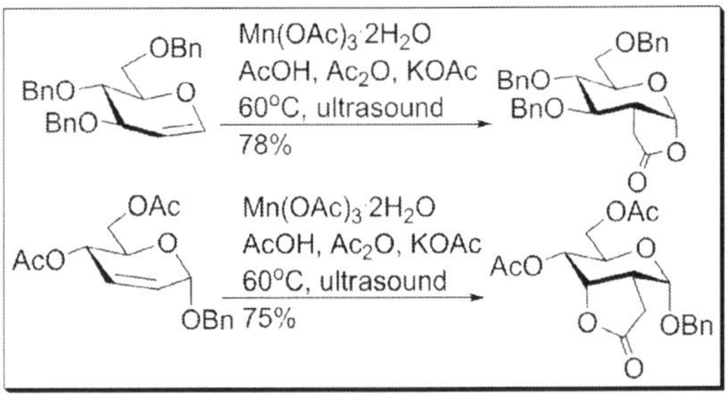

Scheme 93 Preparation of bicyclic carbohydrate γ-lactones under sonication conditions.

15 Transformation of carbohydrates under photoinduced reactions

In search of greener reaction conditions for the chemical transformations of carbohydrates, photoinduced reactions have attracted attention by virtue of their eco-friendliness and versatility such as mild conditions at room temperature, requirement of less anhydrous conditions, sufficient activation by using stoichiometric quantities of reagents, high yield, and low waste production.[199] In general, the photocatalytic reactions are often induced by the presence of a photocatalyst in the system. The interaction between an electronically excited photocatalyst and organic molecules generate diverse classes of reactive intermediates, which are utilized in a variety of reactions to furnish diverse classes of molecules. Although a wide variety of organic transformations have been carried out using photoinduced reaction,[200] their applications in the transformation of carbohydrates or synthesis of oligosaccharide are quite limited.[201] However, in the recent past a few reports have appeared in the literature describing the transformation of carbohydrates in photoinduced reactions.

16 Photoinduced glycosylation

In 1990, Griffin et al.[202] first introduced photochemical preparation of glycosyl cations from thioglycosides. Photoinduced cleavage of anomeric arylthio groups from the thioglycosides in the presence of an electron transfer agent, for example, 1,4-dicyanonaphthalene (DCN) resulted in the formation of glycosyl cations, which are reactive glycosylating agents. A wide range of oligosaccharides have been synthesized using the intermediate thioglycosides of monosaccharides followed by photochemical cleavage of the sulphide groups (Scheme 94).

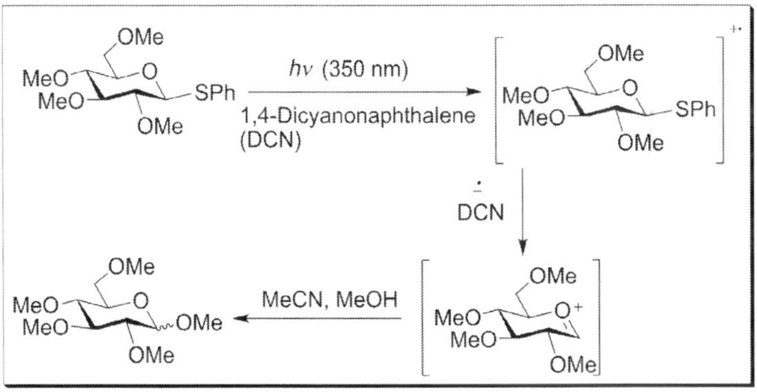

Scheme 94 Photoinduced glycosylation via the formation of glycosyl cation.

Later, Nakanishi et al.[203] reported a reasonably green approach for photochemical glycosylations, which involved photoinduced cleavage of anomeric arylthio groups of unprotected thioglycosides in the presence of 2,3-dichloro-5,6-dicyano-*p*-benzoquinone (DDQ) under long UV irradiation to furnish *O*-glycosides. Chances of the production of self-coupled oligosaccharides or 1,6-anhydro sugar derivatives have been significantly reduced by using boronic acid as a temporary protecting groups (Scheme 95).

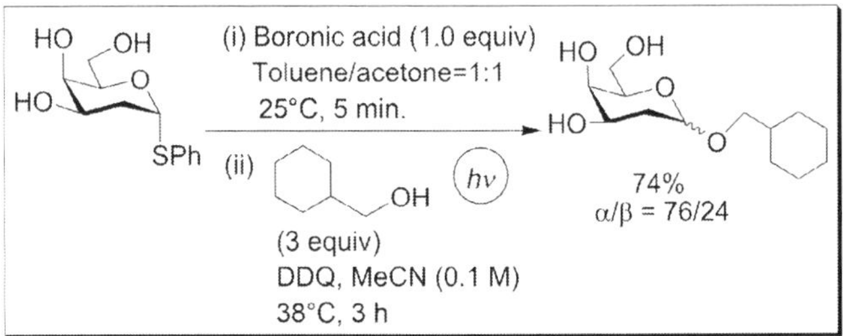

Scheme 95 Photoinduced glycosylation using DDQ as photocatalyst.

In addition to the development of UV light-induced photochemical reactions, attempts were made to develop visible light-induced glycosylations also. In this quest, Ragains and coworkers prepared *O*-glycosides by irradiating 4-(4-methoxyphenyl) but-3-enyl thioglucoside donors under visible light in the presence of Umemoto's reagent[204] (a triflate salt) as a photocatalyst to generate glycosyl cations (Scheme 96).[205] The reaction was found equally effective for the activation of thio- and selenoglycosides under mild user-friendly conditions without the requirement of photosensitizer.[206] Mao et al. reported the activation of thioglycosides using a light-driven strategy via the involvement of a CF_3 radical and subsequent glycosylation with glycosyl acceptors (Scheme 97).[207] Both UV and visible light or even sunlight were used as the source of light. The high efficiency of this light-driven glycosylation protocol has been further extended by the rapid one-pot sequential assembly of oligosaccharides.

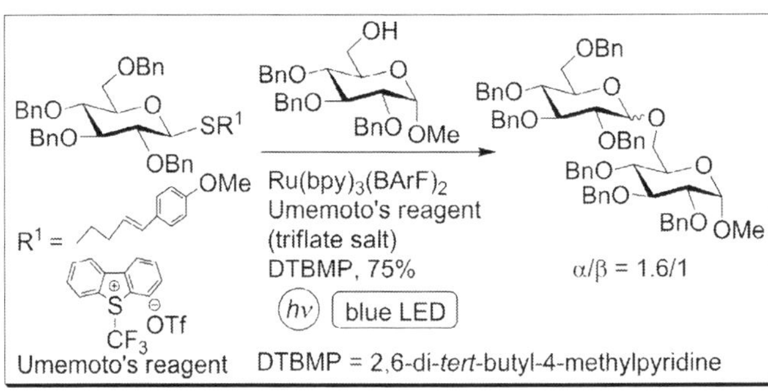

Scheme 96 Photoinduced *O*-glycosylation in the presence of Umemoto's reagent.

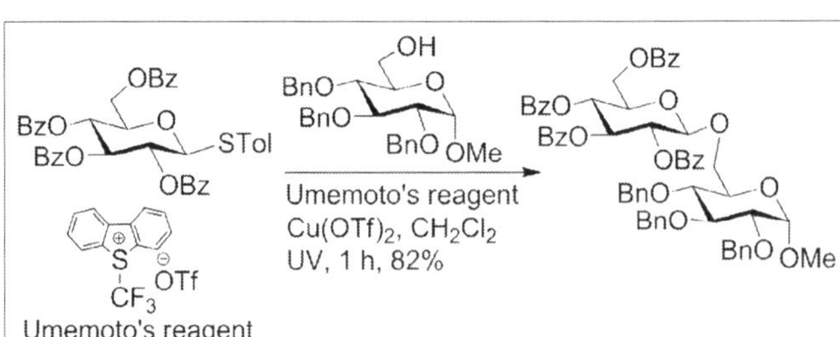

Scheme 97 Photoinduced 1,2-*trans*-*O*-glycosylation in the presence of Umemoto's reagent.

Ye and coworkers[208] demonstrated unique light-driven activation of thioglycosides by thiophilic activator such as $Cu(OTf)_2$ toward the preparation of *O*-glycosides. The cleavage of the C-S bond took place under UV irradiation to form glycosyl radical, which was oxidized to form oxycarbenium ion in the presence of an oxidant and finally *O*-glycoside was formed by nucleophilic addition of an alcohol. This glycosylation protocol differs from the conventional light-mediated glycosylation because of the formation of *S*-radical during the course of the reaction, which was proved by the inhibition of the glycoside formation in the presence of a radical scavenger TEMPO [(2,2,6,6-tetramethylpiperidin-1-yl)oxyl radical]. During the reaction, $Cu(OTf)_2$ acted as an oxidant, which was evident from the reduction of its quantity as the reaction proceeded toward its completion (Scheme 98). Besides the primary hydroxy acceptor, sialic acid glycosylation with a variety of glycosyl acceptors furnished sialylated products in very good yields (72–85%) with significant stereoselectivity.

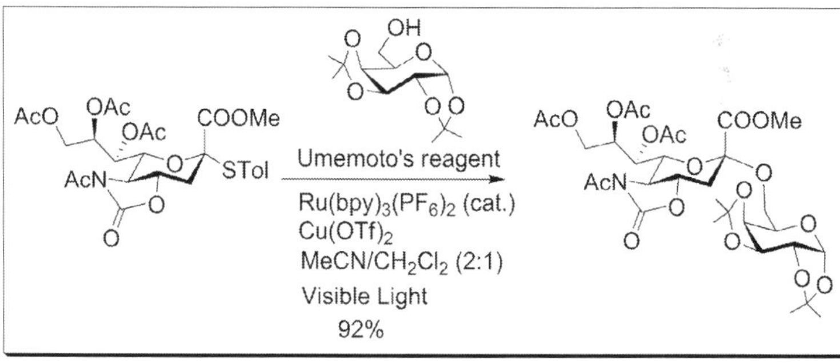

Scheme 98 Strategy for the light-induced activation of thioglycosides and glycosylation.

Bowers and coworkers[209] developed the *O*-glycosylation of thioglycosides activated via visible light. Visible light catalysis permits the efficient formation of single-electron-transfer (SET) redox cycles. Mechanistic studies indicate that the full reaction is light sensitive, and it occurs through a mechanism involving the decomposition of an oxidatively formed sulfur radical cation and propagation via the reduction of the thiol as a side product (Scheme 99). It was demonstrated that the metal-ligand complex is generated by oxidative quenching of an excited state visible light catalyst, such as $Ru(bpy)_3Cl_2$ or $Ir[dF(CF_3)ppy]_2(dtbbpy)PF_6$.

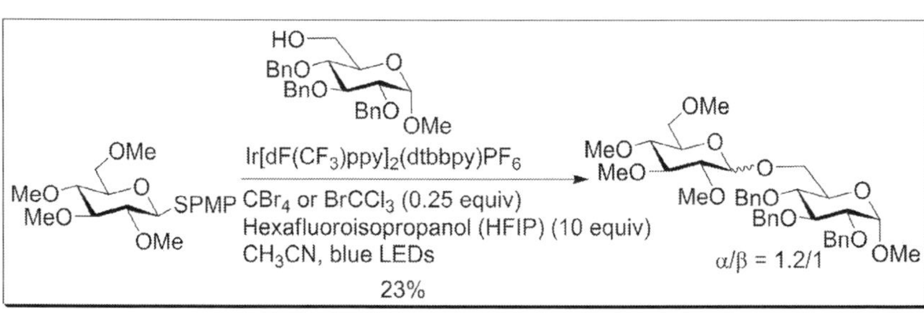

Scheme 99 *O*-Glycosylation of thioglycosides activated by visible light.

Recently, Crich et al.[210] reported photocatalytic formation of glycosidic bonds via the formation of glycosylated 2,2,6,6-tetramethylpiperidin-1-ol (TEMPOL) in the presence of an iridium-based photocatalyst and blue LEDs. The reaction proceeds at room temperature without requirement of any electrophilic additives.

Andrews et al.[211] investigated the scope of photochemical reduction of glucosyl halide to generate glucosyl radicals,[212] which could be reduced to 1-deoxy sugar (anhydro alditol) derivative by a trapping agent such as ^{t}BuSH. A combination of $[Ru(bpy)_3]^{2+}$ and *N,N*-diisopropylethylamine (DIPEA) has been used as the stoichiometric reducing agent to debrominate glucosyl bromide to glucosyl radical (Scheme 100).[213]

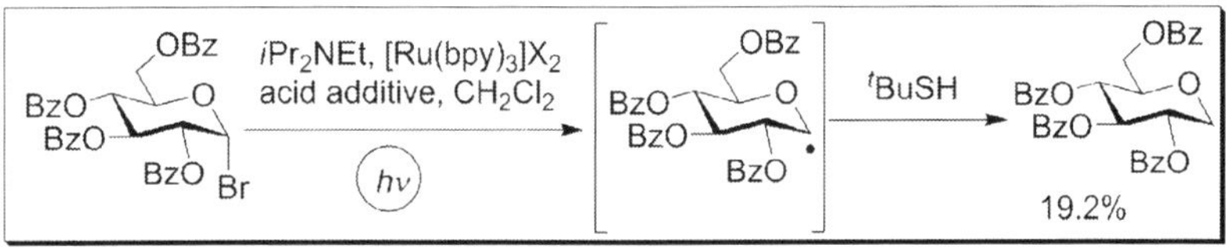

Scheme 100 Photoinduced reduction of glucosyl radical.

Andrews et al.[214] and Giese et al.[215] reported a more efficient, light-mediated conjugate addition of glycosyl radicals to acrolein in the presence of $[Ru(dmb)_3]^{2+}$ as a catalyst (Scheme 101).

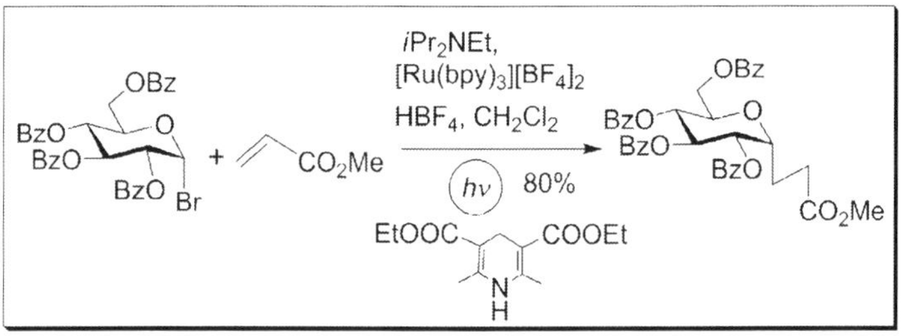

Scheme 101 Continuous photoredox-mediated synthesis of *C*-glycosyl compounds in photoflow reactor.

Recently, photoinduced synthesis of 2-iodo, 2-deoxy, and 2,3-unsaturated glycosides was reported by Mukherjee and coworkers.[216] Treatment of glycals with peroxides in the presence of NIS at room temperature furnished 2-deoxy-2-iodo-α-glycosides in good yields with predominant *trans*-diaxial selectivity. Surprisingly, 2-deoxyglycosides or 2,3-unsaturated-α-glycosides (Ferrier product) were formed as predominant products under photocatalytic conditions in the presence of a catalytic amount of Eosin Y (Scheme 102).

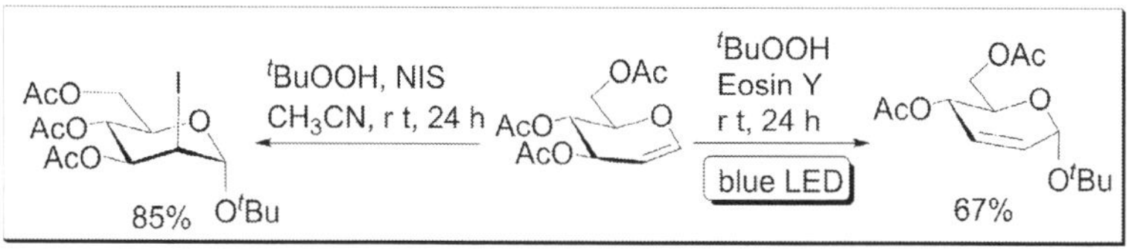

Scheme 102 Photoinduced synthesis of 2-deoxy-2-iodo, 2-deoxy and 2,3-unsaturated glycosides.

Toshima and coworkers[217, 218] reported photoinduced glycosylation of glycosyl trichloroacetimidates catalyzed by organophoto acids. Organophoto acids exhibit a unique property as their acidity increases in the photoexcited state and hence can be used to control glycosylation by light switching. Urea and thiourea derivatives acted as organophoto acids in the photoinduced glycosylation of alcohols with α-D-glucosyl trichloroacetimidate under long wavelength UV irradiation to give satisfactory yields of glycosides. In addition, high β-stereoselectivity was attained using higher concentrations of thiourea as the organophoto acid, whereas high α-stereoselectivity was achieved under low concentrations (Scheme 103).[219]

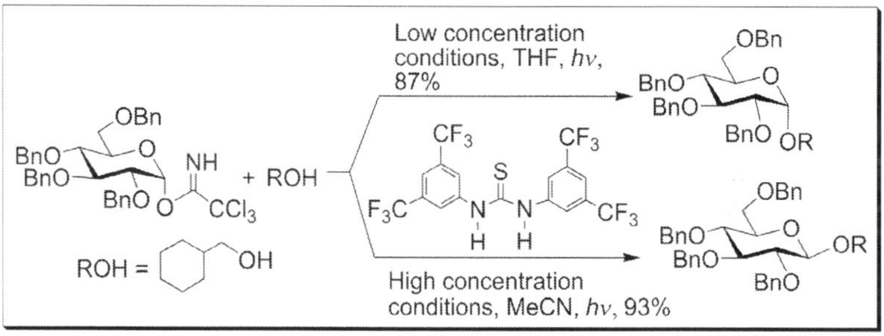

Scheme 103 Stereocontrolled photoinduced glycosylation of alcohols with α-glucosyl trichloroacetimidates using aryl thiourea as an organophoto acid.

17 Photoinduced synthesis of *S*-linked glycoconjugates

Fiore and coworkers[220] conceptualized the photoinduced thiol-ene coupling[221] as a click ligation tool for thiodisaccharide synthesis. A representative reaction was conducted at room temperature, without deoxygenation of the reaction mixture and under irradiation from a UV-visible lamp (λ_{max} = 420 nm, 40 W) in the presence of 2,2-dimethoxy-2-phenylacetophenone (DPAP) as the photoinitiator (Scheme 104).

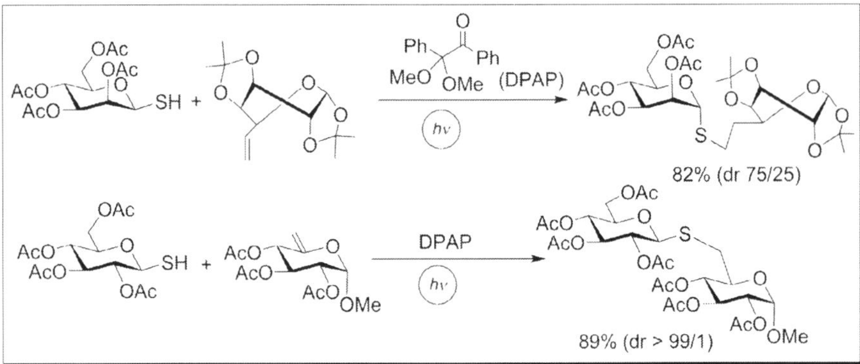

Scheme 104 Optimization of the photoinduced reactions of thiol with alkene.

The synthesis of a new family of 1-deoxy *S*-disaccharides was established by Staderini and coworkers[222] via the free radical hydrothiolation of glycals by sugar thiols (Scheme 105). Davis and coworkers[223] also described the development of a convergent approach for the synthesis of a novel class of *S*-linked glyconjugate proteins through the site-specific ligation of 1-thioglycoses to proteins under photochemical conditions.

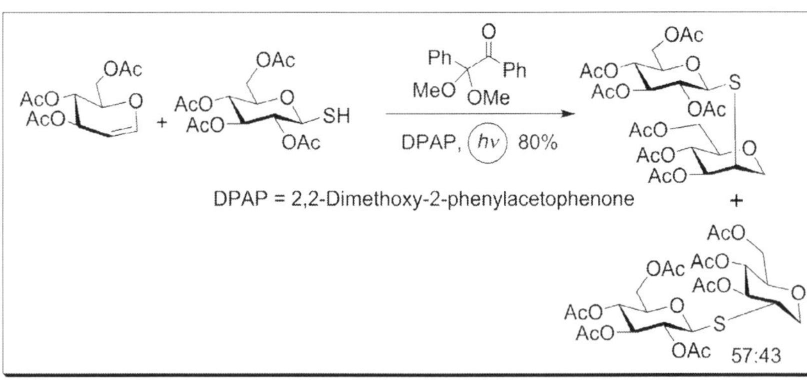

Scheme 105 Synthesis of *S*-disaccharides via the hydrothiolation of glycal.

A different study on the thiol-yne coupling (TYC) reaction was carried out by Massi and coworkers using a glycosylated alkyne and *N*-acetyl-L-cysteine methyl ester) (Scheme 106).[224]

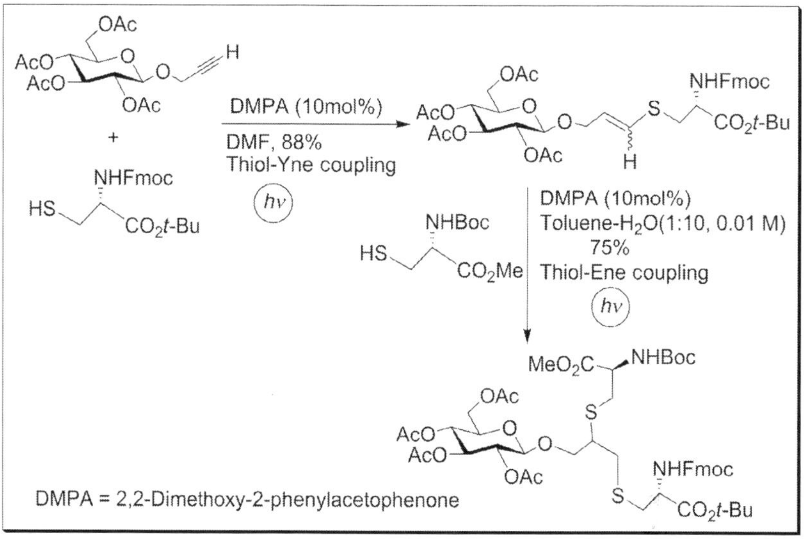

Scheme 106 Photoinduced glycosylation of cysteine-containing peptides.

18 Electrochemical glycosylation

Electrochemistry has also been successfully applied for stereoselective glycosylations.[225] Noyori et al. demonstrated electrochemical glycosylation for the first time in 1986.[226] The electrochemical glycosylations of aryl glycosides have been carried out using protected as well as unprotected glycosyl donors. In 1990, Sinaÿ[227a] and Lubineau[227b] independently reported electrochemical glycosylations using thioglycosides, which have lower oxidation potentials than those of aryl glycosides (Scheme 107).

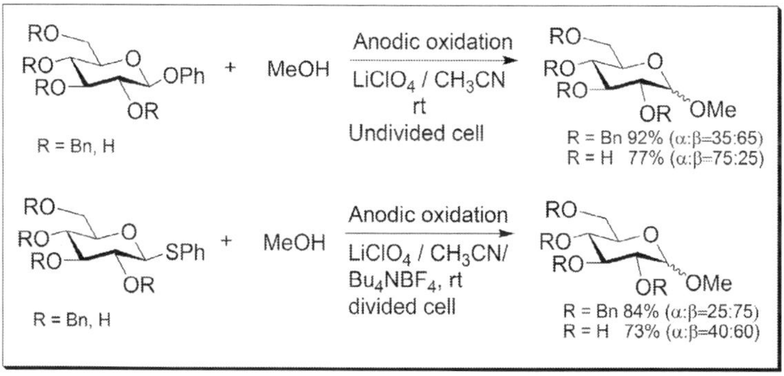

Scheme 107 Electrochemical glycosylations of protected and unprotected aryl glycosides glycosyl donors.

Later, other chalcogenoglycosides, such as seleno- and telluroglycosides, were used as glycosyl donors in the electrochemical glycosylations.[228–232] In 1998, Pinto and coworkers reported chemically induced, radicalcation-initiated glycosylation reactions with phenyl selenoglycosides, using tris(4-bromo phenyl)aminium hexachloroantimonate, ($(p\text{-}BrC_6H_4)_3N^+SbCl_6^-$, BAHA), as the single electron transfer reagent (Scheme 108).[229]

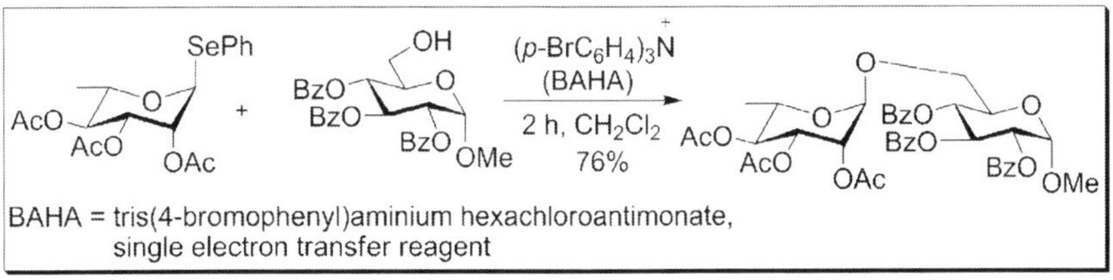

Scheme 108 BAHA-mediated glycosylations with selenoglycoside donors.

Sinaÿ and coworkers[233] have reported similar glycosylations with ethyl thioglycosides (Scheme 109), and this group has subsequently applied the method to the synthesis of oligosaccharides.[234] However, thioglycosides have been more popular than heavy chalcogenoglycosides because of their ease in preparation and handling.[233–242]

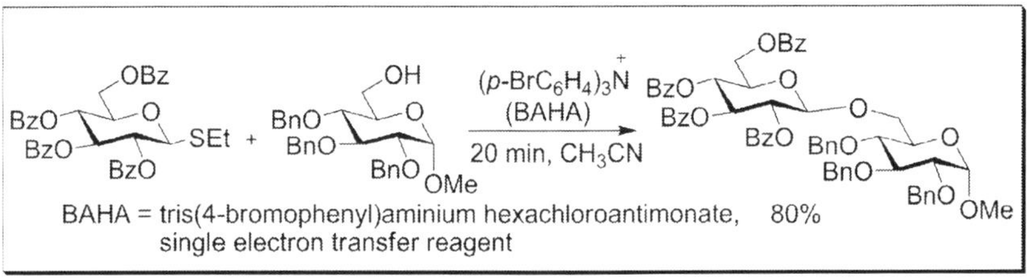

Scheme 109 BAHA-mediated glycosylations with thioglycoside donors.

Despite these revolutionary works, there was little information on the structures and reactivities of the glycosylation intermediates generated by the electrochemical method, and there were only a few examples of oligosaccharide synthesis using such intermediates. Recently reported electrochemically generated glycosylation intermediates and their applications to the oligosaccharide synthesis are presented later.

Nokami et al. performed electrochemical oxidation of thioglycosides in the absence of a glycosyl acceptor using Bu_4NOTf as a supporting electrolyte at low temperature to generate glycosyl triflates, which acts as the donor in the subsequent glycosylations (Scheme 110).[243–247]

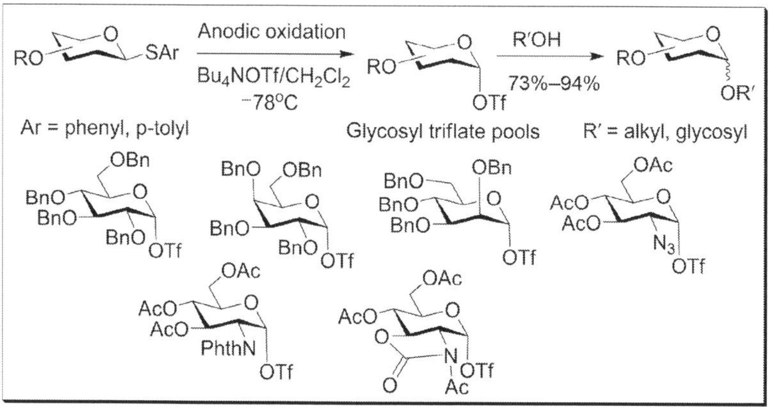

Scheme 110 Electrochemical generation of various types of glycosyl triflates as glycosyl triflate pool.

The successful generation of highly reactive glycosyl triflates by the "glycosyl triflate pool" approach encouraged to use it as a precursor for other glycosylation intermediates. In addition, glycosyl sulfonium ions, which are useful for α-selective glycosylation, could be prepared by the reaction of glycosyl triflates with diorganosulfides (RSR′).[248–250] Indeed, glycosyl triflates were quantitatively converted into the corresponding glycosyl sulfonium ions by the reaction with diorganosulfides. 2-Azido-2-deoxyglycosyl triflate gave a mixture of α and β isomers of the glycosyl sulfonium ion, whereas the β-isomer of glycosyl sulfonium ions were achieved as a single isomer from 2-deoxy-2-phthalimidoglycosyl triflate because of the neighboring group participation of the 2-phthalimido group.

During the course of study, it was observed that the rate of conversion of the α-isomer of the glycosyl sulfonium ion into the corresponding methyl glycoside was faster than that of the β-isomer[248] indicating the α-glycosyl sulfonium ion to be more reactive than its β-isomer.[244] Appearance of the large coupling constants of the axial protons in the ^1H NMR spectra confirmed the β-isomer having conventional $^4C_1$ chair conformation. In contrast, the distorted conformation of the pyranoid ring of the α-isomer is proposed as a major reason for its higher reactivity (Scheme 111).

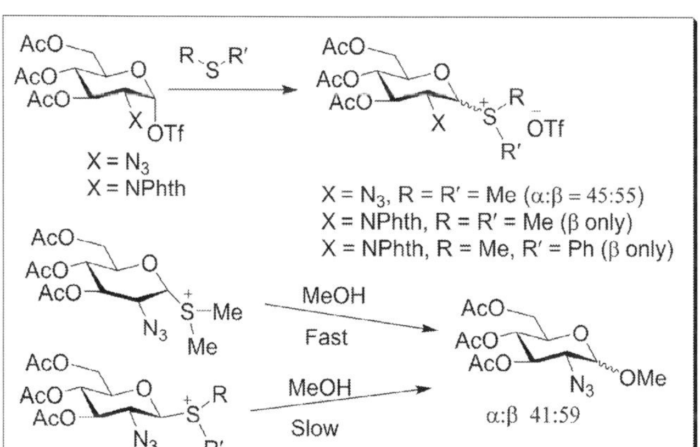

Scheme 111 Conversion of glycosyl triflates into glycosyl sulfonium ions and their reactivities.

Glycosyl sulfonium ions are stable enough to be observed spectroscopically at low temperature and are reasonably reactive at elevated temperature.[249] Therefore, glycosyl sulfonium ion could be stored at low temperature and glycosylations with various acceptors, were performed at – 10°C. One of the benefits of glycosyl sulfonium ions as a glycosyl donor is that a thioglycoside is applicable as a glycosyl acceptor thus the disaccharide obtained is also useful as the next glycosyl donor[248] (Scheme 112).

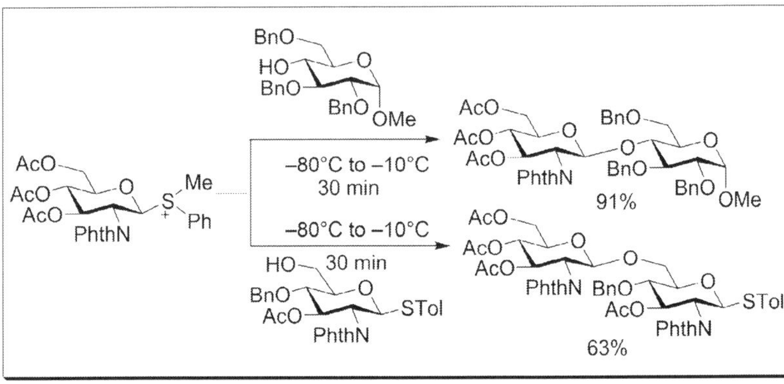

Scheme 112 Application of glycosyl sulfonium ions as glycosyl donors.

To achieve the elongation protocol of oligosaccharides based on the one-pot anodic oxidation (electrochemical glycosylation)-deprotection sequence, Yoshida and coworkers[250]used a glycosyl donor equipped with a temporary protecting group that was stable during the anodic oxidation. Among various protecting groups, 9-fluorenylmethyl carbonate (Fmoc) was found to be appropriate as a temporary protecting group of the glycosyl donor. The Fmoc group was stable under acidic condition and anodic oxidation; however, the Fmoc group was easily cleaved in a one-pot reaction using triethylamine (Et_3N) as a base after completion of the electrochemical glycosylation. Based on this method, synthesis of linear (1→6)-β-oligoglucosides was performed in 86% average yield. The glycosidic bonds were formed with β-selectively because of the neighboring group participation of the 2-*O*-benzoyl group (Scheme 113).

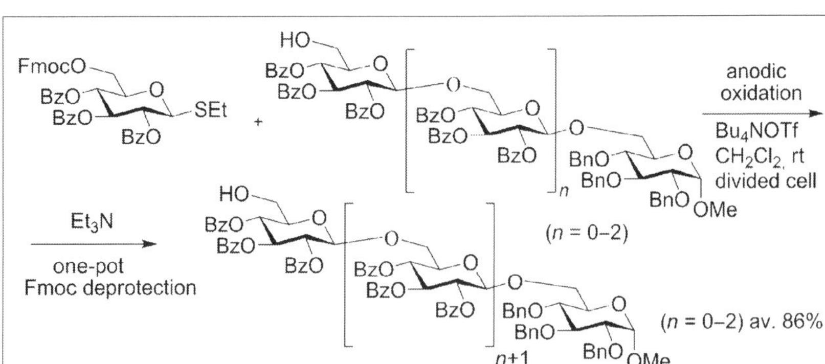

Scheme 113 One-pot electrochemical synthesis of oligosaccharide.

Nishiyama and coworkers[251] developed novel electrochemically induced glycosylation reactions using glucopyranose orthobenzoate derivative as glycosyl donor with a variety of glycosyl acceptors such as sugars, steroids, and adamantanes promoted by an electrochemically generated acid (EGA) produced by the anodic oxidation of alcohol (Scheme 114).

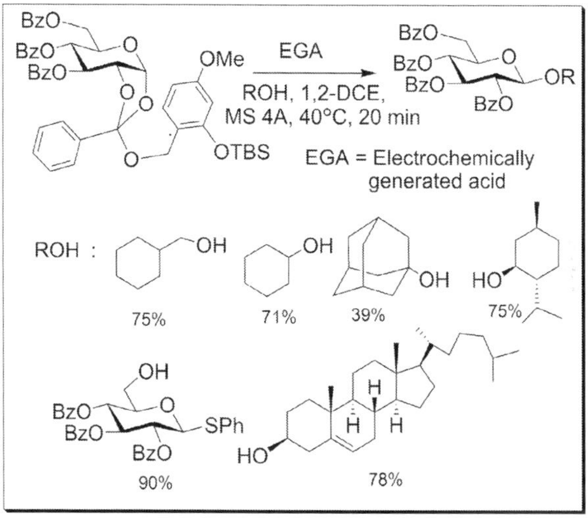

Scheme 114 Glycosylation promoted by electrochemically generated acid (EGA).

Yoshida and coworkers[252] electrochemically generated $ArS(ArSSAr)^+$, which was found to be an effective activator for glycosylation of thioglycosides. It was observed that efficiency of the reaction strongly depends on the nature of the counter anion, and $B(C_6F_5)_4^-$ was the most effective. The method was applicable to various glycosyl donors such as thiogalactosides and thiomannosides, glycosylated with a number of acceptors with secondary hydroxy groups (Scheme 115).

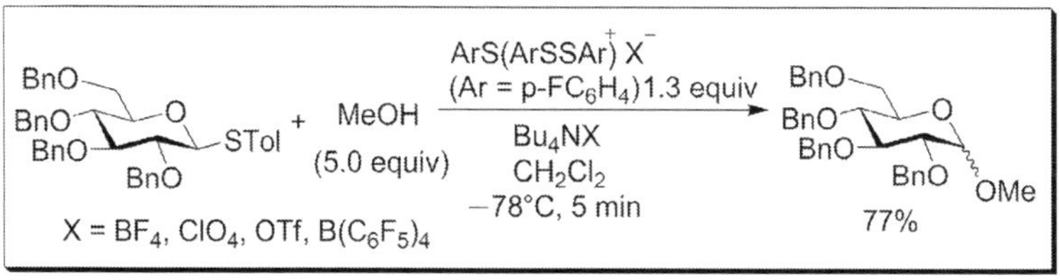

Scheme 115 Electrochemically generated $ArS(ArSSAr)^+B(C_6F_5)_4^-$ as an activator of thioglycosides.

Suzuki et al.[253] generated alkoxycarbenium ions that accumulated as "cation pools" by the low-temperature electrochemical oxidation of α-phenylthioethers. They developed a one-pot method for electrochemical glycosylation, which involves anodic oxidation of thioglycosides to generate glycosyl cation equivalents followed by their reactions with glycosyl acceptors added to the reaction medium (Scheme 116).

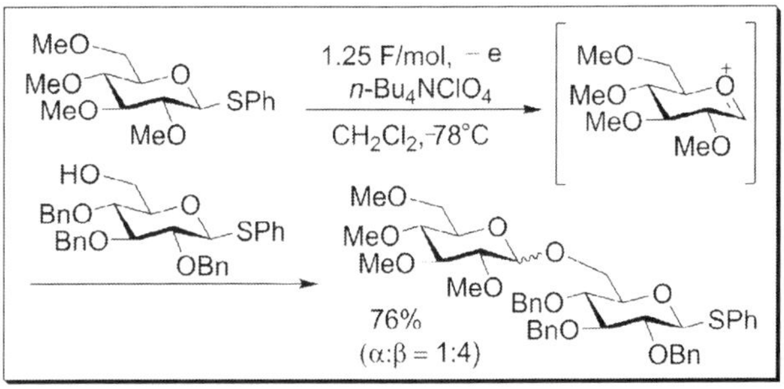

Scheme 116 One-pot electrochemical glycosylation involving anodic oxidation of thioglycosides.

Drouin et al.[254] investigated the roles of protecting groups and the solvent in the electrochemical glycosylation of thiomannosides. They also observed notable differences between electrochemical and chemical glycosylation (Scheme 117).

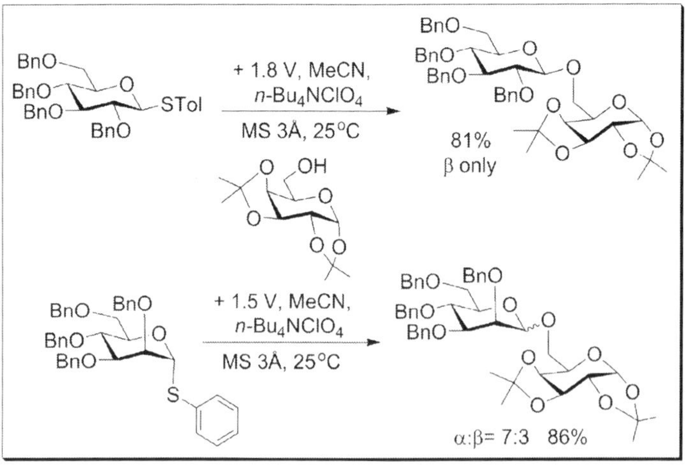

Scheme 117 Stereochemical outcomes in the electrochemical glycosylation of perbenzylated *gluco* and *manno* thioglycosides.

19 Glycosylation under high pressure

The effect of pressure on the rate and direction of organic reactions in solution has long been known and has been an object of extensive physico-chemical studies. There are also some possibilities and usefulness of high-pressure techniques in carbohydrate chemistry. In 1990, Sasaki et al.[255] demonstrated that elevated pressures accelerate the reaction of glucosyl bromide with alcohols in the presence of hindered amines to afford α-glycosides in good yields and high selectivity (Scheme 118).

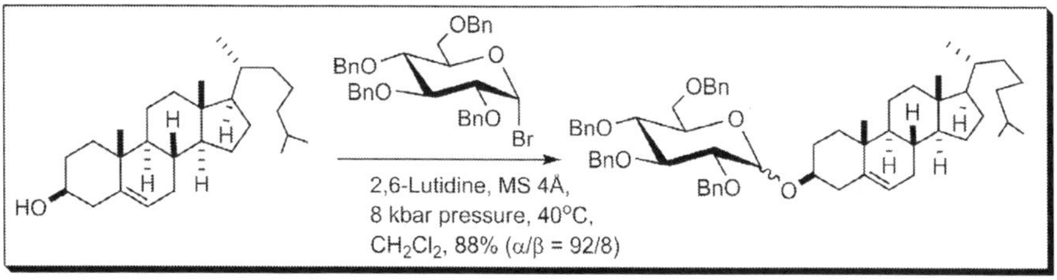

Scheme 118 Glycosylation of cholesterol under high pressure.

Kochetkov et al.[256] reported the glycosylation of trityl ethers by 1,2-*O*-cyanoethylidene derivatives of sugars, and the polycondensation of the monomers in dichloromethane proceeded with absolute stereoselectivity, giving rise to a 1,2 *trans*-glycosidic linkage, at ambient pressure (1.4 GPa) and room temperature (Scheme 119).

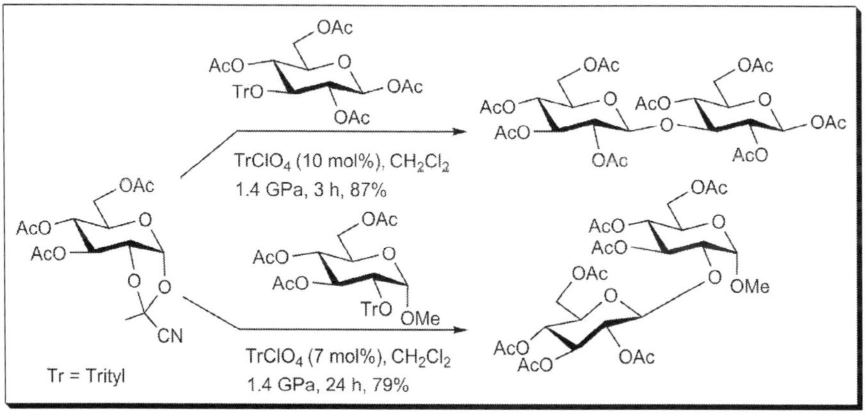

Scheme 119 Stereoselective glycosylation under high pressure.

20 Conclusion

Over the years, significant development has been achieved in the area of chemical transformations of carbohydrates and synthesis of oligosaccharides using environmentally benign greener reaction conditions. Among several green chemical synthetic strategies developed so far, technology-based solvent-free reaction conditions attracted attentions from a large number of researchers. Significant number of reports appeared in the literature in which several chemical transformations of carbohydrate substrates were carried out using green solvents in combination with nonconventional energy sources such as microwave, ultrasound, high pressure, and photoinduced. However, modern, sustainable, environmentally friendly, and automated reaction conditions need to be developed in the future for the preparation of complex oligosaccharides and glycoconjugates for their application in medicinal chemistry.

Acknowledgments

A. Si thanks CSIR, New Delhi, for providing Senior Research Fellowship. The work is supported by SERB, New Delhi (Project No. EMR/2015/000282), Council of Scientific and Industrial research (CSIR), New Delhi (02(0237)/15/EMR-II), and Bose Institute (AKM).

References

1. El Ashry, E. S. H.; Nemr, A. E., Eds. *Synthesis of naturally Occurring Nitrogen Heterocycles From Carbohydrates*; Blackwell Publishing Ltd: Oxford, UK, **2005**.
2. Heterocycles From Carbohydrate Precursors. In *Topics in Heterocyclic Chemistry*; El Ashry, E. S. H., Ed.; vol. 7; Springer, **2007**; pp 133–177.
3. Marco-Contelles, J.; de Opazo, E. *J. Carbohydr. Chem.* **2002**, *21*, 201–218.
4. Madsen, R. *Eur. J. Org. Chem.* **2007**, 399–415.
5. Boutureira, O.; Matheu, M. I.; Diaz, Y.; Castillón, S. Synthesis of C-Nucleosides. In *Chemical Synthesis of Nucleoside Analogues*; Merino, P., Ed.; John Wiley & Sons, Inc.: Hoboken, NJ, USA, 2013.
6. Boutureira, O.; Matheu, M. I.; Diaz, Y.; Castillón, S. *Chem. Soc. Rev.* **2013**, *42*, 5056–5072.
7. Arya, P.; Joseph, R.; Chou, D. T. H. *Chem. Biol.* **2002**, *9*, 145–156.
8. Cao, H.; Hwang, J.; Chen, X. Carbohydrate-Containing Natural Products in Medicinal Chemistry. In *Opportunity, Challenge and Scope of Natural Products in Medicinal Chemistry*; Research Signpost: India, 2011, pp 411.
9. Aplin, J. D.; Wriston, J. C., Jr. *CRC Crit. Rev. Biochem.* **1981**, *10*, 259–306.
10. Ferrier, R. J. *Carbohydrate Chemistry*; vol. 32. Royal Society of Chemistry: Cambridge, UK, **2000**.
11. Pigman, W., Ed.; *The Carbohydrates: Chemistry and Biochemistry Physiology*; Academic Press Inc.: London, **1957**.
12. Kennedy, J. F. *Carbohydrate Chemistry: Macromolecules*; vol. 15. Royal Society of Chemistry: Cambridge, UK, **1986**.
13. Kociensky, P. *Protecting Groups*, 3rd ed.; Georg Thieme Verlag: Stuttgart, **2003**.

14. Wuts, P. G. M.; Greene, T. W. *Greene's Protecting Groups in Organic Synthesis*, 4th ed.; John Wiley & Sons, Inc.: New York, USA, **2007**.
15. Hanessian, S. *Preparative Carbohydrate Chemistry*; Marcel Dekker: New York, **1997**.
16. Anastas, P. T.; Warner, J. C. *Green Chemistry: Theory and Practice*; Oxford University Press: New York, **1998**.
17. (a) Farrán, A.; Cai, C.; Sandoval, M.; Xu, Y.; Liu, J.; Hernáiz, M. J.; Linhardt, R. J. *Chem. Rev.* **2015**, *115*, 6811–6853; (b) Chatterjee, C.; Pong, F.; Sen, A. *Green Chem.* **2015**, *17*, 40–71.
18. Patel, D. D.; Lee, J. M. *Chem. Rev.* **2012**, *12*, 329–355.
19. Tadesse, H.; Luque, R. *Energy Environ. Sci.* **2011**, *4*, 3913–3929.
20. Zhang, Q.; Vigier, D. O. K.; Royer, S.; Jerome, F. *Chem. Soc. Rev.* **2012**, *41*, 7108–7146.
21. Carriazo, D.; Serrano, M. C.; Gutierrez, M. C.; Ferrer, M. L.; del Monte, F. *Chem. Soc. Rev.* **2012**, *41*, 4996–5014.
22. (a) Tuck, C. O.; Perez, E.; Horvath, I. T.; Sheldon, R. A.; Poliakoff, M. *Science* **2012**, *337*, 695–699; (b) Clark, J. H. *J. Chem. Technol. Biotechnol.* **2007**, *82*, 603–609.
23. Brunner, G. A. *Rev. Chem. Biomol. Eng.* **2010**, *1*, 321–342.
24. Zhang, Z.; Du, B.; Zhang, L.-J.; Da, Y.-X.; Quan, Z.-J.; Yang, L.-J.; Wang, X.-C. *RSC Adv.* **2013**, *3*, 9201–9205.
25. Garcia, C.; Hoyos, P.; Hernáiz, M. J. *Biocatal. Biotransform.* **2017**, 131–140.
26. Chao-Jun, L. Green Chemical Syntheses and Processes; ACS Symposium Series; vol. 767. American Chemical Society: Washington, DC, **2000**.
27. Rodrigues, F.; Canac, Y.; Lubineau, A. *Chem. Commun.* **2000**, 2049–2050.
28. Gonzalez, M. A.; Jimenez Requejo, J. L.; Palacios Albarran, J. C.; Galbis Perez, J. A. *Carbohydr. Res.* **1986**, *158*, 53–66.
29. Bragnier, N.; Scherrmann, M.-C. *Synthesis* **2005**, *5*, 814–818.
30. Wang, J.; Li, Q.; Ge, Z.; Li, R. *Tetrahedron* **2012**, *68*, 1315–1320.
31. Norsikian, S.; Zeitouni, J.; Rat, S.; Gérard, S.; Lubineau, A. *Carbohydr. Res.* **2007**, *342*, 2716–2728.
32. (a) Bisht, S. S.; Pandey, J.; Sharma, A.; Tripathi, R. P. *Carbohydr. Res.* **2008**, *343*, 1346–1399; (b) Misra, M.; Sharma, R.; Kant, R.; Maulik, P. R.; Tripathi, R. P. *Tetrahedron* **2011**, *67*, 740–748; (c) Rajaganesh, R.; Das, T. M. *Carbohydr. Res.* **2012**, *357*, 139–142.
33. Garcia Gonzalez, F.; Gomez Sanchez, A. *Adv. Carbohydr. Chem.* **1965**, *20*, 303–355.
34. Misra, A. K.; Agnihotri, G. *Carbohydr. Res.* **2004**, *339*, 1381–1387.
35. Yadav, J. S.; Reddy, B. V. S.; Sreenivas, M.; Satheesh, G. *Synthesis* **2007**, 1712–1716.
36. Sato, S.; Naito, Y.; Aoki, K. *Carbohydr. Res.* **2007**, *342*, 913–918.
37. Carpenter, C. A.; Kenarb, J. A.; Price, N. P. J. *Green Chem.* **2010**, *12*, 2012–2018.
38. Yadav, L. D. S.; Rai, V. K. *Synlett* **2007**, *8*, 1227–1230.
39. Watanabe, M.; Aizawa, Y.; Iida, T.; Aida, T. M.; Levy, C.; Sue, K.; Inomata, H. *Carbohydr. Res.* **2005**, *340*, 1925–1930.
40. Schmid, W.; Whitesides, G. M. *J. Am. Chem. Soc.* **1991**, *113*, 6674–6675.
41. Wu, Q.; Pan, Q.; Zhao, S.; Imker, H.; Serianni, A. S. *J. Org. Chem.* **2007**, *72*, 3081–3084.
42. Aoyama, N.; Kobayashi, S. *Chem. Lett.* **2006**, *35*, 238–239.
43. (a) Tanaka, T.; Nagai, H.; Noguchi, M.; Kobayashi, A.; Shoda, S.-i. *Chem. Commun.* **2009**, 3378–3379; (b) Tanaka, T.; Matsumoto, T.; Noguchi, M.; Kobayashi, A.; Shoda, S.-i. *Chem. Lett.* **2009**, *38*, 458–459; (c) Yoshida, N.; Noguchi, M.; Tanaka, T.; Matsumoto, T.; Aida, N.; Ishihara, M.; Kobayashi, A.; Shoda, S.-i. *Chem. Asian J.* **2011**, *6*, 1876–1885.
44. Komarova, B. S. K.; Maryasina, S. S.; Tsvetkov, Y. E.; Nifantiev, N. E. *Synthesis* **2013**, *45*, 471–478.
45. Zhu, X.; Schmidt, R. R. *Chem. Eur. J.* **2004**, *10*, 875–887.

46. Estrine, B.; Bouquillon, S.; Henin, F.; Muzart, J. *Green Chem.* **2005**, *7*, 219–223.
47. Feng, S.; Li, C. *J. Agric. Food Chem.* **2015**, *63*, 5732–5739.
48. Yu, L.; Li, J.; Ramirez, J.; Chen, D.; Wang, P. G. *J. Org. Chem.* **1997**, *62*, 903–907.
49. (a) Ichikawa, Y.; Matsukawa, Y.; Tamura, M.; Ohara, F.; Isobe, M.; Kotsuki, H. *Chem. Asian J.* **2006**, *1*, 717–723; (b) Ichikawa, Y.; Matsukawa, Y.; Isobe, M. *J. Am. Chem. Soc.* **2006**, *128*, 3934–3938.
50. (a) Micskei, K.; Juhász, Z.; Ratkovic, Z. R.; Somsák, L. *Tetrahedron Lett.* **2006**, *47*, 6117–6120; (b) Kovács, G.; Gyarmati, J.; Somsák, L.; Micskei, K. *Tetrahedron Lett.* **1996**, *37*, 1293–1296; (c) Kovács, G.; Tóth, K.; Dinya, Z.; Somsák, L.; Micskei, K. *Tetrahedron* **1999**, *55*, 5253–5264; (d) Kovács, G.; Micskei, K.; Somsák, L. *Carbohydr. Res.* **2001**, *336*, 225–228.
51. Kumar, R.; Maulik, P. R.; Misra, A. K. *Glycoconj. J.* **2008**, *25*, 595–602.
52. (a) Review articles focusing on the application of CuAAC in carbohydrate chemistry:Dedola, S.; Nepogodiev, S. A.; Field, R. A. *Org. Biomol. Chem.* **2007**, *5*, 1006–1017; (b) Dondoni, A. *Chem. Asian J.* **2007**, *2*, 700–708; (c) Santoyo-Gonzalez, F.; Hernandez-Mateo, F. *Top. Heterocycl. Chem.* **2007**, *7*, 133–178; (d) Aragão-Leoneti, V.; Campo, V. L.; Gomes, A. S.; Field, R. A.; Carvalho, I. *Tetrahedron* **2010**, *66*, 9475–9492; (e) He, X. P.; Zeng, Y. L.; Zang, Y.; Li, J.; Field, R. A.; Chen, G. R. *Carbohydr. Res.* **2016**, *429*, 1–22; (f) Tiwari, V. K.; Mishra, B. B.; Mishra, K. J.; Mishra, N.; Singh, A. S.; Chen, X. *Chem. Rev.* **2016**, *116*, 3086–3240.
53. Pelletier, G.; Zwicker, A.; Allen, C. L.; Schepartz, A.; Miller, S. J. *J. Am. Chem. Soc.* **2016**, *138*, 3175–3182.
54. Wernicke, A.; Belniak, S.; Thévenet, S.; Descotes, G.; Bouchu, A.; Queneau, Y. *J. Chem. Soc. Perkin Trans.* **1998**, *1*, 1179–1182.
55. (a) Scherrmann, M.-C.; Lubineau, A.; Queneau, Y. Functionalization of Carbohydrates in Water. In *Handbook of Green Chemistry*; Wiley-VCH Verlag GmbH & Co. KGaA, **2010**; (b) Queneau, Y.; Pinel, C.; Scherrmann, M.-C. *C. R. Chimie.* **2011**, *14*, pp 688–699.
56. Huddleston, J. G.; Visser, A.; Reichert, W. M.; Willauer, H. D.; Broker, G. A.; Rogers, R. D. *Green Chem.* **2001**, *3*, 156–164.
57. Kragl, U.; Eckstein, M.; Kaftzik, N. *Curr. Opin. Biotechnol.* **2002**, *13*, 565–571.
58. Wilkes, J. S. *Green Chem.* **2002**, *4*, 73–80.
59. MacFarlane, D. R.; Forsyth, M.; Howlett, P. C.; Kar, M.; Passerini, S.; Pringle, J. M.; Ohno, H.; Watanabe, M.; Yan, F.; Zheng, W.; Zhang, S.; Zhang, J. *Nat. Rev. Mater.* **2016**, *1*, 15005–15019.
60. Shi, R.; Wang, Y. *Sci. Rep.* **2016**, *6*, 19644–19655.
61. (a) Vekariya, R. L. *J. Mol. Liq.* **2017**, *227*, 44–60; (b) Miao, W.; Chan, T. H. *Acc. Chem. Res.* **2006**, *39*, 897–908.
62. (a) Baudequin, C.; Bregeon, D.; Levillain, J.; Guillen, F.; Plaquevent, J.-C.; Gaumont, A.-C. *Tetrahedron-Asymmetry* **2005**, *16*, 3921–3945; (b) Poletti, L.; Chiappe, C.; Lay, L.; Pieraccini, D.; Polito, L.; Russo, G. *Green Chem.* **2007**, *9*, 337–341.
63. (a) Murugesan, S.; Linhardt, R. J. *Curr. Org. Synth.* **2005**, *2*, 437–451; (b) El Seoud, O. A.; Koschella, A.; Fidale, L. C.; Dorn, S.; Heinze, T. *Biomacromolecules* **2007**, *8*, 2629–2647; (c) Galan, M. C.; Jones, R. A.; Tran, A.-T. *Carbohydr. Res.* **2013**, *375*, 35–46.
64. Takada, A.; Kadokawa, J.-i. *Biomolecules* **2015**, *5*, 244–262.
65. Forsyth, S. A.; MacFarlane, D. R.; Thomson, R. J.; Itzstein, M. V. *Chem. Commun.* **2002**, *7*, 714–715.
66. Murugesan, S.; Karst, N.; Islam, T.; Wiencek, J. M.; Linhardt, R. J. *Synlett* **2003**, *9*, 1283–1286.
67. Abbott, A. P.; Bell, T. J.; Handa, S.; Stoddart, B. *Green Chem.* **2005**, *7*, 705–707.

68. Zhang, J.; Ragauskas, A. J. *Carbohydr. Res.* **2005**, *340*, 2812–2815.
69. Anas, S.; Sajisha, V. S.; Rajan, R.; Kumaran, R. T.; Radhakrishnan, K. V. *Bull. Chem. Soc. Jpn.* **2007**, *80*, 553–560.
70. (a) Lian, G.; Zhang, X.; Yu, B. *Carbohydr. Res.* **2015**, *403*, 13–22; (b) Oscarson, S. Thioglycosides. In *Carbohydrates in Chemistry and Biology*; Ernst, B., Hart, G. W., Sinaÿ, P., Eds.; Wiley-VCH Verlag GmbH, 2008; pp 93–116; (c) Codée, J. D. C.; Litjens, R.E.J.N.; van den Bos, L. J.; Overkleeft, H. S.; van der Marel, G. A. *Chem. Soc. Rev.* **2005**, *34*, 769–782.
71. (a) Santra, A.; Sau, A.; Misra, A. K. *J. Carbohydr. Chem.* **2011**, *30*, 85–93; (b) Sau, A.; Misra, A. K. *Synlett* **2011**, 1905–1911.
72. Park, T.-J.; Weiwer, M.; Yuan, X.; Baytas, S. N.; Munoz, E. M.; Murugesan, S.; Linhardt, R. J. *Carbohydr. Res.* **2007**, *342*, 614–620.
73. Augé, J.; Sizun, G. *Green Chem.* **2009**, *11*, 1179–1183.
74. Yadav, J. S.; Reddy, B. V. S.; Reddy, J. S. S. *J. Chem. Soc. Perkin Trans.* **2002**, *1*, 2390–2394.
75. Pakulski, Z. *Synthesis* **2003**, 2074–2076.
76. Poletti, L.; Rencurosi, A.; Lay, L.; Russo, G. *Synlett* **2003**, 2297–2300.
77. Rencurosi, A.; Lay, L.; Russo, G.; Caneva, E.; Poletti, L. *J. Org. Chem.* **2005**, *70*, 7765–7768.
78. Sasaki, K.; Nagai, H.; Matsumura, S.; Toshima, K. *Tetrahedron Lett.* **2003**, *44*, 5605–5608.
79. Sasaki, K.; Matsumura, S.; Toshima, K. *Tetrahedron Lett.* **2004**, *45*, 7043–7047.
80. Zhang, J.; Ragauskas, A. *Beilstein J. Org. Chem.* **2006**, *2*, 2–12.
81. (a) Galan, M. C.; Jouvin, K.; Alvarez-Dorta, D. *Carbohydr. Res.* **2010**, *345*, 45–49; (b) Galan, M. C.; Tran, A. T.; Boisson, J.; Benito, D.; Butts, C.; Eastoe, J.; Brown, P. *J. Carbohydr. Chem.* **2011**, *30*, 486–497; (c) Galan, M. C.; Tran, A. T.; Whitaker, S. *Chem. Commun.* **2010**, *46*, 2106–2108.
82. (b) Galan, M. C.; Brunet, C.; Fuensanta, M. *Tetrahedron Lett.* **2009**, *50*, 442–445.
83. Kumar, V.; Talisman, I. J.; Malhotra, S. V. *Eur. J. Org. Chem.* **2010**, 3377–3381.
84. Fevre, M.; Pinaud, J.; Leteneur, A.; Gnanou, Y.; Vignolle, J.; Taton, D.; Miqueu, K.; Sotiropoulos, J. M. *J. Am. Chem. Soc.* **2012**, *134*, 6776–6784.
85. He, X.; Chan, T. H. *Synthesis* **2006**, 1645–1651.
86. Huang, J. Y.; Lei, M.; Wang, Y. G. *Tetrahedron Lett.* **2006**, *47*, 3047–3050.
87. Pathak, A. K.; Yerneni, C. K.; Young, Z.; Pathak, V. *Org. Lett.* **2008**, *10*, 145–148.
88. Yerneni, C. K.; Pathak, V.; Pathak, A. K. *J. Org. Chem.* **2009**, *74*, 6307–6310.
89. Pepin, M.; Hubert-Roux, M.; Martin, C.; Guillen, F.; Lange, C.; Gouhier, G. *Eur. J. Org. Chem.* **2010**, 6366–6371.
90. Sittel, I.; Tran, A. T.; Benito-Alifonso, D.; Galan, M. C. *Chem. Commun.* **2013**, *49*, 4217–4219.
91. Yamada, C.; Sasaki, K.; Matsumura, S.; Toshima, K. *Tetrahedron Lett.* **2007**, *48*, 4223–4227.
92. Galan, M. C.; Tran, A. T.; Bromfield, K.; Rabbani, S.; Ernst, B. *Org. Biomol. Chem.* **2012**, *10*, 7091–7097.
93. (a) Brunner, G. *J. Food Eng.* **2005**, *67*, 21–33; (b) Tan, K. T.; Lee, K. T. *Renew. Sustain. Energy Rev.* **2011**, *15*, 2452–2456.
94. (a) Lang, Q.; Wai, C. M. *Talanta* **2001**, *53*, 771–782; (b) Pourmortazavi, S. M.; Hajimirsadeghi, S. S. *J. Chromatogr. A* **2007**, *1163*, 2–24; (c) Miller, L. *J. Chromatogr. A* **2012**, *1250*, 250.
95. Jessop, P. G.; Ikariya, T.; Noyori, R. *Chem. Rev.* **1999**, *99*, 475–493.
96. Jessop, P. G.; Hsiao, Y.; Ikariya, T.; Noyori, R. *J. Am. Chem. Soc.* **1996**, *118*, 344–355.
97. Oku, T.; Arita, Y.; Tsuneki, H.; Ikariya, T. *J. Am. Chem. Soc.* **2004**, *126*, 7368–7377.

98. Amandi, R.; Licence, P.; Ross, S. K.; Aaltonen, O.; Poliakoff, M. *Org. Process. Res. Dev.* **2005**, *9*, 451–456.
99. Nolen, S. A.; Lu, J.; Brown, J. S.; Pollet, P.; Eason, B. C.; Griffith, K. N.; Gläser, R.; Bush, D.; Lamb, D. R.; Liotta, C. L.; Eckert, C. A.; Thiele, G. F.; Bartels, K. A. *Ind. Eng. Chem. Res.* **2002**, *41*, 316–323.
100. Ikushima, Y.; Hatakeda, K.; Sato, O.; Yokoyama, T.; Arai, M. *Angew. Chem. Int. Ed. Eng.* **1999**, *38*, 2910–2914.
101. (a) Mesiano, A. J.; Beckman, E. J.; Russell, A. J. *Chem. Rev.* **1999**, *99*, 623–633; (b) Mori, T.; Okahata, Y. *Chem. Commun.* **1998**, 2215–2216; (c) Kamat, S. V.; Beckman, E. J.; Russell, A. J. *Crit. Rev. Biotechnol.* **1995**, *15*, 41–71.
102. Li, X.-B.; Ogawa, M.; Monden, T.; Maeda, T.; Yamashita, E.; Naka, M.; Matsuda, M.; Hinou, H.; Nishimura, S.-I. *Angew. Chem. Int. Ed. Eng.* **2006**, *45*, 5652–5655.
103. Cardona, A.; Boutureira, O.; Castillón, S.; Diaz, Y.; Matheu, M. I. *Green Chem.* **2017**, *19*, 2687–2694.
104. Florindo, C.; Oliveira, F. S.; Rebelo, L. P. N.; Fernandes, A. M.; Marrucho, I. M. *ACS Sustain. Chem. Eng.* **2014**, *2*, 2416–2425.
105. Smith, E. L.; Abbott, A. P.; Ryder, K. S. *Chem. Rev.* **2014**, *114*, 11060–11082.
106. Abbott, A. P.; Boothby, D.; Capper, G.; Davies, D. L.; Rasheed, R. K. *J. Am. Chem. Soc.* **2004**, *126*, 9142–9147.
107. Hayyan, A.; Mjalli, F. S.; Alnashef, I. M.; Al-Wahaibi, Y. M.; Al-Wahaibi, T.; Hashim, M. A. *J. Mol. Liq.* **2013**, *178*, 137–141.
108. (a) Liu, P.; Hao, J.-W.; Mo, L.-P.; Zhang, Z.-H. *RSC Adv.* **2015**, *5*, 48675–48704; (b) Alonso, D. A.; Baeza, A.; Chinchilla, R.; Guillena, G.; Pastor, I. M.; Ramón, D. J. *Eur. J. Org. Chem.* **2016**, 612–632.
109. (a) Zhang, L.; Yu, H. *BioResources* **2013**, *8*, 6014–6025; (b) Zhang, L.-X.; Yu, H.; Yu, H.-B.; Chen, Z.; Yang, L. *Chin. Chem. Lett.* **2014**, *25*, 1132–1136.
110. Rus, C.; Luff, C.; Begli, A. H.; Koenig, B. *Synth. Commun.* **2012**, *42*, 3112–3116.
111. Ilgen, F.; Ott, D.; Kralisch, D.; Reil, C.; Palmberger, A.; Konig, B. *Green Chem.* **2009**, *11*, 1948–1954.
112. Rokade, S. M.; Garande, A. M.; Ahmad, N. A. A.; Bhate, P. M. *RSC Adv.* **2015**, *5*, 2281–2284.
113. Poehnlein, M.; Ulrich, J.; Kirschhoefer, F.; Nusser, M.; Muhle-Goll, C.; Kannengiesser, B.; Brenner-Weiβ, G.; Luy, B.; Liese, A.; Syldatk, C.; Hausmann, R. *Eur. J. Lipid Sci. Technol.* **2015**, *117*, 161–166.
114. Rokade, S. M.; Bhate, P. M. *Carbohydr. Res.* **2015**, *416*, 21–23.
115. Rokade, S. M.; Bhate, P. M. *Carbohydr. Res.* **2015**, *415*, 28–30.
116. Horváth, I. T.; Rábai, J. *Science* **1994**, *266*, 72–75.
117. Gladysz, J. A.; Emnet, C. Fluorous Solvents and Related Media. In *Handbook of Fluorous Chemistry*, Gladysz, J. A., Curran, D. P., Horváth, I. T., Eds.; Wiley-VCH Verlag GmbH & Co.: Weinheim, Germany, **2004**.
118. (a) Nakamura, Y.; Takeuchi, S.; Okumura, K.; Ohgo, Y.; Curran, D. P. *Tetrahedron* **2002**, *58*, 3963–3969; (b) Sokeirik, Y. S.; Hoshina, A.; Omote, M.; Sato, K.; Tarui, A.; Kumadaki, I.; Ando, A. *Chem. Asian J.* **2008**, *3*, 1850–1856; (c) Omorodion, H.; Twamley, B.; Platts, J. A.; Baker, R. J. *Cryst. Growth Des.* **2015**, *15*, 2835–2841; (d) Koda, Y.; Terashima, T.; Sawamoto, M. *J. Am. Chem. Soc.* **2014**, *136*, 15742–15748; (e) Tojino, M.; Mori, M.; Kasuya, M. C. Z.; Hatanaka, K.; Kawaguchi, A.; Nagata, K.; Shirai, T.; Mizuno, M. *Bioorg. Med. Chem. Lett.* **2012**, *22*, 1251–1254; (f) Gladysz, J. A.; Curran, D. P. *Tetrahedron* **2002**, *58*, 3823–3825.
119. Studer, A.; Jeger, P.; Wipf, P.; Curran, D. P. *J. Org. Chem.* **1997**, *62*, 2917–2924.

120. Zhang, W. *Chem. Rev.* **2004**, *104*, 2531–2556.
121. (a) Zhang, W. *Tetrahedron* **2003**, *59*, 4475–4489; (b) Dobbs, A. P.; Kimberley, M. R. *J. Fluor. Chem.* **2002**, *118*, 3–17.
122. (a) Zhang, W. *Comb. Chem. High Throughput Screen.* **2007**, *10*, 219–229; (b) Zhang, W. *Top. Curr. Chem.* **2006**, *266*, 145–166.
123. Curran, D. P. *J. Fluor. Chem.* **2008**, *129*, 898–902.
124. (a) Andrade, R. B.; Plante, O. J.; Melean, L. G.; Seeberger, P. H. *Org. Lett.* **1999**, *1*, 1811–1814; (b) Carrel, F. R.; Seeberger, P. H. *J. Org. Chem.* **2008**, *73*, 2058–2065.
125. Zhang, F.; Zhang, W.; Zhang, Y.; Curran, D. P.; Liu, G. *J. Org. Chem.* **2009**, *74*, 2594–2597.
126. Yang, B.; Jing, Y.; Huang, X. *Eur. J. Org. Chem.* **2010**, 1290–1298.
127. Ko, K.-S.; Park, G.; Yu, Y.; Pohl, N. L. *Org. Lett.* **2008**, *10*, 5381–5384.
128. (a) Jaipuri, F. A.; Collet, B. Y. M.; Pohl, N. L. *Angew. Chem. Int. Ed.* **2008**, *47*, 1707–1710; (b) Pohl, N. L. Carbohydrate Microarrays and Fluorous-Phase Synthesis: Interfacing Fluorous-Phase Tags with the Direct Formation of Glycoarrays. In *Current Fluoroorganic Chemistry: New Synthetic Directions, Technologies, Materials, and Biological Applications*, Soloshonok, V. A., Mikami, K., Yamazaki, T., Welch, J. T., Honek, J. F., Eds.; ACS Symposium Seriesvol. 949American Chemical Society: Washington, DC, 2007; pp 261; (c) Ko, K.-S.; Jaipuri, F. A.; Pohl, N. L. *J. Am. Chem. Soc.* **2005**, *127*, 13162–13163.
129. Mamidyala, S. K.; Ko, K.-S.; Jaipuri, F. A.; Park, G.; Pohl, N. L. *J. Fluor. Chem.* **2006**, *127*, 571–579.
130. (a) Chen, G.-S.; Pohl, N. L. *Org. Lett.* **2008**, *10*, 785–788; (b) Park, G.; Ko, K.-S.; Zakharova, A.; Pohl, N. L. *J. Fluor. Chem.* **2008**, *129*, 978–982; (c) Miura, T.; Goto, K.; Waragai, H.; Matsumoto, H.; Hirose, Y.; Ohmae, M.; Ishida, H.-k.; Satoh, A.; Inazu, T. *J. Org. Chem.* **2004**, *69*, 5348–5353.
131. Miura, T.; Hirose, Y.; Ohmae, M.; Inazu, T. *Org. Lett.* **2001**, *3*, 3947–3950.
132. Manzoni, L.; Castelli, R. *Org. Lett.* **2006**, *8*, 955–957.
133. Jaipuri, F. A.; Pohl, N. L. *Org. Biomol. Chem.* **2008**, *6*, 2686–2691.
134. Zong, C.; Venot, A.; Dhamale, O.; Boons, G.-J. *Org. Lett.* **2013**, *15*, 342–345.
135. (a) Carrel, F. R.; Geyer, K.; Codée, J. D. C.; Seeberger, P. H. *Org. Lett.* **2007**, *9*, 2285–2288.
136. Nüchter, M.; Müller, U.; Ondruschka, B.; Tied, A.; Lautenschläger, W. *Chem. Eng. Technol.* **2003**, *26*, 1208–1216.
137. Lidström, P.; Tierney, J.; Wathey, B.; Westman, J. *Tetrahedron* **2001**, *57*, 9225–9283.
138. Caddick, S. *Tetrahedron* **1995**, *51*, 10403–10432.
139. Corsaro, A.; Chiacchio, U.; Pistara, V.; Romeo, G. *Curr. Org. Chem.* **2004**, *8*, 511–538.
140. Gawande, M. B.; Shelke, S. N.; Zboril, R.; Varma, R. S. *Acc. Chem. Res.* **2014**, *47*, 1338–1348.
141. Kappe, C. O.; Stadler, A.; Dallinger, D. In *Microwaves in Organic and Medicinal Chemistry*; Mannhold, R., Kubinyi, H., Folkers, G., Eds.; vol. 52; **2013**; .
142. Leadbeater, N. E. In *Microwave Heating as a Tool for Sustainable Chemistry*, Cann, M. C., Ed.; CRC Press, Taylor & Francis Group: Boca Raton, USA, **2010**.
143. Adam, D. *Nature* **2003**, *421*, 571–572.
144. Toda, F.; Tanaka, K. *Chem. Rev.* **2000**, *100*, 1025–1074.
145. Kappe, C. O. *Angew. Chem. Int. Ed. Eng.* **2004**, *43*, 6250–6284.
146. Bornaghi, L. F.; Poulsen, S.-A. *Tetrahedron Lett.* **2005**, *46*, 3485–3488.
147. Lin, H.-C.; Chang, C.-C.; Chen, J.-Y.; Lin, C.-H. *Tetrahedron-Asymmetry* **2005**, *16*, 297–301.
148. Yoshimura, Y.; Shimizu, H.; Hino, H.; Nishimura, S.-I. *Tetrahedron Lett.* **2005**, *46*, 4701–4704.

149. Rat, S.; Mathiron, D.; Michaud, P.; Kovensky, J. E.; Wadouachi, A. *Tetrahedron* **2007**, *63*, 12424–12428.
150. Shimizu, H.; Yoshimura, Y.; Hinou, H.; Nishimura, S.-I. *Tetrahedron* **2008**, *64*, 10091–10096.
151. Zhang, G.; Liu, Q.; Shi, L.; Wang, J. *Tetrahedron* **2008**, *64*, 339–344.
152. Seo, J.; Michaelian, N.; Owens, S. C.; Dashner, S. T.; Wong, A. J.; Barron, A. E.; Carrasco, M. R. *Org. Lett.* **2009**, *11*, 5210–5213.
153. Ko, Y.-C.; Tsai, C.-F.; Wang, C.-C.; Dhurandhare, V. M.; Hu, P.-L.; Su, T.-Y.; Lico, L. S.; Zulueta, M. M. L.; Hung, S.-C. *J. Am. Chem. Soc.* **2014**, *136*, 14425–14431.
154. Wang, C. C.; Lee, J. C.; Luo, S. Y.; Kulkarni, S. S.; Huang, Y. W.; Lee, C. C.; Chang, K. L.; Hung, S. C. *Nature* **2007**, *446*, 896–899.
155. Wang, C. C.; Kulkarni, S. S.; Lee, J. C.; Luo, S. Y.; Hung, S. C. *Nat. Protoc.* **2008**, *3*, 97–113.
156. Hinou, H.; Saito, N.; Ogawa, M.; Maeda, T.; Nishimura, S.-I. *Int. J. Mol. Sci.* **2009**, *10*, 5285–5295.
157. Limousin, C.; Cléophax, J.; Petit, A.; Loupy, A.; Lukacs, G. *J. Carbohydr. Chem.* **1997**, *16*, 327–342.
158. Achar, T. K.; Bose, A.; Mal, P. *Beilstein J. Org. Chem.* **2017**, *13*, 1907–1931.
159. Do, J.-L.; Friščić, T. *ACS Cent. Sci.* **2017**, *3*, 13–19.
160. Wang, G. W. *Chem. Soc. Rev.* **2013**, *42*, 7668–7700.
161. Stolle, A.; Szuppa, T.; Leonhardt, S. E. S.; Ondruschka, B. *Chem. Soc. Rev.* **2011**, *40*, 2317–2329.
162. Stolle, A.; Schmidt, R.; Jacob, K. *Faraday Discuss.* **2014**, *170*, 267–286.
163. Tyagi, M.; Khurana, D.; Kartha, K. P. R. *Carbohydr. Res.* **2013**, *379*, 55–59.
164. Mugunthan, G.; Kartha, K. P. R. *J. Carbohydr. Chem.* **2008**, *27*, 294–299.
165. Patil, P. R.; Kartha, K. P. R. *Green Chem.* **2009**, *11*, 953–956.
166. Kumar, V.; Yadav, N.; Kartha, K. P. R. *Carbohydr. Res.* **2014**, *397*, 18–26.
167. Kumar, V.; Taxa, N.; Jangir, R.; Bharatam, P. V.; Kartha, K. P. R. *J. Org. Chem.* **2014**, *79*, 3427–3439.
168. Boissou, F.; Sayou, N.; Vigier, K. D. O.; Barakat, A.; Marinkovic, S.; Estrine, B.; Jerome, F. *ChemSusChem* **2015**, *8*, 3263–3269.
169. (a) Sethi, K. P.; Kartha, K. P. R. *Carbohydr. Res.* **2016**, *434*, 132–135; (b) Sarala, G.; Kartha, K. P. R. *Trends Carbohydr. Res.* **2016**, *8*, 19–28.
170. Mason, T. J.; Peters, D. *Practical Sonochemistry: Power Ultrasound Uses and Applications*, 2nd ed.; Horwood Publishing, **2003**.
171. Mason, T. J. *Chem. Soc. Rev.* **1997**, *26*, 443–451.
172. Cella, R.; Stefani, H. A. Ultrasonic Reactions. In *Green Techniques for Organic Synthesis and Medicinal Chemistry*, Zhang, W., Cue, B. W., Eds.; John Wiley & Sons, Ltd.: Chichester, UK, **2012**.
173. (a) Kardos, N.; Luche, J.-L. *Carbohydr. Res.* **2001**, *332*, 115–131; (b) Wang, Y.; McGuigan, C. *Synth. Commun.* **1997**, *27*, 3829–3833.
174. Bera, S.; Mondal, D.; Martin, J. T.; Singh, M. *Carbohydr. Res.* **2015**, *410*, 15–35.
175. Grice, P.; Ley, S. V.; Pietruszka, J.; Osborn, H. M. I.; Priepke, H. W. M.; Warriner, S. L. *Chem. Eur. J.* **1997**, *3*, 431–440.
176. (a) Holý, A.; Soucek, M. *Tetrahedron Lett.* **1971**, *12*, 185–188; (b) Brown, L.; Coreeda, M. *J. Org. Chem.* **1984**, *49*, 3875–3880.
177. Robin, M. J.; Hawrelak, S. D.; Kanai, T.; Siefert, J. M.; Mengel, R. *J. Org. Chem.* **1979**, *44*, 1317–1322.
178. (a) Quaedflieg, P.J.L.M.; Timmers, C. M.; Kal, V. E.; van der Marel, G. A.; Kuyl Yeheskiely, E.; van Boom, J. H. *Tetrahedron Lett.* **1992**, *33*, 3081–3084; (b) Zhu, T.; Boons, G. J. *Tetrahedron-Asymmetry* **2000**, *11*, 199–205.

179. (a) Nicolaou, K. C.; Winssinger, N.; Pastor, J.; de Roose, F. *J. Am. Chem. Soc.* **1997**, *119*, 449–450; (b) Roussel, F.; Knerr, L.; Grathwohl, M.; Schmidt, R. R. *Org. Lett.* **2000**, *2*, 3043–3046.
180. Deng, S.; Gangadharmath, U.; Chang, C.-W.T. *J. Org. Chem.* **2006**, *71*, 5179–5185.
181. Tanifum, C. T.; Zhang, J.; Chang, C.-W.T. *Tetrahedron Lett.* **2010**, *51*, 4323–4327.
182. Gholap, A. R.; Venkatesan, K.; Daniel, T.; Lahoti, R. J.; Srinivasan, K. V. *Green Chem.* **2003**, *5*, 693–696.
183. Fischer, E. *Ber. Dtsch. Chem. Ges.* **1920**, *53*, 1589–1605.
184. Doerschuk, A. P. *J. Am. Chem. Soc.* **1952**, *74*, 4202–4203.
185. Deng, S.; Chang, C.-W.T. *Synlett* **2006**, 756–760.
186. Tatina, M.; Yousuf, S. K.; Mukherjee, D. *Org. Biomol. Chem.* **2012**, *10*, 5357–5360.
187. Zhao, J.; Wei, S.; Yue, A.; Shao, H. *Chin. J. Chem.* **2012**, *30*, 627–633.
188. Benito-Alifonso, D.; Jones, R. A.; Tran, A.; Woodward, H.; Smith, N.; Galan, M. C. *Beilstein J. Org. Chem.* **2013**, *9*, 1867–1872.
189. Tanifum, C. T.; Chang, C.-W.T. *J. Org. Chem.* **2009**, *74*, 634–644.
190. Lu, S. R.; Lai, Y. H.; Chen, J. H.; Liu, C. Y.; Tony Mong, K. K. *Angew. Chem. Int. Ed. Eng.* **2011**, *50*, 7315–7320.
191. Shaikh, N.; Russo, L.; Cipolla, L.; Nicotra, F. *Mol. Divers.* **2011**, *15*, 341–345.
192. Brochette-Lemoine, S.; Trombotto, S.; Joannard, D.; Descotes, G.; Bouchu, A.; Queneau, Y. *Ultrason. Sonochem.* **2000**, *7*, 157–161.
193. Jarikote, D. V.; O'Reilly, C.; Murphy, P. V. *Tetrahedron Lett.* **2010**, *51*, 6776–6778.
194. da Silva, G. B.; Guimarães, B. M.; Assis, S. P. O.; Lima, V. L. M.; de Oliveira, R. N. *J. Braz. Chem. Soc.* **2013**, *24*, 914–921.
195. Wang, G.-N.; André, S.; Gabius, H.-J.; Murphy, P. V. *Org. Biomol. Chem.* **2012**, *10*, 6893–6907.
196. Lee, H. S.; Nguyen, H. M.; Koo, Y. M.; Ha, S. H. *Process Biochem.* **2008**, *43*, 1009–1012.
197. Sridhar, P. R.; Prabhu, K. R.; Chandrasekaran, S. *J. Org. Chem.* **2003**, *68*, 5261–5264.
198. (a) Arasappan, A.; Fraser-Reid, B. *J. Org. Chem.* **1996**, *61*, 2401–2406; (b) Merritt, J. R.; Naisang, E.; Fraser-Reid, B. *J. Org. Chem.* **1994**, *59*, 4443–4449; (c) Yousuf, S. K.; Mukherjee, D.; Mallikharjunrao, L.; Taneja, S. C. *Org. Lett.* **2011**, *13*, 576–579.
199. (a) König, B. *Eur. J. Org. Chem.* **2017**, 1979–1981; (b) Kabra, K.; Chaudhary, R.; Sawhney, R. L. *Ind. Eng. Chem. Res.* **2004**, *43*, 7683–7696.
200. Friedmann, D.; Hakki, A.; Kim, H.; Choi, W.; Bahnemann, D. *Green Chem.* **2016**, *18*, 5391–5411.
201. Sangwan, R.; Mandal, P. K. *RSC Adv.* **2017**, *7*, 26256–26321.
202. Griffin, G. W.; Bandara, N. C.; Clarke, M. A.; Tsang, W. S.; Garegg, P. J.; Oscarson, S.; Silwanis, B. A. *Heterocycles* **1990**, *30*, 939–947.
203. Nakanishi, M.; Takahashi, D.; Toshima, K. *Org. Biomol. Chem.* **2013**, *11*, 5079–5082.
204. Umemoto, T.; Ishihara, S. *J. Am. Chem. Soc.* **1993**, *115*, 2156–2164.
205. Spell, M. L.; Deveaux, K.; Bresnahan, B. C. G.; Bernard, L.; Sheffield, W.; Kumar, R.; Ragains, J. R. *Angew. Chem. Int. Ed. Eng.* **2016**, *55*, 6515–6519.
206. (b) Spell, M.; Wang, X.; Wahba, A. E.; Conner, E.; Ragains, J. *Carbohydr. Res.* **2013**, *369*, 42–47.
207. Mao, R.-Z.; Xiong, D.-C.; Guo, F.; Li, Q.; Duanb, J.; Ye, X.-S. *Org. Chem. Front.* **2016**, *3*, 737–743.
208. (a) Yu, Y.; Xiong, D.-C.; Mao, R.-Z.; Ye, X.-S. *J. Org. Chem.* **2016**, *81*, 7134–7138; (b) Mao, R.-Z.; Guo, F.; Xiong, D.-C.; Li, Q.; Duan, J.; Ye, X.-S. *Org. Lett.* **2015**, *17*, 5606–5609.
209. Wever, W. J.; Cinelli, M. A.; Bowers, A. A. *Org. Lett.* **2013**, *15*, 30–33.

210. Wen, P.; Crich, D. *Org. Lett.* **2017**, *19*, 2402–2405.
211. Andrews, R. S.; Becker, J. J.; Gagne, M. R. *Angew. Chem. Int. Ed. Eng.* **2010**, *49*, 7274–7276.
212. Dupuis, J.; Giese, B.; Ruegge, D.; Fishcer, H.; Korth, H. G.; Sustmann, R. *Angew. Chem.* **1984**, *96*, 887–888.
213. Andrews, R. S.; Becker, J. J.; Gagne, M. R. *Org. Lett.* **2011**, *13*, 2406–2409.
214. Andrews, R. S.; Becker, J. J.; Gagne, M. R. *Angew. Chem. Int. Ed. Eng.* **2012**, *51*, 4140–4143.
215. Giese, B.; Dupuis, J.; Leising, M.; Nix, M.; Linder, H. J. *Carbohydr. Res.* **1987**, *171*, 329–341.
216. Rasool, F.; Bhat, A. H.; Hussain, N.; Mukherjee, D. *ChemistrySelect* **2016**, *1*, 6553–6557.
217. Iwata, R.; Uda, K.; Takahashi, D.; Toshima, K. *Chem. Commun.* **2014**, *50*, 10695–10698.
218. Kimura, T.; Eto, T.; Takahashi, D.; Toshima, K. *Org. Lett.* **2016**, *18*, 3190–3193.
219. Geng, Y.; Kumar, A.; Faidallah, H. M.; Albar, H. A.; Mhkalid, I. A.; Schmidt, R. R. *Angew. Chem. Int. Ed. Eng.* **2013**, *52*, 10089–10092.
220. Fiore, M.; Marra, A.; Dondoni, A. *J. Org. Chem.* **2009**, *74*, 4422–4425.
221. (a) Posner, T. *Ber. Dtsch. Chem. Ges.* **1905**, *38*, 646–657; (b) Dondoni, A.; Marra, A. *Chem. Soc. Rev.* **2012**, *41*, 573–586; (c) McSweeney, L.; Dénès, F.; Scanlan, E. M. *Eur. J. Org. Chem.* **2016**, 2080–2095; (d) Lázár, L.; Borbás, A.; Somsák, L. *Carbohydr. Res.* **2018**, *470*, 8–12; (e) Lázár, L.; Juhász, L.; Batta, G.; Borbás, A.; Somsák, L. *New J. Chem.* **2017**, *41*, 1284–1292; (f) Lázár, L.; Csávás, M.; Tóth, M.; Somsák, L.; Borbás, A. *Chem. Pap.* **2015**, *69*, 889–895; (g) József, J.; Juhász, L.; Illyés, T. Z.; Csávás, M.; Borbás, A.; Somsák, L. *Carbohydr. Res.* **2015**, *413*, 63–69; (h) Lázár, L.; Csávás, M.; Hadházi, Á.; Herczeg, M.; Tóth, M.; Somsák, L.; Barna, T.; Herczegh, P.; Borbás, A. *Org. Biomol. Chem.* **2013**, *11*, 5339–5350; (i) Lázár, L.; Csávás, M.; Herczeg, M.; Herczegh, P.; Borbás, A. *Org. Lett.* **2012**, *14*, 4650–4653; (j) Bege, M.; Bereczki, I.; Herczeg, M.; Kicsák, M.; Eszenyi, D.; Herczegh, P.; Borbás, A. *Org. Biomol. Chem.* **2017**, *15*, 9226–9233.
222. Staderini, S.; Chambery, A.; Marra, A.; Dondoni, A. *Tetrahedron Lett.* **2012**, *53*, 702–704.
223. Floyd, N.; Vijayakrishnan, B.; Koeppe, J. R.; Davis, B. G. *Angew. Chem.* **2009**, *121*, 7938–7942.
224. Minozzi, M.; Monesi, A.; Nanni, D.; Spagnolo, P.; Marchetti, N.; Massi, A. *J. Org. Chem.* **2011**, *76*, 450–459.
225. Marra, A.; Scherrmann, M.-C. *Carbohydr. Chem.* **2014**, *40*, 160–177.
226. Noyori, R.; Kurimoto, I. *J. Org. Chem.* **1986**, *51*, 4320–4322.
227. (a) Amatore, C.; Jutand, A.; Mallet, J.-M.; Meyer, G.; Sinaÿ, P. *J. Chem. Soc. Chem. Commun.* **1990**, 718–719; (b) Balavoine, G.; Gref, A.; Fischer, J.-C.; Lubineau, A. *Tetrahedron Lett.* **1990**, *31*, 5761–5764.
228. Yamago, S.; Kokubo, K.; Yoshida, J. *Chem. Lett.* **1997**, 111–112.
229. Mehta, S.; Pinto, B. M. *Carbohydr. Res.* **1998**, *310*, 43–51.
230. Yamago, S.; Kokubo, K.; Hara, O.; Masuda, S.; Yoshida, J. *J. Org. Chem.* **2002**, *67*, 8584–8592.
231. France, R. R.; Rees, N. V.; Wadhawan, J. D.; Fairbanks, A. J.; Compton, R. G. *Org. Biomol. Chem.* **2004**, *2*, 2188–2194.
232. France, R. R.; Compton, R. G.; Davis, B. G.; Fairbanks, A. J.; Rees, N. V.; Wadhawan, J. D. *Org. Biomol. Chem.* **2004**, *2*, 2195–2202.
233. Marra, A.; Mallet, J.-M.; Amatore, C.; Sinaÿ, P. *Synlett* **1990**, 572–574.
234. Zhang, Y.-M.; Mallet, J.-M.; Sinaÿ, P. *Carbohydr. Res.* **1992**, *236*, 73–88.
235. Mallet, J.-M.; Meyer, G.; Yvelin, F.; Jutand, A.; Amatore, C.; Sinaÿ, P. *Carbohydr. Res.* **1993**, *244*, 237–246.

236. Amatore, C.; Jutand, A.; Meyer, G.; Bourhis, P.; Machetto, F.; Sinaÿ, P.; Tabeur, C.; Zhang, Y.-M. *J. Appl. Electrochem.* **1994**, *24*, 725–729.
237. Balavoine, G.; Berteina, S.; Gref, A.; Fischer, J.-C.; Lubineau, A. *J. Carbohydr. Chem.* **1995**, *14*, 1217–1236.
238. Nokami, J.; Osafune, M.; Ito, Y.; Miyake, F.; Sumida, S.; Torii, S. *Chem. Lett.* **1999**, 1053–1054.
239. Bohé, L.; Crich, D. *Carbohydr. Res.* **2015**, *403*, 48–59.
240. Mitsudo, K.; Kawaguchi, T.; Miyahara, S.; Matsuda, W.; Kuroboshi, M.; Tanaka, H. *Org. Lett.* **2005**, *7*, 4649–4652.
241. Tanaka, N.; Ohnishi, F.; Uchihata, D.; Torii, S.; Nokami, J. *Tetrahedron Lett.* **2007**, *48*, 7383–7387.
242. Drouin, L.; Compton, R. G.; Fairbanks, A. J. *J. Phys. Org. Chem.* **2008**, *21*, 516–522.
243. Nokami, T.; Shibuya, A.; Tsuyama, H.; Bowers, A. A.; Crich, D.; Suga, S.; Yoshida, J. *J. Am. Chem. Soc.* **2007**, *128*, 10922–10928.
244. Nokami, T.; Saito, K.; Yoshida, J. *Carbohydr. Res.* **2012**, *363*, 1–6.
245. Nokami, T.; Shibuya, A.; Yoshida, J. *Trends Glycosci. Glycotechnol.* **2008**, *20*, 175–185.
246. Nokami, T.; Shibuya, A.; Saigusa, Y.; Manabe, S.; Ito, Y.; Beilstein, J. *Beilstein J. Org. Chem.* **2012**, *8*, 456–460.
247. Nokami, T.; Shibuya, A.; Manabe, S.; Ito, Y.; Yoshida, J. *Chem. Eur. J.* **2009**, *15*, 2252–2255.
248. Nokami, T.; Nozaki, Y.; Saigusa, Y.; Shibuya, A.; Manabe, S.; Ito, Y.; Yoshida, J. *Org. Lett.* **2011**, *13*, 1544–1547.
249. Nokami, T. *Trends Glycosci. Glycotechnol.* **2012**, *24*, 203–214.
250. Nokami, T.; Tsuyama, H.; Shibuya, A.; Nakatsutsumi, T.; Yoshida, J. *Chem. Lett.* **2008**, *37*, 942–943.
251. Kawa, K.; Saitoh, T.; Kaji, E.; Nishiyama, S. *Org. Lett.* **2013**, *15*, 5484–5487.
252. Saito, K.; Saigusa, Y.; Nokami, T.; Yoshida, J. *Chem. Lett.* **2011**, *40*, 678–679.
253. Suzuki, S.; Matsumoto, K.; Kawamura, K.; Suga, S.; Yoshida, J. *Org. Lett.* **2004**, *6*, 3755–3758.
254. Drouin, L.; Compton, R.; Fietkau, N.; Fairbanks, A. J. *Synlett* **2007**, *17*, 2711–2717.
255. Sasaki, M.; Gama, Y.; Yasumoto, M.; Ishigami, Y. *Tetrahedron Lett.* **1990**, *31*, 6549–6552.
256. (a) Kochetkov, N. K.; Zhulin, V. M.; Klimov, E. M.; Malysheva, N. N.; Makarova, Z. G.; Ott, A. Y. *Carbohydr. Res.* **1987**, *164*, 241–254; (b) Kochetkov, N. K.; Zhulin, V. M.; Malysheva, N. M.; Klimov, E. M.; Makarova, Z. G. *Carbohydr. Res.* **1987**, *167*, C8–C10.

5-(Hydroxymethyl)furfural and 5-(glucosyloxymethyl)furfural in multicomponent reactions

2

Weigang Fan, Charlie Verrier, Lianjie Wang, Mohammed Ahmar, Jia-Neng Tan, Florence Popowycz, Yves Queneau
Univ Lyon, Institut de Chimie et Biochimie Moléculaires et Supramoléculaires, CNRS, Université Lyon 1, INSA Lyon, CPE Lyon, ICBMS, UMR 5246; Université Claude Bernard, Bâtiment Lederer, Villeurbanne Cedex, France

Chapter outline

1 Introduction

The dehydration of various carbohydrates to furanic aldehydes offers an impressive scope of applications. One side is the "bio-based chemistry" viewpoint, since 5-(hydroxymethyl)furfural (HMF) is a renewable platform molecule able to provide functional intermediates with industrial relevance. Another aspect is directly related to their chemical functionalities, notably the presence of the aldehyde function, which makes them available for a wide range of reactions, among which are the multicomponent reactions (MCRs). This is the focus proposed in this chapter. Furfural, obtained from 5-carbon sugars, has been widely employed in such strategies, but 6-carbon sugar-derived 5-(hydroxymethyl)furfural (HMF) and its glucosylated analog 5-(glucosyloxymethyl)furfural (GMF), obtained from isomaltulose, have been much less employed. In keeping with our standing interest for green and bio-based chemistry,[1–9]

Recent Trends in Carbohydrate Chemistry. https://doi.org/10.1016/B978-0-12-817467-8.00002-5
© 2020 Elsevier Inc. All rights reserved.

we have investigated several uses of HMF and GMF, including in MCR processes. This chapter will review this subject based on our own results and relevant literature from other groups, after a short introduction on each key topic, namely multicomponent reactions, HMF and GMF, in order to provide the full flavor of this combination of synthetic methodology and carbohydrate chemistry.

1.1 Multicomponent reactions

Multicomponent reactions (MCR) are chemical reactions where at least three molecules react to form a single product that retains all or most of the atoms of the starting materials.[10, 11] The concept of MCRs is rather rough: undoubtedly, they belong to the category of one-pot reactions, but it is not clear whether one-pot reactions that involve sequential addition of reactants at different stages of a stepwise process should also be considered as MCRs, or whether this refer only to reactions in which all three (or more) partners are present in the starting mixture.[12, 13] Due to the importance of MCRs in synthetic chemistry, a nomenclature has been established to refer to a MCR involving *n* different components, thus called *n*-CR (Fig. 1). Depending on the reactivity of engaged chemical entities and the reaction conditions, all components in MCRs might assemble in linear form (like the Mannich reaction) or undergo further cyclization (like the Hantzsch dihydropyridine synthesis) to provide complex molecules in a single step from simple building blocks.

MCRs have been known for more than 160 years, and in 1850, Adolph Strecker discovered the reaction of carbonyl compounds with amines and the cyanide anion.[14] Since then, numerous popular MCRs have been reported, such as the Hantzsch dihydropyridine synthesis (1881),[15] the Biginelli reaction (1891),[16, 17] the Mannich reaction (1912),[18] the Passerini reaction (1921),[19] or the Ugi reaction (1959).[20] The fact that all these MCRs are name reactions undoubtedly indicates their importance in organic synthesis.

MCRs possess many apparent advantages over multistep processes, such as high efficiency, atom economy, reduced waste generation, and time and energy economy. These features that comply with the majority of Anastas and Warner's principles of green chemistry make the use of MCRs an ideal synthetic strategy,[21] especially in the perspective of diversity-oriented synthesis. Indeed, MCRs have been widely used in combinatorial chemistry to produce large libraries of small molecules useful for the discovery of lead compounds in the fields of pharmaceuticals and agrochemicals.

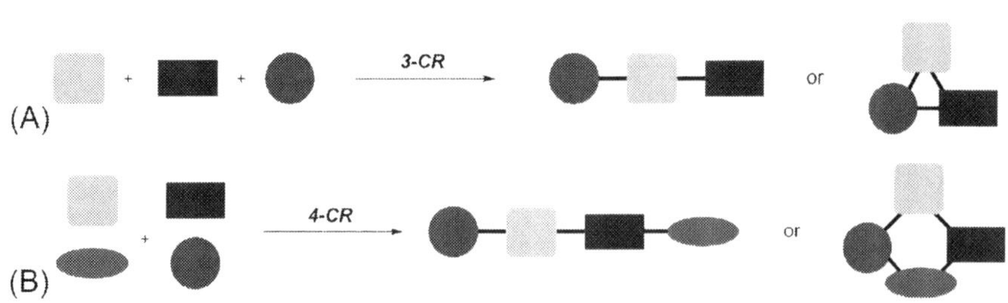

Fig. 1 Schematic representation of MCRs involving different entities in 3-CR (A) or 4-CR (B).

The utilization of MCRs in polymer chemistry has also been revealed as a promising strategy to make highly functionalized polymers.

Aldehydes are fundamental building blocks in organic synthesis and probably the most commonly employed class of chemicals in MCRs. Their inherent electrophilicity allows for the formation, without the need for activation, of reactive species through condensation reactions with amines or carbonyl compounds that can be further reacted with another chemical entity to generate a MCR product. It is, for example, the case in the Strecker, Ugi, Hantzsch, or Mannich reactions.

Most methodological studies on MCRs, focusing mainly on reactivity and conditions, use commercially available substrates for investigating the scope of the reaction, benzaldehyde being the classical reference substrate. In a more sustainable perspective that would also include the preference for renewable resources, the application of MCRs to the valorization of biosourced aldehydes would combine the features of MCRs in terms of atom and energy economy and the renewability of the starting platform molecules.

Furfural is probably the most commonly encountered bio-based aldehyde in multicomponent reactions, as it is highly available and exhibits a predictable reactivity like simple aromatic aldehydes. However, the bio-based HMF and GMF tend to behave differently compared to the unsubstituted furfural in many reactions due to the presence of additional hydroxy groups on the substrates raising solubility, reactivity, and/or stability issues. Furthermore, despite the high attractiveness of such strategies, the moderate availability of HMF before the recent years has limited its use as starting materials toward complex molecules via MCRs. GMF remains much less used due to its lower availability; however, its peculiar glycosylated structure offers interesting opportunities and is therefore an interesting analog.

1.2 5-(Hydroxymethyl)furfural

5-(Hydroxymethyl)furfural (HMF) has become of increasing interest for the scientific community during the last decades, along with the advent of green chemistry principles. Indeed, it has been ranked among the top "10+4" bio-based products to exploit in biorefineries.[22] This biosourced C-6 furanic derivative can be obtained by high temperature, acid-mediated dehydration of hexoses and poly-hexoses such as cellulose or inulin (Fig. 2). Due to its limited stability to both acidic conditions and heat,[23–25] the absence of efficient and industrially relevant methods for its obtention in pure form has been a brake for the development of more sustainable processes using (HMF) until the beginning of the century. However, recent development of numerous modern methods for the preparation and isolation of (HMF) (often from complex mixtures of partially dehydrated sugars, rehydration products, and oligomeric compounds known as humins) drastically increased its attractiveness, because of its new availability.

For (HMF) production, different systems using Brønsted or Lewis acid catalysts have been employed in both homogeneous and heterogeneous conditions, with various efficiencies.[26–31] The key points are the control of the selectivity during the sugar dehydration and the stability of the generated (HMF) in the acidic conditions employed for its formation. Those parameters can be tuned by the nature of the catalyst employed and

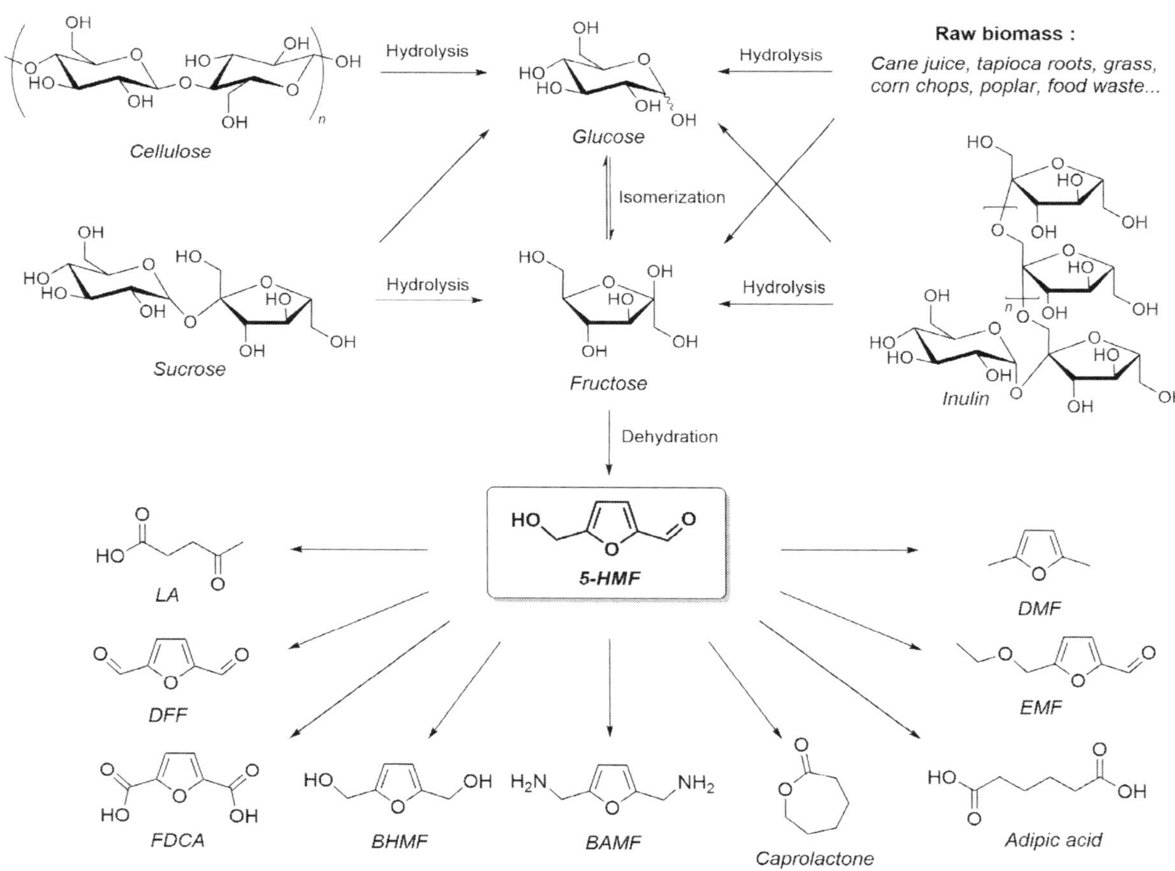

Fig. 2 Overview of HMF production from biosourced materials and its main applications. See text for abbreviations.

the reaction conditions, for example, using a biphasic medium allowing a continuous extraction of the generated furanic aldehyde. In a more applicative perspective, raw biomasses,[32] such as tapioca roots, wood chops, corncob, or even food waste, have also been converted into (HMF), taking advantage of the high carbohydrate content of those natural feedstocks. As a consequence of this increasing availability, (HMF) has rapidly become one of the most attractive bio-based platform molecules due to the presence of three reactive functional groups, an aldehyde, a primary alcohol, and a furan ring, which offers possible derivatizations through a wide range of chemical transformations (Fig. 2).[33–35]

Most established applications of (HMF) concern the production of industrially relevant molecules, with the aim of replacing currently used petroleum-based chemicals, in keeping with the necessary renewal of chemical industry that must comply with the principles of sustainable chemistry. Representative examples of HMF-derived molecules concern the field of polymer chemistry, with the preparation of monomers such as diformylfuran (DFF), furandicarboxylic acid (FDCA), bis(hydroxymethyl)furan (BHMF), bis(aminomethyl)furan (BAMF), levulinic acid (LA), caprolactone, or adipic acid. Bio-based liquid fuel has also been obtained from HMF, among which dimethylfuran (DMF) and ethoxymethylfurfural (EMF) proved of great interest. Those small molecules can be obtained by simple transformations, such as hydration, oxidations, reductions, and etherification of HMF. Besides the aforementioned applications, (HMF) has however recently been used as substrate for more complex transformations, including multicomponent reactions, and this aspect will be detailed throughout this chapter.

1.3 5-(α-D-Glucosyloxymethyl)furfural

Another interesting class of bio-based furanic derivatives is the family of 5-glycosylated analogs of HMF, among which 5-(α-D-glucosyloxymethyl)furfural GMF stands as the most accessible and used molecule.[36] Similar to HMF, GMF and analogs have been identified or isolated in many natural sources such as plants, foods, processed vegetables, or extraction products from these plants. It has been found, for example, in caramel candy.[37] During a caramelization process, glycosylated furans are formed by dehydration of oligosaccharides at high temperature, or by glycosylation between HMF and other sugars. Several GMF-type molecules could also be found in the extracts of some traditional medicinal herbs and fruits, such as *Rehmanniae Radix*, *Prinsepia uniflora*, and *Amelanchier canadensis* (Fig. 3).[38–42]

The most commonly employed method to obtain GMF is the dehydration of isomaltulose [6-*O*-(α-D-glucopyranosyl)-D-fructose] (Scheme 1). This disaccharide, also known as palatinose, has been used in Japan as a sugar alternative in foods since 1985, and is manufactured by enzymatic rearrangement of sucrose on the industrial scale.[43–46] This key step involves the conversion of the α(1→2) linkage between glucose and fructose to a α(1→6) linkage under the action of sucrose isomerase. The production of GMF from isomaltulose requires a selective acid-promoted dehydration of the substrate in strictly anhydrous conditions to avoid undesired cleavage of the glycosidic bond leading to side formation of glucose and HMF. The best result for producing GMF was reported by Lichtenthaler et al., using the acidic resin Dowex 50 WX4 (H^+ form) in DMSO at 120°C, in the presence of molecular sieves for trapping

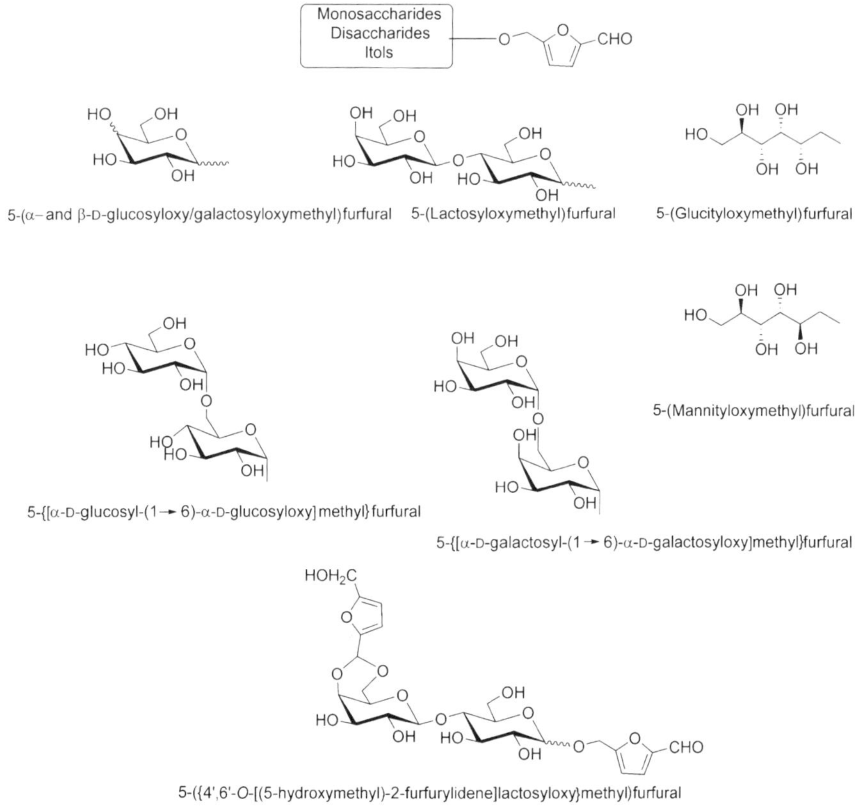

Fig. 3 Examples of glycosylated HMF analogs isolated from plant extracts and food.

the water produced during the reaction, and giving GMF in 70% yield after purification.[47] More recently, Koenig reported that the dehydration of isomaltulose could also be performed in solvent-free conditions, through the formation of a low melting mixture between the carbohydrate and choline chloride (1:1 ratio wt/wt). Heating this mixture at 100°C under reduced pressure in the presence of zinc chloride as catalyst gave up to 52% of the glucosylated furan in only 1 h.[48] This method may be of interest as an environmentally friendly GMF synthesis process. Choline chloride is known as an interesting additive or cosolvent for producing furanic aldehydes from carbohydrates.[49, 50] Analogs of GMF can also be synthesized following a similar approach from other disaccharides. Sodium aluminate-mediated isomerization of glucose-containing disaccharides followed by acidic dehydration toward the furanic derivative was applied to melibiose, gentiobiose, and primeverose that were converted into α-GalMF, β-GMF, and β-xylMF, respectively, with yields between 45% and 50% over two steps (Scheme 1).[51] Alternatively, glycosylation reactions of HMF with glucose or galactose have been described to produce GMF and GalMF.[37, 52]

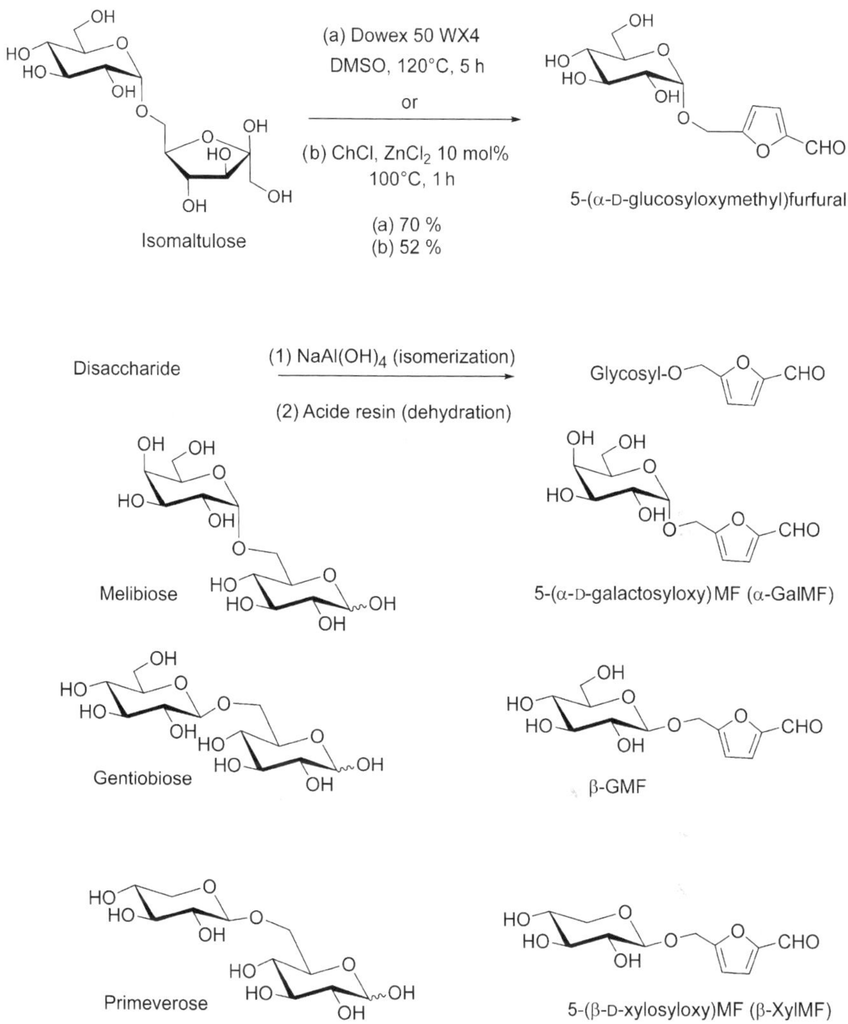

Scheme 1 Isomerization/dehydration of disaccharides toward glycosylated bio-based furans.

In the following sections, examples depicted in Fig. 4 taken from literature including our own recent work will illustrate the remarkable diversity of chemical architectures which can be obtained from HMF or GMF in one unique step. Actually, all examples concern reactions involving the furanic aldehyde and an amine forming in situ an intermediate imine, which can evolve differently depending on the nature of the third (or more) component. The first two sections concern examples of nucleophilic attack of the intermediate imine by enolates (Biginelli, *aza*-Baylis-Hillman), followed

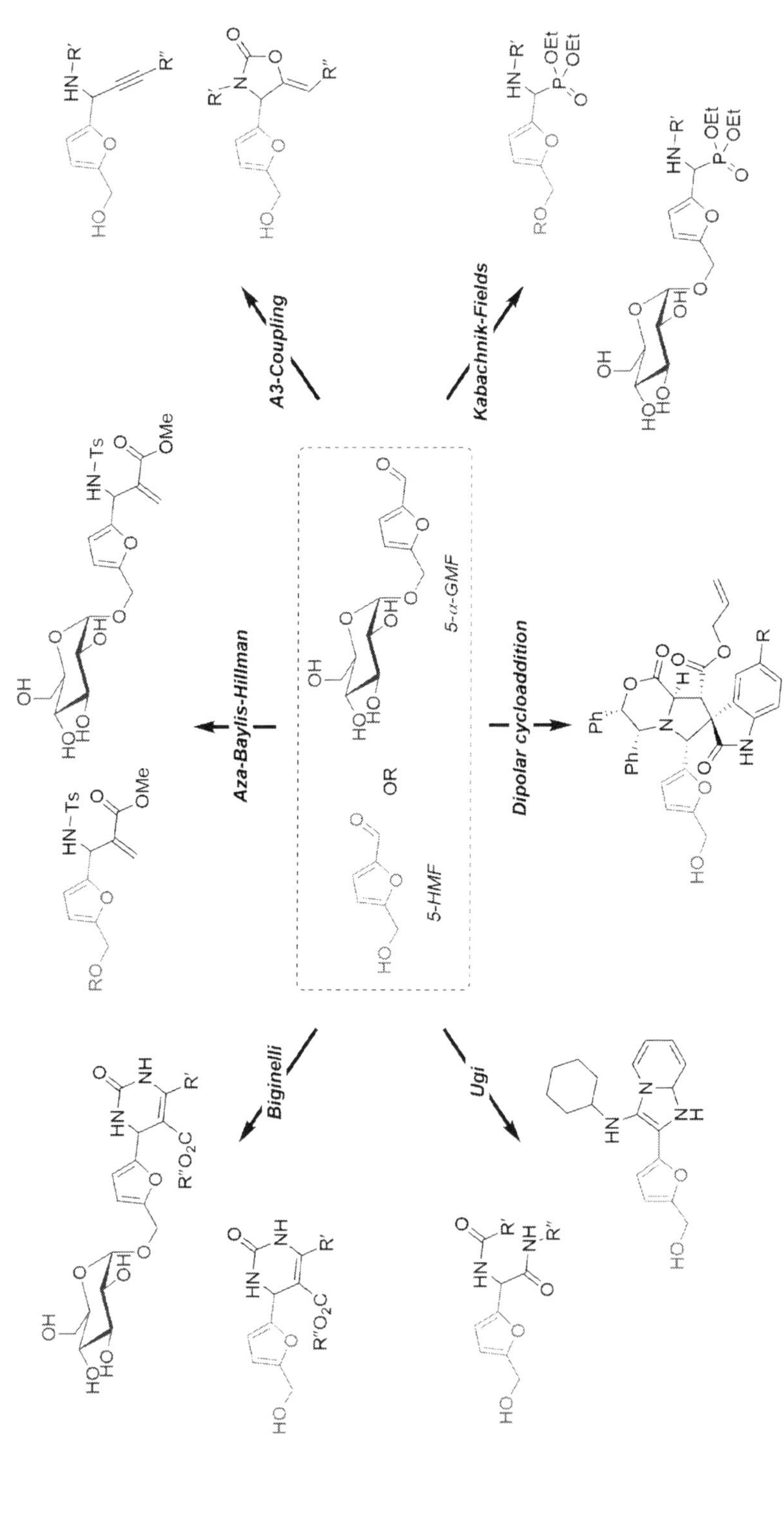

Fig. 4 Multicomponent reactions of HMF and GMF.

by a section dealing with acetylides additions (A^3 reactions), then one about addition of isonitriles (Ugi-type reactions). The last section concerns the addition of a nucleophilic heteroatom, namely a phosphorus atom in the Kabachnik-Fields reaction, and examples of cycloadditions of furanic 1,3-dipoles.

2 Biginelli-type reactions

In 1891, the Italian chemist Pietro Biginelli reported for the first time the reaction of an aldehyde, urea, and a 1,3-dicarbonyl compound, yielding with unprecedented efficiency the 3,4-dihydropyrimidin-2-one (DHPM) scaffold, a heterocyclic motif present in several bioactive molecules.[53, 54] Among mechanistic propositions for this reaction is the formation of a Knoevenagel product, or pathways involving either an iminium or an enamine intermediate (Scheme 2). More recent studies by Kappe, supported by calculations, showed the prevalence of the *N*-acyliminium intermediate formed by condensation of the aldehyde and the urea under acidic activation.[55] This highly electrophilic transient intermediate then undergoes irreversible nucleophilic addition of the dicarbonyl compound generating a new C–C bond. The resulting urea cyclizes through a second condensation on one of the carbonyl group to produce, after tautomerization, the dihydropyrimidine ring.

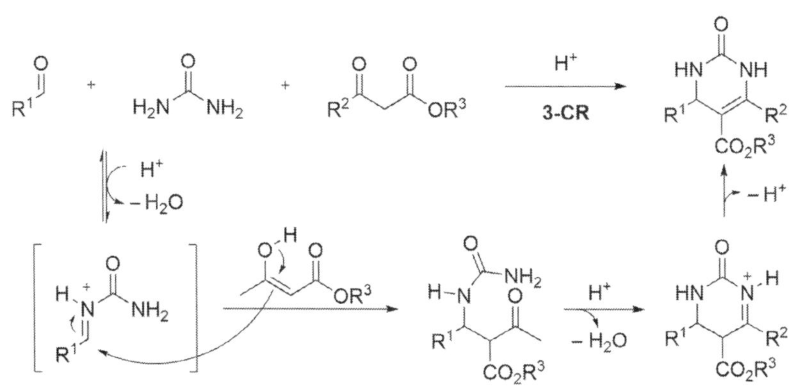

Scheme 2 Mechanism of the Biginelli reaction proposed by Kappe.

The first example of Biginelli-type reaction of HMF was reported by Tulshian and coworkers to prepare adenosine receptor (A2A) antagonists (Scheme 3).[56] The condensation of HMF, indanone **1** and guanidine **2** in refluxing EtOH provided the compound **3**. Though very few information about the conditions and yield were given, satisfyingly, the HMF-adorned analog **3** appeared as a promising active compound, among the different fused pyrimidines tested, as A2A adenosine receptor antagonist with approximately 100-fold selectivity over the A1 receptor.

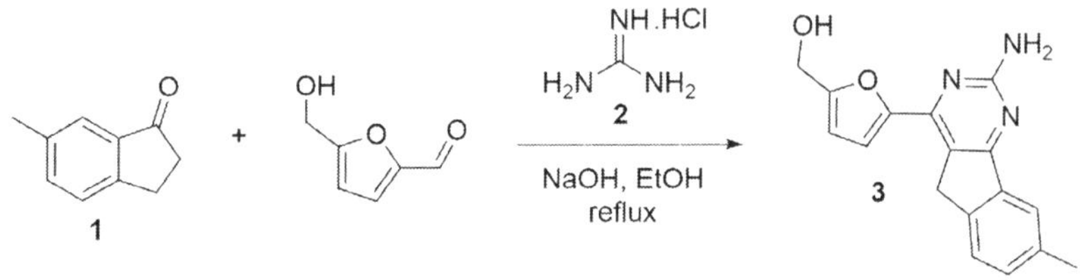

Scheme 3 Preparation of a bioactive pyrimidine through Biginelli reaction of HMF.

As part of our ongoing research on the valorization of bio-based platform chemicals, our group has studied the Biginelli reaction employing HMF as aldehyde component in order to develop a more general method toward HMF-derived DHPMs (Scheme 4).[57] Initial attempts employing simple Brønsted acids such as HCl or trifluoroacetic acid as catalyst for the reaction of HMF with urea and acetylacetone did not prove successful, as no desired heterocycle could be obtained. Even though, in similar conditions, furfural provided the corresponding DHPM with good yield, the use of HMF produced a complex mixture of degradation products, illustrating the significantly different behavior between furfural and HMF, and proving that furfural cannot be considered as a pertinent model for HMF. From our initial results, we logically suspected the strong acidity of the medium to trigger side reactions, and we turned our attention to the use of milder Lewis acids. Among the tested catalysts, zinc(II) chloride gave the best results, both in terms of yield and selectivity toward the DHPM versus the formation of the Knoevenagel adduct between acetylacetone **4** and HMF. An examination of the reaction conditions allowed to run the reaction in neat conditions, using 20 mol% of catalyst at 80°C, and the corresponding DHPM **6a** could be isolated in 86% after 8 h. Changing the aldehyde to the more complex α-GMF, the carbohydrate-adorned DHPM **6b** was obtained in 62% as an equimolar mixture of two diastereoisomers. The scope of the Biginelli reaction was also explored with respect to variations on the 1,3-dicarbonyl compound, using different diketones and ketoesters that reacted with similar efficiency toward **6c-h**. *N*-methyl urea **5a**, thiourea **5b** as well as urea-type heterocyclic partners **5c** and **5d** were also successfully employed in the MCR with HMF.

3 Aza-Morita-Baylis-Hillman reaction

The aza-Morita-Baylis-Hillman (aza-MBH) reaction is a three-component reaction involving an aldehyde, an amine (or a sulfonamide), and an electron-deficient alkene, typically an acrylate-type compound. The reaction is promoted by Lewis bases, such as tertiary amines or phosphines, and proceeds as follows: the nitrogen-containing partner condenses on the aldehyde to form an imine, while the Lewis base reversibly adds on the electron-deficient alkene, resulting in a betaine intermediate (Scheme 5). The latter possesses an enolate-like reactivity and can add on the imine to form another zwitterionic intermediate that can eliminate the Lewis base to regenerate the carbon-carbon double bond. Whereas this reaction was first reported in a two-step

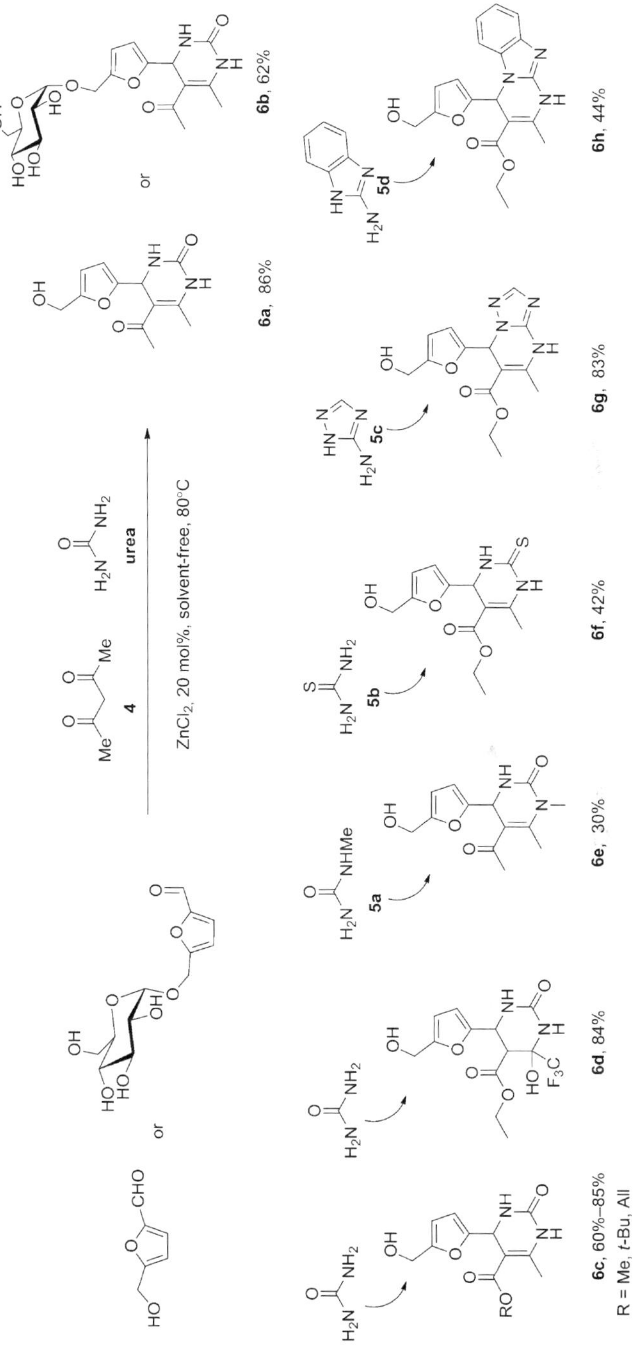

Scheme 4 Solvent-free Biginelli reaction of HMF and GMF.

procedure, requiring the isolation of the imine.[58, 59] Kahn showed later the possibility to combine the three components in a MCR,[60] giving direct access to functionalized allylic amines from simple starting materials.

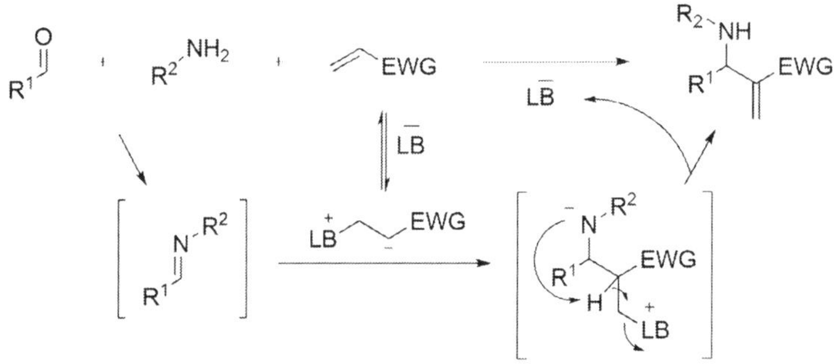

Scheme 5 Mechanism of the aza-MBH reaction. LB = Lewis base.

Considering our previous experiences in MBH reactions of bio-based furanic aldehydes,[61, 62] our group recently studied the possibility to prepare new carbohydrate-containing allylic amines using this strategy from GMF.[63]

We started our study using *p*-toluenesulfonamide as nitrogen-containing partner and methyl acrylate, with various combinations of Lewis acids/Lewis bases to promote, respectively, the imine formation and the carbon-carbon bond-forming reaction. After optimization, we found that the best results were obtained using lanthanum triflate and 3-hydroxyquinuclidine (3-HQD) in catalytic amounts (5 and 20 mol%, respectively) in methanol at room temperature, and the model allylic sulfonamide **9a** could be isolated in 70% yield. We then explored the scope of the reaction using our optimized conditions (Scheme 6). Unsurprisingly, other sulfonamides could be employed with similar efficiency, and other acrylates and acrylonitriles also proved to be reactive partners under our conditions. Finally, it has been possible to change the nature of the link between the furanic aldehyde and the sugar, as well as the sugar itself, as both β-XylMF and β-GMF provided the corresponding allylic amines **9g** and **9h** in good yields.

4 A^3-coupling of bio-based furanic aldehydes

The reaction between an aldehyde, an amine, and an alkyne under copper catalysis is referred to as the A^3-coupling reaction.[64, 65] This MCR constitutes a straightforward access toward propargylamines, highly valuable building blocks in the synthesis of complex amines.

Mechanistically, the A^3-coupling reaction starts with the condensation of a primary or secondary amine on an aldehyde (Scheme 7), generating the corresponding iminium ion. Meanwhile, a terminal alkyne undergoes amine-assisted cupration to form the corresponding copper acetylide, and this catalytically generated organometallic species adds on the transient iminium to produce the final propargylic amine and

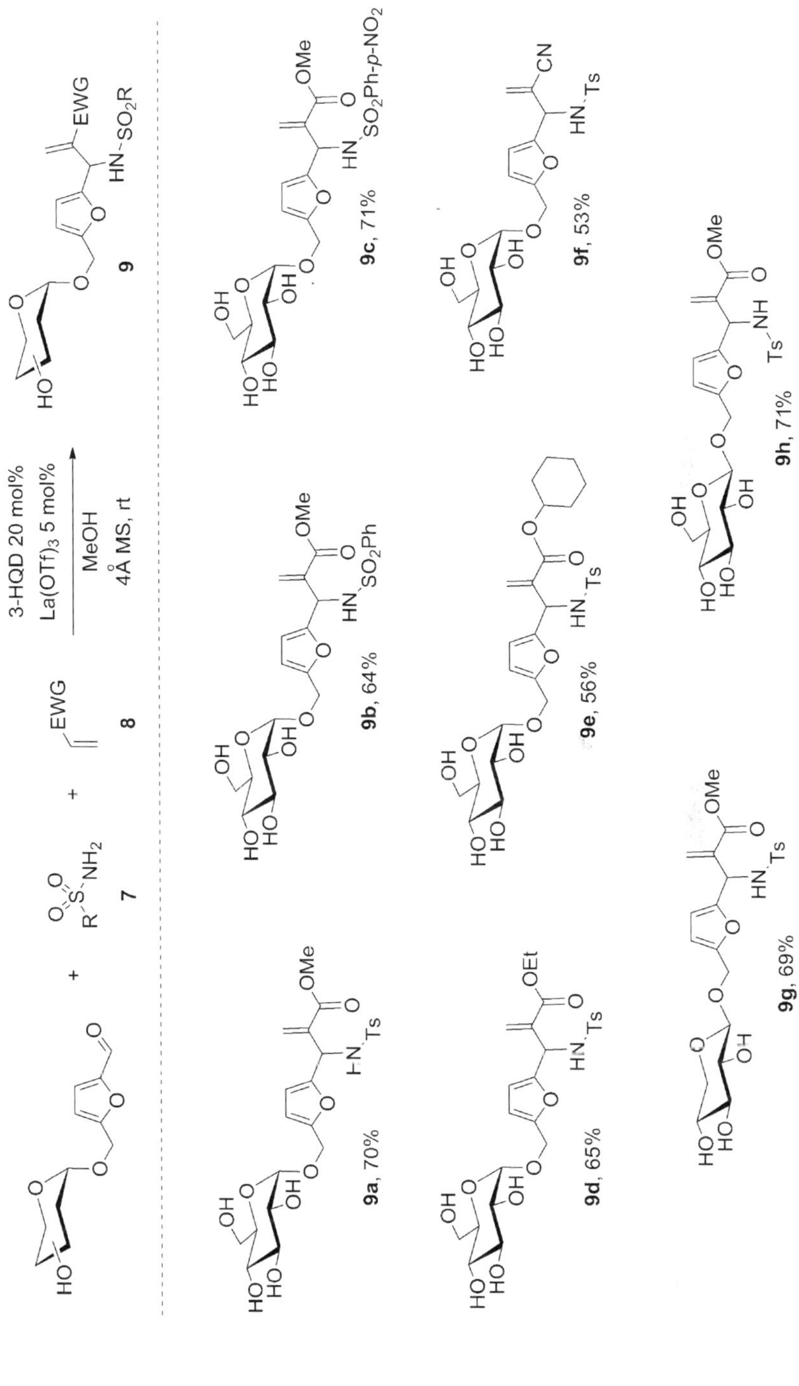

Scheme 6 Scope of the aza-MBH reaction of glycosylated bio-based furfurals.

regenerate the copper(I) catalyst. Other metals have been employed in similar processes, such as ruthenium or gold; however, copper remains the most popular and the only one applied to bio-based furanic aldehydes.

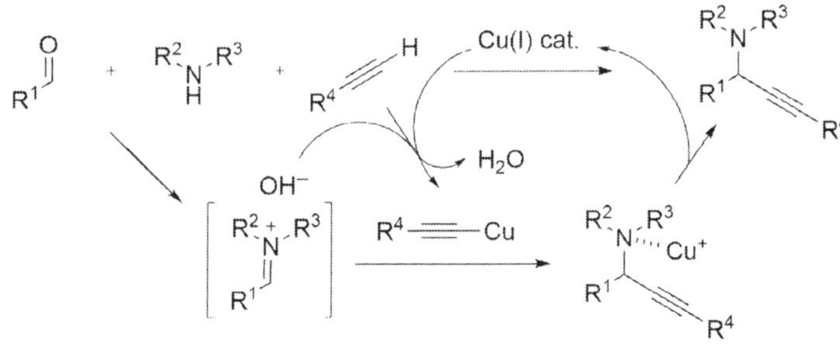

Scheme 7 Mechanism of the A^3-coupling reaction.

The first utilization of HMF in A^3-coupling reaction was reported by Kundu in 2002 (Scheme 8).[66] Strictly speaking, HMF was first immobilized on a Sieber-amide resin using a succinate linker via its primary hydroxy group, prior to being subjected to A^3-coupling conditions. The obtained resin-supported HMF **10** was treated with a variety of *N*-substituted piperazines and terminal alkynes **11** in the presence of two equivalents of copper(I) chloride in dioxane at 85°C. The produced propargylamine **12** could be released from the resin by acidic hydrolysis in the presence of trifluoroacetic acid, and isolated in moderate yields.

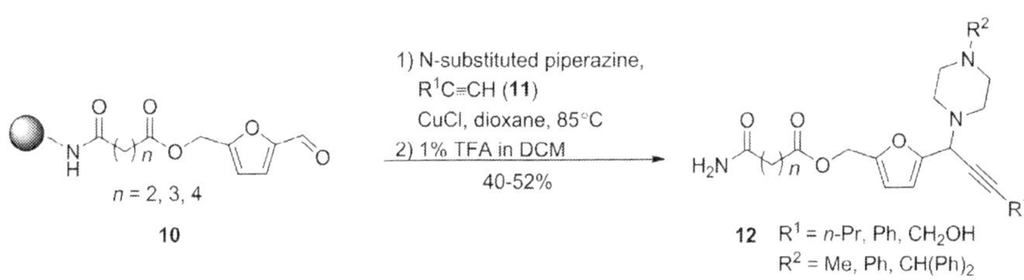

Scheme 8 A^3-coupling of supported HMF.

In 2018, Xu et al. described a modified A^3-coupling of HMF incorporating a new partner in the MCR, namely carbon dioxide, resulting in a four-component reaction that yields exo-methylene oxazolidinones (Scheme 9).[67] The authors proposed a classical A^3-coupling mechanism generating a secondary propargylamine, followed by

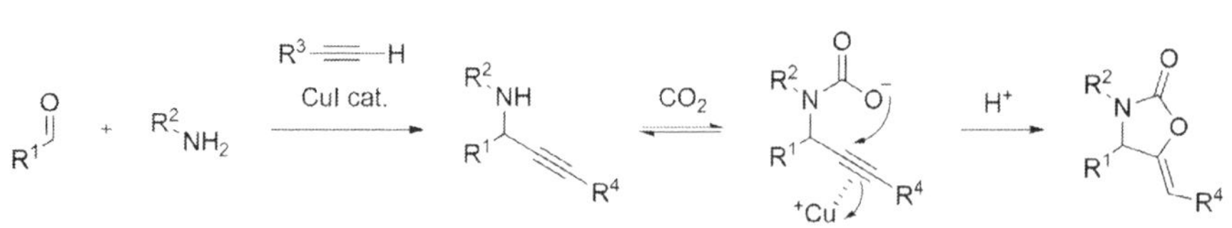

Scheme 9 Mechanistic proposal for the CO_2 incorporation in A^3-coupling reaction.

reversible addition of the amine on carbon dioxide. The resulting carbamate then cyclizes via anti-addition on the copper-coordinated triple bond to form the oxazolidine ring. Alternatively, another mechanism could not be ruled out, where a carbamate salt is formed from the amine and CO_2 and then condenses on the aldehyde to undergo A^3-coupling, the product being able to directly cyclize as previously proposed.

The screening of catalysts showed that copper(I) iodide gave superior reactivity and increasing both the temperature and the pressure of CO_2 facilitated the coupling reaction. Finally, the optimized conditions were established as using 30 mol% of CuI as catalyst at 75°C in ethanol. Various primary aliphatic amines **13** and aromatic alkynes **14** were employed under the optimized conditions, leading to corresponding oxazolidinones **15** in 17%–84% yields (Scheme 10). Notably, a double coupling occurred on 1,3-diethynylbenzene to give bis(oxazolidinone) **15i** in 25% yield. As for the amine component, it was found that aniline was not able to yield the oxazolidinone, perhaps due to its low nucleophilicity on nitrogen atom. In the case of secondary aliphatic amines, such as diethylamine and pyrrolidine, the corresponding propargylamines were obtained without incorporation of CO_2.

5 Ugi-type reactions

The reaction between a carboxylic acid, an aldehyde, an amine, and an isonitrile is one of the most famous MCRs ever reported, commonly known as the Ugi reaction. It allows for the direct preparation of *N*-acylated α-aminoamides from simple building blocks. The admitted mechanism relies on the formation of an imine from an aldehyde and a primary amine, and its protonation by the carboxylic acid triggers the nucleophilic attack of the isonitrile partner on the imine (Scheme 11). The generated nitrilium ion is then attacked by the carboxylate molecule, and a 1,4-acyl-shift provides the final product of the Ugi 4-CR reaction.

Multicomponent reactions in combination with tandem reactions provide an opportunity to rapidly access molecules with high level of structural complexity. This concept has been applied to HMF, as the Ugi reaction has been incorporated in a tandem sequence taking advantage of the reactivity of the furan heterocycle in Diels-Alder cycloaddition, in order to prepare a large library of polycyclic drug-like molecules (Scheme 12).[68] Similar to the A^3-coupling reaction (*vide supra*), the authors first immobilized HMF on a MPEG support thanks to formylacetal linkage on the hydroxymethyl group. In order to design a tandem Ugi/Diels-Alder sequence, the authors used benzyl isocyanide in combination with **16** and diverse amine and fumaric acid monoamide (**17**) partners to generate structural diversity among the library. In this strategy, the Ugi product incorporates an electron-rich diene that is the furan ring as well as the electron-deficient alkene of the fumarate moiety, able to undergo intramolecular 4+2 cycloaddition to generate a polycyclic adduct **18**. Twenty amines and five fumaric acid monoamide building blocks were tested in parallel with benzyl isocyanide and MPEG-supported HMF as constant components in MeOH at 50°C. After the reaction completion, the mixture was purified through gel permeation chromatography (GPC) to remove small molecules. Treatment of obtained immobilized

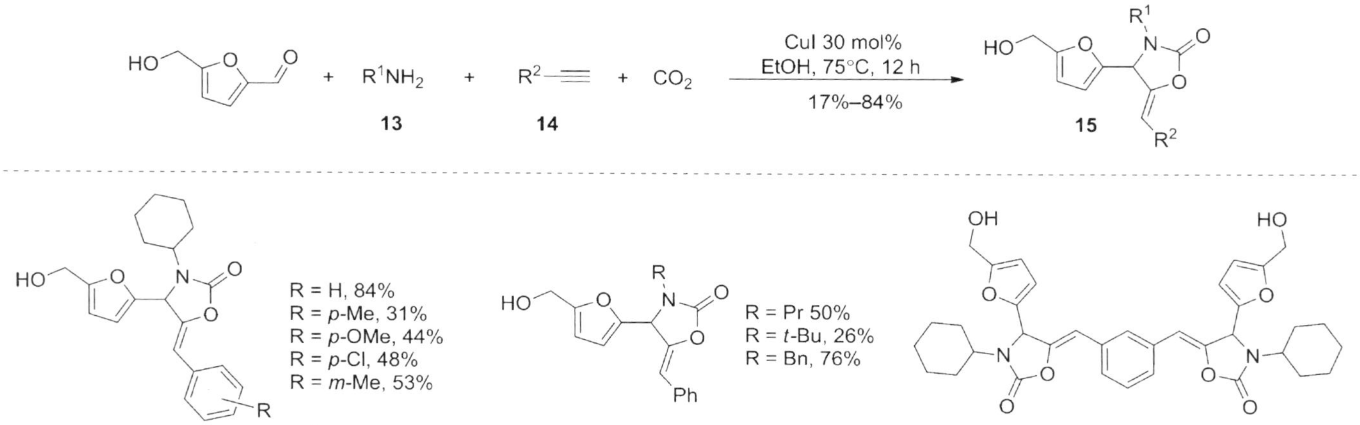

Scheme 10 A^3 coupling reaction of HMF, primary amines, terminal alkynes and CO_2.

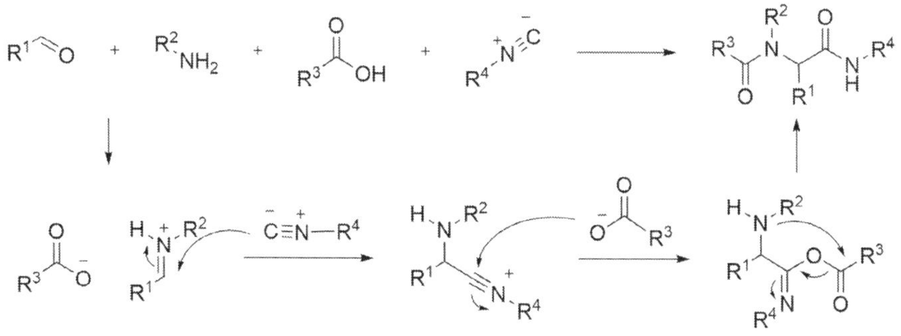

Scheme 11 Mechanism of the Ugi 4-CR.

products with trifluoroacetic acid in dichloromethane at room temperature released the products **19** from their support. Among 100 parallel trials, 96 reactions proceeded successfully, delivering the desired products, among which 68 reactions gave yields of **19** superior to 60%. Finally, it was found that insoluble polymer beads TentaGel S OH resin-supported HMF was also an efficient platform for solid phase synthesis of heterotricycles via Ugi/Diels-Alder reaction.

A variant of the Ugi reaction using aminopyridines, also known as Groebcke-Blackburn-Bienaymé reaction (Scheme 13), has been applied for the production of novel compounds owning bioactivities against the enzyme HIV-1 reverse transcriptase.[69] Using HMF and cyclohexylisonitrile, the furan-substituted imidazolopyridine **20** is formed after reaction at 100°C in the presence of acidic montmorillonite K10.

A last example of Ugi-like 3-CR of HMF, using two amines **21** and **22** and diphenylmethyl isocyanide **23**, has been reported using microfluidic technology, to produce the corresponding Ugi 3-CR adducts **24** and **25** via combinatorial chemistry approach (Scheme 14).[70] However, the products were only detected by mass spectrometry and the yields were assumed to be modest.

6 Kabachnik-Fields reaction

α-Aminophosphonates constitute an important class of organophosphorus compounds because of their wide range of biological properties and structural analogies with α-amino acids, the most commonly used aminophosphonate being the famous glyphosate. The Kabachnik-Fields reaction represents undoubtedly the most straightforward approach to α-aminophosphonates starting from simple aldehydes, amines and dialkylphosphites.[71] Analogically to the A^3-coupling, the transformation proceeds through the formation of a reactive imine that undergoes nucleophilic addition of the phosphorous partner (Scheme 15).

Some HMF-derived α-aminophosphonates had been previously prepared by Skowroński via the Pudovik reaction,[72] which is the stepwise version of the Kabachnik-Fields reaction, requiring the isolation of the imine. In this case, the dialkylphosphite addition was promoted using a catalytic amount of trifluoroacetic acid. However, our group recently reported a more efficient method to directly access HMF-derived

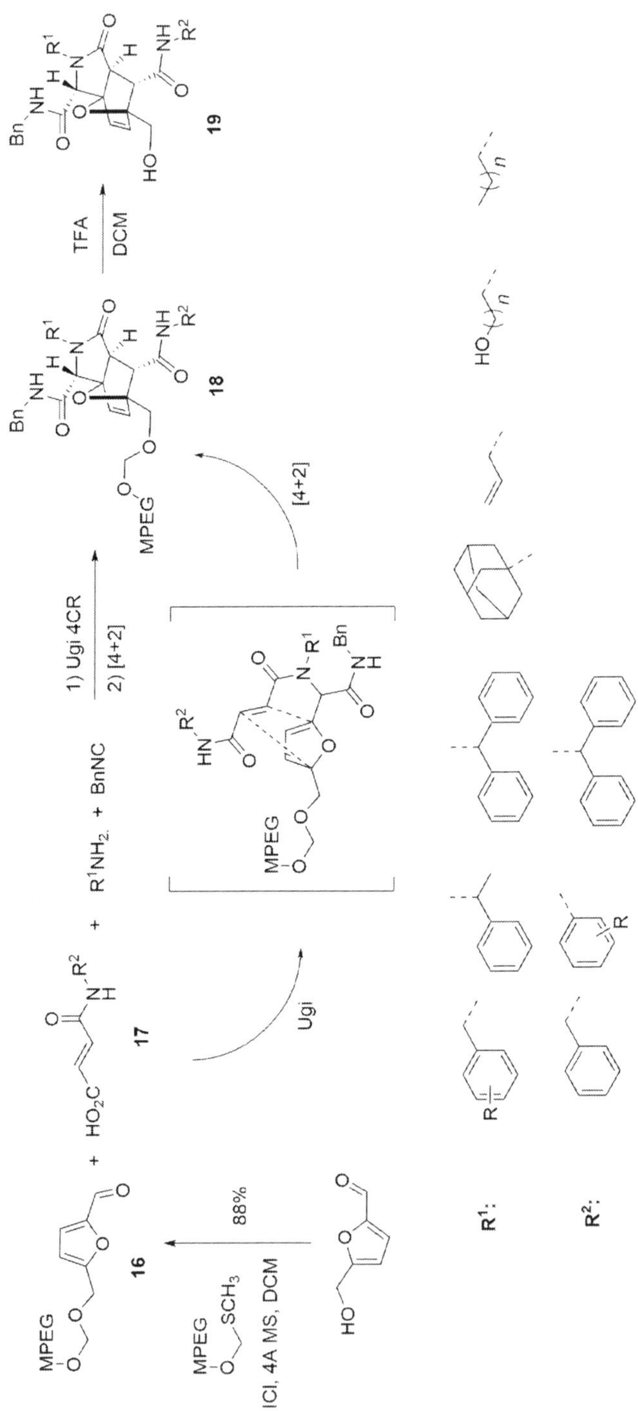

Scheme 12 Design of the Ugi/Diels-Alder tandem sequence.

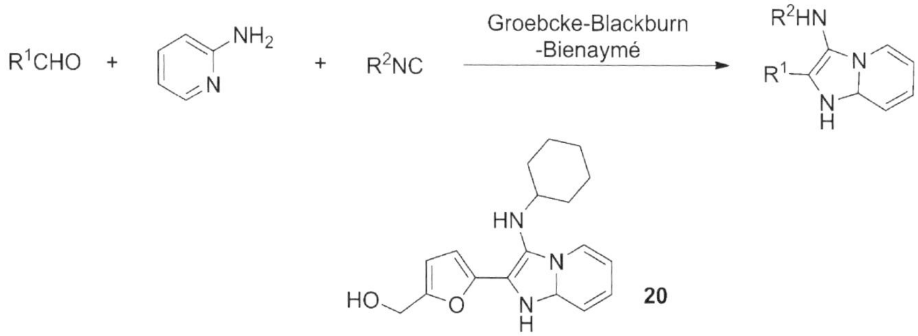

Scheme 13 Groebcke-Blackburn-Bienaymé reaction and its product from HMF.

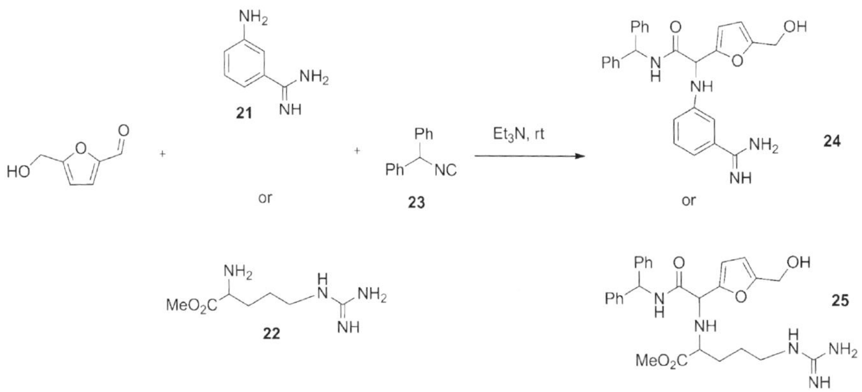

Scheme 14 Ugi-type 3-CR of HMF.

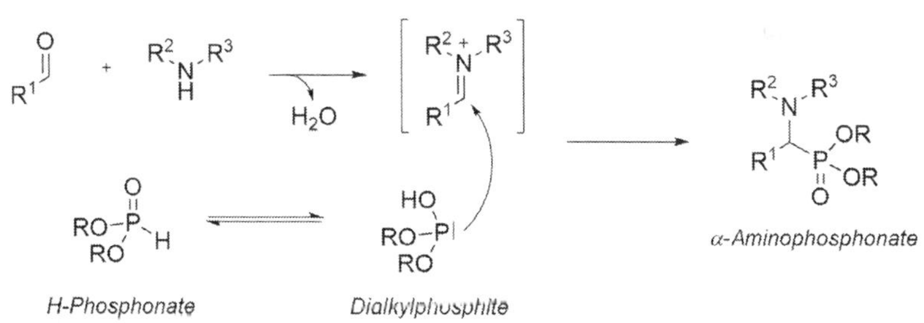

Scheme 15 Mechanism of the Kabachnik-Fields reaction.

aminophosphonates through a Kabachnik-Fields reaction from HMF, amines and dialkylphosphites (Scheme 16). An optimization of the conditions for the MCR led us to run the reaction using 5 mol% of molecular iodine as Lewis acid catalyst in 2-MeTHF.[73] A wide range of amines **26** and dialkylphosphites **27** were investigated, providing the corresponding α-aminophosphonates **28** in fair to excellent yields. For the amine component, primary anilines show better reactivity than benzylamine, aliphatic amines and secondary anilines. The reactions with anilines were performed at room temperature

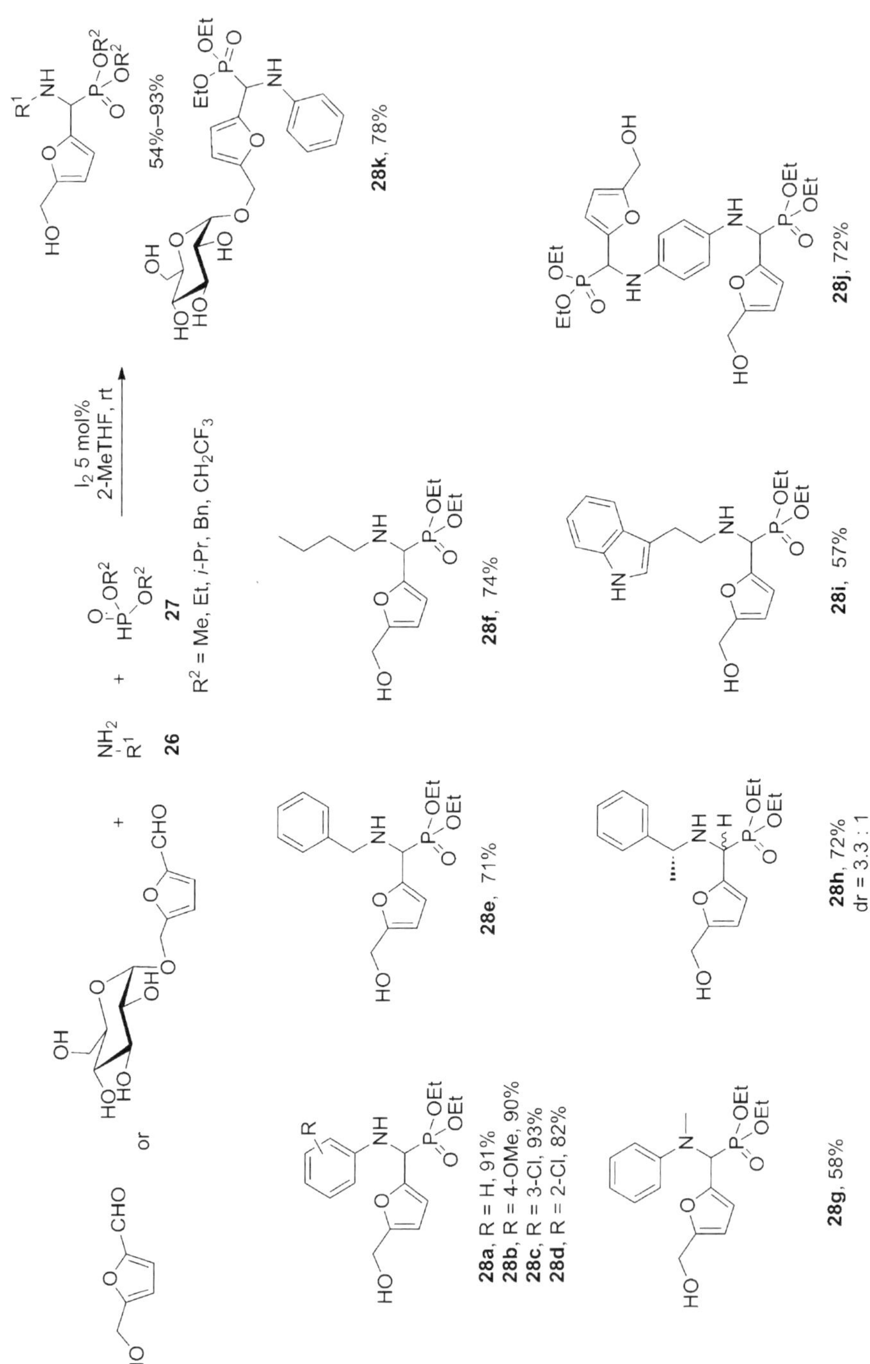

Scheme 16 Kabachnik-Fields reaction involving HMF and α-GMF.

and provided higher yields than that from aliphatic amines. The latter have to be promoted using higher temperature (up to 50°C). It should be noted that in the case of anilines, lower yields were obtained at higher temperatures, probably due to the thermal instability of the product. (*R*)-α-Methylbenzylamine provided **28h** as a mixture of two diastereoisomers in 3.3:1 ratio, whereas *p*-phenylenediamine provided double Kabachnik-Fields product **28j** in 72% yield using two equivalents of HMF. Concerning the phosphite component, dimethyl-, di*iso*propyl- and dibenzyl-phosphites reacted smoothly, delivering the corresponding products in 86–89% yields. Interestingly, tris(2,2,2-trifluoroethyl)phosphite that has two strong electron-withdrawing groups also led to the target product, albeit with a modest yield of 54%. Triethylphosphite was also successfully reacted under the optimal conditions, giving the product **28a** in 74% yield. The reaction with diethylphosphite proceeded as well under solvent-free conditions at room temperature, **28a** being obtained in 91% yield after 8 h. The method could be extended to GMF, leading to the sugar-containing aminophosphonate **28k** in 72% yield without any change in the reaction conditions (5 mol% I_2, 2-MeTHF, rt). Finally, derivatizations on the Kabachnik-Fields product taking advantage of the intact hydroxy group were investigated, such as conversions to aldehyde and azide.[73]

It must be noted that, though slower, the catalyst-free reaction is also possible. Together with the use of the bio-based solvent 2-MeTHF, this offers a particularly interesting example of a clean, atom-economical, and bio-based approach to complex targets.

7 Multicomponent dipolar cycloadditions

Cycloaddition of HMF-derived 1,3 dipoles constitutes a valuable strategy for the rapid obtention of HMF-derived heterocycles. Like the precedent examples of MCRs, the strategy involves the condensation of a nitrogen-centered nucleophile on the aldehyde to form a carbon-nitrogen double bond. However, in order to turn this imine-like electrophile into a 1,3-dipole, the intermediate must undergo deprotonation in the α position of the nitrogen as shown in Scheme 17, before thermal cycloaddition with an olefinic dipolarophile.

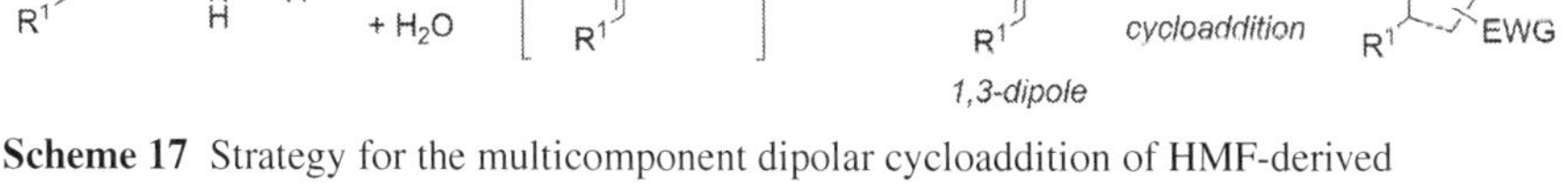

Scheme 17 Strategy for the multicomponent dipolar cycloaddition of HMF-derived 1,3-dipoles.

Schreiber prepared a couple of HMF-based spirooxindole derivatives **33** via the reaction between macrobead-supported HMF **29**, 5,6-diphenylmorpholin-2-one **30** and isatin-derived dipolarophiles **31**.[74] The approach here relies on the condensation of **30** on **29** to form the corresponding iminium-ion. The presence of a carboxylate on the amine scaffold allows for its deprotonation to produce a chiral azomethine ylide **32** that can react with **31** in a highly diastereoselective fashion (Scheme 18).[75]

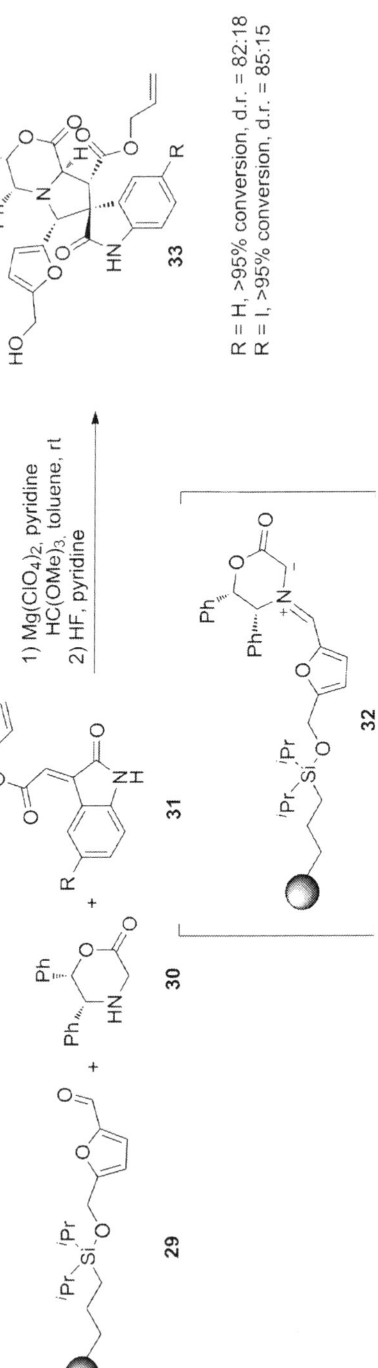

Scheme 18 Construction of spiro-oxindoles from HMF.

Magnesium perchlorate was employed as a mild Lewis acid to promote this reaction. Moreover, pyridine and methylorthoformate were added to minimize the cleavage of the loaded aldehyde from the silicon linker and as a dehydrating agent, respectively. The optimized conditions afforded good conversion (> 95%) and diastereoselectivity (> 82:18). After completion of this 3-CR, the macrobeads were treated with HF pyridine, thus releasing the final compound **33**.

Our group recently explored the possibility to use a similar approach for the production of HMF and GMF-derived isoxazolidines: heterocycles that can be found in the structure of several natural products and biologically active drugs.[76] The strategy is based on the use of *N*-substituted hydroxylamines that can condense on HMF to form a stable nitrone, able to react with electron-deficient olefins via cycloaddition (Scheme 19).

Preliminary results have been obtained using HMF, *N*-methylhydroxylamine, and methyl acrylate as dipolarophile. The hydroxylamine being unstable, they are usually sold as hydrochloride salts **34**, rendering necessary the addition of a base in the medium to promote the condensation with carbonyl compounds. In our hands, sodium hydrogenocarbonate gave the best results, and a screening of the conditions showed a good reactivity in alcoholic medium for both the nitrone **35** formation and its cycloaddition with acrylates. Our best results were obtained when the reaction was carried out in isopropanol (1M) at 80°C, using 2 equivalents of dipolarophile. Following this procedure, different dipolarophiles **36**, namely acrylamides and cyclic enones could be engaged in the MCR, and the corresponding HMF-derived isoxazolidines **37** were obtained in relatively moderate yields. Extension of the method to GMF required to work in more polar ethanol, and the product **37e** as a mixture of diastereo- and regio-isomers of reaction with methyl acrylate was formed in 43% yield (unpublished results).

8 Conclusion

Currently, the research concerning the application of HMF and its glucosylated analogs in MCRs is confined to a few existing MCRs. Extending the scope to other MCRs, and further discovering new MCRs employing HMF as a substrate, remains a challenge considering its relatively unstable nature. Both developing the utilization of HMF in "old" MCRs and in novel MCRs will be useful investigations expected to construct efficiently complex value-added molecules from biomass by straightforward and atom-economical routes. The use of GMF enables the incorporation of a glucosyl moiety in the targets, which can prove useful in terms of biological applications, or more simply, in terms of physicochemical properties, by providing a polar and water-soluble appendage.

Acknowledgments

Financial support from CNRS, UCBL, INSA Lyon, and CPE Lyon is gratefully acknowledged, as well as the ANR for a postdoctoral fellowship to C. V. (FurCab project, ANR-15-CE07-0016) and the China Scholarship Council for PhD grants to WF and LW.

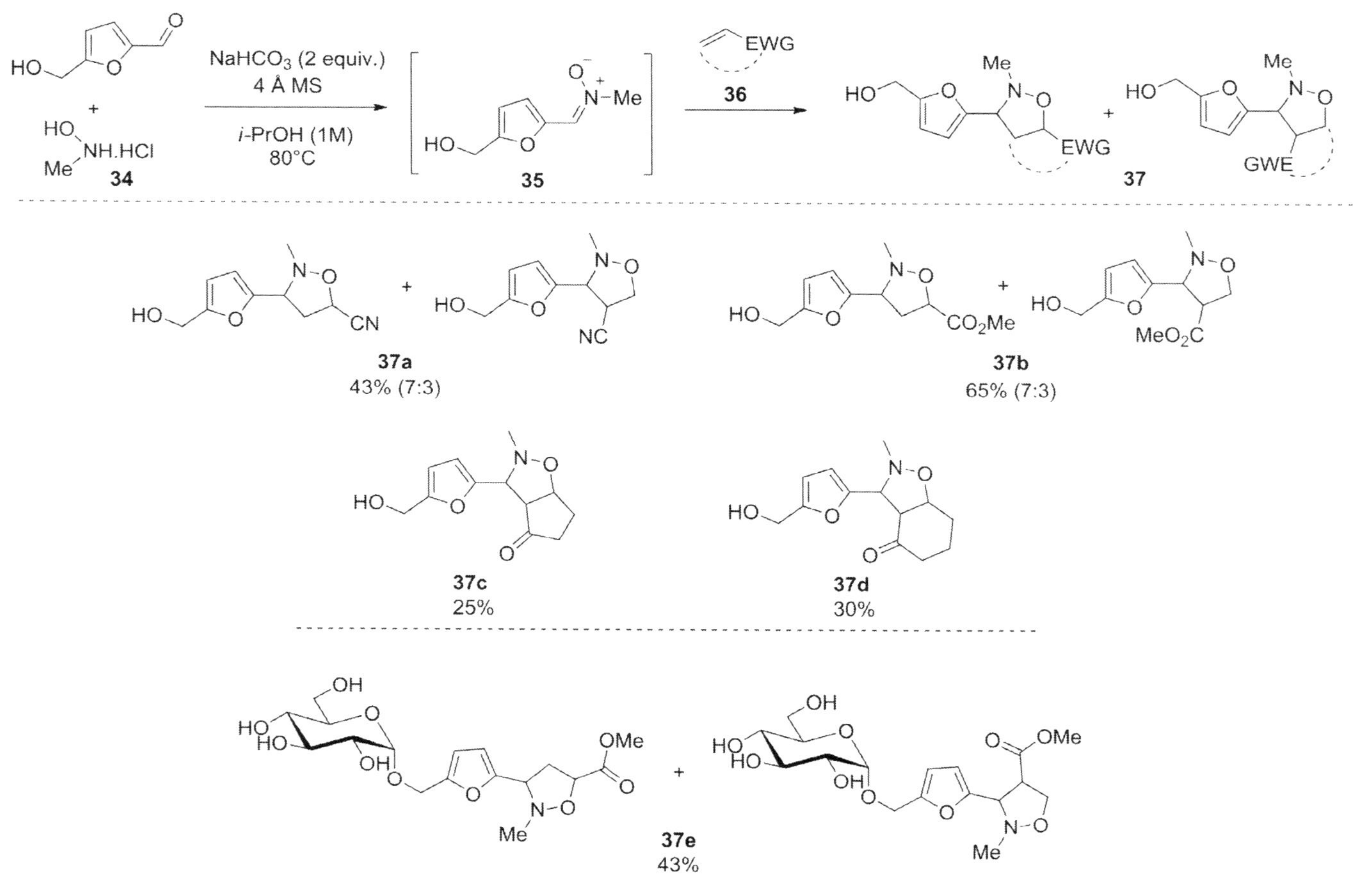

Scheme 19 Multicomponent preparation of isoxazolidines from bio-based aldehydes.

References

1. Rauter, A. P.; Vogel, P.; Queneau, Y., Eds. Carbohydrates in Sustainable Development, Part 1. Renewable Resources for Chemistry and Biotechnology. In *Topics in Current Chemistry*, Springer: Heidelberg, **2010**; Vol. 294.
2. Rauter, A. P.; Vogel, P.; Queneau, Y., Eds. Carbohydrates in Sustainable Development, Part 2. A Mine for New Synthons and Materials. In *Topics in Current Chemistry*, Springer: Heidelberg, **2010**; Vol. 295.
3. Scherrmann, M.-C.; Lubineau, A.; Queneau, Y. Functionalization of Carbohydrates in Water. In *Handbook of Green Chemistry*; Anastas, P., Ed.; Green Solvents. Reactions in water, Li, C. J., Ed.; Wiley, **2010**; Vol. 5, pp 291–330.
4. Wang, L.; Queneau, Y. Carbohydrate-Based Amphiphiles: A Resource for Biobased Surfactants. In *Encyclopedia of Sustainability Science and Technology*, Meyers, R. A., Ed.; Green Chemistry and Chemical Engineering, Han, B., Wu, T., Eds; Springer, **2018**. https://doi.org/10.1007/978-1-4939-2493-6_1009-1.
5. Queneau, Y.; Jarosz, S.; Lewandowski, B.; Fitremann, J. Sucrose Chemistry and Applications of Sucrochemicals. *Adv. Carbohydr. Chem. Biochem.* **2008**, *61*, 217–292.
6. Queneau, Y.; Chambert, S.; Besset, C.; Cheaib, R. Recent Progress in the Synthesis of Carbohydrate-Based Amphiphilic Materials: The Examples of Sucrose and Isomaltulose. *Carbohydr. Res.* **2008**, *343*, 1999–2009.
7. Epoune Lingome, C.; Gadenne, B.; Alfos, C.; Queneau, Y.; Moebs-Sanchez, S. Ring Opening of Epoxidized Methyl or Ethyl Oleate by Alkyl Glycosides. *Eur. J. Lipid Sci. Technol.* **2017**, *119*, 1600413.
8. Derrien, E.; Ahmar, M.; Martin-Sisteron, E.; Raffin, G.; Queneau, Y.; Marion, P.; Beyerle, M.; Pinel, C.; Besson, M. Oxidation of Aldoses Contained in Softwood Hemicellulose Acid Hydrolysates Into Aldaric Acids Under Alkaline or Non-Controlled pH Conditions. *Ind. Eng. Chem. Res.* **2018**, *57*, 4543–4552.
9. Jiang, S.; Verrier, C.; Ahmar, M.; Lai, J.; Ma, C.; Muller, E.; Marion, P.; Queneau, Y.; Pera-Titus, M.; Jérôme, F.; De Oliveira Vigier, K. Unveiling the Role of Choline Chloride in Furfural Synthesis From Highly Concentrated Feeds of Xylose. *Green Chem.* **2018**, *20*, 5104–5110.
10. Zhu, J.; Wang, Q.; Wang, M.-X. *Multicomponent Reactions in Organic Synthesis*; Wiley-VCH Verlag GmbH & Co. KGaA: Weinheim, **2015**.
11. Zhu, J.; Bienaymé, H. *Multicomponent Reactions;* Wiley-VCH Verlag GmbH & Co. KGaA: Weinheim, **2005**.
12. Herrera, R. P.; Marqués-López, E. *Multicomponent Reactions: Concepts and Applications for Design and Synthesis*; John Wiley & Sons, Inc.: Hoboken, New Jersey, **2015**.
13. Levi, L.; Müller, T. J. J. Multicomponent Syntheses of Functional Chromophores. *Chem. Soc. Rev.* **2016**, *45*, 2825–2846.
14. Strecker, A. Uber die künstliche Bildung der Milchsäure und einen neuen, dem Glycocoll homologen Körper. *Justus Liebigs Ann. Chem.* **1850**, *75*, 27–45.
15. Hantzsch, A. Condensationsprodukte aus aldehydammoniak und ketonartigen verbindungen. *Ber. Dtsch. Chem. Ges.* **1881**, *14*, 1637–1638.
16. Biginelli, P. Ueber Aldehyduramide des Acetessigäthers. *Ber. Dtsch. Chem. Ges.* **1891**, *24*, 1317–1319.
17. Biginelli, P. Ueber Aldehyduramide des Acetessigäthers. II. *Ber. Dtsch. Chem. Ges.* **1891**, *24*, 2962–2967.
18. Mannich, C.; Krösche, W. Ueber ein Kondensationsprodukt aus Formaldehyd, Ammoniak und Antipyrin. *Arch. Pharm.* **1912**, *250*, 647–667.

19. Passerini, M.; Simone, L. Sopra gli isonitrili (I). Composto del p-isonitril-azobenzolo con acetone ed acido acetico. *Gazz. Chim. Ital.* **1921**, *51*, 126–129.
20. Ugi, I. K. Versuche mit isonitrilen. *Angew. Chem. Int. Ed.* **1959**, *71*, 386.
21. Cioc, R. C.; Ruijter, E.; Orru, R. V. A. Multicomponent Reactions: Advanced Tools for Sustainable Organic Synthesis. *Green Chem.* **2014**, *16*, 2958–2975.
22. Bozell, J. J.; Petersen, G. R. Technology Development for the Production of Biobased Products From Biorefinery Carbohydrates—The US Department of Energy's "Top 10" Revisited. *Green Chem.* **2010**, *12*, 539–554.
23. Girisuta, B.; Janssen, L. P. B.; Heeres, H. J. A Kinetic Study on the Decomposition of 5-Hydroxymethylfurfural Into Levulinic Acid. *Green Chem.* **2006**, *8*, 701–709.
24. Nikolov, P.; Yaylayan, V. A. Thermal Decomposition of 5-(Hydroxymethyl)-2-Furaldehyde (HMF) and Its Further Transformations in the Presence of Glycine. *J. Agric. Food Chem.* **2011**, *59*, 10104–10113.
25. Galkin, K. I.; Krivodaeva, E. A.; Romashov, L.; Zalesskiy, S. S.; Kachala, V. V.; Burykina, J. V.; Ananikov, V. P. Critical Influence of 5-Hydroxymethylfurfural Aging and Decomposition on the Utility of Biomass Conversion in Organic Synthesis. *Angew. Chem. Int. Ed.* **2016**, *55*, 8338–8342.
26. Yu, I. K. M.; Tsang, D. C. W. Conversion of Biomass to Hydroxymethylfurfural: A Review of Catalytic Systems and Underlying Mechanisms. *Bioresour. Technol.* **2017**, *238*, 716–732.
27. Teong, S. P.; Yi, G.; Zhang, Y. Hydroxymethylfurfural Production From Bioresources: Past, Present and Future. *Green Chem.* **2014**, *16*, 2015–2026.
28. Agarwal, B.; Kailasam, K.; Sangwan, R. S.; Elumalai, S. Traversing the History of Solid Catalysts for Heterogeneous Synthesis of 5-Hydroxymethylfurfural From Carbohydrate Sugars: A Review. *Renew. Sustain. Energy Rev.* **2018**, *82*(Part 3), 2408–2425.
29. Chinnappan, A.; Baskar, C.; Kim, H. Biomass Into Chemicals: Green Chemical Conversion of Carbohydrates Into 5-Hydroxymethylfurfural in Ionic Liquids. *RSC Adv.* **2016**, *6*, 63991–64002.
30. Saha, B.; Abu-Omar, M. M. Advances in 5-Hydroxymethylfurfural Production From Biomass in Biphasic Solvents. *Green Chem.* **2014**, *16*, 24–38.
31. Dashtban, M.; Gilbert, A.; Fatehi, P. Recent Advancements in the Production of Hydroxymethylfurfural. *RSC Adv.* **2014**, *4*, 2037–2050.
32. Menegazzo, F.; Ghedini, E.; Signoretto, M. 5-Hydroxymethylfurfural (HMF) Production From Real Biomasses. *Molecules* **2018**, *23*, 2201.
33. Hu, L.; Lin, L.; Wu, Z.; Zhou, S.; Liu, S. Recent Advances in Catalytic Transformation of Biomass-Derived 5-Hydroxymethylfurfural Into Innovative Fuels and Chemicals. *Renew. Sustain. Energy Rev.* **2017**, *74*, 230–257.
34. Gupta, K.; Rai, R. K.; Singh, S. K. Metal Catalysts for the Efficient Transformation of Biomass-Derived HMF and Furfural to Value Added Chemicals. *ChemCatChem* **2018**, *10*, 2326–2349.
35. Kong, X.; Feng, Z.; Kozinski, J. A.; Butler, I. S.; Xu, L.; Song, H.; Wei, X. Catalytic Conversion of 5-Hydroxymethylfurfural to Some Value-Added Chemicals. *Green Chem.* **2018**, *20*, 3657–3682.
36. Tan, J. N.; Ahmar, M.; Queneau, Y. Glucosyloxymethylfurfural (GMF): A Creative Renewable Scaffold Towards Bioinspired Architectures. *Pure Appl. Chem.* **2015**, *87*, 827–839.
37. Urashima, T.; Suyama, K.; Adachi, S. The Condensation of 5-(Hydroxymethyl)-2-Furfuraldehyde With Some Aldoses on Heating. *Food Chem.* **1988**, *29*, 7–17.
38. Won, T. H.; Liao, L.; Kang, S. S.; Shin, J. Simultaneous Analysis of Furfural Metabolites From Rehmanniae Radix Preparata by HPLC-DAD–ESI-MS. *Food Chem.* **2014**, *142*, 107–113.
39. Lin, A.-S.; Qian, K.; Usami, Y.; Lin, L.; Itokawa, H.; Hsu, C.; Morris-Natschke, S. L.; Lee, K.-H. 5-Hydroxymethyl-2-Furfural, a Clinical Trials Agent for Sickle Cell Anemia, and

Its Mono/Di-Glucosides From Classically Processed Steamed Rehmanniae Radix. *J. Nat. Med.* **2008**, *62*, 164–167.
40. Li, S.-L.; Song, J.-Z.; Qiao, C.-F.; Zhou, Y.; Qian, K.; Lee, K.-H.; Xu, H.-X. A Novel Strategy to Rapidly Explore Potential Chemical Markers for the Discrimination Between Raw and Processed Radix Rehmanniae by UHPLC–TOFMS With Multivariate Statistical Analysis. *J. Pharm. Biomed. Anal.* **2010**, *51*, 812–823.
41. Zhou, H.; Zhao, R.; Yang, J. Two New Alkaloid Galactosides From the Kernel of Prinsepia uniflora. *Nat. Prod. Res.* **2013**, *27*, 687–690.
42. Adhikari, D. P.; Schutzki, R. E.; DeWitt, D. L.; Nair, M. G. Effects of Amelanchier Fruit Isolates on Cyclooxygenase Enzymes and Lipid Peroxidation. *Food Chem.* **2006**, *97*, 56–64.
43. Hagen, R. W.; Lorenz, S. Ein neues bakterielles Umwandlungsprodukt der Saccharose. *Angew. Chem.* **1957**, *69*, 641.
44. Lina, B. A. R.; Jonker, D.; Kozianowski, G. Isomaltulose (Palatinose®): A Review of Biological and Toxicological Studies. *Food Chem. Toxicol.* **2002**, *40*, 1375–1381.
45. Schwengers, D.; Benecke, H. Processes for Producing Leucrose. US4693974A, 1987-09-15, **1987**.
46. Munir, M.; Schneider, B.; Schiweck, H. 1-*O*-α-D-glucopyranosyl-D-fructose: darstellung aus saccharose und ihre reduktion zu 1-*O*-α-D-Glucopyranosyl-D-glucitol. *Carbohydr. Res.* **1987**, *164*, 477–485.
47. Lichtenthaler, F. W.; Martin, D.; Weber, T.; Schiweck, H. Studies on Ketoses, 7–5-(α-D-Glucosyloxymethyl)furfural: Preparation From Isomaltulose and Exploration of Its Ensuing Chemistry. *Liebigs Ann. Chem.* **1993**, (9), 967–974.
48. Ruß, C.; Luff, C.; Begli, A. H.; Koenig, B. Solvent-Free Preparation of 5-(α-D-Glucosyloxymethyl)furfural From Isomaltulose–Choline Chloride Melts. *Synth. Commun.* **2012**, *42*, 3112–3116.
49. Liu, F.; Audemar, M.; De Oliveira Vigier, K.; Cartigny, D.; Clacens, J.-M.; Costa Gomes, M. F.; Pádua, A. A. H.; De Campo, F.; Jérôme, F. Selectivity Enhancement in the Aqueous Acid-Catalyzed Conversion of Glucose to 5-Hydroxymethylfurfural Induced by Choline Chloride. *Green Chem.* **2013**, *15*, 3205–3213.
50. Jérôme, F.; De Oliveira Vigier, K. Catalytic Conversion of Carbohydrates to Furanic Derivatives in the Presence of Choline Chloride. *Catalysts* **2017**, *7*, 218.
51. Martin, D.; Lichtenthaler, F. W. Versatile Building Blocks From Disaccharides: Glycosylated 5-Hydroxymethylfurfurals. *Tetrahedron Asymmetry* **2006**, *17*, 756–762.
52. Cottier, L.; Descotes, G.; Soro, Y. Synthesis of Acetylated Ranunculin Diastereoisomers and δ–Glucosyloxy–γ–oxo Esters From α or β Glucosylmethylfurfural. *J. Carbohydr. Chem.* **2005**, *24*, 55–71.
53. Kappe, O. Recent Advances in the Biginelli Dihydropyrimidine Synthesis: New Tricks From an Old Dog. *Acc. Chem. Res.* **2000**, *33*, 879–888.
54. Nagarajaiah, H.; Mukhopadhyay, A.; Moorthy, J. N. Biginelli Reaction: An Overview. *Tetrahedron Lett.* **2016**, *57*, 5135–5149.
55. Kappe, O. A Reexamination of the Mechanism of the Biginelli Dihydropyrimidine Synthesis. Support for an N-Acyliminium ion Intermediate. *J. Org. Chem.* **1997**, *62*, 7201–7204.
56. Matasi, J. J.; Caldwell, J. P.; Hao, J.; Neustadt, B.; Arik, L.; Foster, C. J.; Lachowicz, J.; Tulshian, D. B. The Discovery and Synthesis of Novel Adenosine Receptor (A2A) Antagonists. *Bioorg. Med. Chem. Lett.* **2005**, *15*, 1333–1336.
57. Fan, W.; Queneau, Y.; Popowycz, F. HMF in Multicomponent Reactions: Utilization of 5-Hydroxymethylfurfural (HMF) in the Biginelli Reaction. *Green Chem.* **2018**, *20*, 485–492.

58. Shi, Y.-L.; Shi, M. Aza-Baylis–Hillman Reactions and Their Synthetic Applications. *Eur. J. Org. Chem.* **2007**, (18), 2905–2916.
59. Declerck, V.; Martinez, J.; Lamaty, F. Aza-Baylis – Hillman Reaction. *Chem. Rev.* **2009**, *109*, 1–48.
60. Bertenshaw, S.; Kahn, M. Phosphine Mediated Synthesis Of 2-Methylidene-3-Amino Esters and Ketones. *Tetrahedron Lett.* **1989**, *30*, 2731–2732.
61. Tan, J.-N.; Ahmar, M.; Queneau, Y. HMF Derivatives as Platform Molecules: Aqueous Baylis-Hillman Reaction of Glucosyloxymethylfurfural Toward New Biobased Acrylates. *RSC Adv.* **2013**, *3*, 17649–17653.
62. Tan, J.-N.; Ahmar, M.; Queneau, Y. Bio-Based Solvents for the Baylis-Hillman Reaction of HMF. *RSC Adv.* **2015**, *5*, 69238–69242.
63. Tan, J.-N.; Ahmar, M.; Queneau, Y. Glycosyloxymethylfurfural (GMF) in Multicomponent *aza*-Morita-Bayliss-Hillman Reaction: Rapid Access to Highly Functionalized Carbohydrate Scaffolds. *Curr. Org. Synth.* **2018**, *15*, 430–435.
64. Peshkov, V. A.; Pereshivko, O. P.; Van der Eycken, E. V. A Walk Around the A3-Coupling. *Chem. Soc. Rev.* **2012**, *41*, 3790–3807.
65. Liu, Y. Recent Advances on Diversity-Oriented Heterocycle Synthesis via Multicomponent Tandem Reactions Based on A3 Coupling. *ARKIVOC* **2014**, *I*, 1–20.
66. Gupta, P.; Singh, S. K.; Pathak, A.; Kundu, B. Template-Directed Approach to Solid-Phase Combinatorial Synthesis of Furan-Based Libraries. *Tetrahedron* **2002**, *58*, 10469–10474.
67. Wu, Q.; Chen, J.; Guo, X.; Xu, Y. Copper(I)-Catalyzed Four-Component Coupling Using Renewable Building Blocks of CO_2 and Biomass-Based Aldehydes. *Eur. J. Org. Chem.* **2018**, *24*, 3105–3113.
68. Oikawa, M.; Ikoma, M.; Sasaki, M. Parallel Synthesis of Tandem Ugi/Diels–Alder Reaction Products on a Soluble Polymer Support Directed Toward Split-Pool Realization of a Small Molecule Library. *Tetrahedron Lett.* **2005**, *46*, 415–418.
69. Bode, M. L.; Rousseau, A.; Gravestock, D.; Moleele, S. S.; Van der Westhuyzen, C. W. Imidazopyridines and Imidazopyrimidines as HIV-1 Reverse Transcriptase Inhibitors and Their Preparation, Pharmaceutical Compositions and Use in the Treatment of HIV Infection. WO2010032195A1, **2010**.
70. Theberge, A. B.; Mayot, E.; El Harrak, A.; Kleinschmidt, F.; Huck, W. T. S.; Griffiths, A. D. Microfluidic Platform for Combinatorial Synthesis in Picolitre Droplets. *Lab Chip* **2012**, *12*, 1320–1326.
71. Keglevich, G.; Bálint, E. The Kabachnik–Fields Reaction: Mechanism and Synthetic Use. *Molecules* **2012**, *17*, 12821.
72. Cottier, L.; Descotes, G.; Lewkowski, J.; Skowroński, R. Synthesis and Its Stereochemistry of Aminophosphonic Acids Derived From 5-Hydroxymethylfurfural. *Phosphorus, Sulfur Silicon Relat. Elem.* **1996**, *116*, 93–100.
73. Fan, W.; Queneau, Y.; Popowycz, F. The Synthesis of HMF-Based α-Amino Phosphonates via One-Pot Kabachnik–Fields Reaction. *RSC Adv.* **2018**, *8*, 31496–31501.
74. Lo, M. M. C.; Neumann, C. S.; Nagayama, S.; Perlstein, E. O.; Schreiber, S. L. Library of Spirooxindoles Based on a Stereoselective Three-Component Coupling Reaction. *J. Am. Chem. Soc.* **2004**, *126*, 16077–16086.
75. Sebahar, P. R.; Williams, R. M. The Asymmetric Total Synthesis of (+)- and (−)-Spirotryprostatin B. *J. Am. Chem. Soc.* **2000**, *122*, 5666–5667.
76. Berthet, M.; Cheviet, T.; Dujardin, G.; Parrot, I.; Martinez, J. Isoxazolidine: A Privileged Scaffold for Organic and Medicinal Chemistry. *Chem. Rev.* **2016**, *116*, 15235–15283.

Alkyne dicobalt complexes in carbohydrates: Synthetic applications

3

Ana M. Gómez, J. Cristóbal López
Bioorganic Chemistry Department, Instituto Química Orgánica General (IQOG-CSIC), Madrid, Spain

Chapter outline

1 Introduction

The complexation of an alkyne to dicobalt hexacarbonyl, $Co_2(CO)_6$, as in compound **2** (Scheme 1), facilitates the (acid-mediated) access to cobalt-stabilized carbocations, for example **3** (Scheme 1), which upon treatment with a given nucleophile, that is, **Nu** (Scheme 1) provides access to differently substituted propargyl systems, for example, **4** (Scheme 1). This transformation, first discovered by Nicholas and Pettit[1, 2] and currently known as the Nicholas reaction,[3] is based on the enhanced stability of carbonium ions α- to an organometallic substituent, which was at the origin of the efficient and regiospecific transformation **3**→**4** (Scheme 1). The Nicholas reaction

Recent Trends in Carbohydrate Chemistry. https://doi.org/10.1016/B978-0-12-817467-8.00003-7
© 2020 Elsevier Inc. All rights reserved.

is compatible with a variety of carbon and heteroatom nucleophiles, thus providing access to a broad range of propargylic derivatives, **5**. Additionally, the overall transformation **1→5** (Scheme 1) benefits from (i) the ready availability of the starting propargyl alcohols by nucleophilic addition of organometallic alkynyl reagents to aldehydes or ketones; (ii) the easy formation of dicobalt hexacarbonyl complexed propargyl derivatives **2**, by reaction of the corresponding alkyne **1**, with dicobalt octacarbonyl; (iii) the use of the cobalt cluster-containing derivatives, for example, **4** (Scheme 1), in subsequent transformations, such as the Pauson-Kand reaction[4]; or (iv) the easy regeneration of the original alkyne bond by a number of existing methods for cobalt decomplexation.

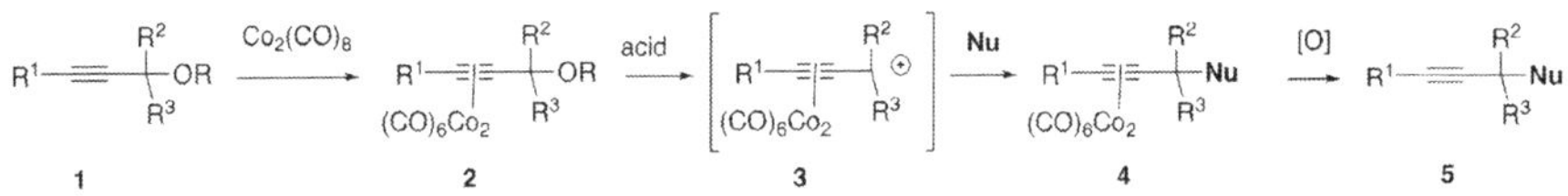

Scheme 1 The Nicholas reaction.

Owing to the previously mentioned virtues, the Nicholas reaction has found ample use in organic chemistry. In this chapter, we intend to provide a brief overview on synthetic transformations mediated by cobalt complexed propargylic derivatives in carbohydrate substrates, including recent contributions by our research group.

At this point, a special mention to the pioneering contributions of Isobe's group to this chemistry is pertinent.[5] Accordingly, studies from Isobe's laboratories have ranged from anomeric equilibration of cobalt complexed propargylic *C*-glycosides, to complex organic synthesis employing carbohydrate-derived dicobalt hexacarbonyl complexes as the starting materials, *vide infra*.

An inspection of literature references dealing with the use of dicobalt hexacarbonyl-stabilized propargylic cations in synthesis will show, at least, four different types of representations (Fig. 1). Thus, even though drawings as **D** (Fig. 1) might provide a more "realistic" picture, by highlighting the bent nature of the dicobalt

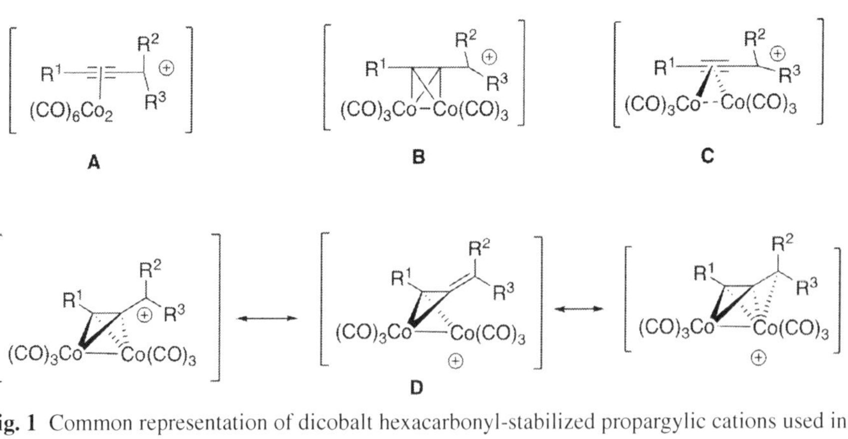

Fig. 1 Common representation of dicobalt hexacarbonyl-stabilized propargylic cations used in the literature.

hexacarbonyl compounds, for the sake of simplicity, we have opted for drawings type **A** to represent dicobalt hexacarbonyl derivatives throughout the chapter.

2 Dicobalt hexacarbonyl-mediated anomerization of alkynyl *C*-glycosides

Isobe and coworkers reported a general methodology for the introduction of different acetylenic moieties onto the anomeric position of pyranoses with the resultant *C*-glycosides having complete α-orientation (Scheme 2).[6] For instance, Ferrier-type alkynylation of tri-*O*-acetyl-D-glycal **6**, with silylacetylenes in a Lewis acid-catalyzed process, allowed the stereoselective introduction of alkynyl groups in almost quantitative yields leading to alkynyl *C*-glycosides **8** (Scheme 2A).[7] The reaction proceeds through the formation of an allylic oxycarbenium cation **7**, which reacts selectively at the anomeric center with the appropriate silylacetylene. The anomeric effect of the ring oxygen and the conformation of the pyranose ring favor the α-attack of the incoming group, and the reaction was shown to be completely selective leading to pseudoaxial α-anomer of *C*-glycosides type **8** (Scheme 2A).

Similarly, *C*-glycosylation of 2-acetoxy-D-glycal **9** produced unstable *C*-glycosidic compounds **10**, which were easily converted to allylic alcohols **11** (Scheme 2B).

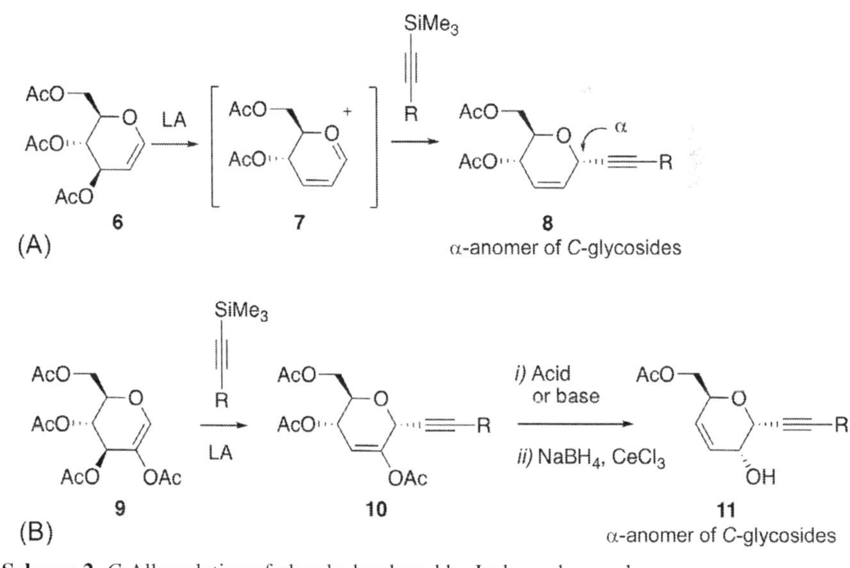

Scheme 2 *C*-Alkynylation of glycals developed by Isobe and coworkers.

In the early 1990s, contributions from Isobe's laboratories also showed that the α-anomer of alkynyl *C*-glycosides could be "anomerized" to the corresponding β-isomers by means of their corresponding dicobalt hexacarbonyl complexes. The overall process for α→β conversion, including cobalt complexation **12**→**13**, acid epimerization **13**→**14**, and oxidative decomplexation **14**→**15**, is illustrated in Scheme 3.[8] After the equilibration reaction, the cobalt complexes can be oxidatively decobaltated with a variety of reagents, such as iodine,[9] amine *N*-oxides,[3] cerium(IV) ammonium nitrate (CAN),[10] ferric nitrate,[1a] tetrabutylammonium fluoride (TBAF), etc.,[11] to unveil the alkyne moiety. Among these oxidants, iodine has shown to react efficiently, in short times, to restore the original triple bond in almost quantitative yield. Isobe and Hosokawa also found reductive conditions for the cobalt decomplexation by the use of tributyltin hydride or triethylsilane.[12] The original acetylene derivatives are then selectively transformed into the corresponding *cis*-olefin or *cis*-vinylsilanes.

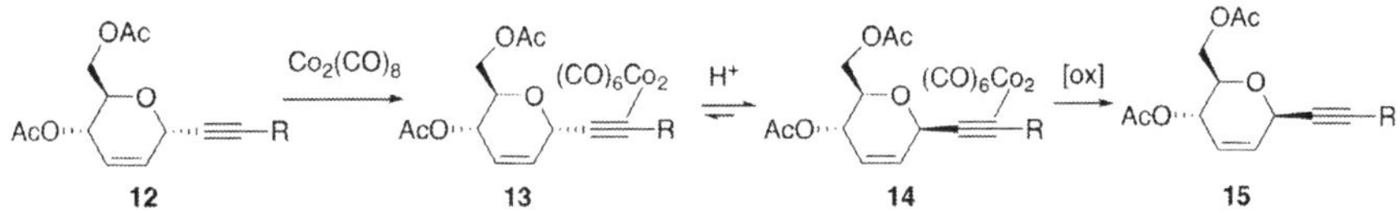

Scheme 3 Isobe's anomerization of alkynyl *C*-glycosides through cobalt complexes.

The crucial acid-catalyzed (α→β) equilibration features a propargylic anomeric oxycarbenium ion stabilized by the cobalt complex, which initiates an opening and closing of the tetrahydropyran ring leading to an equilibrium in which the thermodynamically more stable equatorial isomer will predominate. The ratio of the observed equilibration ranges from moderate to excellent, and the *C*-glycoside (α,β) ratio is controlled by the relative stability of both anomers.

Some illustrative examples are depicted in Scheme 4. In the preferred pyranose conformation of the alkynyl *C*-glycosides' α-anomer the silylacetylene moiety is usually located in an α-axial orientation, for example, **16** (Scheme 4A). However, after formation, the corresponding dicobalt hexacarbonyl complexes, for example, **19** (Scheme 4A), the conformation of the pyranose ring usually changes so that the dicobalt hexacarbonyl complexed alkynyl substituent will adopt an α-equatorial orientation (see also **17**→**20**, **18**→**21**, Scheme 4). Then, in the acid-mediated (α→β) epimerization, the driving force in the formation of the β-isomer in preference to the α-isomer could be explained on the basis of the minimization of 1,3-diaxial steric interactions in the lowest energy conformation of the pyranose ring.[13] More recently, an exhaustive and plausible rationale for the dicobalt hexacarbonyl complexed C1 alkynyl group epimerization has been advanced.[14]

Accordingly, in the case of 2,3-dehydropyranose systems, such as **19**, the ratio of (α→β) equilibration is rather moderate (1:6) accounting for the stability difference due to the disfavored interaction between the axial anomeric proton H-1 and C-6 (Scheme 4A). In the corresponding saturated ring system, that is, **20**, the equilibration provides the β-isomer **23** in a much higher ratio (α/β= 1:19), owing to the three pairs of severe 1,3-diaxial interactions existing in **20** (Scheme 4B). On the other hand, with pyranose systems containing a double bond between C-3 and C-4, such as **21**, it was observed that the α/β ratios increase in accordance with the size of the OH-2-substituent (Scheme 4C). Thus, considering that 1,3-interactions are virtually the same, the effect of the bigger C-2 substituents are then explained on the basis of an additional larger gauche effect between the C-2 substituent and the bulky cobalt-acetylene residue (Scheme 4C).

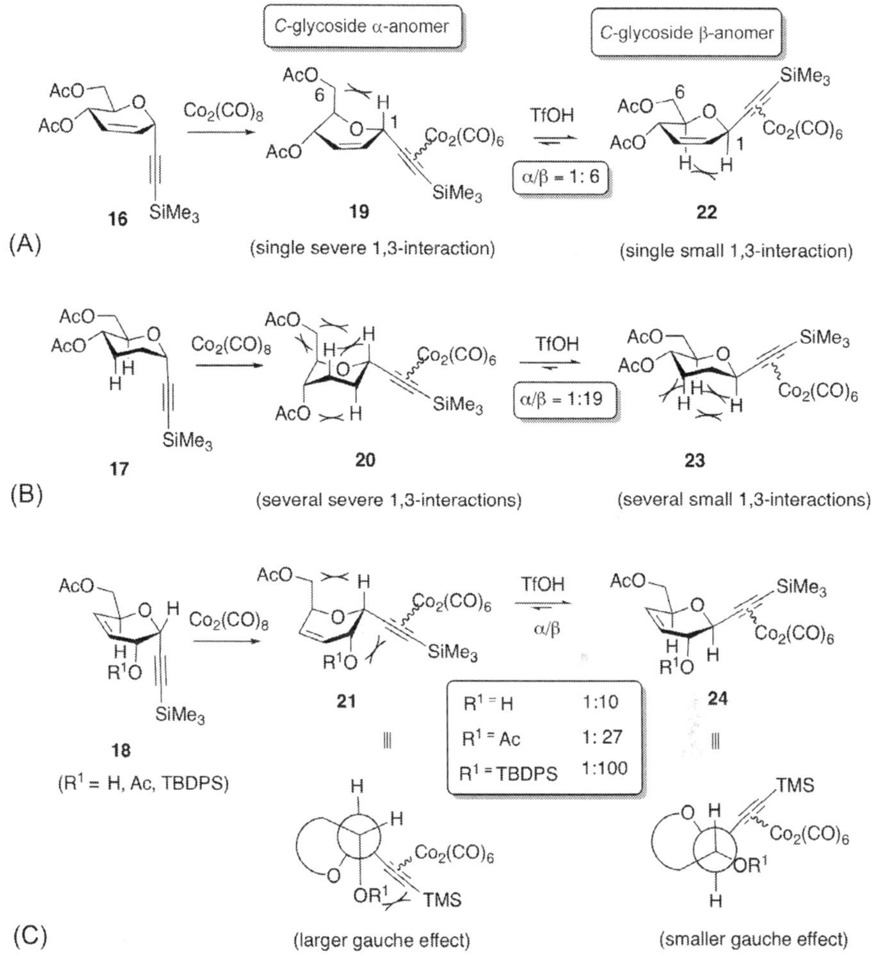

Scheme 4 Anomerization of cobalt complexes in unsaturated (**16**, **18**) and dideoxy (**17**) pyranose derivatives.

The synthesis of "segment C" of the antibiotic *tautomycin* exemplified the application of this methodology to the preparation of natural products.[15] Thus, Isobe's group retrosynthetic analysis of the spiranic segment C of *tautomycin* led to two subsegments C-1 and C-2, which were prepared by application of the epimerization protocol to (phenylthio)alkynyl pyranoses **26** and **27**, readily available from tri-*O*-acetyl-D-glycal **6** and levoglucosenone **25**, respectively (Scheme 5).[16] Thus, complex formation and epimerization of *C*-glycoside **28** afforded a 1:1.1 mixture of α- and β-anomers **28** and **30**, respectively. This epimerization was found to be difficult and not very efficient because of the presence of a β-C2 axial methyl group in the pyranose ring. However, after separation of both anomers, and three additional equilibrations of the recovered α-isomer **28**, the β-anomer **30** could finally be obtained in 65% yield (Scheme 5A). Conversely, the equilibration of *C*-glycoside α-anomer **29** was strongly shifted to its β-isomer **31** because two substituents (methyl and acetoxymethyl) prefer to adopt an equatorial disposition (Scheme 5B).

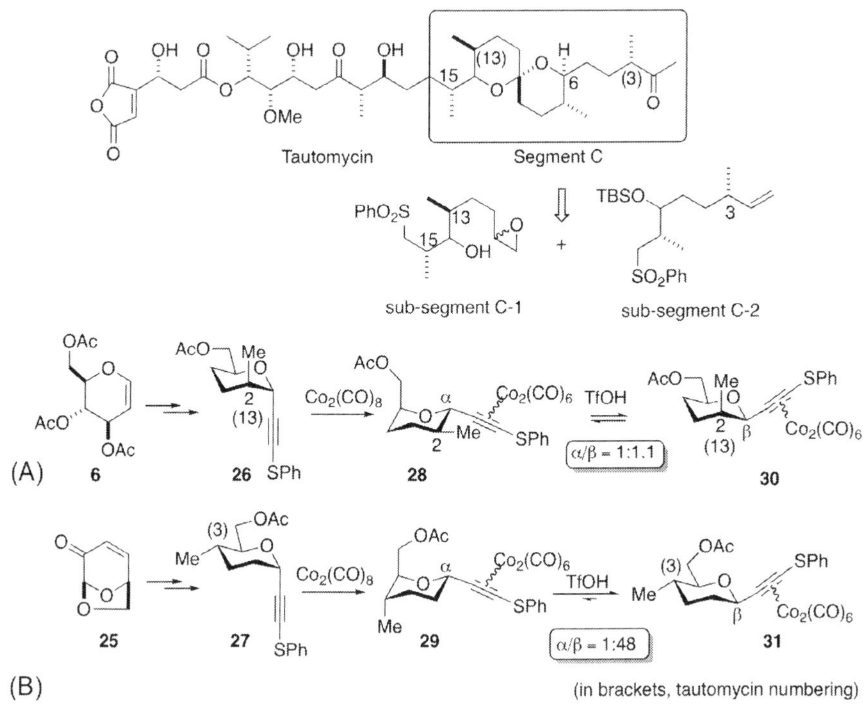

Scheme 5 Anomerization of cobalt complexes **28** and **29** en route to the synthesis of segment C of *Tautomycin*.

Désiré and Veyrières applied the anomerization conditions described by Isobe to "fully oxygenated" alkynyl *C*-glucopyranoside α-anomer **32** (Scheme 6).[17] In this substrate, unlike in previous examples by Isobe's group, the cobaltation step (**32**→**33**) did not cause a conformational change [$^4C_1 \rightarrow {}^1C_4$] in the ensuing pyranoside **33**, which after complexation still exhibited a slightly distorted $^4C_1$ chair conformation with the bulky complexed alkyne in an axial orientation (the alternate $^1C_4$ conformer would have imposed two severe 1,3-diaxial interactions, Scheme 6). Addition of triflic acid to **33** then brought an equilibrium with a low 1:3.3 (α/β) ratio, reflecting the slightly higher thermodynamic stability of **34** when compared to **33**.

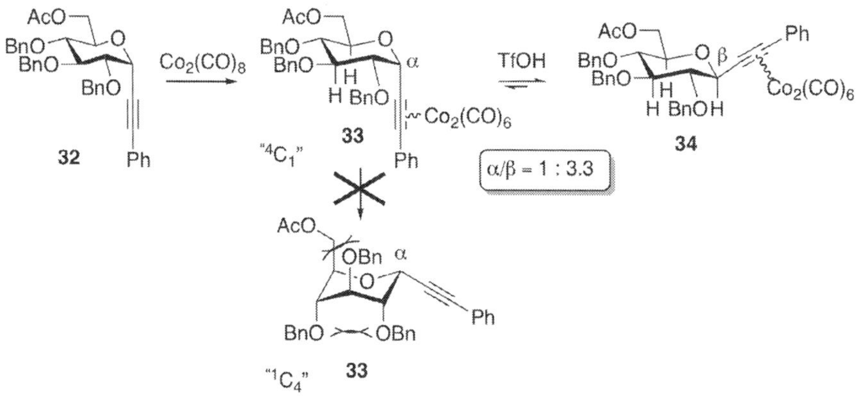

Scheme 6 Equilibrium of cobalt complexes of fully substituted alkynyl α and β *C*-glucopyranosides **33** and **34**, respectively.

More recently, Yu and coworkers have disclosed density functional theory (DFT) calculations to rationalize the remarkable differences observed in the attempted Nicholas epimerization of aziridinyl and allyl *C*-(α-glycosides) **35** and **36**, respectively (Scheme 7).[18] These compounds, owing to the presence of the terminal silyloxy substituent at the lateral chain, possessed two reactive sites for Nicholas-type transformations. Thus, whereas in both cases, upon activation with $BF_3 \cdot OEt_2$ (CH_2Cl_2, AcOH), the terminal silyl group at the lateral chain was replaced by an acetate in an S_N1 Nicholas-type reaction, only the allylic glycoside **36** underwent partial (α→β) anomerization (Scheme 7B), with the α-disposition of aziridinyl derivative **37** remaining unchanged (Scheme 7A).

The observed experimental results were in agreement with the quantum calculations, which had predicted (i) that in the unsaturated glycoside **36**, both Nicholas substitution and Nicholas epimerization would be likely to have similar rates and to

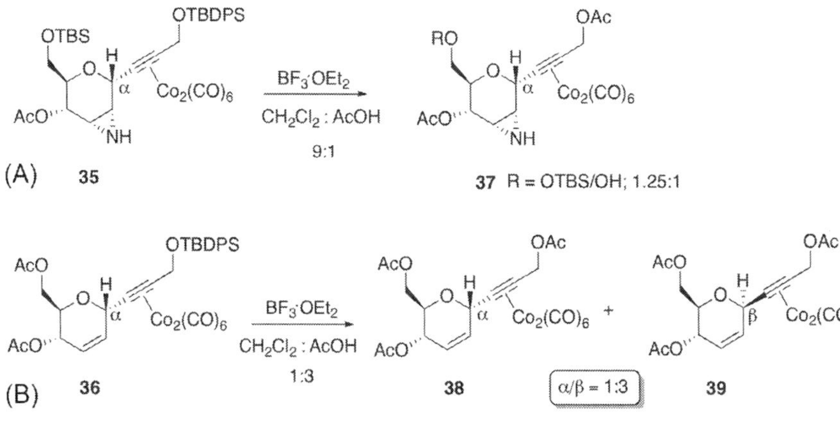

Scheme 7 Attempted epimerization of the aziridinyl and allyl *C*-glycosides α-anomers **35** and **36**, respectively.

happen simultaneously; (ii) that the β-anomer **39** would be thermodynamically preferred over the α-anomer **39** by 1 kcal/mol, value in accordance with the experimentally observed (α/β) rate (1:3); and (iii) that the barriers for the anomerization and substitution reactions of the azidirinyl derivative **35** would differ by 6 kcal/mol so that the latter process will definitely be the preferred reaction pathway. The annulated aziridine ring enhances the rigidity of the glycoside backbone, which in turn slows down the epimerization reaction.

3 Dicobalt hexacarbonyl-mediated ring-opening of alkynyl *C*-glycosides

As an extension of their studies on (dicobalt hexacarbonyl)alkynyl *C*-glycosides, Isobe *et al.* showed that the anomeric Nicholas oxycarbenium ion involved in the aforementioned epimerization could be trapped by external nucleophiles to produce cobalt complexes of ring-opened sugar derivatives, for example, **40–44** (Scheme 8).

Thus, treatment of dicobalt hexacarbonyl complex of alkynyl 2,3-unsaturated *C*-glycoside **19**, with TfOH in the presence of Ac_2O provided open-chain derivative **40**, in good yield. The product was, however, isolated as the most stable *E* olefin (Scheme 8A). The observed *Z*→*E* isomerization of the initial double bond demonstrated that, in this type of systems, not only the cobalt moiety but also the π-electrons of the double bond take part in the stabilization of the intermediate allylic cation.[19]

Furthermore, when the ring-opening reaction was conducted using pivaloyl tetrafluoroborate,[20] a variety of nucleophiles could be used to trap the ring-opened electrophilic intermediate cation to generate differently substituted building blocks, that is, **41–44** (Scheme 8B). In general, *Z*→*E* isomerization of the double bond was

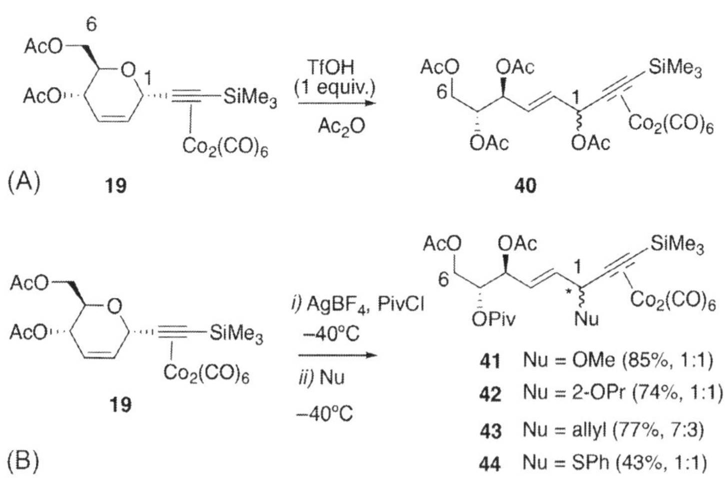

Scheme 8 Ring-opening nucleophile trapping reactions of $\Delta_{2,3}$-unsaturated-α-*C*-dicobalt hexacarbonyl alkynyl glycosides **19**.

always observed in the final products, which were obtained as 1:1 diastereomeric mixtures at the new stereogenic center (C1, carbohydrate numbering, Scheme 8). However, as an exception, reaction with allyl silane as the nucleophile proved to be stereoselective furnishing a 7:3 diastereomeric mixture, **43** (Scheme 8B). All these linear cobalt complexes could be uneventfully demetallated in good yields by the use of iodine.

Additional studies by Isobe's group showed that cobalt complexes arising from $\Delta_{3,4}$-unsaturated systems, that is, **21** and **45** (Scheme 9) required stronger reaction conditions (longer reaction times and additional amount of acid) since they do not have π-electrons to further stabilize the intermediate Nicholas cation. All products (**46–48**) were obtained as diastereomeric mixtures at C-3 (open-chain compound numbering) although good selectivity could be observed when a bulky substituent (i.e., TBDPS ether) was present at C-2 of the sugar ring (C-4, open-chain compound numbering). In these cases, the reaction proceeds to provide 3,4-*syn* diols as the major isomers. The authors provided a rationalization for this stereoselectivity using arguments similar to those initially proposed by Schreiber and coworkers.[21]

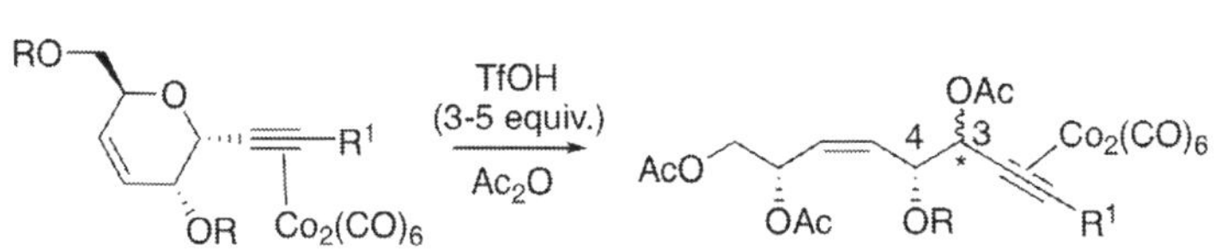

21a R = Ac, R^1 = $SiMe_3$
21b R = TBDPS, R^1 = $SiMe_3$
45 R = TBDPS, R^1 = H

46 R = Ac, R^1 = $SiMe_3$ (54%, 53:47)
47 R = TBDPS, R^1 = $SiMe_3$ (84%, 94:6)
48 R = TBDPS, R^1 = H (71%, 88:12)

Scheme 9 Ring-opening reaction of $\Delta_{3,4}$-unsaturated dicobalt hexacarbonyl alkynyl *C*-glycosides' α-anomers.

4 Dicobalt hexacarbonyl-mediated formation of ether rings from sugar acetylenes

Isobe and coworkers then drove their attention to the application of the intramolecular Nicholas cyclization as a method for the conversion of dicobalt hexacarbonyl alkynyl *C*-glycosides into medium-sized rings.

In this context, oxepane ring formation from six-membered sugar precursors is usually impractical due to thermodynamic constraints.[22] However, Isobe's group found that Nicholas-type recyclization of the aforementioned open-chain derivatives could be used to prepare seven-membered oxepane derivatives (Scheme 10).[23] In this context, initial studies showed that open-chain Nicholas alcohol **49** (readily obtained from $\Delta_{3,4}$-unsaturated derivative **21b**, Scheme 9, via acetyl derivative **47** Scheme 10) underwent triflic acid-mediated cyclization leading to oxepene **50**, as a single isomer (Scheme 10). Apparently, this highly selective cyclization is thermodynamically controlled since the two bulky substituents, cobalt acetylene moiety at C-3 and TBDPS ether function at C-4, prefer to adopt a *trans* relationship to minimize steric strain. Decomplexation using oxidative conditions then originated oxepene **51**, a useful synthetic precursor for subunits of marine *trans*-fused polyether toxins.[24]

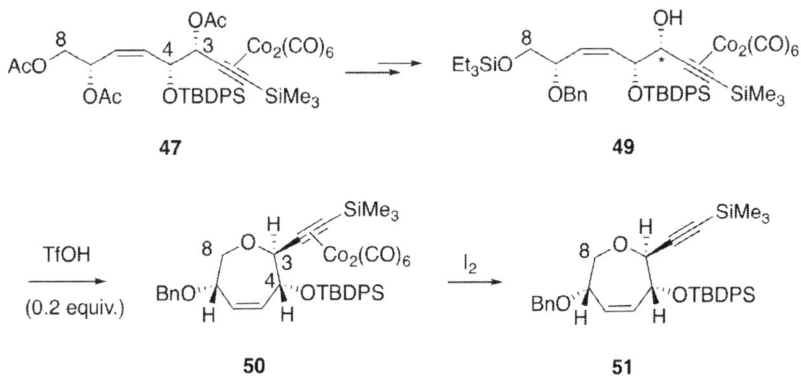

Scheme 10 Application of an intramolecular Nicholas cyclization in the preparation of synthetically useful oxepanes.

Having realized that open-chain *trans*-olefin derivatives resulting from the ring-opening of alkynyl $\Delta_{2,3}$-unsaturated *C*-glycosides, for example, **55** (Scheme 11), would preclude a spontaneous ring-forming reaction, Isobe *et al.* introduced an alternative ring-recyclization strategy to medium-sized 7-, 8-, 9-, and 10-membered rings by positioning the nucleophilic hydroxy group at the, otherwise, terminal site of the acetylene (Scheme 11).[25] In order to induce the Nicholas cyclization, the precursors **54** were prepared by *C*-glycosidation of tri-*O*-acetyl-D-glycal **6**, with trimethylsilylacetylene followed by palladium catalyzed ene-yne coupling reaction with the appropriate vinyl iodide, **53** (Scheme 11). Then, cobalt complexation and acid treatment of the dicobalt hexacarbonyl complexed allylic *C*-glycosides **54** generated the intermediate Nicholas cations **55**, by a process in which an initially formed (C-1) *cis*-allylic cation isomerized to the more stable *trans*-allylic cation. Under these premises, recyclization (onto C5-OH) to regenerate the original pyranose ring will no longer be possible, and cyclization occurs onto the terminal hydroxy group to afford **56**. This protocol was then successfully applied to the formation of cyclic ethers of various sizes, including 7-, 8-, 9-, and 10-membered rings.[26] The resulting cyclic compounds, that is, **56** (Scheme 11), with the dicobalt hexacarbonyl complex located inside the ring system, were found to be inert to the usual oxidative decomplexation conditions, probably due to the high-ring strain imposed by an internal acetylenic moiety in the desired ene-yne cyclic ethers. However, application of high H_2 pressure, in the presence of Rh-C, proved to be effective for the synthesis of the decomplexed dienes **57** and **58** (Scheme 11). Accordingly, the carbon atoms corresponding to the original acetylenes ended up as olefinic carbons in the 8-, 9-, and 10-membered derivatives **58**. However, a double-bond transposition was observed in the case of the seven-membered system, **57**.

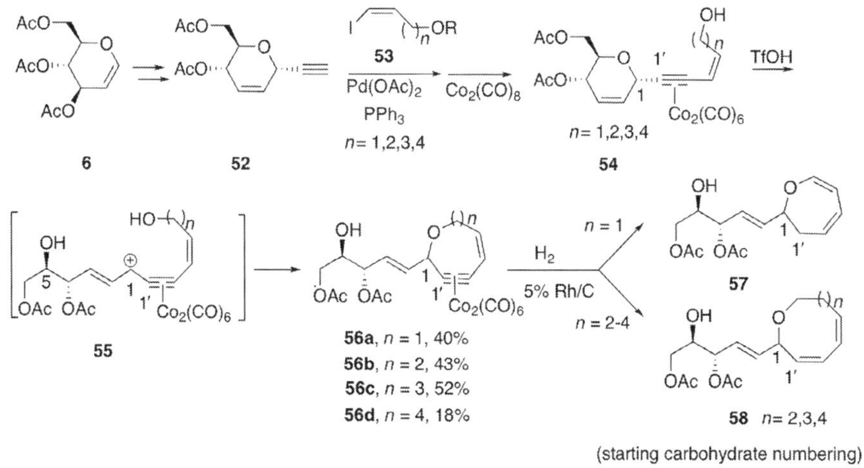

Scheme 11 Medium ether ring formation via dicobalt hexacarbonyl complexes.

The synthesis of *ciguatoxin* and *gambiertoxin* by Isobe and coworkers, which make use of most of the ring-opening and ring-closing methodologies described by his group, provides a remarkable example of the full potential of dicobalt hexacarbonyl alkynyl derivatives in synthesis.[27] In this context, Isobe *et al.* have reviewed their efforts in the synthesis of *ciguatoxin* by application of Nicholas-related processes,[28] and readers are referred to this manuscript for a detailed account. Some selected examples are, however, outlined in Scheme 12.[29] *Trans*-alkenes containing a six-membered ring

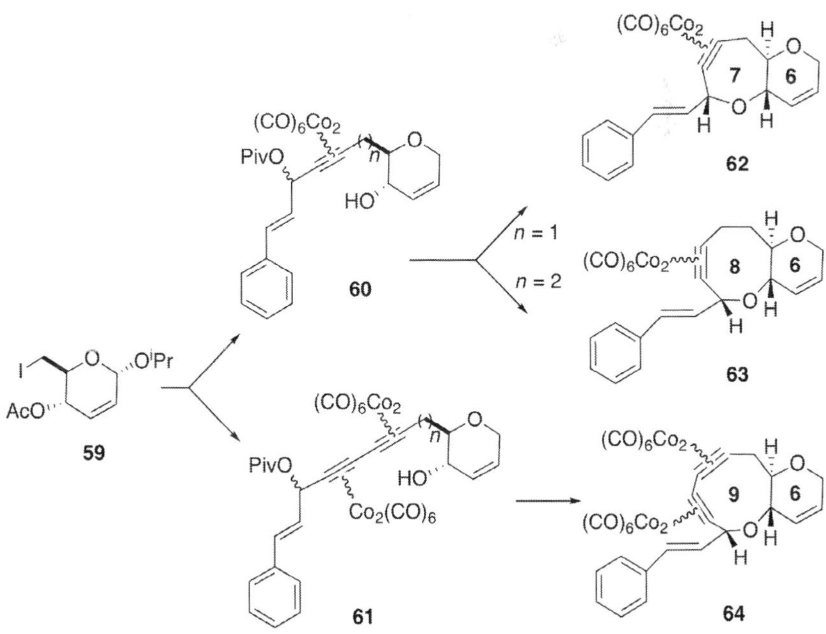

Scheme 12 Fused bicyclic ether formation via dicobalt hexacarbonyl complexes.

alcohol on a dicobalt hexacarbonyl alkyne tether, as in **60** (Scheme 12), were cyclized to give 7–6 and 8–6 bicycles, **62** and **63**, respectively, with predominant or exclusive formation of the *syn* stereoisomer. Likewise, derivative **61** comprising a tether with two complexed alkynes could be cyclized to give 9–6 bicycle **64**.

In 2003, Takase *et al.* developed a concise method for the synthesis of various alkynyl *C*-glycosides type **67**, by use of an intramolecular Nicholas cyclization. These derivatives had been identified as useful building blocks for oligonucleotide synthesis and in antisense DNA strategies.[30]

The approach, depicted in Scheme 13, involved a sequence of (i) complexation, (ii) five-membered ring-formation, and (iii) decomplexation, which could be conducted in a one-pot operation. The cobalt-mediated cyclization was shown to be reversible, then leading to the thermodynamically more stable β-isomers.

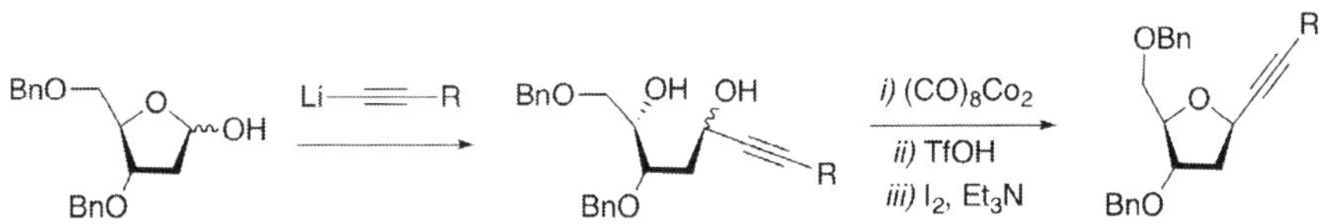

65 66 67

Scheme 13 Stereoselective synthesis of alkynyl *C*-glycosides with a 2-deoxy-β-D-*erythro*-pentofuranosyl moiety via intramolecular Nicholas reaction.

5 Glycosylations based on alkyne dicobalt hexacarbonyl complexes

In 1997, Mukai *et al.* developed an intramolecular glycosylation method for the transformation **68→69**, shown in Scheme 14.[31] The designed reaction pathway involved acid-mediated release of the propargylic glycosyl-acceptor partner **71** to generate a Nicholas cation **70**, which evolved by liberating a nonnucleophilic lactone **73**, and a glycosyl cation **72** that could intercept the aforementioned glycosyl acceptor **71** to give disaccharides **69**. The intramolecular nature of the process was proven via a crossover experiment, and the observed disaccharide yields ranged from moderate to low (77%–37%). The authors studied anomeric mixtures of D-gluco-, D-galacto-, and D-mannopyranosides as glycosyl donors, and excellent stereoselection could be observed in the case of 2-*O*-benzoyl *gluco* and *manno* derivatives.

Very recently, Li and coworkers have reported on the use of cobalt hexacarbonyl propargyl cations for the activation of thioglycosides in intermolecular glycosylation protocols.[32] The glycosylation process is outlined in Scheme 15 and involves the use of a cobalt hexacarbonyl propargyl derivative, for example, **75**, functioning as the glycosyl acceptor and a thioglycoside donor, for example, **74**. As a result of their studies, a plausible reaction pathway was advanced. Thus, interaction of the $Co_2(CO)_6$ propargylated acceptor with TMSOTf (0.1 equiv.) produced the Nicholas cation **76**, liberating a carbohydrate residue that could be silylated to produce **77**, as an activated glycosyl acceptor. The reaction process could then continue with the nucleophilic addition of the anomeric

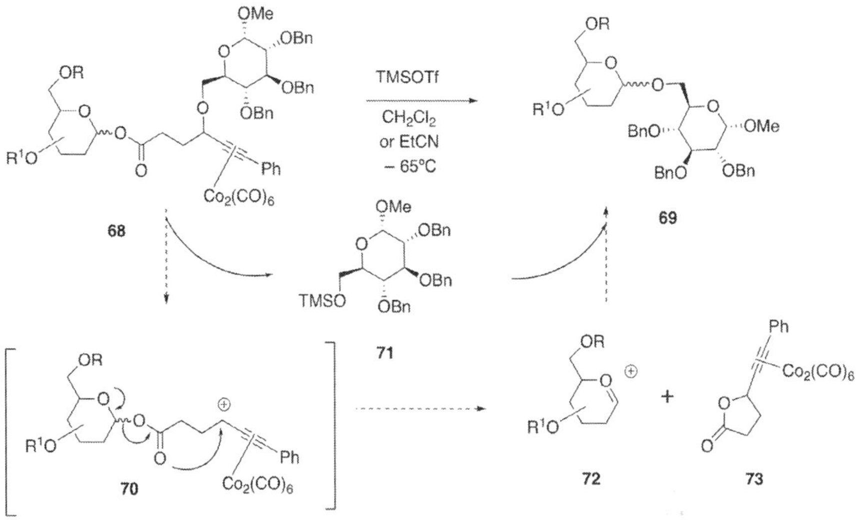

Scheme 14 Intramolecular aglycon delivery glycosylation mediated by an alkyne dicobalt hexacarbonyl complex.

sulfur atom on **74**, to Nicholas cation **76**, causing the cleavage of the C–S bond and resulting in the generation of glycosyl cation **78**, which could be trapped by **77** to provide disaccharide **79**, as well as $Co_2(CO)_6$ complexed propargyl *p*-methylphenylsulfide **80** (Scheme 15). The scope of the method, also studied by the authors, proved to be very broad and the glycosylation could be applied to acceptors with primary and secondary hydroxy groups, providing disaccharides in moderate to good yields (69%–85%).

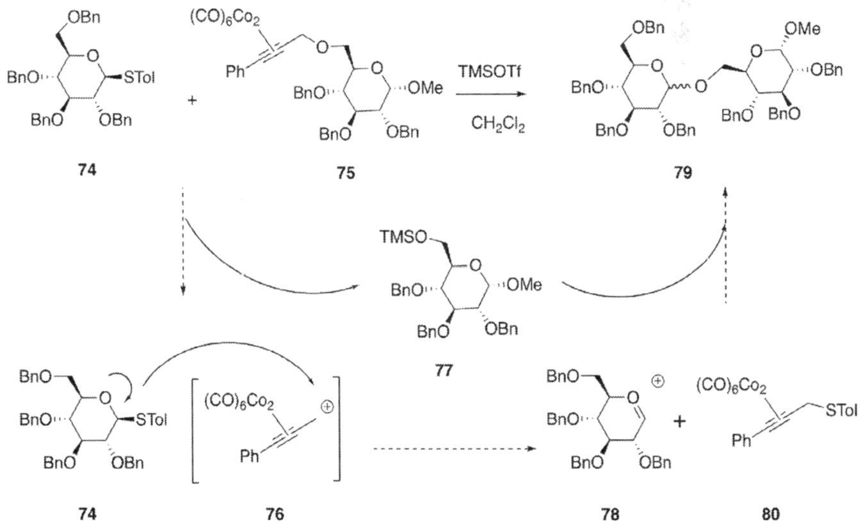

Scheme 15 $Co_2(CO)_6$-propargyl cation-mediated intermolecular glycosylation of thioglycosides.

Nicholas-type cations were also used by Schmidt's group in the preparation of *C*-disaccharides of ketoses.[33] Thus, they had identified *C*-ketosides **81** (Scheme 16) as potential candidates for glycosyl-transferase inhibitor precursors. Their retrosynthetic analysis visualized the formation of the desired *C*-ketosides by cyclization of a Nicholas cation **83** to *C*-ketoside **82**, as the key step (Scheme 16). It is noteworthy that a related acid-triggered cyclization on decobaltated **84** had proven to be unsuccessful. Subsequent synthetic manipulations of the dicobalt hexacarbonyl cluster **82** would then ultimately lead to **81**. Access to the alkynyl precursor **83** was imagined via the complex structure **84** accessible by sequential addition of allylmagnesium bromide and a pyranose-derived lithium acetylide to a gluconolactone derivative, **85**.

The key steps of Schmidt's synthesis are highlighted in Scheme 17. Thus, 2,3,4,6-tetra-*O*-benzyl-D-glucono-1,5-lactone **85a** (R = Bn) was transformed into isomeric 2′-(*S*) and 2′-(*R*) Co_2-$(CO)_6$-alkynyl derivatives, (*S*)-**86** and (*R*)-**86**, respectively (Scheme 17). Additionally, (*S*)-**86** and (*R*)-**86** were composed of two isomeric C-1 derivatives. The Nicholas-type cyclization was next studied in most of these isomeric derivatives. Accordingly, Nicholas-type cyclization of each of the two isomeric C-1 alcohols (*R*)-**86** was carried out separately to provide, in each case, a 1:1 anomeric mixture of (*R*)-**87** ketosides (76% and 83% yield), which after separation of the anomers, and decobaltation with ceric ammonium nitrate (CAN), provided (*R*)-**88**α and (*R*)-**88**β, in 88% and 81% yield, respectively. On the other hand, a single C-1 isomer of cobaltated (*S*)-**86** was cyclized and demetalated (CAN), to yield a 10:1 mixture of α- and β-anomeric derivatives (*R*)-**87**α and (*R*)-**87**β, respectively. Further processing of these derivatives led to all possible isomers of **81**.

6 Dicobalt hexacarbonyl-mediated Ferrier(II)-type carbocyclizations from pyranose derivatives

Several methods for the transformation of dicobalt hexacarbonyl-derived pyranoses into cyclohexanone derivatives have been described by several research groups. These transformations displayed certain resemblance to the so-called Ferrier(II) carbocyclization or Ferrier(II) rearrangement. The Ferrier(II) carbocyclization process **89**→**90** depicted in Scheme 18 was reported by Ferrier in 1979.[34, 35] The transformation involved regiospecific hydroxymercuration of the vinyl ether moiety to an unstable hemiacetal, that is, **89**→**91**, which evolves to a 1,5-dicarbonyl intermediate **92**, that might experience an aldol-like cyclization to give cyclohexanone **90** (Scheme 18).

A report by Harrity and coworkers in 2002[36] was the first to illustrate the value of dicobalt hexacarbonyl clusters in the Ferrier-related construction of cyclohexanones, for example, **93**→**94** (Scheme 19). Thus, the hexacarbonyl dicobalt moiety at C-5 in enyne **93** facilitated the scission of the propargylic C−5−O bond by stabilization of the intermediate reactive carbocationic species **95**, which experienced a completely

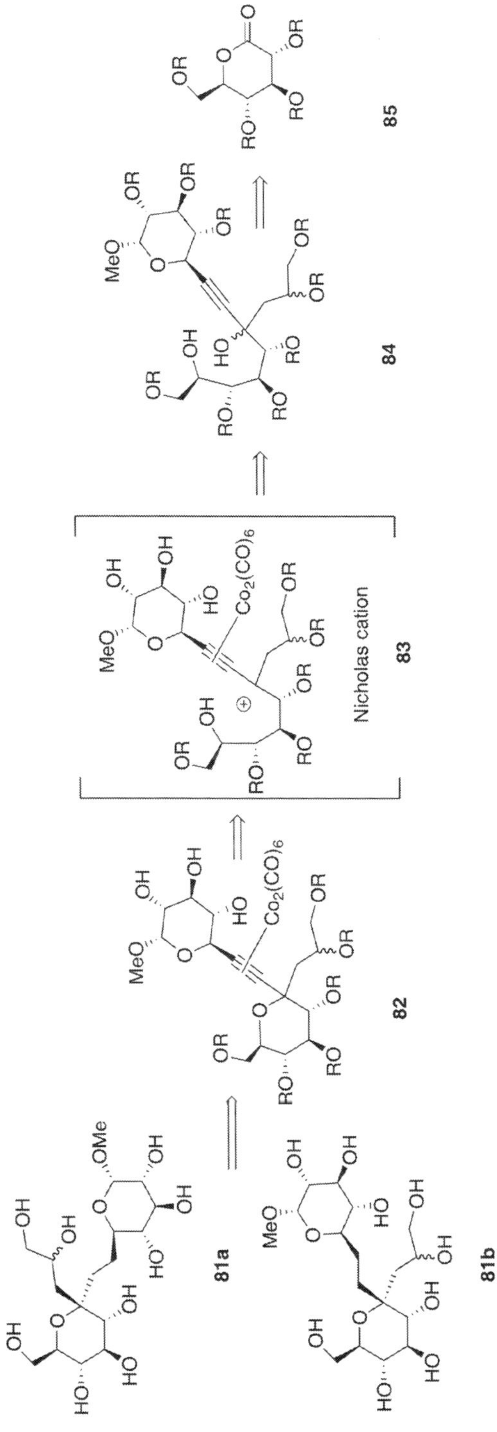

Scheme 16 Retrosynthesis of ketose *C*-disaccharides by Nicholas cyclization.

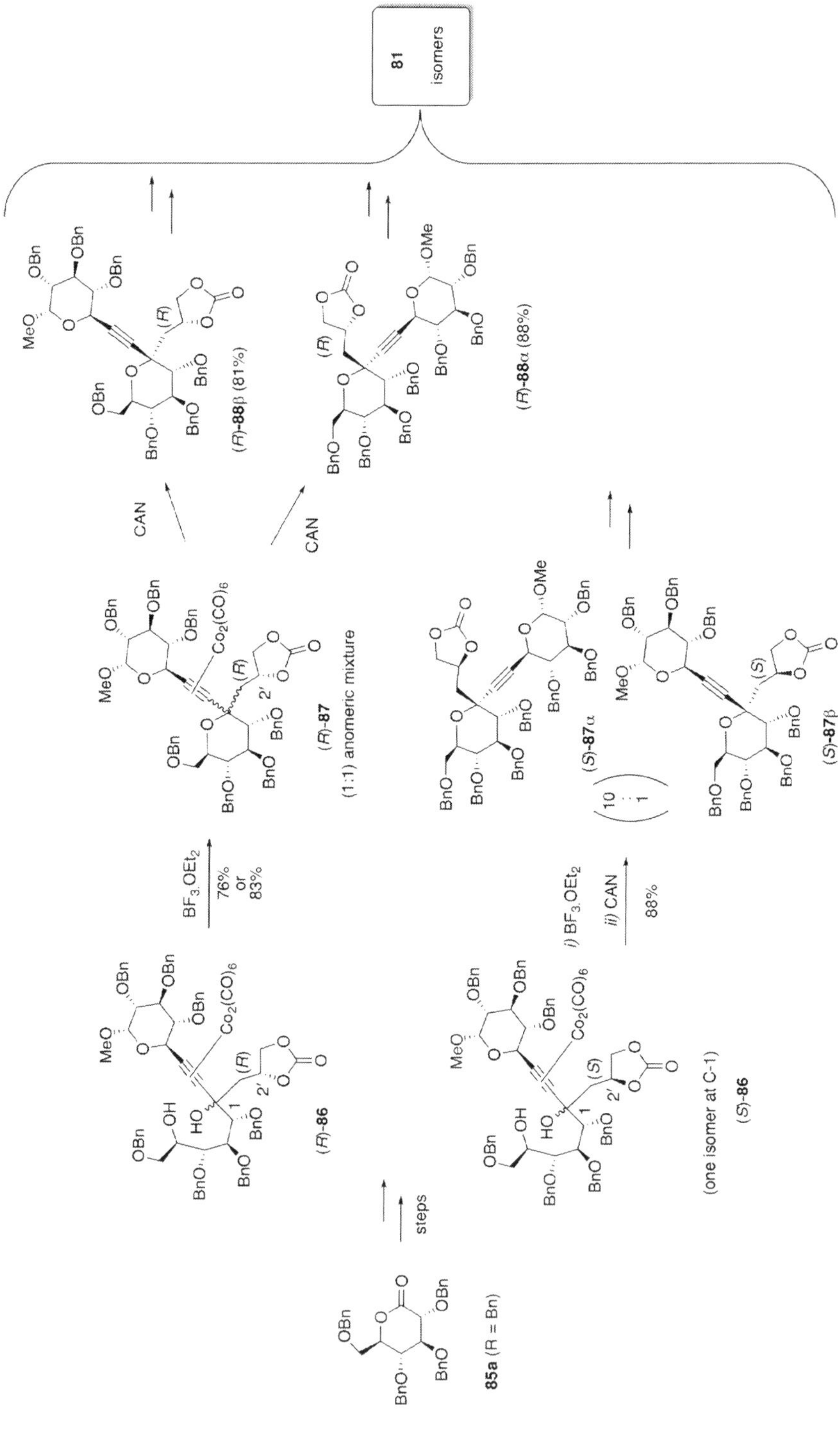

Scheme 17 Schmidt's synthesis of *C*-ketosides by Nicholas-type cyclization.

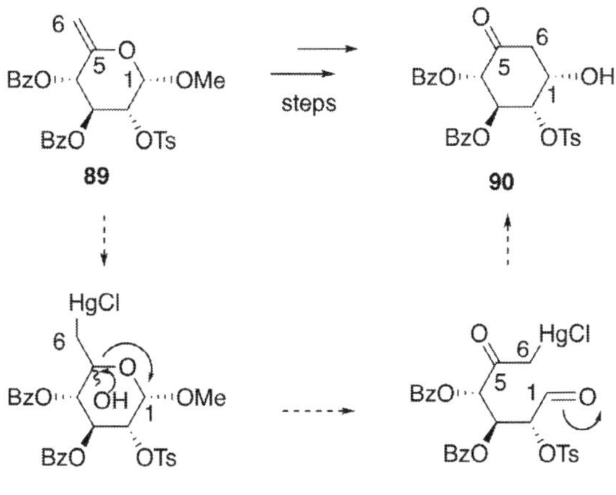

Scheme 18 Ferrier(II) rearrangement or Ferrier carbocyclization.

regioselective addition to the enolate moiety at C-1 to yield cyclohexanone **94**, most likely through a chair-like transition state, for example, **96**.[37] Although initial studies were performed in racemic series, in their work with enantiomerically enriched compounds, Harrity and coworkers were able to establish that minimal racemization occurred at the stereogenic centre during the rearrangement process.[38]

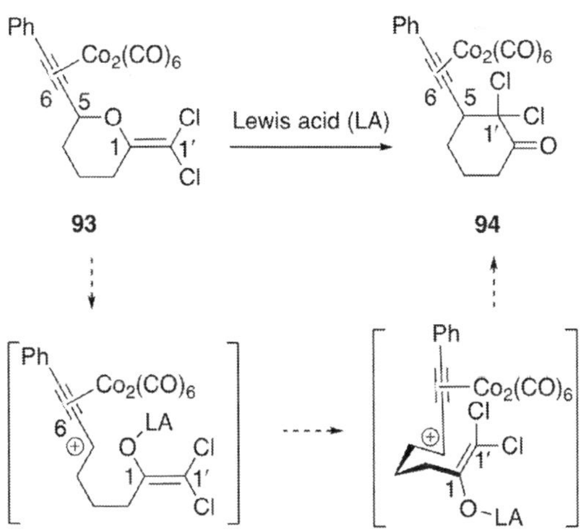

Scheme 19 Harrity's dicobalt hexacarbonyl mediated cyclohexanone formation.

On the other hand, as a continuation of their interest in sugar-to-carbocycle transformations, for example, **97→98** (Scheme 20),[39] Sinaÿ and coworkers reported on the carbocyclic ring closure of phenyl 5-*C*-methylene-1-thio- and -1-selenopyranosides, and the corresponding aryl *C*-glycosides mediated by triisobutylaluminum (TIBAL).[40] In their studies, they acknowledged the key role of an electron-donating group (EDG) by stabilizing their proposed key (C-1)-carbocationic intermediate **99**, in its cyclization process to ketone **100**, the latter subsequently reduced (TIBAL) to the corresponding alcohol **98** (Scheme 20). In this context, Sollogoub, Sinaÿ, and coworkers recognized the potential of the dicobalt hexacarbonyl cluster as an EDG in their general strategy for the preparation of carbocycles from carbohydrate derivatives. As a result, they reported a concise method for the stereocontrolled synthesis of gem-difluorocarba-α- and β-D-glucopyranoses **103**, based on a Nicholas-type cyclization **101**→**105**, followed by synthetic manipulations (Schemes 21 and 22).[41] Accordingly, synthesis of the α-isomer (α-**103**) was carried out by TIBAL-mediated Nicholas cyclization **101**→**102**, followed by synthetic manipulations (CAN, partial hydrogenation, ozonolysis, and reduction) on the alkynyl dicobalt hexacarbonyl group for the installation the C6-OH group (Scheme 21). In this process, the stereoselective reduction of the C-1 carbonyl group in ketone intermediate **105**, to yield the C-1 axial hydroxy group in α-**102**, was explained by the authors via intramolecular hydride-delivery from the isobutyl group of TIBAL on the less-hindered β-face of the molecule. On the contrary, the Ferrier(II)-Nicholas rearrangement of **101**, mediated by $Cl_3TiO\textit{i}Pr$ led to cyclohexanone **105** that, after quenching of the Lewis acid with THF, could be reduced under steric control by super hydride (Et_3BHLi) to yield equatorial alcohol **106**, precursor of β-**103** via a synthetic sequence related to the one mentioned earlier (Scheme 22).

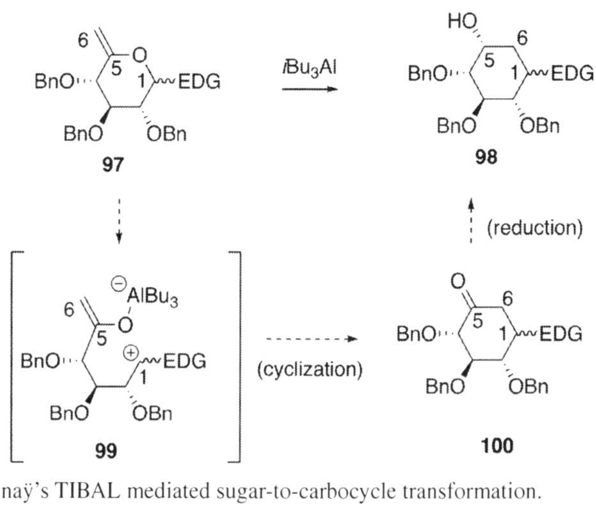

Scheme 20 Sinaÿ's TIBAL mediated sugar-to-carbocycle transformation.

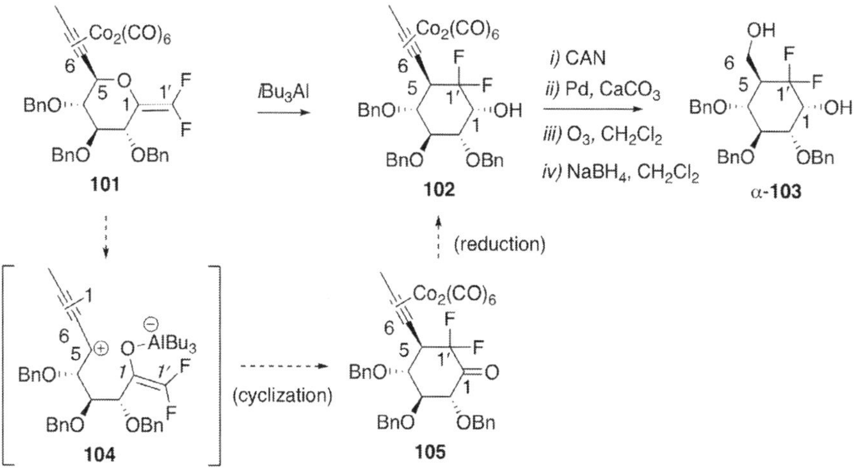

Scheme 21 Sinaÿ and Sollogoub's TIBAL mediated synthesis of 5a-difluoro-α-D-carbaglucopyranose.

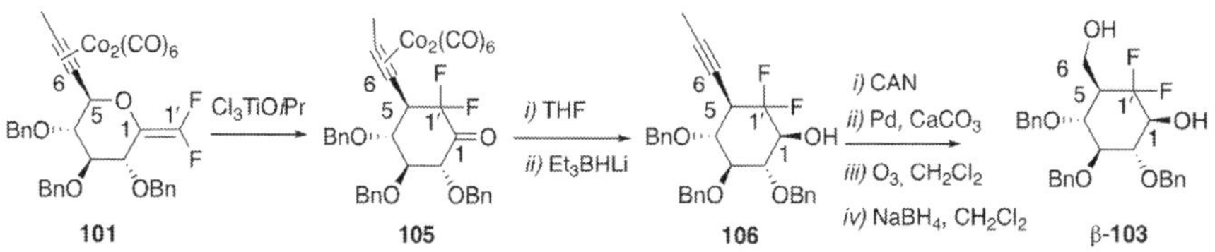

Scheme 22 TIBAL-mediated synthesis of 5a-difluoro-β-D-carbaglucopyranose.

By application of the aforementioned protocol, Sollogoub and coworkers were able to synthesize 5a-difluoro α- and β-carbamanno- (**109**) and 5a-difluoro α- and β-carbagalactopyranoses (**112**), from alkynyl derivatives **107** and **110**, respectively, readily available from the parent monosaccharides (Scheme 23).[42] As mentioned previously, the cobalt clusters arising from **107** and **110** played a dual role in the process: (i) as the electron-donating function stabilizing the propargyl cation and (ii) as a precursor of the CH_2OH group in the desired carbasugars. Unlike their previous example with the D-gluco-derivative [**101**→α-**103**] (Scheme 22), the reduction of the intermediate (C-1) ketones in the *manno* and *galacto* series proved not to be completely stereoselective, leading to α,β-anomeric mixtures **108** and **111**, respectively (Scheme 23).

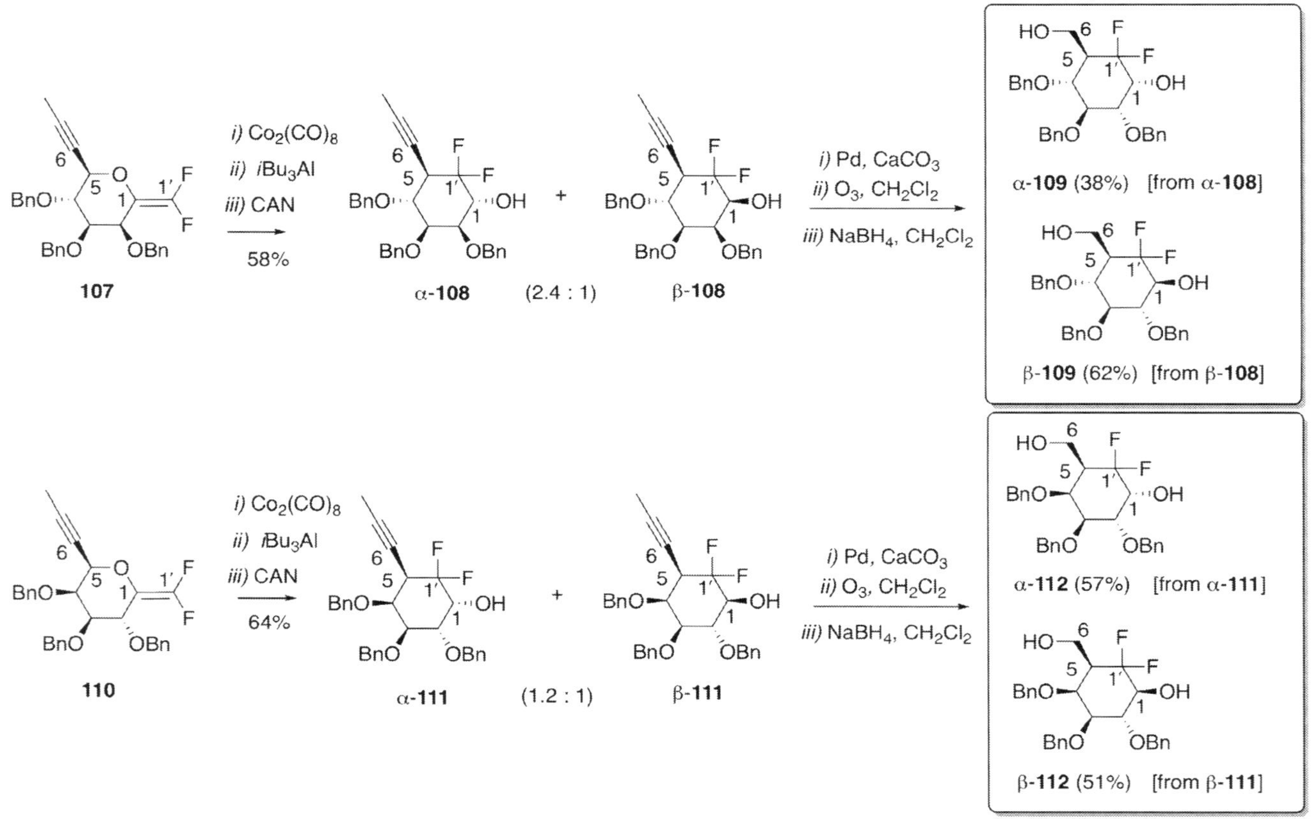

Scheme 23 TIBAL mediated synthesis of 5a-difluoro-α,β-D-carbamanno- and 5a-difluoro-α,β-D-carbagalactopyranoses, **109** and **112**, respectively.

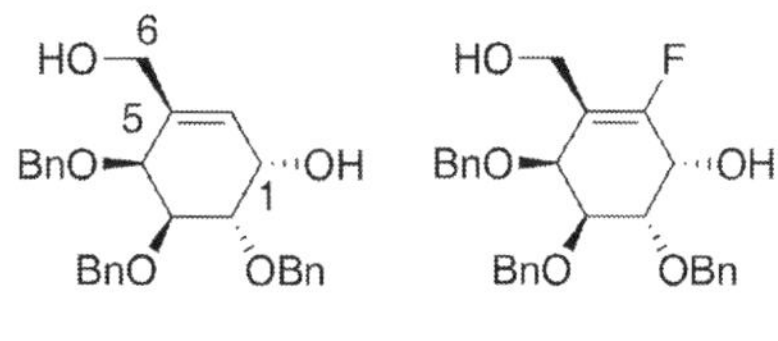

(+)-MK7607 **113**

Fig. 2 (+)-MK7607 and its 5a-fluoro analogue, **113**.

In a subsequent related work, Sardinha *et al.* reported on the synthesis of the 5a-fluoro analogue of (+)-MK7607 (**113**) (Fig. 2), a naturally occurring pseudo-carbasugar with herbicidal activity, from α-**112**.[43]

More recently, Chang and Isobe have reported on an additional diastereoselective dicobalt hexacarbonyl-assisted Nicholas-type cyclization to alkynyl cyclohexanones.[14] Their approach, highlighted in Scheme 24, was largely based in chemistry previously developed by Isobe's group. Thus, a completely α-stereoselective alkynylation on 2-acetoxy-D-glycal triacetate (**9**) led to **114**, which was then transformed into **115a–e**, a series of compounds differing on the substituent at the acetylenic terminal position (Scheme 24A). Cobalt-assisted Ferrier(II)-type cyclization of **115** was successfully mediated by diethylaluminum chloride (Et_2AlCl),[44] to give cyclohexanones **116** (Scheme 24B). Their results showed that smaller substituents at the alkyne terminal position (R^1), as in **115c**, **115e**, resulted in higher stereoselectivities in the cyclization leading to **116** (Scheme 24B).

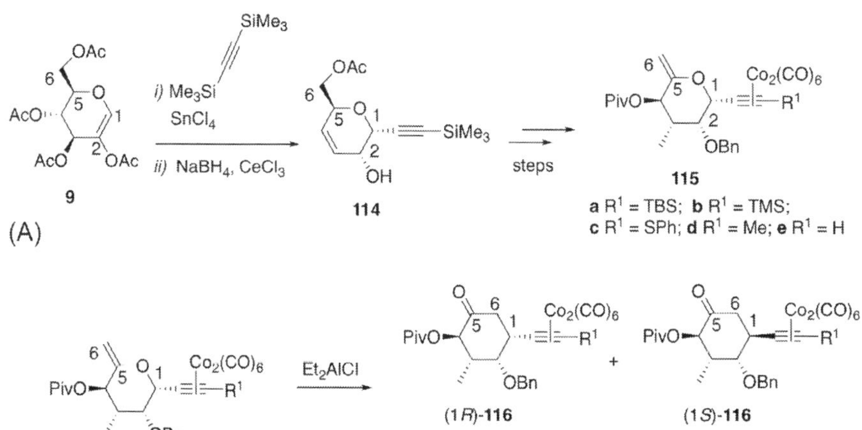

	(1*R*)-116	(1*S*)-116
116a (75%)	60	40
116b (78%)	67	33
116c (62%)	100	—
116d (92%)	83	17
116e (88%)	100	—

Scheme 24 Nicholas-Ferrier(II) type carbocyclization by Isobe's group.

In summary, contributions from Sollogoub's and Isobe's laboratories have served to illustrate the efficiency of the alkynyl dicobalt hexacarbonyl species in the Ferrier(II) carbocyclization, as depicted in Scheme 25. Thus, whereas in Sollogoub's approach the alkynyl cobalt cluster was placed at C-5 (Scheme 25A), in Isobe's protocol the dicobalt hexacarbonyl substituent was located at C-1 (Scheme 25B), and in each case the propargyl cation, stabilized by the dicobalt hexacarbonyl cluster, played a key role in the cyclization.

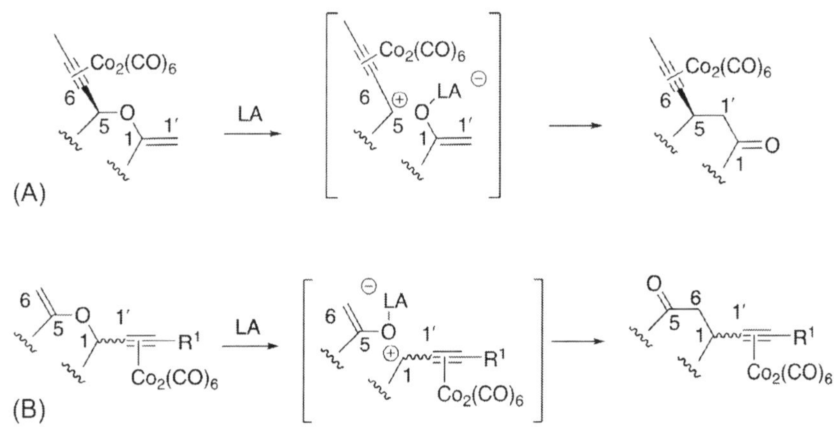

Scheme 25 Approaches to Nicholas-Ferrier(II)-type carbocyclization by Sollogoub's (A) and Isobe's (B) group.

7 Pyranosidic dicobalt hexacarbonyl propargyl oxycarbenium ions versus oxycarbenium ions—Some remarkable features

Some early examples from our laboratories also served to illustrate the idiosyncrasy of anomeric (C-1) dicobalt hexacarbonyl propargyl cations, compared to the parent alkynyl cations. Thus, as part of a synthetic program we required appreciable amount of bis(*C*-pyranosides) type **119**, with one of the C1-substituents being an alkynyl group (Scheme 26).[45, 46] A straightforward retrosynthetic analysis to 1-*C*-alkynyl *C*-pyranosides **119**, depicted in Scheme 26, showed two possibilities based on the *C*-glycosylation of either *C*-alkynyl pyranoses **117** or their dicobalt hexacarbonyl derivatives **118**. At the outset of this work, only one example of the transformation of a *C*-vinyl ketose into a bis(*C*-pyranoside) had been described.[47]

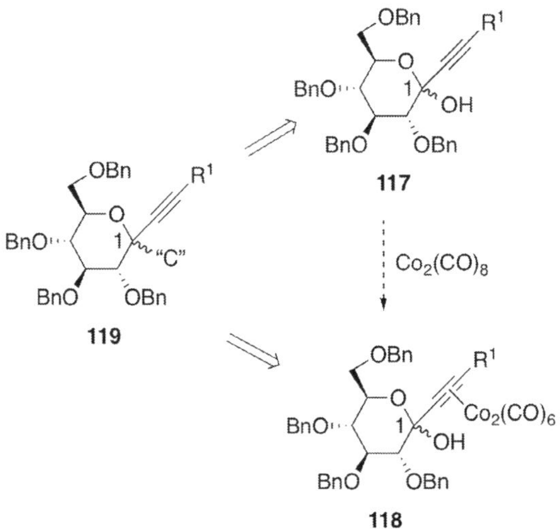

Scheme 26 Retrosynthetic analysis of *C*-pyranosides type **119** from 1-*C*-alkynylpyranoses.

Some representative examples of our studies are displayed in Scheme 27. Thus, the *C*-allylation (allyltrimethylsilane) of "cobalt-free" 1-*C*-alkynylpyranose **117** ($BF_3{\cdot}OEt_2$) took place smoothly to give *C*-glycoside **120**, as a sole stereoisomer, in 67% yield (Scheme 27A). On the other hand, a related reaction of **117** in the presence of 1,3,5-trimethoxybenzene as nucleophile provided open-chain bis-arylated derivative **121** (51% yield, Scheme 27B).

Conversely, the reactions of dicobalt hexacarbonyl analogues **118** and **124** with allyltrimethylsilane and 1,3,5-trimethoxy benzene, respectively, took place to give entirely different compounds (Scheme 27C and D).[48] Thus, allylation of dicobalt hexacarbonyl derivative **118** (allyl trimethylsilane, $BF_3{\cdot}OEt_2$) furnished unsaturated, 1,4-di-*C*-allyl derivative **122**, which after uneventful iodine-mediated decobaltation provided **123**. The incorporation of the second allyl unit had taken place at C-4 with complete stereoselectivity (Scheme 27C). Reaction of **124** with 1,3,5-trimethoxybenzene ($BF_3{\cdot}OEt_2$) provided C3-branched *C*-glycal **125**, which was decobaltated (I_2, THF) to provide **126** (Scheme 27D). Unlike the previous case, only one aryl residue had been incorporated, this time at C-3 with complete stereoselectivity (Scheme 27D).

From the results in Scheme 27, it became clear that the dicobalt hexacarbonyl cluster in 1-*C*-alkynylpyranose complexes **118** and **124** exerted a drastic effect on the chemical behavior of the intermediate anomeric oxycarbenium ions, when compared to "cobalt free" analogue **117** (compare Scheme 27A and Scheme 27C; and Scheme 27B and Scheme 27D).

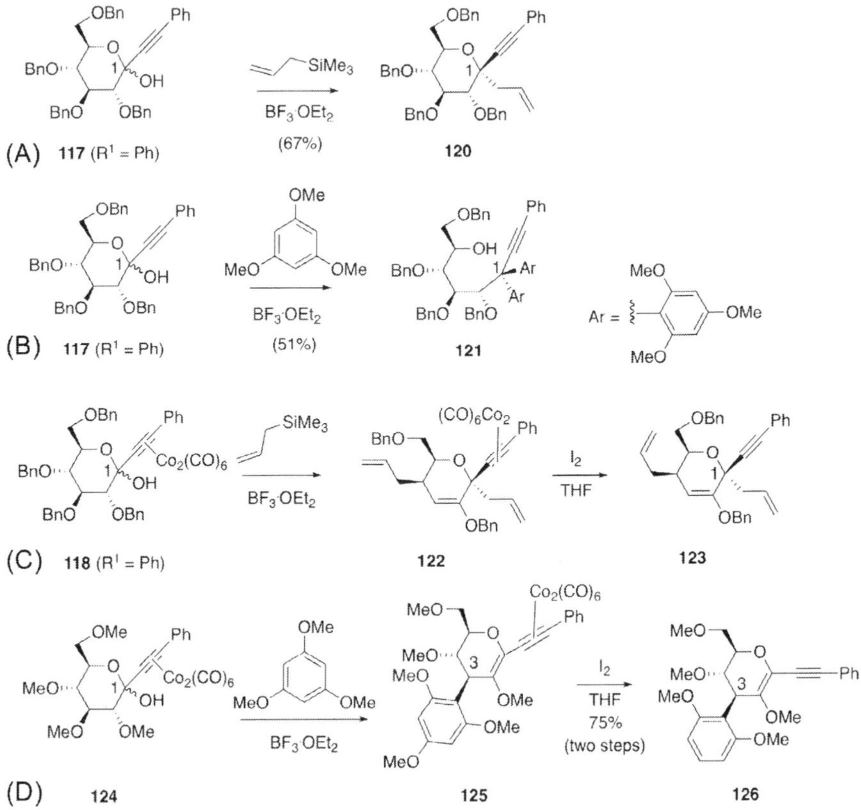

Scheme 27 Lewis-acid mediated reactions of *C*-alkynyl and *C*-dicobalt hexacarbonyl alkynyl pyranoses with allyl trimethylsilane and 1,3,5-trimethoxybenzene.

Our proposed reaction pathways for the aforementioned transformations are depicted in Scheme 28A and B. Thus, the formation of compounds **120** and **121** from **117** could be explained by alkylation or (Friedel-Crafts) arylation of an intermediate 1-*C*-alkynyloxycarbenium ion **127** leading to bis(*C*-glycosides) **120** (Scheme 28A) or **128** (Scheme 28B), respectively. The former could be isolated as the final reaction product, whereas the highly reactive bis(*C*-glycoside) intermediate **128** underwent further acid-catalyzed ring-opening to generate a highly stabilized (benzylic, propargylic) C-1 oxycarbenium ion **129**, able to capture a second nucleophile unit, leading to **121** (Scheme 28B). On the other hand, Nicholas-oxycarbenium ion **130**, arising from Lewis acid activation of dicobalt hexacarbonyl derivatives **118** or **124**, displayed alternate reaction patterns. Thus, the formation of bis-allylated derivative **122** was explained by stereoselective nucleophilic ring-opening of the bicyclic dioxolanyl ion **131** (in equilibrium with **130**), to provide a postulated C4-allyl derivative **132**, followed by elimination to *C*-glycal **133** (Scheme 28C).[1b, 49] The latter behaved as a Ferrier(I)-type substrate, *vide infra*, leading to vinylogous Nicholas cation **134**,[50, 51] able to capture a second allyl nucleophile at C-1 to lead, after decobaltation, to 1-*C*-alkynyl-1-*C*-allyl derivative **123** (Scheme 28C). A direct elimination process from oxycarbenium Nicholas cation **130** to a postulated glycal **135** might have been at the origin of the formation of the 3-*C*-branched glycal **125**, via arylation at C-3 of the vinylogous Nicholas cation **136** (Scheme 28D).

From these results, it became of interest to us the chemical behavior of the postulated vinylogous Nicholas cations **134** and **136** (Scheme 28C and D), and under these premises we initiated a study on the synthesis and reactivity of the *C*-(dicobalt hexacarbonyl)alkynylglycals, for example, **133**, **135**, as precursors of the aforementioned vinylogous pyranosidic Nicholas cations.

8 Dicobalt hexacarbonyl complexes of alkynyl compounds as precursors of pyranosidic Ferrier-Nicholas cations—Synthesis and transformations

8.1 Ferrier rearrangement or Ferrier(I) reaction—Ferrier-Nicholas cations

The transformation of glycals (1,5-anhydrohex-1-enitols, for example, **137**, Scheme 29A) into 2,3-unsaturated glycosyl derivatives, for example **139**, reported by Ferrier in 1962, is currently known as Ferrier rearrangement.[52–54] In this transformation, allylic oxycarbenium ions, for example, **138**, are currently accepted as the reaction intermediates. In comparison with "normal" glycosyl oxycarbenium ions, Ferrier cations **138** are stabilized by the absence of the electron-withdrawing C2-substituent and by additional allylic conjugation.

As previously mentioned, we had postulated the existence of dicobalt hexacarbonyl complexes of 1-*C*-alkynyl **133** and **135** as reaction intermediates and precursors of vinylogous Nicholas cations **134** and **136** (Scheme 28), respectively, which we had previously termed Ferrier-Nicholas cations.[48] In this context, we became interested in the study of Ferrier-Nicholas cations **142** and **143** that, as we visualized, could arise from dicobalt hexacarbonyl substrates **140** and **141**, respectively (Scheme 29B and C).

We had reasoned that "Ferrier-Nicholas" cations, for example **142**, **143**, would enjoy additional Nicholas-type stabilization when compared to Ferrier cations and that the presence of the 1-*C*- or 3-*C*-substituent would result in regioselective transformations.

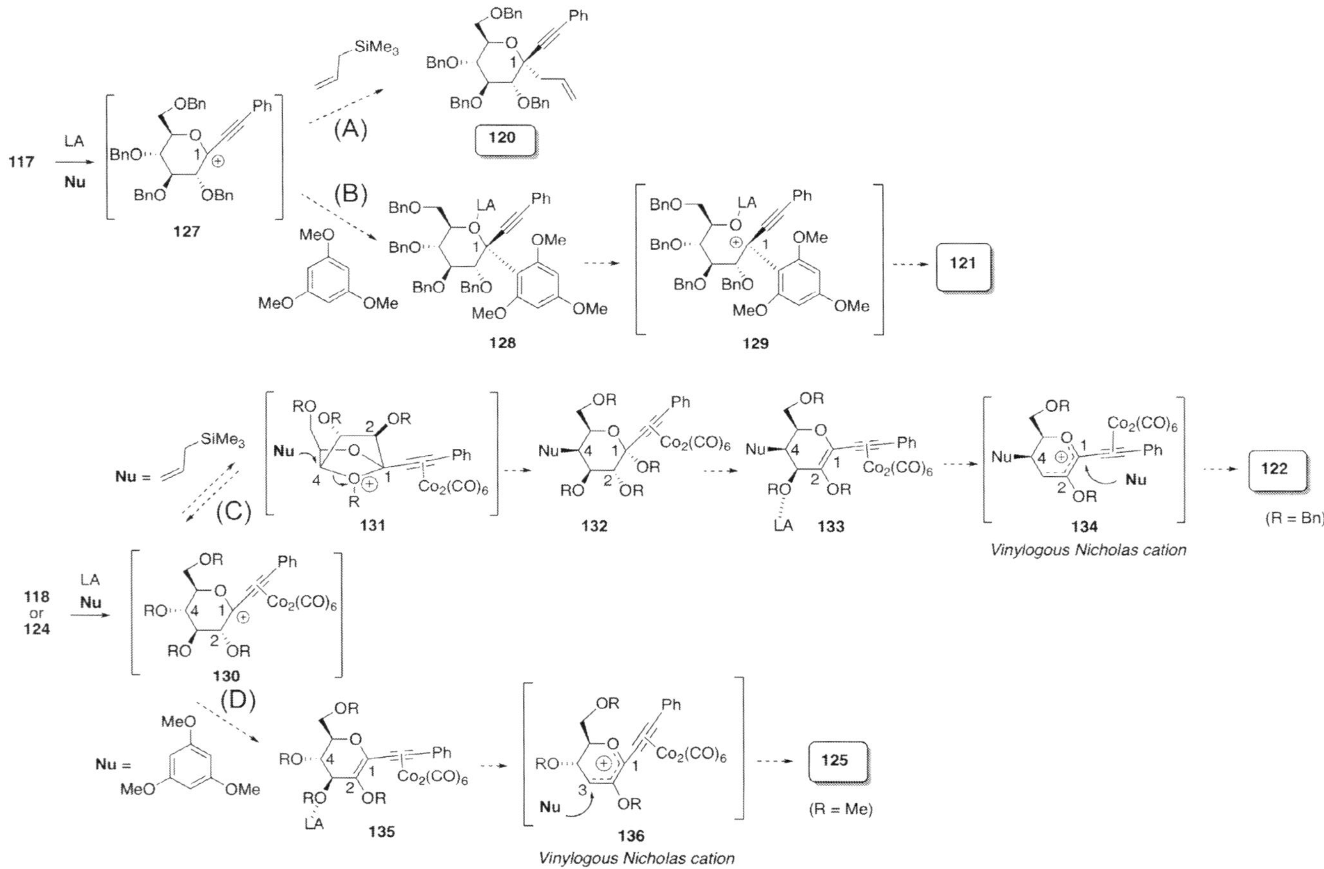

Scheme 28 Proposed reaction pathways in the transformations of *C*-alkynyl (**117**) and *C*-dicobalt hexacarbonyl alkynyl (**118**, **124**) pyranoses.

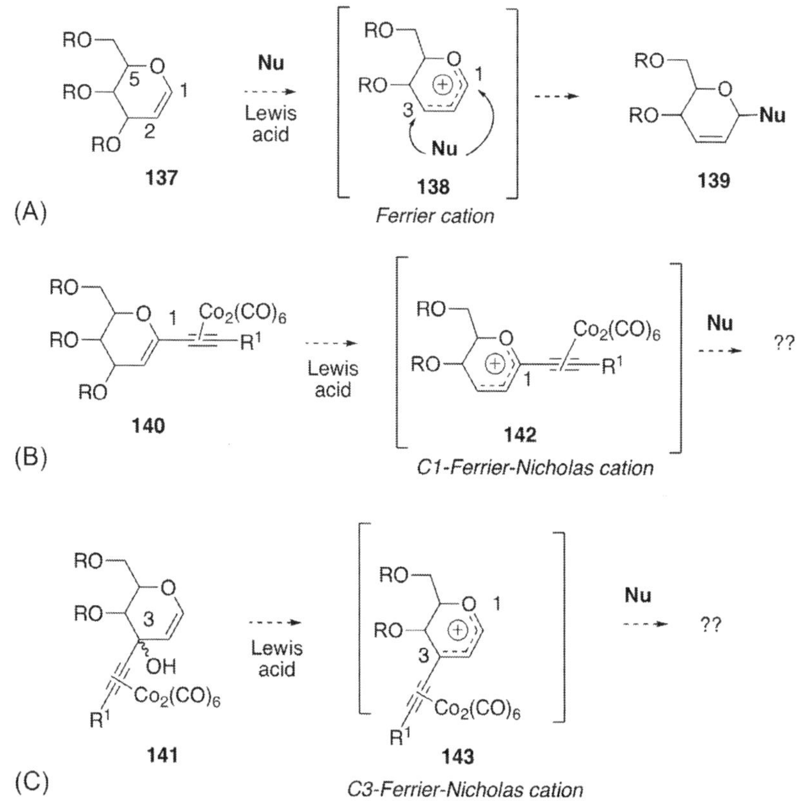

Scheme 29 Ferrier rearrangement, Ferrier cations (A), and Ferrier-Nicholas cations (B,C).

8.2 C1-Ferrier-Nicholas cations

Regarding the study of C1-Ferrier-Nicholas cations, we initially synthesized derivatives **140**, with a dicobalt hexacarbonyl complex of the alkynyl group at C1, as their potential precursors. Thus, compounds **140** (Fig. 3), differing on the C6-substituent, were efficiently prepared from tri-*O*-acetyl-D-glycal **6**, in five- or six-step synthetic sequences.[55, 56]

The chemical behavior of these derivatives, upon treatment with $BF_3{\cdot}OEt_2$, proved to be highly dependent on the nature of the substituent at C6. Thus, 6-*O*-triisopropylsilyl glycal **140a** reacted with a series of nucleophiles, for example furan,

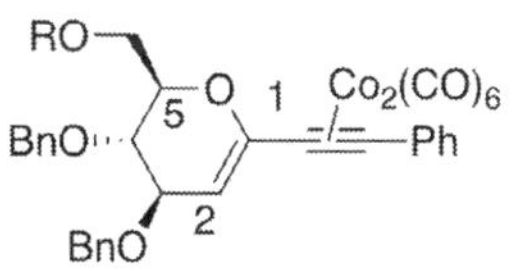

140

a R = TIPS; **b** R = H; **c** R = Bn; **d** R = allyl

Fig. 3 Dicobalt hexacarbonyl complexes of 1-C-alkynylglycals **140a–d**.

allyl trimethylsilane, indole, in the presence of $BF_3{\cdot}OEt_2$, at −20°C, to give stereoselectively 3-*C*-branched glycals **144** in fairly good yield (Scheme 30). Remarkably, the reaction of **140a** with pyrrole, which took place at −78°C, provided access to 3-*C*-pyrrolylglycal **145**, whose highly nucleophilic glycal double bond proved to be able to react with a second nucleophile unit in a two-step electrophilic addition process (A_dE),[57] leading to bis(*C*-glycosides) **146**, in moderate yields (Scheme 29).

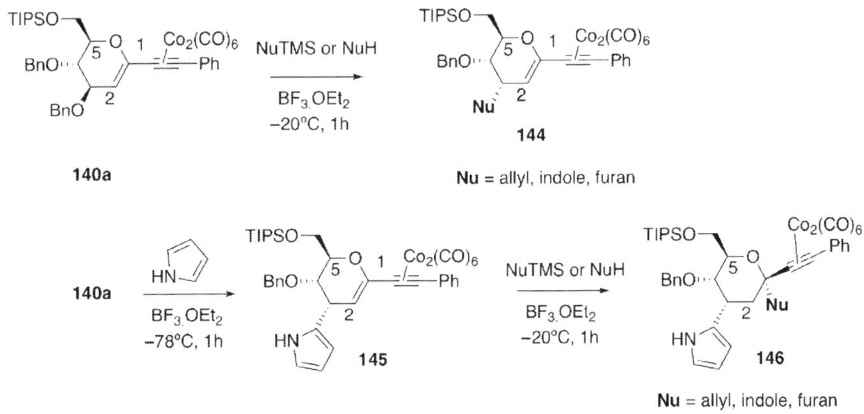

Scheme 30 Reactivity studies on 6-*O*-triisopropyl *C*-glycal **140a**.

6-Hydroxy *C*-glycal **140b**, on the other hand, underwent acid-catalyzed cyclization to a mixture of epimeric 3-OBn-1,6-anhydro derivatives **147** and branched tetrahydrofuran derivative, **148** (Scheme 31). Our rationale for the observed transformation involved the reversible formation of Ferrier-Nicholas cation **149**, which will be at the origin of the observed C3-epimeric mixture in **147**. Alternatively, an intramolecular, two-step electrophilic cyclization process (Ad_E) of the C-6 hydroxy group onto the

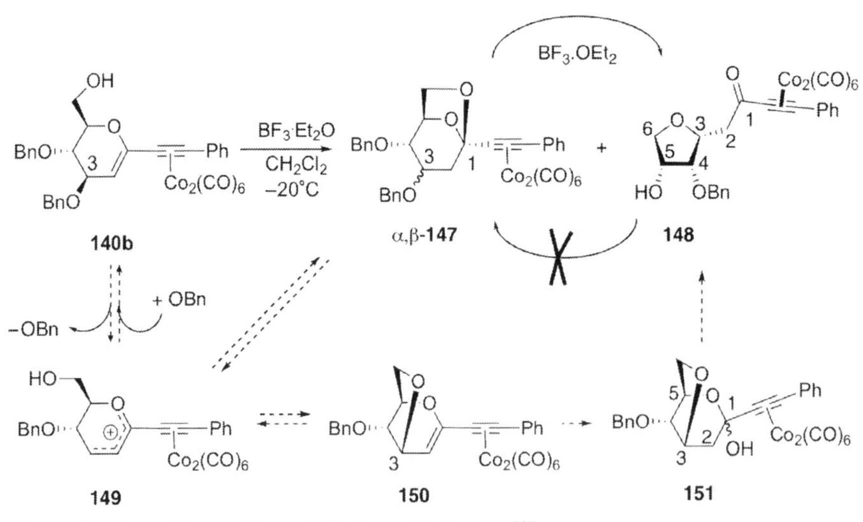

Scheme 31 Reactivity studies on 6-hydroxy *C*-glycal **140b**.

glycal double bond would have led to the formation of the 1,6-anhydro derivative, **147**. On the other hand, regioisomeric intramolecular (C3) Ferrier-type reaction would produce **150**, which could undergo acid catalyzed hydration to hemiketal **151**, whose retroketalization will irreversibly produce branched tetrahydrofuran **148**, as a sole isomer. Along this line, **147** could be transformed to **148** under the same reaction conditions, whereas the opposite did not occur.

Reaction of 6-OBn derivative **140c**, in the presence of $BF_3{\cdot}OEt_2$, also proved to be a remarkable process, providing a mixture of 1,6-anhydro derivative **152** and oxepane **153** (Scheme 32). The proposed reaction pathway invoked a 1,6-hydride transfer from the 6-*O*-benzyl group onto an intermediate Ferrier-Nicholas cation **154**,[58] to generate a benzylic oxycarbenium ion and regenerate the glycal double bond in **155**. The latter is the key intermediate in the formation of **152** and **153**. Thus, hydrolysis of the oxycarbenium ion followed by addition to the glycal double bond produced **152**, whereas an alternate path involving cyclization through attack of the glycal double bond to the oxycarbenium ion produced, stereoselectively, bicyclic Nicholas cation **156**, which would evolve to oxepane **153** by water addition, to form an intermediate ketose, followed by pyranose ring-opening.

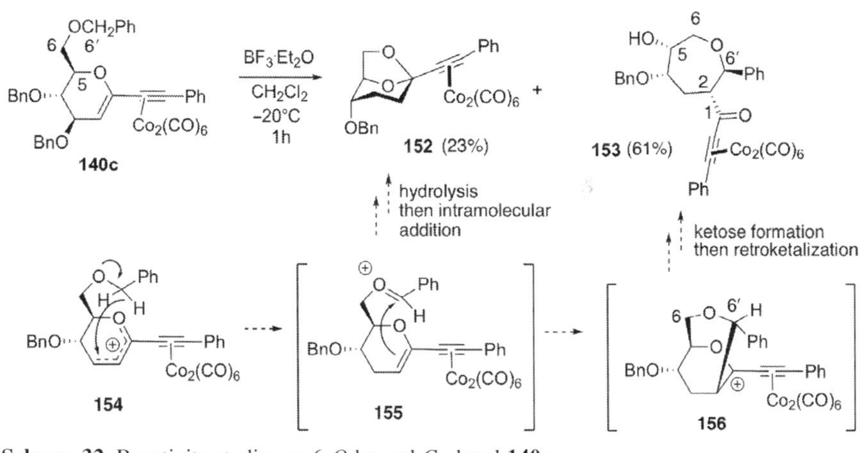

Scheme 32 Reactivity studies on 6-*O*-benzyl *C*-glycal **140c**.

Finally, 6-*O*-allyl derivative **140d** displayed, upon treatment with $BF_3{\cdot}OEt_2$, a behavior similar to that of its 6-OBn analogue **140c**, furnishing a mixture of oxepane and 1,6-anhydro sugar derivatives, **157** and **158**, respectively (Scheme 33). It is noteworthy that replacement of the terminal phenylalkynyl group by H resulted in an improved yield of oxepane over the 1,6-anhydro derivative (compare **157a/158a** versus **157b/158b**). This tendency had also been observed in the cyclization of 6-OBn derivatives.[55] Oxepanes **157** were of special interest to us because they could be used as test substrates for the Pauson-Khand cyclization reaction.[3, 59] Indeed, trimethylamine *N*-oxide (TMANO) treatment of **157a** provided tricyclic derivative **159** in 49% yield (Scheme 33).

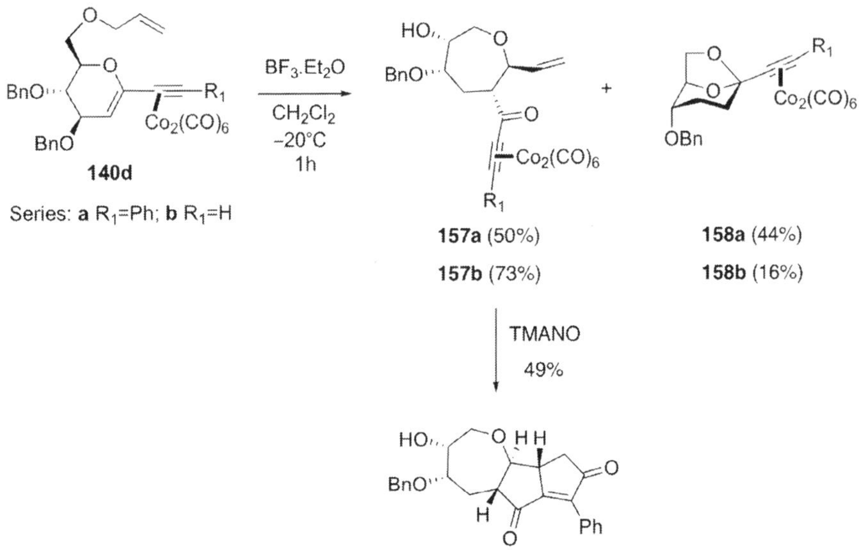

159

Scheme 33 Reactivity studies on 6-*O*-benzyl *C*-glycal **140d**, and Pauson-Khand reaction.

In this context, it is relevant to mention that the first Pauson-Khand reaction on carbohydrate derivatives had been reported by Isobe and Takai in 1999 (Scheme 34).[60] They studied the Pauson-Khand reaction of dicobalt hexacarbonyl complexes of both 2-*O*-allyl-1-*C*-alkynyl anomers **160** to give access to tricyclic derivatives **161** and **162**. The reactions were mediated by *N*-methylmorpholine-*N*-oxide.

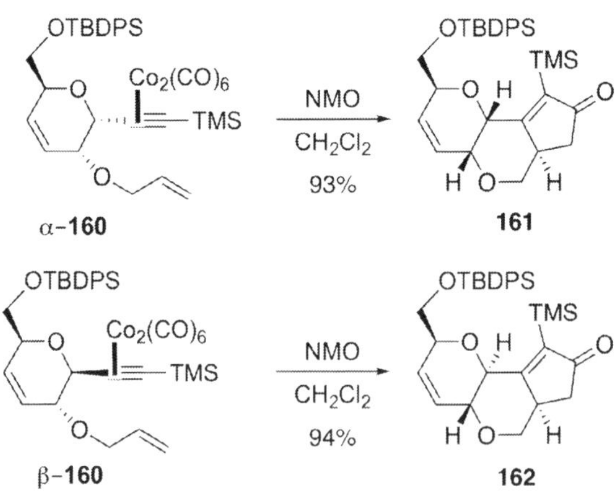

Scheme 34 First Pauson-Khand cyclization on carbohydrate derivatives.

8.3 C3-Ferrier-Nicholas cations

As starting material for the studies on 3-*C*-alkynylglycals and their dicobalt hexacarbonyl derivatives, the 3-*C*-alkynyl derivative **163**, available in a four-step sequence from tri-*O*-acetyl-D-glycal, **6**, and its cobalt complexed counterpart **164**, were used (Scheme 35).[61] Glycosylation reactions of MeOH with **163** and **164** took place at −20°C ($BF_3{\cdot}OEt_2$, CH_2Cl_2) to give methyl glycosides **165** and **166**, respectively. Even though the glycosylation of the dicobalt hexacarbonyl derivative **164** gave better yields (87%) than that of the 3-*C*-alkynyl glycal **165** (63%), the main difference was the observed anomeric selectivity. Thus, whereas **163** produced a 6:1 (α/β) mixture of anomers (**165**), in keeping with literature precedents, glycosylation of dicobalt hexacarbonyl complex of 3-*C*-alkynylglycal, **164**, furnished a 1:3 (α/β) mixture of methyl glycosides (**166**). The dependency of the anomeric selectivity on the presence of the cobalt cluster could be established when **165** [6:1 (α/β) anomeric mixture] was treated with $Co_2(CO)_8$ followed by $BF_3{\cdot}OEt_2$ to yield **166** as a 1:3 (α/β) mixture of methyl glycosides (Scheme 35).

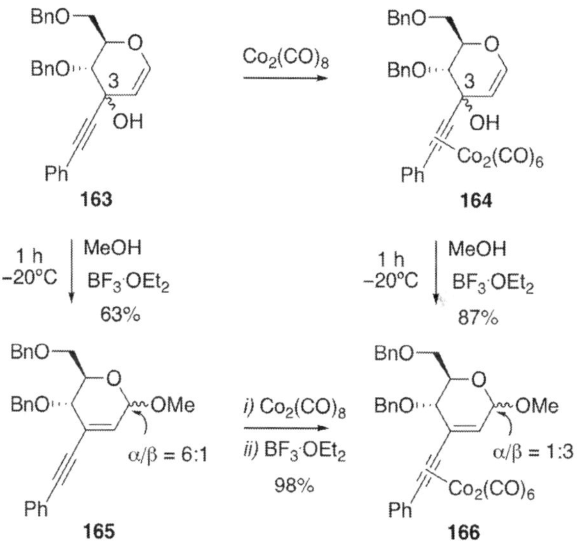

Scheme 35 Glycosylation of methanol with 3-*C*-alkynylglycal **163** and dicobalt hexacarbonyl complex of 3-*C*-alkynylglycal **164**.

The role of the cobalt complex in the reactivity of alkynyl compound **164** could be established when a less reactive glycosyl acceptor, allyltrimethyl silane, was employed (Scheme 36). Thus, *C*-glycosylation of allyltrimethyl silane with "decobaltated" alkynylglycal **163** produced the α-anomer of allyl *C*-glycoside **167** in a scarce 12% yield, whereas *C*-allylation of **164** produced α-anomer of allyl *C*-glycoside **168** in 70% yield (two steps, from **163**). Decobaltation of **168** was uneventfully accomplished by treatment with tetrabutylammonium fluoride (TBAF).

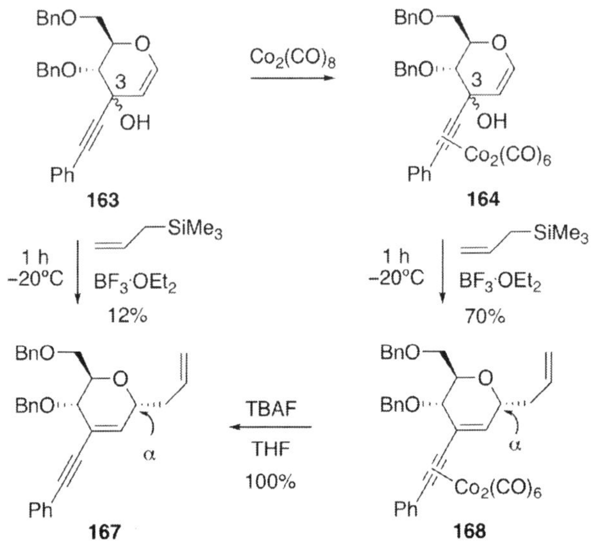

Scheme 36 Glycosylation of allytrimethylsilane with 3-*C*-alkynylglycal and the dicobalt hexacarbonyl complex of 3-*C*-alkynylglycal, **163** and **164**, respectively.

The glycosylation of *N*-methylindole with glycals **163** and **164** leading to *C*-indolyl derivatives **169** and **170** was also studied using different reaction times (Scheme 37). From these experiments at least two conclusions could be drawn: first, the Nicholas activation was critical to obtain synthetically useful yields of heteroaryl *C*-indolyl derivatives **170** (compare 4% yield versus 43% yield, entry *i*, and 12% versus 69% yield, entry *ii*, Scheme 37), and second, longer reaction times favored the formation of β-glycosides over α-glycosides (compare α/β selectivities in entries *i* and *ii*, Scheme 37).

The results in Scheme 37 (variation of the α/β anomeric ratio in 1-*C*-indolyl derivative **170** with the reaction time) seemed to indicate the existence of an equilibrium in favor of the thermodynamically more stable β-anomers, more pronounced in the doubly activated 1-*C*-indolyl derivative β-**170**. The α→β equilibration, in these derivatives, was explained by the authors by invoking the intermediacy of an open-chain vinylogous Nicholas cation, that is, **171** (Scheme 38A). Along this line, the observed reaction of **164** with two equivalents of *N*-methylindole to yield open-chain, bis-indolyl, derivative **172**, could be explained by reaction of open-chain Nicholas cation **171**, with a second indole molecule (Scheme 38B).

8.4 Ferrier-Nicholas systems based on (2-deoxy-2-C-methylenepyranosyl)alkynes

A recent report has identified dicobalt hexacarbonyl complexes of 2-*C*-methyleneketose derivatives, for example, **173**, as precursors of 2′,3-difunctionalized *C*-glycals, for example, **174** (Scheme 39).[62] Thus, the dicobalt hexacarbonyl complex of *C*-ketoside **173** readily obtained from tri-*O*-acetyl-D-glycal, **6**, reacted with heteroaryl nucleophiles in

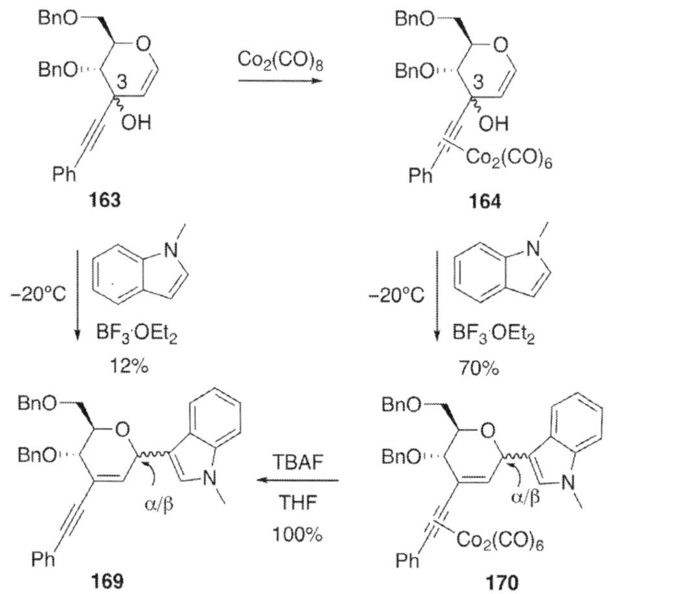

i)	20 min	4% yield	α/β ratio = 2:1	10 min	43% yield	α/β ratio = 3:1
ii)	60 min	12% yield	α/β ratio = 1.5:1	60 min	69% yield	α/β ratio = 1:2

Scheme 37 Glycosylation of *N*-methylindole with 3-*C*-alkynyl- and dicobalt hexacarbonyl complex of 3-*C*-glycals **163** and **164**, respectively. Effect of the reaction time.

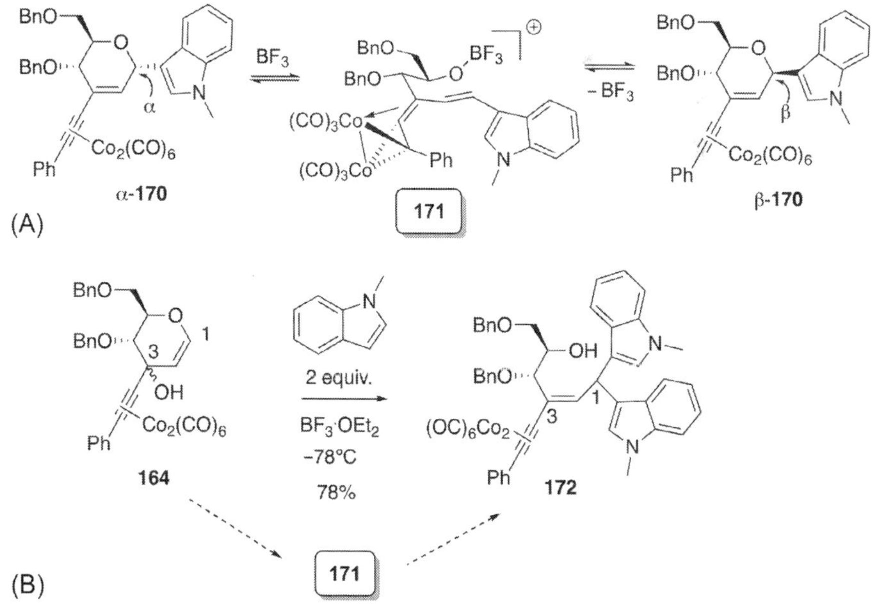

Scheme 38 Proposed rationale for the "α/β" equilibration in *C*-indolyl glycosides. Existence of an open-chain allylic Nicholas cation intermediate **171** (A), and synthesis of di-indolyl open-chain derivative **172** (B).

the presence of Lewis acids ($BF_3{\cdot}OEt_2$, $InBr_3$) to give difunctionalized *C*-glycals. The proposed reaction pathway exemplified in Scheme 39 with indole as the nucleophile, involved the sequential formation of two different Ferrier-Nicholas, vinylogous cations, that is, **175** and **177**. Each (vinylogous) cation possessed two reactive sites (α, γ), and in each case reacted at the distal γ-position, rather than at the anomeric (α-) site, to yield the bis-functionalized species **172**. The reactions took place with moderate yields affording the α-anomer of 3-*C*-functionalized derivative as the sole isomer when $BF_3{\cdot}OEt_2$ was used as catalyst, or as major isomers when $InBr_3$ was used as promoter.

The reaction was also extended to the use of pyrrole, *N*-methylindole, or thiophenol as nucleophiles, with similar results.

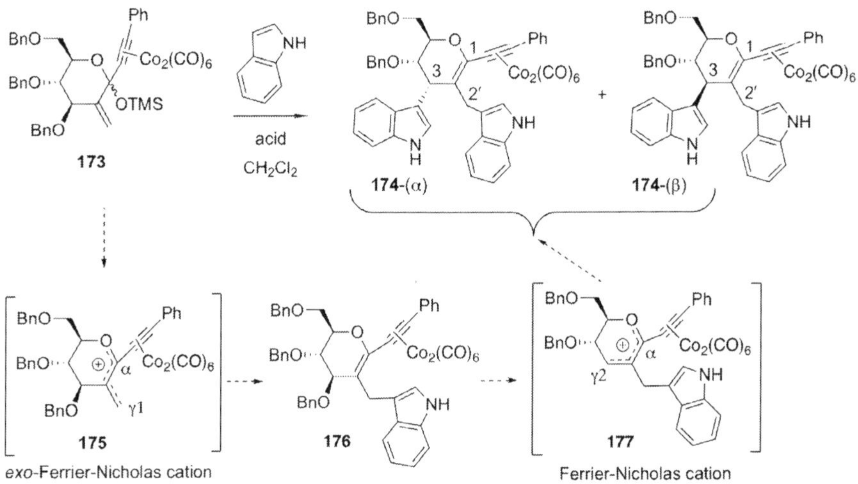

Scheme 39 Reaction of dicobalt hexacarbonyl complex of 1-*C*-alkynyl-2-deoxy-2-*C*-methylenepyranoside with indole.

9 Conclusion

In summary, we have illustrated the rich chemistry associated to carbohydrate-derived dicobalt hexacarbonyl complexes. In fact, the incorporation of dicobalt hexacarbonyl clusters to carbohydrate derivatives, pioneered by Isobe's group seminal contributions, has enabled the development of a variety of novel synthetically useful transformations. In most of these processes, the anomeric position of the carbohydrate plays an active role. For example, the incorporation of a cobalt cluster at the anomeric position facilitates the α→β equilibration of *C*-alkynyl glycosides, the formation of *C*-ketosides and the formation of open-chain derivatives. Finally, the additional stabilization provided by the Nicholas' dicobalt hexacarbonyl cluster to pyranosidic carbenium ions and allylic oxycarbenium ions has been key in the development of novel Ferrier(II)-Nicholas, and Ferrier(I)-Nicholas transformations, respectively.

Acknowledgments

We gratefully acknowledge the Spanish Ministerio de Economía y Competitividad (MINECO) and Fondo Europeo de Desarrollo Regional (FEDER) (grant CTQ2015-66702-R, MINECO/FEDER, UE), Ministerio de Ciencia, Innovación y Universidades (MCIU), Agencia Estatal de Investigación (AEI), and Fondo Europeo de Desarrollo Regional (FEDER) (grant RTI2018-094862-B-I00, MCIU/AEI/FEDER, UE), for financial support.

References

1. (a) Nicholas, K. M.; Pettit, R. *Tetrahedron Lett.* **1971**, *27*, 3475–3478; (b) Nicholas, K. M.; Pettit, R. *J. Organomet. Chem.* **1972**, *44*, C21–C24; (c) Connor, R. E.; Nicholas, K. M. *J. Organomet. Chem.* **1977**, *125*, C45–C48; (d) Lockwood, R. F.; Nicholas, K. M. *Tetrahedron Lett.* **1977**, *33*, 4163–4166; (e) Nicholas, K. M. *Acc. Chem. Res.* **1987**, *20*, 207–214; (f) Nicholas, K. M. *J. Org. Chem.* **2015**, *80*, 6943–6950.
2. Reviews: (a) B. J. Teobald, Tetrahedron 58 (2002) 4133–4170; (b) Röse, P.; Hilt, G. *Synthesis* **2016**, *48*, 463–492; (c) Green, J.R. *Synlett* **2012**, 1271–1282; (d) Diaz, D.D.; Betancort, J.M.; Martin, V.S. *Synlett* **2007**, 343–359; (e) Müller, T.J.J. *Eur. J. Org. Chem.* **2001**, 2021–2033; (f) Bromfield, K.M.; Graden, H.; Ljungdahl, N.; Kahn, N. Dalton Trans. **2009**, 5051–5061; (g) Kahn, N. *Curr. Org. Chem.* **2012**, *16*, 322–334; (h) Green, J.R. *Curr. Org. Chem.* **2001**, *5*, 809–826; (i) Zweig, J.E.; Kim, D.E.; Newhouse, T.R. *Chem. Rev.* **2017**, *117*, 11680–11752; (j) Omae, I. *Appl. Organomet. Chem.* **2007**, *21*, 318–355
3. Schreiber, S. L.; Sammakia, T.; Crowe, W. E. *J. Am. Chem. Soc.* **1986**, *108*, 3128–3130.
4. Pauson, P. L.; Khand, I. U. *Ann. N. Y. Acad. Sci.* **1977**, *295*, 2–14.
5. (a) Isobe, M.; Nishizawa, R.; Hosokawa, S.; Nishikawa, T. *Chem. Commun.* **1998**, 2665–2676; (b) Saeeng, R.; Isobe, M. *Chem. Lett.* **2006**, *35*, 552–557.
6. Ichikawa, Y.; Isobe, M.; Konobe, M.; Goto, T. *Carbohydr. Res.* **1987**, *171*, 193–199.
7. Tsukiyama, T.; Isobe, M. *Tetrahedron Lett.* **1992**, *33*, 7911–7914.
8. Tanaka, S.; Tsukiyama, T.; Isobe, M. *Tetrahedron Lett.* **1993**, *34*, 5757–5760.
9. Magnus, P.; Becker, D. P. *J. Chem. Soc. Chem. Commun.* **1985**, 640–642.
10. Seyferth, D.; Wehman, A. T. *J. Am. Chem. Soc.* **1970**, *92*, 5520–5522.
11. Jones, G. B.; Wright, J. M.; Rush, T. M.; Plourde, G. W., II; Kelton, T. F.; Mathews, J. E.; Huber, R. S.; Davidson, J. P. *J. Org. Chem.* **1997**, *62*, 9379–9381.
12. Hosokawa, S.; Isobe, M. *Tetrahedron Lett.* **1998**, *39*, 2609–2612.
13. Tanaka, S.; Isobe, M. *Tetrahedron* **1994**, *50*, 5633–5644.
14. Chang, W.-C.; Isobe, M. *Tetrahedron* **2014**, *70*, 8324–8333.
15. (a) Ichikawa, Y.; Tsuboi, K.; Jiang, Y.; Naganawa, A.; Isobe, M. *Tetrahedron Lett.* **1995**, *36*, 7101–7104; (b) Jiang, Y.; Isobe, M. *Tetrahedron* **1996**, *52*, 2877–2892; (c) Tsuboi, K.; Ichikawa, Y.; Jiang, Y.; Naganawa, A.; Isobe, M. *Tetrahedron* **1997**, *53*, 5123–5142.
16. (a) Jiang, Y.; Ichikawa, Y.; Isobe, M. *Synlett* **1995**, 285–288; (b) Jiang, Y.; Ichikawa, Y.; Isobe, M. *Tetrahedron* **1997**, *53*, 5103–5122.
17. Desire, J.; Veyrieres, A. *Carbohydr. Res.* **1995**, *268*, 177–186.
18. Tsou, P.-K.; Lee, Y.-C.; Lankau, T.; Isobe, M.; Yu, C.-H. *J. Phys. Org. Chem.* **2018**, *31*, e3780.
19. Tanaka, S.; Tatsuta, N.; Yamashita, O.; Isobe, M. *Tetrahedron* **1994**, *50*, 12883–12894.
20. Tanaka, S.; Isobe, M. *Synthesis* **1995**, 859–862.
21. Schreiber, S. L.; Klimas, M. T.; Sammakia, T. *J. Am. Chem. Soc.* **1987**, *109*, 5749–5759.

22. Dey, S.; Ganesh, N. V.; Jayaraman, N. Ring Expansion Methodologies of Pyranosides to Septanosides and Structures of Septanosides. In *Domino and Intramolecular Rearrangement Reactions as Advanced Synthetic Methods in Glycoscience*, Witczak, Z. J., Bielski, R., Eds.; 2016.
23. (a) Tanaka, S.; Isobe, M. *Tetrahedron Lett.* **1994**, *35*, 7801–7804; (b) Tanaka, S.; Tatsuta, N.; Yamashita, O.; Isobe, M. *Tetrahedron* **1994**, *40*, 12883–12894. Repetida ref 16.
24. Yasumoto, T.; Murata, M. *Chem. Rev.* **1993**, *93*, 1897–1909.
25. Isobe, M.; Yenjai, C.; Tanaka, S. *Synlett* **1994**, 916–918.
26. Yenjai, C.; Isobe, M. *Tetrahedron* **1998**, *54*, 2509–2520.
27. (a) Hosokawa, S.; Isobe, M. *Synlett* **1995**, 1179–1180. (b) Hosokawa, S.; Isobe, M. *Synlett* **1996**, 351–352; (c) Hosokawa, S.; Kirschbaum, B.; Isobe, M. *Tetrahedron Lett.* **1998**, *39*, 1917–1920; (d) Hosokawa, S.; Isobe, M. *J. Org. Chem.* **1999**, *64*, 37–48; (e) Saeeng, R.; Isobe, M. *Tetrahedron Lett.* **1999**, *40*, 1911–1914; (f) Saeeng, R.; Isobe, M. *Heterocycles* **2001**, *54*, 789–798; (g) Kira, K.; Isobe, M. *Tetrahedron Lett.* **2000**, *41*, 5951–5955; (h) Kira, K.; Isobe, M. *Tetrahedron Lett.* **2001**, *42*, 2821–2824; (i) Kira, K.; Hamajima, A.; Isobe, M. *Tetrahedron* **2002**, *58*, 1875–1888; (j) Takai, S.; Isobe, M. *Org. Lett.* **2002**, *4*, 1183–1186; (k) Takai, S.; Sawada, N.; Isobe, M. *J. Org. Chem.* **2003**, *68*, 3225–3231; (l) Baba, T.; Huang, G.; Isobe, M. *Tetrahedron* **2003**, *59*, 6851–6872; (m) Liu, T.-Z.; Isobe, M. *Synlett* **2000**, 266–268; (n) Liu, T.-Z.; Kirschbaum, B.; Isobe, M. *Synlett* **2000**, 587–590; (o) Liu, T.-Z.; Li, J.-M.; Isobe, M. *Tetrahedron* **2000**, *56*, 10209–10219; (p) Liu, T. Z.; Isobe, M. *Tetrahedron* **2000**, *56*, 5391–5404; (q) Nonoyama, A.; Hamajima, A.; Isobe, M. *Tetrahedron* **2007**, *63*, 5886–5894; (r) Hamajima, A.; Isobe, M. *Org. Lett.* **2006**, *8*, 1205–1208; (s) Ichikawa, Y.; Isobe, M.; Masaki, H.; Kawai, T.; Goto, T.; Katayama, C. *Tetrahedron* **1987**, *43*, 4759–4766; (t) Ichikawa, Y.; Isobe, M.; Goto, T. *Tetrahedron* **1987**, *43*, 4749–4758; (u) Ohyabu, N.; Nishikawa, T.; Isobe, M. *J. Am. Chem. Soc.* **2003**, *125*, 8798–8805; (v) Nishikawa, T.; Urabe, D.; Isobe, M. *Angew. Chem. Int. Ed.* **2004**, *43*, 4782–4785.
28. Isobe, M.; Hamajima, A. *Nat. Prod. Rep.* **2010**, *27*, 1204–1226.
29. Isobe, M.; Hosokawa, S.; Kira, K. *Chem. Lett.* **1996**, 473–474.
30. Takase, M.; Morikawa, T.; Abe, H.; Inouye, M. *Org. Lett.* **2003**, *5*, 625–628.
31. Mukai, C.; Itoh, T.; Hanaoka, M. *Tetrahedron Lett.* **1997**, *38*, 4595–4598.
32. Xia, M.-J.; Yao, W.; Meng, X.-B.; Lou, Q.-H.; Li, Z.-J. *Tetrahedron Lett.* **2017**, *58*, 2389–2392.
33. Streicher, H.; Geyer, A.; Schmidt, R. R. *Chem. Eur. J.* **1996**, *2*, 502–510.
34. Ferrier, R. J. *J. Chem. Soc. Perkin Trans.* **1979**, *1*, 1455–1458.
35. Ferrier, R. J.; Middleton, S. *Chem. Rev.* **1993**, *93*, 2779–2831.
36. Carbery, D. R.; Reignier, S.; Myatt, J. W.; Miller, N. D.; Harrity, J. P. A. *Angew. Chem. Int. Ed.* **2002**, *41*, 2584–2587.
37. Meek, S. J.; Harrity, J. P. A. *Tetrahedron* **2007**, *63*, 3081–3092.
38. Carbery, D. R.; Reignier, S.; Miller, N. D.; Adams, H.; Harrity, J. P. A. *J. Org. Chem.* **2003**, *68*, 4392–4399.
39. (a) Das, S. K.; Mallet, J.-M.; Sinaÿ, P. *Angew. Chem. Int. Ed.* **1997**, *36*, 493–496.
    (b) Sollogoub, M.; Mallet, J.-M.; Sinaÿ, P. *Tetrahedron Lett.* **1998**, *39*, 3471–3472.
40. Sollogoub, M.; Mallet, J.-M.; Sinaÿ, P. *Angew. Chem. Int. Ed.* **2000**, *39*, 362–364.
41. Deleuze, A.; Menozzi, C.; Sollogoub, M.; Sinaÿ, P. *Angew. Chem. Int. Ed.* **2004**, *43*, 6680–6683.
42. Sardinha, J.; Guieu, S.; Deleuze, A.; Fernandez-Alonso, M. C.; Rauter, A. P.; Sinaÿ, P.; Marrot, J.; Jimenez-Barbero, J.; Sollogoub, M. *Carbohydr. Res.* **2007**, *342*, 1689–1703.
43. Sardinha, J.; Rauter, A. P.; Sollogoub, M. *Carbohydr. Res.* **2008**, *49*, 5548–5550.

44. Meek, S. J.; Pradaux, F.; Demont, E. H.; Harrity, J. P. A. *J. Org. Chem.* **2007**, *72*, 3467–3477.
45. Gomez, A. M.; Uriel, C.; Valverde, S.; Jarosz, S.; Lopez, J. C. *Tetrahedron Lett.* **2002**, *43*, 8935–8940.
46. Gomez, A. M.; Uriel, C.; Valverde, S.; Jarosz, S.; Lopez, J. C. *Eur. J. Org. Chem.* **2003**, 4830–4837.
47. Oishi, T.; Nagumo, Y.; Hirama, M. *Chem. Commun.* **1998**, 1041–1042.
48. Gomez, A. M.; Uriel, C.; Valverde, S.; Lopez, J. C. *Org. Lett.* **2006**, *8*, 3187–3190.
49. (a) Saksena, A. K.; Green, M. J.; Mangiaracina, P.; Wong, J.; Kreutner, W.; Gulbenkian, A. R. *Tetrahedron Lett.* **1985**, *26*, 6423–6426; (b) Melykian, G. G.; Mineif, A.; Vostrowsky, O.; Bestmnn, H. J. *Synthesis* **1991**, 633–636.
50. Vinylogous Nicholas cations are generated from dicobalt complexes arising from propargyl-allyl ether or hydroxyl derivatives.
51. There are relatively few examples of vinylogous Nicholas reactions (VNR). For intramolecular VNR, see: (a) I. Kolodziej, J. R. Green, *Synlett* (2011) 2397–2401; (a) I. Kolodziej, J. R. Green, *Synlett* (2011) 2397–2401; (b) Kolodziej, I.; Green, J.R. *Org. Biomol. Chem.* **2015**, *13*, 10852–10864; for intermolecular VNR, see: (c) E. Alvaro, M. C. de la Torre, M. A. Sierra, *Org. Lett.* 5 (2003) 2381–2384; (d) Mahmood, A.; Ngenzi, R.; Penner, P.M.; Green, J.R. *Synlett* **2016**, *27*, 1245–1250
52. The terms Ferrier rearrangement or Ferrier(I) reaction: Ferrier, R. J.; Overend, W. G.; Ryan, A. E. *J. Chem. Soc. C* **1962**, 3667–3670. are used to differentiate this process from the previously mentioned transformation of hex-5-enopyranosyl derivatives into functionalized cyclohexanones, currently termed Ferrier carbocyclization, or Ferrier (II) reaction.
53. (a) Ferrier, R. J.; Hoberg, J. O. *Adv. Carbohydr. Chem. Biochem.* **2003**, *58*, 55–119; (b) Ferrier, R. J.; Zubkov, O. A. *Org. React.* **2003**, *62*, 569–736; (c) Ferrier, R. J. *Top. Curr. Chem.* **2001**, *215*, 153–175.
54. (a) Gomez, A. M.; Miranda, S.; Lopez, J. C. *Carbohydr. Chem.* **2017**, *42*, 210–247; (b) Gomez, A. M.; Lobo, F.; Uriel, C.; Lopez, J. C. *Eur. J. Org. Chem.* **2013**, 7221–7262; (c) Ansari, A. A.; Lahiri, R.; Vankar, Y. D. *ARKIVOC* **2013**, *2*, 316–362.
55. Gomez, A. M.; Lobo, F.; Miranda, S.; Lopez, J. C. *Chem. Eur. J.* **2014**, *20*, 10492–10502.
56. Gomez, A. M.; Lobo, F.; Perez de las Vacas, D.; Valverde, S.; Lopez, J. C. *Chem. Commun.* **2010**, *46*, 6159–6161.
57. (a) Smit, W. A.; Caple, R.; Smoliakova, I. P. *Chem. Rev.* **1994**, *94*, 2359–2382; (b) Smit, W. A.; Schegolev, A. A.; Gybin, A. S.; Mikaelian, G. S.; Caple, R. *Synthesis* **1984**, 887–890.
58. Byerley, A. L. J.; Kenwright, A. M.; Lehmann, C. W.; MacBride, J. A. H.; Steel, P. G. *J. Org. Chem.* **1998**, *63*, 193–194.
59. (a) Closser, K. D.; Quintal, M. M.; Shea, K. M. *J. Org. Chem.* **2009**, *74*, 3680–3688; (b) Lebold, T. P.; Carson, C. A.; Kerr, M. A. *Synlett* **2006**, 364–368; (c) Quintal, M. M.; Closser, K. D.; Shea, K. M. *Org. Lett.* **2004**, *6*, 4949–4952; (d) Jamison, T. F.; Shambayati, S.; Crowe, W. E.; Schreiber, S. L. *J. Am. Chem. Soc.* **1987**, *119*, 4353–4363.
60. Isobe, M.; Takai, S. *J. Organomet. Chem.* **1999**, *589*, 122–125.
61. Miranda, S.; Lobo, F.; Gomez, A. M.; Lopez, J. C. *Eur. J. Org. Chem.* **2017**, 2501–2511.
62. Miranda, S.; Gomez, A. M.; Lopez, J. C. *Eur. J. Org. Chem.* **2018**, 5355–5374.

Gold-catalyzed methodologies in carbohydrate syntheses

4

Srinivas Hotha
Department of Chemistry, Indian Institute of Science Education & Research, Pune, India

Chapter outline

1 History

Gold intimately fascinated ever since the evolution of humans started and hence has been rightly considered "the King of Metals." Gold played a pivotal role in the evolution of civilizations, and still today continues to influence economics, politics, arts, religion, and technology and, hence, touched every part of the human life cycle. Mesmerizing shine and unreactive nature of gold toward routinely used chemicals in day-to-day life coupled with scarcity made this a precious commodity. Gold is available all over the globe and mining is more or less a global phenomenon with more than 90 countries involved in the extraction using various processes such as cyanidation, flotation, gravity separation, smelting, and chemical extraction. The jewelry industry accounts for the utilization of the majority of the mined gold with ~60% share, while roughly 20% is held as monetary reserves in banks. Remarkable insensitivity to corrosion and oxidation, malleability, ductility, and mechanical robustness along with high thermal ($317\,W\,m^{-1}\,K^{-1}$) and electrical conductivities ($45.2\,S\,m^{-1}$) of metallic gold

Recent Trends in Carbohydrate Chemistry. https://doi.org/10.1016/B978-0-12-817467-8.00004-9
© 2020 Elsevier Inc. All rights reserved.

were exploited by the burgeoning electronics industry. However, the gold rush did not catch enough attention from synthetic organic chemistry community until the end of the past century.[1a]

Chemically, gold is represented by the symbol Au, which is derived from the Latin word *aurum*, and is a transition metal of electronic configuration [Xe] $4f^{14}5d^{10}6s^{1}$ with atomic number 79. The atomic weight of metallic [Au] is 196.9665, which is at the cross section of the group 11 and period 6. Au is positioned in between platinum (Pt) and mercury (Hg) in the periodic table of elements. Gold is considered a noble metal as it is noncorrosive and does not undergo oxidation under normal conditions. Metallic gold dissolves in aqua regia or in a solution of cyanide salts. Nonetheless, various oxidation states of gold are known, ranging from -1 to $+5$; but Au^{0}, Au^{1+}, and Au^{3+} are more versatile species for many applications in chemistry. Au intermediates undergo protodeauration that is not observed with mercury complexes though they are in the same period. These electronic properties have pushed gold catalysis to emerge as a powerful synthetic tool for a variety of functional group transformations.[1b]

2 Introduction

Organic reactions facilitated by gold complexes have received very little attention in the past owing to the preconceived notion that gold is expensive and inert; however, the discovery that compounds of gold can be prepared in catalytically active forms has spurred on great interest in investigating the chemical and catalytic properties of gold.[2] Of the many different oxidation states, Au(0), Au(I), and Au(III) are relevant in the presence of organic substrates. Au(I) instantaneously disproportionates to Au(III) and Au(0) in the absence of any stabilizing ligands. Theoretical investigations demonstrated that R_3PAu^+ can react more specifically with alkenes and is isolobal to H^+ and LHg^{2+}, while Au(III) is more alkynophilic.[3] Relativistic effects attain maximum in the periodic table at gold, and as a consequence the size of gold(I) is smaller than that of silver(I).[4] In addition, Au(I) and Au(III) are diamagnetic, and hence monitoring of reactions and understanding the mechanism by NMR is facilitated.

The majority of gold-catalyzed reactions involve gold species **1–3** as intermediates depending on the nature of nucleophile and the substrate (Scheme 1).

Mechanistically, a gold-catalyst can form a linear two-coordinate π-complex with unsaturated carbons (alkenes, alkynes, allenes) for subsequent nucleophilic attack.[2] The metal then "slips" to one end of the multiple bonds generating the positive charge at the other, which is available for the nucleophilic attack.[5] Loss of proton followed by the protodeauration can regenerate the cationic gold species for the fresh catalytic action (Scheme 2).

Addition of water and alcohols to alkynes delivering the Markovnikov product in case of terminal alkynes by Thomas et al. in 1976 is one of the very first beginnings of the gold-catalysis in organic reactions (Scheme 2).[6] However, the field did not gain

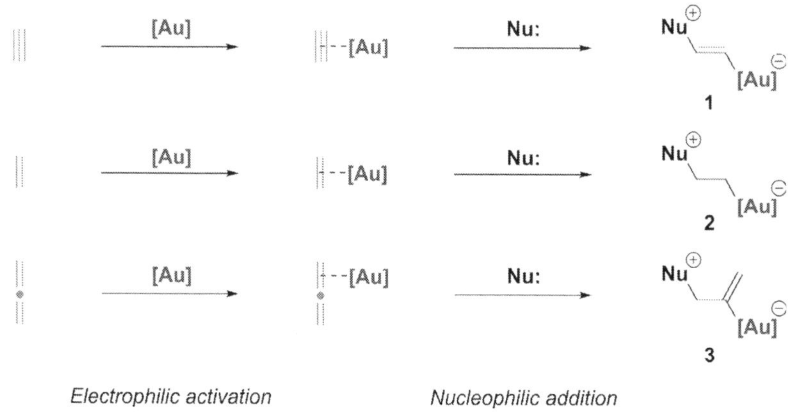

Scheme 1 Common modes of activation in the gold-catalyzed activation of unsaturated compounds.

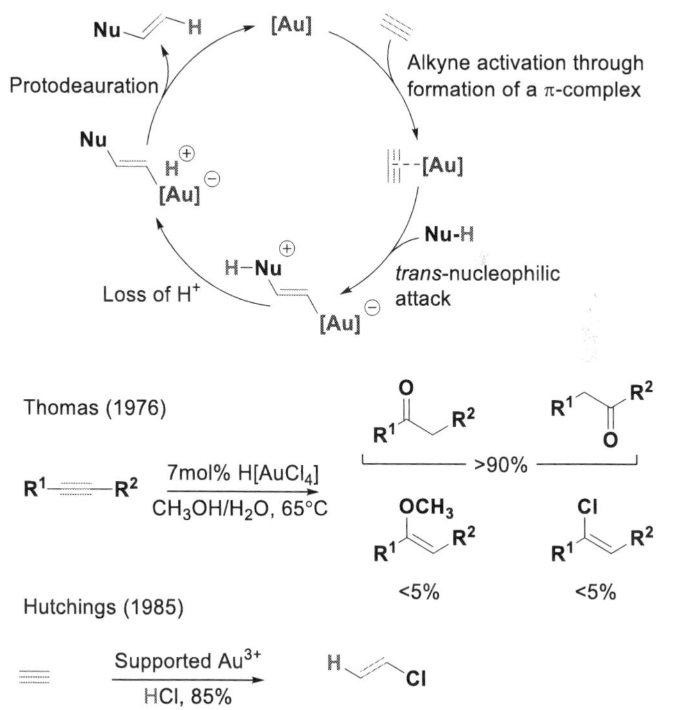

Scheme 2 General catalytic cycle of gold-catalyzed alkyne activation and early gold-catalyzed reactions.

much momentum until the next pioneering effort by Hutchings who has developed a mercury-free heterogeneous process for the conversion of acetylene to an industrially important vinyl halide[7] in the presence of Au(III) supported on charcoal and HCl.[7] In 2000, Hashmi's group showed a highly interesting gold-catalyzed synthesis of arenes **5** by selective cross cycloisomerization/dimerization of compound **4**. Later studies proved that 2 mol% of $AuCl_3$ in acetonitrile transformed furans **6** to hydroxyarene **5**, which has spurred great deal of interest among synthetic organic chemistry community (Scheme 3).[8]

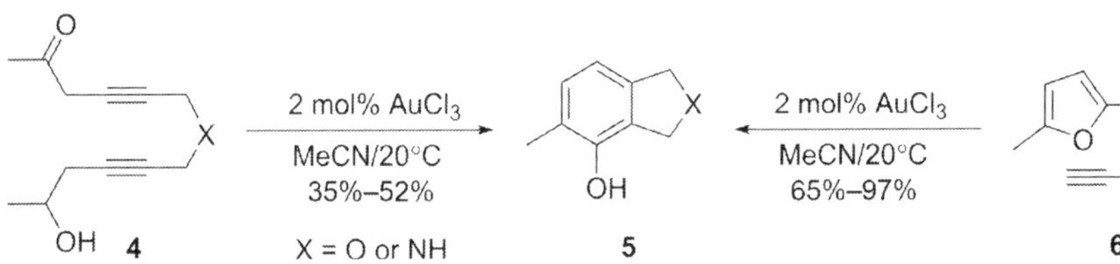

Scheme 3 Gold-catalyzed synthesis of arenes.

3 Gold catalysis in carbohydrate chemistry

The gold rush in carbohydrate chemistry started much before organic chemists appreciated the versatility of the chemistry of gold. Interestingly, one of the very early approved treatments for rheumatoid arthritis was Auranofin (**7**, Fig. 1), an orally administered drug, that is a hybrid molecule containing tetra-*O*-acetyl-1-thio-β-D-glucopyranose and gold. Auranofin has been in clinical trials as a drug for HIV, amebiasis, tuberculosis, and ovarian cancer.[9]

4 Heterogeneous oxidation of carbohydrates using gold-catalysts

The oxidation of polyols is a demanding and industrially relevant target, which is traditionally carried out by platinum or palladium nanoparticles. Rossi and Prati have shown that gold nanoparticles can be effective catalysts for the oxidation of alcohols and quite recently they have identified synergistic effects of Pd or Pt with the Au-carbon catalysts for the oxidation of D-sorbitol (**8**) to D-gluconic acid (**9**) and D-gulonic acid (**10**) (Scheme 4).[10a,b]

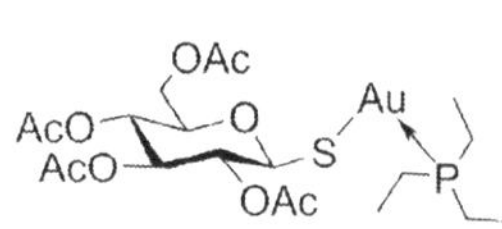

7

Fig. 1 Structure of Auranofin (approved as Ridaura in 1985).

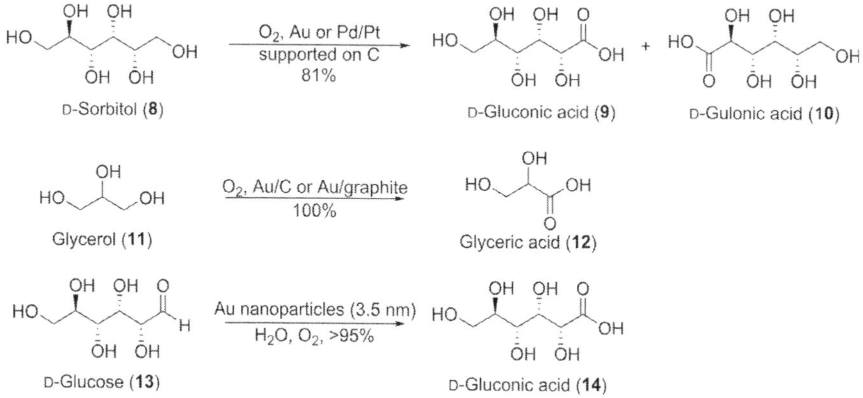

Scheme 4 Oxidation of polyols using gold.

Hutchings in 2002 reported a highly efficient oxidation of glycerol (**11**) to glyceric acid (**12**) with 100% selectivity using 1% Au on either charcoal or graphite as catalyst at 60°C in aqueous media and subsequently utilized the similar protocol for the oxidation of D-glucose (**13**) to D-gluconic acid (**14**) (Scheme 4).[11] Microwaves are noticed to improve the oxidation of free carbohydrates to the corresponding aldonates under gold-catalysis conditions.[10d]

5 Homogeneous gold-catalysis in carbohydrate chemistry

Homogeneous gold-catalysis in carbohydrate chemistry started with the application of Hashmi's arene synthesis on monosaccharide building blocks. Diversity-oriented approach using carbohydrate templates for the combinatorial libraries that are chiral, polycyclic, oxygen-rich, and natural product-like prompted Hotha and his coworkers to exploit diversity available within glycal or 1,2:5,6-di-*O*-isopropylidene glucose-based scaffolds.[12a] In one of the efforts, Hashmi's reaction was attempted on the glucose-derived substrate **15** to obtain the isobenzofuran annulated monosaccharide **16** in very high yields (90%–95%) along with the formation of the alcohol **17** in <5% yield (Scheme 5).[12a]

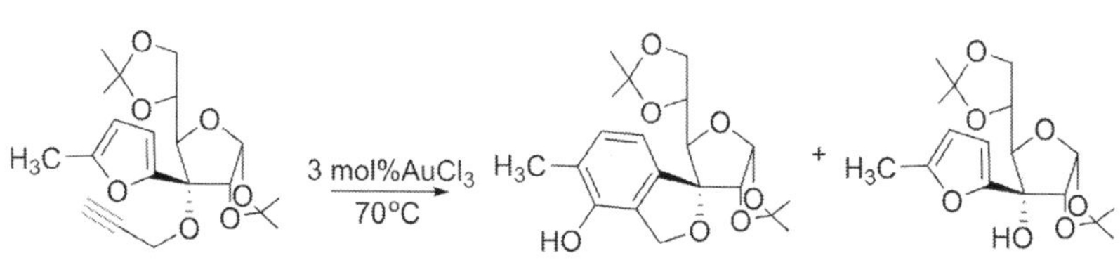

15 16 (95%) 17 (5%)

Scheme 5 Gold-catalyzed synthesis of isobenzofuran annulated monosaccharide.

Subsequently, Hotha and Kashyap identified a Ferrier-like reaction with 3-*O*-propargyl derivative **18** to afford **19** instead of the anticipated product (**20**) that will be obtained if intramolecular trapping of the Au-alkyne with strategically positioned olefin occurred (Scheme 6).[12b] Around the same time, they have also noticed formation of C-2 methylene glycosides instead of anticipated spirocyclic glycosides.[12c] These observations nicely corroborated the fact that a propargyloxy moiety could become a leaving group in the presence of $AuCl_3$ in acetonitrile. Later on gold(III) chloride and phenylacetylene catalytic system has been developed by Vankar's group for the Ferrier rearrangement, and *O*-glycosylation of anomeric acetates as glycosyl donors.[12d] Galan's group noticed that catalytic quantities of Au(I) and AgOTf provide direct access to α-glycosides from glycals at room temperature.[12e]

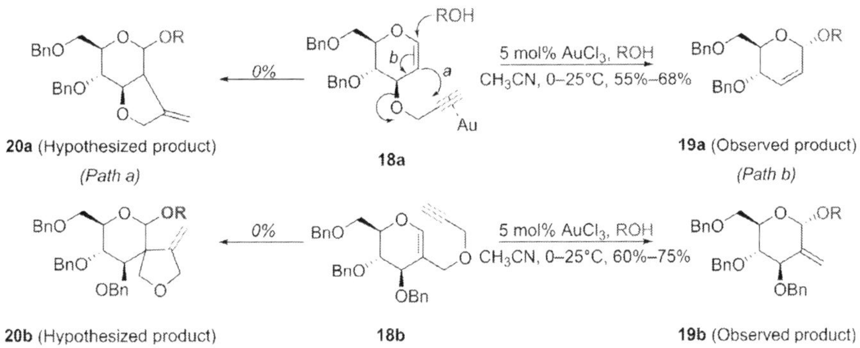

Scheme 6 Unexpected Ferrier-like product during the gold catalyzed reaction on glycal template.

The gold-catalyzed activation of propargyl ethers was soon followed by the activation of propargyl glycosides by $AuCl_3$ catalysis (Scheme 7).[13] It was noticed that the per-*O*-benzylated propargyl glucoside **21** reacted with the acceptor **22** to give the disaccharide **23** in very good yield.[13] The reaction was initially hypothesized to form cyclopropanone **25** beside the oxycarbenium ion intermediate **24**. Similarly, galactosyl and mannosyl donors were also activated by 3 mol% $AuCl_3$ in acetonitrile at 60°C for 2–15 h. The yields obtained with simple acceptors and carbohydrate-derived primary alcohols were good, but low yields were noticed with secondary alcohols, and glycosides of tertiary alcohols were not reported in 2006.[13] The presence of a nonparticipating benzyl ether at the C-2 position resulted in low anomeric selectivity and the participating benzoates at C-2 did not react at all under gold-catalyzed glycosidation conditions. Later studies also proved that one can activate even methyl glycosides in the presence of gold-halides at elevated temperature.[14]

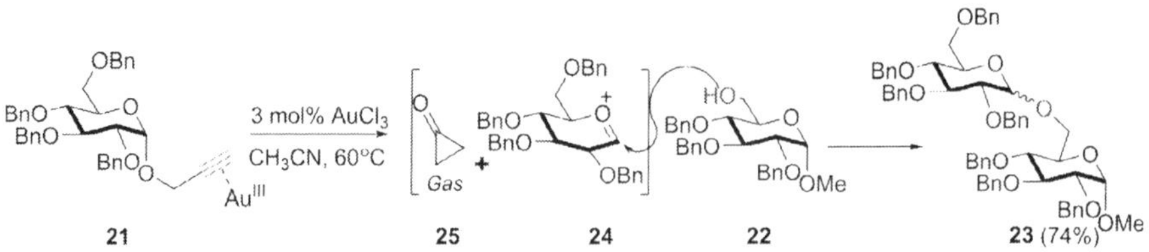

Scheme 7 Activation of propargyl glucosides under gold-catalysis.

The protocol has been successfully applied for the synthesis of 1,6-anhydrosugars,[15a] amino acid glycoconjugates,[15b] disaccharides,[15c] and trisaccharides.[15] The mechanism of the gold-catalyzed activation of propargyl glycosides was initially hypothesized by the formation of the entropically favored highly volatile cyclopropanone **25**, which was later proved to be otherwise.[16b] However, poor anomeric selectivity in the glycosidation, nonreactivity of O-2 benzoyl glycosides in the gold-catalyzed glycosidation, and hydrolysis[16a] of the interglycosidic bond due to the high temperature prompted Hotha's group to synthesize a combinatorial library of many glycosyl donors (**26a–l**) with alkynylated appendages at the C-1 position.[16b]

Results from that study showed *gem*-dialkylated appendages containing alkynes to be more reactive under gold-catalyzed glycosidation conditions with acceptor **27** to give the disaccharide **29**; the glycosidations could be conducted at 25°C with *gem*-dialkyl substituted appendages. Ethynylcyclohexyl glycosides **26k** were selected over others as ethynylcyclohexanol precursors (**26h–j**, **26l**) for the glycosyl donor preparation are cheap and easily available; the remaining donors **26a–g** either did not react or gave unsatisfactory yields. Though the reaction could be carried out at 25°C, the C-2 benzoate ester-containing donors were observed to be resistant to the glycosidation conditions even at higher temperature (Scheme 8). The mechanism was proved by GC-MS studies wherein Hotha and Kayastha noticed that ethynyl cyclohexanol was formed in the reaction mixture. Generally, yields of this reaction were noticed to be higher compared to the activation by propargyl glycosides; primary alcohol acceptors gave excellent conversions, and secondary alcohols resulted in very good yields. Increased reactivity of the *gem*-dialkyl substituents has been attributed to the well-studied Thorpe-Ingold effect.[16c]

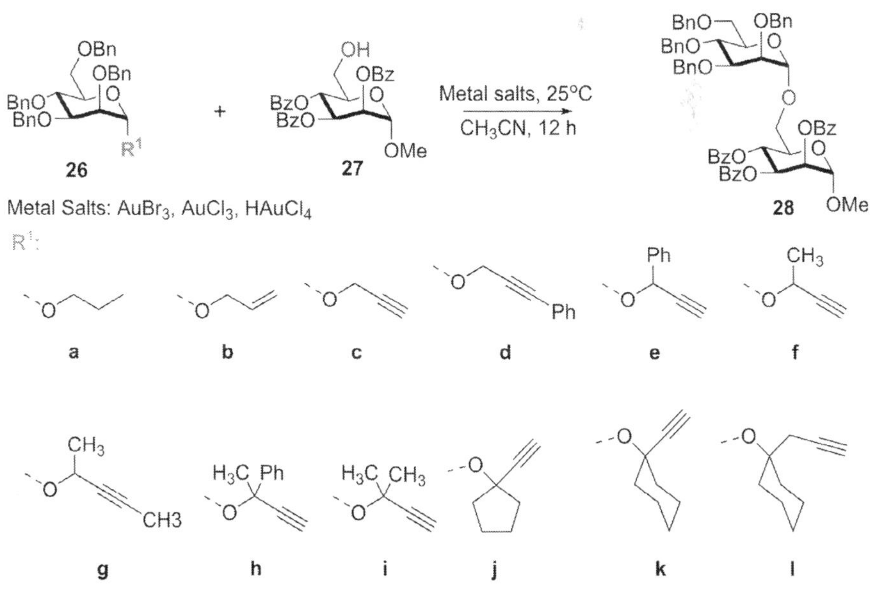

Optimzed conditions: R_1 = **26k**, Metal salt = 5 mol% each of $AuCl_3$ and $AgSbF_6$, 4Å MS, 25°C

Scheme 8 Optimization of the alkynyl-leaving groups.

Besides advancement of increased reactivity, the scope of the glycosidation was still limited to reactive donors with 2-*O*-benzyl protected groups, and hence the control of anomeric stereochemistry through neighboring group participation could not be achieved. Increasing the temperature of the glycosidation resulted in the hydrolysis of the inter-glycosidic bonds and formation of 1,6-anhydro sugars.[15d]

Nevertheless, these new glycosyl donors have spurred good deal of interest among other carbohydrate chemists to improve further.[17a] Mamidyala and Finn reported $AuCl_3$-catalyzed glycosidation using 2-butynyl glycosides.[17b] Another report from the Kunz group[16c] showed that gold-catalysts are suitable for activation of glycosyl imidates, which was recently investigated by the Schmidt group through NMR studies.[17e] Roy *et al.* have reported the use of $AuCl_3$ alone or $AuCl_3$ combined with phenylacetylene as novel catalysts for trichloroacetimidate activation.[17d,e] The glycosylation procedure involving equimolar quantities of $AuCl_3$ and phenylacetylene was applied with many acceptors and donors.[17d] Sureshan *et al.* claimed that tolyl thioglycosides could be activated by catalytic amount of $AuCl_3$ for glycosidation; however, it was later corrected that a sub-stoichiometric amount of $AuCl_3$ was required for the glycosylation.[17f] Glycosyl sulfoxides could be utilized as glycosyl donors under the catalysis of $AuCl_3$ and AgOTf as reported by Vankar and his co-workers.[17g] Furthermore, Boc-pyranones were also activated under $AuCl_3$ conditions.[17h]

6 2,2-Dimethylbut-3-ynyl thioglycosides as glycosyl donors

The Zhu group prepared 2,2-Dimethylbut-3-ynyl thioglycosides containing the alkyne functionality and investigated them as glycosyl donors (Scheme 9).[18] The donor **29** showed some advantages compared to the original propargyl glycosides; however, donor synthesis was less straightforward. Nevertheless, they have identified that their glycosyl donor can be activated using 10 mol% $AuCl_3$ and 30 mol% AgOTf to undergo glycosylation using acceptor **30** to afford the disaccharide **31** and the by-product **32** from the leaving group (Scheme 9).[18a,b] Higher reactivity of these donors was attributed to the Thorpe–Ingold effect due to the presence of *gem*-dimethyl groups. In a subsequent and recent study, a gold-catalyzed procedure for the highly diastereoselective synthesis of α-D-glucosides was accomplished by the activation of anomeric *o*-ethynylphenyl β-D-thioglucoside donors.[18c]

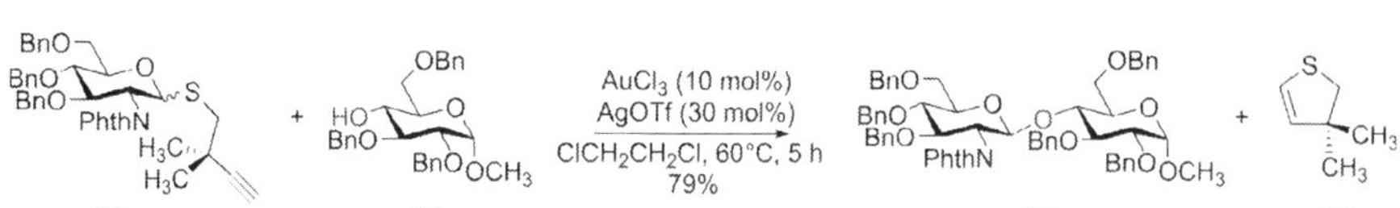

Scheme 9 Activation of thioglycosides containing alkyne-bearing leaving groups.

The utility of this type of glycosyl donor was demonstrated by the successful synthesis of many disaccharides containing benzyls, benzoyls, azides, phthalimides, etc. The selectivity is mainly 1,2-*trans* through the participating protecting groups at the C-2 position.

7 Activation of orthoesters

Often, the 1,2-*trans* stereoselectivity in carbohydrate chemistry is achieved through the neighboring group participation of protecting groups at the C-2 position such as benzoates, acetates, or other esters. However, propargyl glycosides could not be successfully activated when esters were present at the C-2 position (Scheme 10).[13]

To improve the 1,2-*trans* selectivity for the gold-catalyzed glycosidations, Sureshkumar and Hotha synthesized propargyl 1,2-orthoesters (e.g., **33**) to afford various glycosides and disaccharides (**34–42**) easily.[19] In these reactions, $AuBr_3$ was found to be superior compared to $AuCl_3$ due to the balanced Lewis and Brønsted acidity of $AuBr_3$ over $AuCl_3$.[19a] AuCl and Au_2O_3 did not activate the propargyl orthoesters demonstrating the importance of gold and alkyne the overall catalysis. The Au(III) halide and phenylacetylene combination also has been explored for the activation of pyranosyl 1,2-orthoesters by Vankar's group.[19c]

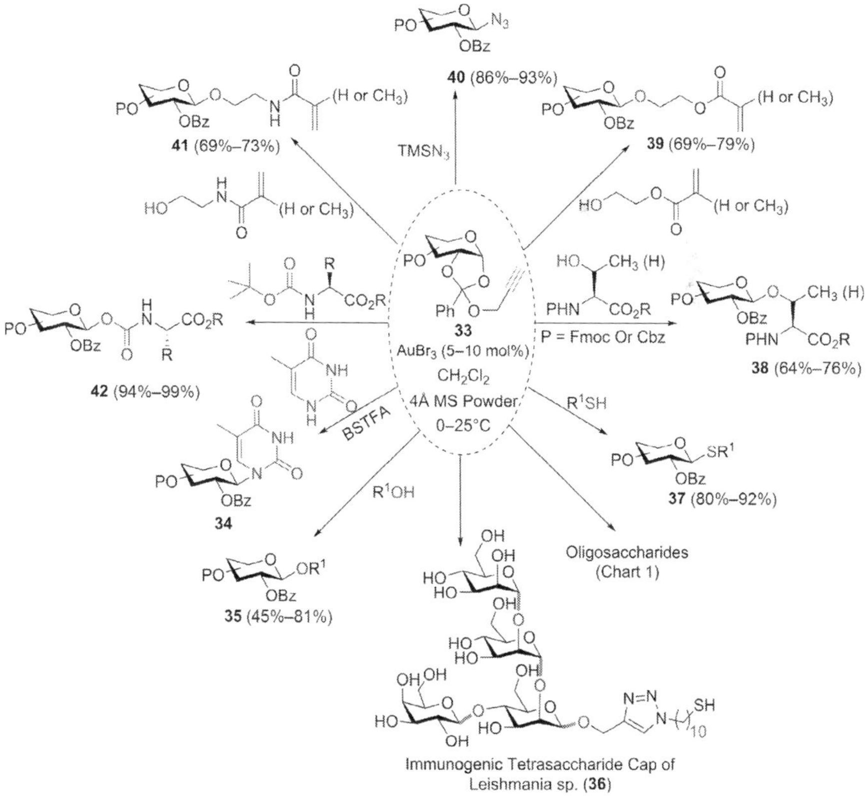

Scheme 10 Utility of 1,2-orthoesters for glycoside syntheses.

8 Activation of glycosyl esters

Soon after the report of gold-catalyzed activation of propargyl glycosides, Yu identified glycosyl *ortho*-alkynylbenzoate donors (e.g., **43**) using $Ph_3PAuOTf$ or $Ph_3PAuNTf_2$ with acceptor **30** to afford disaccharide **44** and the benzopyrenone **45**.[20] These glycosyl donors are also shelf-stable, and they can be activated under mild conditions.[20a,b] Thorough studies have been performed to understand the reaction mechanism, catalyst development, and variations on the leaving group.[20d] *ortho*-Alkynylbenzoate donor chemistry has been demonstrated to be efficient for the synthesis of phosphates, oximes, purine, and pyrimidine nucleosides, *C*-glycosides,[20e] etc.[20f] Additionally, Au(I) salts were found to be better suited for the activation of 1,2-anhydro sugar derivatives than conventional Lewis acid promoters.[20f, 20g]

In a later study, Dutta *et al.* showed that the glycosylation between *ortho*-alkynylbenzoates (e.g., **46**) and acceptor **22** can also occur at room temperature to give disaccharide **47** with molecular iodine as the mild, easily available, and cheap promoter; additionally, iodine-based activation can chemoselectively activate *ortho*-alkynylbenzoates in the presence of thioglycosides also (Scheme 11).[20h] In a very recent study, phenylpropiolate glycosyl donors (**49**) were also activated using 10 mol% of $AuCl_3$ to afford many *O*/*N*-glycosides (e.g., **50**) along with the phenylpropiolic acid **51** in high yields (Scheme 11).[20i]

Excellent functional group compatibility coupled with mild activation procedure for the activation glycosyl alkynyl esters was reported for the stereoselective synthesis of challenging β-mannopyranoses,[20j] β-rhamnopyranosides,[20k] α-linked glucosamine derivatives,[20l] β-Kdo glycosides,[20m] and altropyranosides.[20n]

9 Activation of glycosyl carbonate

After thoroughly screening many reagents, Mishra et al. found that 1-ethynylcyclohexyl glycosyl carbonates can be activated in the presence of gold(I)-phosphite catalyst (**52**) and AgOTf.[21] The model glycosylation reaction carried out between the glucosyl carbonate **51a** and the alcohol **30** was complete in 15 min by the addition of 8 mol% each of gold-phosphite (**52**) and AgOTf (Scheme 12) affording the disaccharide **53** in excellent yield. Gold trihalides were found to convert ethynylcyclohexyl glycosyl carbonates in sub-stoichiometric (0.3 eq.) quantities at 25°C; however, the efficiency of the reaction dropped as the amount of gold trihalide decreased.[21]

The Au(I)-phosphite and AgOTf combination was found to be the best for the activation of glycosyl carbonates within 15 min at 25°C. The surprising synergistic effect of Au-phosphite and AgOTf was rationalized by the ability of Ag ion in scavenging the chloride ion to form AgCl and *in situ* gold-triflate, which is more reactive. Alternatively, AgOTf also avoids the formation of less reactive halide-bridged dimeric gold complex. Cyclic olefin **54** is the sole by-product formed in the reaction, which does not compete with the nucleophile, and no CO_2 liberation was noticed. The versatility of this reaction was demonstrated by the synthesis of pyrimidine nucleoside **55**, phenyl glucoside **56**, serine glucoside **57**, and glycosyl azide **58** (Scheme 13).[21]

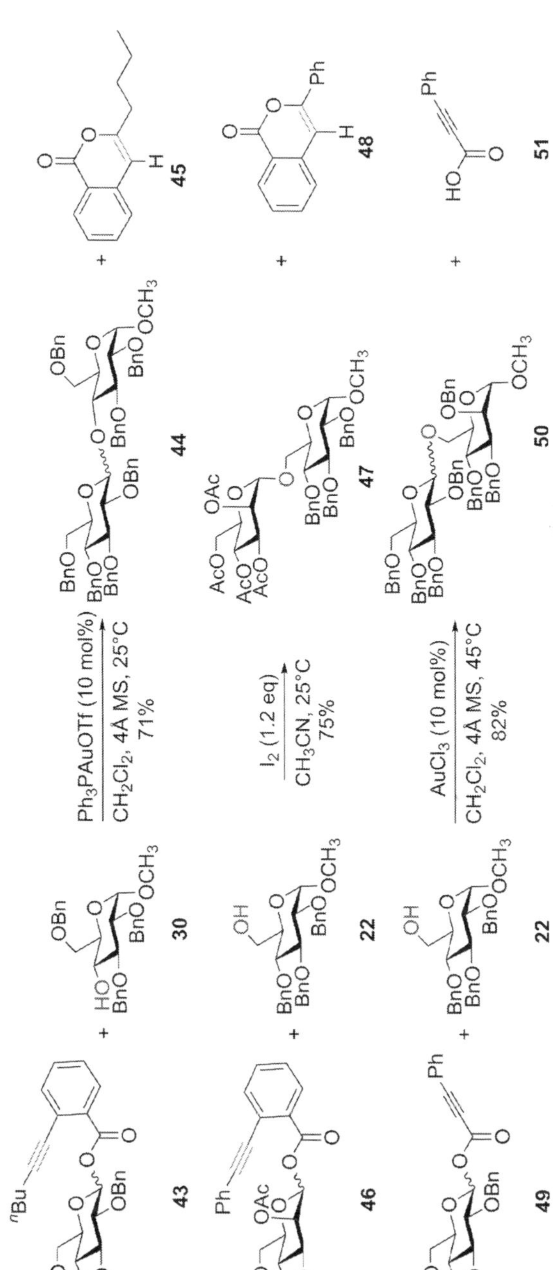

Scheme 11 Gold-catalyzed activation of anomeric esters.

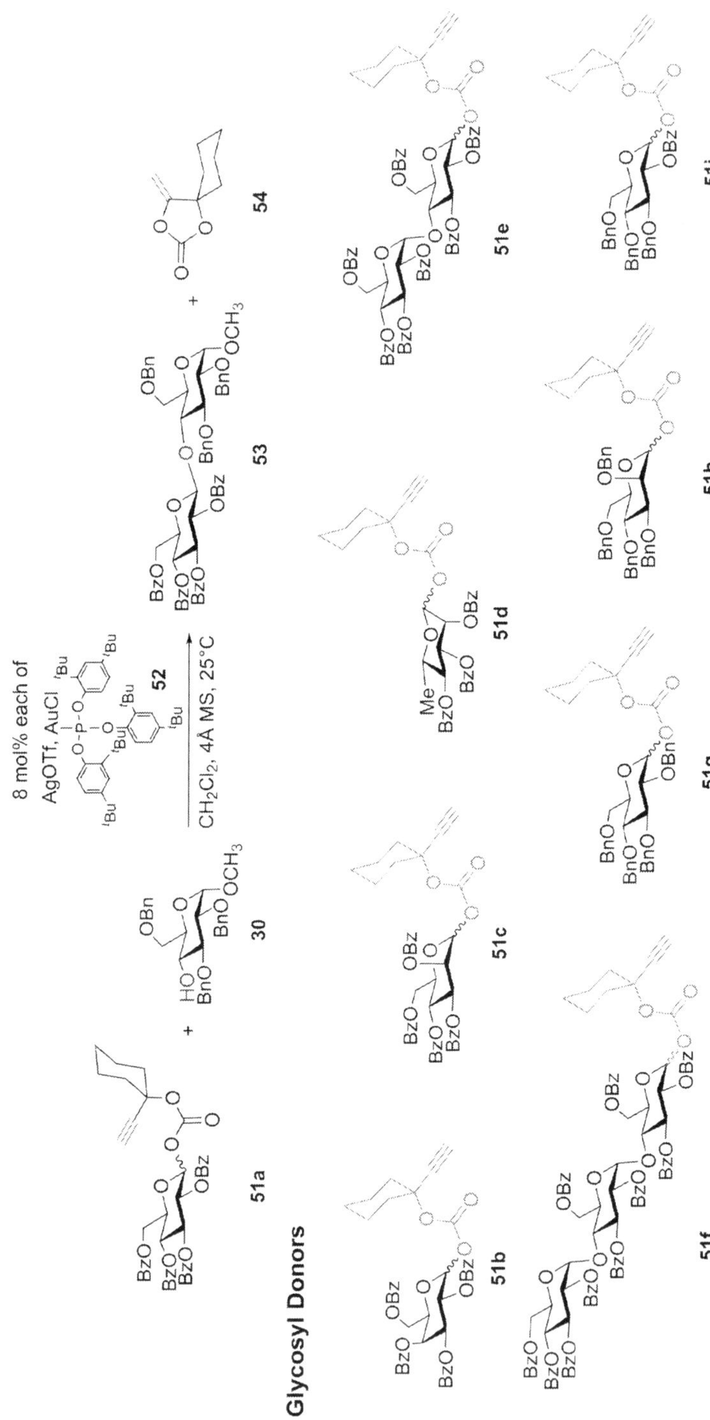

Scheme 12 Alkynyl glycosyl carbonates as glycosyl donors.

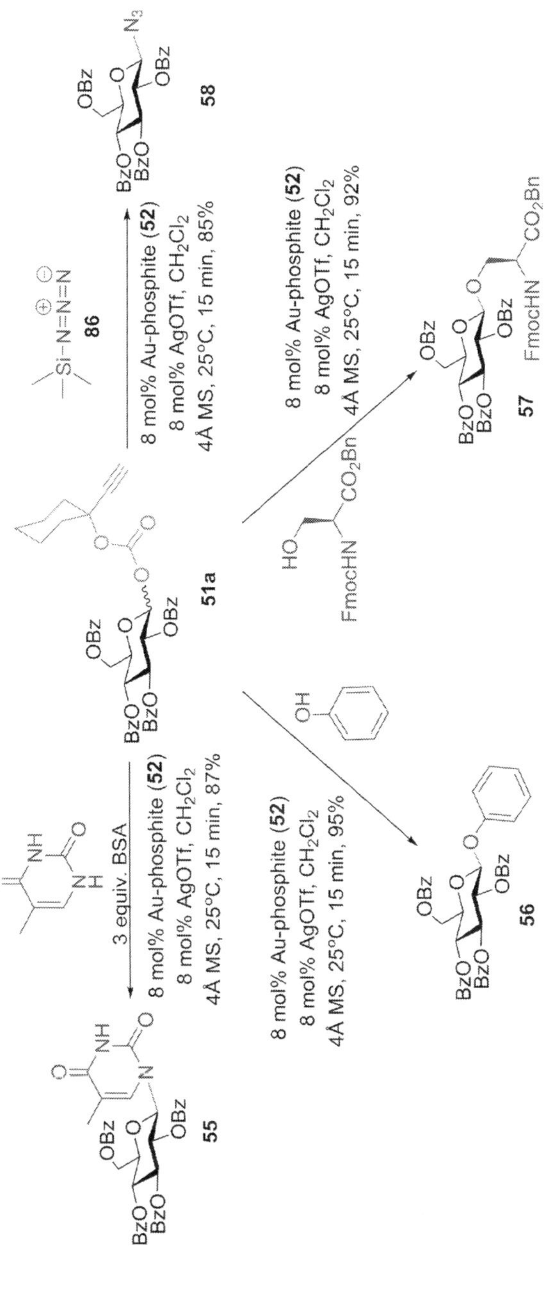

Scheme 13 Synthesis of various glycosides using alkynyl glycosyl carbonates.

The strength of the method was demonstrated by synthesizing a tridecasaccharide mannose-capped arabinan motif reminiscent of the cell wall lipoarabinomannan. The reaction tolerated most of the protecting groups routinely employed in carbohydrate chemistry, namely esters, ethers, silyl ethers, Nap-ethers, azides, benzylidenes, and isopropylidenes.[21]

10 Synthesis of oligosaccharides

Gold-catalysis has significantly contributed to the development of strategies for the synthesis of oligosaccharides. The phenomenal applications to the oligosaccharide syntheses can be attributed to the mild, catalytic, and highly reactive nature of the donor chemistry and noncompetitive leaving group of the glycosyl donor. There are many glycosyl donors available in the synthetic carbohydrate chemists' arsenal to synthesize monosaccharides and disaccharides. However, only a limited number of glycosyl donors found applications for the synthesis of higher oligosaccharides so far. Many disaccharides, trisaccharides, and oligosaccharides have been synthesized in a facile manner using salient features of gold-catalyzed donor chemistry that evolved over the last one decade. Some of the synthesized oligosaccharide structures are listed in Charts 1 and 2.

So far, glycosyl 1,2-orthoesters,[22a] alkynyl glycosyl carbonates,[21, 22b–d] and *o*-alkynylbenzoyl glycosyl esters[20e,22e–i] are reported for the syntheses of oligosaccharides. For example, Hotha's group continued to dominate the synthesis of oligofuranosides from the *Mycobacterium tuberculosis* cell wall.[22a–d] They have demonstrated the synthesis of a heneicosafuranoside **59** (19 Ara*f*s and 2 Gal*f*s)[22a] and a nonadecaarabinofuranoside **60** (19 Ara*f*s) of the arabinogalactan,[22b] a heneicosasaccharide **61** (15 Ara*f*s and 6 Man*p*s) portion of the lipoarabinomannan,[22c] and a pentacosasaccharide **62** (23 Ara*f*s and 2 Gal*f*s) moiety of the arabinogalactan.[22d]

11 Gold-catalyzed synthesis of glycolipids

The Yu group accomplished total synthesis of many saponins containing a steroidal moiety attached to a glycan. Among these, the synthesis of TMG-chitotriomycin (**63**),[22e] periploside (**64**),[22f] astrosterioside (**65**),[22g] digitoxin (**66**),[22h] and goniopectenoside B (**67**)[22i] are notable due to the size and complexity of the compounds (Chart 2). Gold-catalyzed activation of *o*-alkynylbenzoate glycosyl donors provided excellent yields for the glycosidation of digitoxin compared to the NIS/AgOTf-mediated tolylthio donor activation while preparing 3-aminosugar analogues of digitoxin and digoxin.[22j]

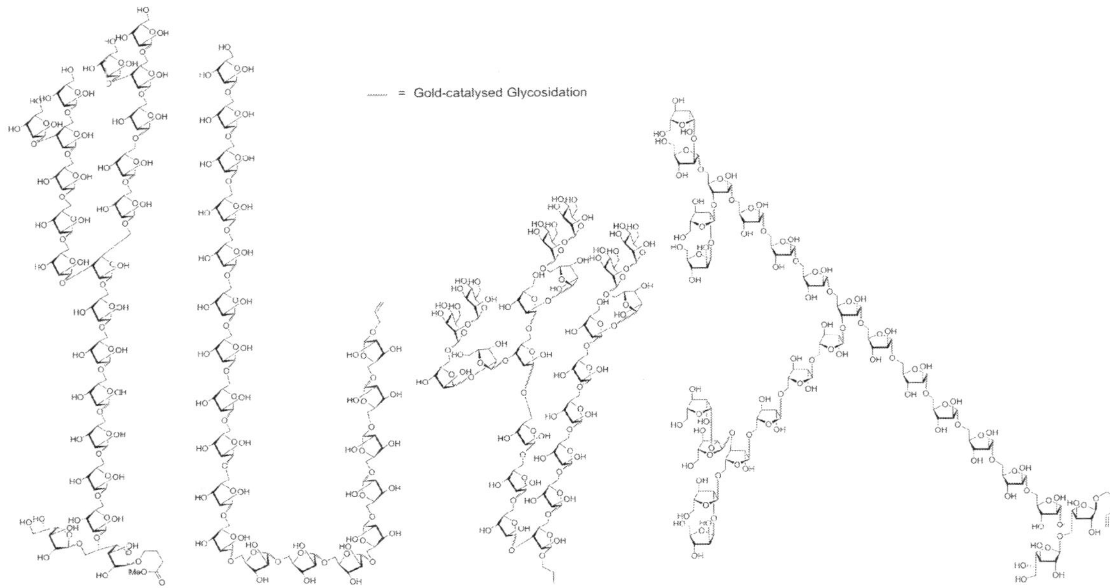

59 60 61 62

Description	59	60	61	62
Leaving group	Orthoesters	Carbonate	Carbonate	Carbonate
Activation by	7 mol% $AuCl_3$	8 mol% each of gold-phosphite **52** and AgOTf		
Number of sugar units	21	19	21	25
Overall yield	0.09% in 47 steps	6.4% in 23 steps	0.016% in 56 steps	0.0012% in 77 steps
Year of publication	2017[22a]	2017[22b]	2017[22c]	2018[22d]

Chart 1 Gold-catalyzed methods for the synthesis of oligosaccharides from Hotha's group.

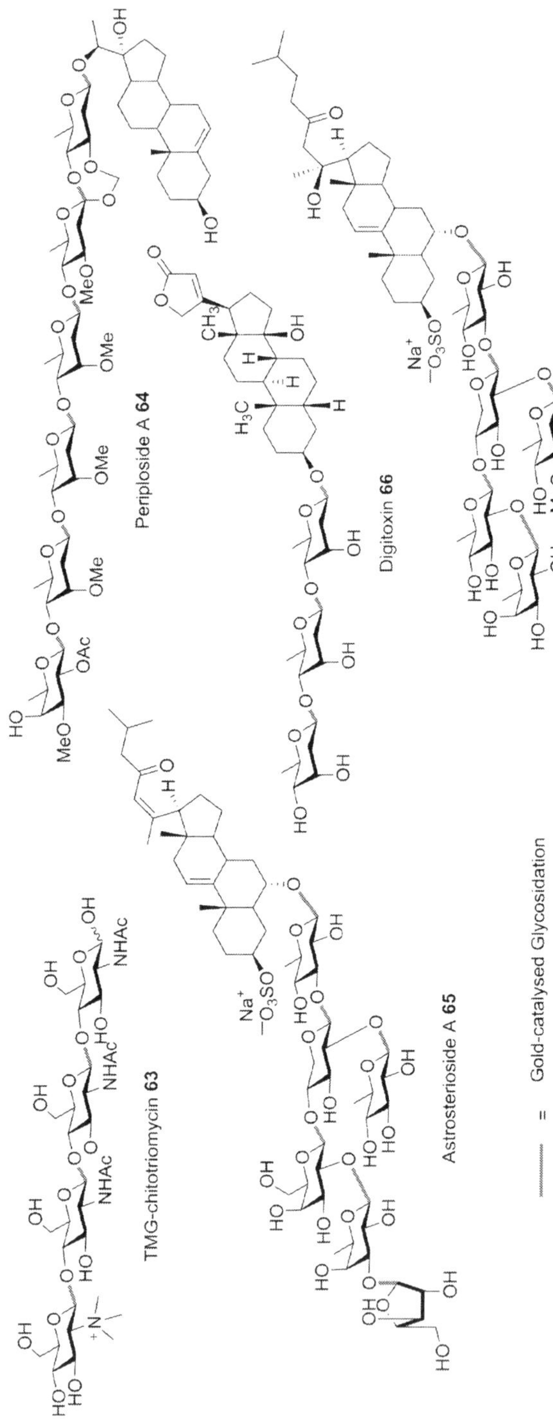

Chart 2 Gold-catalyzed methods for the synthesis of oligosaccharides from the Yu group.

12 Gold-catalysis on carbohydrates for the total synthesis of biologically significant molecules

Total synthesis of natural products is one of the fascinating fields of organic chemistry. Many natural products have been synthesized exploiting contemporary reactions. The gold-catalysis has shown its mettle in the total synthesis field as well. The chiron approach wherein chiral starting materials are used for the synthesis of complex molecules is quite useful. In this context, the utilization of saccharide-based synthons is exploited as carbohydrate scaffolds are easily accessible and several asymmetric reactions can be carried out with precise control of chirality. For example, a simple protocol for the synthesis of (1→1)- and (1→6)-linked *C*-disaccharides has been developed by a one-pot [Au]-catalyzed alkynol cycloisomerization followed by $BF_3{\cdot}OEt_2$, Et_3SiH reduction to obtain pyran-furan (**71**), pyran-pyran (**72**,**73**), and furan-furan (**74**) from the corresponding monosaccharides (**68–70**) (Scheme 14).[23a] A novel etherification protocol was developed using *o*-alkynylbenzoates and gold catalysis while providing access to diosgenyl glycoconjugates.[23b]

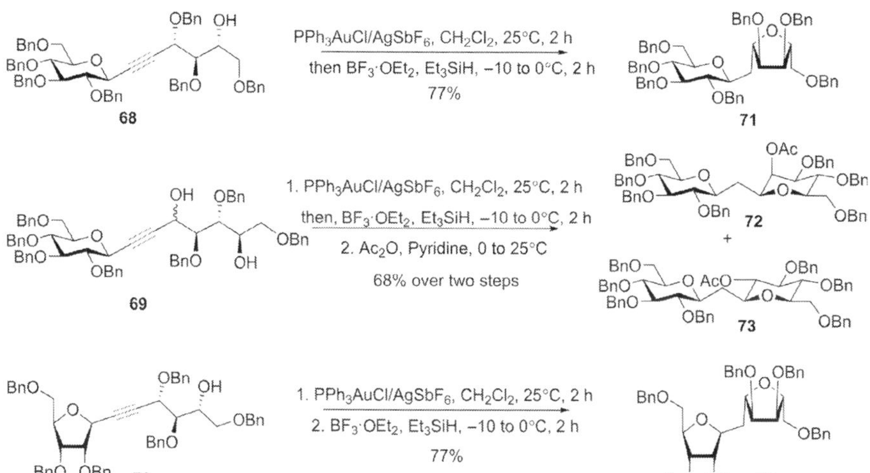

Scheme 14 Gold catalysis for *C*-saccharides.

Gold catalysis on carbohydrate scaffolds has been explored for the synthesis of some natural product-like scaffolds and hybrid molecules (Scheme 15).[24] For example, glycals **75** are converted to *C*-glycosides **77** in the presence of catalytic amounts of Ph_3PAuCl and AgOTf at room temperature by reacting with propargylic acetates **76**.[24a] In another direction, alkyne present in the enyne **78** was activated by $AuCl_3$ followed by hydration and host of interesting annulations resulted in the formation of highly substituted furan derivatives **79**.[24b] Interestingly, a Sonogashira-type coupling was accomplished in the absence of Cu(I) salt using catalytic quantities of $PdCl_2(PPh_3)_2$ and AuI for the synthesis of nonnatural amino acid derivatives **82** from alkyne **80** and aryl iodide **81** precursors.[24c] Ring annulation on glucose-derived monosaccharide **84** was noticed in the presence of catalytic amounts of Au and Ag salts from anilines **83** to obtain a fused multi-cyclic framework **85**.[24d] In yet another facet, $AuBr_3$ was found to be a suitable catalyst for the stereoselective cyclization of 2-*C*-malonyl carbohydrate derivatives at the anomeric center affording a bicyclic lactone with the retention of one of the ester groups. Further, the ester has been opened in order to get various glycosides using TMSOTf. Similarly, 2-*C*-branched methyl glycosides are found to react with the glycosyl acceptors under gold catalysis to afford transglycosylated products in a facile manner.[24e–f]

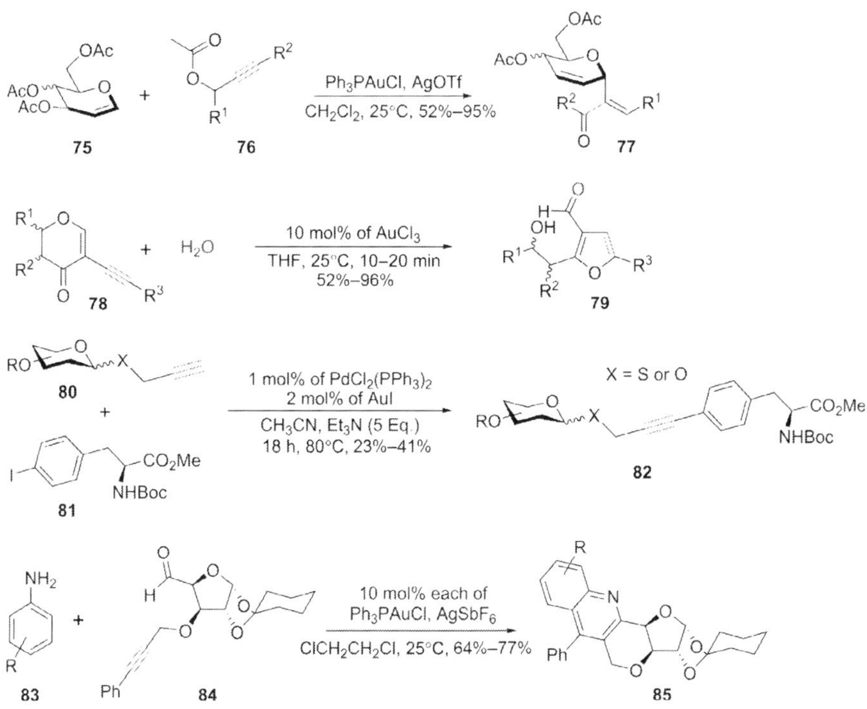

Scheme 15 Gold catalysis for the synthesis of nonnatural glycosidic linkages and hybrid scaffolds.

13 Conclusion

Slowly and steadily, the chemistry of gold has transformed from Alchemy to all chemists. The carbohydrate chemistry also found great advantages in the chemistry of gold. Gold-catalyzed glycosidations have become popular among glycosylation chemists due to the unique reactivity, mild and catalytic activation of alkynyl-bearing leaving groups. Gold catalysis on carbohydrate scaffolds has been dominated by the chemistry of glycosidations. Conventional glycosyl donors (e.g., 1,2-anhydro sugars, glycals, trichloroacetimidates, and esters) are also found to be activated under gold-catalyzed conditions. All these reactions stand on the Lewis acidity, high alkynophilicity, and the ability to produce Brønsted acid when gold halides react with alcohols. Prominent advantages of gold-catalyzed glycosylations are (i) ease of preparation of the glycosyl donors, (ii) catalytic activation using noble metal salts, (iii) large substrate scope due to the mild nature of the catalytic system, (iv) non-competing by-products from the leaving group, (v) operational convenience, and (vi) orthogonality with other conventional glycosyl donors. These unique features of gold-catalysis will increase the interest of glycochemists in exploiting salient features of gold-catalysis in the future.

Acknowledgments

Srinivas Hotha acknowledges the DST-SERB (EMR/2017/003823) for offering financial support. He also acknowledges all the hard work, dedication, patience, and perseverance of all the past and present members of the Gold-catalyzed Glycosidation for Glycoconjugates (3G) Group.

References

1. (a) Kramer, S.; Gagosz, F. In *Gold Catalysis: An homogeneous Approach*, Toste, F. D., Michelt, V., Eds.; Imperial College Press: London, **2014**; pp 1–50; (b) Bond, G.; Louis, C.; Louis, C.; Thompson, T. *Catal. Sci. Ser.* **2006**, *6*, 1–71.
2. (a) Hashmi, A. S. K. *Chem. Rev.* **2007**, *107*, 3180–3211; (b) Hashmi, A. S. K. *Angew. Chem. Int. Ed.* **2005**, *44*, 6990–6993.
3. Raubenheimer, H. G.; Schmidbaur, H. *Organometallics* **2012**, *31*, 2507–2522.
4. Schwerdtfeger, P.; Hermann, H. L.; Schmidbaur, H. *Inorg. Chem.* **2003**, *42*, 1334–1342.
5. (a) Gorin, D. J.; Toste, F. D. *Nature* **2007**, *446*, 395–403; (b) Füstner, A.; Davies, P. W. *Angew. Chem. Int. Ed.* **2007**, *46*, 3410–3449.
6. Norman, R. O. C.; Parr, W. J. E.; Thomas, C. B. *J. Chem. Sci. Perkin Trans.* **1976**, *1*, 1983–1987.
7. Hutchings, G. J. *J. Catal.* **1985**, *96*, 292–295.
8. (a) Hashmi, A. S. K.; Frost, T. M.; Bats, J. W. *J. Am. Chem. Soc.* **2000**, *122*, 11553–11554; (b) Hashmi, A. S. K.; Schwarz, L.; Choi, J.-H.; Frost, T. M. *Angew. Chem. Int. Ed. Eng.* **2000**, *39*, 2285–2288.
9. Kean, W. F.; Hart, L.; Buchanan, W. W. *British J. Rheum.* **1997**, *36*, 560–572.
10. (a) Dimitratos, N.; Porta, F.; Prati, L.; Villab, A. *Catal. Lett.* **2005**, *99*, 181–185; (b) Beltrame, P.; Comotti, M.; Pina, C. D.; Rossi, M. *Appl. Catal. A: Gen.* **2006**, *297*, 1–7; (c) Comotti, M.; Pina, C. D.; Matarrese, R.; Rossi, M.; Siani, A. *Appl. Catal. A: Gen.* **2005**, *291*, 204–209; (d) Omri, M.; Pourceau, G.; Becuwe, M.; Wadouachi, A. *ACS Sustain. Chem. Eng.* **2016**, *4*, 2432–2438.
11. (a) Carrettin, S.; McMorn, P.; Johnston, P.; Griffin, K.; Kiely, C. J.; Attard, G. A.; Hutchings, G. J. *Top. Catal.* **2004**, *27*, 131–136; (b) Carrettin, S.; McMorn, P.; Johnston, P.; Griffin, K.; Kiely, C. J.; Hutchings, G. J. *Phys. Chem. Chem. Phys.* **2003**, *5*, 1329–1336; (c) Carrettin, S.; McMorn, P.; Johnston, P.; Griffin, K.; Hutchings, G. J. *Chem. Commun.* **2002**, 696–697.

12. (a) Maurya, S. K.; Hotha, S. *Tetrahedron Lett.* **2006**, *47*, 3307–3310; (b) Kashyap, S.; Hotha, S. *Tetrahedron Lett.* **2006**, *47*, 2021–2023; (c) Kashyap, S.; Vidadala, S. R.; Hotha, S. *Tetrahedron Lett.* **2007**, *48*, 8960–8962; (d) Roy, R.; Rajasekaran, P.; Mallick, A.; Vankar, Y. D. *Eur. J. Org. Chem.* **2014**, 5564–5573; (e) Palo-Nieto, C.; Sau, A.; Galan, M. C. *J. Am. Chem. Soc.* **2017**, *139*, 14041–14044.
13. Hotha, S.; Kashyap, S. *J. Am. Chem. Soc.* **2006**, *128*, 9620–9621.
14. Vidadala, S. R.; Hotha, S. *Chem. Commun.* **2009**, 2505–2507.
15. (a) Thadke, S. A.; Hotha, S. *Tetrahedron Lett.* **2010**, *51*, 5912–5914; (b) Shaikh, A. Y.; Sureshkumar, G.; Pati, D.; Gupta, S. S.; Hotha, S. *Org. Biomol. Chem.* **2011**, *9*, 5951–5959; (c) Vidadala, S. R.; Gayatri, G.; Sastry, G. N.; Hotha, S. *Chem. Commun.* **2011**, *47*, 9906–9908; (d) Kayastha, A. K.; Hotha, S. *Beilstein J. Org. Chem.* **2013**, *9*, 2147–2155.
16. (a) Kayastha, A. K.; Hotha, S. *Tetrahedron Lett.* **2010**, *51*, 5269–5272; (b) Kayastha, A. K.; Hotha, S. *Chem. Commun.* **2012**, *48*, 7161–7163; (c) Dang, T. T.; Boeck, F.; Hintermann, L. *J. Org. Chem.* **2011**, *76*, 9353–9361.
17. (a) Nielsen, M. M.; Pedersen, C. M. *Chem. Rev.* **2018**, *118*, 8285–8358; (b) Mamidyala, S. K.; Finn, M. G. *J. Org. Chem.* **2009**, *74*, 8417–8420; (c) Götze, S.; Fitzner, R.; Kunz, H. *Synlett* **2009**, *2009*, 3346–3348; (d) Roy, R.; Palanivel, A. K.; Mallick, A.; Vankar, Y. D. *Eur. J. Org. Chem.* **2015**, 4000–4005; (e) Peng, P.; Schmidt, R. R. *J. Am. Chem. Soc.* **2015**, *137*, 12653–12659; (f) Vibhute, A. M.; Dhaka, A.; Athiyarath, V.; Sureshan, K. M. *Chem. Sci.* **2016**, *7*, 4259–4263; (g) Palanivel, A.; Chennaiah, A.; Dubbu, S.; Mallick, A.; Vankar, Y. D. *Carbohydr. Res.* **2017**, *437*, 43–49; (h) Liu, W.; Chen, Q.; Liang, J.; Du, Z.; Zhang, K.; Zheng, X.; O'Doherty, G. A. *Synlett* **2015**, 1683–1686.
18. (a) Adhikari, S.; Baryal, K. N.; Zhu, D.; Li, X.; Zhu, J. *ACS Catal.* **2013**, *3*, 57–60; (b) Adhikari, S.; Li, X.; Zhu, J. *J. Carbohydr. Chem.* **2013**, *32*, 336–359; (c) Zheng, Z.; Zhang, L. *Carbohydr. Res.* **2019**, *471*, 56–63.
19. (a) Sureshkumar, G.; Hotha, S. *Tetrahedron Lett.* **2007**, *48*, 6564–6568; (b) Sureshkumar, G.; Hotha, S. *Chem. Commun.* **2008**, 4282–4284; (c) Mallick, A.; Mallikharjunarao, Y.; Rajasekaran, P.; Roy, R.; Vankar, Y. D. *Eur. J. Org. Chem.* **2016**, 579–588.
20. (a) Li, Y.; Yang, Y.; Yu, B. *Tetrahedron Lett.* **2008**, *49*, 3604–3608; (b) Li, Y.; Yang, X.; Liu, Y.; Zhu, C.; Yang, Y.; Yu, B. *Chem. Eur. J.* **2010**, *16*, 1871–1882; (c) Zhu, Y.; Yu, B. *Angew. Chem. Int. Ed.* **2011**, *50*, 8329–8332; (d) Tang, Y.; Li, J.; Zhu, Y.; Li, Y.; Yu, B. *J. Am. Chem. Soc.* **2013**, *135*, 18396–18405; (e) Chen, X.; Wang, Q.; Yu, B. *Chem. Commun.* **2016**, *52*, 12183–12186; (f) Li, W.; Yu, B. *Chem. Soc. Rev.* **2018**, *47*, 7954–7984; (g) Yu, B. *Acc. Chem. Res.* **2018**, *51*, 507–516; (h) Dutta, S.; Sarkar, S.; Gupta, S. J.; Sen, A. K. *Tetrahedron Lett.* **2013**, *54*, 865–870; (i) Shaw, M.; Thakur, R.; Kumar, A. *J. Org. Chem.* **2019**, *84*, 589–605; (j) Sun, P.; Wang, P.; Zhang, Y.; Zhang, X.; Wang, C.; Liu, S.; Lu, J.; Li, M. *J. Org. Chem.* **2015**, *80*, 4164–4175; (k) Zhu, Y.; Shen, Z.; Li, W.; Yu, B. *Org. Biomol. Chem.* **2016**, *14*, 1536–1539; (l) Li, J.; Dai, Y.; Li, W.; Laval, S.; Xu, P.; Yu, B. *Asian J. Org. Chem.* **2015**, *4*, 756–762; (m) Mi, X.; Lou, Q.; Fan, W.; Zhuang, L.; Yang, Y. *Carbohydr. Res.* **2017**, *448*, 161–165; (n) Shen, Z.; Mobarak, H.; Li, W.; Widmalm, G.; Yu, B. *J. Org. Chem.* **2017**, *82*, 3062–3071.
21. Mishra, B.; Neralkar, M.; Hotha, S. *Angew. Chem. Int. Ed.* **2016**, *55*, 7786–7791.
22. (a) Thadke, S. A.; Mishra, B.; Islam, M.; Pasari, S.; Manmode, S.; Rao, B. V.; Neralkar, M.; Shinde, G. P.; Walke, G.; Hotha, S. *Nat. Commun.* **2017**, *8*, 14019; (b) Mishra, B.; Manmode, S.; Panda, R. R. A.; Hotha, S. *Eur. J. Org. Chem.* **2017**, 4794–4802; (c) Islam, M.; Shinde, G. P.; Hotha, S. *Chem. Sci.* **2017**, *8*, 2033–2039; (d) Pasari, S.; Manmode, S.; Walke, G.; Hotha, S. *Chem. Eur. J.* **2018**, *24*, 1128–1139; (e) Yang, Y.; Li, Y.; Yu, B. *J. Am. Chem. Soc.* **2009**, *131*, 12076–12077; (f) Zhang, X.; Zhou, Y.; Zuo, J.; Yu, B. *Nat. Commun.* **2015**, *6*, 5879; (g) Dai, Y.; Yu, B. *Chem. Commun.* **2015**, *51*, 13826–13829; (h) Ma, Y.; Li, Z.;

Shi, H.; Zhang, J.; Yu, B. *J. Org. Chem.* **2011**, *76*, 9748–9756; (i) Xiao, G.; Yu, B. *Chem. Eur. J.* **2013**, *19*, 7708–7712; (j) Zeng, J.; Sun, G.; Wang, R.; Zhang, S.; Teng, S.; Liao, Z.; Meng, L.; Wan, Q. *Org. Chem. Front.* **2017**, *4*, 2450–2454.

23. (a) Narute, S. B.; Rout, J. K.; Ramana, C. V. *Chem. Eur. J.* **2013**, *19*, 15109–15114; (b) Zhang, L.; Li, L.; Bai, S.; Zhou, X.; Wang, P.; Li, M. *Org. Lett.* **2016**, *18*, 6030–6033.
24. (a) Huang, N.; Liao, H.; Yao, H.; Xie, T.; Zhang, S.; Zou, K.; Liu, X.-W. *Org. Lett.* **2018**, *20*, 16–19; (b) Mal, K.; Sharma, A.; Das, I. *Chem. Eur. J.* **2014**, *20*, 11932–11945; (c) Jeon, M. H.; Mathew, B. P.; Kuram, M. R.; Myung, K.; Hong, S. Y. *Org. Biomol. Chem.* **2016**, *4*, 11518–11524; (d) Reddy, B. V. S.; Majumder, N.; Sridhar, B. *Tetrahedron Lett.* **2014**, *55*, 6081–6084; (e) Vidadala, S. R.; Pimpalapalle, T. M.; Linker, T.; Hotha, S. *Eur. J. Org. Chem.* **2011**, 2426–2430; (f) Pimpalpalle, T. M.; Vidadala, S. R.; Hotha, S.; Linker, T. *Chem. Commun.* **2011**, *47*, 10434–10436.

Glycomimetics with unnatural glycosidic linkages

5

Viktor Kelemen, Anikó Borbás
Department of Pharmaceutical Chemistry, University of Debrecen, Debrecen, Hungary

Chapter outline

1 Introduction

Carbohydrates, in the form of oligosaccharides, polysaccharides, and glycoconjugates, are ubiquitous in all living organisms and play crucial roles in a wide range of biological processes, including intercellular adhesion, signaling, trafficking, disease progression, pathogen detection, and immune responses.[1–7] Translation of the numerous biological activities of carbohydrates into therapeutic applications is, however, challenging due to their structural complexicity, hydrophilicity, and weak binding properties. The sensitivity of the glycosidic bond to enzymatic hydrolysis has further hindered the application of glycosides as biological probes or drug candidates. To address these issues, a broad variety of carbohydrate mimetics have been prepared.

One major class of carbohydrate mimetics results from linkage modifications. Replacement of the glycosidic oxygen by a methylene group, sulfur, selenium, or nitrogen is the most frequently used tactic to produce derivatives with increased stability, while maintaining the biological properties of the natural compounds (Fig. 1A). Over the last decade, a variety of other synthetic linkage modes comprising two or three bridging atoms in place of a native *O*-glycosidic bond have been designed (Fig. 1B and C). The extended interglycosidic bridges provide specific conformational properties and, in some cases, advantageous binding capabilities to carbohydrate mimetics. Beyond the naturally occurring head-to-head and head-to-tail linkages involving

Recent Trends in Carbohydrate Chemistry. https://doi.org/10.1016/B978-0-12-817467-8.00005-0
© 2020 Elsevier Inc. All rights reserved.

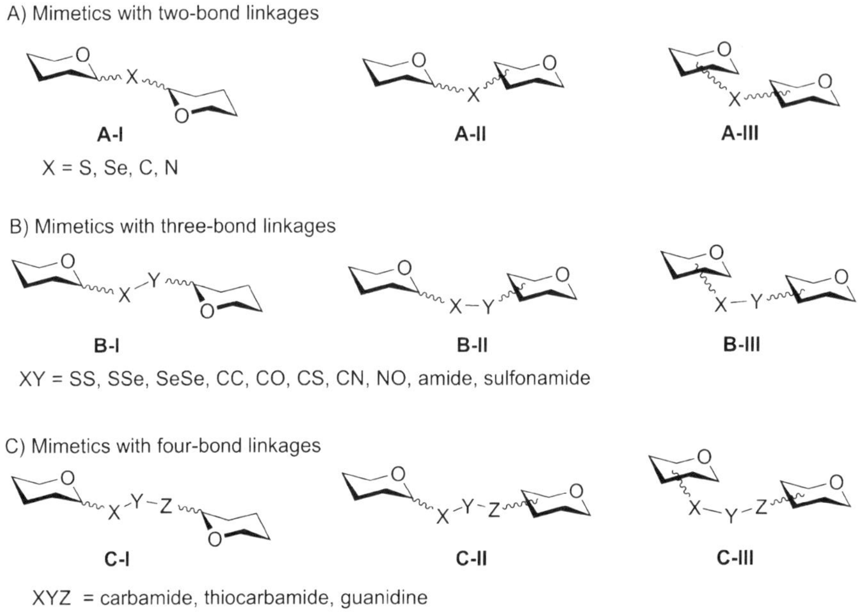

Fig. 1 Various linking modes of carbohydrate mimetics represented on pyranoside structures: (A) mimetics with two-bond linkages, (B) mimetics with three-bond linkages, (C) mimetics with four-bond linkages; (I) head-to-head, (II) head-to-tail, and (III) tail-to-tail connections.

at least one anomeric center (Fig. 1I and II), diglycose derivatives, also known as neoglycosides or pseudodisaccharides, consisting of two monosaccharides linked at nonanomeric positions by various two-, three-, and four-bond linkages (Fig. 1III) have also been prepared.

The main objective of this chapter is to provide an overview of the broad variety of synthetic interglycosidic linking modes developed so far. We focus on the key synthetic steps leading to the formation of disaccharide mimetics. Fig. 1 summarizes the linkage modifications on pyranoid disaccharide structures. Similar mimetics consisting of furanose rings as well as construction of higher oligomers with modified linking modes are also discussed. Due to the numerous literature results and excellent review articles available,[8–13] we mainly focus on the synthetic methods published in the last 10–15 years.

2 Carbohydrate mimetics with two-bond interglycosidic linkages

Substitution of the interglycosidic oxygen between two monosaccharide units by another heteroatom (S, Se, N) or carbon is the most obvious way to produce carbohydrate mimetics.

2.1 Thioglycosides

Thioglycosides, in which the glycosidic oxygen is replaced by a sulfur atom, are among the most important glycomimetics reported to date due to their stability and close structural similarity to the natural *O*-glycosides. As *S*-glycosides are more stable than their *O*-glycoside analogues against both chemical and enzymatic degradations, thiodisaccharides and thiooligosaccharides are convenient probes for enzyme inhibition studies as competitive inhibitors of glycoside hydrolases.[14]

A variety of methods are employed for synthesis of the thioglycosides[15] including glycosylation of thio acceptors with activated glycosyl donors,[16, 17] S_N2-like displacement of a good leaving group with 1-thioaldoses,[18, 19] Michael or Michael-type additions of 1-thioaldose to unsaturated acceptors,[20] ring-opening of sugar epoxides[21] and thiiranes[22] by a 1-thioaldose nucleophile, and Ferrier reactions between glycals and sulfur-containing coupling partners.[23] Recently, it has been demonstrated that the efficient formation of thiodisaccharides is possible by photoinduced free-radical addition of sugar thiols to unsaturated carbohydrate rings.[24–27]

Excellent reviews on the synthesis of thiooligosaccharides, which cover the earlier literature, have already been issued.[15, 18, 28, 29] Herein, we summarize the advances of the past decade.

2.1.1 Introduction of sulfur to the anomeric position

The nonreducing disaccharide trehalose, consisting of two glucose molecules linked by an α,α-1,1-linkage, has important cell-protective activities in plants, fungi, and various microorganisms. Although many efforts have been devoted to synthesize 1-thiotrehalose analogues as potential therapeutic agents, the introduction of α,α-anomeric linkages with high stereocontrol proved to be particularly challenging.

In 2012, Xin and Zhu reported a highly stereoselective, five-step synthesis of α,α-1-thiotrehalose **7** from 1,6-anhydro-D-glucose derivative **1** (Scheme 1).[30] First, α-glycosyl thiol **2** was prepared by trimethylsilyl triflate-mediated nucleophilic ring-opening of **1** with bis(trimethylsilyl)sulfide.[31, 32] After diacetylation followed by selective *S*-deacetylation, the α-glucosylthiol acceptor **3** was glycosylated with glucosyl thioimidate **4β** or glucosyl trichloroacetimidate **5α** producing the protected 1-thiotrehalose **6** with excellent α-stereoselectivity. In the key coupling step, Schmidt's inverse procedure[33] with donor **5α** provided the highest yield.[34] Final deprotection of **6** gave 1-thiotrehalose **7** in 90% yield. Zhu and coworkers utilized the protected **6** as an intermediate for the synthesis of **8** as a thioglycoside analogue of the diacyl trehalose maradolipid.[34]

By modifying the Zhu method mentioned previously, Compain and coworkers developed a very elegant two-step access to 1-thiotrehalose (Scheme 2).[35] In the key TMSOTf-mediated double thioglycosylation step, the 1,6-anhydro-tri-*O*-benzyl-D-glucose **1** behaved both as a glycosyl donor and as a precursor of glycosyl acceptor **9**. In this convergent process, two carbon-sulfur bonds were created at the anomeric positions with a high level of stereocontrol affording the protected thiotrehalose in various efficacy depending on the stoichiometry of the reactants. Applying 0.5 equiv. of $(TMS)_2S$ in the presence of 0.9 equiv. of TMSOTf resulted in **10** with the highest 56% yield.

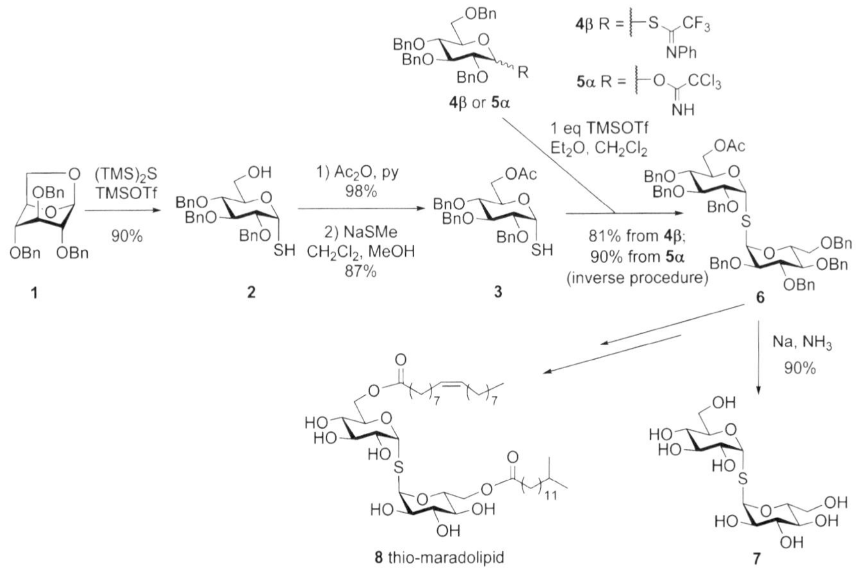

Scheme 1 Stereoselective synthesis of 1-thiotrehalose **7** and thio-maradolipid **8**, starting from 1,6-anhydro-D-glucose derivative **1**, developed by Xin and Zhu.[30]

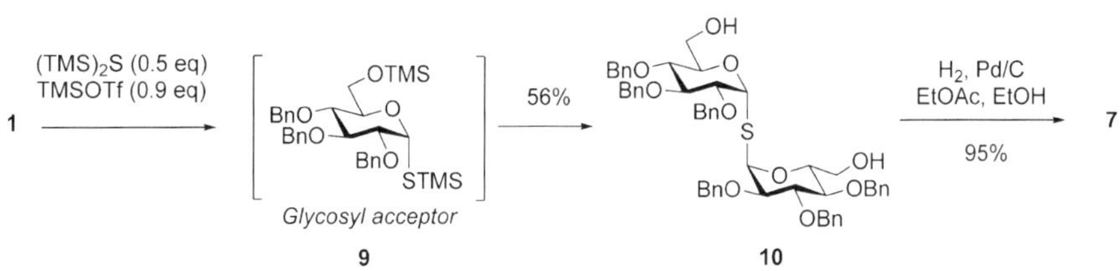

Scheme 2 Compain's two-step route to 1-thiotrehalose **7**.

As a demonstration of the synthetic utility of the highly stereoselective synthesis of 1,2-*cis* α-1-thioglucose **2**, Zhu and Wang reported the synthesis of α-*S*-(1→6)-linked glucosyl oligomers up to pentasaccharide.[36] The sulfur atom was introduced to α-anomeric position in the very first step by the earlier discussed ring-opening reaction of 1,6-anhydroglucose **1** to produce **2**, then monosaccharide building blocks **11** and **14** were prepared by routine transformations (Scheme 3). The anomeric sulfanyl group in **14** was masked with monomethoxytrityl (MMTr) group. Thioglycosylation of 6-iodinated **14** with α-glycosylthiol **15** in the presence of the phase transfer catalyst tetrabutylammonium hydrogensulfate (TBAHS) gave disaccharide building block **16** in 68% yield. The further synthetic steps involved thioglycosylation of 6-iodinated sugars with α-glycosylthiols (**11**+**14**, **17**+**14**, and **18**+**16**) under phase transfer conditions followed by removal of the MMTr group using trifluoroacetic acid (TFA) in the presence of triethylsilane. In each coupling step, the thiolate anion was generated in situ using $NaHCO_3$, and the α-*S*-disaccharide was then readily formed by nucleophilic displacement of the iodo substituent. As it was expected, the α-stereochemical integrity at the

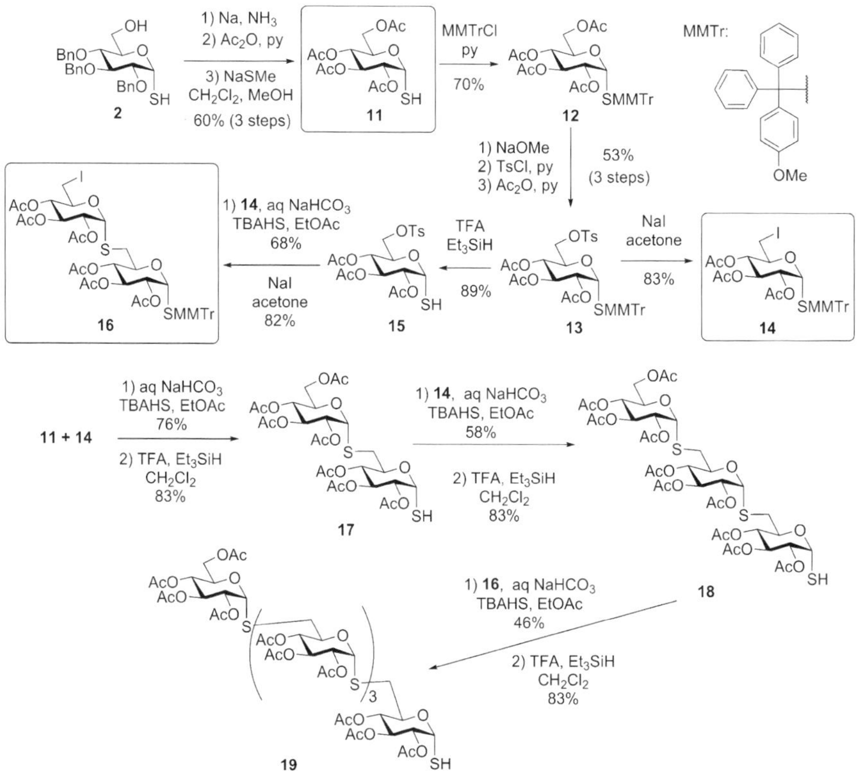

Scheme 3 Zhu's nucleophilic approach to the synthesis of α-*S*-(1→6)-linked tri- and pentaglucosylthiols **18** and **19**.[36]

anomeric center of glycosylthiols was maintained during the coupling reactions due to the intrinsic configurational stability of glycosylthiols under basic conditions.[37]

A similar nucleophilic method was developed by Belz and Williams for synthesizing *S*-linked α-1,6-oligomannosides (Scheme 4). The readily available 1,2-*trans* mannosyl isothiouronium bromide **20** and anomeric thioacetate **24** were used as configurationally homogeneous α-sulfanyl precursors and the interglycosidic sulfur linkages were generated through coupling α-thiolates with 6-deoxy-6-iodo sugars. The reducing end methyl thiomannobioside **25** was formed by the condensation of thioacetate **24** with 6-iodomannoside **21** in the presence of Et_2NH. Although transformation of **25** to a 6′-iodide proved to be problematic, a successful 6′-iodination was accomplished through a five-step protocol. Next, coupling of disaccharide thioacetate **23** with the 6′-iodo disaccharide **26**, followed by activation of the resulting tetrasaccharide **27** to 6‴-iodide **28**, and iterative coupling with **23** afforded the *S*-linked hexasaccharide **29**.[38]

In 2009, Falconer and coworkers reported the facile preparation of a series of 1,1′-thiodisaccharides by $BF_3{\cdot}Et_2O$-catalyzed couplings of sugar per-*O*-acetates with 1-thiosugars.[39] Reaction of glucose and galactose per-*O*-acetates **30** and **31** with tetra-*O*-acetylated 1-thioglucose **32** provided the desired 1,2-*trans*-linked

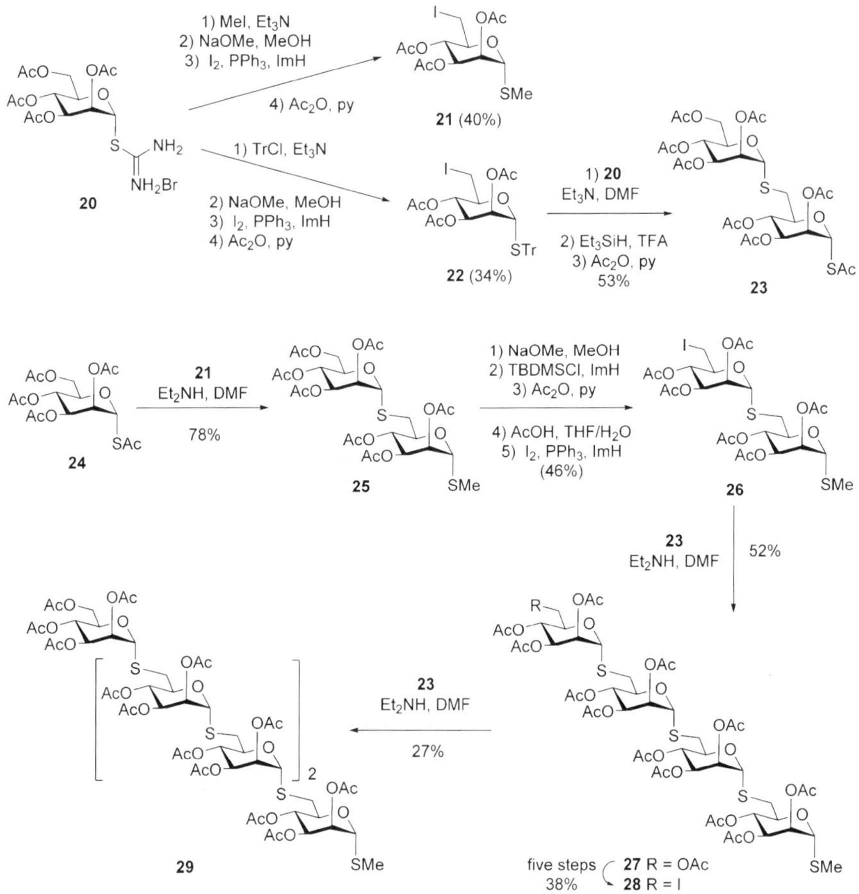

Scheme 4 Nucleophilic block synthesis route to an α-*S*-(1→6)-linked methyl hexamannoside developed by Belz and Williams.[38]

1,1′-thiodisaccharides **33** and **34** in good yields with minimal by-products (Scheme 5). The galactose donor **31** was more reactive, 2 h reaction time was sufficient to complete its consumption, while complete conversion of the glucose donor **30** required a longer, 18 h reaction. The study was extended to investigate the reaction of fully acetylated D-mannose (**35α,β**), L-fucose (**36**), and L-rhamnose (**37**) with acetylated 1-thio-D-glucose **32**, -galactose **38**, -mannose **39**, and L-fucose **40** (Scheme 5). Single anomer peracetylated glycosyl donors yielded an anomeric mixture of thiodisaccharides (at the donor anomeric carbon), with the dominance of the 1,2-*trans*-linked product (Table 1). In the *gluco* and *galacto* cases, very high 1,2-*trans*-β-selectivity was observed. With mannose as donor, the 1,2-*trans* α-anomer was favored in a 3:1 ratio, and a 1,2-*trans*:1,2-*cis* ratio of ca. 2:1 was obtained with the acetylated 6-deoxy L-sugar donors.

Jana and Misra described a fast and stereoselective synthesis of β-glycosylthiols, including D- and L-sugars and disaccharides (e.g., **32**, **40**, and **51–55**), and their

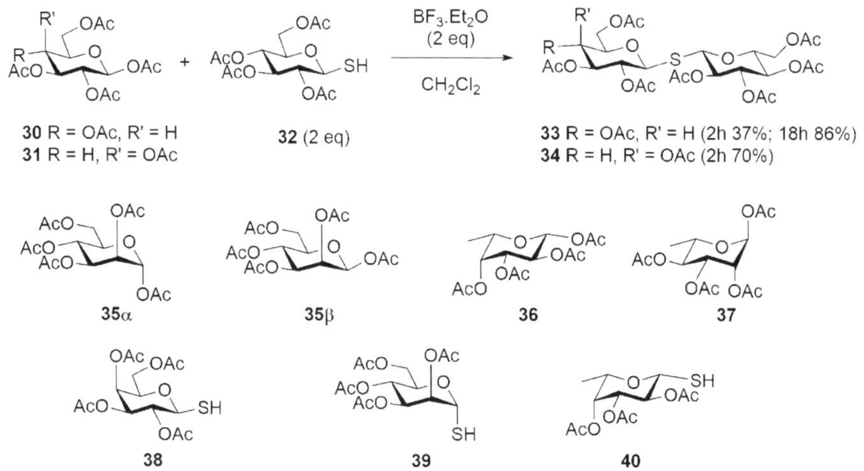

Scheme 5 Lewis acid-mediated reactions of glucose and galactose per-*O*-acetates **20** and **21** with thiosugar **22** and structure of further sugar reactants involved in the study of Falconer's group.[39]

application in the synthesis of (1,1′)-β-thio-linked di- and trisaccharides such as **47** and **57** (Scheme 6). The exclusive formation of β-glycosylthiols was explained by the S_N2 displacement of the α-bromide ion by the carbonotrithioate ion derived from the reaction of carbon disulfide and sodium sulfide, followed by elimination of CS_2.[40]

Over the last decade, the radical-mediated thiol-ene coupling (also called thio-click reaction), where a reactive thiyl radical undergoes an addition reaction onto an alkene to furnish a thioether linkage, has emerged as a mild and efficient method for the preparation of sulfur-containing glycoconjugates and thiosugars.[24–27] Our group reported that photoinitiated thiol-ene coupling reactions of 2-substituted glycals represent a generally applicable strategy for the stereoselective formation of the challenging 1,2-*cis* α-thioglycosidic linkage.[41–43] We demonstrated on the exemplary case of 2-acetoxy-D-glycal **58** that the UV-light-initiated addition of a range of thiols (e.g., **32** and **59–63**) including amino acid, peptide, and various sugar thiols in the presence of the photoinitiator 2,2-dimethoxy-2-phenylacetophenone (DPAP) occurs with good to high yields and complete stereoselectivity providing an easy access to 1,2-*cis* α-linked thiodisaccharides (**66–69**) and thioglycoconjugates (**64** and **65**) (Table 2).[41] The reactions required mild conditions (irradiation at λ_{max} 365 nm for 3×15 min at room temperature) and showed tolerance to air, water, and a broad range of functional groups. We have found that in the reactions with thiols **59–63**, the conversion of glycal **58** varied between 60% and 90%, depending on the thiol reactivity, and the conversion was not increased by prolonged exposure to UV light or using higher thiol excess. We extended the study to thiol-ene reactions of benzyl-protected 2-hydroxyglycal **70**.[42] As expected, compound **70**, possessing a more electron-rich double bond than that of **58**, showed higher reactivity toward the electrophilic thiyl radicals. With thiol **63**, for instance, complete conversion of **70** was observed; however, the isolated yield of **71** was only 59% because decomposition also occurred due to lability of the benzyl protecting groups under the conditions of the radical addition (Table 2).

Table 1 Synthesis of β,β- and α,β-(1,1)-thiodisaccharides by $BF_3 \cdot Et_2O$-mediated couplings of sugar per-*O*-acetates with 1-thiosugars.

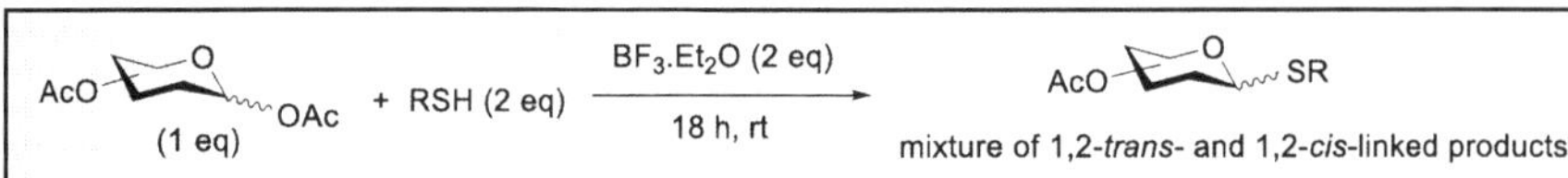

Per-*O*AC-sugar	Thiosugar RSH	1,2-*trans*-linked product A	1,2-*cis*-linked product B	Ratio A:B[a]
30	**38**	**34** (80%)	Not isolated	9:1
31	**38**	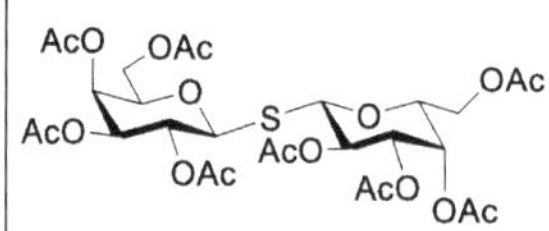 **41** (85%)[b]	Not isolated	9:1
31	**39**	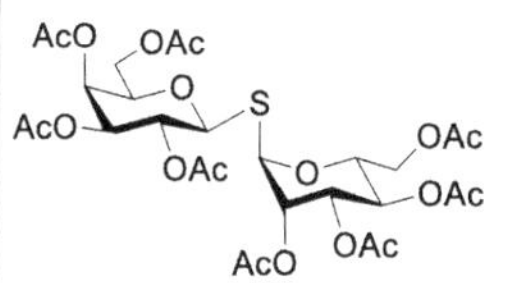 **42** (70%)[b]	Not isolated	9:1
31	**40**	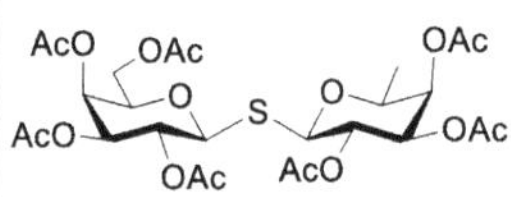 **43** (68%)[b]	Not isolated	9:1
35α or **35β**	**32**	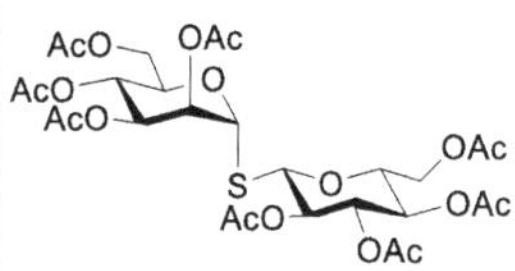 **44** (**35α** 39%, **35β** 32%)	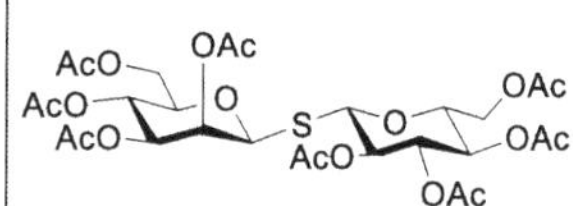 **45** (**35α** 14%, **35β**)	**35α** 2.8:1 **35β** 2.9:1
35β	**38**	**42** (42%)	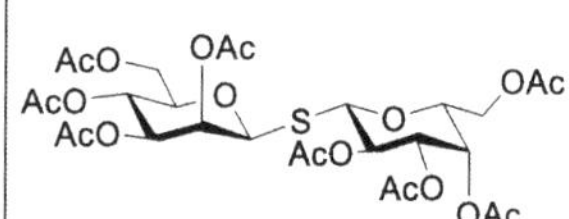 **46** (15%)	2.8:1
36	**32**	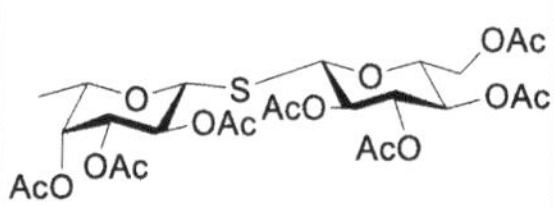 **47** (48%)	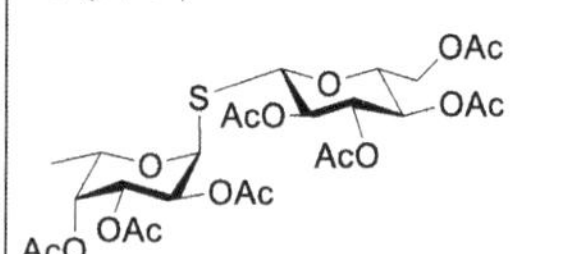 **48** (26%)	1.8:1
37	**32**	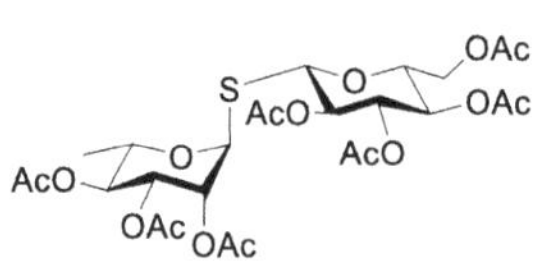 **49** (50%)	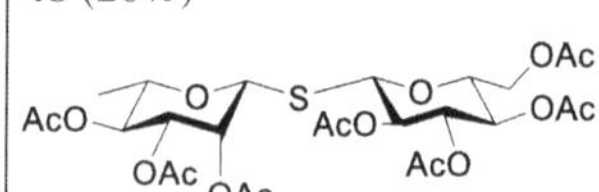 **50** (27%)	1.9:1

[a] Determined by NMR.
[b] 2h.

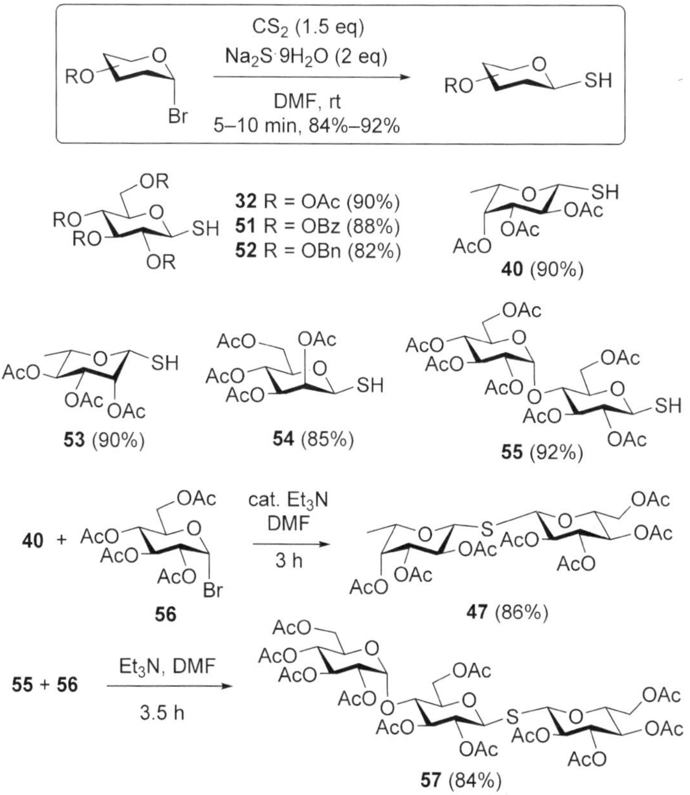

Scheme 6 Misra's stereoselective synthesis of β-glycosylthiols and their application in the formation of β-thio-linked oligosaccharides.[40]

The free-radical hydrothiolation of 2-substituted D-glycals occurs, presumably, exclusively via the $^4C_1$ conformer of the carbon-centered radical **58-R**, which is formed from the $^4H_5$ half chair conformer of the glycal upon bottom-face attack of the thiyl radical in the propagation step (Scheme 7). The exquisite stability of the $^4C_1$ conformation of D-hexopyranoses, the kinetically preferred axial thiyl addition and the kinetically preferred axial H-abstraction steps of cyclic alkenes[44] together ensure the exclusive α-stereoselectivity of the reaction.[43]

Following the investigation of the thiol-ene couplings of further 2-substituted glycals including the L-fucose-derived **72**, the 2-acetamido-D-glycal obtained from glucose **76**, and 2-acetoxy-D-glycals obtained from galactose (**73**), allose (**74**), and gulose (**75**), it has been found that the reaction temperature profoundly influences the efficacy of the reactions by controlling the equilibrium of the reversible propagation step of the radical chain process.[43] While all glycals reacted with full 1,2-*cis* α-selectivity, their reactivity was varying in a configuration-dependent manner, and with enoses bearing axial acetoxy substituents low conversions and, thereby, low yields were obtained at room temperature (Table 3). Interestingly, our study revealed that the reaction progress could be promoted by cooling, while it was inhibited by heating.

Table 2 Borbás's thio-click approach for the stereoselective synthesis of 1,2-*cis* α-linked thioglycosides from 2-hydroxyglycals **58** and **70**.

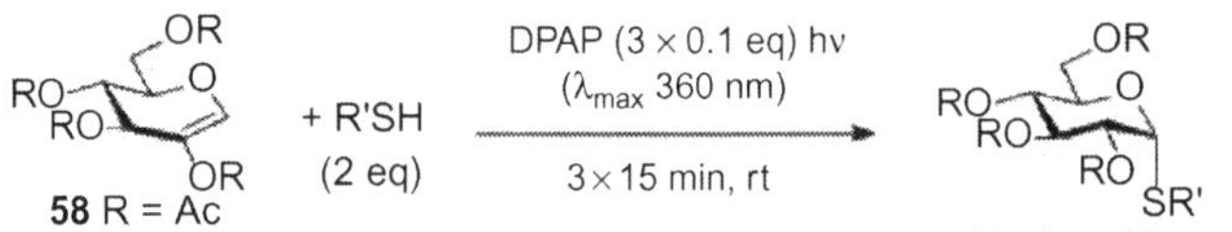

58 R = Ac
70 R = Bn

α-thioglucosides

Thiol	Product (yield)	Solvent
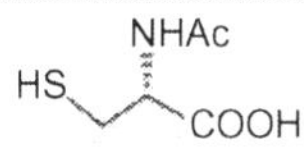 **59**	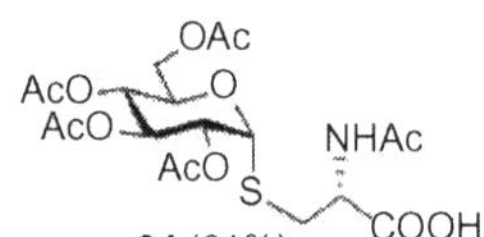 **64** (61%)	toluene-MeOH
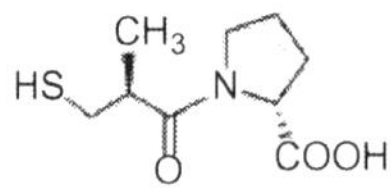 **60**	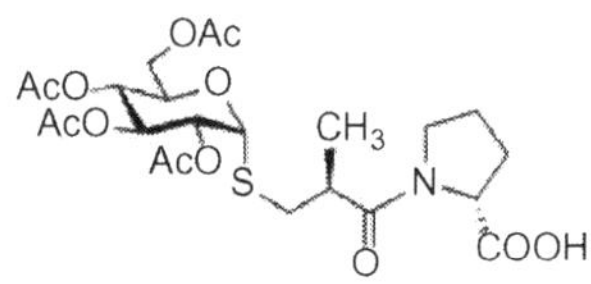 **65** (88%)	MeOH
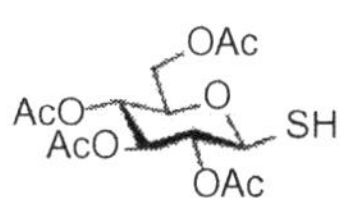 **32**	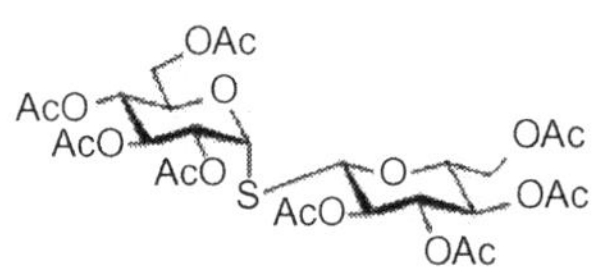 **66** (87% with 1.2 eq thiol)	toluene
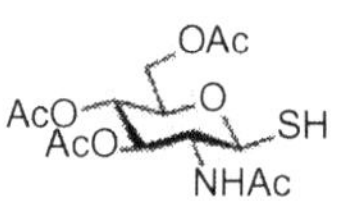 **61**	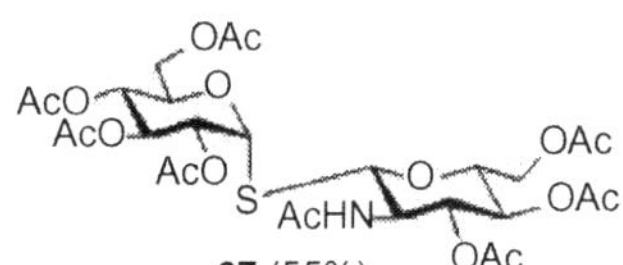 **67** (55%)	toluene-MeOH
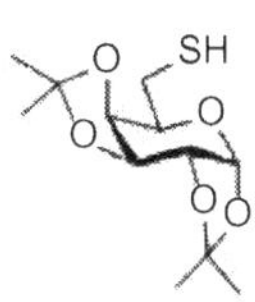 **62**	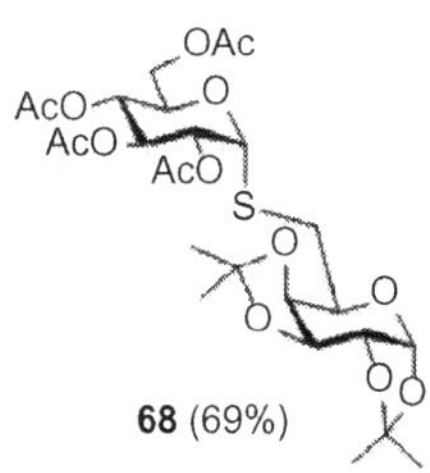 **68** (69%)	toluene
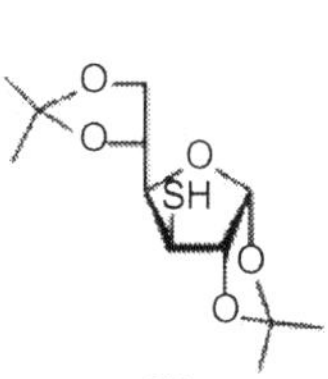 **63**	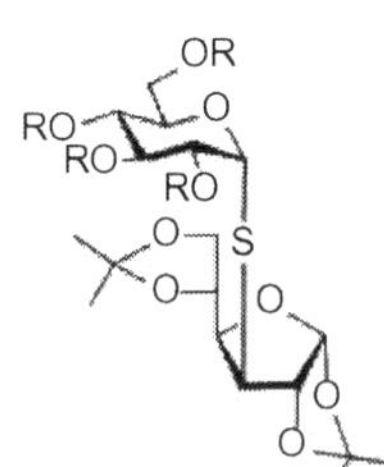 **69** R = Ac (~60% conversion of **58**, 59% yield) **71** R = Bn (full conversion of **70**, 59% yield)	toluene

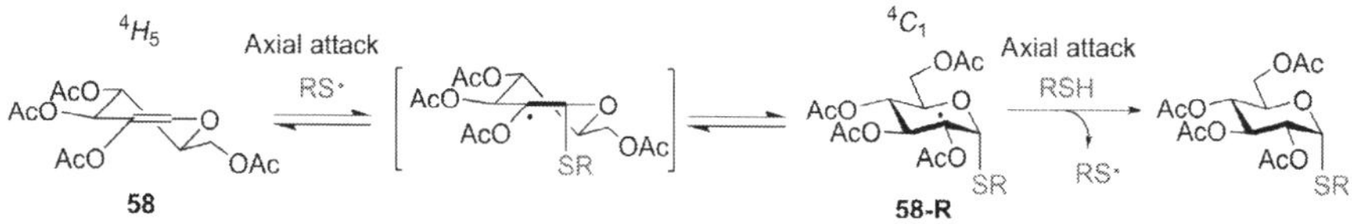

Scheme 7 Plausible mechanistic pathway of free-radical addition of thiols to 2-acetoxyglycal **58** (**58**→**58-R**: propagation step).

Table 3 Configuration- and temperature-dependent reactivity of 2-substituted glycals in thiol-ene coupling reactions.

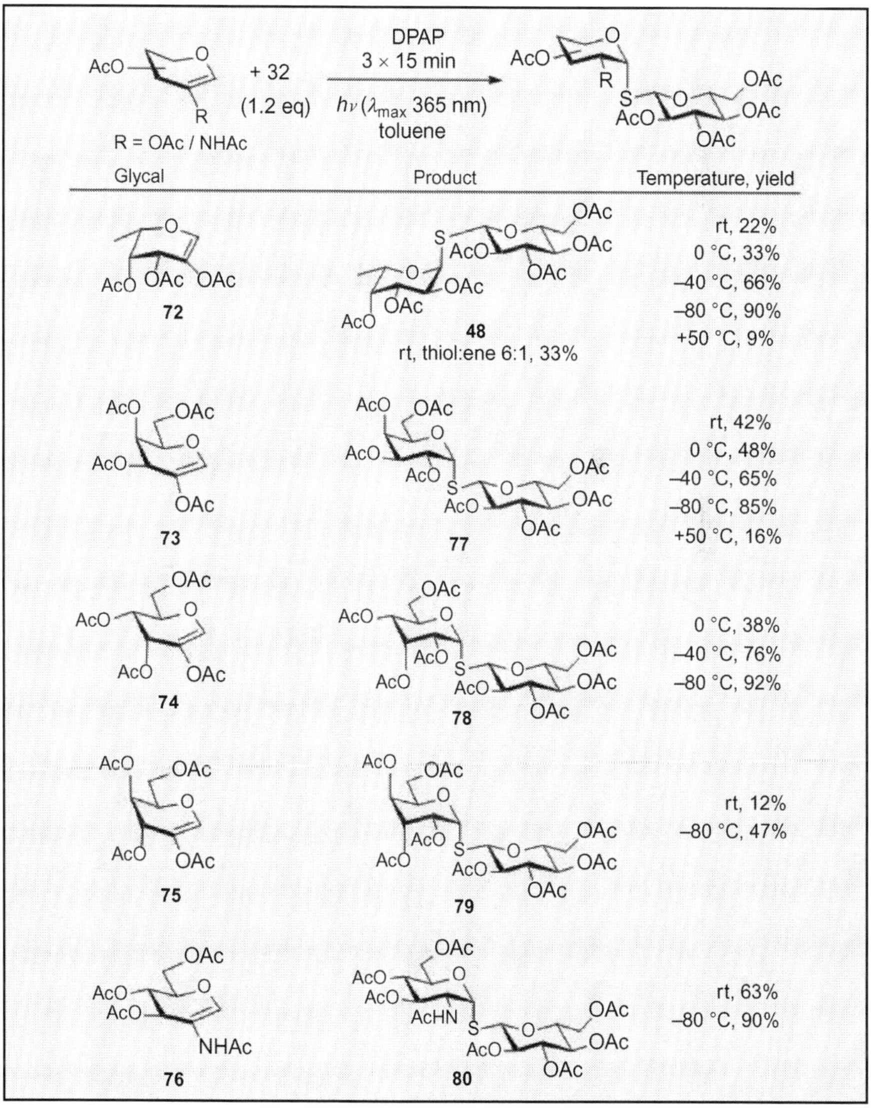

At −80°C, the equilibrium of the rapidly reversible addition of the thiyl radical to alkenes is shifted sufficiently toward products, leading to efficient addition reactions. The heating exerted an opposite effect on the conversions. Running the reactions of **72** or **73** with **32** at +50°C, low conversion of the glycals was observed that resulted in a significant decrease in the yield of products **48** and **77**. The insufficient reactivity of the 2-acetamido derivative **76** at room temperature, which is probably caused by electronic effects, could also be increased by cooling.

The low-temperature thiol-ene coupling was exploited for the expeditious synthesis of a series of 1,2-*cis* α-thio-linked L-fucosides, -D-galactosides, and D-GlcNAc disaccharide mimetics.[43] As it is illustrated with the reactions of 2-acetoxy-L-glycal **72**, the thiol-ene couplings worked efficiently with a very low, 1.2 equiv. thiol excess and in different solvents to provide the α-thiofucosides **82–85** with full stereoselectivity in good to excellent yields (Table 4).

On the basis of the low-temperature photoinitiated thio-click reaction Lázár et al. developed an iterative 1,2-*cis*-thioglycosylation protocol for the synthe-

Table 4 Efficient thiol-ene coupling reactions of 2-acetoxy-L-glycal **72** under Borbás conditions.

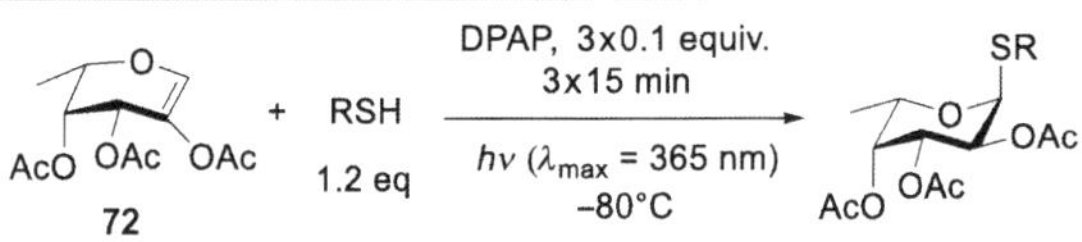

Thiol	Solvent	Product (yield)
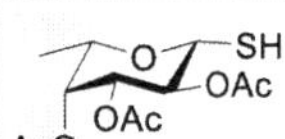 **40**	Toluene	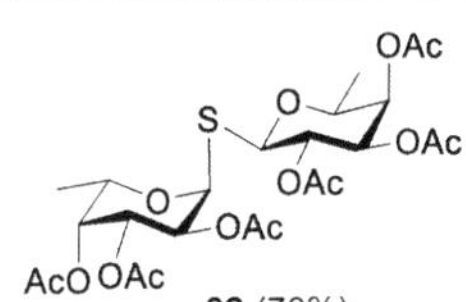 **82** (73%)
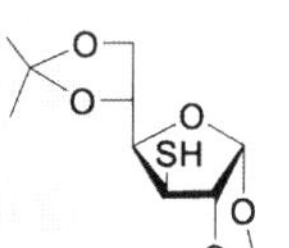 **63**	Toluene	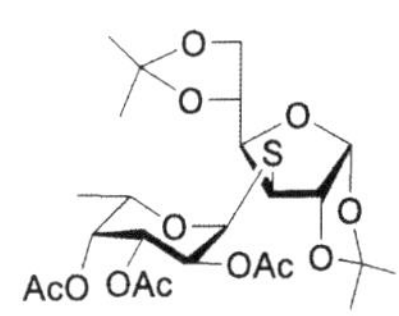 **83** (66%)
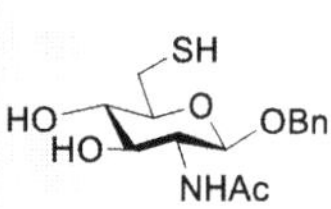 **81**	Toluene:MeOH 1:1	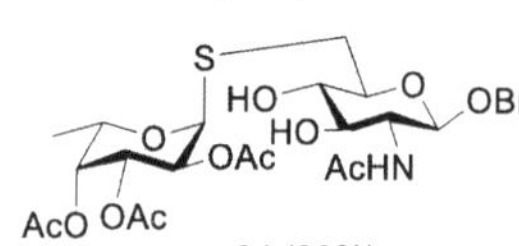 **84** (90%)
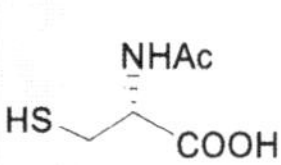 **59**	DMF:toluene:MeOH 1:4:6	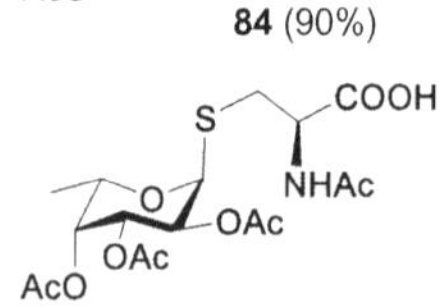 **85**

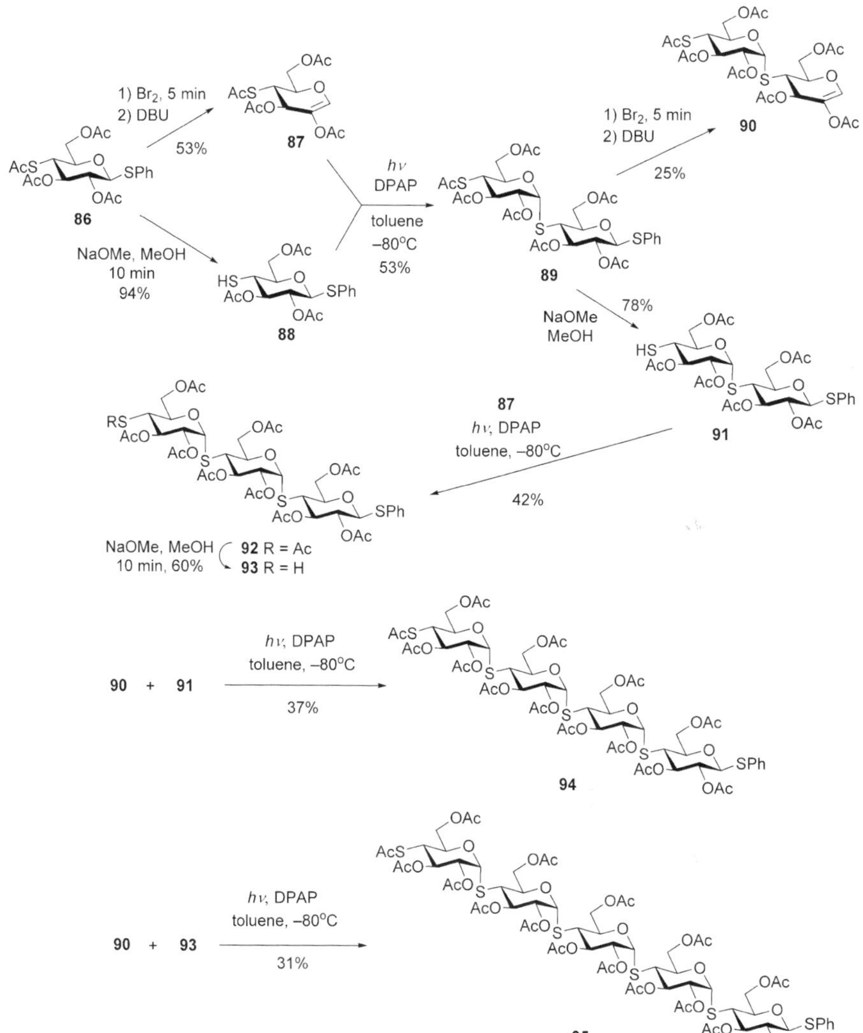

Scheme 8 Thio-click approach to the synthesis of α-**S**-(1→4)-linked maltooligomers developed by Lázár et al.[45]

sis of thio-linked maltooligomers up to pentasaccharide (Scheme 8).[45] The starting 1,4-dithioglucopyranoside **86** was converted to both glycal **87**, via bromination followed by DBU-mediated elimination, and thiol **88**, by selective *S*-deacetylation. In contrast to Zhu's nucleophilic approach to α-*S*-(1→6)-linked glucosyl oligomers (Scheme 3), the 1,2-*cis* α-configured anomeric sulfur atom was introduced in the coupling step with full stereoselectivity to produce trithiomaltose **89** as the only product. Toward higher oligomers **92**, **94**, and **95**, the necessary 2-acetoxyglycals and thiols were obtained directly from the coupling products **89** and **92** without further protecting group manipulations.

The temperature did not affect the conversion significantly, however, decreasing the temperature decreased the disulfide formation, and thus less thiol was needed. This method represents a simple alternative to the known syntheses of similar products.[46, 47]

Pellegrini-Moïse, Chapleur, and coworkers investigated the additions of various thiols to trisubstituted activated furanoid and pyranoid *exo*-glycals under basic and radical conditions.[48] The hydrothiolation reaction took place with full regio- and high stereoselectivity affording various tertiary *S*-glycosides, including adducts of (*E*)- and (*Z*)-*exo*-glycals with aromatic and aliphatic thiols, amino acids, and a 6-thiosugar. Reactions of the D-*gulo*-configured furanoglycosylidene **96**-*E* with unprotected primary thiols (e.g., **97**) under basic conditions using 1,8-diazabicyclo[5.4.0]undec-7-ene (DBU) went to completion in 2–3 h at room temperature with high yields (Scheme 9). The radical-mediated thiol-ene couplings under thermal activation also worked, albeit with a lower efficacy.

Varela and coworkers have described successful approaches to prepare thiooligosaccharides possessing pento- and hexofuranose sugars as constituents, and many of these mimetics have been evaluated as enzyme inhibitors.[22, 49–52] They studied a MoO_2Cl_2-promoted glycosidation route to *S*-linked thiodisaccharides with a furanose unit as nonreducing end (Scheme 10). From 1-*O*-acetyl-2,3,5-tri-*O*-benzoyl-β-D-ribofuranose **99**, the MoO_2Cl_2-catalyzed glycosidation reaction with thiol **100** gave compound **101**, whereas glycosylation was unsuccessful when $SnCl_4$ was employed as the catalyst.[49] The thioglycosylation took place under diasterocontrol of the 2-*O*-benzoyl group leading to selective formation of the 1,2-*trans*-thioglycosidic linkage. The method was extended to arabino- and galactofuranoses to produce thiodisaccharides as potential

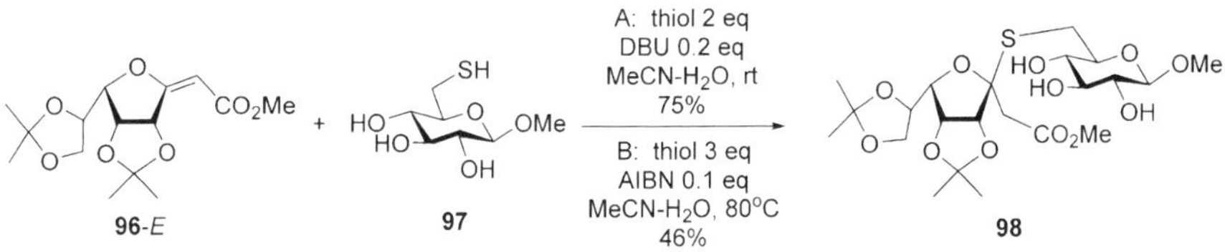

96-*E* 97 98

Scheme 9 Stereoselective synthesis of tertiary *S*-glycoside **98** by base- and radical-mediated additions of thiol **97** to push-pull exoglycal **96**.

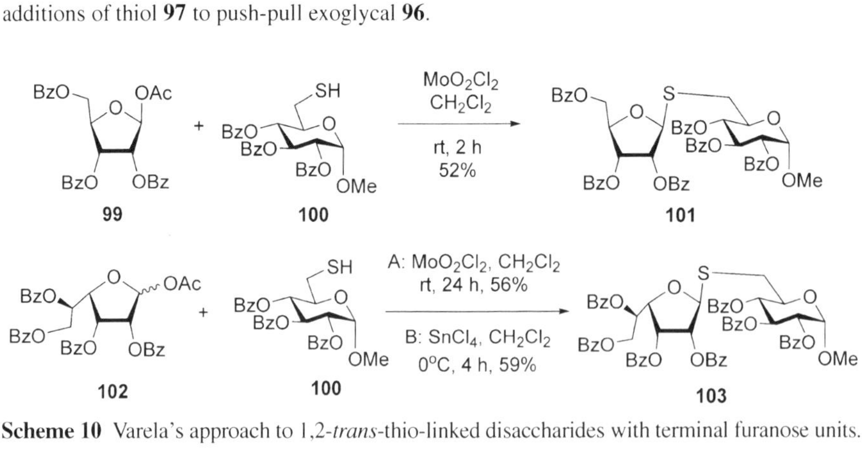

99 100 101

102 100 103

Scheme 10 Varela's approach to 1,2-*trans*-thio-linked disaccharides with terminal furanose units.

enzyme inhibitors. Noteworthy, in these cases $SnCl_4$ also promotes the glycosylations successfully.[50] Reaction of the 1-*O*-acetyl-per-*O*-benzoyl-D-galactofuranose **102** with **100**, catalyzed by either $SnCl_4$ or MoO_2Cl_2, gave thiodisaccharide **103** in a stereoselective manner (Scheme 10).

Varela's group described the synthesis of furanosyl thiodisaccharide analogues of β-D-Gal*f*-(1→5)-D-Gal*f*, a rather common motif in polyfuranosides of pathogenic microorganisms (Scheme 11).[52] To get access to the 6-deoxy-5-thio-α-L-*altro* disaccharide mimetic **109**, the key 5-thiosugar **107** was prepared from the conveniently protected methyl β-D-galactofuranoside **104** via the D-*galacto*-configured epoxide **105**. Reaction of **105** with thiourea in MeOH afforded the 5,6-epithio-α-L-*altro* derivative **106** and its subsequent $LiAlH_4$ reduction gave the expected 6-deoxy-5-thiofuranose **107** with good yield. In the coupling step, the sulfanyl group of **107** was glycosylated using the per-*O*-benzoyl-Gal*f* trichloroacetimidate **108** as glycosyl donor to afford, after a two-step deprotection, glycomimetic **109**. The synthesis of the D-*galacto* analogue of **107** via the corresponding L-*altro*-configured epoxide and its coupling with **108** to produce a 6-deoxy-5-thio-β-D-*galacto* analogue of β-D-Gal*f*-(1→5)-D-Gal*f* was also described.[52]

The Ferrier rearrangement[53–55] using a sugar thiol nucleophile is a viable method of synthesizing disaccharide mimetics, as found by Osborn and coworkers (Scheme 12).[23] They prepared a range of (1→4)- and (1→6)-*S*-linked disaccharides using glycals derived from glucose and galactose, and sulfur-containing coupling partners derived from methyl α-D-glucopyranoside. From the reactions of 3,4,6-tri-*O*-acetyl-D-glycal **110** with acetyl- or benzyl-protected 6-thio-α-D-glucopyranosides **111a** and **111b**, using $LiBF_4$ as Lewis acid, the 1,2-*S*-linked disaccharides **112a** and **112b** were obtained with high yields and a 10:1 ratio of the α and β anomers. Reactions of 2-acetoxyglycal **58** under $LiBF_4$ activation showed low efficacy. When the Lewis acid was exchanged for $BF_3{\cdot}Et_2O$, decreased reaction times, improved yields, and exclusive α-stereoselectivity were observed with both types of glycals using either primary (**111b**)

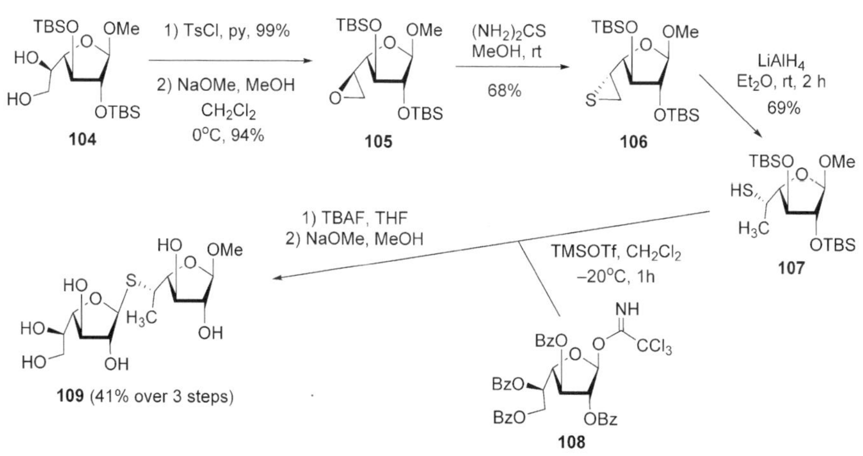

Scheme 11 Synthesis of 6-deoxy-5-thio-L-altrofuranose **107** and its use as an *S*-nucleophile to prepare β-D-galactofuranosyl-(1→5)-*S*-α-L-altrofuranoside **109**.

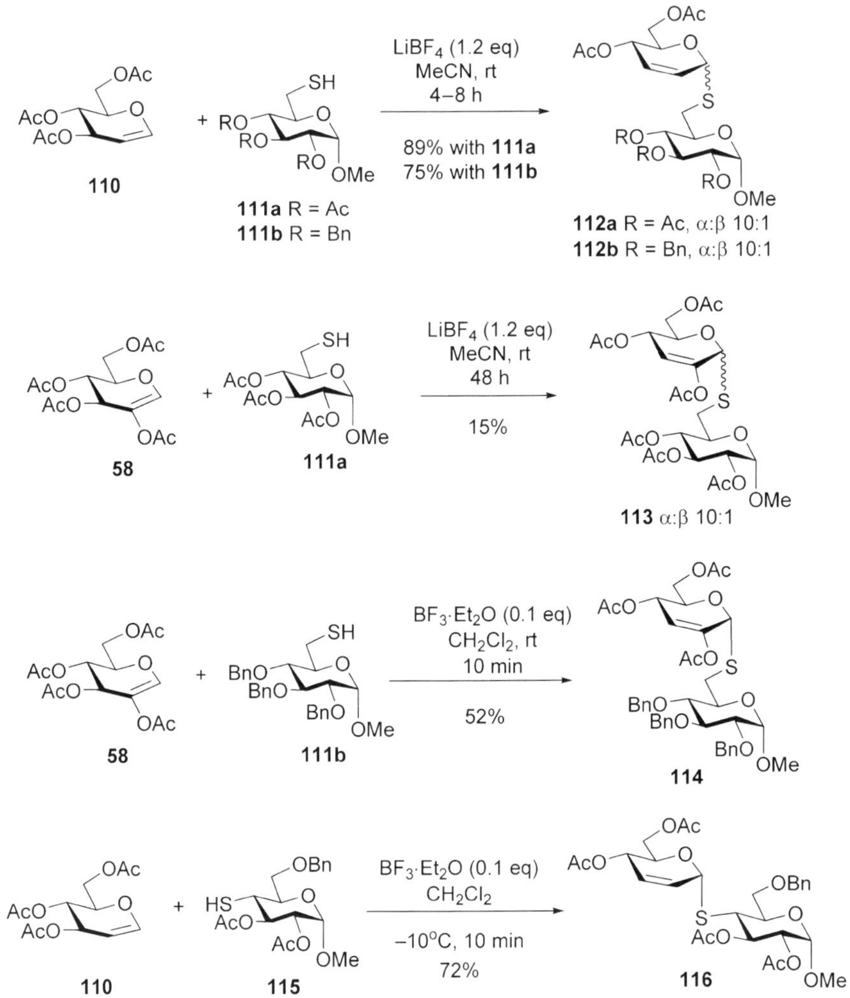

Scheme 12 Osborn's stereoselective synthesis of *S*-linked disaccharide mimetics *via* Ferrier reactions.

or secondary thiol (**115**) coupling partners. Functionalization of the 2,3-double bond within the disaccharides via OsO_4-mediated dihydroxylation was also reported.[23]

2.1.2 Introduction of the thio-linkage to a non-anomeric position

The photoinduced thiol-ene additions of unsubstituted and 1-substituted glycals have been applied to the synthesis of 2-thio-linked saccharide mimetics (Scheme 13). Dondoni and coworkers investigated the hydrothiolation of glycals with 1-thiosugars as a regioselective method for the preparation of (1→2)-*S*-linked disaccharides.[56] The efficiency and the diastereoselectivity of the reaction highly depended on the configuration of the glycals. Hydrothiolations of glycals **110** and **119**, derived from glucose and galactose, respectively, occurred with low levels of diastereoselectivity. For the

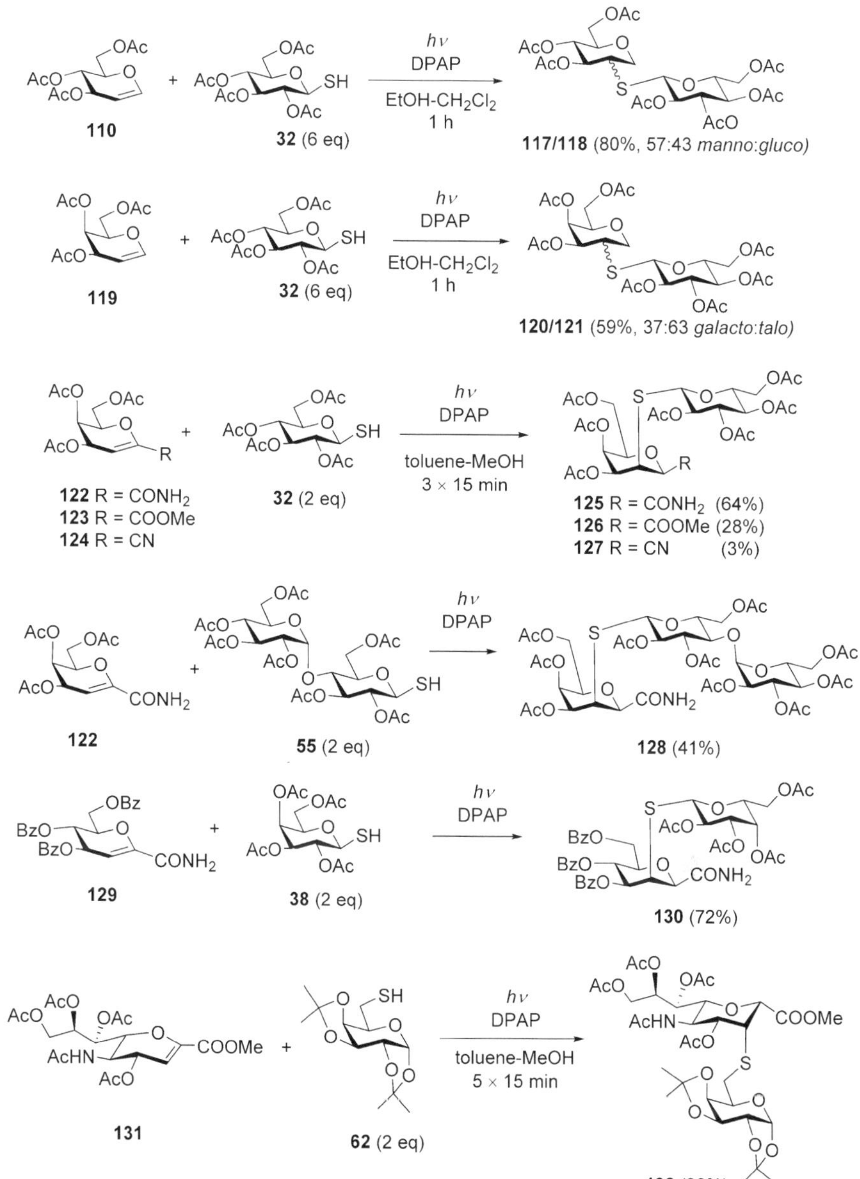

Scheme 13 Introduction of thio-substituents to C2 position of the pyranose ring by photoinitiated additions of thiols to unsubstituted and 1-substituted glycals.

addition of 1-thiosugars to glycals derived from D-allose and D-gulose derivatives, complete stereoselectivity was observed, albeit in low yields. Somsák and coworkers investigated the free-radical hydrothiolation of amide-, methoxycarbonyl-, and cyano-substituted glycal derivatives (e.g., **122–124** and **129**) with a range of sugar thiols.[57] The coupling reactions took place with full regio- and stereoselectivity due to the preferential axial attack of the thiyl radical to the less-substituted C2 position of the pyranose ring and the preferred axial attack on the C1 radical in the hydrogen abstraction step. The glycal configuration (**122** vs **129**) had no effect on the stereochemical outcome of the reaction. However, the efficiency of the additions to **122–124** was highly variable depending on the different electron-withdrawing capabilities of the 1-C substituents (CN>COOMe>$CONH_2$).

We studied the hydrothiolation of a 2,3-unsaturated *N*-acetylneuraminic acid derivative **131** with 6-thiogalactose **62** as a method for the synthesis of an *S*-linked mimetic of the natural α-(2→6)-linked sialyl-galactoside structure. The reaction provided **132** as the main product bearing an axial thioether linkage at C2 position of the Neu5Ac ring. The low efficiency of the addition can be explained by the low electron density of the double bond owing to the electron-withdrawing carboxyl moiety on Neu5Ac-2-ene.[42]

Dondoni's group have also studied the photoinduced addition of 1-thiosugars to 4-exomethylene pentofuranoside and 5-exomethylene hexopyranoside derivatives for the introduction of the *S*-interglycosidic linkage to the primary position of the acceptor substrates.[58] This method, yielding the Gly-*S*-CH_2-Gly structure, was further explored by others for the synthesis of C-S-bridged glycomimetics, which will be discussed later in this chapter.

Photoinduced thiol-ene couplings of 2,3-unsaturated glycosides represent a further strategy for introduction of a thio-linkage to C2-position of glycopyranosides.[41, 59] Studying the reaction of **133** with a range of glycosylthiols, we have found that the thiyl radical added axially to position C2 of the sugar ring in all cases (Scheme 14). With thiols **32** and **39**, the corresponding 3-deoxy-2-thiodisaccharides **134** and **135** were formed rapidly and in high yields. The synthetic utility of the thiol-ene coupling of 2,3-unsaturated glycosides was demonstrated by the simple and efficient synthesis of **138**, an α-(1→2)-thio-linked mannobioside mimetic bearing an azide functionalized aglycon, which was converted to multivalent mannoclusters for lectin-binding studies.[59, 60]

Messaoudi and coworkers reported an efficient synthetic strategy to (1→2)-*S*-linked di- and oligosaccharide mimetics by Pd-catalyzed Buchwald-Hartwig-Migita cross-coupling of 1-thiosugars with 2-iodoglycals (Scheme 15).[61] After the appropriate solvent, reaction temperature, and reaction time have been carefully selected, the coupling reaction of β-1-thioglucose **32** with 2-iodoglycal **139** using a palladium-G3-XantPhos palladacycle precatalyst furnished the desired β-(1→2)-*S*-linked disaccharide **141** in 86% yield (Scheme 15). The cross-coupling reaction was successfully extended to 2-iodoglycal **140** and various 1-thiols including 1-thio-α-D-glucose **11** and 1-thiomaltotriose **144** to give the corresponding (1→2)-*S*-linked di- and oligosaccharides in 55%–80% yields.

Witczak and coworkers developed a conjugated addition approach for stereoselective synthesis of thiodisaccharides by using levoglucosenone or isolevoglucosenone as reactive Michael addition acceptors.[20, 62] Thioglycosylation of levoglucosenone **146** with 1-thiogalactose **38** proceeded with complete stereoselectivity, affording the 3-deoxy-4-thio-linked **147** in high yield (Scheme 16). Stereoselective reduction of its C-2 keto function with L-selectride, and subsequent acetylation and acetolytic

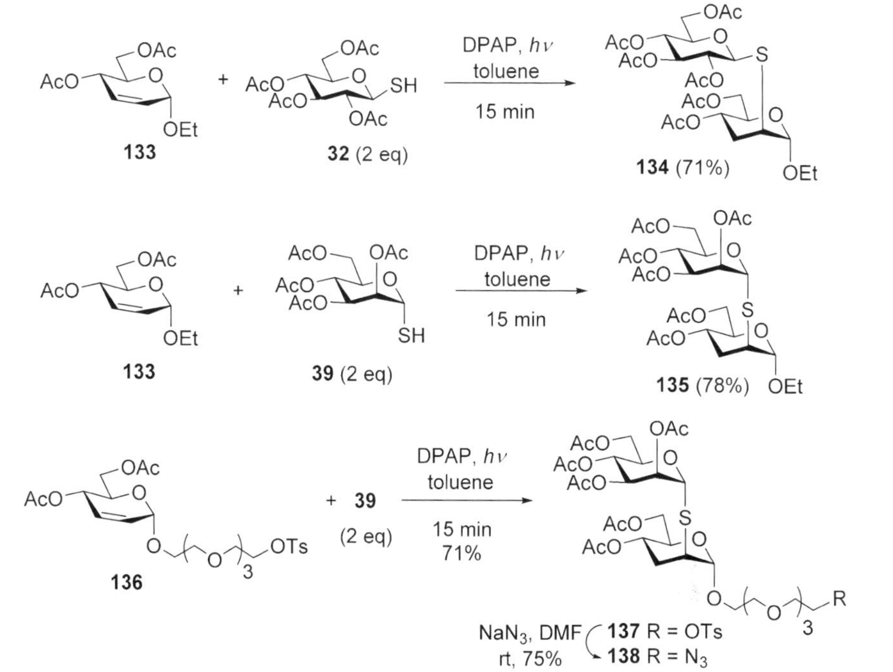

Scheme 14 Stereoselective synthesis of 1→2-linked thiodisaccharides including spacer-armed α-(1→2)-thio-linked mannobioside **138**.

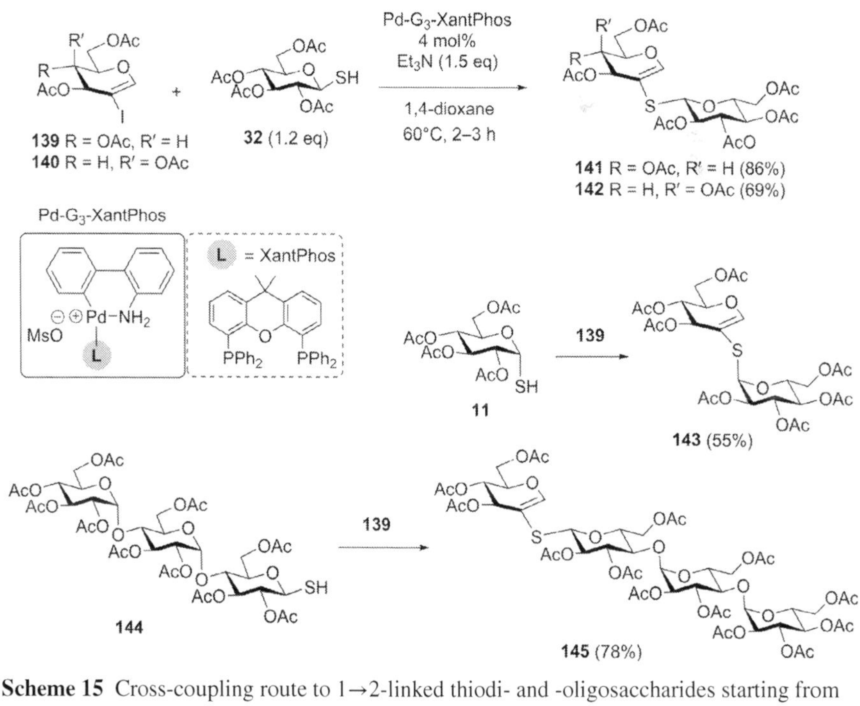

Scheme 15 Cross-coupling route to 1→2-linked thiodi- and -oligosaccharides starting from 2-iodoglycals.

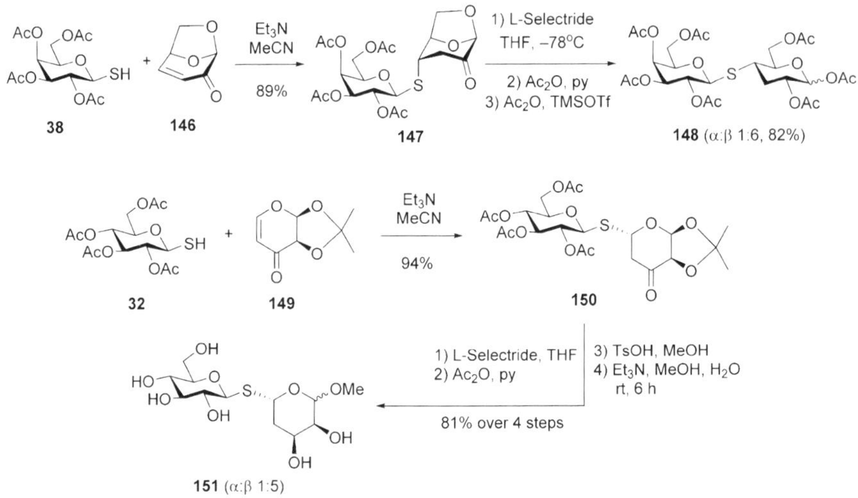

Scheme 16 Witczak's method for stereoselective synthesis of 3-deoxy-(1→4)-4- and 4-deoxy-(1→5)-5-thiodisaccharides by conjugate addition of 1-thiosugars to reactive enones.

cleavage of the 1,6-anhydro ring gave thiodisaccharide **148** as an anomeric mixture. The base-catalyzed addition of thiol **32** to enone **149** and subsequent reduction of the C-3 keto function of the obtained adduct **150** proceeded also in a stereoselective manner to produce, after deprotection, the corresponding 4-deoxy-5-thio-linked disaccharide **151**. The Michael adduct glycomimetics, in both protected and unprotected forms, were used in antibacterial and antineoplastic studies.[63, 64]

Varela and coworkers extensively studied the base-catalyzed Michael addition of thiols to the α,β-unsaturated carbonyl system of dihydropyran-2-ones, which are readily available from 2-acetoxyglycals, for synthesizing disaccharide mimetics with thioglycosidic or thioether linkages.[65–68] Base-catalyzed addition of 6-thioglucose derivative **100** to hex-3-enopyranosid-2-ulose **152**, obtained by allylic rearrangement of 2-acetoxy-tri-*O*-acetyl-D-glycal obtained from the protected galactose precursor, afforded the 4,6′-thioether-linked disaccharides **153** and **154** in 93% combined yield (Scheme 17). The observed high facial selectivity in favor of the 4-thio-α-D-*threo* isomer in **153** was explained by the directing effect of the axially oriented anomeric substituent. Reduction of the carbonyl group of the major product **153** with $NaBH_4$ in methanol gave disaccharides with the α-D-*xylo* (**155**) and the α-D-*lyxo* (**156**) epimers, in a 2:1 ratio.[66] Varela's group explored the method for the synthesis of 3-deoxy-4-thiodisaccharides with a pentopyranose unit as reducing or nonreducing end.[67] Addition of 1-thioxylose **157** to enone **152** gave thiodisaccharide **158** bearing the pentopyranosyl ring at the nonreducing end. Conducting the reaction at −18°C, complete stereoselectivity was observed providing **158** as the only product. It was found that the stereochemical outcome during the reduction of ulose **158** was highly variable depending on the reaction temperature, reducing agent, and solvent. Using $NaBH_4$ in a MeOH-THF mixture at rt afforded disaccharide **159**, bearing the α-D-*xylo* configured moiety in 70% yield as the major product, while reduction with an excess of $NaBH_4$

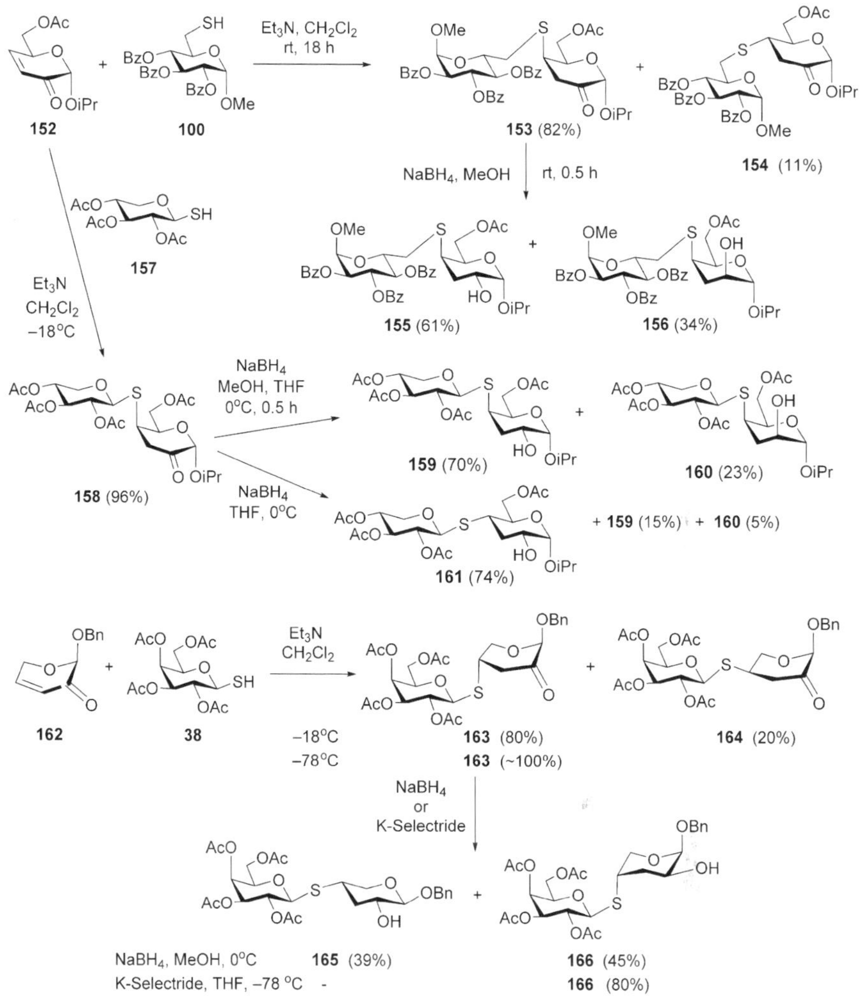

Scheme 17 Varela's stereoselective synthesis of 3-deoxy-4-glycosylthio-hexo- and pentopyranosides by conjugate addition of sugar thiols to dihydropyran-2-ones.

in the nonprotic solvent THF led to isomerization of C-4 producing product **161** in 74% yield. Thioglycosylation of pentose-derived ulose **162**, prepared from benzyl β-D-arabinopyranoside, resulted in, at −18°C, the formation of thiodisaccharides **163** and **164** in a 4:1 ratio. The stereoselectivity of the Michael addition could be increased by lowering the temperature, and at −78°C, **163** formed as the only product in quantitative yield. In this case, reduction of thioulose **163** with K-selectride in THF at −78°C afforded **166**, that contains the β-D-*threo* configured residue, as the only product.

On the basis of the earlier results, bivalent thiolactose derivatives were synthesized by Uhrig and coworkers for enzyme inhibitory and lectin-binding studies. The Michael acceptor **167** bearing a propargyl aglycon was prepared starting from

2-acetoxy-3,4,6-tri-*O*-acetyl-D-glycal, obtained from the protected galactose precursor, using Ferrier rearrangement. Michael addition of 2,3,4,6-tetra-*O*-acetyl-1-thio-β-D-galactopyranose **38** to enone **167** at −70°C yielded the 4-thio-α-D-*threo* configured **168** as the only product (Scheme 18). The key step of the synthesis was the isomerization-reduction of ulose **168** using excess $NaBH_4$ in THF to produce the 4-thiolactose mimetic **169**. This compound and its analogues were then subjected to azide-alkyne cycloaddition reactions to give glycoclusters such as **171**.[68]

Varela's group extended the Michael addition approach to the synthesis of 2-acetamido-2,3-dideoxy-(1→4)-thiodisaccharides by addition of thiogalactose **38** to hexose- or pentose-derived acetyl oximes.[69] The α,β-unsaturated ketoximes **172*E*** and **172*Z*** were obtained from the corresponding enone in two steps (Scheme 19). Addition of 1-thiogalactose **38** to the ketoximes took place with complete facial selectivity directed by the β-configured aglycon, and, under optimized conditions, the oxime configuration could also be controlled to produce either **173*E*** or **173*Z*** in a diastereoselective manner. The role of 1,4-dithiothreitol (DTT) in the conjugate addition step was to prevent the disulfide formation from **38**. Stereoselective reduction of the acetoxime group was achieved with $LiAlH_4$ in THF and the obtained thiodisaccharide, after deacetylation, showed promising β-galactosidase inhibitory activity.

Varela and coworkers described a direct route of synthesizing sulfur-linked analogues of numerous naturally occurring and nonoccurring disaccharides by ring-opening reactions of sugar oxiranes and thiiranes (Scheme 20).[21, 22, 70] Toward synthesis of the sugar oxiranes the starting enone **152** was converted into the 2-OH derivative **175** using Luche-reduction.[71] This free hydroxy group directed the attack of *m*-CPBA from the α-face of the compound in the next step, giving the *allo*-configured epoxide **176** as major product, which was then coupled with 1-thioglucose **32** and 1-thiogalactose **38** to give the regioisomeric (1→3)- and (1→4)-linked thiodisaccharides **178–181** with D-*gulo* and D-*gluco* configuration at the reducing end. The disadvantage of this method was the lack of regioselectivity in the coupling step.

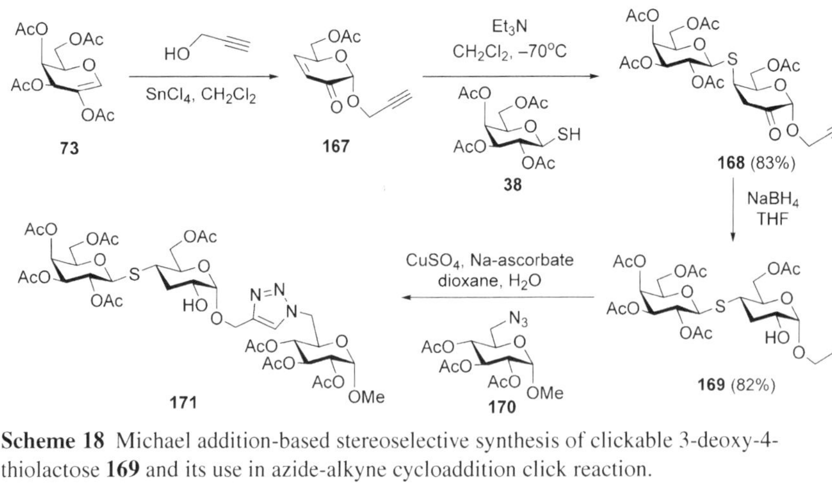

Scheme 18 Michael addition-based stereoselective synthesis of clickable 3-deoxy-4-thiolactose **169** and its use in azide-alkyne cycloaddition click reaction.

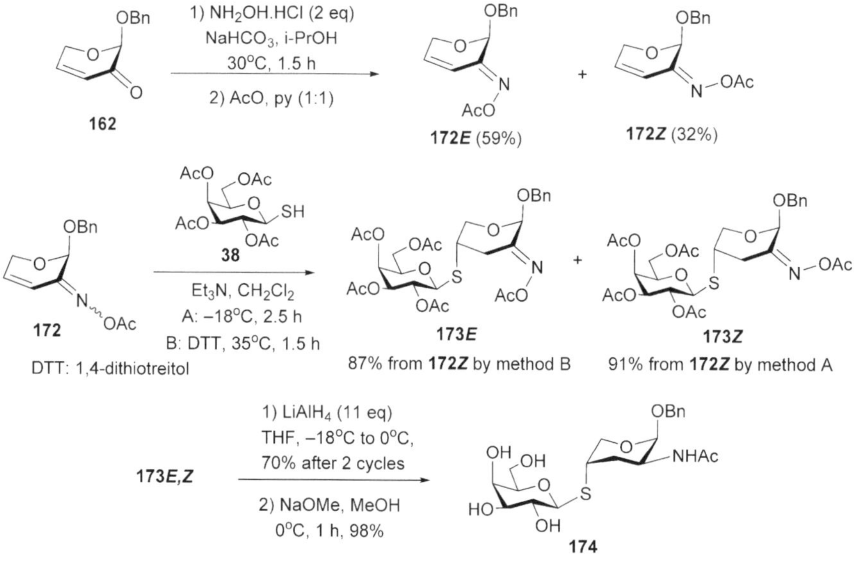

Scheme 19 Synthesis of 2-acetamido-2,3-dideoxy-4-thiodisaccharide **174** by Michael addition of 1-thiogalactose to (*E*)- and (*Z*)-acetyloximes.

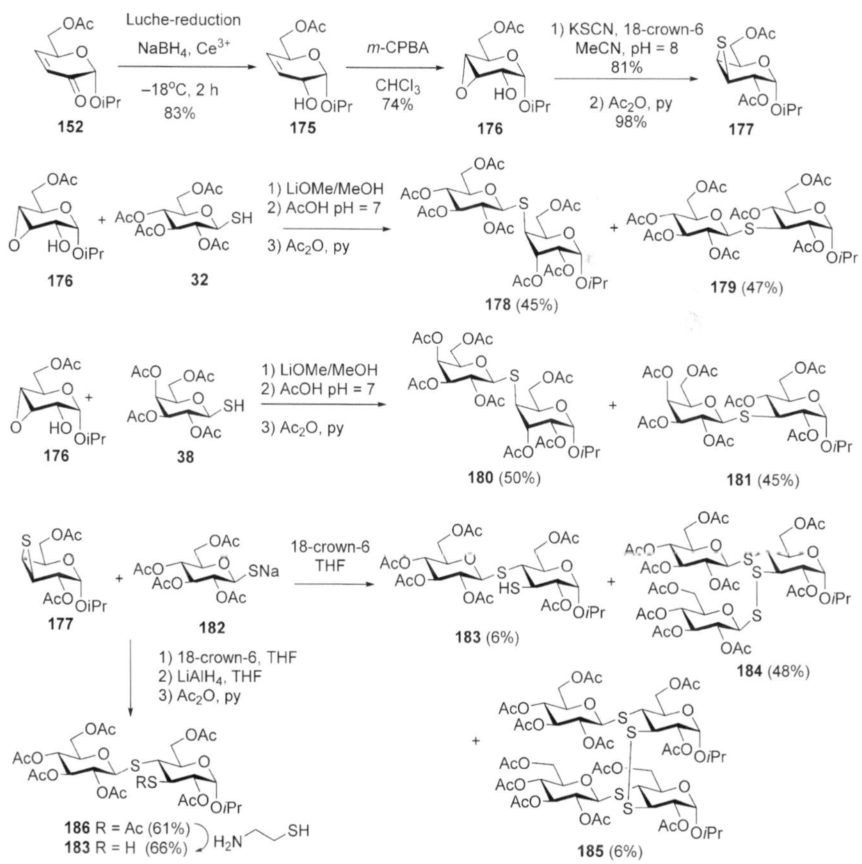

Scheme 20 Synthesis of (1→3)- and (1→4)-*S*-linked disaccharides by ring-opening reactions of sugar oxiranes and thiiranes.

To extend the scope of the ring-opening reactions, epoxide **176** and its protected derivatives were used to synthesize thiiranes such as **177**. Nucleophilic opening of the thiirane ring of **177** with sodium thiolate **182**, prepared in situ from the corresponding thiol with NaH, afforded the expected dithiodisaccharide **183** along with its disulfides **184** and **185** formed by oxidative coupling of thiols that are present in the reaction mixture. The yield of the desired **183** was successfully increased by $LiAlH_4$ reduction of the crude mixture of reaction of **177** with **182** followed by acetylation and selective *S*-deacetylation of the obtained **186** by cysteamine.[22] 3,4-Dithiolactose was prepared in an analogous way by ring-opening of **177** with the *galacto* analogue of **182**.[70]

Glycosylation of the sulfanyl group of **183** with furanosyl and pyranosyl trichloroacetimidates (e.g., with **187**) provided the corresponding branched dithiotrisaccharides such as **188**, albeit with relatively low yields (Scheme 21).[22]

Varela and coworkers studied the oxidation of thiodisaccharides to the corresponding sulfoxides or sulfones because of the potential biological activities of such molecules. The sulfur atom in compound **179** could be selectively oxidized into sulfoxide using excess *m*-CPBA as oxidizing agent, and the obtained diastereoisomers could be separated by column chromatography (Scheme 22).[72]

Cumpstey et al. studied the synthesis of thioether-linked pseudodisaccharides using S_N2 displacement of carbohydrate sulfonates by carbohydrate thiol nucleophiles (Scheme 23). First, carbohydrate sulfonates were prepared (e.g., **192** and **195**), next carbohydrate thioesters (**193** and **196**) were synthesized from these sulfonates by S_N2 substitution with thioacetate. Using the sulfonates and the in situ-generated thiolates as coupling partners, nonsymmetric (**197**) and symmetric thioether pseudodisaccharides (**198**) were obtained in a stereoselective manner, with inversion of the configuration at the sulfonate-bearing carbon. The synthesis of the symmetric thioether-linked products could also be accomplished via dimerization of the sulfonates using a divalent sulfur anion as nucleophile (**195**→**198**). Deprotection of the pseudodisaccharides and oxidation of the bridging sulfur to sulfoxide or sulfon with *m*-CPBA were also studied.[73]

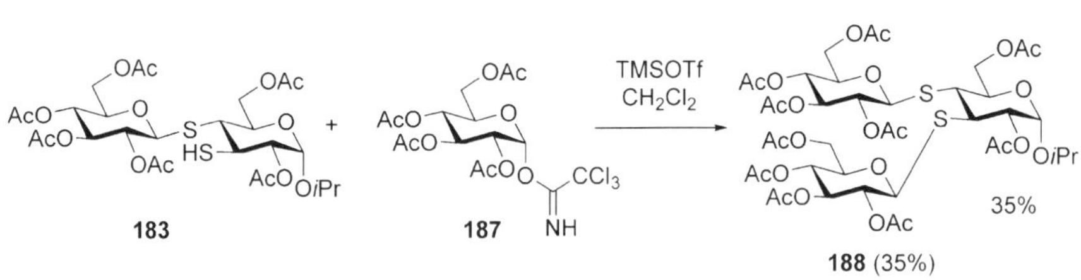

Scheme 21 Synthesis of branched dithiotrisaccharide **188**.

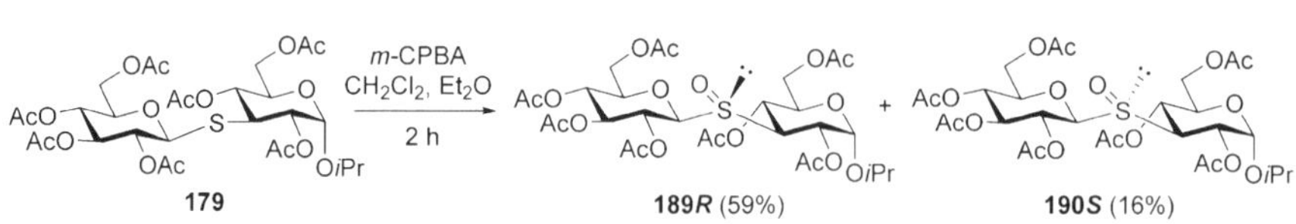

Scheme 22 Oxidation of a thiodisaccharide to diastereoisomeric sulfoxides.

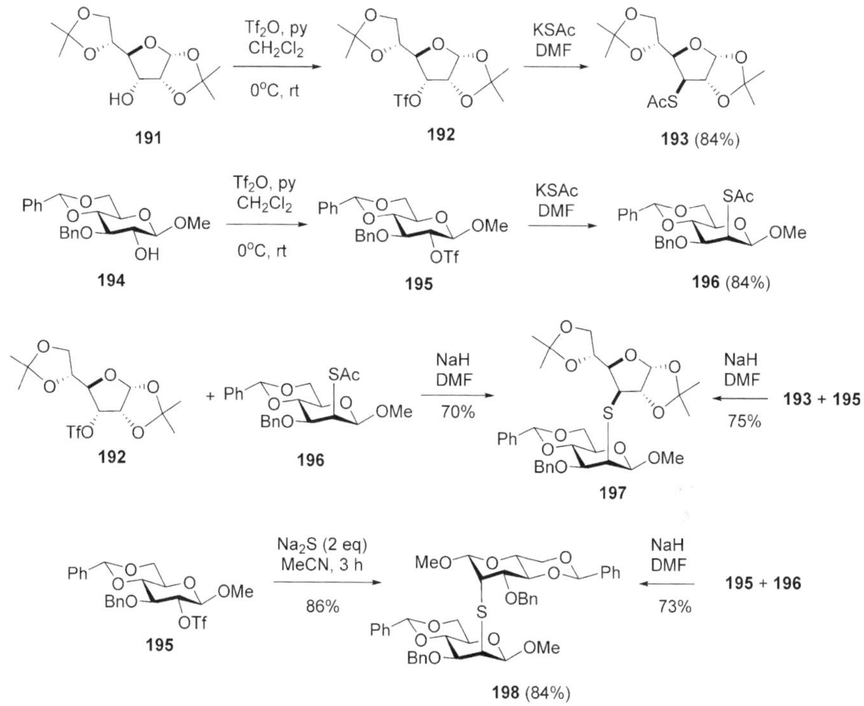

Scheme 23 Cumpstey's S_N2 approach for the synthesis of tail-to-tail thiodisaccharides.[73]

2.2 Selenoglycosides

The synthesis of selenium-linked glycoconjugates and disaccharides has received increasing attention over the last decade, because these derivatives can be used in protein-carbohydrate interactions studies as hydrolytically stable glycomimetics.[74, 75]

One early synthesis of selenodisaccharides, reported by Czernecki's group,[76] was based on in situ selenolate anion formation by the reduction of symmetric diglycosyl diselenide and subsequent reactions of this selenolate with tosylated glycosyl units. Another pioneering method was described by Pinto and coworkers in which a selenolglycosyl acceptor was glycosylated with a glycosyl trichloroacetimidate donor in the presence of a catalytic amount of TMSOTf, producing a (1→4)-linked selenodiglucoside.[77]

The main challenge during the preparation of selenodisaccharides is the easy and fast oxidation of sugar selenols into symmetric diselenides. Therefore, recent methods are based on in situ formation of a nucleophilic selenolate in the presence of an electrophilic coupling partner. On the other hand, anomeric selenolates, similarly to anomeric thiolates, show high configurational stability.[37, 78, 79] Consequently, introduction of the selenium into the anomeric position before the coupling step is a useful tactic for the stereoselective synthesis of glycosidically linked selenodisaccharides.

Ando, Ishihara, and coworkers developed a stereoselective selenoglycosylation approach using either α- or β-4-methylbenzoyl selenoglycosides as the glycosyl donors (Scheme 24).[78, 79] The 1-(*p*-methylbenzoyl)seleno-β-D-gluco- (**202β**) and galacto- (**203β**) pyranosides were prepared with complete stereoselectivity by reacting α-glycosyl bromides (**199**, **200**) with potassium *p*-methylselenobenzoate **201** under phase transfer conditions.[78] Interestingly, analogous reactions of β-glycosyl chlorides with **201** in the presence of Bu_4NHSO_4 failed to give the desired α-selenoglycosides. The 18-crown-6-mediated reactions of **204** and **205** yielded anomeric mixtures from which the α-configured selenoglycoses **202α** and **203α** were isolated, respectively, by silica gel chromatography.[79] In the key step, the α- or β-configured acyl selenoglycose was activated by the reaction with Cs_2CO_3 and a secondary amine to produce an anomeric selenolate anion, which reacted in situ with *O*-triflyl sugars to yield selenoglycosides (e.g., **207**, **208** and **210**) while retaining the anomeric configuration. The reaction occurred by an S_N2 mechanism leading to inversion of configuration on the *O*-triflyl sugars. In the case of the 6-*O*-triflylgalactose **206**, its use in high excess was critical to suppress the formation of the corresponding heterodiselenide sugar as an inseparable byproduct.[78]

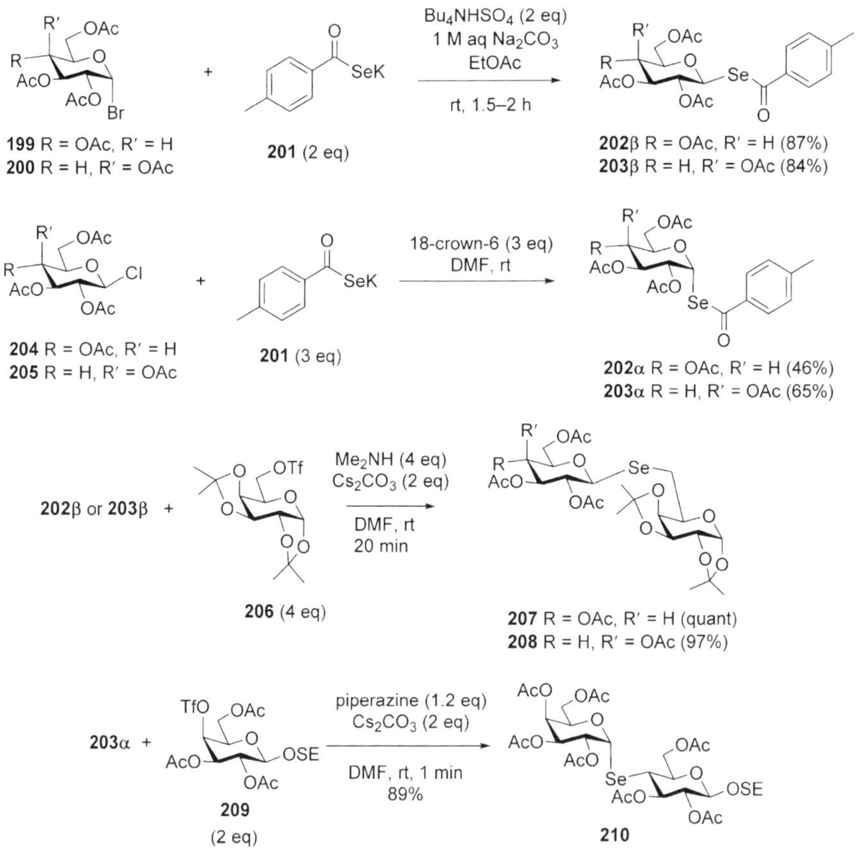

Scheme 24 Nucleophilic selenoglycosylation of primary and secondary triflates with α- and β-selenolate anions generated in situ from *p*-methylbenzoyl selenoglycoses [SE: 2-(trimethylsilyl)ethyl].

Ando's group reported the synthesis of selenolactose **211** and related compounds by using *p*-methylselenobenzoic anhydride, instead of potassium *p*-methylselenobenzoate, as a novel selenating reagent for introduction of selenium into mono- or disaccharides.[74] The same group developed a facile route to selenoglycosides by transacetalization reaction between a selenoacetal and a glycosyl imidate. This reaction was utilized in the synthesis of 2-(trimethylsilyl)ethyl selenogalactoside **213**, which, upon treatment with TBAF in the presence of 4-*O*-triflate **209** as the electrophile, was transformed into selenolactose **211**.[80] After deprotection, the obtained Se-lactoses such as **212** were studied as ligands for human galectin-9 N-terminal carbohydrate recognition domain (Scheme 25).[74]

Szilágyi and coworkers reported that (1→2)-*trans* β-selenium-linked heterodisaccharides could efficiently be synthesized under mild conditions via reactions of glycosyl-isoselenuronium salts with monosaccharide derivatives bearing primary or secondary triflate groups (Scheme 26). Salts **215** and **216** were prepared from acetobromo and chloro sugars with commercially available odorless selenourea in high

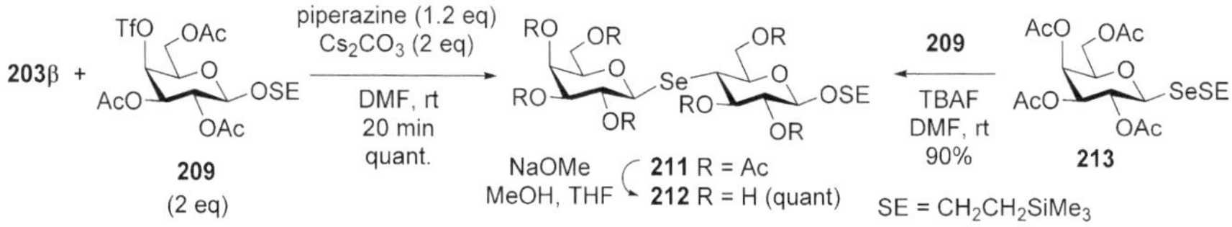

Scheme 25 Ando's different approaches for the synthesis of selenolactose.

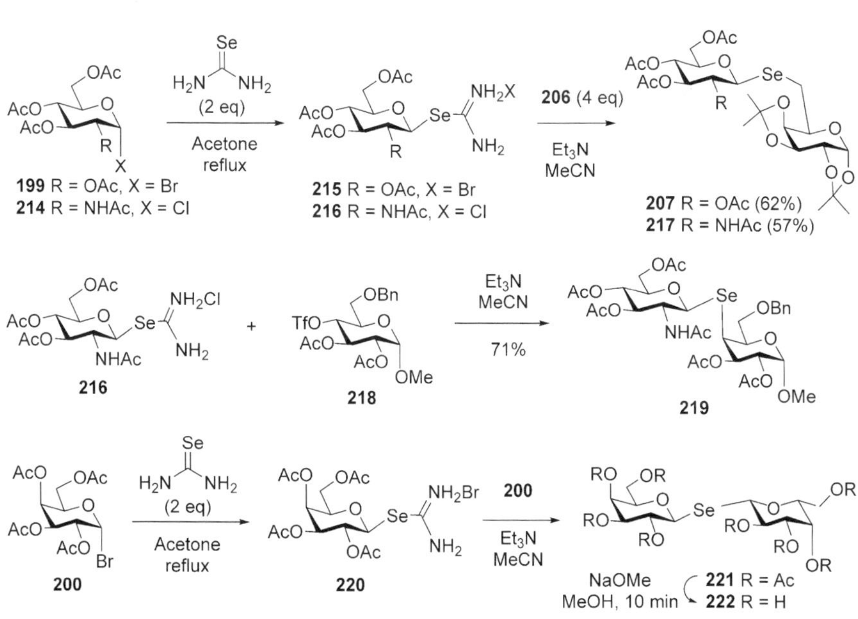

Scheme 26 Facile route to (1→2)-*trans* β-selenoglycosides using isoselenuronium salts developed by Szilágyi's group.

yields. Reactions of these glycosyl-isoselenuronium salts with various electrophiles, including glycosyl 6- or -4-*O*-triflates **206** and **218**, in the presence of Et_3N yielded the corresponding β-selenoglycosides in good to high yields, while retaining the anomeric configuration.[81]

Symmetric selenodigalactoside **221** was synthesized by this method and, after deprotection, **222** was tested as inhibitor in human galectin-binding assays.[75]

Fournière and Cumpstey reported the synthesis of selenoether-linked symmetric and unsymmetric disaccharides of furanoses and pyranoses.[82] Their method involved introduction of selenium by displacement of a carbohydrate triflate (**192**) or mesylate (**226**) by a selenobenzoate anion nucleophile, conversion of the obtained selenoesters into diselenides by methanolysis and aerial oxidation, reduction of the diselenides (**224**, **228**) to reveal a selenol, and in situ displacement of a second carbohydrate sulfonate in an S_N2 reaction to give selenoethers such as **225** and **229** (Scheme 27). The selenoether formation worked well with primary and secondary selenol coupling partners using 1.2–1.3 equiv. of triflate with respect to selenolate. The elaborated route was successfully applied for straightforward synthesis of **233** without protection of the glucose-building unit **230**. The symmetric selenoether-bridged diallofuranose **225**, after deprotection, was used in lectin-binding studies.[73]

Lüdtke and coworkers elaborated a straightforward route to selenoether-linked disaccharides coupled through the primary positions (Scheme 28). The key transformation was the one-step synthesis of symmetric diglycose diselenides (e.g., **235** and **240**) by the reaction of sugar primary tosylates with lithium diselenide, prepared by

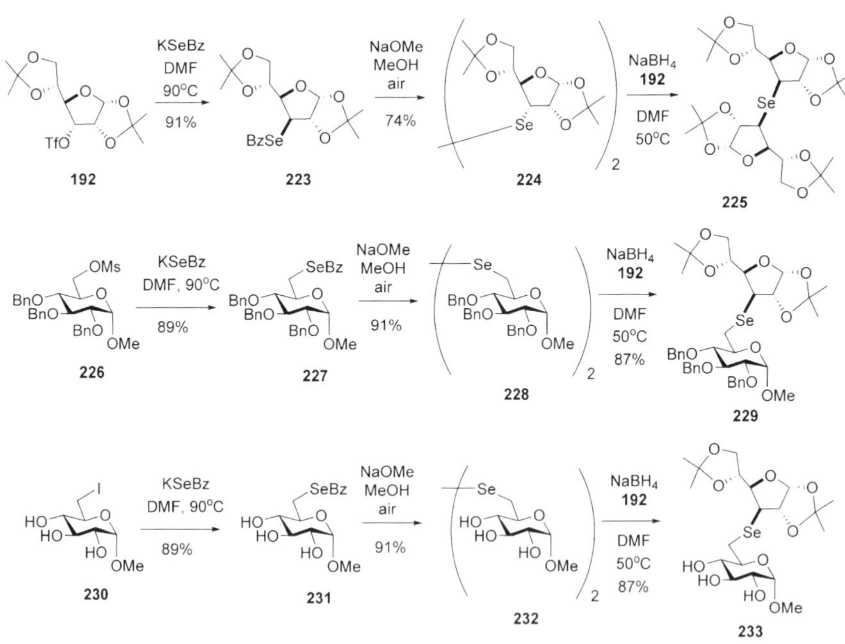

Scheme 27 Syntheses of (3→3)- and (3→6)-linked selenoether pseudodisaccharides by in situ formation of the selenol coupling partners from symmetric diselenides.

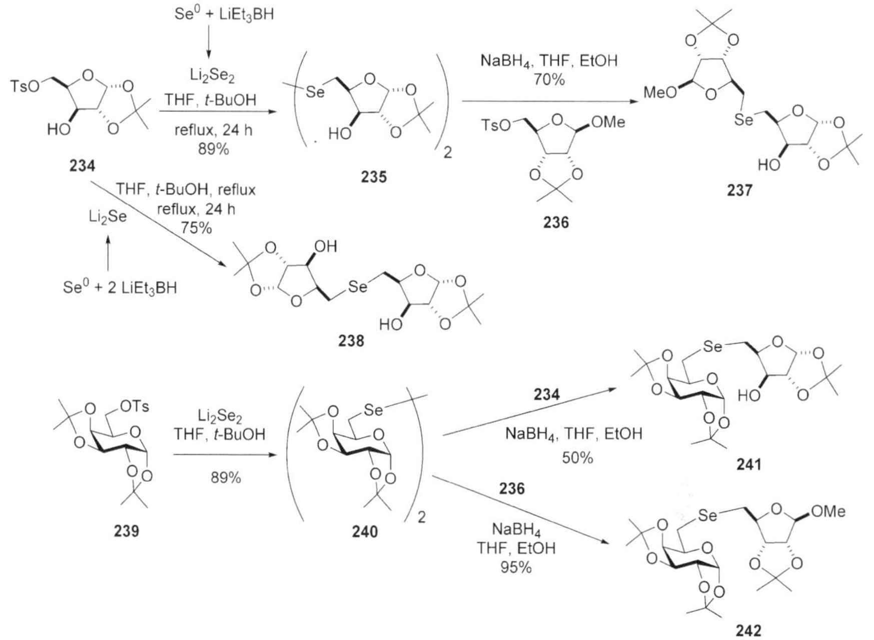

Scheme 28 Straightforward synthesis of selenoether pseudodisaccharides coupled through the primary positions.

the reaction of elemental selenium with $LiEt_3BH$. Reduction of the diselenide bond with $NaBH_4$ released the selenium nucleophiles, which were reacted in situ with a sugar tosylate to produce the nonglycosidic selenium-linked heterodisaccharides **237**, **241**, and **242**.[83] It was also demonstrated that changing the stoichiometry between the elemental selenium and $LiEt_3BH$, lithium selenide was generated and its reaction with 5-*O*-tosylxylofuranose **234** directly furnished the symmetric selenoether disaccharide **238** in 73% yield.[84]

2.3 *N-Glycosides and neoglycosides*

N-Glycosides in which the anomeric oxygen is replaced by nitrogen are essential constituents of valuable sugar-related derivatives including *N*-glycoproteins and nucleic acids. *N*-Glycosyl mimics of glycosides[12] including aminooxylated carbohydrates[85] are of great pharmaceutical interest, for example, as anti-diabetic or anti-cancer agents.[86, 87] Importantly, like sugars, *N*-glycosyl compounds exhibit mutarotation, which makes the stereoselective *N*-glycosylation difficult.

Surprisingly, synthetic methods for the creation of *N*-glycosidic bonds between two monosaccharide units are scarce. More efforts have been made to prepare pseudodisaccharide derivatives of valienamine, a highly functionalized cyclohexene that is found in nature as a substructure of acarbose or the validamycin family of antibiotics.[88] Cumpstey et al. synthesized various valienamine analogue pseudodisaccharides, from which some proved to exhibit inhibitory activity against α-glucosidase II (Scheme 29). From the protected valienol C-1 bromide **243** as carbaglycosyl donor and protected 3-aminoglucoside **244**, using Hünig's base and heating, the desired pseudodisaccharide was obtained with moderate yield but with good stereoselectivity (α:β=7:1).[89] Another method described by Cumpstey et al. utilized valienol **246**[90] or analogues as electrophile and methyl 6-amino-2,3,4-tri-*O*-benzyl-6-deoxy-6-*N*-nosyl-α-D-glucopyranoside **247** as activated nucleophile. The α-configured **246** reacted readily with **247** under Mitsunobu conditions in THF to afford the β-linked pseudodisaccharide **248** with a good 64% yield. Interestingly, the β-analogue of **246** failed to give the corresponding α-linked pseudodisaccharide.[91]

Inspired by Shing Pd-catalyzed allylic amination[92, 93] Cumpstey et al. synthesized trichloroacetimidate derivatives as a new type of leaving group for this reaction and used C-6 amine **250** as the nucleophile. They observed that under Shing's conditions, using bis(dibenzylideneacetone)palladium(0) and trimethylolpropane phosphite (TMPP), the α-imidate gave no clean reaction; however, under the same conditions, using the β-imidate **249**, the β-linked pseudodisaccharide **251** was obtained with high yield and complete stereoselectivity.[91]

Synthesis of tail-to-tail amine-linked neodisaccharides has been investigated by various approaches including reductive amination, aziridine opening, epoxide opening, and Mitsunobu reaction.[94–97] Cumpstey and coworkers developed a mild method to connect two monosaccharides at nonanomeric position by a nitrogen

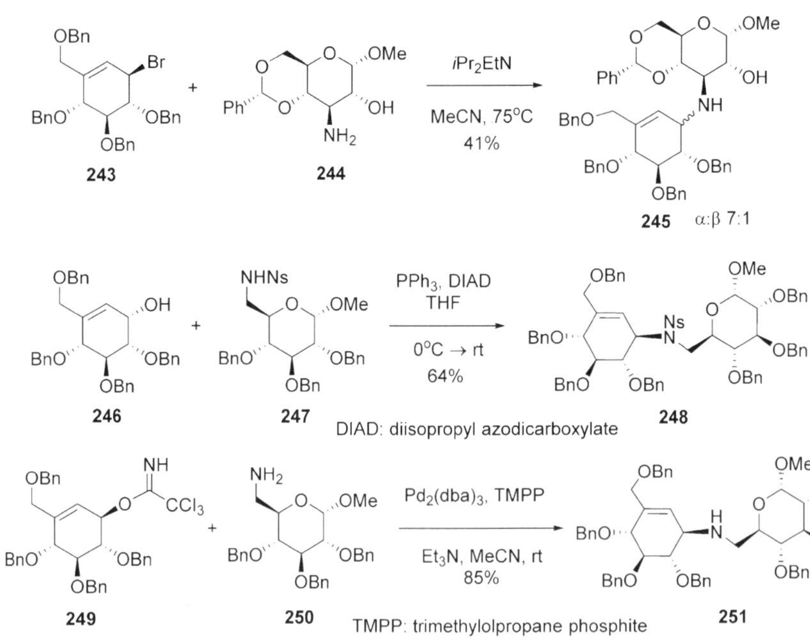

Scheme 29 Cumpstey's different approaches to (1→3)- and (1→6)-linked valienamine pseudodisaccharides.

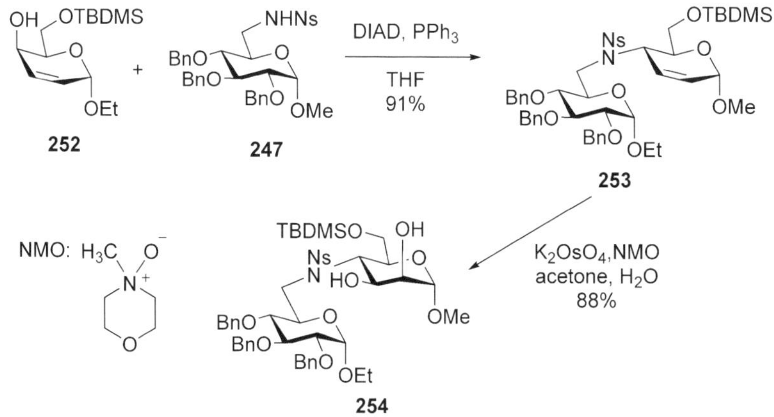

Scheme 30 Synthesis of (3→6)-amino-linked neodisaccharide by Mitsunobu coupling of an allylic electrophile.

atom. Initially, they have found that attempted installation of a bridging nitrogen into primary-secondary or secondary-secondary positions by Mitsunobu coupling failed.[97] Therefore, in order to facilitate the Mitsunobu reaction, an allylic alcohol of enhanced electrophilicity was used as the electrophilic coupling partner (Scheme 30). The 2,3-unsaturated-4-OH derivative **252** was prepared from tri-*O*-acetyl-D-glycal, obtained from protected galactose precursor, by means of Ferrier rearrangement. Mitsunobu coupling between the newly formed reactive electrophile and nosylamides (e.g., **247**), resulted in the desired 4-aminoneodisaccharides such as **253** with excellent stereoselectivity. The refunctionalisation of the double bond of **253** was successfully performed *via* osmium tetroxide catalyzed dihydroxylation, using *N*-methylmorpholine *N*-oxide (NMO) as the cooxidant, to yield the *manno*-configured diol **254** with full stereoselectivity.[98]

Chemoenzymatic synthesis of di- and trisaccharide mimetics with nonglycosidic amino bridges has also been reported by Thiem and coworkers (Scheme 32).[99] The formation of the amino bridge occurred via Dess-Martin oxidation of the primary position of protected or unprotected sugars (e.g., **255**) and subsequent reductive amination of the obtained dialdoses using an amino sugar coupling partner (e.g., **257**). *N*-Acetylglucosamine-based disaccharides, obtained by enzymatic glycosylation, was also subjected to the oxidation-reductive amination protocol to produce (6→6)- and (6→2)-*N*-linked glycosylated lactosamine mimetics (Scheme 31).

Peri and coworkers described the efficient synthesis of a lipid-A-disaccharide mimetic containing a β-*N*(OMe) interglycosidic linkage by a Koenigs-Knorr condensation of glycosyl bromide donor **259** with the 6-methoxyaminoglucoside aglycon **260** (Scheme 32). Both glycosyl building blocks were equipped with lipid chains before glycosylation. Disaccharide **261** was formed quantitatively as a 1:2 mixture of α- and β-anomers. Deprotection of the major anomer provided **262** which proved to be an efficient antagonist of lipid A on MT2 macrophages.[100]

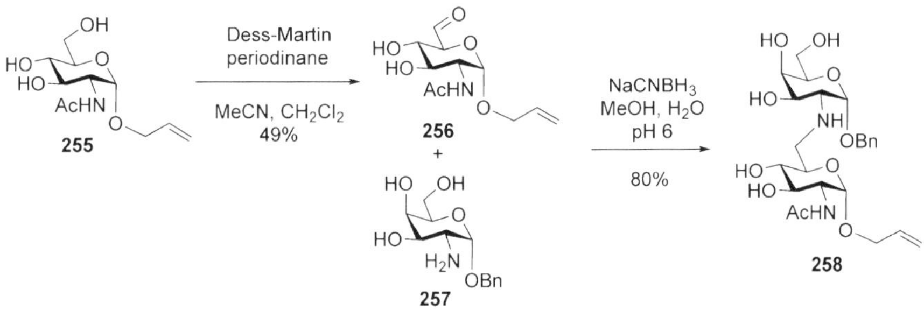

Scheme 31 Rapid synthesis of a (2→6)-amino-bridged disaccharide by reductive amination.

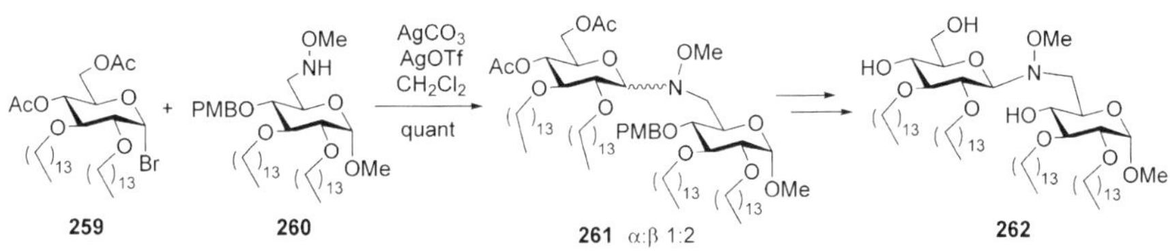

Scheme 32 Synthesis of a β-*N*(OMe)-linked lipid A mimetic by Koenigs-Knorr glycosylation of a C-6 methoxyaminated acceptor.

2.4 C-Glycosides

C-Glycosyl derivatives, in which the interglycosidic oxygen is replaced by a methylene unit, are stable to enzymatic and chemical hydrolysis. In this chapter, we use the term C-glycosides, which is the trivial name of these compounds, although it is a misnomer according to the IUPAC carbohydrate nomenclature. Due to the high stability, *C*-glycosides found broad application in medicinal chemistry and drug design. Strategies toward stereoselective synthesis of *C*-disaccharides include anionic methods, cationic methods, free-radical approaches, cyclization strategies, methods based on sigmatropic rearrangement, and transition-metal-mediated cross-coupling reactions. Over the last decades, many review articles have been published on the synthesis of *C*-glycosides,[101–103] including *C*-oligosaccharides,[104, 105] and excellent comprehensive reviews on this topic were published in 2017 and 2018.[106–108] Therefore, we only provide a short review on the results of the past five years.

In 2013 Ramana and coworkers described different approaches to synthesize (1→1) and (1→6)-linked *C*-disaccharides via alkynol cycloisomerization and reductive deoxygenation, carried out in one pot (Scheme 33). In the first route, the key steps were to form a hemiacetal (e.g., **265**) via coupling the starting lactone (**263**) to a sugar-derived, linear, lithiated alkyne (**264**), and reducing them in the same pot, thus obtaining the desired *C*-glycosides (**266**). The substrate's flexibility can be manipulated by changing the size of the ring, be it furanoid or pyranoid structures. Afterwards, different catalysts were used for alkynol cycloisomerization to yield **268**, among which $AuCl(PPh_3)$/$AgSbF_6$ proved to yield the best results. Having completed the cycloisomerization, the product was reduced with $BF_3{\cdot}Et_2O$ and Et_3SiH to obtain the desired *C*-disaccharide

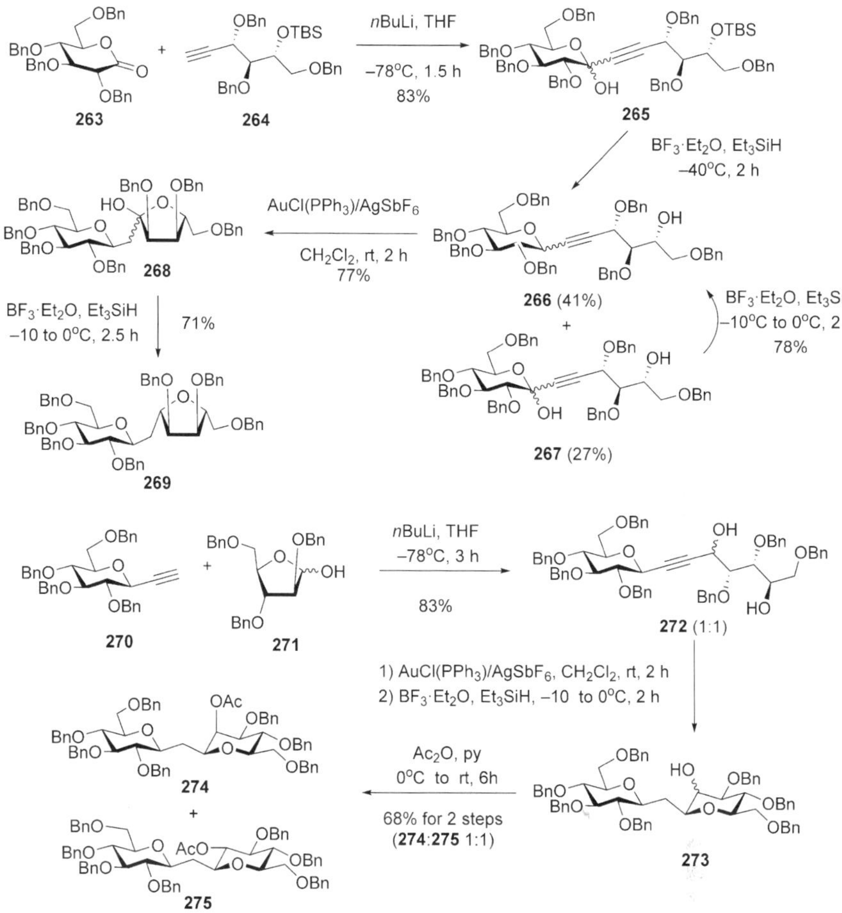

Scheme 33 Synthesis of (1→1)-linked *C*-disaccharides *via* addition of lithiated alkynes to sugar electrophiles followed by one-pot alkynol cycloisomerization-deoxygenation.

269. It is worth noting that during the first reduction step, a large amount of diol (**267**) was formed, but this could be readily reduced into the desired product.

In an alternative approach, instead of linear alkynes, lithiated C-1 ethynyl glycoside and a benzylated furanose lactol derivative were used. Although the formation of the C-linked trehalose precursor **272**, starting from alkyne **270** and arabinofuranose **271**, was not stereoselective, the synthesized products could be isolated with an excellent 83% combined yield. Applying the optimized conditions in the one-pot cycloisomerization and the reduction steps, followed by acetylation, the epimeric methylene-linked trehalose mimetics **274** and **275** were isolated in good overall yield. Pyranosyl and furanosyl (1–6)-*C*-disaccharides were obtained by using a furanosyl 5,6-epoxide as the electrophile in the coupling step.[109]

A Stille cross-coupling approach was developed by Werz and coworkers for efficient synthesis of (1→2)-, (1→3)- and (1→4)-*C*-disaccharides.[110] Reaction of stannylated glycal

276 with glucose-derived exocyclic bromoolefins **277** or **280** under the standard Stille coupling conditions produced the expected (1→2)- and (1→3)-linked *C*-disaccharides (**278** and **281**) in high yields. The pseudodisaccharidic dienes could be either completely reduced to 2-deoxy *C*-disaccharides (**279**) or refunctionalized in an oxidative-reductive manner to produce the corresponding 2′-OH disaccharides (e.g., **282**). Under similar conditions, (1→4)-*C*-disaccharides were obtained successfully (Scheme 34).

An efficient way of synthesizing (1→2)- and (1→6)-*C*-linked pseudodisaccharides was reported by Oroszova and coworkers.[111] Having prepared the pinacol boronate **283** from D-glucal (not recommended trivial name), it was coupled to methyl 2,3,4-tri-*O*-benzyl-6-deoxy-6-iodo-α-D-glucopyranoside **284** under various conditions, among which the best one was the Pd-catalyzed cross-coupling with $[HP(t\text{-}Bu)_2Me]BF_4$, giving the desired (1→6)-*C*-glycoside **285** in 72% yield. Using **286** as the alkyl iodide coupling partner, (1→2)-*C*-disaccharide **287** was also synthesized with the aid of $NiBr_2$·diglyme, 4,4′-di-*t*-butyl-2,2′-dipyridyl and *t*-BuOK in *tert*-amyl alcohol. The synthesized glycal conjugates were refunctionalized using various oxidation-reduction methods giving α- and β-linked D-*gluco*- (**288** and **289**) and D-*manno*-configured pseudodisaccharides.[111]

Mukherjee and coworkers developed a strategy based on cross-dehydrogenative coupling of sugar enol-ethers to furanose-derived terminal alkenes (e.g., **291**), resulting in pseudodisaccharides like **292** and **294** (Scheme 35). Using $Pd(OAc)_2$ as catalyst and AgOAc as oxidating agent, the desired *C*-glycosides were obtained in excellent yields. Afterwards, their attention focused on the less costly $Cu(OAc)_2$ as oxidating agent; however, in those cases, the yield was unsatisfactory. Changing the catalyst to $Pd(TFA)_2$ or $PdCl_2$ resulted in lower yields as well, just as using pyridine-dioxane as solvent. The reactions were fully *E*-stereoselective in all cases and took place efficiently with both pyranoid and furanoid (e.g., **290** and **293**) enol-ethers (Scheme 36).[112]

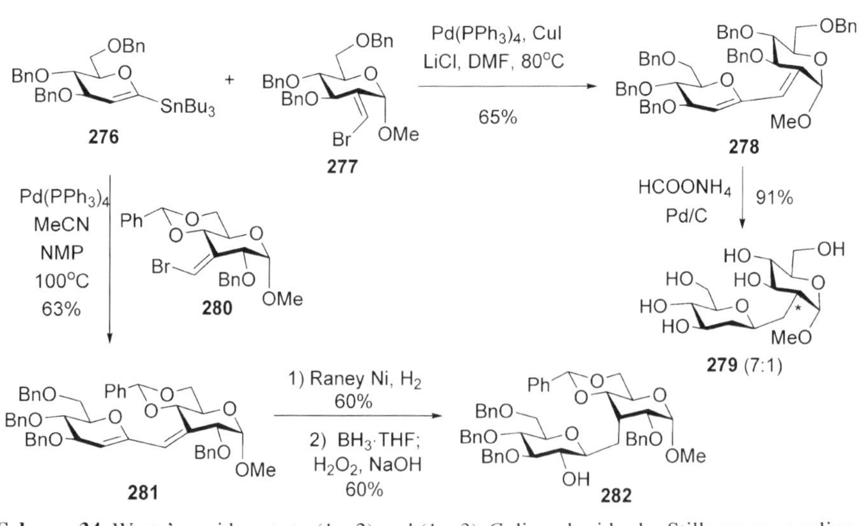

Scheme 34 Wertz's rapid route to (1→2) and (1→3)-*C*-disaccharides by Stille cross-coupling.

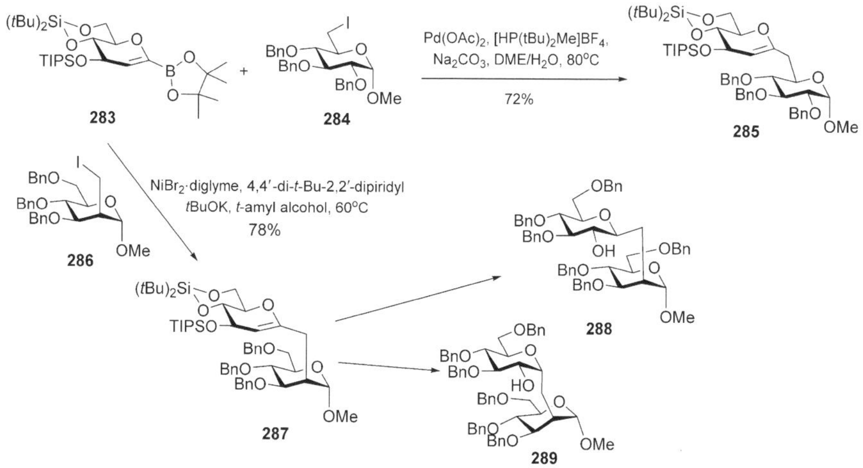

Scheme 35 Synthesis of (1→6) and (1→2)-*C*-disaccharides by sp^2-sp^3 Suzuki-Miyaura reaction.

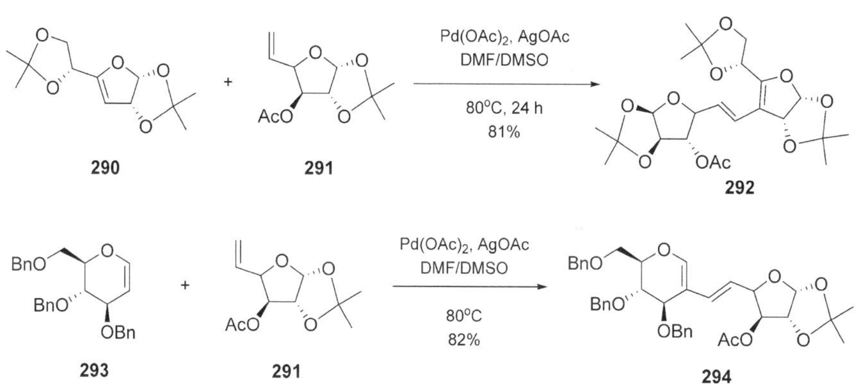

Scheme 36 Cross-dehydrogenative sp^2-sp^2 coupling route to *C*-disaccharide mimetics linked at nonanomeric positions.

3 Mimetics with three-bond interglycosidic connections

Carbohydrate structures containing a three-bond intersaccharidic linkage with two heteroatoms, although scarcely found in nature, display important biological activity. The oligosaccharide unit of enedyine antibiotics of potent antitumor activity contain a C-O (namenamicin)[113, 114] or an N-O (calicheamicins and esperamicins)[115] interglycosidic moiety, and the hydroxylamin linkage of calicheamicin participates in the antitumor action. On the other hand, the disulfide bridge is an abundant three-bond motif in nature and plays an essential role in the stability of proteins, peptides, and many other biologically important products.[116–118]

Inspired by the potential biological activity, many research efforts have been devoted to produce disaccharide mimetics with novel linking modes, including S-S, Se-S, Se-Se, C-S, N-O, and C-O, interglycosidic bonds. In this field, the synthesis and application of aminooxylated carbohydrates[85, 119] have recently been summarized in-depth in excellent review articles.

3.1 S-S, Se-S, Se-Se-linked mimetics

In recent years, diglycosyl disulfides, particularly the unsymmetric ones, have become important novel carbohydrate scaffolds of potential biological activity.[120–122]

Iadinosi's group reported a straightforward route to symmetric glycosyl disulfides starting from per-*O*-acetylated sugars (e.g., **30**, **31**, and **35**), applying a sequential approach including anomeric iodide generation (**295**), isothiuronium salt formation (**296**) and thiolate oxidation (Scheme 37). In this way, the glucose- and galactose-derived disulfides **297** and **298** could be obtained with high β,β-selectivity, while in the *manno* case disulfide **299** was formed as a 1:1 mixture of the α,α- and α,β-anomers. The use of a catalytic amount of phenyl diselenide proved to be crucial for the fast and efficient conversion of the glycosyl thiolate, formed in situ by the action of Et_3N, to the corresponding disulfide. Noteworthy, the whole synthesis took place efficiently within a few hours without isolation of any intermediates.[123]

Iodine-NIS-mediated direct oxidation of an *S*-acetyl group, located at any position of a sugar skeleton, was reported as another straightforward method for the synthesis of symmetric disulfide disaccharides (Scheme 38).[124] Using excess I_2 in the presence of 0.5 equiv. of NIS in acetonitrile at room temperature, the starting *S*-acetyl-containing sugars were completely converted to the corresponding disulfides. The formation of sugar disulfides (e.g., **300–306**) typically occurred within 1 h, except for GlcNAc disulfide **302**, which required 7 h. Efficient reduction of disulfides into

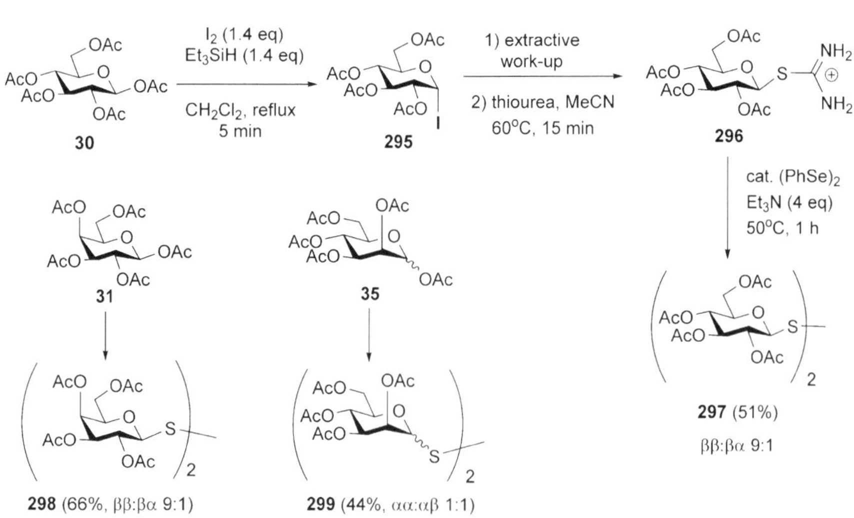

Scheme 37 Rapid access to symmetric glycosyl disulfides from sugar peracetates.

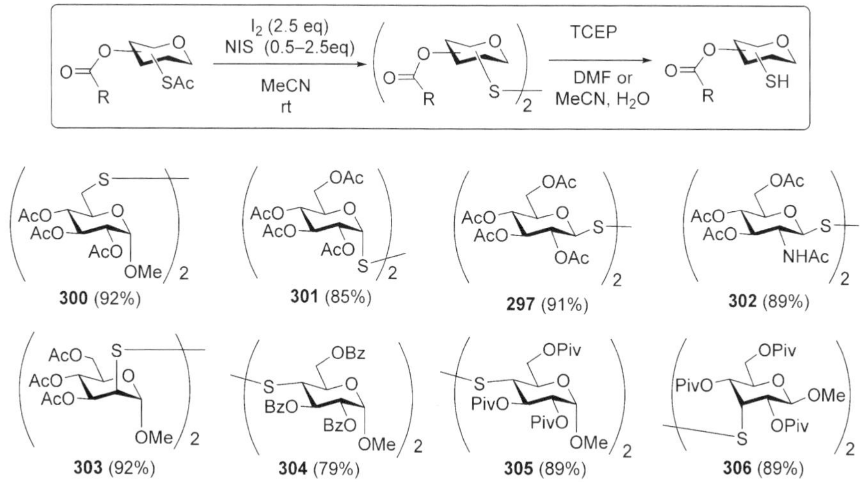

Scheme 38 Straightforward synthesis of symmetric disulfides by I_2-NIS-mediated oxidation of thioacetates, developed by Dong.

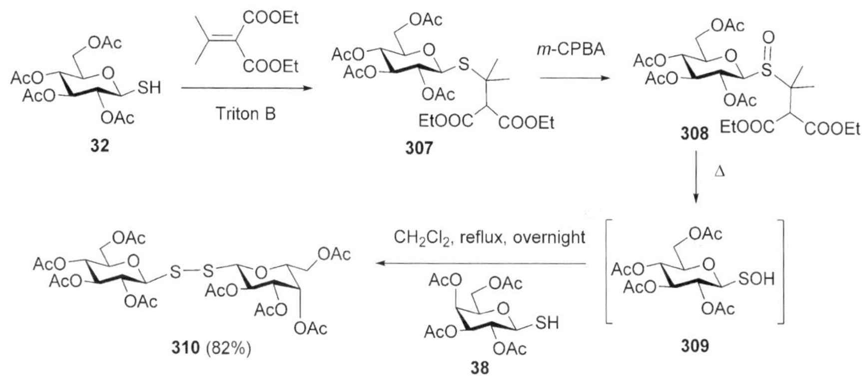

Scheme 39 Access to unsymmetric disulfide **310** through **in situ** generated sulfenic acid **309**.

the corresponding sugar thiols by tris(2-carboxyethyl)phosphane (TCEP) was also described.

The first synthesis of the challenging unsymmetric diglycosyl disulfides, established by Szilágyi, Herczegh, and coworkers, was based on nucleophilic substitution of a glycosyl methanethiosulfonate ester with glycosly thiolate.[125] Mixed diglycosyl disulfides were also prepared directly from two different thiols in a one-pot manner.[126]

Italian authors reported an efficient route to mixed disulfides using in situ-generated glucosyl sulfenic acid **309** as a novel glycosylsulfenyl-transfer reagent (Scheme 39).[127] The sulfinyl **308**, prepared through synthesis of sulfide **307** followed by oxidation with *m*CPBA, was converted to the transient sulfenic acid by thermolysis. In situ coupling of **309** with 1-thiogalactose **38** gave the mixed disulfide **310** in a high yield.

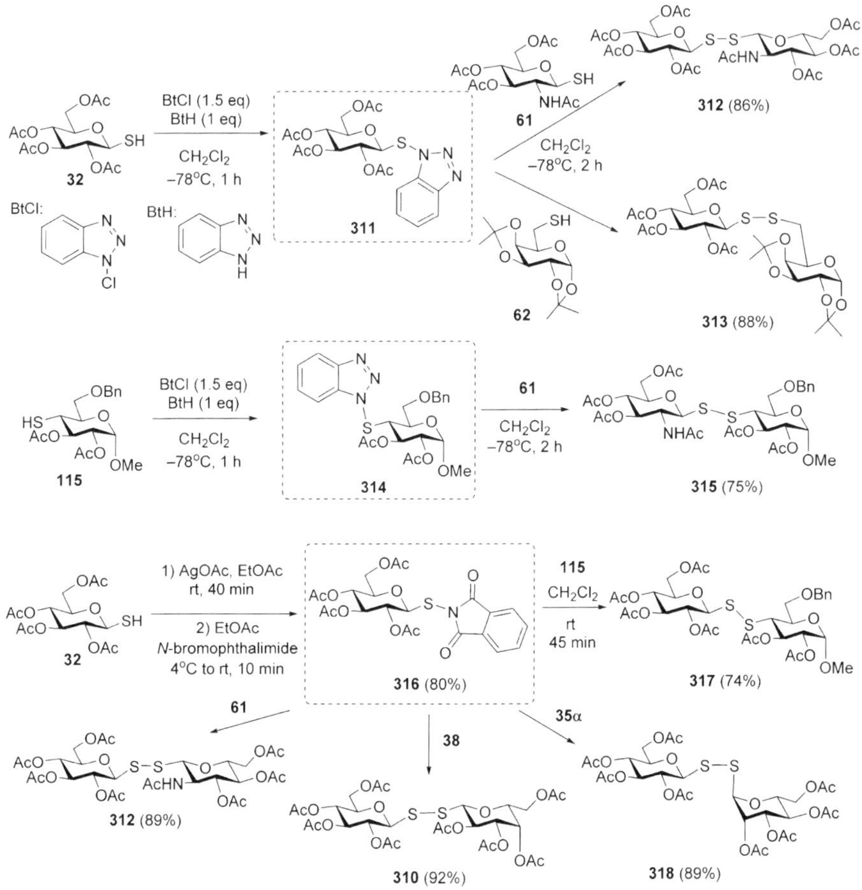

Scheme 40 Different approaches to unsymmetric disulfides linked at both anomeric and non-anomeric positions from Szilágyi's and Hunter's groups.

Szilágyi and coworkers reported the synthesis of a large number of unsymmetric glycosyl disulfides by using glycosylthio-benzothiazol[128] or -phthalimide derivatives[129] as reactive sulfenylating species (Scheme 40). An efficient one-pot synthesis of mixed disulfides such as **312**, **313**, and **315** was achieved via the corresponding benzothiazolylsulfenyl derivatives **311** and **314**. The (glycosylthio)phtalimides (e.g., **316**), obtained from the silver salts of 1-thioaldoses proved to be convenient sulfenyl-transfer reagents and reacted readily with glycosylthiols to yield the corresponding mixed disulfides (**310**, **312**, **317**, and **318**) with high yields.

Inspired by a novel method for the synthesis of unsymmetric disulfides of aliphatic and aromatic thiols,[130] Zhu and coworkers described direct transformation of a 1:1 mixture of two different 1-thiosugars into mixed (1-1)-disulfide using DDQ-mediated oxidation (Scheme 41).[131] The reactions gave the corresponding α,β-heterocoupled products **319–322** in good yields. Noteworthy, β,β-disulfide side products were formed in all cases by homocoupling of the β-thiols of higher nucleophilicity.

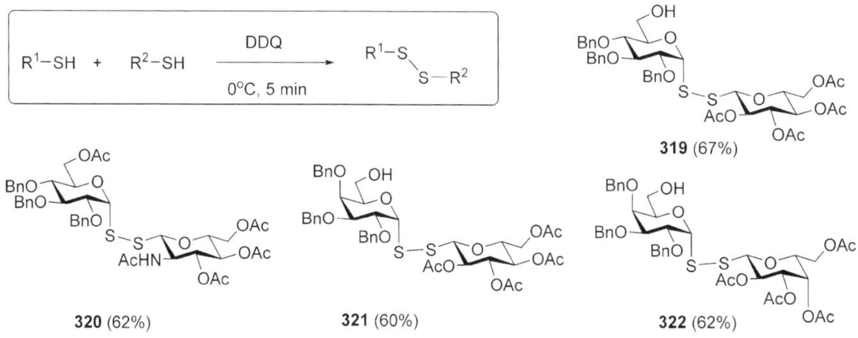

Scheme 41 Synthesis of unsymmetric α,β-(1→1) glycosyl disulfides using DDQ as oxidant.

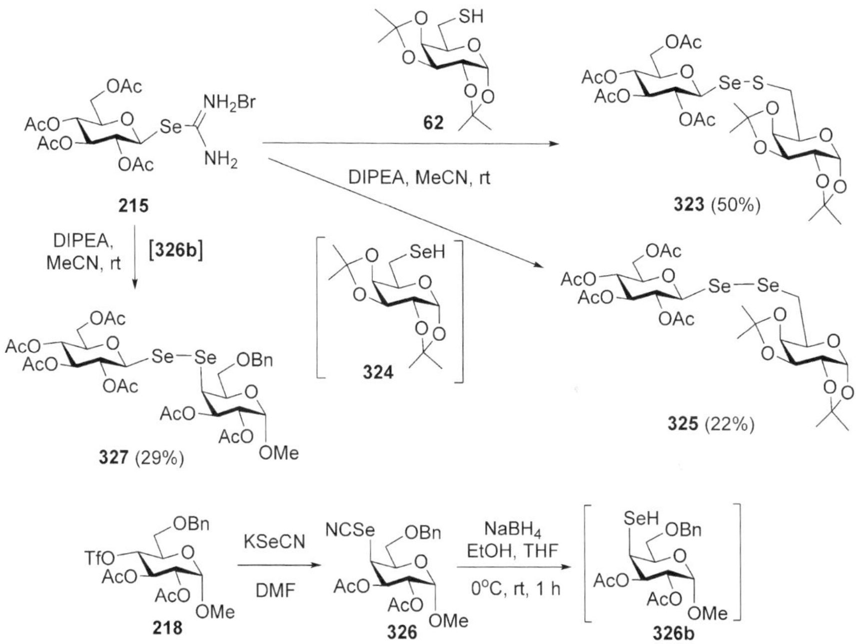

Scheme 42 Synthesis of selenium-sulfur and selenium-selenium-linked disaccharide mimetics.

Szilágyi and coworkers reported the synthesis of diglycosyl selenosulfides and unsymmetric diglycosyl diselenides by coupling of glycosyl isoselenuronium salts with sugar thiols or sugar selenols (Scheme 42).[132] Reaction of **215** with thiol **62** in the presence of excess Hünig's base (DIPEA) afforded the expected selenosulfide **323** in 50% yield. Coupling of the isoselenuronium salt **215** with in situ-generated selenols **324** and **326b** yielded the corresponding (1→4)- and (1→6)-linked mixed diglycosyl diselenides **325** and **327** in moderate yields. Formation of symmetric diglycosyl diselenide side products from isoselenuronium salt **215** and selenols **324** and **326b** was observed in these reactions.

3.2 C-S, C-N, N-O, C-O, SO_2-N-linked disaccharides

Dondoni and coworkers have reported the stereoselective synthesis of glycomimetics displaying the Gly-S-CH_2-Gly scaffolds via photoinduced hydrothiolation of 4-exomethylene-ribofuranoside and 5-exomethylene glucopyranoside derivatives with 1-thioaldoses.[58] As an extension of this method, our group and Somsák's group investigated the thiol-ene reactions of *exo*-glycals and furanoid exomethylene derivatives to produce C-S-bridged glycomimetics (Scheme 43).[42, 133, 134] Addition reactions to *exo*-glycals of both D-*gluco* **328** and D-*galacto* **330** configuration furnished the β-C-S-bridged disaccharide products **329** and **331**, respectively, with complete regio- and stereoselectivity.[133, 134] Reaction of 3-*exo*-methylene-xylofuranose derivative **332** with 1-thiomannose **35α** produced compound **333** as a single diastereoisomer with D-*allo*-configuration at the furanose unit.[133] The stereoselectivity of the additions was governed by the preferred axial H-abstraction by the most populated conformation of the intermediate carbon-centered radical formed in the thiyl addition step.

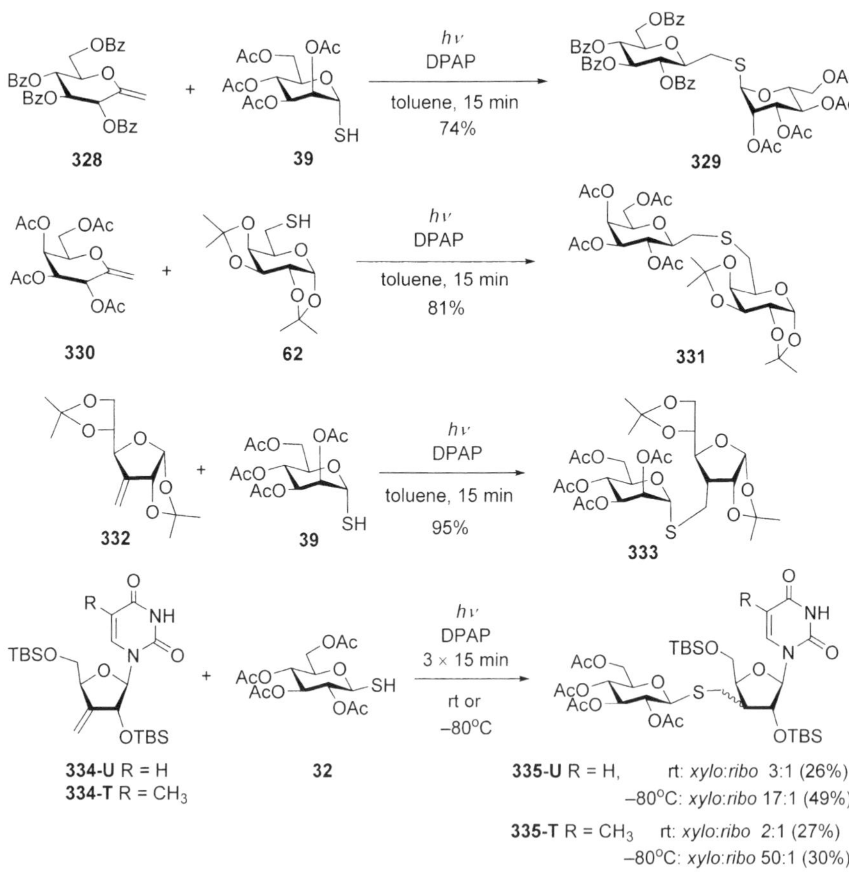

Scheme 43 Thio-click-based synthesis of C-S-linked disaccharide mimetics and sugar modified nucleosides.

This method was also utilized for the synthesis of sugar-modified nucleosides.[135] It was found that the nucleobase aglycon influenced the stereochemical outcome of the reaction, which could also be controlled by the reaction temperature. Addition of thiols to the 3-exomethylene bond of uridine **334-U** and ribothymidine **334-T** derivatives at -80°C gave the D-*xylo* nucleosides with high stereoselectivity.

The application of *S*-glycopyranosyl-*N*-monoalkyl dithiocarbamates (DTCs) as masked 1-glycosylthiols for the synthesis of *S*-linked carbohydrate mimetics was studied by Santoyo-Gonzalez and coworkers.[136] The novel dithiocarbamates such as **337** were obtained *via* nucleophilic displacement of 1-chloro or -bromosugars with dithiocarbamate salts of amines (Scheme 44). Having studied the effects of stoichiometry of the reagents, it has been observed that using higher molar ratio of the dithiocarbamate salts along with shorter reaction times yields the highest amount of glycosyl dithiocarbamates, whereas using equimolar amounts and extended reaction time results in the formation of symmetric diglycosyldisulfides. These novel compounds were used to synthesize thioglycosyl vinylsulfonates and various *S*-linked homo- and heterodivalent neoglycoconjugates. Coupling the unmasked thiolate obtained from the basic treatment of the carbohydrate dithiocarbamate **337** to electrophilic cyclic sulfate derivative of glucose (**338**) resulted in the formation of heterodivalent neoglycoconjugate **339**.[136]

Paton and coworkers described a nitrile oxide-based approach for the synthesis of (1→6)- and (1→1)-amidoxime-linked disaccharides, which involved coupling of aminomethylene pyranoses via 1,3-nucleophilic addition reaction to pyranosyl nitrile oxides, thus forming amidoxime interglycosidic linkages (Scheme 45). The glucopyranosyl nitrile oxide coupling partner **342** was generated in situ by dehydrochlorination

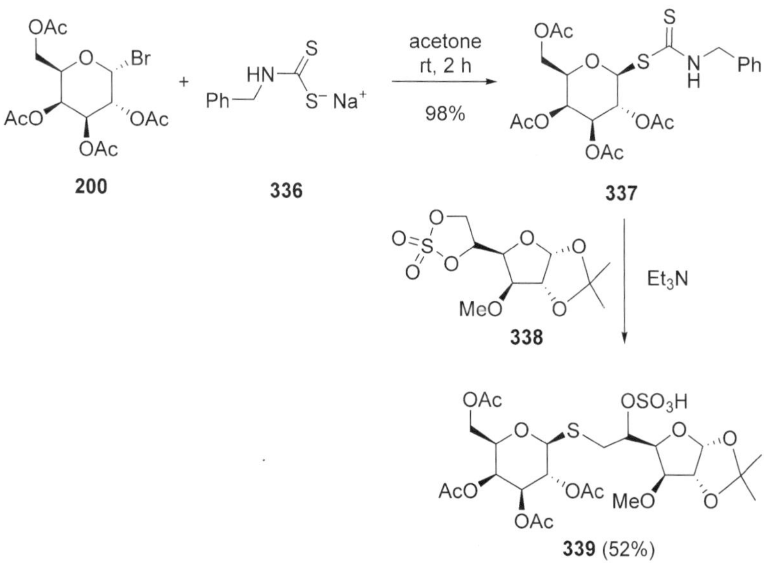

Scheme 44 Synthesis and nucleophilic coupling of *S*-glycopyranosyl-*N*-monoalkyl dithiocarbamate **337** to cyclic sulfate derivative **338**

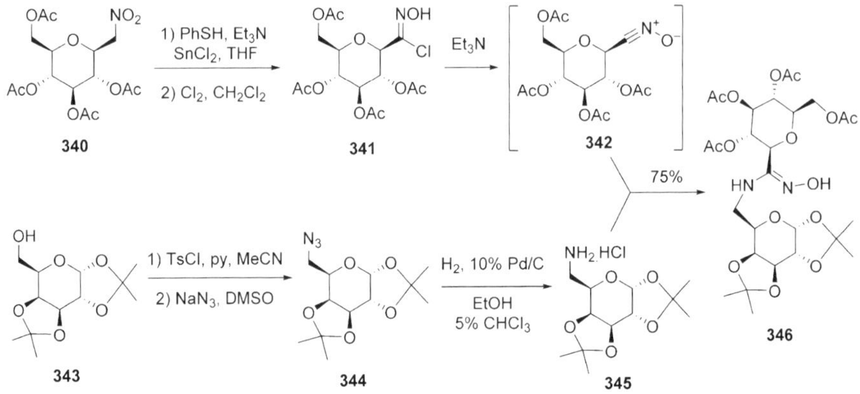

Scheme 45 Nitril-oxide coupling route to (1→6)-amidoxime-linked pseudodisaccharide **346**.

of hydroximoyl chloride **341** and reacted with the aminogalactose component **345**, prepared from **343** by 6-azidation and a subsequent catalytic hydrogenation, to furnish the D-Glc-D-Gal (1→6)-amidoxime disaccharide **346** in a good yield. This approach may prove to be an efficient way of synthesizing a wide array of similar pseudodisaccharides.[137]

Uronamidoxy-linked oligosaccharides as new glycosyl amino acid mimetics were synthesized by Xie and coworkers from properly protected D-ribofuranoid glycoaminoxy acid building blocks (Scheme 46). Having obtained the starting protected momomeric building units **347** and **348** from di-*O*-isopropylidene-D-glucofuranose in several steps, the desired pseudodisaccharide **349** was prepared with good yield using different methods. Using 1-ethyl-3-(3-dimethylaminopropyl)carbodiimide (EDC) and HOBt in DMF-dichloromethane gave the best, 81% yield, whereas using diethyl cyanophosphonate and triethylamine in THF gave 77% and diphenylphosphoryl azide, DIPEA, and $NaHCO_3$ in DMF gave 78%. The synthesized compounds were then converted into building blocks again, either by ester deprotection with trifluoroacetic acid or by hydrazinolysis of the phtalimido moiety, to continue the elongation up to hexamer.[138]

Knowing that iminoalditols are strong glycosidase inhibitors, Blériot et al. synthesized 1,2-*cis*-homoiminosugars utilizing a β-aminoalcohol rearrangement of seven-membered iminosugars. The obtained piperidine iminosugar **353** was treated with diphenyl phosphoryl azide (dppa), DEAD, and PPh_3 in THF, to produce the α-configured 2-azidopiperidine **354** (Scheme 47). This compound, while being a GlcNAc analogue, is also able to incorporate aglycons at the pseudoanomeric position and can form stable iminoglycoconjugates, as shown by the synthesis of its pseudodisaccharide derivatives **356** and **357**.[139]

In 2011, Poulsen and coworkers described a general method by which the anomeric center of one carbohydrate can be linked to virtually any position of a second carbohydrate through a sulfonamide bridge.[140] Peracetylated 1-thioglucopyranose **358** was coupled to a wide array of pyranoid and furanoid aminosugars, in the presence of diethyl bromomalonate. The method is exemplified by the reactions of **358** with amines

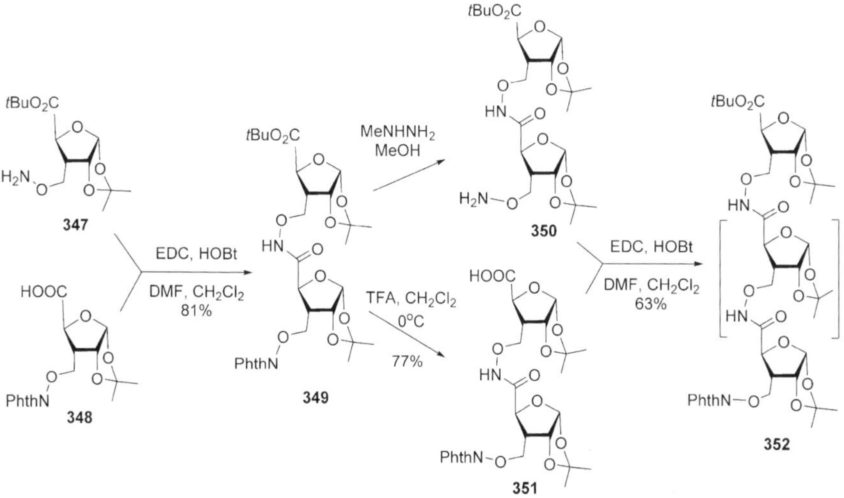

Scheme 46 Synthesis of furanoid uronohydroxamate di- and tetramers.

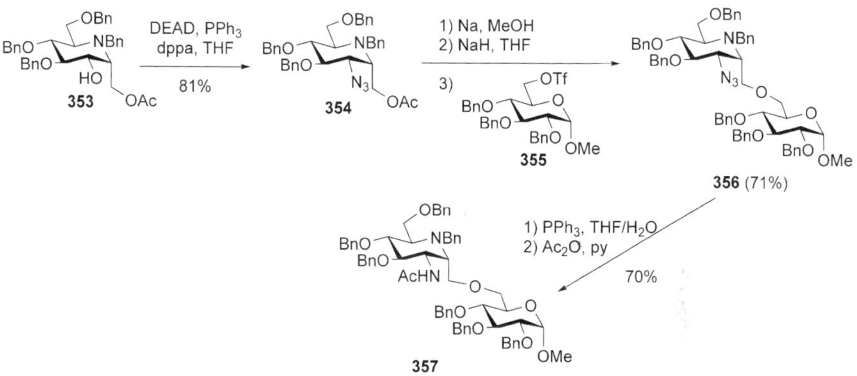

Scheme 47 Synthesis of pseudodisaccharide **357** containing a GlcNAc-analogue iminosugar unit.

359 and **362** to yield the corresponding sulfenamides **360** and **363** in moderate to good yields. Having successfully synthesized the sulfenamide bond between the two carbohydrate moieties, presumably via a sulfenyl bromide intermediate, *m*CPBA was used as oxidizing agent to obtain the desired sulfonamide-linked pseudodisaccharides such as **361** and **364**, which were later deprotected via Zemplén deacetylation (Scheme 48). The coupling processes showed excellent stereoselectivity with retention of configuration at the anomeric centers.

The scope of the reaction was extended to larger thioacetate (e.g., **365** and **368**) and glycosylamine (e.g., **369**) substrates, forming glycomimetics up to octasaccharide (Scheme 49). The yields were overall excellent, however, in cases where low yields were perceived the phenomenon could be attributed to formation of disulfide and diglycosylamine side products, which proved difficult to separate from the products by chromatography due to their similar physical properties. Reaction of 1-thiolactose

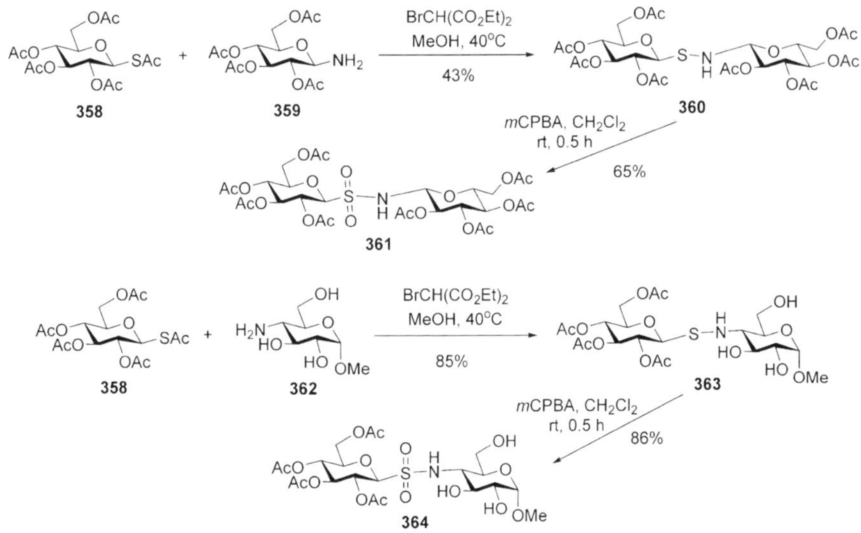

Scheme 48 Poulsen's approach for the synthesis of (1→1) and (1→4) sulfonamide-bridged disaccharides.[140]

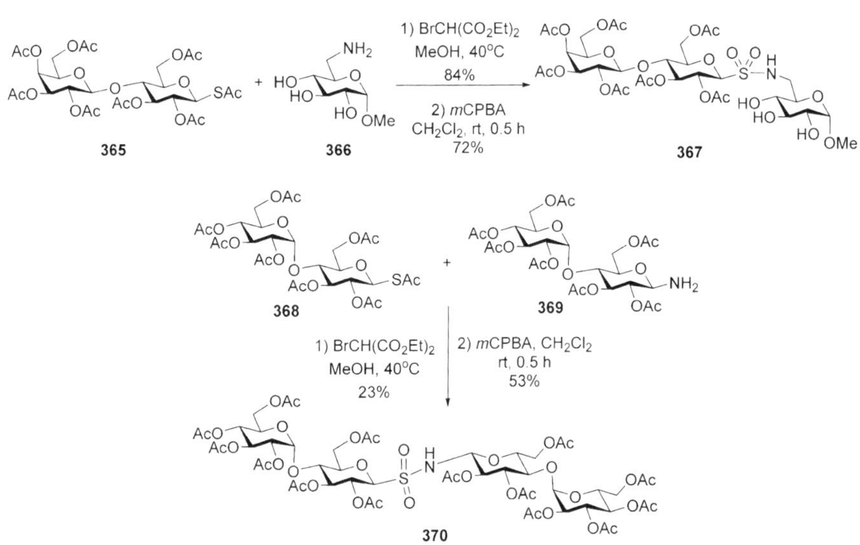

Scheme 49 Synthesis of sulfonamide-linked tri- and tetrasaccharides.

peracetate **365** with 6-amino-6-deoxyglucoside **366**, for example, gave the trisaccharide mimetic **367** with high yields in both steps. However, the 2+2 coupling between 1-thiomaltose peracetate **368** and maltosyl amine **369** to form sulfenamide proceeds with the formation of the corresponding disulfide and diglycosylamine side products, thereby significantly decreasing the overall yield of the sulfonamide tetrasaccharide mimetic **370**.

4 Mimetics linked by four-bond bridges

Oligosaccharide mimetics built up with urea, thiourea, and guanidine interglycosidic bridges can possess phosphate-binding capability, and thereby can serve as artificial receptors for nucleic acids.[141–143]

Prosperi et al. developed a general method to yield intersaccharidic ureido linkages. They synthesized pseudodisaccharides with urea bridges between positions 1→2, 1→4, and 1→6 by connecting aminosugars to in situ generated glycosyl isocyanates. For the synthesis of the (1→4)-urea-bridged disaccharide **375**, the 4-aminoglucoside **373**, obtained from galactoside **371** via the 4-azidoglucoside **372**, was used as the acceptor and glucosyl isocyanide **374** was applied as a donor precursor. Compound **374** was converted to the corresponding isocyanate in the coupling mixture with the aid of pyridine *N*-oxide (Scheme 50). The authors found no difficulty in connecting the isocyanate coupling partner to primary and secondary positions, not even axial amino groups proposed problems. In the future, these compounds are suitable to become building blocks in oligosaccharide synthesis.[144]

García Fernández's group developed a method to synthesize various pseudodi- and trisaccharides, connected in 1→6 β position by urea, thiourea and guanidine linkages.[142] Starting from glycosyl isothiocyanate **376** and 6-aminosugar **366**, the desired thiourea-bridged derivative **377** was obtained with excellent yield (Scheme 51). After converting **377** into carbodiimide **379** using desulfurization reaction, it can be transformed into urea (**380**) and guanidine (**381**) derivatives also with very good yields. Pseudotrisaccharides with alternating pseudoamide-type (urea, thiourea, and guanidine) intersaccharide bridges were also prepared to study their phosphate binding capabilities in water.[142]

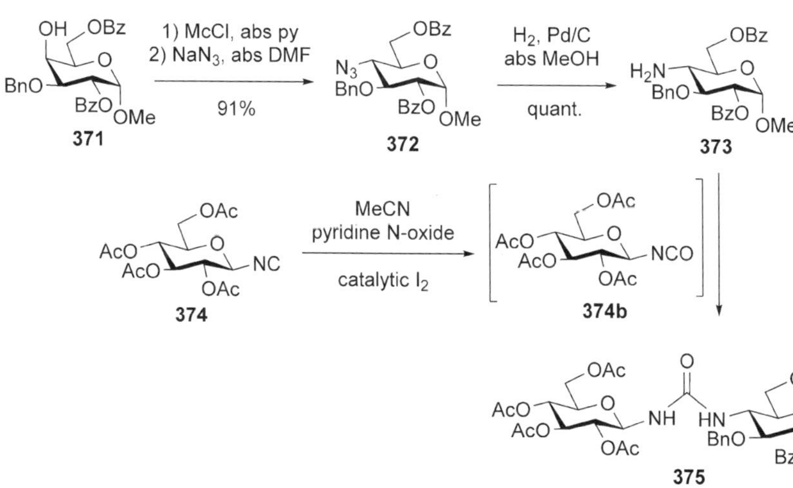

Scheme 50 Synthesis of (1→4)-urea-bridged disaccharide **375** (Mc: monochloromesyl).

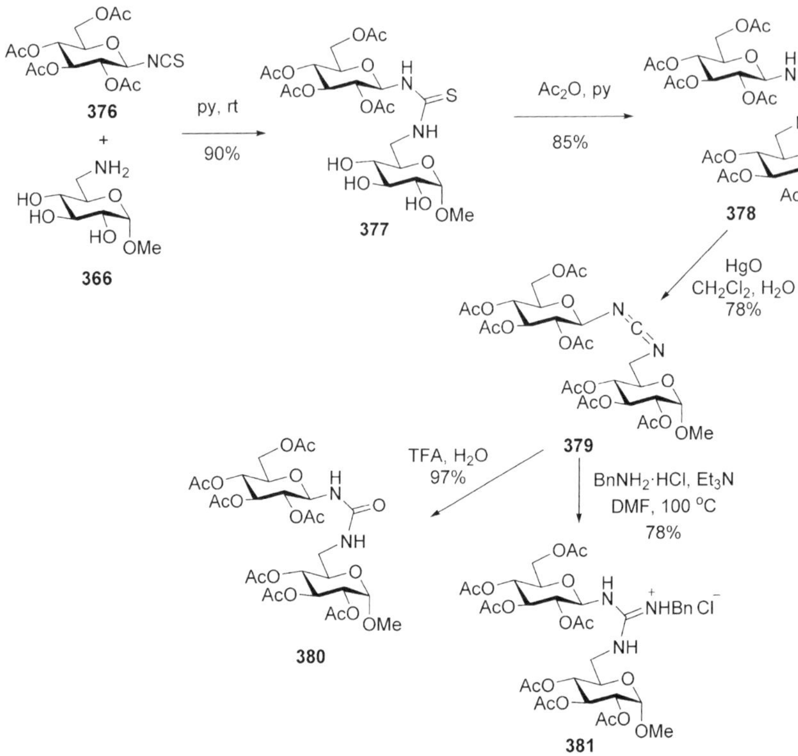

Scheme 51 Synthesis of urea-, thiourea-, and guanidine-linked disaccharide mimetics.

5 Conclusion

In the past 10–15 years numerous, diverse and inspiring studies described the synthesis of novel carbohydrate mimetics bearing linkage modifications. One of the most common derivatives incorporated a sulfur atom in the place of the glycosidic oxygen, mainly because of the two atoms' inherent similarity. This results in the molecule being biologically the most akin to the original compound and still become more stable against enzymatic modification. A wide variety of reactions were applied during the syntheses, such as photoinitiated thiol-ene addition (thio-click reaction), Michael addition, Ferrier rearrangement, and various S_N2 methods. Recently, studies focused on synthesizing selenium-linked disaccharides as well, which compounds can be used in protein-carbohydrate interaction studies as hydrolytically stable glycomimetics. Syntheses of *N*-glycosides, which are essential constituents of sugar-related derivatives such as *N*-glycoproteins and nucleic acids, and pharmaceutically valuable as anti-diabetic and anti-cancer agents are also discussed. Interestingly, a notable amount of valienamine derivatives emerged as pseudodisaccharides. *C*-glycosides, where the interglycosidic oxygen is replaced by a methylene group, are applied in a wide variety in pharmaceutic design and medicinal chemistry. Strategies toward stereoselective synthesis of *C*-disaccharides included anionic methods, cationic methods, free-radical approaches, cyclization strategies, methods based on sigmatropic rearrangement, and transition-metal-mediated cross-coupling reactions. Finally, compounds linked via three-bond or four-bond interglycosidic linkages were discussed. Although aside from disulfides most of them are scarce in nature, these compounds are valuable targets of drug design because of their inherent flexibility, stability and altered steric and electronic properties.

In a short time period, a wide variety of linkage-modified carbohydrate mimetics have been formed. Not only are these compounds potential drug molecules, but their challenging syntheses requiring the application of state-of-the-art synthetic methodologies may provide inspiration for future studies, further broadening the spectrum of medical compounds.

References

1. Bertozzi, C. R.; Kiessling, L. L. Chemical Glycobiology. *Science* **2001**, *291*, 2357–2364.
2. Rudd, P. M.; Elliott, T.; Cresswell, P.; Wilson, I. A.; Dwek, R. A. Glycosylation and the Immune System. *Science* **2001**, *291*, 2370–2376.
3. Helenius, A.; Aebi, M. Intracellular Functions of N-Linked Glycans. *Science* **2001**, *291*, 2364–2369.
4. Varki, A. Glycan-Based Interactions Involving Vertebrate Sialic-Acid-Recognizing Proteins. *Nature* **2007**, *446*, 1023.
5. Varki, A.; Cummings, R. D.; Esco, J. D.; Freeze, H. H.; Stanley, P.; Bertozzi, C. R.; Hart, C. W.; Etzler, M. E. Essentials of Glycobiology; Cold Spring Harbor Laboratory Press: Cold Spring Harbor, N.Y., 2009.
6. Jefferis, R. Glycosylation as a Strategy to Improve Antibody-Based Therapeutics. *Nat. Rev. Drug Discov.* **2009**, *8*, 226.
7. Varki, A. Biological Roles of Glycans. *Glycobiology* **2017**, *27*, 3–49.
8. Sears, P.; Wong, C.-H. Carbohydrate Mimetics: A New Strategy for Tackling the Problem of Carbohydrate-Mediated Biological Recognition. *Angew. Chem. Int. Ed.* **1999**, *38*, 2300–2324.
9. Compain, P.; Martin, O. R. Carbohydrate Mimetics-Based Glycosyltransferase Inhibitors. *Bioorg. Med. Chem.* **2001**, *9*, 3077–3092.
10. Compain, P.; Martin, O. R. Design, Synthesis and Biological Evaluation of Iminosugar-Based Glycosyltransferase Inhibitors. *Curr. Top. Med. Chem.* **2003**, *3*, 541–560.
11. Ernst, B.; Magnani, J. L. From Carbohydrate Leads to Glycomimetic Drugs. *Nat. Rev. Drug Discov.* **2009**, *8*, 661.
12. Koester, D. C.; Holkenbrink, A.; Werz, D. B. Recent Advances in the Synthesis of Carbohydrate Mimetics. *Synthesis* **2010**, *2010*, 3217–3242.
13. Cipolla, L.; Araújo, A. C.; Bini, D.; Gabrielli, L.; Russo, L.; Shaikh, N. Discovery and Design of Carbohydrate-Based Therapeutics. *Expert Opin. Drug Discovery* **2010**, *5*, 721–737.
14. Driguez, H. Thiooligosaccharides as Tools for Structural Biology. *Chembiochem* **2001**, *2*, 311–318.
15. Pachamuthu, K.; Schmidt, R. R. Synthetic Routes to Thiooligosaccharides and Thioglycopeptides. *Chem. Rev.* **2006**, *106*, 160–187.

16. Andrews, J. S.; Mario Pinto, B. Synthesis of a thio Analogue of n-Propyl Kojibioside, a Potential Glucosidase Inhibitor. *Carbohydr. Res.* **1995**, *270*, 51–62.
17. Blanc-Muesser, M.; Defaye, J.; Driguez, H. Syntheses Stereoselectives de 1-Thioglycosides. *Carbohydr. Res.* **1978**, *67*, 305–328.
18. Szilágyi, L.; Varela, O. Non-conventional Glycosidic Linkages: Syntheses and Structures of Thiooligosaccharides and Carbohydrates with Three-Bond Glycosidic Connections. *Curr. Org. Chem.* **2006**, *10*, 1745–1770.
19. Rye, C. S.; Withers, S. G. The Synthesis of a Novel Thio-Linked Disaccharide of Chondroitin as a Potential Inhibitor of Polysaccharide Lyases. *Carbohydr. Res.* **2004**, *339*, 699–703.
20. Witczak, Z. J.; Lorchak, D.; Nguyen, N. A Click Chemistry Approach to Glycomimetics: Michael Addition of 2,3,4,6-tetra-O-acetyl-1-thio-β-D-Glucopyranose to 4-deoxy-1,2-O-Isopropylidene-l-glycero-pent-4-Enopyranos-3-ulose—A Convenient Route to Novel 4-deoxy-(1→5)-5-C-Thiodisaccharides. *Carbohydr. Res.* **2007**, *342*, 1929–1933.
21. Manzano, V. E.; Uhrig, M. L.; Varela, O. Straightforward Synthesis of Thiodisaccharides by Ring-Opening of Sugar Epoxides. *J. Org. Chem.* **2008**, *73*, 7224–7235.
22. Repetto, E.; Manzano, V. E.; Uhrig, M. L.; Varela, O. Synthesis of Branched Dithiotrisaccharides via Ring-Opening Reaction of Sugar Thiiranes. *J. Org. Chem.* **2012**, *77*, 253–265.
23. Ellis, D.; Norman, S. E.; Osborn, H. M. I. Synthesis of S-Linked Carbohydrate Analogues via a Ferrier Reaction. *Tetrahedron* **2008**, *64*, 2832–2854.
24. Dénès, F.; Pichowicz, M.; Povie, G.; Renaud, P. Thiyl Radicals in Organic Synthesis. *Chem. Rev.* **2014**, *114*, 2587–2693.
25. Dondoni, A. The Emergence of Thiol–Ene Coupling as a Click Process for Materials and Bioorganic Chemistry. *Angew. Chem. Int. Ed.* **2008**, *47*, 8995–8997.
26. Dondoni, A.; Marra, A. Recent Applications of Thiol–Ene Coupling as a Click Process for Glycoconjugation. *Chem. Soc. Rev.* **2012**, *41*, 573–586.
27. McSweeney, L.; Dénès, F.; Scanlan, E. M. Thiyl-Radical Reactions in CarboHydrate Chemistry: From Thiosugars to Glycoconjugate Synthesis. *Eur. J. Org. Chem.* **2016**, *2016*, 2080–2095.
28. Driguez, H. Thiooligosaccharides in Glycobiology. *Top. Curr. Chem.* **1997**, *187*, 85–116.
29. Witczak, Z. J. Thio Sugars: Biological Relevance as Potential New Therapeutics. *Curr. Med. Chem.* **1999**, *6*, 165–178.
30. Xin, G.; Zhu, X. A Facile and Highly Stereoselective Synthesis of 1-Thiotrehalose. *Tetrahedron Lett.* **2012**, *53*, 4309–4312.
31. Dere, R. T.; Wang, Y.; Zhu, X. A Direct and Stereospecific Approach to the Synthesis of α-Glycosyl Thiols. *Org. Biomol. Chem.* **2008**, *6*, 2061–2063.
32. Zhu, X.; Dere, R. T.; Jiang, J.; Zhang, L.; Wang, X. Synthesis of α-Glycosyl Thiols by Stereospecific Ring-Opening of 1,6-Anhydrosugars. *J. Org. Chem.* **2011**, *76*, 10187–10197.
33. Schmidt, R. R.; Michel, J. Facile Synthesis of α- and β-O-Glycosyl Imidates; Preparation of Glycosides and Disaccharides. *Angew. Chem. Int. Ed. Engl.* **1980**, *19*, 731–732.
34. Zeng, X.; Smith, R.; Zhu, X. Synthesis of Thioglycoside Analogues of Maradolipid. *J. Org. Chem.* **2013**, *78*, 4165–4170.
35. Tardieu, D.; Céspedes Dávila, M. F.; Hazelard, D.; Compain, P. An Expeditious Synthesis of 1-Thiotrehalose. *Synthesis* **2018**, *50*, 3927–3930.
36. Wang, H.; Zhu, X. Expedient Synthesis of an α-S-(1→6)-Linked Pentaglucosyl Thiol. *Org. Biomol. Chem.* **2014**, *12*, 7119–7126.

37. Caraballo, R.; Deng, L.; Amorim, L.; Brinck, T.; Ramström, O. pH-Dependent Mutarotation of 1-Thioaldoses in Water. Unexpected Behavior of (2S)-d-Aldopyranoses. *J. Org. Chem.* **2010**, *75*, 6115–6121.
38. Belz, T.; Williams, S. J. A Building Block Approach to the Synthesis of a Family of S-Linked α-1,6-Oligomannosides. *Carbohydr. Res.* **2016**, *429*, 38–47.
39. Morais, G. R.; Humphrey, A. J.; Falconer, R. A. A Facile Preparation of Trehalose Analogues: 1,1-Thiodisaccharides. *Carbohydr. Res.* **2009**, *344*, 1039–1045.
40. Jana, M.; Misra, A. K. Stereoselective Synthesis of β-Glycosyl Thiols and Their Synthetic Applications. *J. Org. Chem.* **2013**, *78*, 2680–2686.
41. Lázár, L.; Csávás, M.; Herczeg, M.; Herczegh, P.; Borbás, A. Synthesis of S-Linked Glycoconjugates and S-Disaccharides by Thiol–Ene Coupling Reaction of Enoses. *Org. Lett.* **2012**, *14*, 4650–4653.
42. Lázár, L.; Csávás, M.; Hadházi, Á.; Herczeg, M.; Tóth, M.; Somsák, L.; Barna, T.; Herczegh, P.; Borbás, A. Systematic Study on Free Radical Hydrothiolation of Unsaturated Monosaccharide Derivatives With Exo- and Endocyclic Double Bonds. *Org. Biomol. Chem.* **2013**, *11*, 5339–5350.
43. Eszenyi, D.; Kelemen, V.; Balogh, F.; Bege, M.; Csávás, M.; Herczegh, P.; Borbás, A. Promotion of a Reaction by Cooling: Stereoselective 1,2-cis-α-Thioglycoconjugation by Thiol-Ene Coupling at −80°C. *Chem. Eur. J.* **2018**, *24*, 4532–4536.
44. LeBel, N. A.; Czaja, R. F.; DeBoer, A. Stereochemistry of Free-Radical Additions of Thiols to Substituted Cyclohexenes. *J. Org. Chem.* **1969**, *34*, 3112–3126.
45. Lázár, L.; Borbás, A.; Somsák, L. Synthesis of Thiomaltooligosaccharides by a Thio-Click Approach. *Carbohydr. Res.* **2018**, *470*, 8–12.
46. Blanc-Muesser, M.; Vigne, L.; Driguez, H.; Lehmann, J.; Steck, J.; Urbahns, K. Spacer-Modified Disaccharide and Pseudo-Trisaccharide Methyl Glycosides That Mimic Maltotriose, as Competitive Inhibitors for Pancreatic Alpha-Amylase: A Demonstration of the "Clustering Effect". *Carbohydr. Res.* **1992**, *224*, 59–71.
47. Ratajczak, F.; Greffe, L.; Cottaz, S.; Driguez, H. Convergent Synthesis of 4-Thiomaltooligosaccharides. *Synlett* **2003**, 1253–1254.
48. Richard, M.; Didierjean, C.; Chapleur, Y.; Pellegrini-Moïse, N. Base- and Radical-Mediated Regio- and Stereoselective Additions of Thiols, Thio-Sugars, and Thiol-Containing Peptides to Trisubstituted Activated exo-Glycals. *Eur. J. Org. Chem.* **2015**, *2015*, 2632–2645.
49. Repetto, E.; Marino, C.; Uhrig, M. L.; Varela, O. Two Straightforward Strategies for the Synthesis of Thiodisaccharides with a Furanose Unit as the Nonreducing End. *Eur. J. Org. Chem.* **2008**, *2008*, 540–547.
50. Repetto, E.; Marino, C.; Uhrig, M. L.; Varela, O. Thiodisaccharides With Galactofuranose or Arabinofuranose as Terminal Units: Synthesis and Inhibitory Activity of an exo beta-D-Galactofuranosidase. *Bioorg. Med. Chem.* **2009**, *17*, 2703–2711.
51. Repetto, E.; Marino, C.; Varela, O. Synthesis of the (1→6)-Linked Thiodisaccharide of Galactofuranose: Inhibitory Activity Against a β-Galactofuranosidase. *Bioorg. Med. Chem.* **2013**, *21*, 3327–3333.
52. Lo Fiego, M. J.; Marino, C.; Varela, O. Synthesis of Galactofuranosyl-(1→5)-Thiodisaccharide Glycomimetics as Inhibitors of a β-d-Galactofuranosidase. *RSC Adv.* **2015**, *5*, 45631–45640.
53. Ferrier, R. J. Substitution-With-Allylic-Rearrangement Reactions of Glycal Derivatives. *Top. Curr. Chem.* **2001**, *215*, 153–176.
54. Ferrier, R. J.; Zubkov, O. A. Transformation of Glycals Into 2,3-Unsaturated Glycosyl Derivatives. In: Organic Reactions; 2004.

55. Gómez, A. M.; Lobo, F.; Uriel, C.; López, J. C. Recent Developments in the Ferrier Rearrangement. *Eur. J. Org. Chem.* **2013**, *2013*, 7221–7262.
56. Staderini, S.; Chambery, A.; Marra, A.; Dondoni, A. Free-Radical Hydrothiolation of Glycals: a Thiol-ene-Based Synthesis of S-Disaccharides. *Tetrahedron Lett.* **2012**, *53*, 702–704.
57. Lázár, L.; Juhász, L.; Batta, G.; Borbás, A.; Somsák, L. Unprecedented β-Manno Type Thiodisaccharides With a C-Glycosylic Function by Photoinitiated Hydrothiolation of 1-C-Substituted Glycals. *New J. Chem.* **2017**, *41*, 1284–1292.
58. Fiore, M.; Marra, A.; Dondoni, A. Photoinduced Thiol–Ene Coupling as a Click Ligation Tool for Thiodisaccharide Synthesis. *J. Org. Chem.* **2009**, *74*, 4422–4425.
59. Csávás, M.; Demeter, T.; Herczeg, M.; Timári, I.; Kövér, K. E.; Herczegh, P.; Borbás, A. Rapid Synthesis of Self-Assembling 1,2-Thiomannobioside Glycoconjugates as Potential Multivalent Ligands of Mannose-Binding Lectins. *Tetrahedron Lett.* **2014**, *55*, 6983–6986.
60. Csávás, M.; Malinovská, L.; Perret, F.; Gyurkó, M.; Illyés, Z. T.; Wimmerová, M.; Borbás, A. Tri- and Tetravalent Mannoclusters Cross-Link and Aggregate BC2L-A Lectin From Burkholderia cenocepacia. *Carbohydr. Res.* **2017**, *437*, 1–8.
61. Al-Shuaeeb, R. A. A.; Montoir, D.; Alami, M.; Messaoudi, S. Synthesis of (1→2)-S-Linked Saccharides and S-Linked Glycoconjugates via a Palladium-G3-XantPhos Precatalyst Catalysis. *J. Org. Chem.* **2017**, *82*, 6720–6728.
62. Witczak, Z. J.; Kaplon, P.; Markus Dey, P. Thio-Sugars VII. Effect of 3-deoxy-4-S-(β-d-gluco- and β-d-Galactopyranosyl)-4-Thiodisaccharides and Their Sulfoxides and Sulfones on the Viability and Growth of Selected Murine and Human Tumor Cell Lines. *Carbohydr. Res.* **2003**, *338*, 11–18.
63. Witczak, Z. J.; Sarnik, J.; Czubatka, A.; Forma, E.; Poplawski, T. Thio-Sugar Motif of Functional CARB-Pharmacophore for Antineoplastic Activity. Part 2. *Bioorg. Med. Chem. Lett.* **2014**, *24*, 5606–5611.
64. Korycka-Machała, M.; Brzostek, A.; Dziadek, B.; Kawka, M.; Popławski, T.; Witczak, Z.; Dziadek, J. Evaluation of the Mycobactericidal Effect of Thio-Functionalized Carbohydrate Derivatives. *Molecules* **2017**, *22*, 812.
65. Uhrig, M. L.; Manzano, V. E.; Varela, O. Stereoselective Synthesis of 3-Deoxy-4-S-(1→4)-Thiodisaccharides and Their Inhibitory Activities Towards β-Glycoside Hydrolases. *Eur. J. Org. Chem.* **2005**, *2006*, 162–168.
66. Uhrig, M. L.; Szilágyi, L.; Kövér, K. E.; Varela, O. Synthesis of Non-Glycosidic 4,6′-Thioether-Linked Disaccharides as Hydrolytically Stable Glycomimetics. *Carbohydr. Res.* **2007**, *342*, 1841–1849.
67. Cagnoni, A. J.; Uhrig, M. L.; Varela, O. Synthesis of Pentopyranosyl-Containing Thiodisaccharides. Inhibitory Activity Against β-Glycosidases. *Bioorg. Med. Chem.* **2009**, *17*, 6203–6212.
68. Cagnoni, A. J.; Varela, O.; Kovensky, J.; Uhrig, M. L. Synthesis and Biological Activity of Divalent Ligands Based on 3-deoxy-4-Thiolactose, an Isosteric Analogue of Lactose. *Org. Biomol. Chem.* **2013**, *11*, 5500–5511.
69. Dada, L.; Manzano, V. E.; Varela, O. Design and Synthesis of 2-Acetamido-2,3-dideoxythiodisaccharides via Diastereoselective Conjugate Addition to Sugar Enone O-Acetyl Oximes. Galactosidase Inhibition Studies. *Org. Lett.* **2018**, *20*, 6225–6228.
70. Manzano, V. E.; Uhrig, M. L.; Varela, O. Convenient Synthesis of 4-Thiolactose, 3,4-Dithiolactose and Related Thiooligosaccharides and Disulfides. Inhibitory Activity of the Glycomimetics Against a β-Galactosidase. *Org. Biomol. Chem.* **2012**, *10*, 8884–8894.

71. Luche, J. L. Lanthanides in Organic Chemistry. 1. Selective 1,2 Reductions of Conjugated Ketones. *J. Am. Chem. Soc.* **1978**, *100*, 2226–2227.
72. Colomer, J. P.; Manzano, V. E.; Varela, O. Synthesis of Thiodisaccharide Sulfoxides and Sulfones—Determination of the Configuration of the Sulfur Stereocentre. *Eur. J. Org. Chem.* **2013**, *2013*, 7343–7353.
73. Cumpstey, I.; Ramstadius, C.; Akhtar, T.; Goldstein, I. J.; Winter, H. C. Non-Glycosidically Linked Pseudodisaccharides: Thioethers, Sulfoxides, Sulfones, Ethers, Selenoethers, and Their Binding to Lectins. *Eur. J. Org. Chem.* **2010**, *2010*, 1951–1970.
74. Suzuki, T.; Makyio, H.; Ando, H.; Komura, N.; Menjo, M.; Yamada, Y.; Imamura, A.; Ishida, H.; Wakatsuki, S.; Kato, R.; Kiso, M. Expanded Potential of Seleno-Carbohydrates as a Molecular Tool for X-Ray Structural Determination of a Carbohydrate–Protein Complex With Single/Multi-Wavelength Anomalous Dispersion Phasing. *Bioorg. Med. Chem.* **2014**, *22*, 2090–2101.
75. André, S.; Kövér, K. E.; Gabius, H.-J.; Szilágyi, L. Thio- and Selenoglycosides as Ligands for Biomedically Relevant Lectins: Valency–Activity Correlations for Benzene-Based Dithiogalactoside Clusters and First Assessment for (di)Selenodigalactosides. *Bioorg. Med. Chem. Lett.* **2015**, *25*, 931–935.
76. Czernecki, S.; Randriamandimby, D. Selenoglycosides 5.1 Stereocontrolled Synthesis of Seleno-Disaccharides. *J. Carbohydr. Chem.* **1996**, *15*, 183–190.
77. Mehta, S.; Andrews, J. S.; Svensson, B.; Pinto, B. M. Synthesis and Enzymic Activity of Novel Glycosidase Inhibitors Containing Sulfur and Selenium. *J. Am. Chem. Soc.* **1995**, *117*, 9783–9790.
78. Kawai, Y.; Ando, H.; Ozeki, H.; Koketsu, M.; Ishihara, H. A Facile Method for β-Selenoglycoside Synthesis Using β-p-Methylbenzoyl Selenoglycoside as the Selenating Unit. *Org. Lett.* **2005**, *7*, 4653–4656.
79. Nanami, M.; Ando, H.; Kawai, Y.; Koketsu, M.; Ishihara, H. Stereoselective Synthesis of Various α-Selenoglycosides Using In Situ Production of α-Selenolate Anion. *Tetrahedron Lett.* **2007**, *48*, 1113–1116.
80. Suzuki, T.; Komura, N.; Imamura, A.; Ando, H.; Ishida, H.; Kiso, M. A Facile Method for Synthesizing Selenoglycosides Based on Selenium-Transfer to Glycosyl Imidate. *Tetrahedron Lett.* **2014**, *55*, 1920–1923.
81. Kumar, A. A.; Illyés, T. Z.; Kövér, K. E.; Szilágyi, L. Convenient Syntheses of 1,2-trans Selenoglycosides Using Isoselenuronium Salts as Glycosylselenenyl Transfer Reagents. *Carbohydr. Res.* **2012**, *360*, 8–18.
82. Fournière, V.; Cumpstey, I. Synthesis of Non-Glycosidically Linked Selenoether Pseudodisaccharides. *Tetrahedron Lett.* **2010**, *51*, 2127–2129.
83. Affeldt, R. F.; Braga, H. C.; Baldassari, L. L.; Lüdtke, D. S. Synthesis of Selenium-Linked Neoglycoconjugates and Pseudodisaccharides. *Tetrahedron* **2012**, *68*, 10470–10475.
84. Braga, H. C.; Stefani, H. A.; Paixão, M. W.; Santos, F. W.; Lüdtke, D. S. Synthesis of 5′-Seleno-Xylofuranosides. *Tetrahedron* **2010**, *66*, 3441–3446.
85. Pifferi, C.; Daskhan, G. C.; Fiore, M.; Shiao, T. C.; Roy, R.; Renaudet, O. Aminooxylated Carbohydrates: Synthesis and Applications. *Chem. Rev.* **2017**, *117*, 9839–9873.
86. Borges de Melo, E.; da Silveira Gomes, A.; Carvalho, I. α- and β-Glucosidase inhibitors: Chemical Structure and Biological Activity. *Tetrahedron* **2006**, *62*, 10277–10302.
87. Asano, N. Glycosidase Inhibitors: Update and Perspectives on Practical Use. *Glycobiology* **2003**, *13*, 93R–104R.
88. Chen, X.; Fan, Y.; Zheng, Y.; Shen, Y. Properties and Production of Valienamine and Its Related Analogues. *Chem. Rev.* **2003**, *103*, 1955–1978.

89. Cumpstey, I.; Ramstadius, C.; Borbas, K. E.; Alonzi, D. S.; Butters, T. D. Synthesis and α-Glucosidase II Inhibitory Activity of Valienamine Pseudodisaccharides Relevant to N-Glycan Biosynthesis. *Bioorg. Med. Chem. Lett.* **2011**, *21*, 5219–5223.
90. Cumpstey, I.; Gehrke, S.; Erfan, S.; Cribiu, R. Studies on the Synthesis of Valienamine and 1-epi-Valienamine Starting From d-Glucose or l-Sorbose. *Carbohydr. Res.* **2008**, *343*, 1675–1692.
91. Cumpstey, I.; Ramstadius, C.; Borbas, K. E. Investigation of Coupling Reactions for the Synthesis of Valienamine Pseudodisaccharides. *Synlett* **2011**, *2011*, 1701–1704.
92. Shing, T. K. M.; Cheng, H. M.; Wong, W. F.; Kwong, C. S. K.; Li, J.; Lau, C. B. S.; Leung, P. S.; Cheng, C. H. K. Enantiospecific Synthesis of Pseudoacarviosin as a Potential Antidiabetic Agent. *Org. Lett.* **2008**, *10*, 3145–3148.
93. Shing, T. K. M.; Cheng, H. M. An Alternative Synthesis of 1,1′-Bis-Valienamine From d-Glucose. *J. Org. Chem.* **2010**, *75*, 3522–3525.
94. Neimert-Andersson, K.; Sauer, S.; Panknin, O.; Borg, T.; Söderlind, E.; Somfai, P. Synthesis of New Sugar-Based Surfactants and Evaluation of Their Hemolytic Activities. *J. Org. Chem.* **2006**, *71*, 3623–3626.
95. Coxon, B. A Karplus Equation for 3JHCCN in Amino Sugar Derivatives. *Carbohydr. Res.* **2007**, *342*, 1044–1054.
96. Kroutil, J.; Buděšínský, M. Preparation of Diamino Pseudodisaccharide Derivatives From 1,6-Anhydro-β-d-Hexopyranoses via Aziridine-Ring Cleavage. *Carbohydr. Res.* **2007**, *342*, 147–153.
97. Akhtar, T.; Cumpstey, I. Investigations Into the Synthesis of Amine-Linked Neodisaccharides. *Tetrahedron Lett.* **2007**, *48*, 8673–8677.
98. Cumpstey, I.; Frigell, J.; Pershagen, E.; Akhtar, T.; Moreno-Clavijo, E.; Robina, I.; Alonzi, D. S.; Butters, T. D. Amine-Linked Diglycosides: Synthesis Facilitated by the Enhanced Reactivity of Allylic Electrophiles, and Glycosidase Inhibition Assays. *Beilstein J. Org. Chem.* **2011**, *7*, 1115–1123.
99. Neumann, J.; Weingarten, S.; Thiem, J. Synthesis of Novel Di- and Trisaccharide Mimetics with Non-Glycosidic Amino Bridges. *Eur. J. Org. Chem.* **2007**, *2007*, 1130–1144.
100. Peri, F.; Marinzi, C.; Barath, M.; Granucci, F.; Urbano, M.; Nicotra, F. Synthesis and Biological Evaluation of Novel Lipid A Antagonists. *Bioorg. Med. Chem.* **2006**, *14*, 190–199.
101. Merino, P.; Tejero, T.; Marca, E.; Gomollón-Bel, F.; Delso, I.; Matute, R. Cross-Coupling Reactions for the Synthesis of C-Glycosides and Related Compounds. *Heterocycles* **2012**, *86*, 791–820.
102. Liao, H.; Ma, J.; Yao, H.; Liu, X.-W. Recent Progress of C-Glycosylation Methods in the Total Synthesis of Natural Products and Pharmaceuticals. *Org. Biomol. Chem.* **2018**, *16*, 1791–1806.
103. Choumane, M.; Banchet, A.; Probst, N.; Gérard, S.; Plé, K.; Haudrechy, A. The Synthesis of D-C-Mannopyranosides. *C. R. Chim.* **2011**, *14*, 235–273.
104. Yuan, X.; Linhardt, R. J. Recent Advances in the Synthesis of C-Oligosaccharides. *Curr. Top. Med. Chem.* **2005**, *5*, 1393–1430.
105. Lei, L.; Michael, M.; Maarten, H. D. P. Synthesis of C-Saccharides and Higher Congeners. *Curr. Org. Chem.* **2001**, *5*, 1133–1167.
106. Yang, Y.; Yu, B. Recent Advances in the Chemical Synthesis of C-Glycosides. *Chem. Rev.* **2017**, *117*, 12281–12356.
107. Kitamura, K.; Ando, Y.; Matsumoto, T.; Suzuki, K. Total Synthesis of Aryl C-Glycoside Natural Products: Strategies and Tactics. *Chem. Rev.* **2018**, *118*, 1495–1598.

108. Bokor, É.; Kun, S.; Goyard, D.; Tóth, M.; Praly, J.-P.; Vidal, S.; Somsák, L. C-Glycopyranosyl Arenes and Hetarenes: Synthetic Methods and Bioactivity Focused on Antidiabetic Potential. *Chem. Rev.* **2017**, *117*, 1687–1764.
109. Narute, S. B.; Rout, J. K.; Ramana, C. V. Synthesis of C-Disaccharides Through a One-Pot Alkynol Cycloisomerization-Reductive Deoxygenation. *Chem. Eur. J.* **2013**, *19*, 15109–15114.
110. Koester, D. C.; Kriemen, E.; Werz, D. B. Flexible Synthesis of 2-Deoxy-C-Glycosides and (1→2)-, (1→3)-, and (1→4)-Linked C-Glycosides. *Angew. Chem. Int. Ed.* **2013**, *52*, 2985–2989.
111. Oroszova, B.; Choutka, J.; Pohl, R.; Parkan, K. Modular Stereoselective Synthesis of (1→2)-C-Glycosides Based on the sp2–sp3 Suzuki–Miyaura Reaction. *Chem. Eur. J.* **2015**, *21*, 7043–7047.
112. Hussain, N.; Babu Tatina, M.; Mukherjee, D. Cross Dehydrogenative Coupling of Sugar Enol Ethers With Terminal Alkenes in the Synthesis of Pseudo-Disaccharides, Chiral Oxadecalins and a Conjugated Triene. *Org. Biomol. Chem.* **2018**, *16*, 2666–2677.
113. Weinstein, D. S.; Nicolaou, C. K. Synthesis of the Namenamicin A–C Disaccharide: Towards the Total Synthesis of Namenamicin. *J. Chem. Soc. Perkin Trans.* **1999**, *1*, 545–558.
114. Nicolaou, K. C.; Li, R.; Lu, Z.; Pitsinos, E. N.; Alemany, L. B. Total Synthesis and Full Structural Assignment of Namenamicin. *J. Am. Chem. Soc.* **2018**, *140*, 8091–8095.
115. Smith, A. L.; Nicolaou, K. C. The Enediyne Antibiotics. *J. Med. Chem.* **1996**, *39*, 2103–2117.
116. Scharf, D. H.; Habel, A.; Heinekamp, T.; Brakhage, A. A.; Hertweck, C. Opposed Effects of Enzymatic Gliotoxin N- and S-Methylations. *J. Am. Chem. Soc.* **2014**, *136*, 11674–11679.
117. Bang, E.-K.; Gasparini, G.; Molinard, G.; Roux, A.; Sakai, N.; Matile, S. Substrate-Initiated Synthesis of Cell-Penetrating Poly(disulfide)s. *J. Am. Chem. Soc.* **2013**, *135*, 2088–2091.
118. Jiang, C.-S.; Müller, W. E. G.; Schröder, H. C.; Guo, Y.-W. Disulfide- and Multisulfide-Containing Metabolites From Marine Organisms. *Chem. Rev.* **2012**, *112*, 2179–2207.
119. Chen, N.; Xie, J. N–O Linkage in Carbohydrates and Glycoconjugates. *Org. Biomol. Chem.* **2016**, *14*, 11028–11047.
120. Murthy, B. N.; Sinha, S.; Surolia, A.; Jayaraman, N.; Szilágyi, L.; Szabó, I.; Kövér, K. E. Interactions of Aromatic Mannosyl Disulfide Derivatives With Concanavalin A: Synthesis, Thermodynamic and NMR Spectroscopy Studies. *Carbohydr. Res.* **2009**, *344*, 1758–1763.
121. André, S.; Pei, Z.; Siebert, H.-C.; Ramström, O.; Gabius, H.-J. Glycosyldisulfides From Dynamic Combinatorial Libraries as O-Glycoside Mimetics for Plant and Endogenous Lectins: Their Reactivities in Solid-Phase and Cell Assays and Conformational Analysis by Molecular Dynamics Simulations. *Bioorg. Med. Chem.* **2006**, *14*, 6314–6326.
122. Caraballo, R.; Sakulsombat, M.; Ramström, O. Phosphine-Mediated Disulfide Metathesis in Aqueous Media. *Chem. Commun.* **2010**, *46*, 8469–8471.
123. Adinolfi, M.; Capasso, D.; Di Gaetano, S.; Iadonisi, A.; Leone, L.; Pastore, A. A Straightforward Synthetic Access to Symmetrical Glycosyl Disulfides and Biological Evaluation Thereof. *Org. Biomol. Chem.* **2011**, *9*, 6278–6283.
124. Ge, J.-T.; Zhou, L.; Zhao, F.-L.; Dong, H. Straightforward S–S Bond Formation via the Oxidation of S-Acetyl by Iodine in the Presence of N-Iodosuccinimide. *J. Org. Chem.* **2017**, *82*, 12613–12623.

125. Szilágyi, L.; Illyés, T.-Z.; Herczegh, P. Elaboration of a Novel Type of Interglycosidic Linkage: Syntheses of Disulfide Disaccharides. *Tetrahedron Lett.* **2001**, *42*, 3901–3903.
126. Ribeiro Morais, G.; Falconer, R. A. Efficient One-Pot Synthesis of Glycosyl Disulfides. *Tetrahedron Lett.* **2007**, *48*, 7637–7641.
127. Aversa, M. C.; Barattucci, A.; Bonaccorsi, P. Efficient Synthesis of Unsymmetrical Disulfides Through Sulfenic Acids. *Eur. J. Org. Chem.* **2009**, *2009*, 6355–6359.
128. Stellenboom, N.; Hunter, R.; Caira, M. R.; Szilágyi, L. A High-Yielding, One-Pot Preparation of Unsymmetrical Glycosyl Disulfides Using 1-Chlorobenzotriazole as an In Situ Trapping/Oxidizing Agent. *Tetrahedron Lett.* **2010**, *51*, 5309–5312.
129. Illyés, T.-Z.; Szabó, T.; Szilágyi, L. Glycosylation via Mixed Disulfide Formation Using Glycosylthio-Phthalimides and -Succinimides as Glycosylsulfenyl-Transfer Reagents. *Carbohydr. Res.* **2011**, *346*, 1622–1627.
130. Vandavasi, J. K.; Hu, W.-P.; Chen, C.-Y.; Wang, J.-J. Efficient Synthesis of Unsymmetrical Disulfides. *Tetrahedron* **2011**, *67*, 8895–8901.
131. Smith, R.; Zeng, X.; Müller-Bunz, H.; Zhu, X. Synthesis of Glycosyl Disulfides Containing an α-Glycosidic Linkage. *Tetrahedron Lett.* **2013**, *54*, 5348–5350.
132. Illyés, T.-Z.; Balla, S.; Bényei, A.; Kumar, A. A.; Timári, I.; Kövér, K. E.; Szilágyi, L. Exploring the Syntheses of Novel Glycomimetics. Carbohydrate Derivatives With Se-S- or Se-Se- Glycosidic Linkages. *Chem. Sel.* **2016**, *1*, 2383–2388.
133. Lázár, L.; Csávás, M.; Tóth, M.; Somsák, L.; Borbás, A. Thio-Click Approach to the Synthesis of Stable Glycomimetics. *Chem. Pap.* **2015**, *69*, 889–895.
134. József, J.; Juhász, L.; Illyés, T. Z.; Csávás, M.; Borbás, A.; Somsák, L. Photoinitiated Hydrothiolation of Pyranoid Exo-Glycals: The D-Galacto and D-Xylo Cases. *Carbohydr. Res.* **2015**, *413*, 63–69.
135. Bege, M.; Bereczki, I.; Herczeg, M.; Kicsák, M.; Eszenyi, D.; Herczegh, P.; Borbás, A. A Low-Temperature, Photoinduced Thiol–Ene Click Reaction: A Mild and Efficient Method for the Synthesis of Sugar-Modified Nucleosides. *Org. Biomol. Chem.* **2017**, *15*, 9226–9233.
136. Megia-Fernandez, A.; de la Torre-Gonzalez, D.; Parada-Aliste, J.; Lopez-Jaramillo, F. J.; Hernandez-Mateo, F.; Santoyo-Gonzalez, F. Masked Thiol Sugars: Chemical Behavior and Synthetic Applications of S-Glycopyranosyl-N-Monoalkyl Dithiocarbamates. *Chem. Asian J.* **2013**, *9*, 620–631.
137. Smellie, I. A. S.; Moggach, S. A.; Paton, R. M. Synthesis of Novel Amidoxime-Linked Pseudodisaccharides. *Tetrahedron Lett.* **2011**, *52*, 95–97.
138. Gong, Y.; Sun, H.; Xie, J. Synthesis of Oligosaccharide Mimetics With Glycoaminoxy Acids. *Eur. J. Org. Chem.* **2009**, *2009*, 6027–6033.
139. Blériot, Y.; Auberger, N.; Jagadeesh, Y.; Gauthier, C.; Prencipe, G.; Tran, A. T.; Marrot, J.; Désiré, J.; Yamamoto, A.; Kato, A.; Sollogoub, M. Synthesis of 1,2-cis-Homoiminosugars Derived from GlcNAc and GalNAc Exploiting a β-Amino Alcohol Skeletal Rearrangement. *Org. Lett.* **2014**, *16*, 5512–5515.
140. Lopez, M.; Bornaghi, L. F.; Driguez, H.; Poulsen, S.-A. Synthesis of Sulfonamide-Bridged Glycomimetics. *J. Org. Chem.* **2011**, *76*, 2965–2975.
141. Jiménez Blanco, J. L.; Ortega-Caballero, F.; Ortiz Mellet, C.; García Fernández, J. M. (Pseudo)Amide-Linked Oligosaccharide Mimetics: Molecular Recognition and Supramolecular Properties. *Beilstein J. Org. Chem.* **2010**, *6*(20).
142. Jiménez Blanco, J. L.; Bootello, P.; Benito, J. M.; Ortiz Mellet, C.; García Fernández, J. M. Urea-, Thiourea-, and Guanidine-Linked Glycooligomers as Phosphate Binders in Water. *J. Org. Chem.* **2006**, *71*, 5136–5143.

143. Wessel, H. P.; Lucas, S. D. Oligosaccharide Mimetics. In: Glycoscience: Chemistry and Chemical Biology; Fraser-Reid, B. O., Tatsuta, K., Thiem, J., Eds.; Springer: Berlin, Heidelberg, 2008; pp 2079–2112.
144. Prosperi, D.; Ronchi, S.; Lay, L.; Rencurosi, A.; Russo, G. Efficient Synthesis of Unsymmetrical Ureido-Linked Disaccharides. *Eur. J. Org. Chem.* **2004**, *2004*, 395–405.

Advancements in synthetic and structural studies of septanoside sugars

6

Supriya Dey[a], *Gopal Ch Samanta*, *N. Jayaraman*
Department of Organic Chemistry, Indian Institute of Science, Bangalore, India

Chapter outline

1 Introduction

The family of sugars having the seven-membered oxepane ring system is termed as septanoses and septanosides. Naturally occurring sugars generally adopt a five-membered furanose or six-membered pyranose structural forms, rather than acyclic or higher homologous structural forms, even when such alternate structural forms are possible. Pyranose or furanose cyclic structures are thermodynamically favored[1] and all naturally occurring sugars adopt one of these two forms largely. The notion that a hexose sugar can adopt a structure over and above the commonly referred acyclic, furanose and pyranose forms was reported decades ago by Micheel and coworkers,[2] by allowing the formation of the lactol involving the primary hydroxy group and the aldehyde functionality. This first known septanose sugar was synthesized as the corresponding *O*-peracetate in the early 1930s. Involving the primary hydroxy group to react with the anomeric carbon is also the early known levoglucosan formation,[3]

[a] Current address: Department of Chemistry, The Scripps Research Institute, La Jolla, CA, United States.

Recent Trends in Carbohydrate Chemistry. https://doi.org/10.1016/B978-0-12-817467-8.00006-2
© 2020 Elsevier Inc. All rights reserved.

producing a bicyclic anhydrosugar, constituted of six- and seven-membered ring systems. However, the relevance of isolating monocyclic septanoses was sought-after for reasons that the lactol formation through primary hydroxy group was thought as a distinct possibility. Report of such a success was the methylation reaction of D-galactose, which afforded methyl 2,3,4,5-tetra-*O*-methyl-α-D-galactoseptanoside, in a considerable yield.[4] Substituting the C-6 hydroxy group of a galactopyranose with a sulfanyl group, followed by reaction under acidic treatment, led to a septanose derivative containing an endocyclic sulfur, thereby forming a thioseptanose derivative.[5] 2,3,4,5-Tetra-*O*-methyl-hexoses have a high propensity to provide septanosides under acidic conditions in the presence of an alcohol, as demonstrated early by Anet.[6] This study also attempted to rationalize the conformation of a resulting septanoside as that of a skew-chair conformation, in a close comparison to a less complex cycloheptane structure. Cyclization of a hexose to a septanose was thus recognized as a fruitful approach, to derive sugar structures beyond furanose and pyranose ring structures. The approach was even more elegantly exploited by Stevens and coworkers, in order to secure septanosides and to establish the conformations. Acid treatment of glucose **1** with acetone afforded D-glucoseptanoses **2** and **3**, as diisopropylidene acetals (Scheme 1A).[7] When this reaction was conducted in a kilogram-quantity, it led to a practical method to prepare an anomeric mixture of acetonide protected septanosides **2α** (64 g, 4.2%) and **2β** (43 g, 2.8%). The studies expanded further to establish the conformations of the septanosides in solution and in the solid-state systematically.[8] Acid-mediated ring-opening of a glycopyranoside and ring-closing cyclization was also developed to secure septanosides, as reported by Gelas and coworkers.[9] Thus, treatment of methyl 4,6-*O*-isopropylidene-α-D-glucopyranoside **4** with pyridinium chloride afforded methyl 2,3:4,5-di-*O*-isopropylidene-α- and β-D-glucoseptanoside **5**, in 43% yield and the reaction was conducted without requiring a solvent (Scheme 1B). Diacetonide protection in **5** from monoacetonide-protected **4** occurred through the presumed release of acetone during the course of the reaction.

Whereas the above early developments relied on the reactivity of primary hydroxy group at C-6 carbon with the aldehyde to demonstrate that septanose formation was also a mode of intramolecular cyclization for a hexose, apart from furanose and pyranose forms, dedicated efforts to develop synthetic methods targeting synthesis of septanoses and their derivatives have evolved during the last couple of decades through innovative synthetic approaches. The interest in developing synthetic methods is elevated further due to the oxepane ring system to which septanoses belong to. The following sections discuss features of these newer methods to synthesize functionalized septanoses.

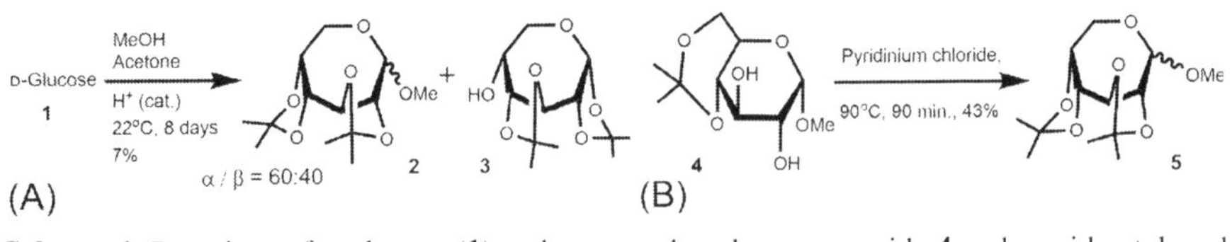

Scheme 1 Reactions of D-glucose (**1**) and protected D-glucopyranoside **4** under acid-catalyzed conditions to afford a septanose **3**, septanosides **2** and **5**.

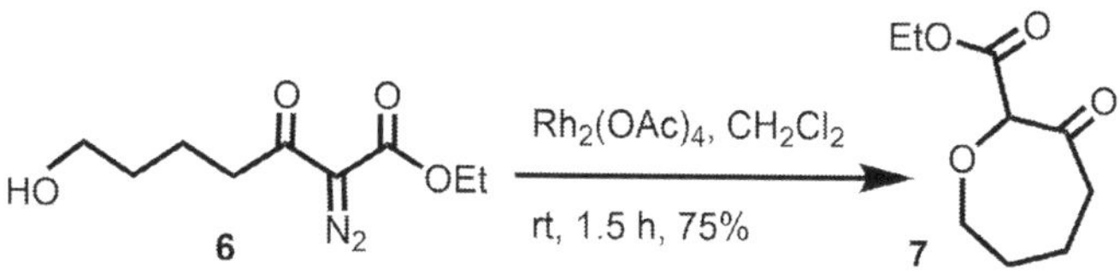

Scheme 2 Reaction of a diazo-synthon **6** to afford oxepane scaffold **7**.

Oxepane scaffolds are seen commonly as a part of multiple fused ring systems, devoid of one or more hydroxy groups, a classic example is the polycyclic ether brevetoxin, which is a marine neurotoxin. Marine and other bioactive natural products present oxepane as a core structure around which multiple modifications occur, resulting in heavily functionalized natural products that exhibit a range of beneficial biological effects.[10, 11] Intriguing structures of these natural products provided opportunities and challenges associated with their synthesis, with a particular emphasis on the assembly of the oxepane scaffold.[12] Cyclization of a preformed linear precursor to the corresponding oxepane system is one of the fruitful approaches to the assembly, an example is the cyclization through rhodium carbene–OH insertion of an appropriately positioned diazo–OH synthon **6**, which afforded oxepane scaffold **7**, in a good yield when conducted on a several grams scale (Scheme 2).[13] Whereas natural product structures provide impetus to assemble the oxepane scaffold, reactions that permit synthesis of septanosides and their derivatives through synthetic methods have been an intense topic of investigation in heptose sugars. Important approaches to devise methods for selective septanose formation rely predominantly on ring-closing reactions of linear precursors and one atom homologation through ring expansion of six-membered ring functionalized precursors. Periodic reviews of the progress in septanoside syntheses appear routinely.[14–17]

2 Ring-closing reactions of linear precursors to septanoses

A linear sugar precursor undergoing cyclization was recognized as one of the earliest methods to derive a septanoside. The equilibration of free sugar to cyclic structures, wherein septanose is one of the forms, was recognized early. When studying the reaction of glucose with acetone, in the presence of catalytic amount of H_2SO_4, Stevens observed the cyclization to 1,2:3,4-di-*O*-isopropylidene-α-D-glucoseptanose, in 1.9% yield, the rest being glucofuranose derivative.[8] The Knoevenagel condensation of a dialdehyde with nitromethane, secured through periodate oxidation of methyl 4,6-*O*-benzylidene-α-D-glucopyranoside **8**, afforded septanoside **9** (Scheme 3).[18] This reaction is an early example for the synthesis of septanosides through cyclization involving non-anomeric carbon. Such carbanion condensation reaction with dialdehyde was also utilized to prepare septanosides of the type as **10**.[19] Other dialdehydes **12** and **15** were taken up for further exploitation to conduct nitroaldol condensation reaction so as to afford 4-nitroseptanosides **13** and **16**, initiated from pyranosides **11** and **14**,

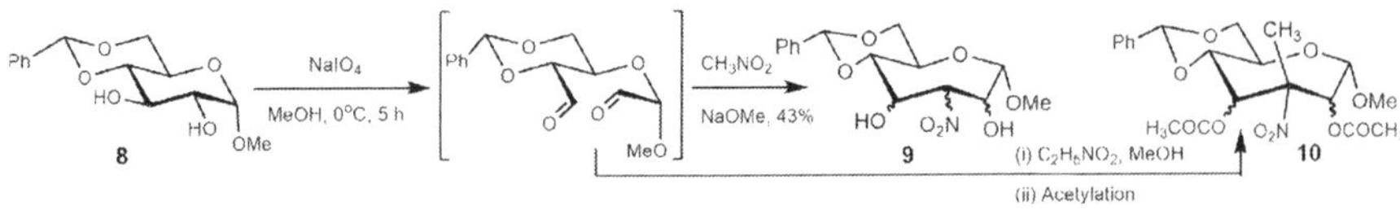

Scheme 3 Knoevenagel reaction of pyranoside derivative **8** to afford septanosides **9** and **10**.

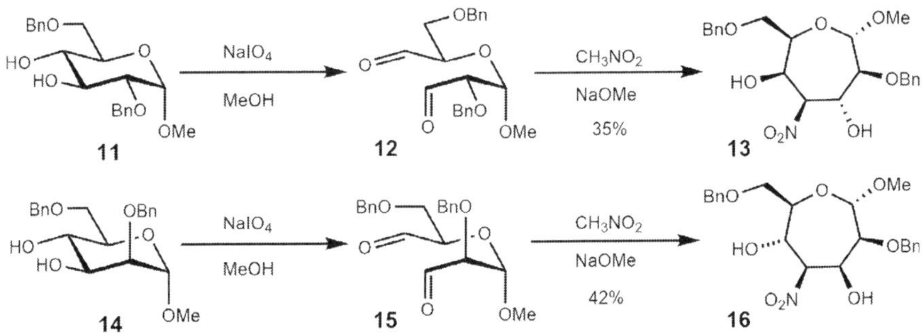

Scheme 4 Reactions of diols **11** and **14** to afford septanosides **13** and **16**, through dialdehyde intermediates **12** and **15**.

respectively (Scheme 4).[20] The dialdehyde precursor for the carbanion condensation still required partially protected diol sugar synthons, as unprotected pyranosides underwent oxidative cleavage at multiple hydroxymethines, their subsequent nitroaldol condensation led to cyclization products with a smaller ring size.

Innovative cyclizations involving 1,3-dipolar cycloaddition reactions emerge to be an interesting approach to prepare functionalized oxepanes and septanosides. Reissig and coworkers developed nitrones as dipoles reacting with an allene dipolarophile, followed by acid-induced cycloetherification between the olefin and distal hydroxy group, to afford functionalized septanosides. A [3+3] cycloaddition, leading to an oxazine, is the key intermediate which acts as synthon for the subsequent cycloetherification reaction. The use of oxazine also ensures functionalization of the resulting septanosides with an amino functionality at C-2 carbon.[21] The scheme of reactions begins from aldopyranose-derived *N*-glycosylhydroxylamine **17a** undergoing (i) [3+3] cyclization to afford **18**; (ii) acid-catalyzed cycloetherification to **19,** and (iii) SmI_2-mediated reduction of 1,2-oxazine to 2-amino-2-deoxyseptanoside **20** (Scheme 5).

The linear precursor **17** is synthesized from 2,3,4-tri-*O*-benzyl-protected aldopyranose sugars, namely, L-arabinose, D-ribose, and D-xylose, upon the reaction of the corresponding lactols with *N*-benzylhydroxylamine. Although *N*-glycosylhydroxylamine **17a** equilibrates in the cyclic and linear forms, the reaction with *tert*-butyldiphenylsilyl (TBDPS) chloride, in the presence of a base, affords *O*-TBDPS-protected acyclic nitrone **17b**.[22] A facile [3+3]-cycloaddition reaction ensues when lithiated *O*-methoxyallene is treated with the nitrones, via the *O*-allenyl intermediate, leading to the 1,2-oxazines, as in **18**. Whereas high diastereomeric purity results with xylose and arabinose synthons, the cycloaddition reaction with ribose derivative affords 2:1 *syn*/*anti* configuration at the newly formed C–N bond. Treatment of 1,2-oxazine with

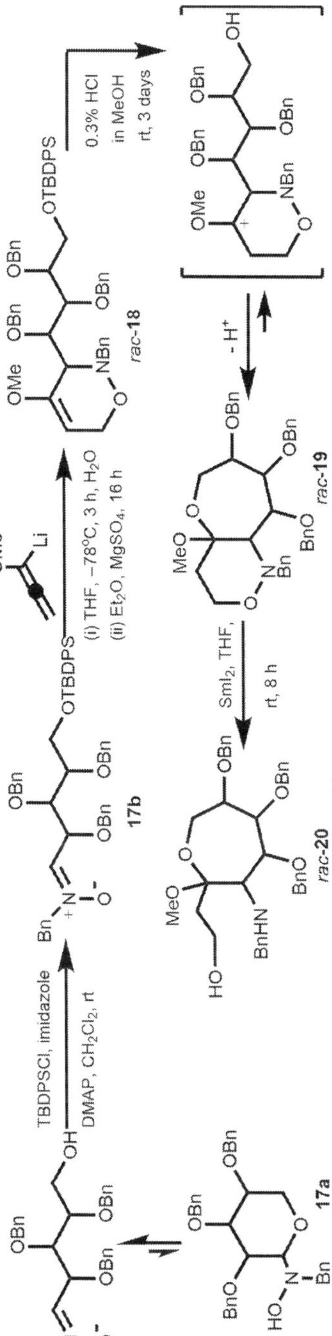

Scheme 5 *N*-glycosylhydroxylamine **17** as a synthon to prepare a septanoside *rac*-**20**.

an acid promotes a cycloetherification involving primary hydroxy group with the endocyclic vinyl ether, resulting in the formation of 6–7 rings fused systems as **19**, with a mixed *cis* and *trans*-ring junctures. The final step of judiciously chosen SmI_2 reagent for the reduction of the 1,2-azine moiety led to novel 2-amino-2-deoxyseptanosides **20–24** (Scheme 5 and Fig. 1).

Endo-cycloetherification of ε,ω-unsaturated alcohol is a fruitful approach to synthesize a septanose ring, from an appropriately functionalized linear precursor. This approach, developed by Castillón and coworkers, required Wittig-Horner reaction of a protected sugar lactol **25** to the corresponding sulfanylalkene derivative **26** (Scheme 6).[23] When this derivative was subjected to electrophile-induced cycloetherification reaction, septanose **27** formation occurred in a moderate yield. The presence of isopropylidene protecting group was found to be necessary in order for the *endo*-cyclization to occur.

The usefulness of the nitrone-alkene cyclization was demonstrated earlier for the formation of oxepanes, from furanose-derived synthons. Upon installation of *O*-allyl moiety at C-3 (**28**) and nitrone moiety at C-4 (**29**) of a furanose derivative, Bhattacharjya and coworkers demonstrated oxepane **32** as a major product of [3+2] cycloaddition (Scheme 7).[24] The substitution at nitrone was observed to influence the cycloaddition, leading to either 6-5 fused (**30**) or 7-5 fused (**31**) bicyclic ring system. The authors surmised a better overlap of frontier orbitals with *N*-phenyl substituent on nitrone and such overlap might not occur with *N*-methyl/*N*-benzyl substituent at the nitrone moiety in the formation of derivative **31**. Further reactions of deprotection,

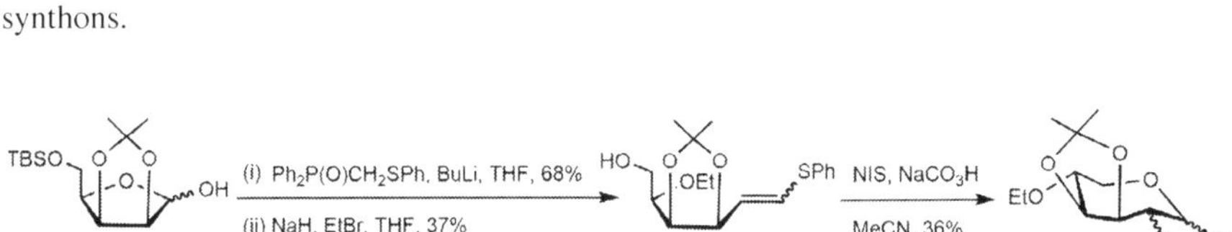

Fig. 1 Molecular structures of septanosides **21–24**, prepared from *N*-glycosylhydroxylamine synthons.

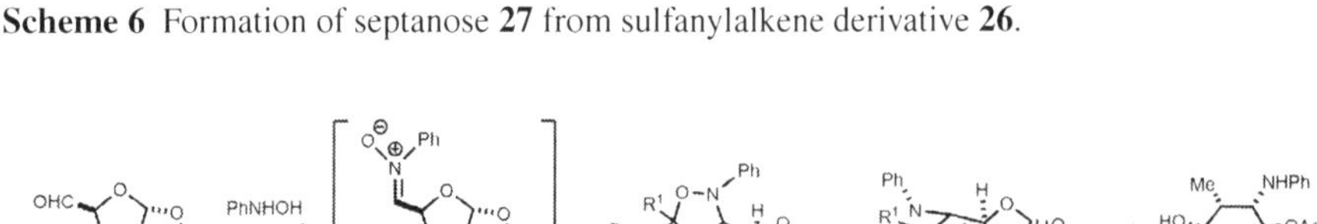

Scheme 6 Formation of septanose **27** from sulfanylalkene derivative **26**.

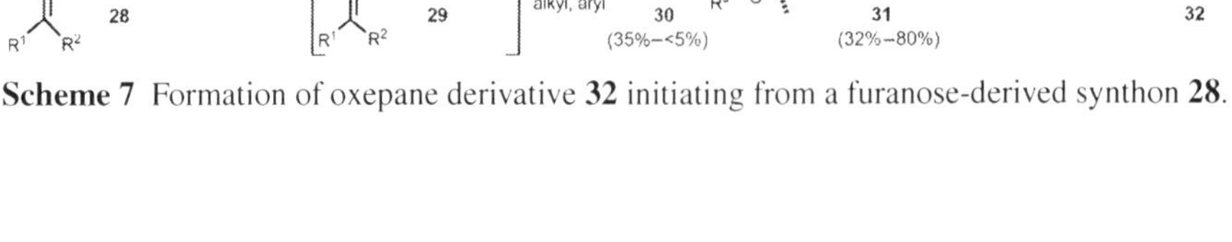

Scheme 7 Formation of oxepane derivative **32** initiating from a furanose-derived synthon **28**.

oxidative cleavage of diol and $Mo(CO)_6$-mediated N-O bond cleavage on **31** afforded amine-substituted oxepane **32**. Cycloaddition reaction through nitrone intermediate thus offered a possibility to generate not only the oxepane ring system but also installation of an amine moiety in the oxepane scaffold.

3 5-Exomethylene pyranoside precursors to septanoses

C-5–C-6-Unsaturated pyranosides are elegant examples of newer synthetic methods to prepare septanosides. An early demonstration was reported by Murphy and coworkers, wherein **33** was subjected to epoxidation, nucleophilic epoxide opening to form 1,5-dicarbonyl derivative **34**. Further equilibration and reaction of the hemiacetal with the internal ketone moiety afforded 1,6-anhydrosugar derivative **35** (Scheme 8).[25] Treatment of **35** with Ac_2O afforded not only 5-*O*-acetate of the 1,6-anhydrosugar **36**, but also 1-*O*-acetylated septanose **37**. Reduction of the carbonyl group in **37** with $NaBH_4$ led to the corresponding alcohol. The work exemplified 5-exomethylene sugar as a precursor for septanose synthesis, wherein generation of a carbonyl group at C-5 carbon of septanose characterized the synthetic approach.

As opposed to epoxidation-nucleophilic ring-opening as shown in Scheme 8, the approach of Chapleur and coworkers involved dihydroxylation of 5-exomethylene sugars originating from differentially protected methyl glucopyranosides, which provided bicyclic 1,6-anhydrosugar derivatives, such as **38**.[26] Reaction of **38** with phenyl thionochloroformate, in the presence of pyridine, led to the formation of septanosyl chloride **41** (Scheme 9), apart from carbonate and thionocarbonate. The propensity of the 1,6-anhydro derivative to equilibrate with a seven-membered septanose system (**39**), subsequent thionocarbonate formation (**40**) and chloride substitution rationalized the septanosyl chloride (**41**) formation. An acetolysis of **38** in the presence of a Lewis acid was shown to cause C-5–O-5 bond cleavage to the septanose derivative **42**. The consequence of acetolysis reaction was the replacement of 4-*O*-benzyl with a 4-*O*-acetate protecting group in **43** (Scheme 10).

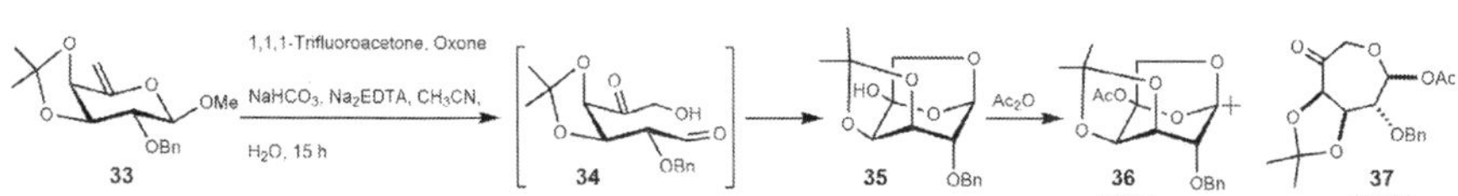

Scheme 8 Conversion of a 5-exomethylene **33** to septanose **37**, through ring-opening and ring-closing reactions.

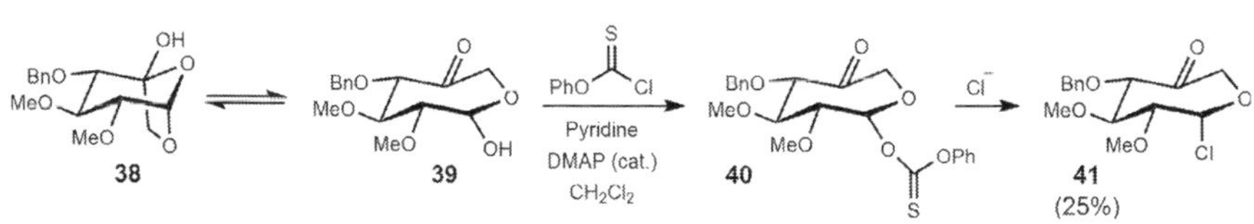

Scheme 9 Reaction of 1,6-anhydrosugar **38** with phenylthionochloroformate to form septanosyl chloride **41**.

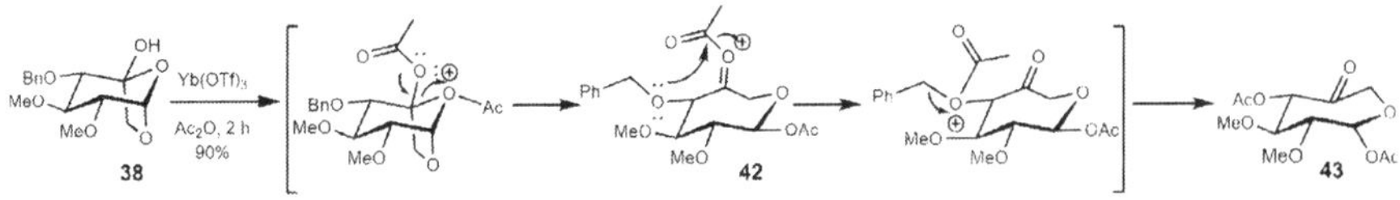

Scheme 10 Acetolysis reaction of 1,6-anhydrosugar **38** to form septanosyl acetate **43**.

Bicyclic sugar intermediates provide innovative means to achieve oxepanes through intramolecular rearrangements. A case in point is the application of Ferrier-Nicholas cation formation on a glycal derivative and further judicious reactions, leading to oxepane formation. Scheme 11 illustrates such a reaction advanced by Gómez and López,[27] wherein the key synthon towards formation of Ferrier-Nicholas cation is the ene-yne **46** from glycal **44**, which was secured through (a) oxidation to **45**; (b) lithium acetylide addition and (c) halide addition using a halogenating agent, and (d) base-mediated halide elimination. The ene-yne derivative **46** undergoes facile complexation with $Co_2(CO)_8$ to **47**, which upon activation with a Lewis acid, initiates a cascade of transformations that results in a functionalized oxepane **48**. The transformations are led by a Ferrier type I reaction, followed by a hydride transfer of benzylic methylene at *O*-6 atom, nucleophilic attack by water and hemiacetal ring-opening. This newer type of reaction opens up the possibility to secure variedly functionalized oxepanes. The carbon-carbon bond formation at C-1, which might be considered as *C*-septanosyl derivatives, offers a further advantage to the reaction.

4 Cyclization of linear precursors with terminal diene functionalities

A significant rise in the studies of the synthesis of septanoses is traced back to the first report of the synthesis of 2,3-unsaturated oxepines, namely, 4,5,6,7-tetrahydrooxepine, and septanosides, through the novel ring-closing metathesis (RCM) reaction, demonstrated early by van Boom and coworkers.[28] The synthetic plan was to secure a 1,8-diene precursor from a furanose **49** through (i) a Wittig reaction to afford **50** and (ii) 5-*O*-allylation to afford dienes **51** and **52**. The RCM reaction enabled the cyclization of the dienes **51** and **52** to afford either 2,3-unsaturated septanoside **53** or oxepine **54**, respectively, depending on substitution at the allyl moiety (Scheme 12). The unhindered nature of varied protecting groups is an added advantage of this approach, thereby enabling the synthesis of oxepine and septanose systems in a predictable manner.

A further development of the reaction in Scheme 12 is the demonstration by Schmidt and Biernet on the tandem RCM-isomerization protocol wherein 2,3-unsaturated oxepine **54** is subjected to double bond isomerization to 1,2-unsaturated oxepine **55**.[29]

Further innovative expansion of the RCM method is the series of work on the synthesis of septanoses and septanosides by Peczuh and coworkers. In an early demonstration, 1,2-unsaturated 1,2-dideoxyseptanoses **59** were synthesized in a three-step procedure involving (i) one-carbon homologation of latent aldehyde **56** to alkene **57**;

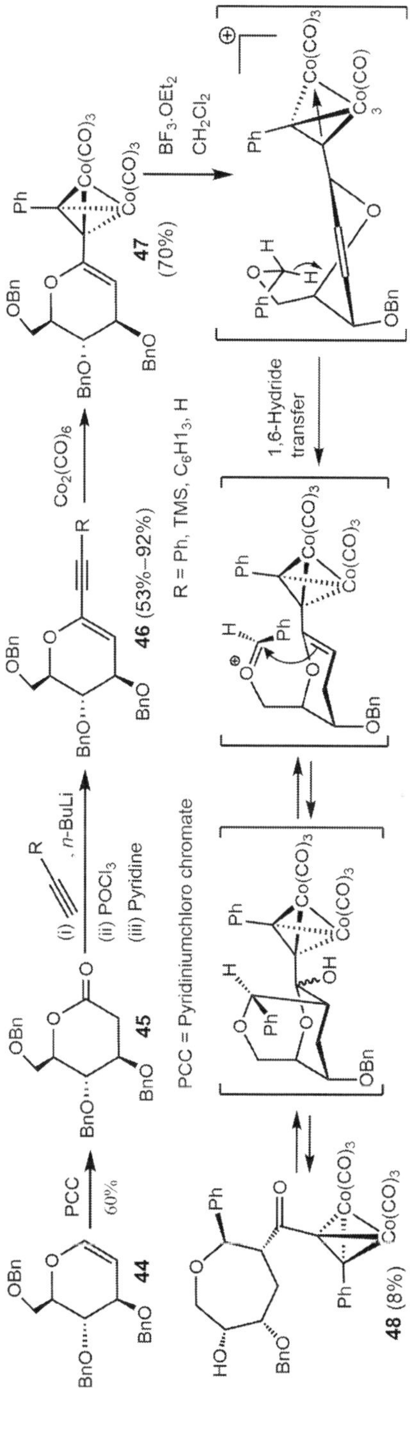

Scheme 11 Ferrier-Nicholas cation-mediated reaction of ene-yne **46** to afford oxepane derivative **48**.

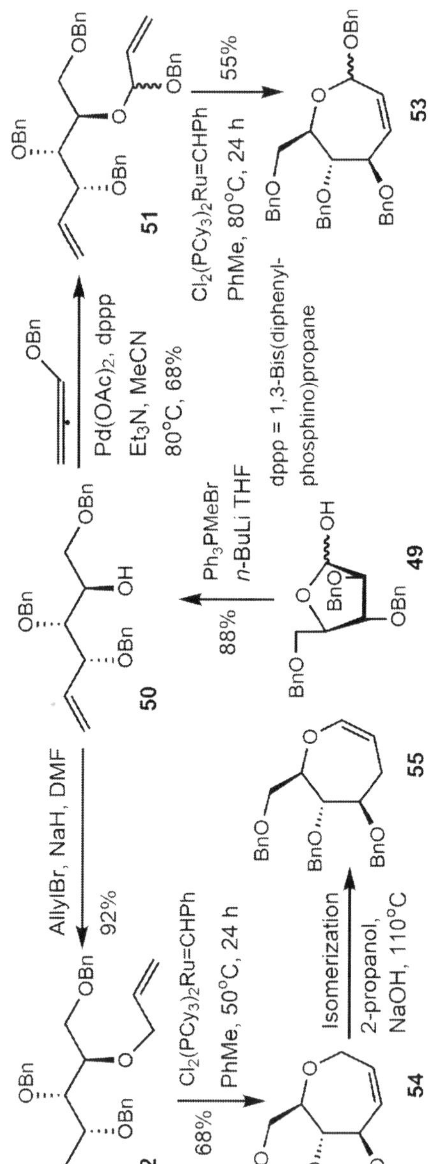

Scheme 12 RCM route to secure oxepine **55** and septanoside **53** from diene linear precursors **51** and **52**.

(ii) vinylation of C-5 alcohol to **58** and (iii) an RCM reaction of the resulting diene to **59** (Scheme 13).[30] Sustained studies brought several innovative aspects related to synthesis, functionalization, and structures of the septanosides.[31] Realizing the use of expensive catalysts for RCM reaction, low efficiency, and scope of protecting groups, alternative cyclization method of linear precursors was also developed, as shown in Scheme 14.[32] This method relied on vinylation of C-1 aldehyde of *O*-benzyl-protected glucose **60** to the corresponding allyl alcohol derivative, oxidative cleavage of alkene moiety to secure α-hydroxyaldehyde **61** and intramolecular acetalization of the resulting aldehyde with C-6 hydroxy group to afford septanose **62**, as a mixture of anomers and epimers. Scalability of the method was tested in the gram-scale formation of final septanose product **63** in the *O*-acetylated form. The reactions were also found to be prone to rearrangement, wherein the newly formed septanose, such as **I** (Fig. 2), preponderantly underwent a ring-opening and subsequent ring-closing to afford a C-5 modified pyranose **III**, as observed in the *O*-debenzylation of **I** to **III**. Instead of vinylation and oxidation, a Wittig methylenation at C-1 aldehyde of a suitably protected glucopyranose afforded derivative **64.** This derivative was converted to aldehyde **68**, through (a) a dihydroxylation to **65**; (b) *O*-benzylidene protection to **66**; (c) regioselective ring-opening and protection of the resulting primary alcohol with TES-moiety to derivative **67,** and (d) oxidation to aldehyde **68**. Intramolecular acetalization, in the presence of acidic alcohol, afforded septanosides **69** and **70** depending on the alcohol used for acetalization (Scheme 15).[33] *O*-Peracetylated septanoses prepared through the above routes were transformed to glycosyl donors upon their conversion to septanosyl halide, for example, **71**, the reaction of which with various acceptors afforded the corresponding septanosides **72–76** shown in Fig. 3.[34]

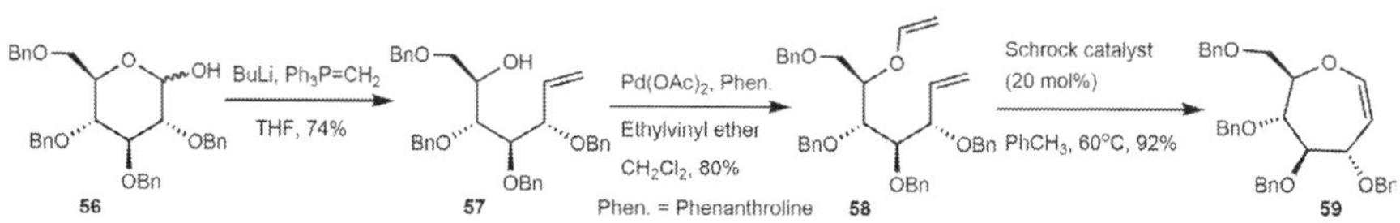

Scheme 13 RCM reaction of diene **58** to afford 1,2-unsaturated oxepine **59**.

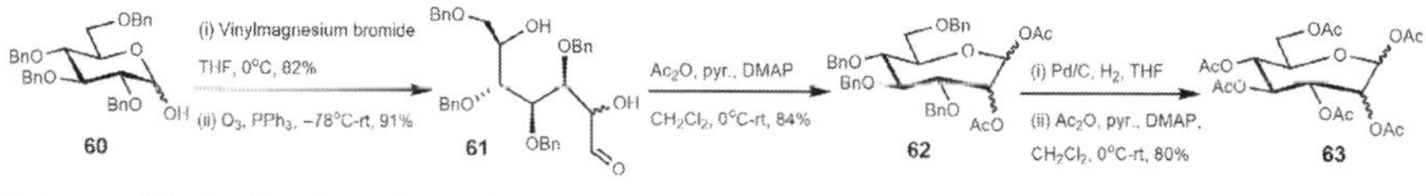

Scheme 14 Cyclization of 2,6-dihydroxyaldehyde **61** to form septanose **62**.

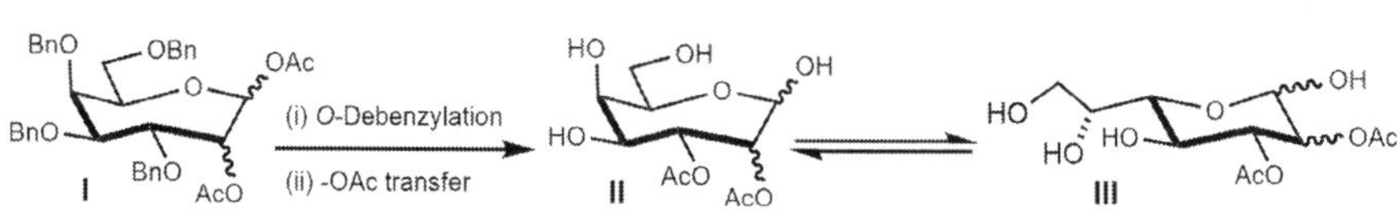

Fig. 2 Equilibration of septanose **I** to a heptopyranose **III**.

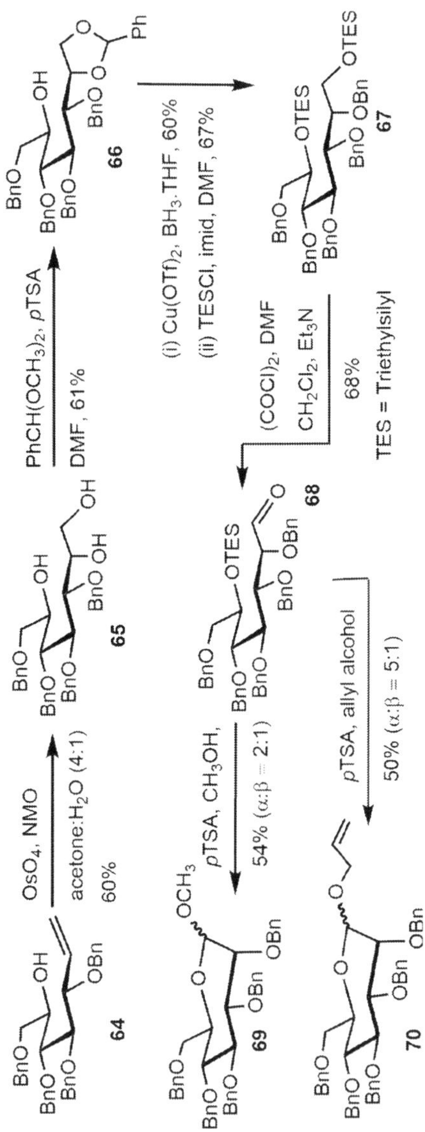

Scheme 15 Conversion of synthon **64** to septanosides **69** and **70**, through intramolecular acetalization.

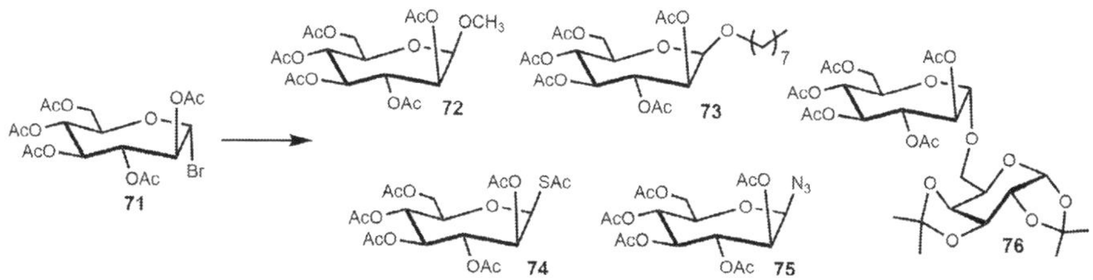

Fig. 3 Molecular structures of varied septanose products **72–76**, derived from septanosyl bromide **71**.

5 Septanose formation through fragment assembly of non-sugar precursors

Whereas septanose formation in the above works is initiated from a sugar precursor, assembly of a septanose using non-sugar fragments was also demonstrated. An important development of such an approach is the report of McDonald and coworkers.[35] Initiating from but-2-ene-1,4-diol **77**, a series of transformations, covering the formation of intermediates **78–81**, was implemented to secure precursor **82**. The key reaction was the metal-catalyzed cycloisomerization of **82**, which afforded 1,2-unsaturated 1,2-dideoxyseptanose **83**, in good yield, and with complete stereochemical control in each step of the reaction (Scheme 16). Having secured the septanose glycal **83**, it was exploited further as glycosyl donor, through conversion to a thioglycoside. Reaction with septanoside acceptors permitted to synthesize septanoside di- and trisaccharides.[35]

As mentioned earlier, synthesis of septanoses over and above that observed as an equilibrium product of the hexose caught the attention, combined further with the increased realization of the importance of oxepane ring system in many natural products. A direction of rich development in septanose synthesis is in one-carbon intramolecular homologation of pyranoses.[36]

6 Cyclopropanated pyrans as one-carbon intramolecular homologation synthons

Bicyclic systems provide an immediate opportunity for transformations involving both of the rings. Pertaining to oxepane and septanose synthesis, this opportunity is exploited through cyclopropanation of glycals with carbenes, followed by judicious bond cleavage of the ring juncture. Cyclopropanation of a glycal was reported early in mid-1960s by Brimacombe and coworkers, wherein the 3,4,6-tri-*O*-methylglycal derived from glucose was subjected to reaction with ethyl trichloroacetate and NaOMe, leading to the formation of corresponding dichlorocarbene-glycal adduct.[37] This work stands out the early example of reactions on 1,2-unsaturated sugars. However, such cyclopropanated glycal adducts were not explored further, particularly as valuable synthons for

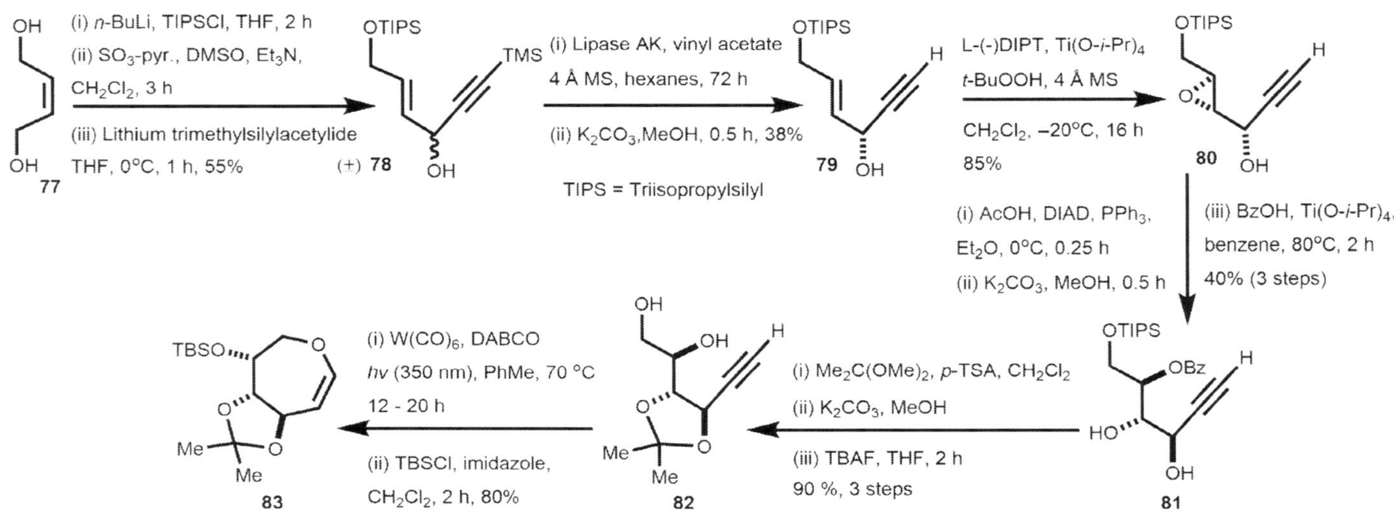

Scheme 16 Formation of alkyne-ol linear precursor **82** and metal-catalyzed cycloisomerization to 1,2-unsaturated septanose **83**.

ring expansions. The possibility to utilize the cyclopropanated adduct as a synthon for ring expansion was reported first by Hoberg and Bozell.[38] A Simmons-Smith reaction of glycal **84** afforded cyclopropane **85**. The strain in the bicyclic ring system, along with the endocyclic oxygen and C-3 acetoxy functionality, was thought to assist an acid-catalyzed cyclopropane ring-opening, concomitant nucleophilic addition at C-1 carbon and elimination of the functionality at C-3 carbon. This expectation was experimentally observed, thus, reaction of **85** with TMSCN as nucleophile and TMSOTf as the acid afforded the 2,3,6,7-tetrahydrooxepine derivative **86** (Scheme 17). The reaction could be considered as a type of Ferrier reaction, with the loss of C-3 functionality, addition of the nucleophile at the anomeric carbon occurring with the C-1–C-2 σ-bond cleavage to afford a 2,3,6,7-tetrahydrooxepine, as opposed to a π-bond migration in the known Ferrier reactions that afford 2,3-unsaturated pyranosides.[39] The presence of C-3–C-4 unsaturation allowed further addition reactions, so as to secure, for example, dibromide **88**, bromohydrin **89,** and epoxide **90** from **87** (Scheme 18). Oxepine **87** was, in turn, prepared by following the ring expansion reaction as in Scheme 17, the precursor being 4,6-*O*-di-(*tert*-butyldimethylsilyl)glycal derived from galactose.[38] Excellent diastereoselectivities of the addition reactions were observed, as a result of the bicyclic structural nature of the synthon and further stabilization of intermediates through, for example, "halogen" bonding in the case of halide product formation.[40]

C-1–C-2-Cyclopropanated pyranoses occupy the most interest in order to readily realize the ring expanded septanose product formation.[36] An early example is the reaction of adduct **91** with MeOH in the presence of K_2CO_3. The reaction was reported to afford bromo-oxepine **92**, in good yield (Scheme 19), due to the formation of a stabilized cation **IV**, involving endocyclic oxygen of the ring expanded septanosyl intermediate.[41] Although the reaction appeared to afford ring expanded product, alternate

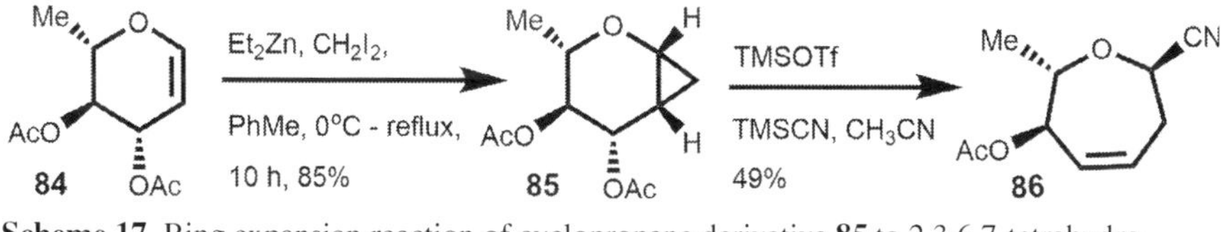

Scheme 17 Ring expansion reaction of cyclopropane derivative **85** to 2,3,6,7-tetrahydro-oxepine derivative **86**.

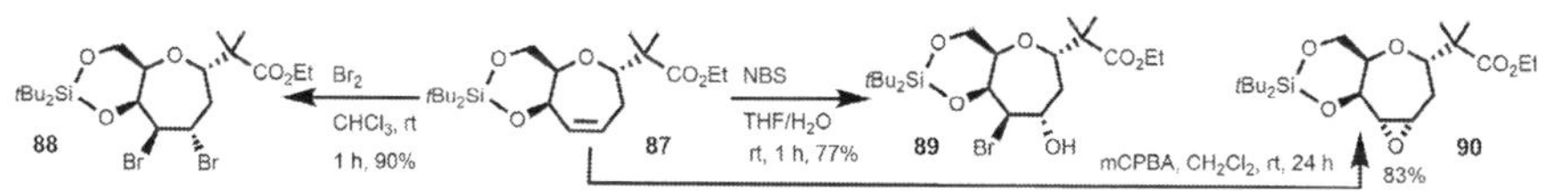

Scheme 18 Reactions of oxepine **87** to addition products **88–90**.

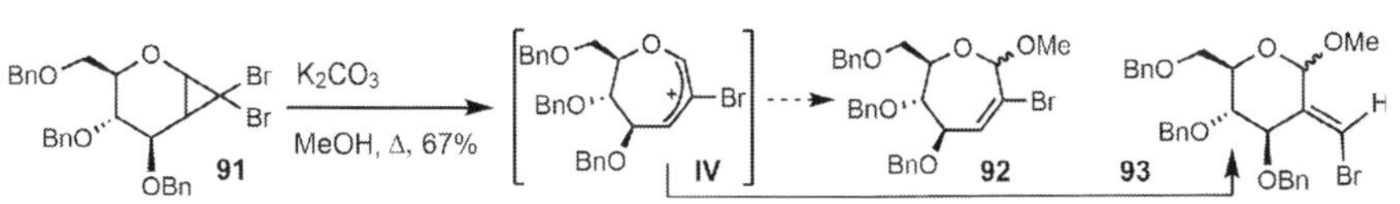

Scheme 19 Reaction of *gem*-dihalocyclopropane derivative **92** under basic condition.

possibilities of base-mediated reactions were not considered, leading to a correction of the reaction product later, as **93**.[42a] The correction of the structure revealed the product to be a 2-*exo*-bromomethylene-2-deoxypyranoside **93**, rather than the originally suggested bromooxepine derivative **92**.

On the other hand, a successful approach to open up the C-1–C-2 ring juncture of the bicyclic system **91** is to conduct the reaction with MeOH, in the presence of Ag(I) under reflux conditions (Scheme 20), to afford bromo-oxepine derivative **92**, as revealed by the work of Harvey and coworkers.[42b] An anomeric mixture of septanose and septanoside was formed in moderate yields, with major anomer being the α-anomer. A further observation was the instability of the oxepine derivative at continued heating of the reaction mixture, which afforded *C*-furanosides.[42b] Scheme 21 depicts the reaction of *gem*-dihalocyclopropanated derivative **91** with allyl alcohol, with the presence of the silver(I) salt, under reflux condition. The reaction led to not only unsaturated septanoside **94**, but also *C*-furanosides **95a** and **95b**, as a function of the reaction duration. Interestingly, only **95a** (12%) and **95b** (45%) formed upon heating **94** with allyl alcohol, in the absence of silver(I) salt. The prevailing acidic condition during the reaction facilitated the ring-opening of the initially formed septanoside to an acyclic derivative, which underwent an intramolecular cycloetherification of C-6 alcohol moiety with the olefin to form the observed *C*-furanoside.

In the case of 1,2-cyclopropanated pyranose derivative, the Lewis acid-mediated ring expansion to a C-3–C-4-unsaturated septanose relied on the labile acyloxy-functionality at C-3 carbon, via endocyclic C-1–C-2 bond cleavage.[38] This approach is still feasible when the C-3 carbon is a keto-moiety, which would lead to an enolether upon ring expansion. The work of Sridhar and Venukumar demonstrated this approach using C-1–C-2-cyclopropanated C-3-ulose sugar derivative **96** which, upon reaction with sugar acceptor **97** in the presence of a Lewis acid, afforded the ring expanded septanoside at the nonreducing end of the disaccharide **98** in good yield, predominantly as α-anomer (Scheme 22**A**).[43] The reduction of the C-3 keto-functionality of

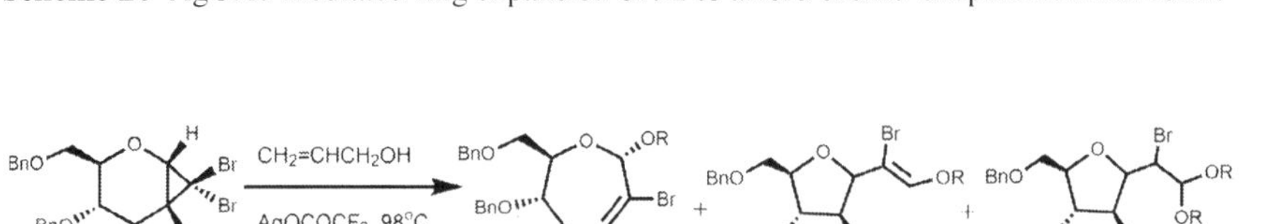

Scheme 20 AgOAc-mediated ring expansion of **91** to afford bromo-oxepine derivatives **92**.

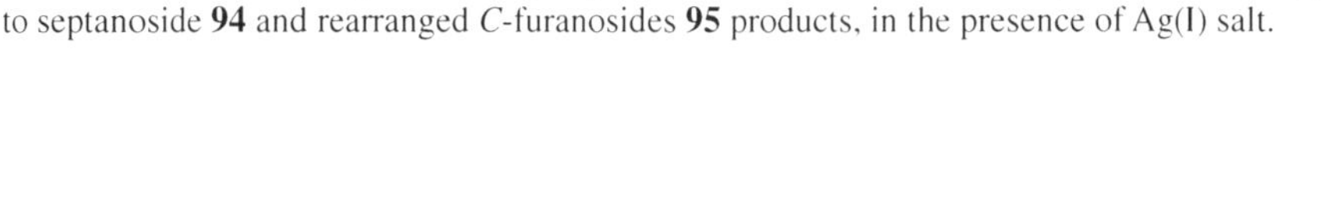

Scheme 21 Reaction of *gem*-dihalocyclopropanated derivative **91** with allyl alcohol, leading to septanoside **94** and rearranged *C*-furanosides **95** products, in the presence of Ag(I) salt.

98 to the corresponding alcohol **99**, followed by utilizing the same as a nucleophile in the reaction with **96** afforded trisaccharide **100** (Scheme 22**B**). It was demonstrated further that the reaction of the ulose **96** with a thiol reagent, in the presence of a Lewis acid, would afford rather an acyclic dithioacetal **101**. The dithioacetal **101**, in turn, was shown to undergo iodonium ion-promoted cyclization and in the presence of acceptor **97** afforded a septanosyl-pyranose/pyranoside disaccharide product **98** (Scheme 23). Interestingly, an additional product **102** of the reaction was also observed by the authors, which was suggested to result from an intramolecular reaction of the carbonyl oxygen with the oxycarbenium ion to form intermediate **V**.[44]

In addition to an ulose functionality at C-3, a further development is the incorporation of an exomethylene moiety at C-3 of the glycal, so as to derive vinyl cyclopropane and subsequent ring expansion reaction.[45] This C-3 exomethylene synthon is equally found to undergo an electrophile-activation-mediated bond cleavage at the ring juncture of cyclopropanated pyran **104**, secured from derivative **103**, to afford the septanoside **105** (Scheme 24). Chloramine-T and phenyltrimethylammonium tribromide were found to mediate the reaction, leading to the installation of an aminomethyl moiety to the C-4 carbon of the newly formed septanoside derivative.

Among 1,2-unsaturated sugars, derivatives that contain the oxygen functionality at C-2, namely, 2-hydroxyglycals, afford a seamless route to prepare septanoses, through cyclopropanation and ring expansion.[46–51] 2-Hydroxyglycals, in the form of protected vinyl ethers, are analogues of 2-uloses,[52] preparation of which proceeds through a mild base-mediated elimination of protected pyranosyl halides. Such vinyl ethers are even

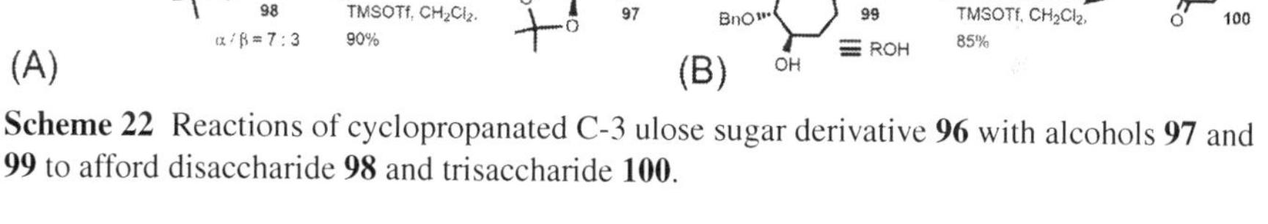

Scheme 22 Reactions of cyclopropanated C-3 ulose sugar derivative **96** with alcohols **97** and **99** to afford disaccharide **98** and trisaccharide **100**.

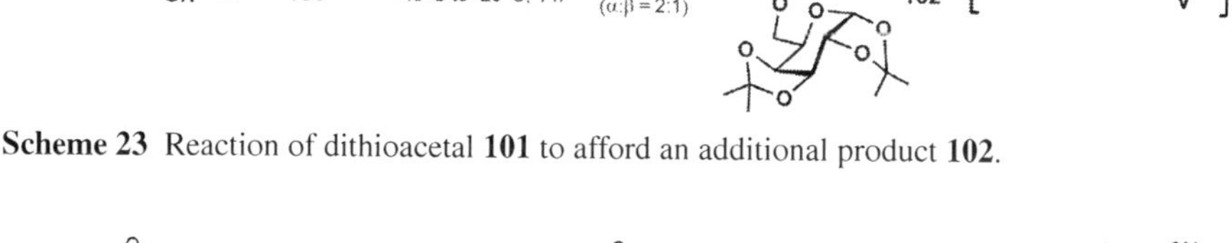

Scheme 23 Reaction of dithioacetal **101** to afford an additional product **102**.

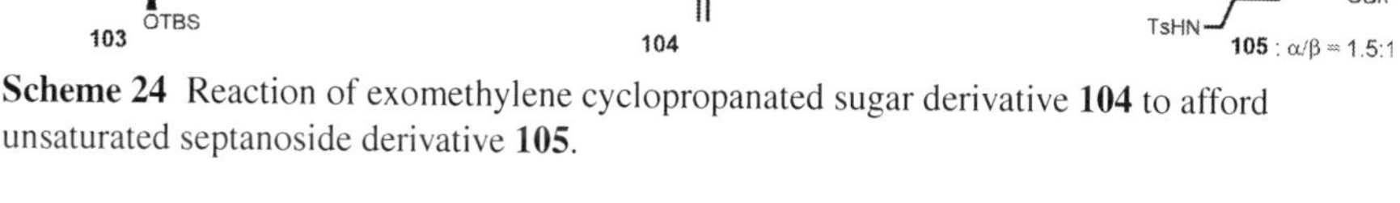

Scheme 24 Reaction of exomethylene cyclopropanated sugar derivative **104** to afford unsaturated septanoside derivative **105**.

more facile synthons for the electrophilic additions and, in the case of carbenes, the addition occurs readily to afford cyclopropanated glycopyranoses, with a retained oxygen functionality at C-2. *Gem*-dihalo-substituted cyclopropanes find interest in these instances also, in order to promote subsequent nucleophilic substitution reactions. The general strategy for the ring expansion of oxyglycal synthon **VI** to septanose **X** is presented in Fig. 4. Nucleophile induced ring-opening of synthon **VII** ensures not only the formation of the seven-membered ring, but also the corresponding septanoside as the glycoside product **VIII**. The resulting septanoside is a C-2–C-3 disubstituted halo-oxepine amenable for further reactions. Oxidation of **VIII** to a 1,2-diketone **IX** and a cursory reduction of the ketone moieties would result in a septanoside **X**, with all carbon sites carrying the hydroxy groups. In practice, this approach is seen to be facile and a number of septanosides have been synthesized through this route. An example is shown in Scheme 25 for the synthesis of septanoside **108** from glycal derivative **106** involving chlorooxepine **107** as the intermediate.[46] Few general features of this approach are that (i) the C-2 functionality in the glycal ensures retaining the oxygen functionality at C-3 of the product septanoside; (ii) ring-opening reaction is amenable under both acid- and base-catalysis; (iii) the intermediate halooxepine presents the unsaturation at fully substituted carbons, thereby preventing possible isomerizations, as observed in glycals without oxygen functionality at C-2 and (iv) the halo-oxepines in the form vinyl halides are synthons for further functionalization at the C-2 carbon of the septanoside.

A range of alkoxides afforded the corresponding septanosides **109–114** as shown in Fig. 5.[47] Further, reactions leading to structures shown in Fig. 6, involving phenoxide nucleophiles, show that these nucleophiles afford septanosides **115–120**, as anomeric mixtures (α/β=1:1). Oxidation of the halo-oxepines to diketones and the reduction reveal a change in the disposition of the diol at C-2 and C-3 carbons. The axial orientation of the phenyl moiety at the anomeric carbon affords a *trans*-disposition of the diol, as in **115α-119α**, whereas the equatorial orientation of the phenyl moiety leads to the diol as a C-3 epimeric mixture as in **115β–118β**, signifying that the facial addition of hydride during reduction depends on the anomeric configuration.

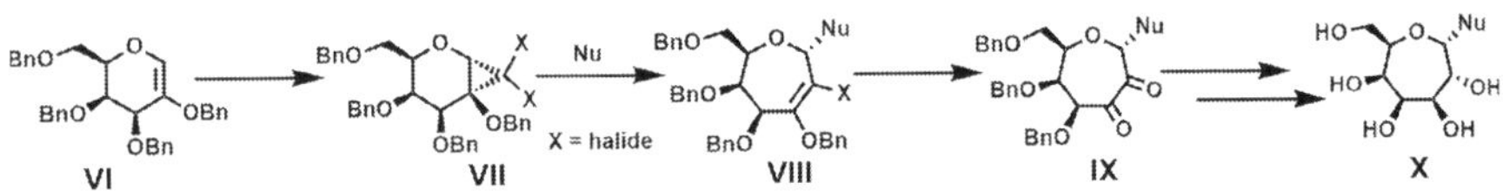

Fig. 4 2-Hydroxyglycal ether **VI** as synthon for cyclopropanation and the ring expansion reaction to form septanose **X**.

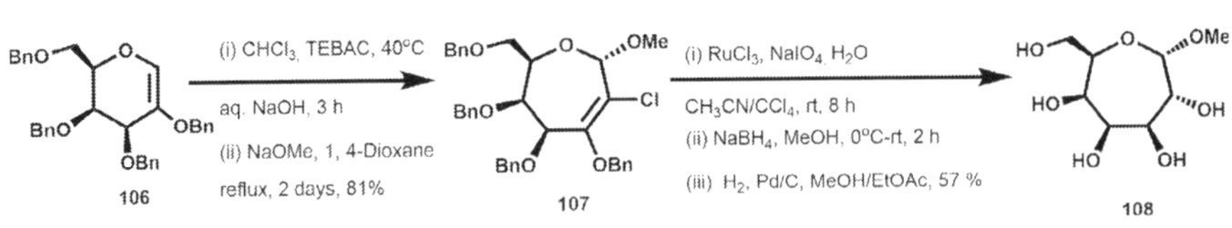

Scheme 25 Cyclopropanation of 2-hydroxyglycal ether **106** and subsequent ring expansion reaction to septanoside **108**.

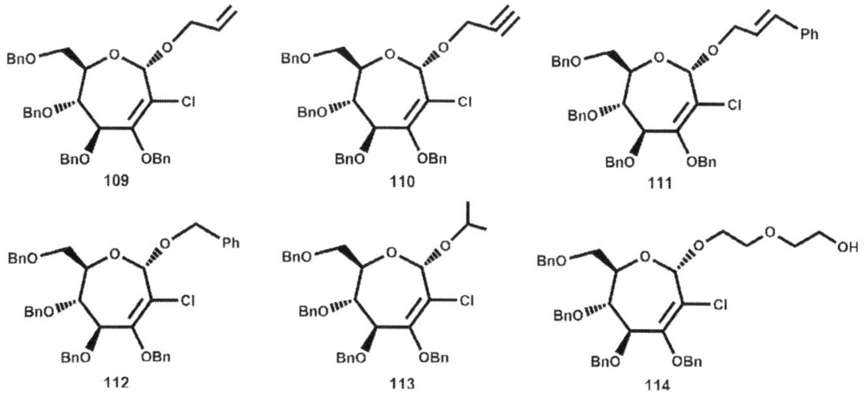

Fig. 5 Molecular structures of septanosides formed from 2-hydroxyglycal synthon with varied acceptor alcohols.

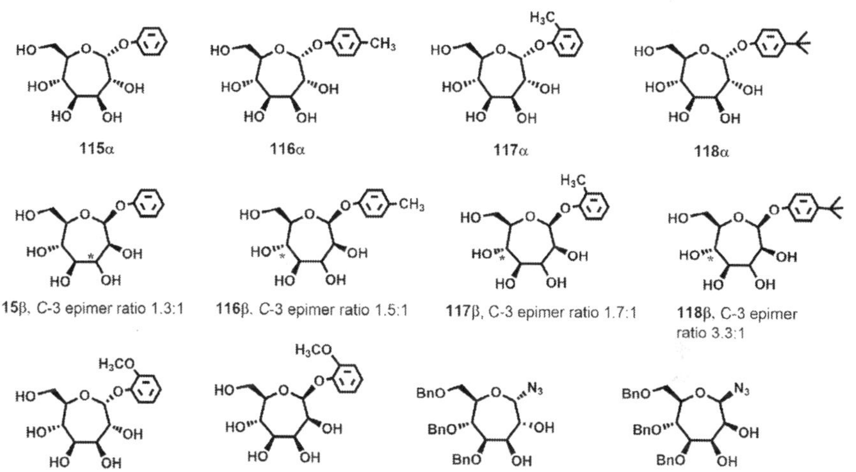

Fig. 6 Molecular structures of septanosides formed from 2-hydroxyglycal synthon with varied phenols and azide as nucleophiles.

These observations were also similar when azide was used as the nucleophile, so as to secure the corresponding septanosyl azides **121**, as an anomeric mixture. However, in this case, the oxidation-reduction sequence of reactions led to diastereoselective *trans*- and *cis*-diol formation with α- and β-anomers of the halooxepines, respectively. There is thus a differing reactivity pattern of halooxepines with varied anomeric substituent. The α-azidoseptanose derivative was taken through reduction further, so as to secure septanosyl amine.

When the alkoxide nucleophiles originate from sugar derivatives, their ring-opening reaction with the cyclopropanated oxyglycals leads to disaccharides **122**, **123** with a septanoside at the nonreducing end.[48] Further, when the oxyglycal moiety originates from a disaccharide, cyclopropanation, followed by ring expansion reactions afford trisaccharides **124**, **125** (Fig. 7). The reactions show that 2-hydroxyglycals of disaccharides are also viable synthons for subsequent dihalocarbene addition and nucleophilic ring expansion. Oxidation-reduction sequence led to diol functionalities at C-2 and C-3 carbons of septanosides with complete diastereoselectivity.

The seamless nature of the reaction could be continued further, in order to prepare di- and triseptanosides.[51] Scheme 26 shows the sequence of reactions leading to the diseptanoside disaccharide. The key reactions of (i) nucleophilic ring-opening of derivative **126** by sugar alkoxide **127**, which is also a 2-hydroxyglycal; (ii) ability to conduct cyclopropanation of the resulting disaccharide to **128**; (iii) alkoxide-mediated ring expansion at the cyclopropanated reducing end of the disaccharide to afford **129**; and (iv) simultaneous oxidation to afford **130,** reduction and final deprotection occurred in good to excellent yields and the diseptanoside **131** could thus be realized.

Whereas the above ring expansion reactions are performed under base-mediated reaction conditions, further studies demonstrate that the reaction can also be conducted under Lewis acid-mediated conditions.[51] Silver acetate was found to be suitable to promote ring expansion reactions in the presence of varied alcohols to afford the corresponding septanosides in good to moderate yields. In addition, 1-*O*-acetyl chlorooxepine also formed as a minor product in these reactions. The later acetate addition product formation becomes even more facile with unreactive acceptor alcohols, such as ^{t}BuOH. When the reaction was conducted in the presence of AgOAc and ^{t}BuOH under reflux, a halooxepine product with an acetate moiety at the C-1 carbon was obtained as the major product. The resulting halooxepines obtained with varied alcohols were found to be stable under prolonged boiling, without undergoing further rearrangements. This observation contrasted with a rearrangement reaction of the newly formed halooxepine, as shown in Scheme 21. The presence of the functionality at C-2 of the 2-hydroxyglycal and the absence of such a moiety in glycals led to differences in the subsequent reactivity patterns of the ring-expanded septanose product.

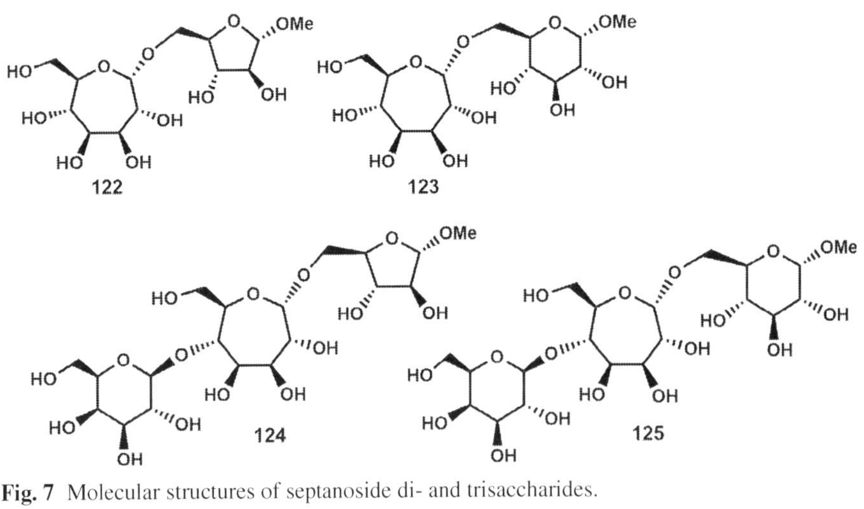

Fig. 7 Molecular structures of septanoside di- and trisaccharides.

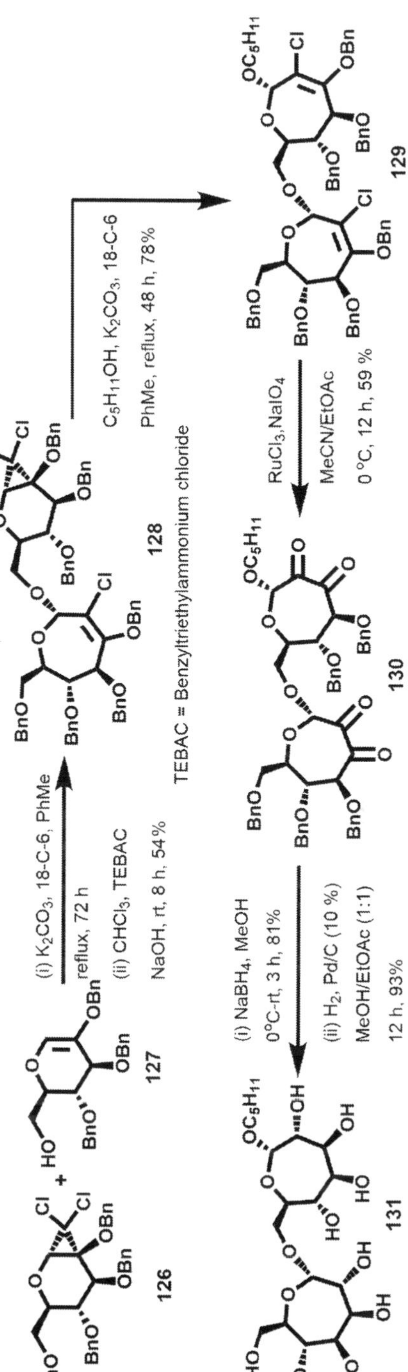

Scheme 26 Reactions leading to the formation of a septanoside disaccharide **131**.

Harvey attributed the rearrangement of septanoside **94**, leading to the formation of *C*-furanoside final product **95a** and **95b** (Scheme 27).[42b] Such a rearrangement is not encountered when cyclopropanated oxyglycal is involved with the ring expansion reaction (Fig. 4).

The halooxepines resulting from ring expansion of cyclopropanated oxyglycal provide further opportunity for effective application of organometallic reactions requiring vinyl halides as synthons. A series of organometallic reactions on halo-oxepines shows formation of varied types of 2-deoxy-2-*C*-alkyl/aryl septanosides **132–136** (Fig. 8).[49] The application of Sonogashira coupling reaction involving acetylenes as substrates with vinyl halide **137** is shown in Scheme 28, the resulting septanoside products **138** are also highly functionalized ene-ynes. On the other hand, reaction of the halooxepine with *n*-BuLi to perform halide-lithium exchange reaction, followed by reaction of the lithiated intermediate with MeI and D_2O afford methyl **135** and deuterium substituted oxepines **136**, respectively.[49]

Cyclopropanated pyranosides were studied as synthons for ring expansion reactions. A rare example is the observation of C-4–C-5-cyclopropanated mannopyranoside derivative **139** undergoing a radical reaction to form 6-exomethylene septanoside **141** (Scheme 29),[53] in addition to the formation of 5-exomethylene **140** and aldehyde **142**.

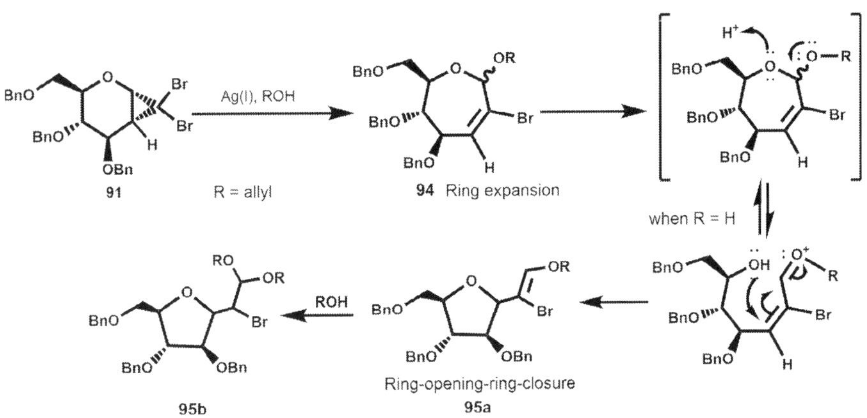

Scheme 27 Plausible mechanism of the formation of *C*-furanosides **95** from cyclopropanated sugar derivative **91**.

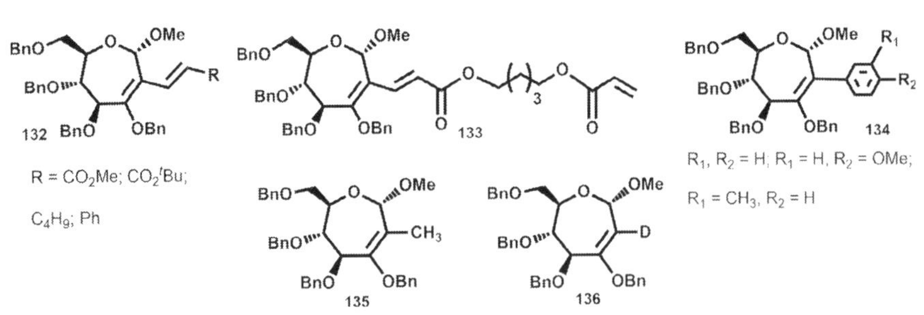

Fig. 8 Molecular structures of C-2 modified septanoside derivatives.

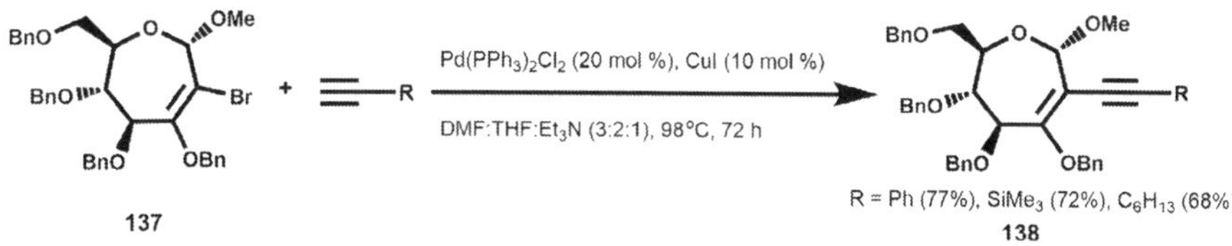

Scheme 28 Sonogashira coupling of vinyl halide **137** to form ene-yne derivative **138**.

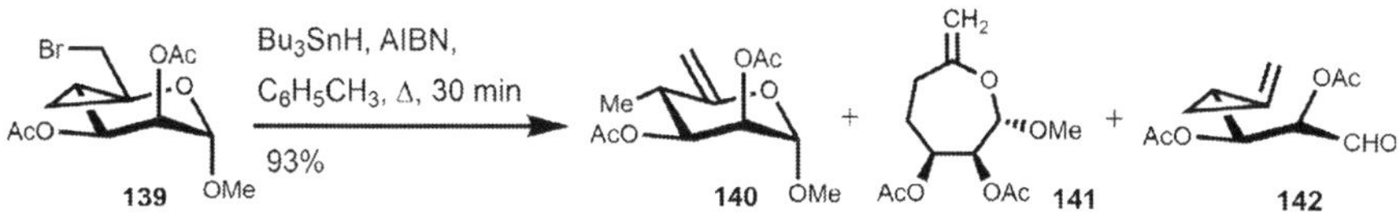

Scheme 29 Formation of 6-exomethylene septanoside **141** from C-4–C-5-cyclopropanated mannopyranoside **139**.

7 Hydrolytic stability of the septanoside glycosidic bond

Assessment of the hydrolytic stability of the glycosidic bond in septanosides provides a clue to stabilities of septanosides, for example, in comparison to pyranosides. Sugar hydroxy group disposition and the conformation are known to play a major role in the hydrolytic stability of pyranosides.[54] For pyranosides, it is well-established that one or more axial orientations of the hydroxy groups provide impetus to faster hydrolysis. For example, methyl α-D-galactopyranoside undergoes faster hydrolysis, under acidic treatment, than methyl α-D-glucopyranoside. The rationale for this difference in glycosidic bond stabilities arising from hydroxy group orientation at remote carbons include qualitative steric influence to involved effects, such as hyperconjugation and the relative ease of the formation of oxycarbenium ion intermediate with altered conformation.[55–57] A similar query to septanosides is thus pertinent. In an early work, Peczuh and coworkers studied the rate of acidic hydrolysis (0.5 M DCl in D_2O) at 50°C on compounds **143–145**. These septanosides were chosen for reasons that **143–145α** share 1,2-*cis*-stereochemistry and a change in configuration occurred at only one carbon in these septanosides. The rate of hydrolysis followed the trend **143>144>145** (Fig. 9). The hydrolysis rates were faster in comparison to pyranoside analogues. Between the anomers of **145**, the α-anomer underwent faster hydrolysis than the β-anomer. A computational study found that the release of MeOH from the α-anomer is favored by 2.2 kcal/mol over the release of MeOH from the β-anomer.[58, 59]

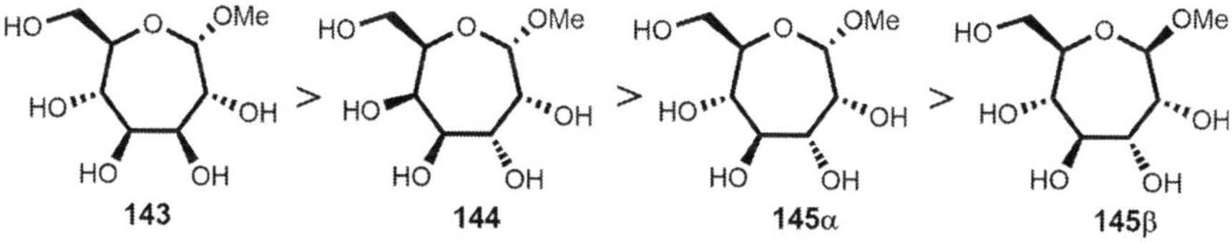

Fig. 9 Glycosidic bond stabilities of septanosides with varied configurations.

A study of three different types of septanosides **146–148**, shown in Fig. 10, was undertaken to assess the hydrolytic stability of the glycosidic bonds.[51] Septanoside **146** is a monosaccharide with *n*-pentyl substituent as the aglycon moiety, septanoside **147** is a diseptanoside with the same aglycon moiety at the reducing end and septanoside **148** presents an interesting situation of having chlorine at C-2 carbon of the septanoside. The hydrolysis was monitored by distinct chemical shift differences in a given proton of the *n*-pentyl moiety before and after hydrolysis by ^{1}H NMR spectroscopy, from which the rate of hydrolysis was determined. The hydrolysis rate followed first-order kinetics in all three derivatives. The observed rate of hydrolysis and the associated thermodynamic parameters are given in Table 1. When conducted at two different temperatures, derivatives **146** and **147** showed that the monosaccharide underwent hydrolysis faster than the disaccharide. Within the disaccharide **147**, the nonreducing end glycosidic bond was harder to hydrolyze than that at the reducing end. Derivative **148** was the slowest of these derivatives, glycosidic bond hydrolysis in this case required temperatures as high as 85–90°C, even at these elevated temperatures, the hydrolysis was still slower.

The rate of hydrolysis depends on the formation and stability of the oxycarbenium ion. In the case of pyranosides, a higher propensity to form the boat conformation stabilizes the oxycarbenium ion, the impetus for the formation of this charged species is influenced by the orientation of the exocyclic hydroxymethyl substituent, which acts as a dipole.[56, 57] Further, field effects induced by an electron-withdrawing substituent at C-2 carbon affects the oxycarbenium ion formation.[60] Such charged species evolution in septanoside might encounter these combined effects affecting the conformation and stability. Computational studies show that *gg* orientation of exocyclic

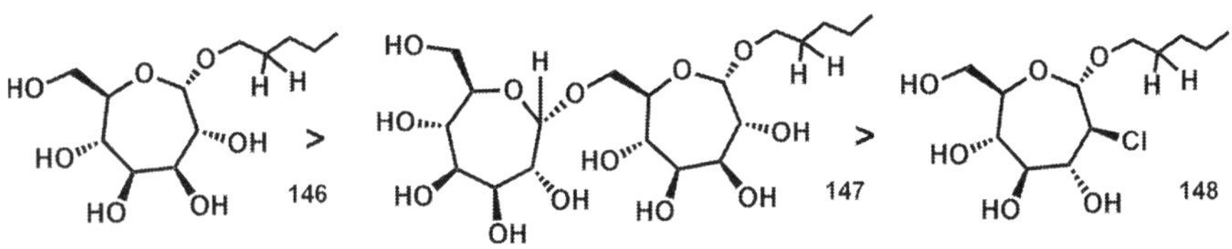

Fig. 10 Molecular structures of septanosides and trends of their glycosidic bond stabilities.

Table 1 Reaction rates and activation parameters of hydrolysis of septanosides **146–148**.

Compound	Rate of hydrolysis (k_{obs}) ($10^4 s^{-1}$) 35°C	45°C	85°C	90°C	ΔH# (kcal/mol)	ΔS# (cal/mol K)	ΔG# (kcal/mol)
146	2.34	5.61	–	–	12.95	−32.88	23.26
147	1.54	3.31	–	–	13.18	−33.17	23.56
	–	1.36					
148	–	–	0.418	2.12	72.44	−123.60	27.90

hydroxymethyl moiety, wherein O-7 is situated *gauche* to C-5 and O-6 of the septanoside, is stable by 2.1 kcal/mol and stabilizes the oxycarbenium ion. In the case of the chloro derivative **148**, the local minima corresponded to *tg*-conformation, wherein O-7 is situated *gauche* to C-5 and *trans* to O-6. In this conformation, the C-7–O-7 dipole is oriented antiperiplanar to the developing positive charge and destabilizes the oxycarbenium ion formation. This conformation along with inductive effect arising from chloride substitution at C-2 appeared to be responsible for the inability of derivative **148** to undergo acidic hydrolysis.

8 Solid-state structures of septanoses

The larger ring size of septanoses leads to larger variation in the conformations of the septanose skeleton, as compared to furanoses and pyranoses.[15] As a result, the wheel of conformational itineraries to the seven-membered septanose envelops larger spatiotemporal distribution. Fig. 11 shows the wheels of conformational itineraries for the chair-twist-chair pseudorotations and boat-twist-boat pseudorotations.[61–63] Assessing the conformations of the septanosides can be traced back to early work of Jackobs and Sundaralingam, who determined the single-crystal X-ray structure of a septanose, namely, 5-*O*-chloroacetyl-1,2:3,4-di-*O*-isopropylidene-α-D-glucoseptanose, from which a conformation adopting chair-to-twist chair was determined.[61] A majority of structures determined thus far, by solid-state and solution-phase techniques, adhere to twist-chair conformation as the predominant one. Structural elucidation of septanoses was pioneered by Stevens in a series of works, in both solid-state and solution-phase determinations. A compilation of the structures elucidated on septanoses by single-crystal X-ray structural determination is presented in Table 2.

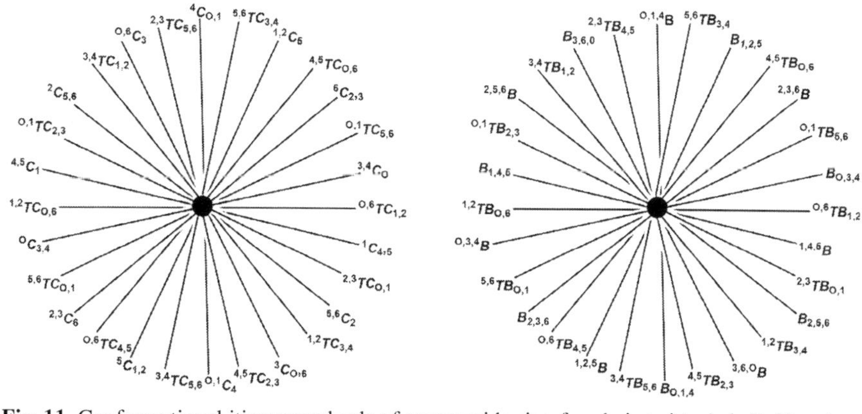

Fig. 11 Conformational itinerary wheels of septanoside ring for chair-twist-chair (left) and boat-twist-boat (right) displacements.

Table 2 Conformation of septanose derivatives determined by solid-state structural analysis

Entry	Septanose	Chemical structure	Conformation
1	5-*O*-Chloroacetyl-1,2:3,4-di-*O*-isopropylidene-α-D-glucoseptanose[61]	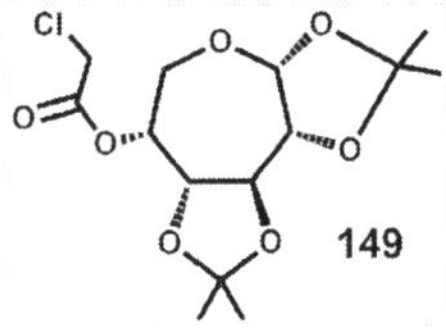 149	In between chair and twist-chair conformation
2	3-*O*-Acetyl-1,2:4,5-di-*O*-isopropylidene-α-D-glucoseptanose[64]	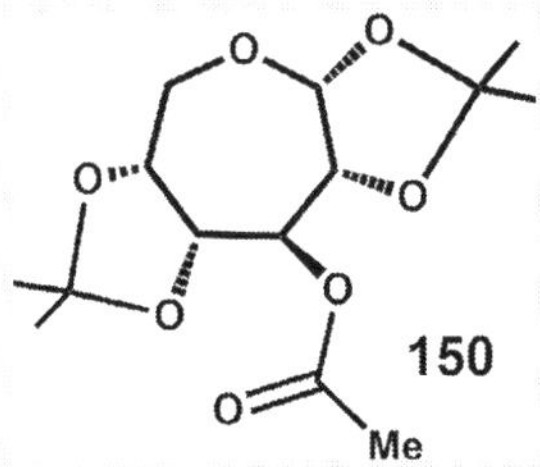 150	Distorted chair conformation
3	5-*O*-Acetyl-1,2:3,4-di-*O*-isopropylidene-α-D-glucoseptanose[65]	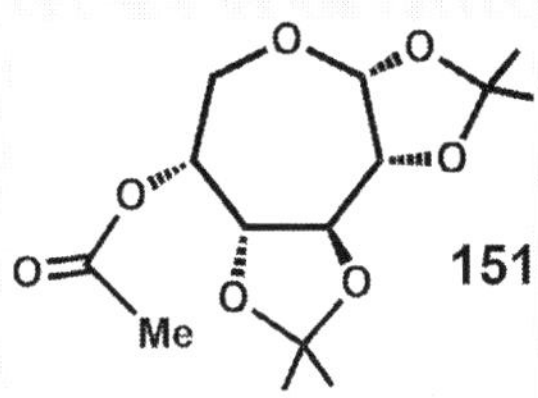 151	$^{4,5}TC_{6,O}$
4	Methyl 2,3,4,5-tetra-*O*-acetyl-α-D-gluco-septanoside[66]	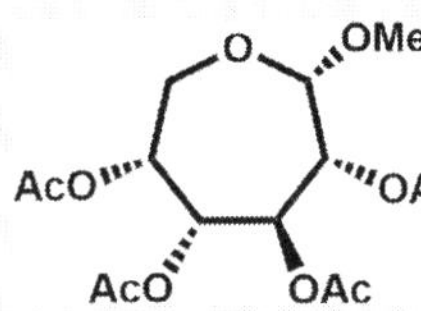 152	$^{4,5}TC_{6,O}$
5	Methyl 2,3,4,5-tetra-*O*-acetyl-β-D-gluco-septanoside[67]	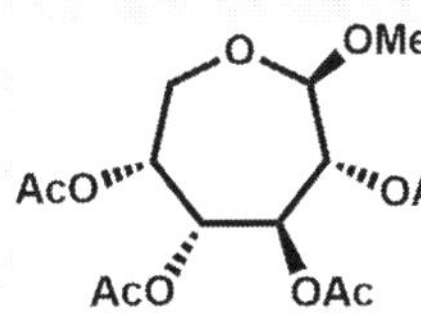 153	$^{1,2}TC_{5,O}$
6	Methyl β-D-glucoseptanoside[68]	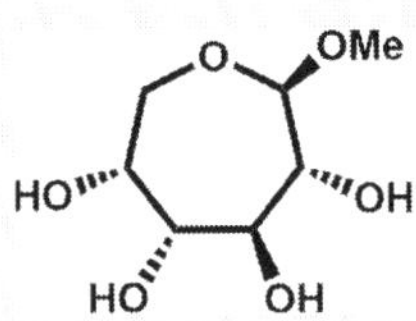 154	$^{2}C_{5,6}$
7	Pentyl 2-chloro-2-deoxy-α-D-*manno*-heptosept-3-uloside[69]	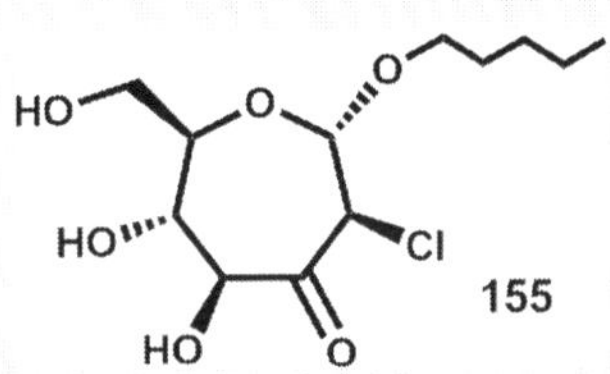 155	$^{5,6}TC_{3,4}$

Table 2 Conformation of septanose derivatives determined by solid-state structural analysis—cont'd

Entry	Septanose	Chemical structure	Conformation
8	*p*-Bromophenyl 4,5,7-tri-*O*-benzyl-β-D-*glycero*-D-*talo*-septanoside[70]	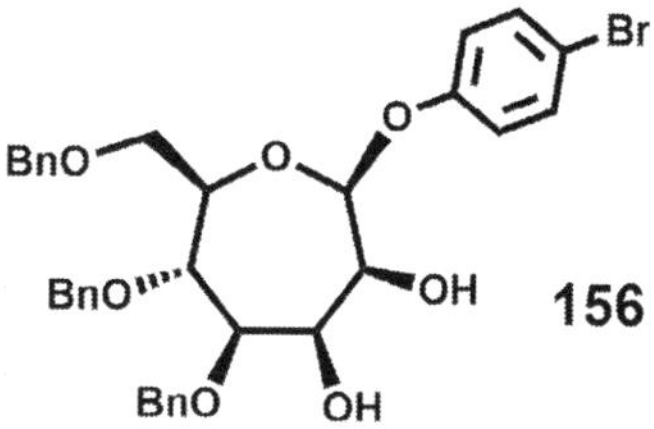 156	$^{0,1}TC_{2,3}$
9	Ethyl 2,3:4,5-di-*O*-isopropylidene-1-thio-α-D-glucoseptanoside[71]	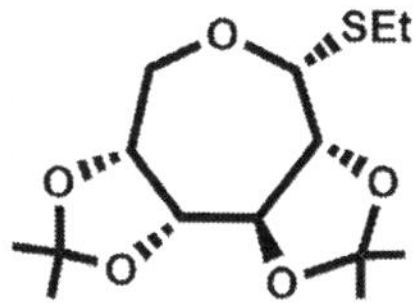 157	$^{5,6}TC_{3,4}$
10	β-D-Glucoseptanose pentaacetate[72]	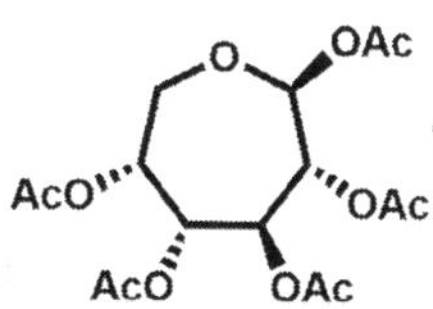 158	$^{5,6}TC_{3,4}$
11	Methyl 5-*O*-acetyl-2-*O*-benzoyl-3,4-*O*-isopropylidene-β-L-idoseptanoside[73]	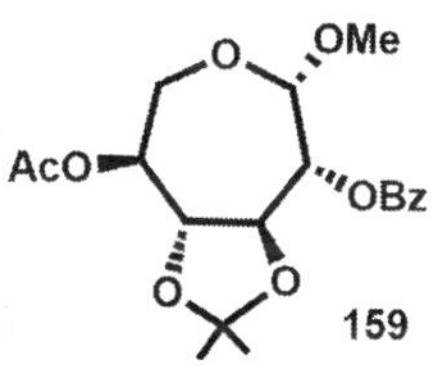 159	$^{0,1}TC_{2,3}$
12	1,6-Anhydro-3,4,5-tri-*O*-benzyl-2-deoxy-D-*xylo*-sept-1-enitol[74]	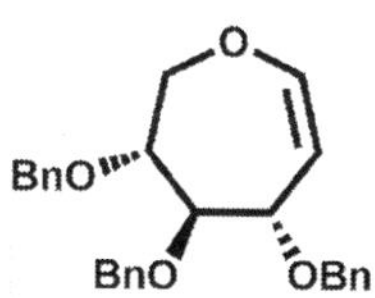 160	$^{6,0}TC_{4,5}$
13	5-*O*-Acetyl-1,2:3,4-di-*O*-isopropylidene-α-D-idoseptanose[75]	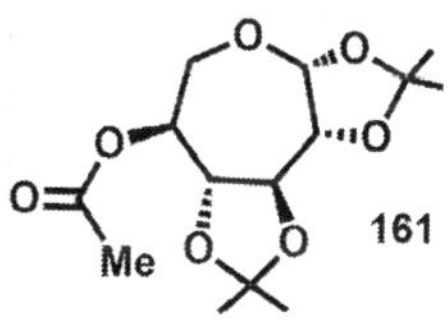 161	$^{1,2,5}B$
14	1,2:3,4-Di-*O*-isopropylidene-4-*C*-[(methylsulfanyl)methoxy]-β-L-*arabino*-hexoseptanos-5-ulose[76]	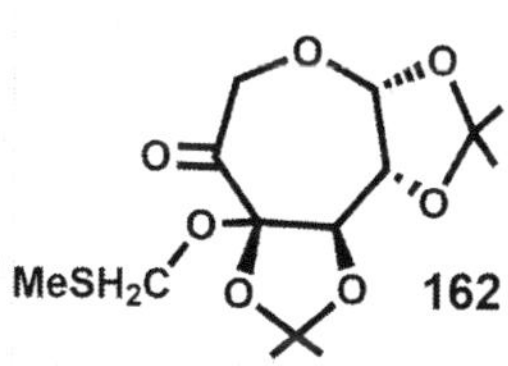 162	$^{3,4,0}B$(L)

9 Solution-phase structural studies of septanoses

The solution-phase structural determinations through NMR spectroscopy have been performed consistently on septanoses, in addition to solid-state structural determinations.[15] Key to solution-phase studies is the assessment of coupling constants and the intramolecular dipolar interactions. An expanded conformational space is feasible in the solution phase and interestingly, in instances, the interconverting nature of the conformations could also be elucidated. As observed through solid-state studies, solution-phase structural studies also show the twist-chair as the preferred conformation of septanoses. Table 3 summarizes conformations as determined in the solution-phase NMR techniques.

Table 3 Conformation of septanose derivatives, as determined through solution-phase NMR spectral studies.

S. No	Septanose	Chemical structure	Conformation
1	Methyl 2-*O*-benzoyl-3,4-*O*-isopropylidene-α-D-glucoseptanoside[77]	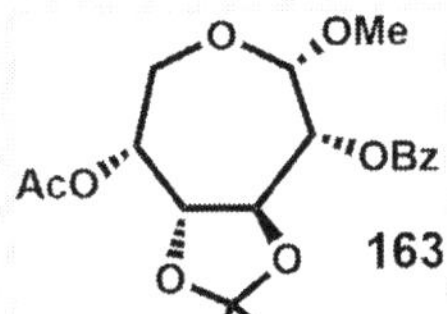 **163**	${}^{2}C_{5,6}$
2	5-*O*-Acetyl-2-*O*-benzoyl-3,4-*O*-isopropylidene-β-L-idoseptanoside[78]	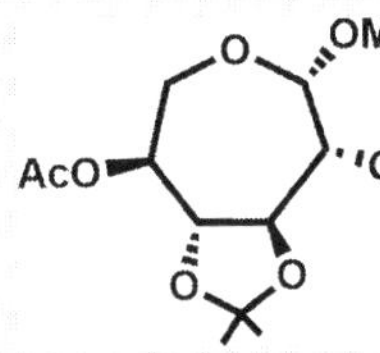 **164**	${}^{0,1}TC_{2,3}$
3	Methyl 2-*O*-benzoyl-5-deoxy-3,4-*O*-isopropylidene-β-L-*threo*-hex-4-enoseptanoside[79]	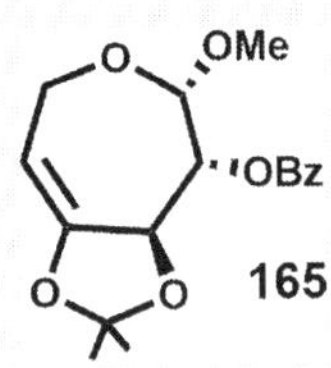 **165**	${}^{4,5}C_{1}$
4	Methyl 2,3:4,5-di-*O*-isopropylidene-β-L-idoseptanoside[80]	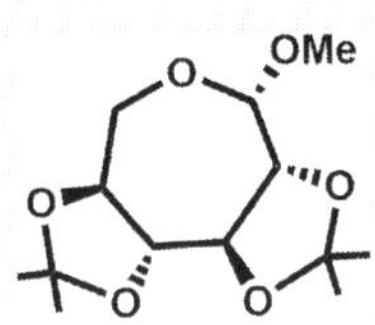 **166**	${}^{6,0}C_{3}$

Table 3 Conformation of septanose derivatives, as determined through solution-phase NMR spectral studies—cont'd

S. No	Septanose	Chemical structure	Conformation
5	Methyl 4,5-di-*O*-acetyl-2,3-*O*-isopropylidene-β-L-idoseptanoside[80]	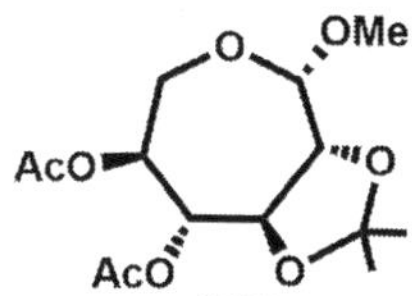 **167**	$^{1,2}TC_{6,0}$
6	Methyl 2,3-di-*O*-acetyl-4,5-*O*-isopropylidene-β-L-idoseptanoside[80]	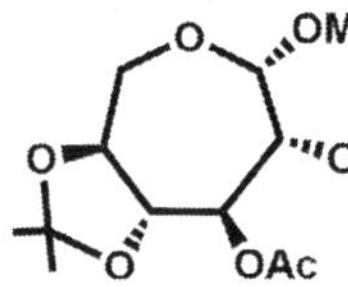 **168**	$^{3,4}TC_{1,2}$
7	Methyl 4,5-di-*O*-acetyl-2,3-di-*O*-isopropylidene-α-D-glucoseptanoside[81]	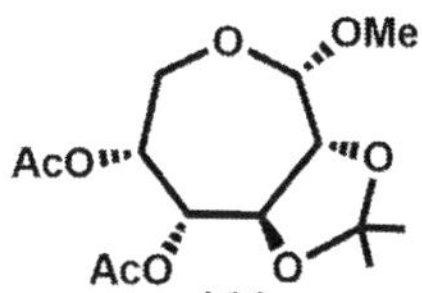 **169**	$^{1,2}TC_{6,0}$
8	Methyl 2,5-di-*O*-acetyl-3,4-*O*-isopropylidene-α-D-glucoseptanoside[81]	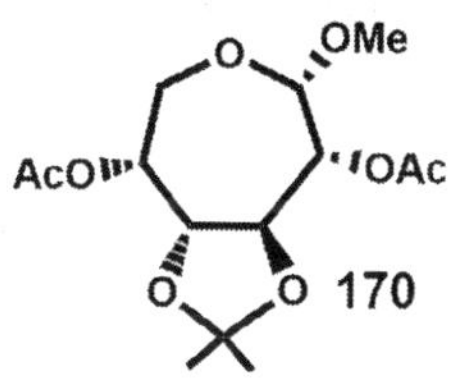 **170**	$^{4,5}TC_{6,0}$
9	3,4,5-Tri-*O*-acetyl-1,2-*O*-isopropylidene-α-D-galactoseptanose[81]	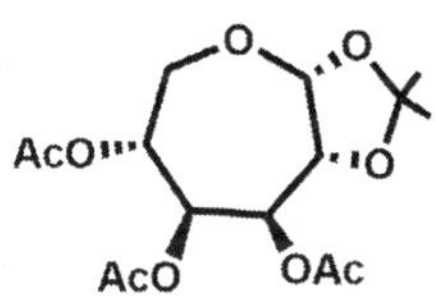 **171**	$^{4,5}TC_{6,0}$
10	5-*O*-Acetyl-1,2:3,4-di-*O*-isopropylidene-α-D-galactoseptanose[82]	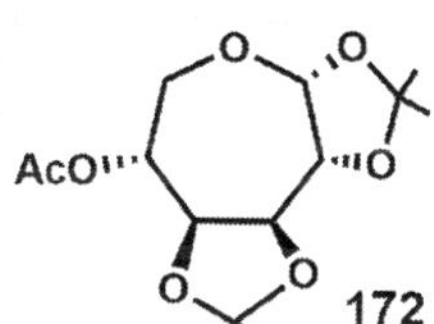 **172**	$^{1,2,5}B$
11	3-*O*-Acetyl-1,2:4,5-di-*O*-isopropylidene-α-D-galactoseptanose[83]	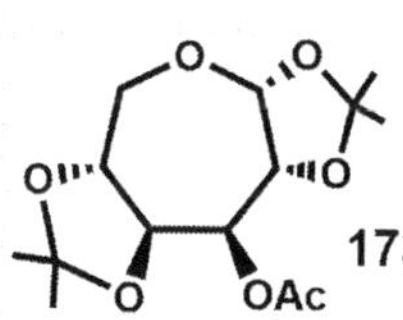 **173**	$^{6,0}TC_{4,5}$

Continued

Table 3 Conformation of septanose derivatives, as determined through solution-phase NMR spectral studies—cont'd

S. No	Septanose	Chemical structure	Conformation
12	4,5-Di-*O*-acetyl-1,2-*O*-isopropylidene-3-*O*-methyl-α-D-glucoseptanose[8]	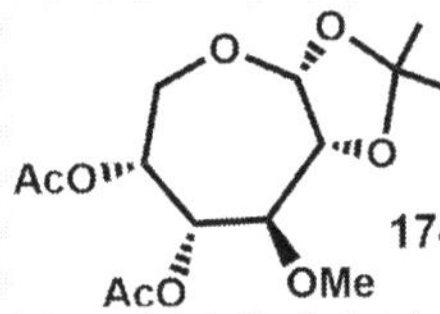 174	${}^{5}C_{1,2}$
13	5-*O*-Acetyl-1,2:3,4-di-*O*-isopropylidene-β-L-altroseptanose[81]	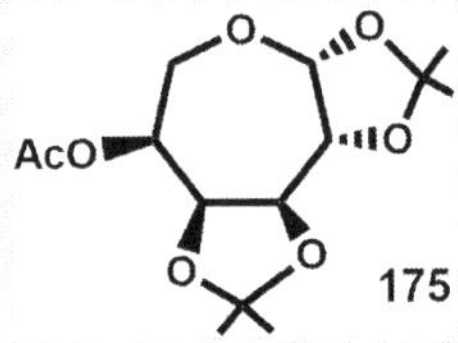 175	${}^{1,2}TB_{3,4}$
14	Methyl 4,5-di-*O*-acetyl-2,3-*O*-isopropylidene-β-D-glucoseptanoside[8]	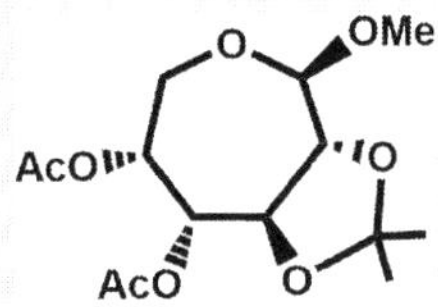 176	${}^{5,6}TC_{3,4}$
15	Methyl 2,3-di-*O*-acetyl-4,5-*O*-isopropylidene-β-D-glucoseptanoside[8]	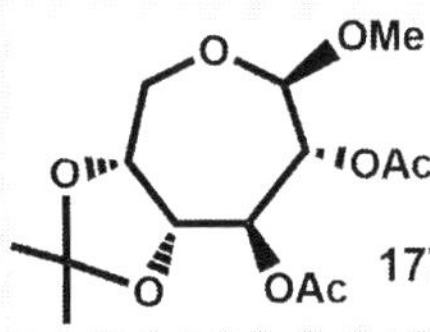 177	${}^{4,5}TC_{6,O}$
16	Methyl 2,5-di-*O*-acetyl-3,4-*O*-isopropylidene-β-D-glucoseptanoside[8]	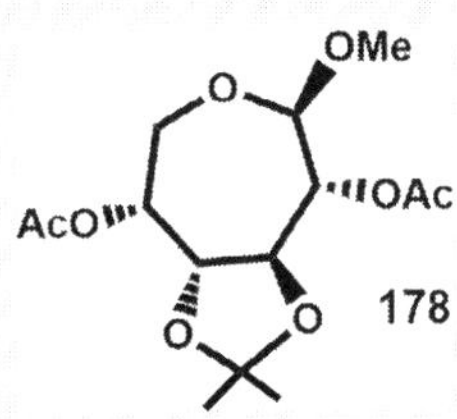 178	${}^{5,6}TC_{3,4}$
17	Methyl 4-*O*-acetyl-2,3-*O*-isopropylidene-β-D-*xylo*-hexoseptanosid-5-ulose[8]	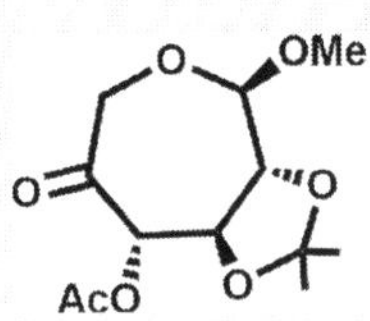 179	${}^{5,6}TC_{3,4}$
18	4,5-Di-*O*-acetyl-2,3-*O*-isopropylidene-α-L-idoseptanoside[8]	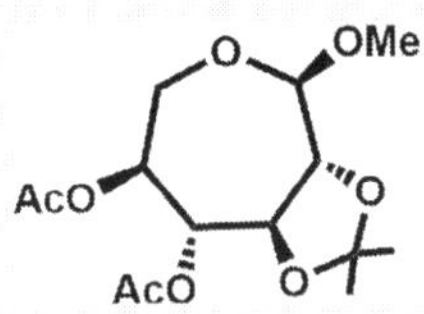 180	${}^{5,6}TC_{3,4}$

Table 3 Conformation of septanose derivatives, as determined through solution-phase NMR spectral studies—cont'd

S. No	Septanose	Chemical structure	Conformation
19	Methyl 2-*O*-benzoyl-3,4-*O*-isopropylidene-β-D *xylo*-hexoseptanosid-5-ulose[8]	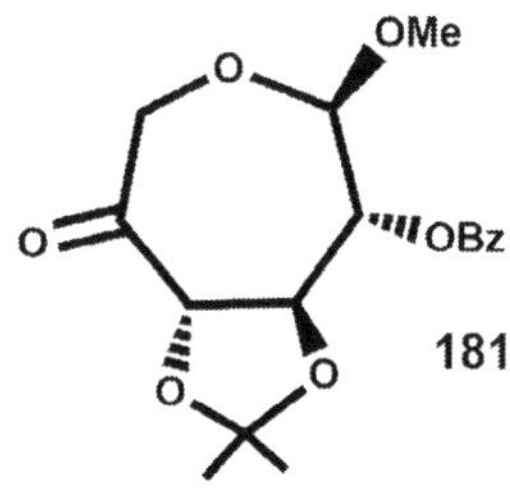 181	${}^{2}C_{5,6}$
20	Methyl 2,5-di-*O*-acetyl-α-L-idoseptanoside[8]	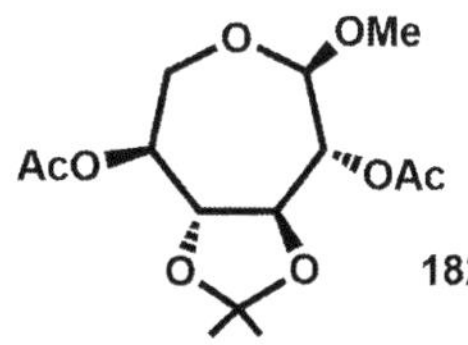 182	${}^{0,1}TC_{2,3} \rightleftharpoons {}^{2}C_{5,6} \rightleftharpoons {}^{3,4}TC_{1,2}$
21	Methyl 2,3,4,5-tetra-*O*-acetyl-α-L-idoseptanoside[8]	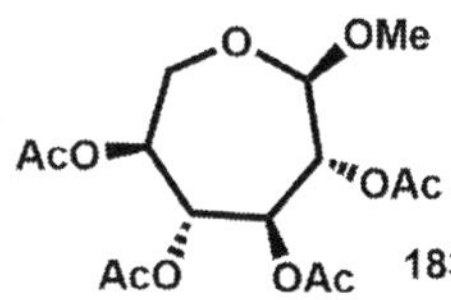 183	${}^{5,6}TC_{3,4}$
22	Methyl 2,3:4,5-di-*O*-isopropylidene-α-L-idoseptanoside[84]	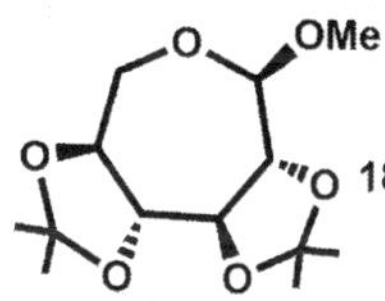 184	${}^{2,3}TC_{4,5}$
23	3-*O*-Acetyl-1,2:4,5-di-*O*-isopropylidene-β-L-idoseptanoside[84]	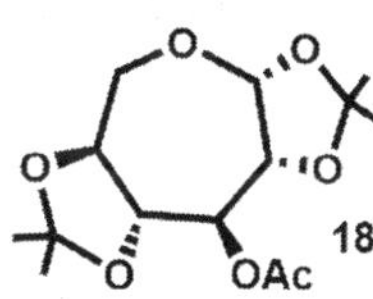 185	${}^{4,5}TC_{6,0}$
24	1,2:3,4-Di-*O*-isopropylidene-5-*C*-methyl-α-D-glucoseptanoside[84]	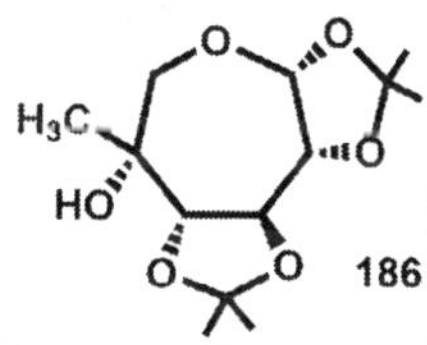 186	${}^{5,6}TC_{3,4}$
25	1,2:3,4-Di-*O*-isopropylidene-5-*C*-methyl-β-L-idoseptanose[84]	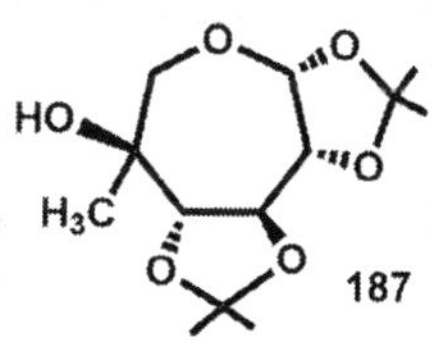 187	${}^{5,6}TC_{3,4}$

Continued

Table 3 Conformation of septanose derivatives, as determined through solution-phase NMR spectral studies—cont'd

S. No	Septanose	Chemical structure	Conformation
26	Methyl 5-*O*-methyl-α-D-*glycero*-D-*ido*-septanoside[85]	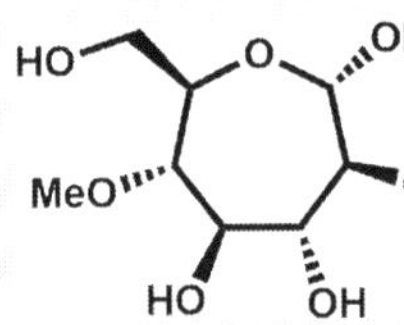 188	$^{3,4}TC_{5,6}$
27	Methyl 5-*O*-methyl-β-D-*glycero*-D-*gulo*-septanoside[85]	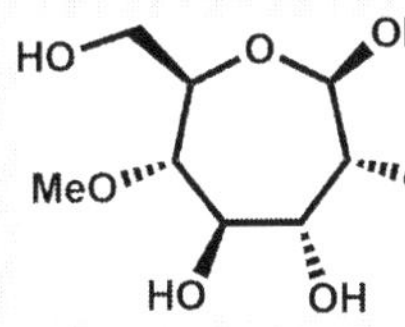 189	$^{6,O}TC_{4,5}$ and $^{3,4}TC_{5,6}$

10 Conclusion

Synthesis and studies of septanose sugars have advanced to a considerable extent, beyond one of the equilibration structural forms of a hexose sugar and one of the important structural forms for conformational elucidation. Many synthetic approaches to septanoses have emerged with a level of ingenuity, not only due to successes that allow implementing well-known organic reactions on to the sugar scaffolds but also to the septanose ring system being a member of larger oxepine structural scaffolds. The oxepine scaffolds themselves form constituent elements of many natural products. On the other hand, the beneficial effects of septanose scaffolds in varied types of functions remain largely unknown. Recent studies by Ernst and coworkers show that FimH lectin binding by a *manno*-septanoside is still weaker, in comparison to a mannopyranoside, as a result of larger conformational flexibility and associated entropic penalty affecting the binding affinity.[86] The work demonstrates that ligand conformational flexibility is also a selection factor for a protein binding to a putative ligand, thereby opening up a horizon to explore the septanoses as ligands to proteins. A structural similarity of septanosides with the corresponding pyranosides was considered as a possibility to verify the inhibition activities of certain glycosidase enzymes by Stütz and coworkers.[87] As surmised, an inhibition of the hydrolysis of chosen *p*-nitrophenyl septanosides by glycosidases was observed although the inhibition was 3–5 orders of magnitude weaker in comparison to the highly potent glycopyranosides. Importantly, the work demonstrated that septanosides exhibit specificities to enzymes, even when these inhibitors might not act as transition state mimics. Material properties of septanosides are largely unknown, whether the additional hydroxymethine *endo*-cyclic moiety, as compared to a pyanoside, modulate the material properties remain to be seen in literature at large. One can anticipate expanding frontiers of septanosides in manifold studies in the future.

In this chapter, the naming is adopted as follows. When the configuration of substituents present at C-2–C-5 is related to a hexose system and the configuration at C-6 to the glycerose system, the septanose is referred to as "*glycero*-septanose." Thus the septanose with a hydroxymethyl substituent at C-6 is referred to as "*glycero*-septanose" with seven carbons in the system. Without a substituent at C-6, the system remains to be "septanose." The naming of compounds is nonuniform in literature, a "*glycero*-septanose" being referred to as "7-hydroxymethyl *glycero*-septanose,"[20] a "heptoseptanose,"[19] and a "septanose" being referred to as "hexoseptanose."[25, 76] With the presence of varied functionalities, such as, an unsaturation, the naming varies as appropriate, without recourse to the configurations at C-2–C-5 substituents and C-6 substituent, example, compound **53** is a benzyl 4,5,7-tri-*O*-benzyl-D-*xylo*-hept-2-enoseptanoside.[28] Further, the naming also adheres to that assigned in the respective literature.

Acknowledgments

Authors thank the Department of Science and Technology (DST), New Delhi, and Council of Scientific and Industrial Research (CSIR), New Delhi, for a generous financial support. NJ thanks DST-SERB for JC Bose National Fellowship. SD and GCH thank CSIR for research fellowships.

References

1. Streitwieser, A.; Heathcock, C. H. Introduction to Organic Chemistry; 3rd ed.; MacMillan Publishing: New York, 1985. Chapters 14 and 27.
2. (a) Micheel, F.; Suckfüll, F. *Annalen* **1933**, *502*, 85–98. (b) Micheel, F.; Spruck, W. *Annalen* **1934**, *67*, 1665–1667.
3. Peat, S. *Adv. Carbohydr. Chem.* **1946**, *2*, 37–77.
4. Bayer, E.; Widder, R. *Justus Liebigs Ann. Chem.* **1965**, *686*, 181–197.
5. Cox, J. M.; Owen, L. N. *J. Chem. Soc. C* **1967**, 1121–1130.
6. Anet, E.F.L.J. *Carbohydr. Res.* **1968**, *8*, 164–174.
7. Stevens, J. D. *Carbohydr. Res.* **1972**, *21*, 490–492.
8. Stevens, J. D. *Aust. J. Chem.* **1975**, *28*, 525–557.
9. Contour, M.-O.; Fayet, C.; Gelas, J. *Carbohydr. Res.* **1990**, *201*, 150–152.
10. Barbero, H.; Díez-Poza, C.; Barbero, A. *Mar. Drugs* **2017**, *15*, 361. (1–34).
11. Basu, S.; Ellinger, B.; Rizzoa, S.; Deraevea, C.; Schürmann, M.; Preut, H.; Arndt, H.-D.; Waldmann, H. *Proc. Natl. Acad. Sci. U. S. A.* **2011**, *108*, 6805–6810.
12. Kleinke, A. S.; Webb, D.; Jamison, T. F. *Tetrahedron* **2012**, *68*, 6999–7018.
13. Nortcliffe, A.; Moody, C. J. *Bioorg. Med. Chem.* **2015**, *23*, 2730–2735.
14. Pakulski, Z. *Pol. J. Chem.* **2006**, *80*, 1293–1326.
15. Dey, S.; Jayaraman, N. *Pure Appl. Chem.* **2014**, *86*, 1401–1419.
16. Vannam, R.; Peczuh, M. W. *Eur. J. Org. Chem.* **2016**, 1800–1812.
17. Dey, S.; Ganesh, N. V.; Jayaraman, N. In: Domino and Intramolecular Rearrangement Reactions as Advanced Synthetic Methods in Glycoscience; Witczak, Z. J., Bielski, R., Eds.; John-Wiley & Sons, 2016; pp 256–278.

18. (a) Baschang, G. *Annalen* **1963**, *663*, 167–173. (b) Wolfrom, M. L.; Nayak, G. E.; Radford, T. *Proc. Natl. Acad. Sci. U. S. A.* **1967**, *58*, 1848–1851.
19. Butcher, M. E.; Ireson, J. C.; Lee, J. B.; Tyler, M. J. *Tetrahedron* **1977**, *33*, 1501–1507.
20. Osborn, H. M. I.; Turkson, A. *Tetrahedron Asymmetry* **2009**, *20*, 2162–2166.
21. Jasiński, M.; Utecht, G.; Fruziński, A.; Reissig, H.-U. *Synthesis* **2016**, *48*, 893–905.
22. Maciaszczyk, J.; Jasiński, M. *Tetrahedron Asymmetry* **2015**, *26*, 510–515.
23. Köver, A.; Matheu, M. I.; Díaz, Y.; Castillón, S. *ARKIVOC* **2007**, 364–379.
24. Datta, S.; Nath, M.; Chowdhury, S.; Bhattacharjya, A. *Tetrahedron Lett.* **2008**, *49*, 2935–2938.
25. Enright, P. M.; Tosin, M.; Nieuwenhuyzen, M.; Cronin, L.; Murphy, P. V. *J. Org. Chem.* **2002**, *67*, 3733–3741.
26. Jackowski, O.; Chrétien, F.; Didierjean, C.; Chapleur, Y. *Carbohydr. Res.* **2012**, *356*, 93–103.
27. Lobo, F.; Gómez, A. M.; Miranda, S.; López, J. C. *Chem. Eur. J.* **2014**, *20*, 10492–10502.
28. Ovaa, H.; Leeuwenburgh, M. A.; Overkleeft, H. S.; van der Marel, G. A.; van Boom, J. H. *Tetrahedron Lett.* **1998**, *39*, 3025–3028.
29. Schmidt, B.; Biernet, A. *Org. Lett.* **2008**, *10*, 105–108.
30. Peczuh, M. W.; Snyder, N. L. *Tetrahedron Lett.* **2003**, *44*, 4057–4061.
31. Saha, J.; Peczuh, M. W. *Adv. Carbohydr. Chem. Biochem.* **2011**, *66*, 121–186.
32. Vannam, R.; Peczuh, M. W. *Org. Biomol. Chem.* **2016**, *14*, 3989–3996.
33. Ramsubhag, R. R.; Peczuh, M. W. *ARKIVOC* **2011**, 92–101.
34. Pote, A. R.; Vannam, R.; Richard, A.; Gascón, J.; Peczuh, M. W. *Eur. J. Org. Chem.* **2018**, 1709–1719.
35. Boone, M. A.; McDonald, F. E.; Lichter, J.; Lutz, S.; Cao, R.; Hardcastle, K. I. *Org. Lett.* **2009**, *11*, 851–854.
36. Harvey, J. E.; Hewitt, R. J.; Moore, P. W.; Somarathne, K. K. *Pure Appl. Chem.* **2014**, *86*, 1377–1399.
37. Brimacombe, J. S.; Evans, M. E.; Forbes, E. J.; Foster, A. B.; Webber, J. M. *Carbohydr. Res.* **1967**, *4*, 239–243.
38. (a) Hoberg, J. O.; Bozell, J. J. *Tetrahedron Lett.* **1995**, *36*, 6831–6834. (b) Batchelor, R.; Hoberg, J. O. *Tetrahedron Lett.* **2003**, *44*, 9043–9045.
39. Ferrier, R. J.; Prasad, N. *J. Chem. Soc. C* **1969**, 570–575.
40. Batchelor, R.; Harvey, J. E.; Northcote, P. T.; Teesdale-Spittle, P.; Hoberg, J. O. *J. Org. Chem.* **2009**, *74*, 7627–7632.
41. Ramana, C. V.; Murali, R.; Nagarajan, M. *J. Org. Chem.* **1997**, *62*, 7694–7703.
42a. Hewitt, R. J.; Harvey, J. E. *Chem. Commun.* **2011**, *47*, 421–423.
42b. Hewitt, R. J.; Harvey, J. E. *J. Org. Chem.* **2010**, *75*, 955–958.
43. Sridhar, P. R.; Venukumar, P. *Org. Lett.* **2012**, *14*, 5558–5561.
44. Venukumar, P.; Sudharani, C.; Sridhar, P. R. *Chem. Commun.* **2014**, *50*, 2218–2221.
45. Ganesh, V.; Kundu, T.; Chandrasekaran, S. *Tetrahedron* **2014**, *70*, 7268–7282.
46. Vijayaganesh, N.; Jayaraman, N. *J. Org. Chem.* **2007**, *72*, 5500–5504.
47. Ganesh, N. V.; Jayaraman, N. *J. Org. Chem.* **2009**, *74*, 739–746.
48. Vijayaganesh, N.; Raghothama, S.; Sonti, R.; Jayaraman, N. *J. Org. Chem.* **2010**, *75*, 215–218.
49. Dey, S.; Jayaraman, N. *Beilstein J. Org. Chem.* **2012**, *8*, 522–527.
50. Dey, S.; Jayaraman, N. *Carbohydr. Res.* **2014**, *389*, 66–71.
51. Dey, S.; Jayaraman, N. *Carbohydr. Res.* **2014**, *399*, 49–56.
52. Lichtenthaler, F. W. *Chem. Rev.* **2011**, *111*, 5569–5609.
53. Gurjar, M. K.; Chakrabarti, A.; Rao, B. V.; Kumar, P. *Tetrahedron Lett.* **1997**, *38*, 6885–6888.

54. Deslongchamps, P. Stereoelectronic Effects in Organic Chemistry. Pergamon Press: Oxford, 1983.30–35.
55. Edward, J. T. *Chem. Ind.* **1955**, 1102–1104.
56. Jensen, H. H.; Nordstrøm, L. U.; Bols, M. *J. Am. Chem. Soc.* **2004**, *126*, 9205–9213.
57. Moumé-Pymbock, M.; Furukawa, T.; Mondal, S.; Crich, D. *J. Am. Chem. Soc.* **2013**, *135*, 14249–14255.
58. Markad, S. D.; Miller, S. M.; Morton, M.; Peczuh, M. W. *Tetrahedron Lett.* **2010**, *51*, 1209–1212.
59. Beck, J. M.; Miller, S. M.; Peczuh, M. W.; Hadad, C. M. *J. Org. Chem.* **2012**, *77*, 4242–4251.
60. Namchuk, M. N.; McCarter, J. D.; Becalski, A.; Andrews, T.; Withers, S. G. *J. Am. Chem. Soc.* **2000**, *122*, 1270–1277.
61. Jackobs, J.; Sundaralingam, M. *J. Chem. Soc. Chem. Commun.* **1970**, 157–158.
62. Cremer, D.; Pople, J. A. *J. Am. Chem. Soc.* **1975**, *97*, 1354–1358.
63. Stoddart, J. F. Stereochemistry of Carbohydrates; Wiley: New York, 1971.
64. Pallister, E. T.; Stephenson, N. C.; Stevens, J. D. *J. Chem. Soc. Chem. Commun.* **1972**, 98–99.
65. Wood, R. A.; James, V. J.; Rae, A. D.; Stevens, J. D.; Moore, F. H. *Aust. J. Chem.* **1983**, *36*, 2269–2277.
66. McConnell, J. F.; Stevens, J. D. *J. Chem. Soc., Perkin Trans. 2* **1974**, *4*, 345–348.
67. Beale, J. P.; Stephenson, N. C.; Stevens, J. D. *J. Chem. Soc. Chem. Commun.* **1971**, 25.
68. Foster, S. J.; James, V. J.; Stevens, J. D. *Acta Crystallogr.* **1983**, *C39*, 610–612.
69. Dey, S.; Basuroy, K.; Jayaraman, N. *Carbohydr. Res.* **2014**, *393*, 37–42.
70. Dey, S.; Basuroy, K.; Jayaraman, N. *Carbohydr. Res.* **2015**, *410*, 9–14.
71. Beale, J. P.; Stephenson, N. C.; Stevens, J. D. *Acta Crystallogr.* **1972**, *B28*, 3115–3121.
72. Bhadbhade, M.; Craig, D. C.; Ng, C. J.; Stevens, J. D. *Carbohydr. Res.* **2012**, *353*, 86–91.
73. Bailey, C. J.; Craig, D. C.; Grainger, C. T.; James, V. J.; Stevens, J. D. *Carbohydr. Res.* **1996**, *284*, 265–270.
74. Peczuh, M. W.; Snyder, N. L.; Fyvie, W. S. *Carbohydr. Res.* **2004**, *339*, 1163–1171.
75. James, V. J.; Stevens, J. D. *Carbohydr. Res.* **1980**, *82*, 167–174.
76. Craig, D. C.; James, V. J.; Stevens, J. D. *Aust. J. Chem.* **1990**, *43*, 2083–2086.
77. Cookson, R. C.; Crabb, T. A.; Frankel, J. J.; Hudec, J. *Tetrahedron* **1966**, *22*, 355–390.
78. Murcko, M. A.; Dipaola, R. A. *J. Am. Chem. Soc.* **1992**, *114*, 10010–10018.
79. Shah, R. H.; Bahl, O. P. *Carbohydr. Res.* **1978**, *65*, 47–55.
80. Ng, C. J.; Craig, D. C.; Stevens, J. D. *Carbohydr. Res.* **1996**, *284*, 249–263.
81. Driver, G. E.; Stevens, J. D. *Carbohydr. Res.* **2001**, *334*, 81–89.
82. James, V. J.; Stevens, J. D. *Cryst. Struct. Commun.* **1982**, *11*, 1933–1938.
83. Driver, G. E.; Stevens, J. D. *Aust. J. Chem.* **1990**, *43*, 2063–2081.
84. Tran, T. Q.; Stevens, J. D. *Aust. J. Chem.* **2002**, *55*, 171–178.
85. Dematteo, M. P.; Mei, S.; Fenton, R.; Morton, M.; Baldisseri, D. M.; Hadad, C. M.; Peczuh, M. W. *Carbohydr. Res.* **2006**, *341*, 2927–2945.
86. Sager, C. P.; Fiege, B.; Zihlmann, P.; Vannam, R.; Rabbani, S.; Jakob, R. P.; Preston, R. C.; Zalewski, A.; Maier, T.; Peczuh, M. W.; Ernst, B. *Chem. Sci.* **2018**, *9*, 646–654.
87. Tauss, A.; Steiner, A. J.; Stütz, A. E.; Tarling, C. A.; Withers, S. G.; Wrodnigg, T. M. *Tetrahedron Asymmetry* **2006**, *17*, 234–239.

N- and *C*-Glycopyranosyl heterocycles as glycogen phosphorylase inhibitors

7

Éva Bokor
Department of Organic Chemistry, University of Debrecen, Debrecen, Hungary

Chapter outline

1 Introduction

Diabetes mellitus is a serious metabolic disorder with the main characteristic of abnormally elevated blood glucose levels (hyperglycemia). Diabetes is an endemic disease with a growing number of patients that is increasing from 366 million in 2011[1] to 425 million in 2017, and a further rise to 629 million is predicted for 2045.[2] Globally, diabetes also entails significant healthcare expenditures, with an annual cost of USD 727 billion in 2017.[2] Due to the long-term complications (such as neuropathy, nephropathy, and particularly cardiovascular damages), diabetes is among the leading causes of death, as well. In 2017, 4 million people died due to this disease.[2]

Type 2 diabetes (T2DM, or noninsulin-dependent diabetes mellitus: NIDDM) is the most frequent form of the disease representing around 90% of all cases.[2] The hyperglycemic state in T2DM is the result of insufficient insulin secretion by the pancreas and/or insensibility of the peripheral tissues toward insulin-stimulated glucose uptake.

Recent Trends in Carbohydrate Chemistry. https://doi.org/10.1016/B978-0-12-817467-8.00007-4
© 2020 Elsevier Inc. All rights reserved.

Current therapy of T2DM is symptomatic, aimed at adjusting the blood sugar concentration close to the normoglycemic state (3.5–6 mM) by a combination of diet, exercise, and administration of oral anti-hyperglycaemic agents. Although several types of antidiabetics acting on different molecular targets (e.g., biguanides, thiazolidindiones, sulfonylureas, glinides, α-glucosidase, and SGLT2 inhibitors) have been developed,[3–5] treatments with these drugs do not provide adequate glucose control for a great proportion of patients.[6]

In T2DM patients, an enhanced glucose release in the liver is perceived, which also contributes to the elevation of serum glucose levels. Based on this, an opportunity of controlling the blood glucose levels is foreseen by an intervention in the carbohydrate metabolic pathways.[7] Glycogen phosphorylase (GP) is the rate-limiting enzyme of the glycogenolysis that catalyses the phosphorolytic breakdown of the storage polysaccharide glycogen. Due to the key role of GP in hepatic glucose production, inhibition of this enzyme has emerged as a therapeutic possibility against T2DM.[8] Additionally, GP inhibition is also regarded as a possible tool for combating other diseased conditions, such as myocardial[9, 10] and cerebral ischemias,[11, 12] and tumors.[13–18]

A large collection of small molecules has been found to inhibit the enzyme through binding to one (or in some cases more) of the seven binding sites of GP (catalytic, allosteric and new allosteric, inhibitor, glycogen storage, benzimidazole, and quercetin sites) identified so far.[19–25] The most widely investigated group of GP inhibitors (GPIs) is that of the derivatives of D-glucose (one of the physiological regulators of the enzyme), which competitively inhibit the enzyme by binding to the catalytic site.[26, 27] Extensive studies with this compound class have revealed four glucose-derived lead structures [*N*-acyl-*N*′-(β-D-glucopyranosyl) ureas, glucopyranosylidene-spiro-heterocycles, and *C*- and *N*-(β-D-glucopyranosyl) heterocycles] which most potent representatives exhibit submicromolar inhibitory activities against GP.[23, 28]

Antihyperglycaemic potential of the glucose derivatives has also been demonstrated by ex vivo and in vivo experiments in several cases,[29–31] and some members have been found to exert unprecedented influence on hepatic metabolism[32] and improve pancreatic β-cell function.[33]

This survey will focus on the *C*- and *N*-(β-D-glycopyranosyl) heterocycles, among which the most efficient derivatives represent the first low nanomolar glucose analog inhibitors of GP. The aim of this overview is to provide the reader with a comprehensive insight into the syntheses and the structure-activity relationships of these very promising inhibitor subclasses, also involving the latest results, which were not, or only partially included in the last related review articles.[23, 28, 34]

2 Syntheses

2.1 Five-membered N-(β-D-glucopyranosyl) heterocycles

2.1.1 Imidazoles

For the synthesis of *N*-glycopyranosyl imidazoles glycosylation reactions of the corresponding imidazole derivatives (e.g., base-mediated[35] or metal-ion-promoted[36] reactions of glycosyl bromides with 1*H*-imidazoles, and Vorbrüggen type condensations of glycosyl acetates with silylated imidazoles[37]) are the most frequently used procedures.

The preparation of *O*-peracetylated 1-glucopyranosyl-4-substituted imidazole **2** was effected by a $Hg(CN)_2$ promoted regioselective substitution reaction of acetobromoglucose **1** with 4-phenylimidazole (Scheme 1). *O*-Deprotection of **2** by the Zemplén protocol gave the test compound **3** in good yield.[38]

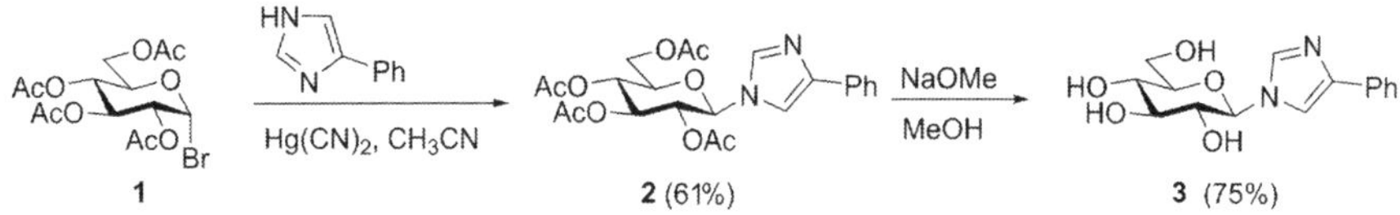

Scheme 1 Synthesis of an *N*-(β-D-glucopyranosyl)imidazole.

2.1.2 1,2,3-Triazoles

1-Glycosyl-1,2,3-triazoles were extensively synthesized for various purposes by the well-known copper-catalyzed azide acetylene cycloaddition (CuAAC) from glycosyl azides.[39,40]

For GP inhibition, a set of 1-(β-D-glucopyranosyl)-4-substituted 1,2,3-triazoles **4** was synthesized by Cu(I)-catalyzed cycloaddition of *O*-peracetylated glucosyl azide **6** with terminal alkynes (Scheme 2).[41–44] On the other hand, starting from the same reactants, 5-halogeno-1,2,3-triazoles **7** were also obtained by using a twofold excess of Cu(I)-halide (Scheme 2).[44, 45] In each series, removal of the *O*-acyl protecting groups was carried out by transesterification to give the unprotected derivatives **5** and **8** in high yields.

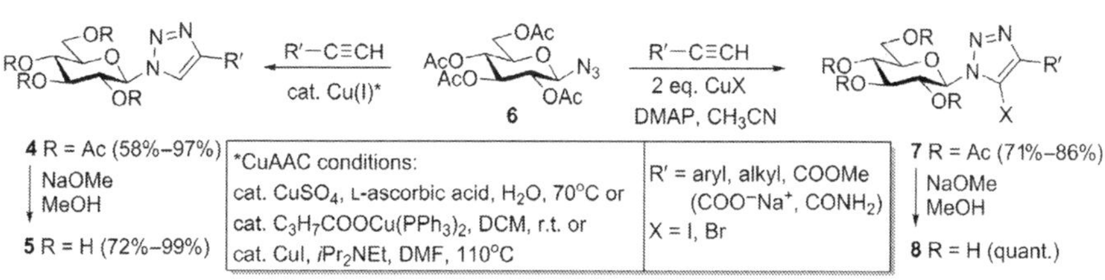

Scheme 2 Synthesis of *N*-(β-D glucopyranosyl) 1,2,3-triazoles.

CuAAC was also used to get further sugar conjugates such as compounds **9-23** having an *N*-glucopyranosyl 1,2,3-triazole moiety (Fig. 1) to test their inhibitory effects against the glycogen phosphorylase enzyme.[42, 44–48]

2.1.3 Tetrazoles

For the synthesis of *N*-glycosyl tetrazoles, the most frequently used method is the reaction of an *O*-peracylated glycosyl bromide with a tetrazole salt (usually obtained by deprotonation with KOH or K_2CO_3) to give mixtures of 1- and 2-glycosylated tetrazoles.[49–51] 1-Glycosyltetrazoles can be obtained from glycosyl isocyanides in

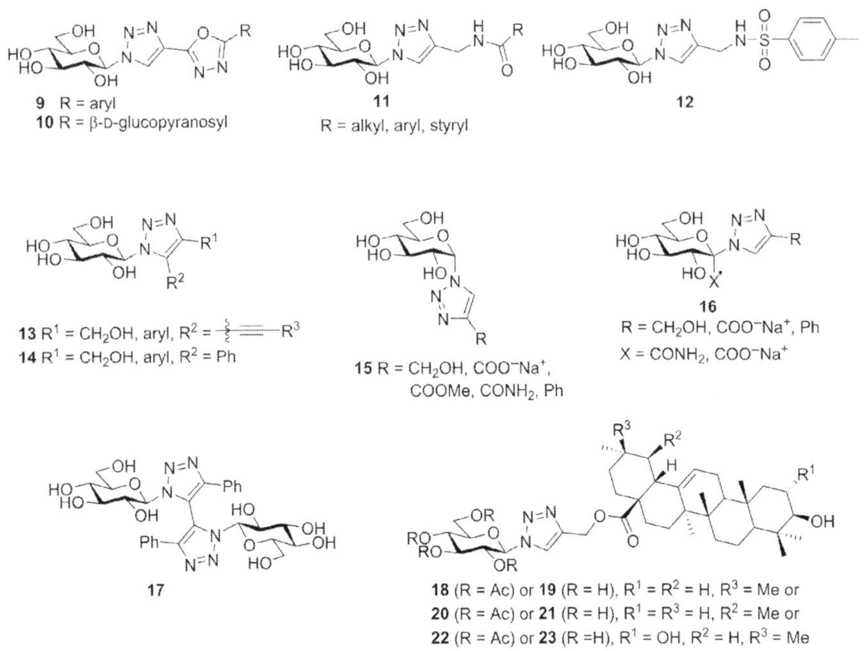

Fig. 1 Further representatives of *N*-glucopyranosyl 1,2,3-triazole derivatives.

multicomponent reactions,[52, 53] while 2-glycosyltetrazoles can be prepared by cycloaddition of glycosyl azides with nitriles.[54]

For GP inhibition studies, the synthesis of *N*-(β-D-glucopyranosyl)-5-substituted tetrazoles **24** was accomplished by glycosylation of 5-substituted tetrazoles with acetobromoglucose **1** in the presence of K_2CO_3 (Scheme 3).[38] The reaction of 5-methyltetrazole with **1** afforded the 2,5-disubstituted derivative **24** (R′=CH_3) along with an undesirable elimination by-product 2-acetoxy-D-glycal **26**. Treatment of **1** with 5-phenyltetrazole resulted in 2,5- and 1,5-disubstituted-tetrazoles[50] (**24** and **27**, respectively), with high preference for the formation of the sterically less crowded regioisomer **24**. *O*-Deacetylation of **24** was carried out by transesterification to give the unprotected derivatives **25** in high yields.

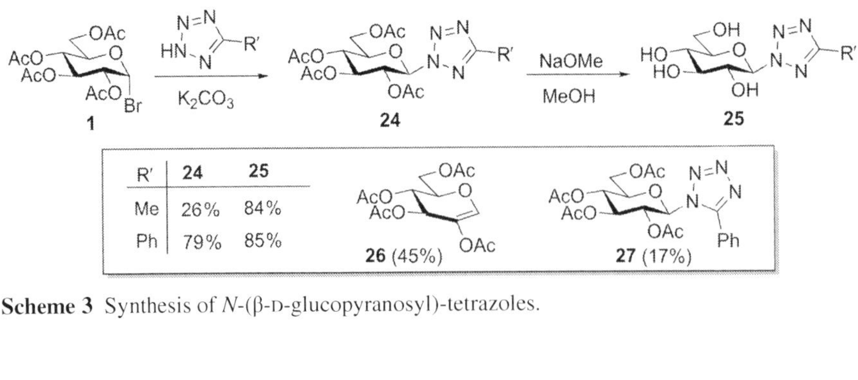

Scheme 3 Synthesis of *N*-(β-D-glucopyranosyl)-tetrazoles.

2.2 Five-membered C-(β-D-glucopyranosyl) heterocycles

2.2.1 2,6-Anhydroaldonic acid derivatives as precursors

For the construction of C-glycopyranosyl heteroarenes, (a) the direct C-glycosylations of electron-rich heterocycles (or those of more reactive metalated derivatives) with electrophilic glycosyl donors, (b) the additions of metalated heterocycles to glyconolactones followed by dehydroxylations of the resulting hemiketals, and (c) the cyclizations of pre-formed C-glycosyl precursors are the most widely applied methods.[28] Since a great number of C-glucopyranosyl hetarene type GP inhibitors were obtained by methods (c) and especially by ring-closures of 2,6-anhydroaldonic acid derivatives, the preparation of the most frequently used members of this compound class is summarized in Scheme 4.

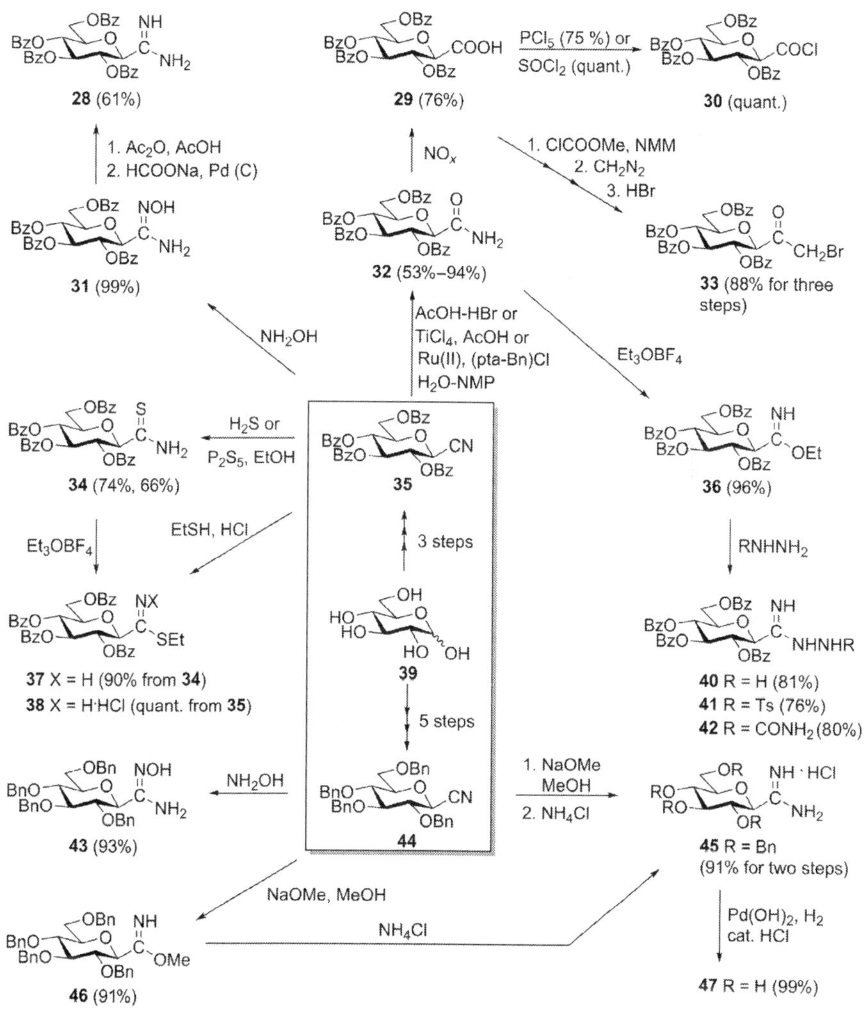

Scheme 4 Syntheses of 2,6-anhydroaldonic acid derivatives as precursors.

O-Perbenzoylated β-D-glucopyranosyl cyanide[55] **35**, easily obtained from D-glucose (**39**) in three steps (including *O*-perbenzoylation of D-glucose followed by treatment with HBr-AcOH and subsequent substitution of the resulting glucosyl bromide with $Hg(CN)_2$), served as the most common starting material. Addition of hydroxylamine to the nitrile group furnished amidoxime **31**.[56, 57] *O*-Acetylation of **31** followed by reductive cleavage of the resulting N–O bond yielded amidine **28**.[58] Partial hydrolysis of cyanide **35** afforded amide[55, 59] **32** from which the anhydro-aldonic acid **29** was obtained by nitrosation.[60] Acid **29** was converted to acid chloride **30**,[58, 60] and to bromomethyl glucosyl ketone **33**.[61] Thioamide **34** was formed by the treatment of cyanide **35** with H_2S[62] or P_2S_5/EtOH[63] whose transformation into the thioimidate **37** was effected upon reaction with Et_3OBF_4.[63] Hydrochloric salt of the previously mentioned iminoester **38** was also obtained by an acid catalyzed reaction of **35** with EtSH.[64] Treatment of amide **32** with Meerwein's salt resulted in imidate[63] **36**, which was converted to various amidrazones **40–42** with hydrazine derivatives.[58, 65, 66] In recent years, some *O*-perbenzylated analogues of the previously mentioned precursors have also been made available. Although the *O*-perbenzylated β-D-glucopyranosyl cyanide **44** had been known for a long time,[67–69] a large-scale preparation of this compound has only recently been accomplished.[70] Amidoxime **43** was prepared in the reaction of **44** with hydroxylamine.[56] Treatment of **44** with MeO^- ions gave imidate[70] **46** whose reaction with NH_4Cl furnished the amidine salt[70] **45**. The synthesis of the unprotected amidine **47** was also performed by reductive cleavage of the *O*-benzyl groups of compound **45**.[71]

2.2.2 Pyrroles

For the formation of 1,2-*trans*-configured *C*-glycopyranosyl pyrroles, *C*-glycosylations of pyrroles with *O*-peracylated glycosyl donors (e.g., glycosyl trichloroacetimidates or halides) in the presence of a Lewis acid are generally applied methods.[28] Following this convenient route, some 2-(β-D-glucopyranosyl)pyrroles were synthesized by the reactions of 1*H*-pyrroles with *O*-peracetylated β-D-glucopyranosyl trichloroacetimidate **50** (Scheme 5).[72] Treatment of the unsubstituted- and the 2-arylpyrroles with **50** in the presence of $BF_3{\cdot}Et_2O$ provided the corresponding 2-mono- (**48**, R′=H) and 2,5-disubstituted pyrroles (**48**, R′=aryl), respectively, in moderate to good yields. Surprisingly, 3-arylpyrroles, when reacted with **50**, gave the 2,3-disubstituted pyrroles **51** instead of the expected, sterically less crowded 2,4-disubstituted regioisomers (not shown). Removal of the *O*-acetyl groups of **48** and **51** was performed by the Zemplén method to get **49** and **52**, respectively, in good yields.

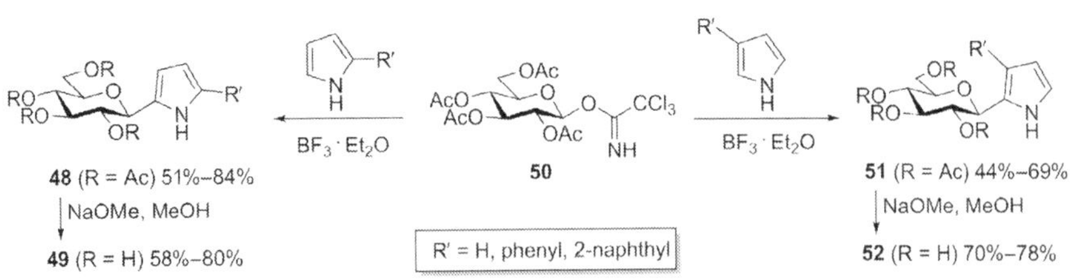

Scheme 5 Synthesis of 2-(β-D-glucopyranosyl)pyrroles.

2.2.3 Isoxazoles

For the construction of 3-glucopyranosyl derivatives of 1,2-azoles a glucosylated phenylethynyl ketone **53** as precursor was prepared by a Pd-catalyzed cross-coupling reaction of acid chloride **30** with $PhC{\equiv}CSnBu_3$ (Scheme 6).[73] Cyclization of **53** with hydroxylamine and subsequent *O*-debenzoylation by the Zemplén protocol furnished the 5-phenylisoxazole **55**.[73] Beside the previously mentioned cyclocondensation reaction, the 1,3-dipolar cycloaddition of alkynes with sugar-based nitrile oxides, obtained from *C*-glycosylated oximes, provided another synthetic possibility to get 3-glycopyranosylisoxazoles.[74–76]

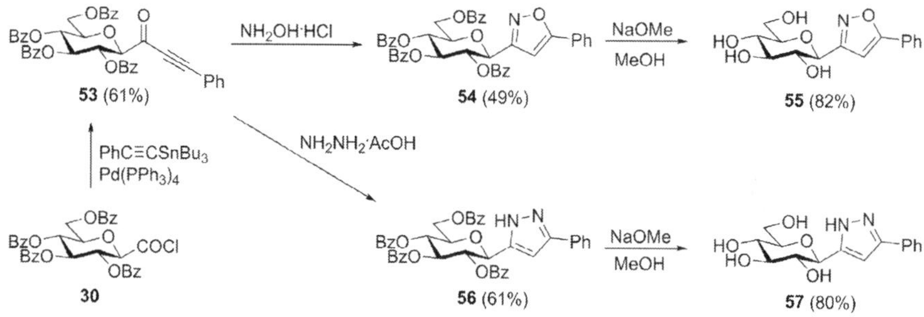

Scheme 6 Synthesis of 3-(β-D-glucopyranosyl)-5-phenylisoxazole and -pyrazole.

2.2.4 Pyrazoles

The 5-phenylpyrazole **57** was obtained by ring closure of ynone **53** with hydrazine acetate followed by *O*-deprotection of **56** (Scheme 6).[73]

2.2.5 Thiazoles

Some 4-aryl-2-glucopyranosyl-thiazoles **59** were synthesized by cyclization of *O*-perbenzoylated *C*-glucosyl thioformamide **34** with α-haloketones, followed by removal of the *O*-benzoyl protecting groups of **58** (Scheme 7).[73]

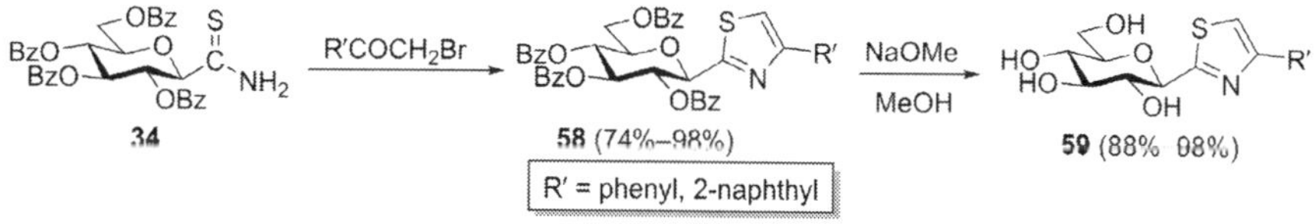

Scheme 7 Synthesis of 2-(β-D-glucopyranosyl)-4-substituted thiazoles.

Based on the previously mentioned Hantzsch-type cyclocondensation other 2-glycopyranosyl-5-substituted thiazoles were also synthesized[77] as well as further analogues were also obtained either by ring-closure of glycopyranosyl cyanides with cystein derivatives followed by oxidation[77] or by the addition of 2-lithiothiazole to aldonolactones[78] followed by functionalization of the C-4 position of the thiazole ring.[79]

2.2.6 Imidazoles

Before the preparation of 2-(β-D-glucopyranosyl)imidazoles for GP inhibition, only one unsubstituted 2-glucosylimidazole could be found in the literature, obtained by the addition of lithiated imidazole to *O*-perbenzylated D-glucono-1,5-lactone followed by reductive cleavage of the hemiacetalic OH.[80]

For the formation of 4(5)-aryl-2-glucopyranosylimidazoles **62,** four synthetic routes were elaborated (Scheme 8). The first synthesis was based on the cyclocondensation of *O*-perbenzoylated *C*-glucosyl formamidine **28** with α-haloketones, however, due to the necessarily applied alkaline conditions the desired heterocycles **60** could be obtained only in low yields.[73] The ring-closure of *O*-perbenzoylated *C*-glucosyl formimidate **36** with α-aminoketones required less basic conditions and furnished the same heterocycles **60** in somewhat better yields.[72] An analogous transformation of the *O*-perbenzylated imidate **46** resulted in similar yields of **61**.[70] Cyclization of the *O*-perbenzylated formamidine **45** with α-haloketones proved to be the most efficient procedure providing imidazoles **61** in good yields.[70] Removal of the protecting groups of **60** by the Zemplén method and those of **61** by catalytic hydrogenation or by EtSH/$BF_3{\cdot}Et_2O$ gave the final products **62** in good to high yields.[70, 73]

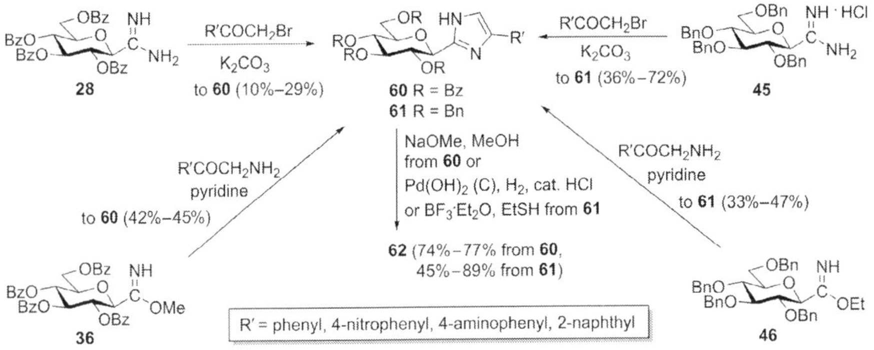

Scheme 8 Synthesis of 4(5)-aryl-2-(β-D-glucopyranosyl)imidazoles.

2.2.7 Oxadiazoles

C-Glucopyranosyl derivatives of oxadiazoles were achieved in three relevant constitutions of the heterocycle. 3-Glucosyl-5-substituted 1,2,4-oxadiazoles **64** were produced by acylation of amidoxime **31** with acid chlorides followed by intramolecular cyclization of the resulting *O*-acylamidoximes **63** (Scheme 9).[56] 3-Glycosyl-5-(mono- or dialkyl)amino-1,2,4-oxadiazoles **66** (Scheme 9) were also obtained by the reaction of the same amidoxime **31** with carbodiimides or Vilsmeyer salts.[81] Protecting group cleavage of **64** and **66** by the Zemplén method gave the test compounds **65** and **67**, respectively, in moderate to excellent yields.

The formation of isomeric 5-glucosyl-3-substituted 1,2,4-oxadiazoles **69** was effected via **68** obtained by 1,3-dipolar cycloaddition of *O*-perbenzoylated glucosyl cyanide **35** with nitrile oxides, generated in situ from hydroximoyl chlorides, followed by *O*-deprotection (Scheme 10).[82, 83]

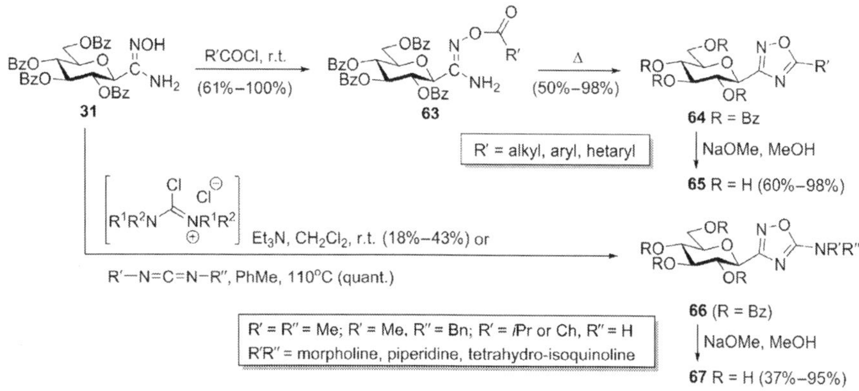

Scheme 9 Synthesis of 3-(β-D-glucopyranosyl)-5-substituted 1,2,4-oxadiazoles.

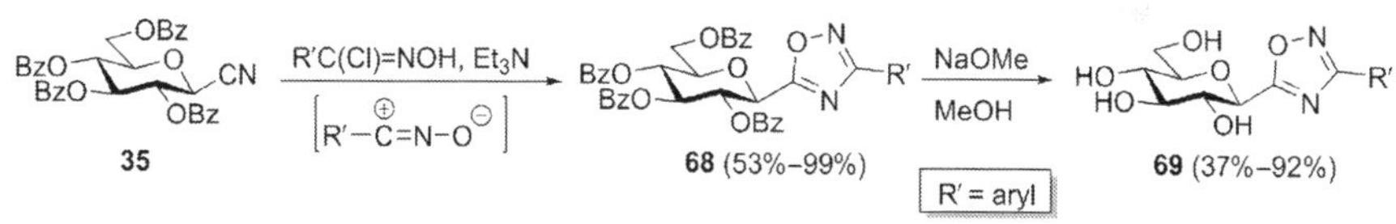

Scheme 10 Synthesis of 5-(β-D-glucopyranosyl)-3-substituted 1,2,4-oxadiazoles.

For the construction of 2-glucosyl-5-substituted 1,3,4-oxadiazoles **72**, two synthetic methods were elaborated (Scheme 11). In the first route, 5-glucosyl-tetrazoles **70** and **71** were prepared by 1,3-dipolar cycloadditions of glycosyl cyanides, **35** and **35a**, respectively, with azide ion.[46, 64] Acylation of these heterocyles accompanied by spontaneous ring-transformation, resulted in the desired 2,5-disubstituted 1,3,4-oxadiazoles **72** in moderate to high yields.[64, 83] In another route, reductive transformation of cyanide **35** in the presence of benzhydrazide furnished the 2,6-anhydroaldose benzoylhydrazone **74** whose intramolecular cyclization under oxidative conditions yielded the corresponding 1,3,4-oxadiazole **72** (R′=Ph).[83] Treatment of cyanide **35** with 4-phenyl- and unsubstituted semicarbazide in the presence of Raney-Ni was also carried out to obtain 2,6-anhydroaldose semicarbazones **75** and **78**, respectively. These compounds proved to be suitable precursors of 2-amino-1,3,4-oxadiazole derivatives.[84] Oxidative ring-closure of 4-phenylsemicarbazone **75** resulted in the 2-phenylamino-1,3,4-oxadiazole **76**, while acylation of semicarbazone **78** and subsequent cyclization of the resulting 4-acylsemicarbazones **79** in the presence of PIDA gave 2-acylamino-1,3,4-oxadiazoles **80**. The ester-protecting groups of **72**, **76,** and **80** were also removed to get the unprotected test compounds **73**, **77,** and **81**, respectively (Scheme 11).

C-Glucosyl-oxadiazolyl-1,2,3-triazoles (Fig. 2) linked by a direct C–C bond (**82**) or by alkyl bridges (**83** and **84**) were also synthesized by CuAAC reactions of suitable ethynylated *C*-glucosyl oxadiazoles.[46, 57]

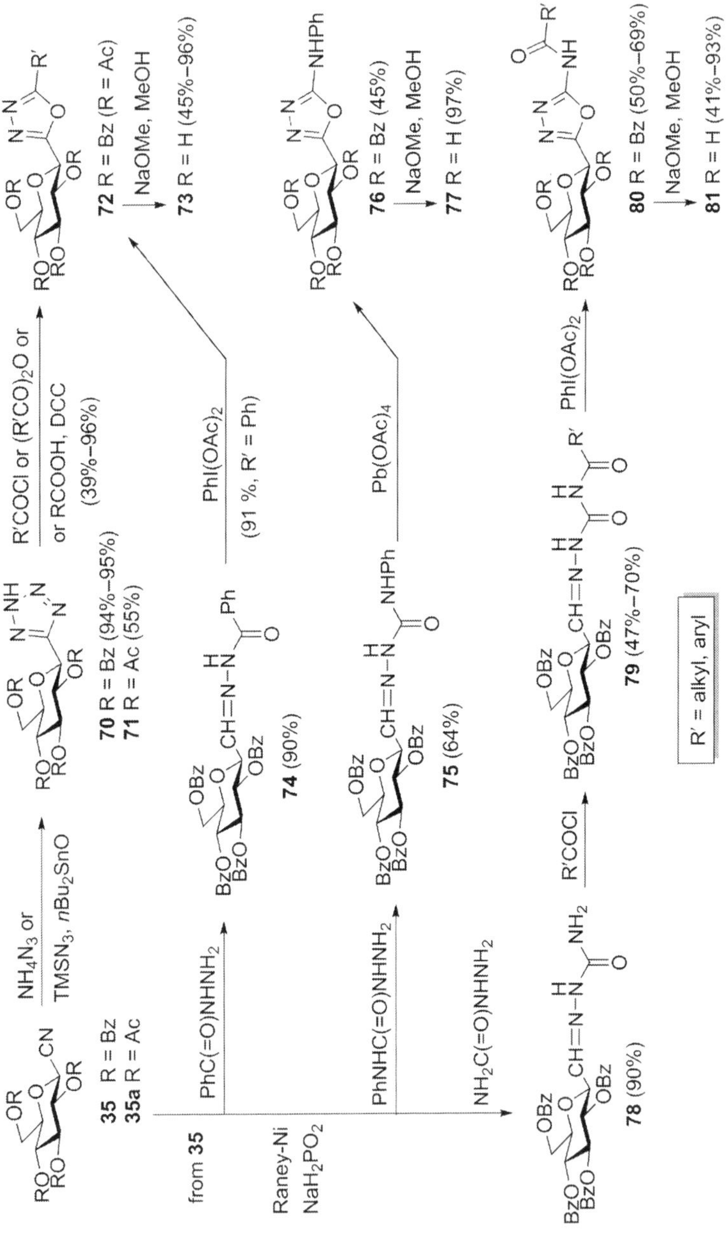

Scheme 11 Syntheses of *C*-(β-D-glucopyranosyl) 1,3,4-oxadiazoles.

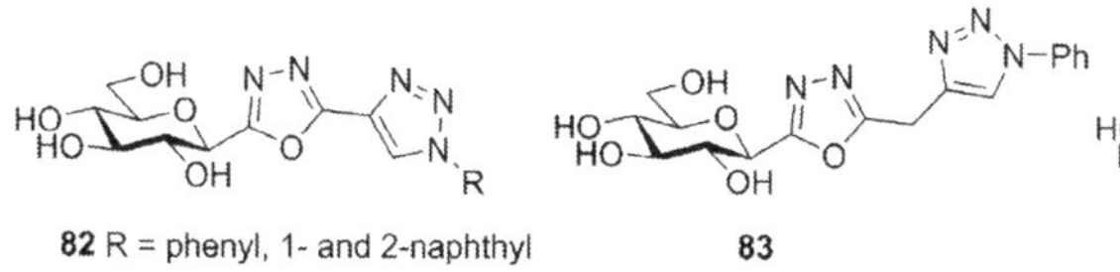

Fig. 2 1,2,3-Triazole-containing glucosyloxadiazoles.

Starting from *C*-glucosyl formamidoxime **31,** the synthesis of two types of trivalent heterocyclic *C*-glucopyranosyl derivatives was also effected (Scheme 12).[57] *O*-Acylation of **31** with 1,3,5-benzenetricarbonyl trichloride followed by microwave-assisted cyclodehydration with TBAF yielded the expected tris-oxadiazole **86**. Treatment of **31** with pent-4-ynoic acid in the presence of EDCI/HOBt and subsequent ring-closure of the resulting *O*-acylamidoxime afforded the alkynylated 1,2,4-oxadiazole **85**. Copper-catalyzed reaction of **85** with 1,3,5-tris(azidomethyl) benzene resulted in the tris-cycloadduct **88**. *O*-Debenzoylation of the obtained homotrivalent molecules **86** and **88** by standard transesterification provided the unprotected test compounds **87** and **89**, respectively, in excellent yields.

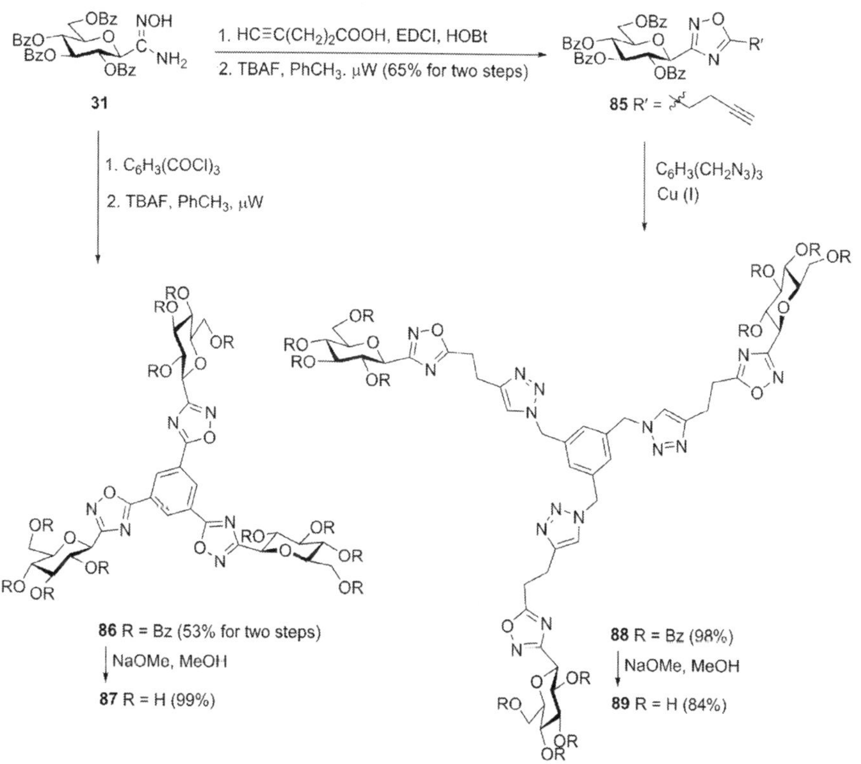

Scheme 12 Synthesis of homotrivalent heterocyclic glucose derivatives.

2.2.8 1,3,4-Thiadiazoles

Similar to the preparation of 2-acylamino-1,3,4-oxadiazoles **81** (Scheme 11), the formation of analogous 1,3,4-thiadiazoles **93** was also examined (Scheme 13).[84] Acid-catalyzed transimination of semicarbazone **78** with thiosemicarbazide furnished thiosemicarbazone **90**. Acylation of **90** resulted in diastereomeric mixtures of 4-acyl-2-acylamino-5-glucosyl-Δ^2-1,3,4-thiadiazolines **91,** which were reacted with PIDA to get the desired 2-acylamino-1,3,4-thiadiazoles **92**. *O*-Debenzoylation of **92** by the Zemplén protocol afforded the final products **93** in good yields.

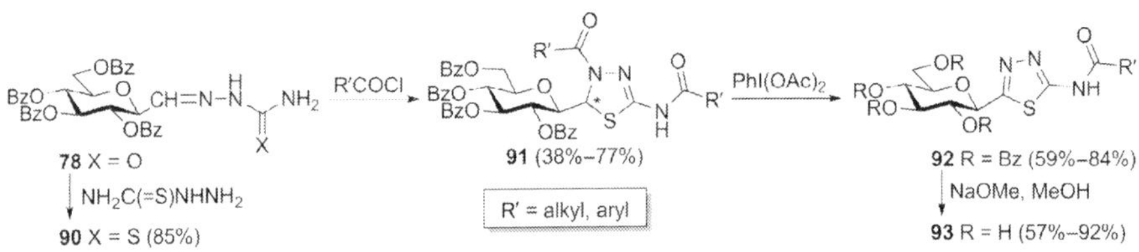

Scheme 13 Synthesis of 2-acylamino-5-(β-D-glucopyranosyl) 1,3,4-thiadiazoles.

2.2.9 Triazoles

1,2,3-Triazoles

The Cu(I)-catalyzed azide-alkyne cycloaddition has been widely applied for constructing 1-substituted-4-glycosyl 1,2,3-triazoles.[28]

For the synthesis of 1-aryl-4-(β-D-glucopyranosyl) 1,2,3-triazoles, CuAAC reactions of *C*-glucopyranosyl ethynes **94** and **95** with pre-formed or in situ generated aryl azides, obtained from areneboronic acids, were carried out (Scheme 14).[38] Removal of the protecting groups of **96** and **97** by standard methods gave the test compounds **98** in excellent yields. In case of the naphthyl-substituted 1,2,3-triazoles, the exchange of the *O*-benzyl protecting groups to acetate esters in routes **94**→**95**→**97** or **94**→**96**→**97** was necessary to avoid the partial saturation of the aromatic rings in the final *O*-deprotection step.[38]

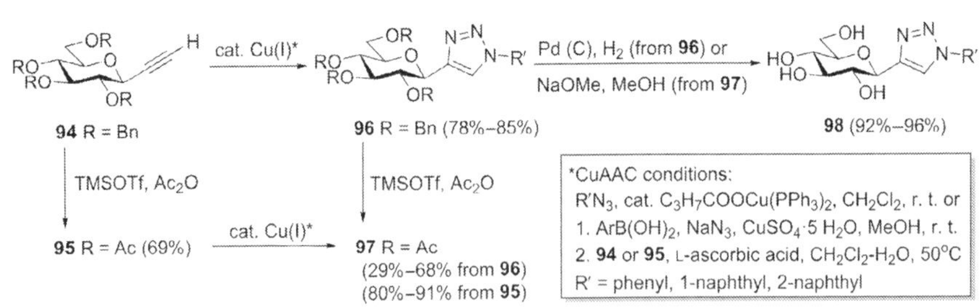

Scheme 14 Synthesis of 4-(β-D-glucopyranosyl)-1-substituted 1,2,3-triazoles.

1,2,4-Triazoles

Some *C*-glycopyranosyl 1,2,4-triazoles were obtained by the reaction of glycopyranosyl cyanides with 1-aza-2-azoniaallene salts[85] or with hydrazonoyl chlorides in the presence of $Yb(OTf)_3$,[86] however, these methods were limited for 1,3,5-trisubstituted 1,2,4-triazoles.

For GP inhibition studies, the desired 3-(β-D-glucopyranosyl)-5-substituted 1,2,4-triazoles **106** were achieved by cyclocondensation reactions (Scheme 15). Thus, (a) the reactions of carboxamidrazones with acid chloride **30** and amidine **28** followed by the intramolecular cyclizations of the resulting N^1-acyl carboxamidrazone type intermediates **99** and **100**, respectively[58]; (b) the treatments of acylthioamides **103** and **104** prepared from acid chloride **30** and thioamide **34**, respectively, with hydrazine[87, 88]; and

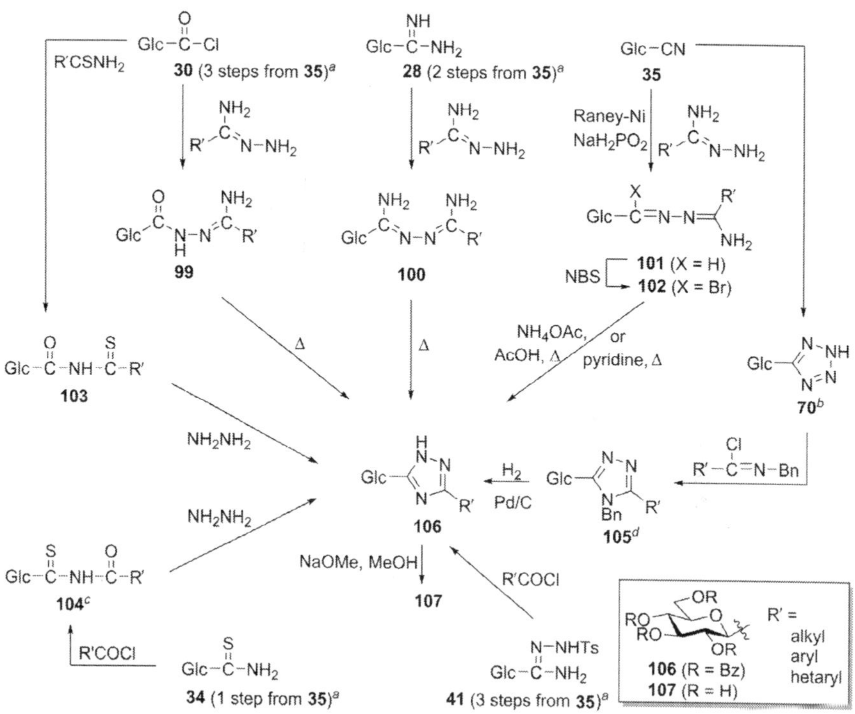

Route	Synthetic steps	Overall yields
35→→→30^a→99→106	5	31%–42%
35→→28^a→100→106	4	41%–43%
35→→→41^a→106	4	38%–47%
35→101→102→106	3	17%–21%
35→→→30^a→103→106	5	18%–24%
35→34^a→104^c→106	3	37%–62%
35→70^b→105→106^d	3	37%–49%

^{a,b} For synthetic details see Scheme 4^a and Scheme 11^b.
^{c} Synthesis of **104** is limited for compounds with R′ = alkyl substituents.
^{d} The order of cleavage of the *N*-benzyl and *O*-benzoyl protecting groups of **105** to get **107** are exchangable.

Scheme 15 Syntheses of 3-(β-D-glucopyranosyl)-5-substituted 1,2,4-triazoles.

(c) the ring-closure of tosylamidrazone **41** with acid chloride followed by spontaneous or TBAF-assisted *N*-detosylation[58, 65, 88] furnished the desired 3,5-disubstituted 1,2,4-triazoles **106**. On the other hand, bromination of 2,6-anhydroaldose imidoylhydrazones **101** obtained from cyanide **35** and carboxamidrazones under reductive conditions and subsequent base-induced cyclization of the bromo derivatives **102** led to the formation of the same heterocycles **106**.[89] Beside the cyclization methods, ring-transformation of tetrazole **70** with *N*-benzyl-carboximidoyl chlorides was also carried out to get *N*-benzylated 1,2,4-triazoles **105** whose debenzylation by catalytic hydrogenolysis afforded the target 3,5-disubstituted derivatives **106**.[88, 90, 91]

Taking into account the length of the different synthetic sequences (starting from the common starting material cyanide **35**) as well as the efficiency of the individual reaction steps, the latter method (route **35→70→105→106**) proved to be the most effective one to get compounds **106**.

It is to be noted that by applying routes **35→→→30→103→106**, **35→34→104→106** and **35→70→105→106**, each isomeric variant of trisubstituted *C*-β-D-glucopyranosyl 1,2,4-triazoles could also be obtained.[87, 90]

Beside 3-glucosyl-1,2,4-triazoles, the related 1,2,4-triazol-5(4*H*)-ones were also synthesized (Scheme 16).[66] Cyclization of N^1-tosyl amidrazone **41** with ethyl chloroformate afforded the 1-tosylated triazolone **111**, while intramolecular ring-closure of N^1-carbamoyl amidrazone **42** yielded the unsubstituted heterocycles **113**. 1-Aryl-1*H*-1,2,4-triazole-5(4*H*)-ones **115** were achieved either by intramolecular cyclizations of N^4-glucopyranosylcarbonyl semicarbazides **109** and **110** prepared from amide **32** via glucopyranosylcarbonyl isocyanate **108** or by a Cu(II) catalyzed coupling of triazolone **113** with areneboronic acid.

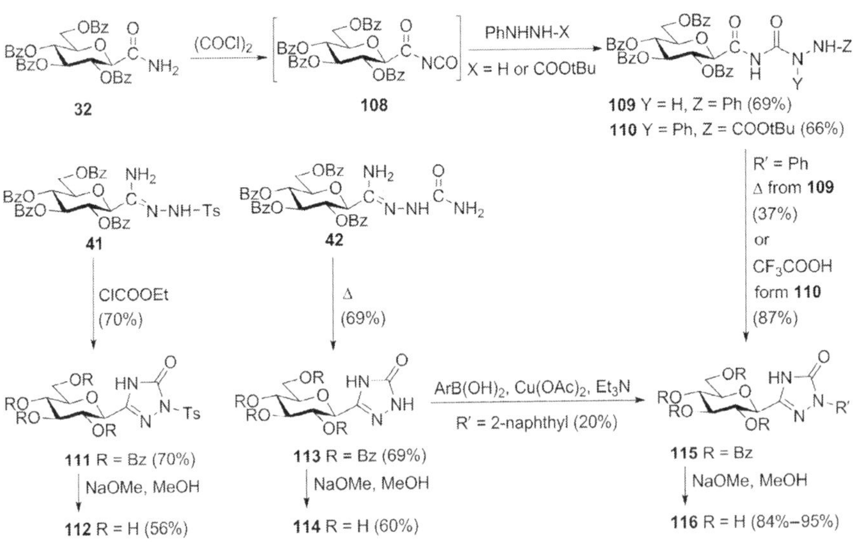

Scheme 16 Synthesis of 3-(β-D-glucopyranosyl)-1-substituted 1*H*-1,2,4-triazol-5(4*H*)-ones.

2.2.10 Tetrazoles

O-Peracylated 5-(β-D-glucopyranosyl)tetrazoles **70**[46, 64] and **71**[64] were prepared from the corresponding glucosyl cyanides as described in Section 2.2.7 (Scheme 11). Treatment of **71** with MeO^- ions afforded the unprotected derivative **117** in quantitative yield (Scheme 17).[64] Compound **70** was used for the formation of *N*-substituted derivatives (Scheme 17).[38] While *N*-methylation of **70** by diazomethane resulted in two regioisomers **119** (R′=Me) and **121** in 1:1 ratio, a copper catalyzed coupling of **70** with benzeneboronic acid provided exclusively the 5-glucosyl-2-phenyltetrazole **119** (R′=Ph) in excellent yield. The synthesis of the latter compound was also accomplished by the reaction of tosylhydrazone[92, 93] **118** with benzenediazonium tetrafluoroborate, albeit in lower yield.[38] The ester-protecting groups of **119** were cleaved by the Zemplén method to give **120** in high yields.

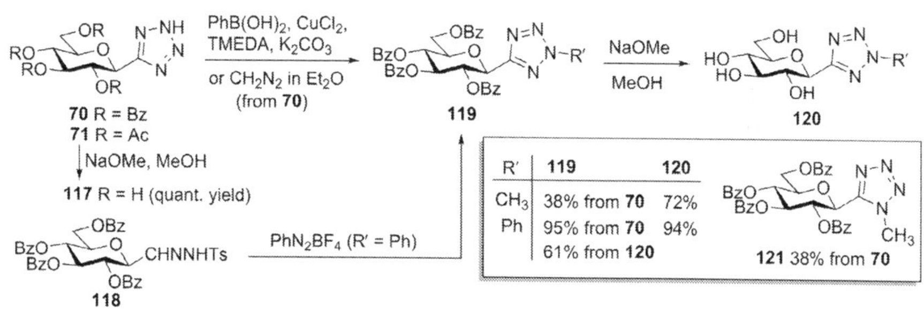

Scheme 17 5-(β-D-Glucopyranosyl)-*N*-substituted tetrazoles.

2.3 Annulated N-(β-D-glucopyranosyl) azoles

N-(β-D-Glucopyranosyl) derivatives of purine bases were synthesized starting from 1,2,3,4,6-penta-*O*-acetyl-D-glucose **122** (Scheme 18).[34] Direct treatment of adenine with **122** in the presence of $SnCl_4$ followed by *O*-deprotection yielded the 9-glucosyladenine **123**, which was converted to the corresponding hypoxantine derivative **125** by an oxidative deamination. Bromination of adenine **123** gave the 8-bromo derivative **124**, from which the 8-oxoadenine **126** was obtained by acetolysis and subsequent removal of the *O*-acetyl group. A set of differently substituted 9- and 7-(β-D-glucopyranosyl) purines **127-134** was achieved by Vorbrüggen type *N*-glycosylation of silylated purines with **122**, followed by *O*-deprotection upon ammonolysis. Treatment of the silylated purine, 6-chloro-, and 2-amino-6-chloropurine with **122** resulted in both N^7 and N^9 regioisomers (the *O*-peracetylated derivatives of **128** and **129**, **131** and **133**, and **132** and **134**, respectively), while analogous reactions of the corresponding 6-bromo- and 2,6-diacetamidopurine gave exclusively the N^9 derivatives (**127** and **130**, respectively, in acetylated forms). The preparation of 9- and 7-glucopyranosylguanines **135** and **136** was also accomplished by a three-step reaction sequence consisting of the Vorbrüggen type condensation of the silylated guanine and acetate **122**, the protection of the O^6-group of the resulting nucleoside analogues (thereby making the two regioisomers separable from each other), and *O*-deprotection.

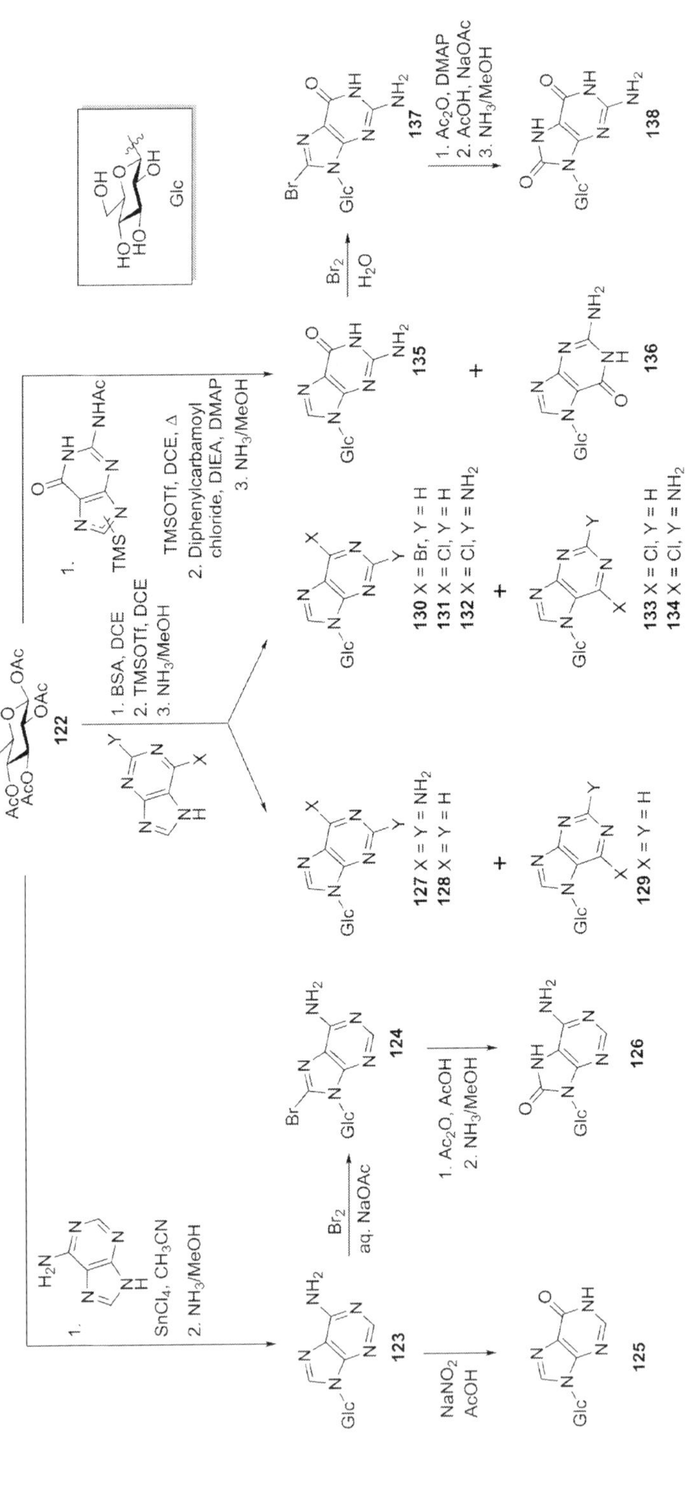

Scheme 18 Synthesis of *N*-(β-D-glucopyranosyl) purines.

The bromination of guanine **135** afforded the 8-bromoguanine **137**. Since the 7-glucosylguanine **136** remained intact under the applied bromination conditions, the preparation of **137** could be effected directly from the mixture of regioisomers **135** and **136**. 8-Oxoguanine **138** was formed starting from **137** by *O*-acetylation, followed by acetolysis of the resulting protected bromide and *O*-deprotection.

2.4 Annulated C-(β-D-glucopyranosyl) azoles

2.4.1 Indoles

To get 2-(β-D-glucopyranosyl)indole **140**, an *N*-tosylated derivative[94] **139** was prepared first by a Pd-catalyzed Sonogashira cross-coupling reaction of *O*-perbenzylated glucosyl ethyne **94** with *N*-tosyl-2-iodoaniline followed by a spontaneous intramolecular ring-closure (Scheme 19). Cleavage of the tosyl group of **139** by Bu_4NF and subsequent hydrogenolytic *O*-debenzylation resulted in the target compound **140**.[72]

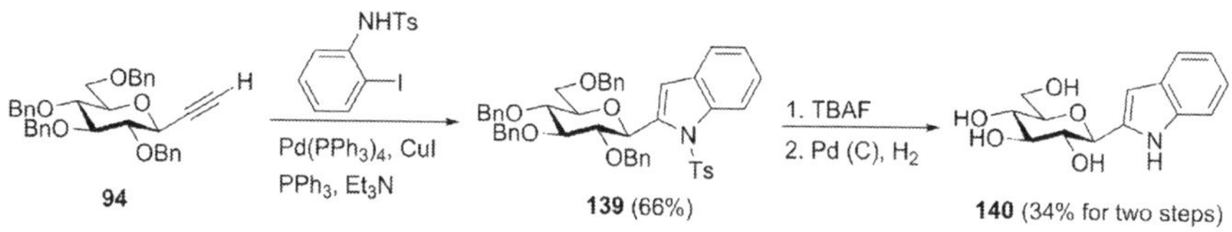

Scheme 19 Synthesis of 2-(β-D-glucopyranosyl)indole.

2.4.2 Benzothiazoles

O-Peracylated 2-glucopyranosylbenzothiazole **141** was synthesized by heterocyclization of the corresponding cyanide **35a** with 2-aminothiophenol (Scheme 20).[64] Compound **141** was also obtained by cyclization of glucosyl nitrile-oxide, generated from the appropriate hydroximoyl chloride, with 2-aminothiophenol,[95] while an *O*-perbenzylated analogue was produced from gluconolactone and 2-lithiobenzothiazole.[96] *O*-Deacetylation of **141** was effected by the Zemplén method to obtain the unprotected derivative **142** in good yield.

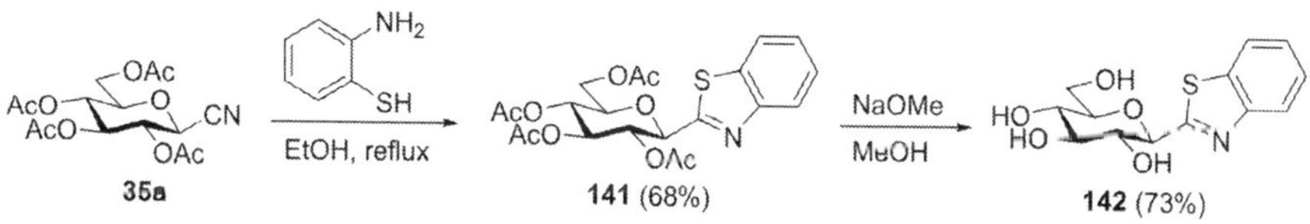

Scheme 20 Synthesis of 2-(β-D-glucopyranosyl)benzothiazole.

2.4.3 Annulated imidazoles

Benzimidazoles and related compounds

For the synthesis of 2-glucopyranosylbenzimidazoles, several ring-closing methods were worked out (Scheme 21). Cyclization of acid **29** with 1,2-diaminobenzenes (DABs) in the presence of triphenylphosphite furnished some benzimidazole

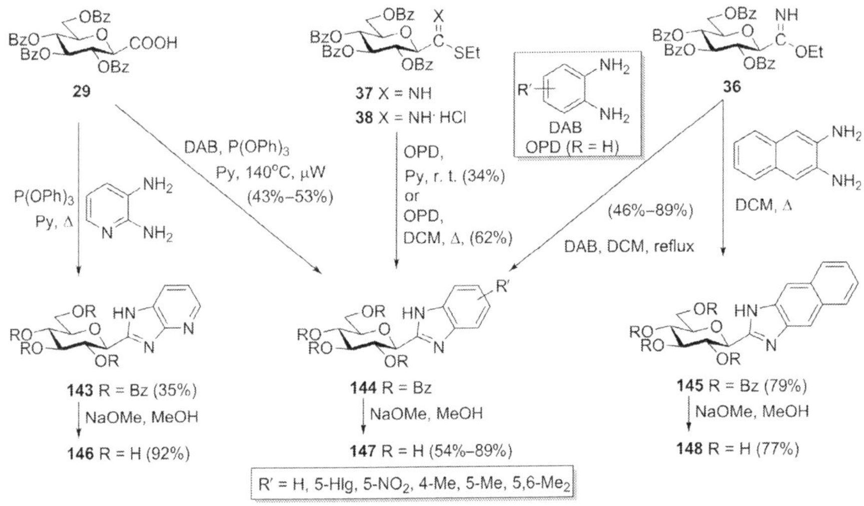

Scheme 21 Synthesis of 2-(β-D-glucopyranosyl)benzimidazoles and related compounds.

derivatives **144** in acceptable yields.[63] Recently, an analogous transformation of **29** with 2,3-diaminopyridine was also effected to produce imidazo[4,5-*b*]pyridine **143**.[61] Among *C*-glucosylated iminoesters, the ethyl-thioimidate hydrochloride **38**, treated with *o*-phenylenediamine (OPD), provided the unsubstituted benzimidazole **144** (R′=H) in low yield.[64] Thioimidate **37**, and especially imidate **36**, when reacted with OPD, furnished **144** (R′=H) in much better yields (62% from **37** and 89% from **36**, respectively).[63] By the reaction of **36** with variously substituted 1,2-diaminobenzenes further derivatives of **144** were also obtained.[63] Additionally, the naphtho[2,3-*d*]imidazole **145** was formed by ring-closing of **36** with 2,3-diaminonaphthalene.[63] *O*-Debenzoylation of the obtained heterocyclic derivatives **143–145** was performed by transesterification to get the final products **146–148**. Besides the previously mentioned synthetic ways, two additional procedures (a) the reaction of *O*-peracetylated *C*-glucopyranosyl formaldoximoyl chloride (actually via an intermediate nitrile oxide) with OPD, and subsequent *O*-deacetylation,[95] and (b) the acid-catalyzed cyclization of unprotected *C*-glucopyranosylmethanal (obtained from the appropriate dimethyl acetal) with OPD, followed by spontaneous dehydrogenation of the resulting benzimidazoline,[97] were developed for the construction of benzimidazole **147** (R′=H).

Further imidazo-fused heterocycles

Very recently, a new series of *C*-glucopyranosyl derivatives of annulated imidazoles was also synthesized (Scheme 22).[61] Ring-closure of *O*-perbenzoylated bromomethyl glucosyl ketone **33** with a set of five- and six-membered aminoheterocycles resulted in the corresponding fused heterocycles, such as imidazo[1,2-*a*]pyridine **149**, imidazo[1,2-*a*]pyrimidine **150**, imidazo[2,1-*b*]thiazole **151**, imidazo[2,1-*b*][1,3,4]thiadiazole

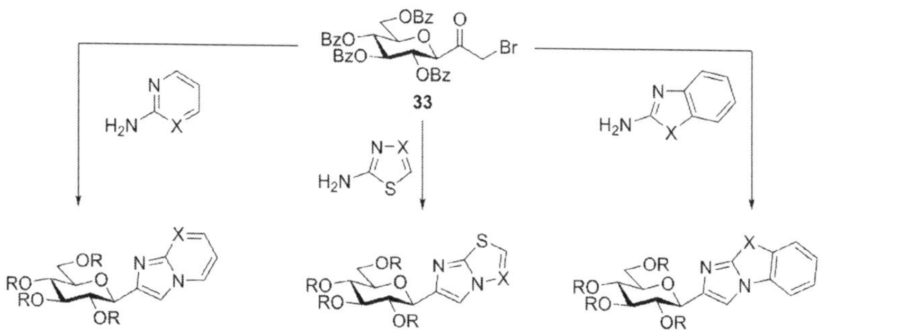

149 R = Bz, X = CH (48%)
150 R = Bz, X = N (48%)
↓ NaOMe, MeOH
156 R = H, X = CH (79%)
157 R = H, X = N (76%)

151 R = Bz, X = CH (66%)
152 R = Bz, X = N (40%)
↓ NaOMe, MeOH
158 R = H, X = CH (93%)
159 R = H, X = N (49%)

153 R = Bz, X = S (58%)
154, 155 R = Bz, X = NEt, NBn (37%–42%)
↓ NaOMe, MeOH
160 R = H, X = S (85%)
161, 162 R = H, X = NEt, NBn (72%–83%)
↓ $Pd(OH)_2/C$, H_2 (from **162**, X = NBn)
163 R = H, X = NH (73%)

Scheme 22 Synthesis of further *C*-(β-D-glucopyranosyl) derivatives of fused imidazoles

152, benzo[*d*]imidazo[2,1-*b*]thiazole **153**, and 9-ethyl- and 9-benzyl-9*H*-benzo[*d*] imidazo[1,2-*a*]imidazoles **154** and **155**, respectively. Subsequent removal of the *O*-benzoyl protecting groups of **149–155** by the Zemplén protocol afforded the test compounds **156–162**, respectively, in good to excellent yields. Hydrogenolytic cleavage of the *N*-benzyl group of **162** was also effected to obtain the fully unprotected derivative **163** in high yield.

2.5 *Six-membered* N-*(β-D-glucopyranosyl) heterocycles*

A set of six-membered *N*-(β-D-glucopyranosyl) heterocycles (including uracil **164**, **166**, **168–171**, thymine **165**, cytosine **167**, 6-azauracil **172**, isocyanuric acid **173**, and phthalazine **174**, **175** derivatives) was obtained by Vorbrüggen type *N*-glycosylation reactions (Scheme 23).[34, 98]

The *O*-peracetylated *N*-glucosyl uracil **176** and thymine **177** were used for the formation of further heterocyclic derivatives (Scheme 24). Tetrazolo-pyrimidinones **178** and **179** were obtained upon reactions of **176** and **177**, respectively, with phosphorus oxychloride and sodium azide. Treatment of **178** and **179** with ammonia in MeOH yielded the unprotected derivatives **182** and **183**, respectively.[34] Photolysis of **178** in aq. acetonitrile followed by *O*-deacetylation resulted in the triazepane derivative **184**.[34] Uracil **176** was converted to the 4-(1,2,4-triazolyl) pyrimidin-2-one[99] **181** which was used as a precursor of N^4-substituted-(β-D-glucopyranosyl)cytosines.[99] Among others, N^4-[2-acridin-9(10*H*)-onyl]cytosine **185** was synthesized by a substitution reaction of **181** with 2-aminoacridone followed by *O*-deprotection.[100]

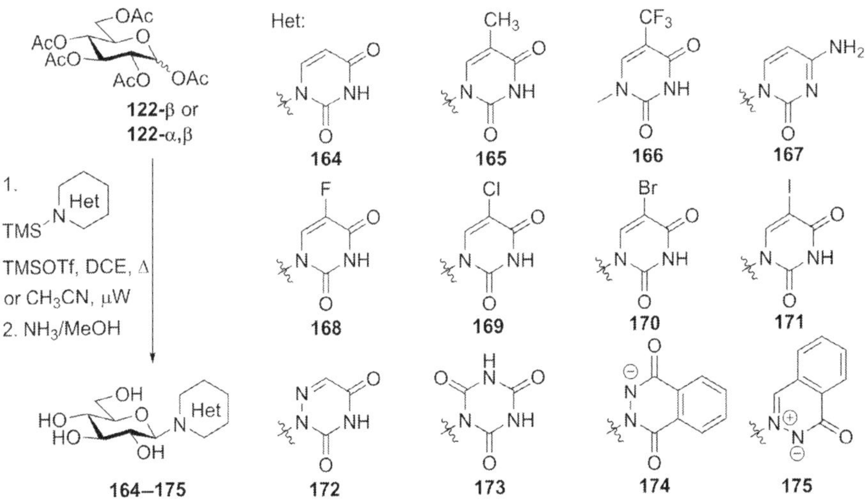

Scheme 23 Synthesis of six-membered *N*-(β-D-glucopyranosyl) heterocycles.

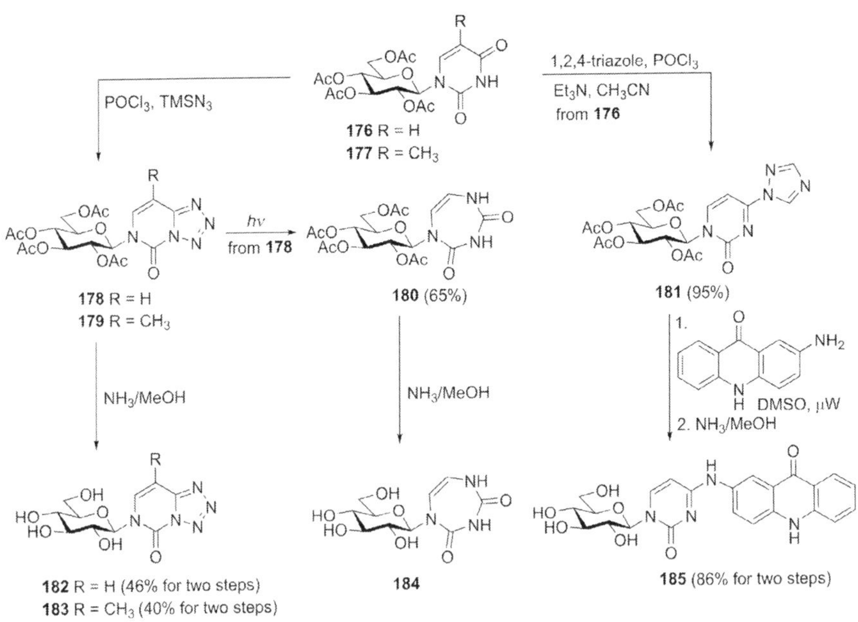

Scheme 24 Transformations of *N*-(β-D-glucopyranosyl) uracil and thymine.

The Pd-catalyzed cross-coupling of 1-glucosyl-5-iodouracil **171** with terminal alkynes resulted in 5-alkynyluracil derivatives **186**, from which alkylfurano[2,3-*d*] pyrimidinones **187** were obtained by an $AgNO_3$-catalyzed 5-*endo*-dig electrophilic cyclization (Scheme 25).[101]

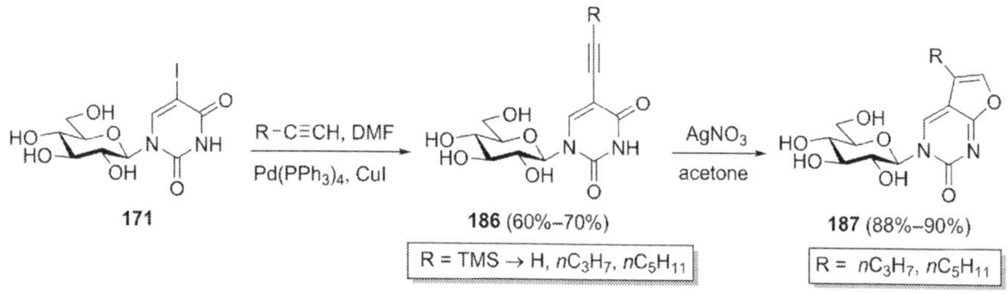

Scheme 25 Transformation of 1-(β-D-glucopyranosyl)-5-iodouracil.

2.6 Six-membered C-(β-D-glucopyranosyl) heterocycles

While *C*-glycopyranosyl derivatives of five-membered heterocycles have received great attention in recent years not only for GP inhibition but also in other fields,[28] the six-membered *C*-glycopyranosyl heterocycles are generally scarcely represented.[28]

Very recently, an extensive study was performed on the preparation of 2-(β-D-glucopyranosyl)pyrimidines (Table 1).[71] Pinner type cyclization of the *O*-perbenzylated glucosyl formamidine **45** with β-ketoesters, dimethyl malonate, and α,β-unsaturated

Table 1 Synthesis of 2-(β-D-glucopyranosyl)pyrimidines.

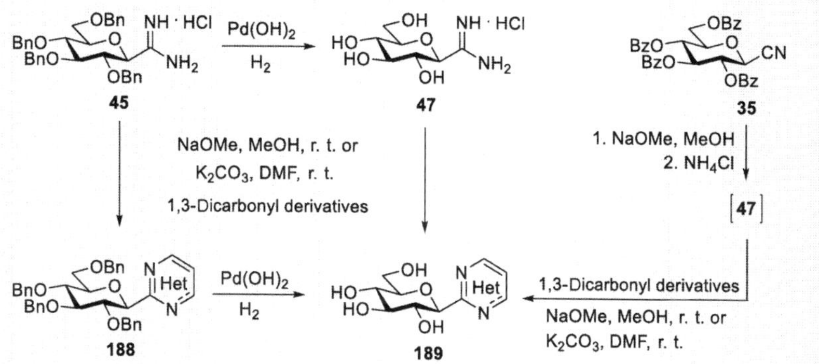

1,3-Dicarbonyl derivatives	Het	Route	Overall yields (%) for **189**
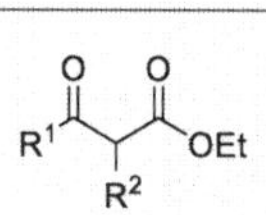	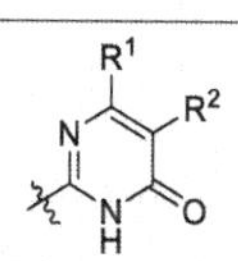	**45→47→189**	64%–87%
		45→188→189	33%–49%
$R^1 = H, R^2 = CH_3, CH_2Cl, Ph; R^1 = CH_3, R^2 = Cl$		**35→[47]→189**	25%–43 %
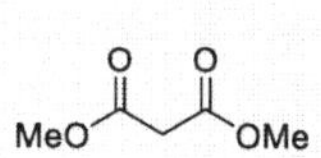	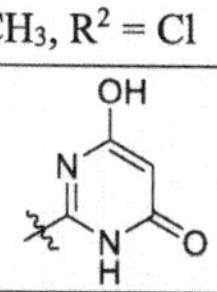	**45→47→189**	70%
		45→188→189	39%
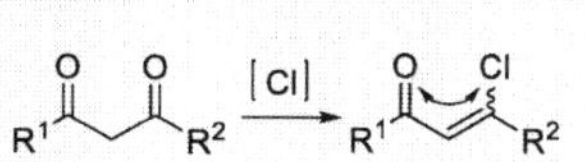	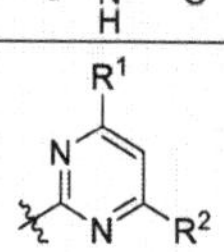	**45→47→189**	61%–74%
		45→188→189	14%–63%
$R^1 = CH_3, R^2 = CH_3, CF_3, R^1 = Ph, R^2 = CH_3, CF_3$		**35→[47]→189**	30%

β-chloroketones derived from the appropriate 1,3-diketones resulted in the desired, diversely substituted pyrimidines **188**. Cleavage of the *O*-benzyl-protecting groups of **188** by catalytic hydrogenation over $Pd(OH)_2$ provided the final products **189**. Heterocycles **189** were also achieved by the reaction of the *O*-deprotected amidine **47** obtained from **45** with the aforementioned 1,3-dielectrophiles. A comparison of the two reaction sequences **45→188→189** vs. **45→47→189** showed the latter route to be more efficient in terms of overall yields. On the other hand, a continuous three-step procedure starting from *O*-perbenzoylated β-D-glycopyranosyl cyanide **35** (via the intermediate amidine **47**) was also worked out to give some pyrimidines **189** in acceptable overall yields.

The preparation of a *C*-glucopyranosyl tetrahydropyrimidin-2-one **192** was carried out starting from the *C*-glucosylated malononitrile **190** in four steps (Scheme 26).[102] The reduction of the nitrile groups of **190** afforded the diamino derivative **191** in quantitative yield. Subsequent treatment of **191** with 1,1′-carbonyldiimidazole followed by *O*-deacetylation under mild basic conditions resulted in the target product **192**.

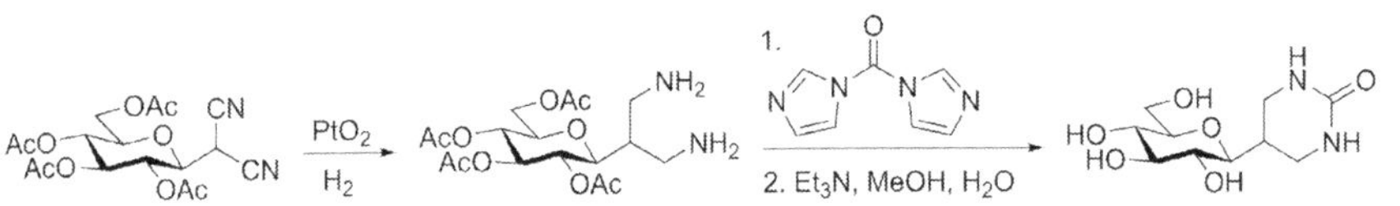

190 **191** (99%) **192** (38% for two steps)

Scheme 26 Synthesis of 5-(β-D-glucopyranosyl)tetrahydropyrimidin-2(1*H*)-one.

2.7 N- and C-Glycopyranosyl heterocycles with modified sugar units

Beside the glucose-based heterocycles, further *N*- and *C*-glycopyranosyl analogues having a sugar unit other than glucose were designed and synthesized in order to investigate the effect of such kind of structural modification on the glycogen phoshorylase inhibitory activity.

β-D-Xylopyranosyl and β-D-glucosaminyl derivatives of some azoles (Scheme 27), such as 1,2,3-triazoles (**195**,[103] **199**[104]), imidazoles (**196**,[105] **200**[104]), 1,2,4-triazoles (**197**,[106] **201**[104]), and benzimidazoles (**198**,[106] **202**[104]) were obtained starting from the appropriate glycosyl azides (**193**, **203**) and cyanides (**194**, **204**) by adaptation of the synthetic methods used for the glucose derivatives (for details see Schemes 2, 8, 15 and 21).

A series of 1-(5-thio-β-D-xylopyranosyl)-1,2,3-triazoles **207** was obtained by copper-catalyzed [3+2] cycloaddition of the 5-thio-xylosyl azide **205** with terminal alkynes followed by *O*-deprotection (Scheme 28).[103] The intermediate *O*-peracetylated 5-thioxylose derivatives **206** were converted to the corresponding sulfoxides **208** and sulfones **210** by the oxidations of the endocyclic sulfur atom (Scheme 28). *O*-Deacetylation of compounds **208** and **210** was effected by acid-catalyzed transesterification to obtain the test compounds **209** and **211**, respectively.[103]

Three series of 1-*C*-oxadiazolyl D-glycals were synthesized[107] in two main synthetic routes: (a) by the construction of the heterocyclic ring on glycal-based

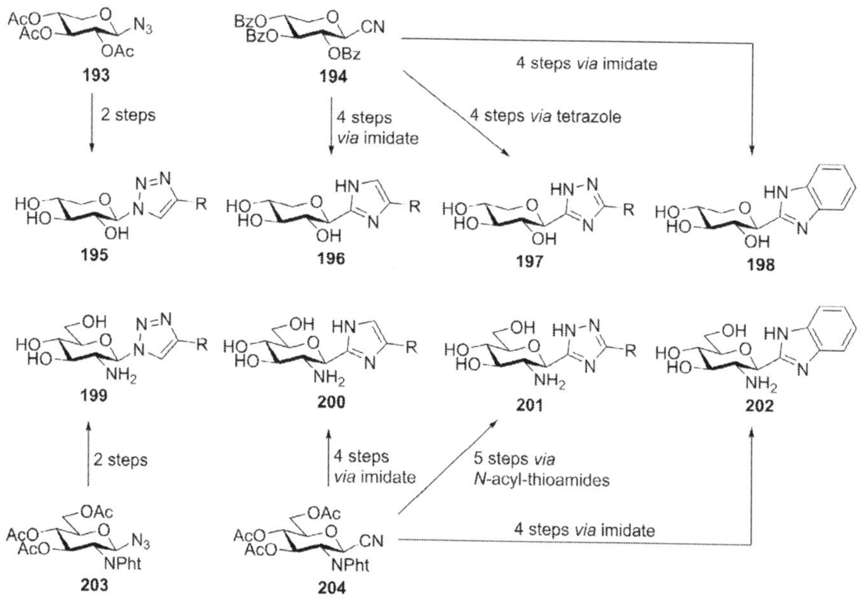

Scheme 27 Syntheses of β-D-xylopyranosyl and β-D-glucosaminyl azoles.

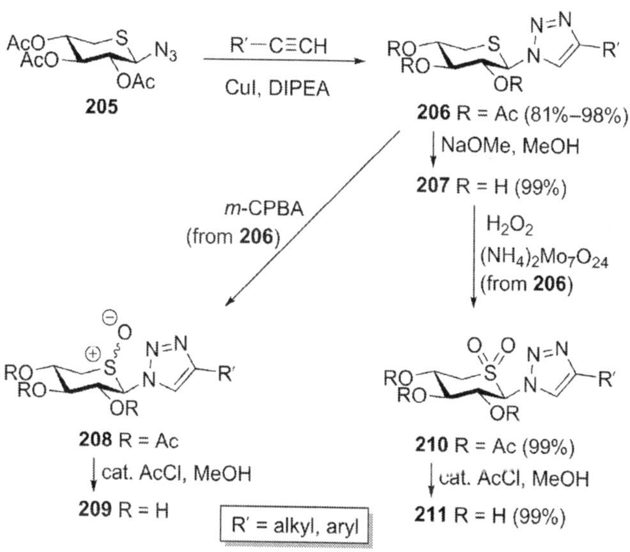

Scheme 28 Synthesis of 1-(5-thio-β-D-xylopyranosyl)-1,2,3-triazoles and related derivatives.

1-substituted 1*H*-1,2,4-triazol-5(4*H*)-ones anhydroaldonic acid type precursors, and (b) by the introduction of the double bond into the sugar unit of the pre-formed *C*-(β-D-glucopyranosyl) heterocycles (Scheme 29). Bromination of the glucopyranosyl cyanide **35** to **212** followed by a Zn-mediated reductive elimination furnished the 1-cyanoglycal **213**. This compound was also obtained directly from **35** by a base-induced elimination, albeit in lower yield. Treatment of **213** with hydroxylamine and subsequent acylation of the resulting amidoxime (not shown) followed by intramolecular cyclization afforded the 3-glucosyl-5-substituted oxadiazoles **215**. The reversed 5-glucosyl-3-substituted 1,2,4-oxadiazoles **216** were prepared by the reaction of cyanide **213** with in situ generated nitrile oxides as well as by the treatment of the corresponding glucopyranosyl heterocyles **68** with DBU. The 1,3,4-oxadiazoles **217** were obtained starting from the appropriate saturated glucosyl heterocycles **72** by radical-mediated bromination followed by reductive elimination of **214**. In the final step, *O*-debenzoylation of compounds **215**–**217** was accomplished by the Zemplén method to give the unprotected derivatives **218**–**220**, respectively, in high yields.

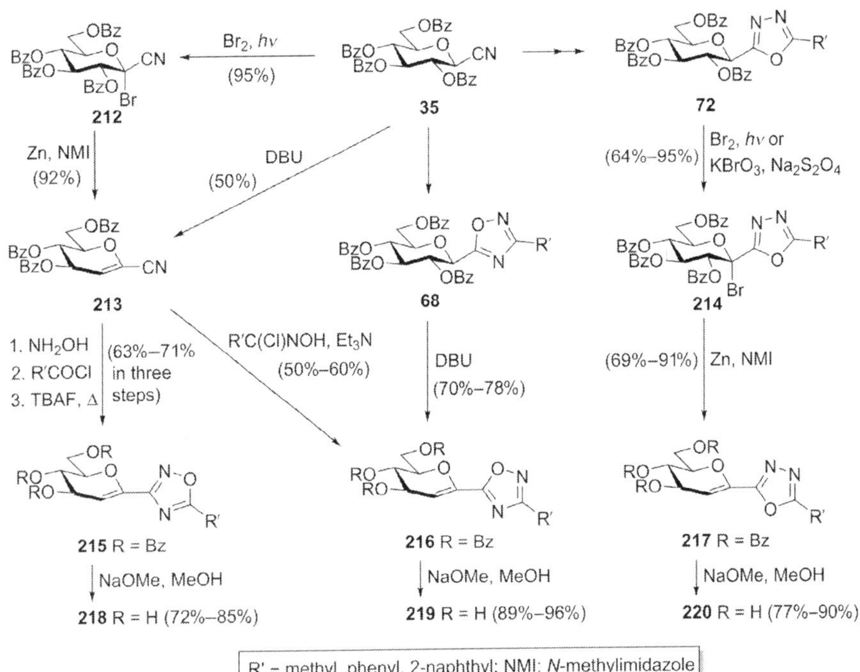

Scheme 29 Syntheses of *C*-(1,2-didehydro-1,2-dideoxy-D-glucopyranosyl)oxadiazoles.

Some (3-deoxy-3-fluoro-β-D-glucopyranosyl)pyrimidine derivatives were synthesized by Vorbrüggen type *N*-glycosylation (Scheme 30). To this end, the desired *O*-peracetylated glucosyl donor **222** was formed starting from 3-deoxy-3-fluoro-1,2:5,6-di-*O*-isopropylidene-α-D-glucofuranose **221** in an acid-catalyzed *O*-deprotection and base-mediated *O*-acetylation sequence.[108] The condensation of acetate **222** with the silylated nucleobases and subsequent *O*-deprotection resulted in the target compounds **223**–**226**.[109–111] For a comparative biological study, synthesis of a 2,3-dideoxy-3-fluoro-glucopyranosyl derivative **227** was also performed.[111]

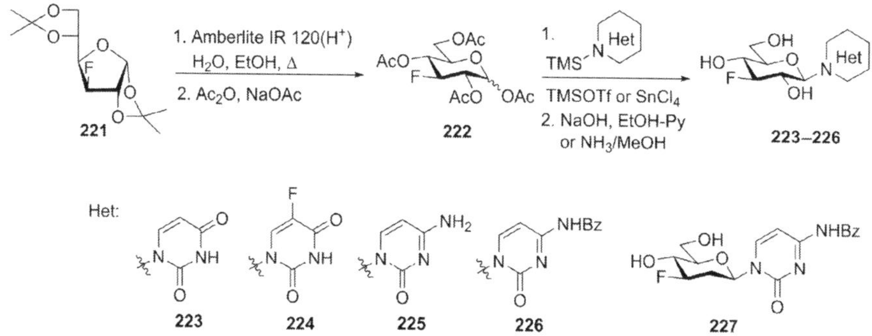

Scheme 30 Synthesis of *N*-(2-deoxy-2-fluoro-β-D-glucopyranosyl)pyrimidines.

By applying a similar synthetic strategy, the preparation of 1-(3′-*C*-hydroxymethyl-β-D-glucopyranosyl)-5-fluorouracil **230** was also effected via **229** starting from the corresponding 3-*C*-acetoxymethyl-1,2:5,6-di-*O*-isopropylidene-α-D-glucofuranose **228** (Scheme 31).[112]

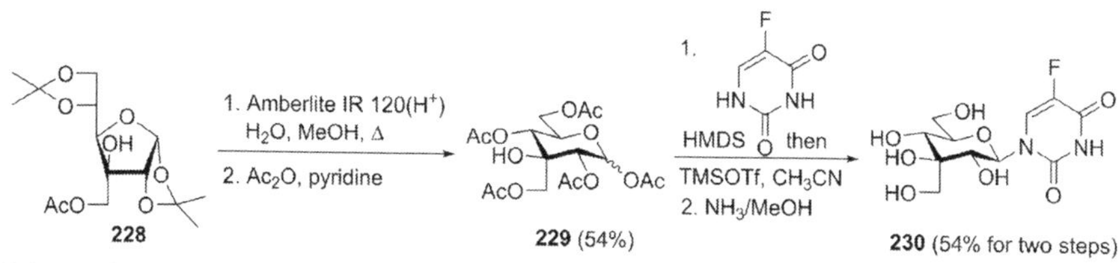

Scheme 31 Synthesis of 1-(3-*C*-hydroxymethyl-β-D-glucopyranosyl)-5-fluorouracil.

3 Glycogen phosphorylase inhibition

The inhibitory potency of the synthesized glycopyranosyl heterocycles was mainly determined against rabbit muscle glycogen phoshorylase b (RMGPb), the "prototype" of GP.[113] This isoform of GP is well-characterized and shows high resemblance (~80% identity in overall, and 100% for the catalytic site) to the pharmacologically relevant human liver glycogen phoshorylase (HLGP).[113] On the other hand, RMGPb is readily available even in crystallized form thereby providing opportunity for assembling extensive structure-activity relationships (SAR) by means of protein crystallography.

3.1 Five-membered N- and C-(β-D-glucopyranosyl) heterocycles

The *N*-acyl-β-D-glucopyranosylamines[114–116] **231** (Table 2) were among the first potent glucose-derived inhibitors of GP and chosen frequently as lead structures for further inhibitor design.

With the aim to replace the NHCO unit of **231** by nonclassical five-membered heterocyclic bioisosters many types of *N*- and *C*-(β-D-glucopyranosyl) azoles have been synthesized as detailed in the previous sections and evaluated as inhibitors of GP.

Table 2 Inhibitory potency (K_i [μM]) of selected *N*-acyl-*N*-(β-D-glucopyranosyl)amines and *N*-(β-D-glucopyranosyl) azoles against RMGPb.

Compound	R′: –CH_2OH	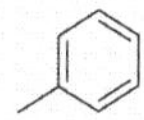	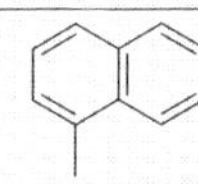	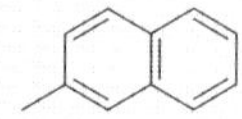
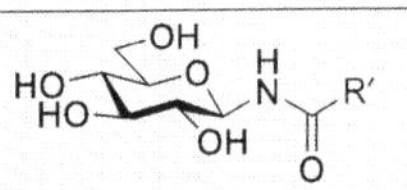 **231**	20[42] 18[41]	81[114] 144[115]	444[116]	10[116] 13[41]
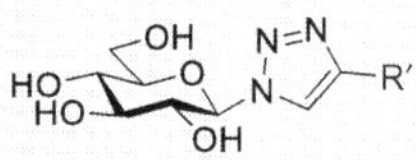 **5**	26[42] 14[41]	162[42] 151[41]	136[41]	36[42] 16[41]
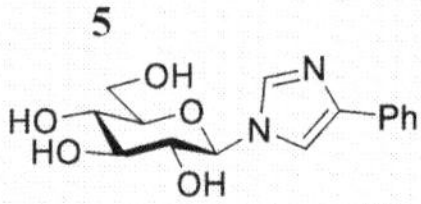 **3** No inh. at 625 μM[38]			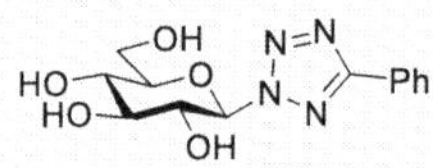 **25** 327[a] [38]	

[a] Calculated from the IC_{50} value (600 μM).[38]

Among *N*-glucopyranosyl heterocycles (Table 2), the 1,2,3-triazoles[41, 42] **5** proved to be inhibitors equipotent with the respective amides **231**. In addition, X-ray crystallographic analysis of the RMGPb-inhibitor complexes also revealed a high similarity between the binding features of the comparable pairs of triazoles **5** and amides **231**.[41]

Insertion of an imidazole ring in the place of the NHCO moiety of *N*-benzoyl-*N*-(β-D-glucopyranosyl)amine **231** (R′=Ph) resulted in noninhibitory compound **3**.[38] The 5-phenyltetrazole[38] **25** exhibited inhibition against RMGPb, however, showed ~2 to 4 times weaker effect than the corresponding 1,2,3-triazole **5** and amide **231**.

Two other types of 1-glucosyl-1,2,3-triazoles, such as **15** and **16** (see Fig. 1) were also tested, however, these compounds displayed weak or no effects against RMGPb.[42]

Among 1-(β-D-glucopyranosyl)-4,5-disubstituted 1,2,3-triazoles[44] (**8**, **13** and **14**), the best inhibitors shown in Fig. 3 exhibited moderate inhibition.

The enzyme kinetic results of the synthesized *C*-β-D-glucopyranosyl azoles with the most frequently investigated R′ substitutents are collected in Table 3. 5-Arylpyrroles[72] **49** displayed no inhibition against RMGPb. Among 1,2-azoles, the 5-phenylisoxazole[73] **55** was inactive at 625 μM concentration, while the analogous pyrazole[73] **57** showed modest inhibitory effect, which is presumably due to the H-bond donor capacity of **57**. A more traceable difference in the inhibitory potency could be observed for the 1,3-azoles. While thiazoles[73] **59** exhibited inhibition in the upper micromolar range, their imidazole counterparts[72, 73] **62** showed ~1000–5000-fold better effects. X-Ray crystallographic analysis of the binding characteristics of imidazoles **62** revealed that the H-bond forming capability of the heterocyclic N-atoms highly contributed to the strong binding.[72] A prominent role was ascribed to the NH-group of the imidazole ring that formed a direct H-bridge with the main chain carbonyl group of the His377 protein residue.[72] A similar interaction was also identified, among others, in the *N*-acyl-β-D-glucopyranosylamine-RMGPb complexes.[21, 114, 118]

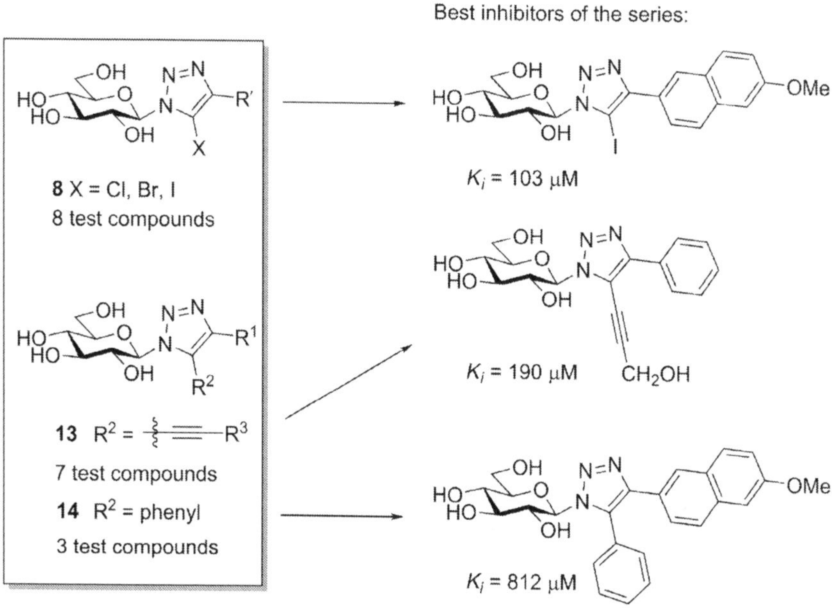

Fig. 3 Inhibitory effect of 1-(β-D-glucopyranosyl)-4,5-disubstituted 1,2,3-triazoles against RMGPb.[44]

Among isomeric oxadiazoles, the 5-glucopyranosyl-3-substituted-1,2,4-oxadiazoles[82, 83] **69** proved to be more effective than the 3-glucopyranosyl-5-substituted analogues[56, 83] **65**, and the inhibitor constants of the former ones (except that of the 1-naphthyl derivative) were comparable to those of the *N*-acyl-β-D-glucopyranosylamines **231** (data shown in Table 2). Surprisingly, 1,3,4-oxadiazoles[83] **73** with bulky aromatic groups were practically inactive, and only the methyl-substituted derivative[64, 117] had a weak inhibitory effect.

C-Glucopyranosyl 1,2,3-triazoles[38] **98** proved to be noninhibitory, while their *N*-glucopyranosyl analogues **5**, as it was shown in Table 2, were good micromolar inhibitors.

The 1,2,4-triazoles[65, 90] **107** in which the position of the heteroatoms is identical to those of oxadiazoles **69** showed even stronger effects. The increased binding affinity of **107** as compared to **69** was the result of the specific NH-His377(CO) H-bond as evidenced by X-ray crystallography.[21, 91]

Introduction of a carbonyl group into the 1,2,4-triazole ring to form an amide function within the heterocycle resulted in moderate inhibitors **116**.[66]

Similar to the 1,2,3-triazoles, exchanging the substituents of the tetrazole ring to get *C*-glucosyl analogues **120** of *N*-glucosyl tetrazole **25** led to complete loss of activity.[38]

This short overview allows the general conclusion that the number and the nature of the heteroatoms as well as the constitution of the heteroring have strong bearings on the efficacy of β-D-glucopyranosyl azoles. On the other hand, the nature of the substituent attached to the azole ring proved to be another decisive factor. For instance, in case of a properly oriented heterocycle, the presence of an aromatic appendage had a positive impact on the binding strength by exploiting hydrophobic contacts in the so-called "β-channel" of the enzyme, as it was evidenced by X-ray crystallography,

Table 3 Inhibitory potency (K_i [μM]) of selected *C*-β-D-glucopyranosyl azoles against RMGPb.

	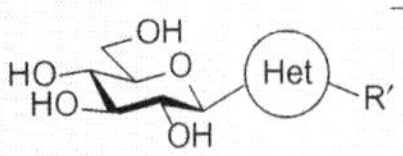	R′				
	Het	$-CH_3$	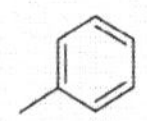	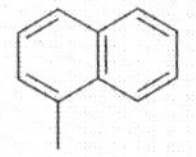	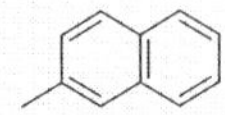	Ref.
49	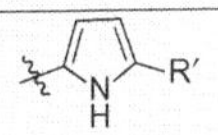	–	NI[a]	–	NI[a]	72
55	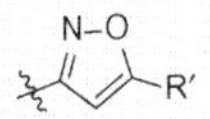	–	NI[a]	–	NI[a]	73
57	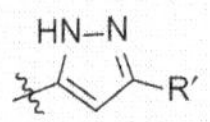	–	400	–	–	73
59	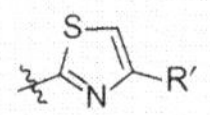	–	310	–	158	73
62	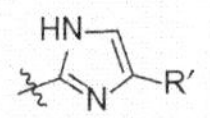	–	0.28	–	0.031	72,73
65	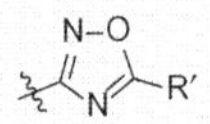	NI[a]	10%[b]	NI[a]	38	56,83
69	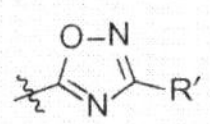	–	64	19	11.6 2.4[c]	82,83
73	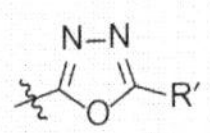	212 145	10%[b]	10%[b]	10%[b]	64,83,118
98	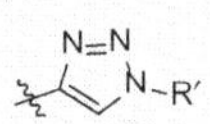	–	NI[a]	NI[a]	NI[a]	38
107	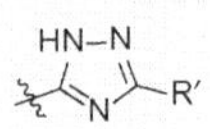	499	7	11.5	0.41	65,88,90
116	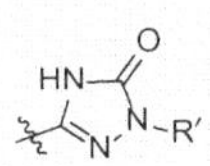	–	191[d]	–	80	66
120	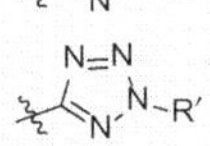	NI[a]	NI[a]	–	–	38

[a] No inhibition at 625 μM.
[b] Inhibition at 625 μM.
[c] The K_i value was measured by Oikonomakos et al. and cited as an unpublished result in Ref. 83.
[d] K_i calculated from the IC_{50} value (350 μM).

for example, for compounds **5**, **62**, **69**, and **107**.[21, 41, 72, 91] In addition, fitting of a 2-naphthyl substituent into the β-cavity proved to be superior to that of phenyl and 1-naphthyl groups, as manifested by changes of the K_i values (compare aryl-substituted derivatives of **5**, **59**, **62**, **65**, **69**, **107**, and **116** in Tables 2 and 3).

Imidazoles **62** and 1,2,4-triazoles **107** possessing several beneficial structural features (e.g., well-positioned H-bond donor and acceptor heteroatoms within the azole ring, bulky aromatic substituent with proper orientation toward the β-channel) have emerged as the most efficient representatives of this inhibitor class. The 2-naphthylimidazole **62** with its unique inhibitory effect is the best glucose analogue inhibitor of GP discovered so far.

These inhibitors were also modified beyond the heterocyclic parts by different types of appending substituents to examine their effect on the inhibition. The most extensive investigations including enzyme kinetic, protein crystallographic and molecular docking studies were implemented with 1,2,4-triazoles **107**.[65, 88, 90, 91, 119] The collection of these compounds with their inhibitory potency against RMGPb is shown in Table 4.

Compounds having an alkyl substituent[65, 90] showed weak or no inhibition, but switching to a phenyl group brought about significant improvement in the inhibiton (K_i=7 μM for R′=Ph).[65, 90] As evidenced by X-ray crystallography,[91] van der Waals interactions of the phenyl group in the β-pocket of the enzyme largely contributed to the strong binding of the latter compound.

Substitution of the phenyl ring in the 4-position by different groups of nonpolar, polar, acidic, and basic character resulted in distinct changes in the inhibition (Table 4): while Me, OH, and OMe groups slightly increased the potency, the *t*-Bu, CF_3, NO_2, COOH, and SO_2NH_2 substituents proved to be detrimental.[65, 88, 90] The 4-aminophenyl derivative[90] emerged as the most effective compound of this set exhibiting nanomolar inhibition.

The inhibitory effect of the previously mentioned compounds pointed out that decisive feature of these endgroups for the inhibition was their size rather than their chemical nature, as it was also confirmed by protein crystallography.[119]

Replacement of the phenyl group by a 2-pyridyl moiety resulted in a poor inhibitor.[90]

Within compounds having multiple substituted phenyl groups,[90] the 3,5-diaminophenyl-1,2,4-triazole was the most potent one, however, showed a twofold weaker effect than the parent phenyl derivative.

Elongation of the aromatic part by an additional benzene ring, such as in the 4-biphenyl derivative, slightly strengthened the binding as compared to the phenyl-1,2,4-triazole. Another biphenyl derivative, bearing the second benzene ring in meta position, displayed a much weaker effect.[88]

The 2-naphthyl derivative[65, 90] with its nanomolar K_i value proved to be the best inhibitor of the whole series. The increased inhibitory potency of this compound as compared to the phenyl derivative was the result of more extensive van der Waals interactions of the 2-naphthyl group in the β-channel of the enzyme.[91] The 1-naphthyl-1,2,4-triazole showed 28-fold weaker effect than the 2-naphthyl derivative, which was due to the less favorable orientation of the 1-naphthyl group in the β-pocket, as evidenced by X-ray crystallography.[88] Attachment of a carboxyl group to the 6-position of the 2-naphthyl ring decreased the inhibition by a factor of ~ 10.[88] Partial saturation of the 2-naphthyl ring resulted in slight weakening of the inhibitory potency.[91] Replacement of the 2-naphthyl group by fluorenyl or anthracenyl moieties resulted in weaker inhibitors, while a further increase of the size of the aromatic part by a pyrenyl residue led to a practically inactive compound.[88]

Beyond this short overview, more detailed SAR explanations, supported by protein crystallographic and computational studies can be found in the cited publications, therefore, the reader is kindly referred to those articles.[88, 91, 119]

The possibility of the bioisosteric replacement of the amide units was also studied on *N*-aryl-**232** and *N*-acyl-*N*′-(β-D-glucopyranosyl) urea type **233** GP inhibitors (Table 5). Substitution of the first NHCO unit of the urea derivatives **232** and **233** by oxadiazole[81, 84] and thiadiazole[84] rings resulted in noninhibitory *C*-glucopyranosyl heterocycles **67**, **77**, **81,** and **93**. It is to be noted that replacement of the second or third NHCO moiety

Table 4 Inhibitory potency (K_i [μM]) of 3-(β-D-glucopyranosyl)-5-substituted 1,2,4-triazoles against RMGPb.[65, 88, 90, 91]

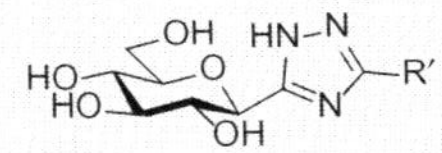

107

R′	R^1	K_i [μM]
$-CH_3$	–	499
$-C(CH_3)_3$	–	NI
$-CH_2OH$	–	105
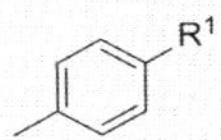	-H	7
	$-CH_3$	1.7
	$-C(CH_3)_3$	778
	$-CF_3$	111
	-OH	2.9
	$-OCH_3$	1.9
	$-NO_2$	33.5
	$-NH_2$	0.67
	-COOH	NI
	$-SO_2NH_2$	98.2
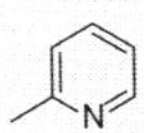	–	707

R′	R^1	K_i [μM]
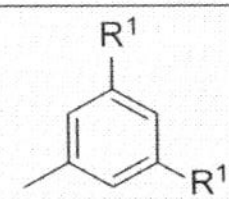	$-CH_3$	39.7
	$-NO_2$	NI
	$-NH_2$	14
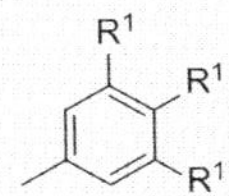	$-OCH_3$	518
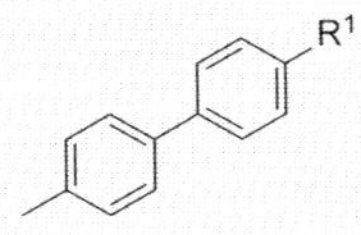	-H	2.38
	-COOH	4.42
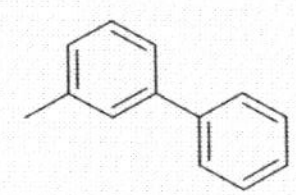	–	68.2

R′	R^1	K_i [μM]
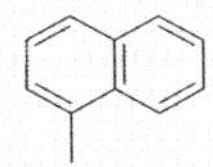	–	11.5
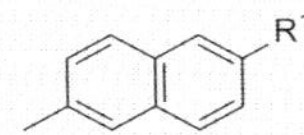	H	0.41
	COOH	4.42
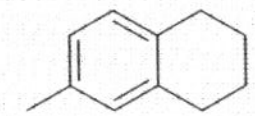	–	1.6
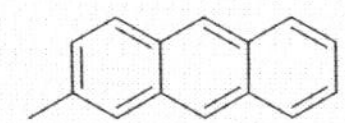	–	20
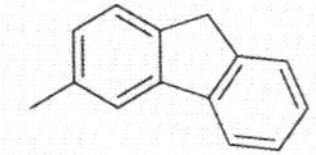	–	1.19
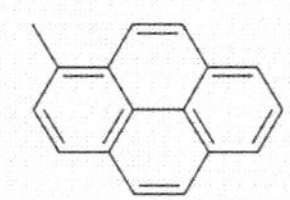	–	>1500

Table 5 Inhibition of RMGPb (K_i [μM]) by *C*-(β-D-glucopyranosyl) azoles derived from *N*-β-D-glucopyranosyl urea type molecules.

	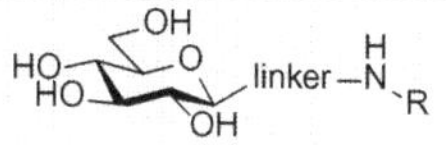	R		
	Linker	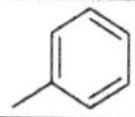	Others	
232	NHCO	18[20]	–	
67	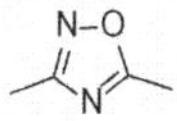	–	-isopropyl and -cyclohexyl No inh. at 625 μM[81]	
77	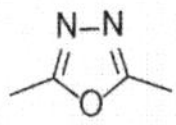	No inh. at 625 μM[84]	–	
	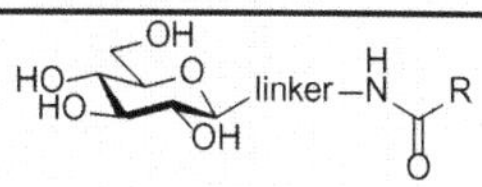	R		
	Linker	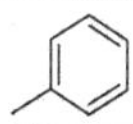	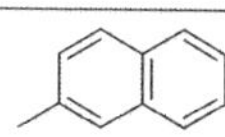	-CH₃
233	NHCO	4.6[124]	0.35[20]	305[124]
81	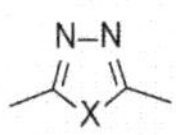 X = O	No inh. at 625 μM[84]		
93	X = S			
	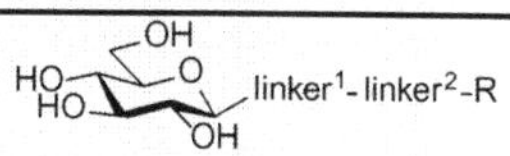	R		
	Linker[1]-Linker[2]	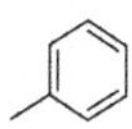	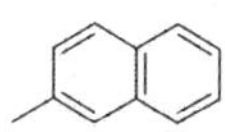	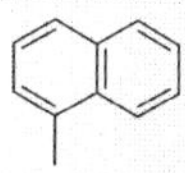
9	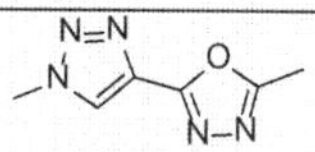	854[46]	No inh. at 625 μM[46]	No inh. at 625 μM[46]
82	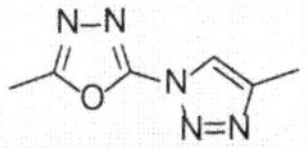	31 % inh. at 1 mM[46]	745[46]	1318[46]
	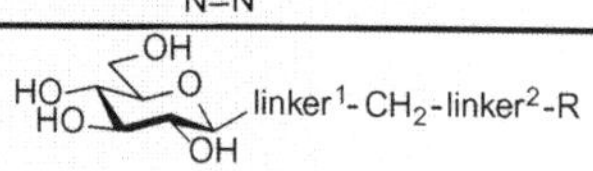	R		
	Linker[1]-CH₂-Linker[2]	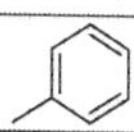	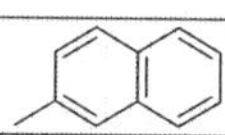	Others
11	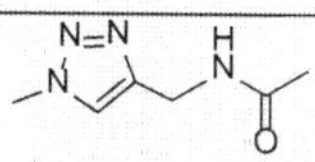	1400[47]	650[47]	-CH₃ 40 % inh. at 5 mM[47]
83	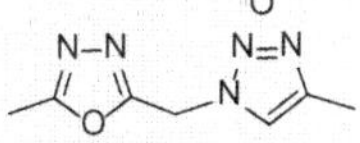	No inh. at 625 μM[46]	–	-CH₂OH 60 % inh. at 1 mM[46]

of **232** and **233** by various heterocylic linkers proved to be more efficient, affording (β-D-glucopyranosyl)aminoazole[121] and (β-D-glucopyranosyl)azolecarboxamide[122–124] type molecules (not shown) with inhibitory activities in the micromolar range.

Further *N*- and *C*-β-D-glucopyranosyl heterocycles (e.g., **9**[46] and **82**[46]), formally derived from acyl ureas by substitution of both NHCO units as well as compounds containing a flexible methylene spacer between linkers 1 and 2, such as **11**[47] and **83,**[46] showed weak or no inhibiton (Table 5).

In order to study the effect of multivalency on the inhibition of glycogen phosphorylase, some heterocyclic derivatives with multiple glucose units were also synthesized (Table 6). Among bis-glucopyranosyl heterocycles, the dimeric 1-(β-D-glucopyranosyl)-1,2,3-triazole[44] **17** showed inhibition in the upper micromolar range, while the 1,2,3-triazolyl-1,3,4-oxadiazole[46] derivative **10** was practically inactive.

The trivalent glucose derivatives[57] **87** and **89**, having three identical *C*-glucopyranosyl heterocyclic units, displayed very weak inhibition, while some members of the tethered D-glucose-triterpene heteroconjugates[48] **18-23** proved low micromolar inhibitors (Table 6).

Table 6 Inhibition of [a]RMGPb or [b]RMGPa by multicomponent heterocyclic glucose derivatives.

Diglucopyranosyl heterocyclic derivatives[44,46]

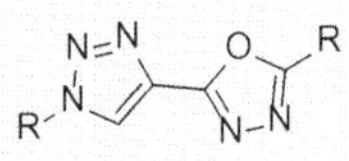

R =

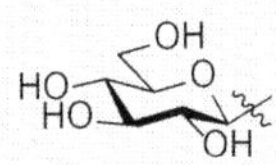

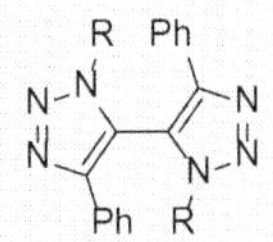

10 19 % inh. at 1 mM[a]

17 K_i = 798 μM[a]

Homotrivalent glucose derivatives[57]

R =

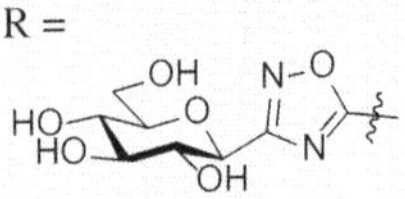

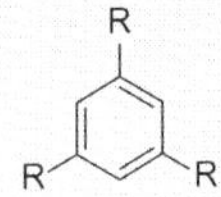

R =

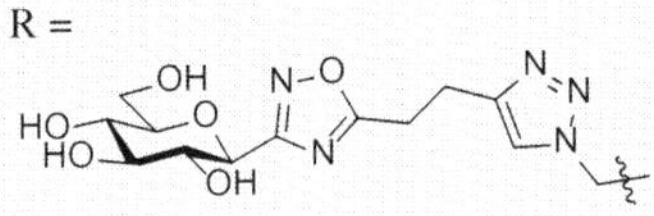

87 62 % inh. at 2.5 mM[a]

89 56 % inh. at 2.5 mM[a]

Heterobivalent triterpene-glucose derivatives[48]

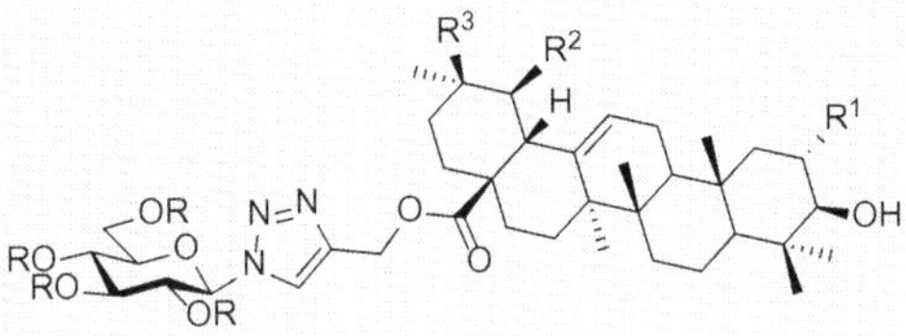

Oleanolic acid (OA) derivatives:
18 (R = Ac) or **19** (R = H), $R^1 = R^2$ = H, R^3 = Me

Ursolic acid (UA) derivatives:
20 (R = Ac) or **21** (R = H), $R^1 = R^3$ = H, R^2 = Me

Maslinic acid (MA) derivatives:
22 (R = Ac) or **23** (R =H), R^1 = OH, R^2 = H, R^3 = Me

	Acid	R	IC_{50} [μM][b]
18	OA	Ac	337
19	OA	H	26
20	UA	Ac	51
21	UA	H	45
22	MA	Ac	NI
23	MA	H	779

The purine-type nucleobases and related nucleosides and nucleotides are known as allosteric inhibitors of GP, which primarily bind to the inhibitor site of the enzyme.[19] Based on this, the inhibitory and binding properties of *N*-β-D-glucopyranosyl purines were also investigated. Selected members of the series are shown in Table 7.

Protein crystallographic studies revealed that the preferred binding site of these compounds was the inhibitor site, and only few of them entered the catalytic site of the enzyme.[34] The enzyme kinetic results showed that both N^9 and N^7 isomers of the unsubstituted *N*-glucopyranosyl purines (**128** and **129**, respectively) were poor inhibitors. The presence of an amino or a halogen substituent at the C-6 position of the N^9 regioisomer proved to be beneficial (**128** vs. **123** and **131**, respectively), while introduction of an additional amino group at C-2 resulted in significant weakening of the inhibition (**123** vs. **127**, and **131** vs. **132**, respectively). The best inhibitor of the series was the 7-(β-D-glucopyranosyl) guanine **136** whose isomeric pair **135** remained practically inactive.[34]

A set of annulated *C*-glucopyranosyl azoles was also tested against RMGPb (Table 8). Among benzocondensed heterocycles, the indole derivative **140** exhibited low inhibitory activity.[72] Kinetic investigations revealed benzothiazole **142** and benzimidazole **147** (R′=H) as micromolar inhibitors with the latter compound being more efficient by approximately one order of magnitude.[64, 117] As evidenced by X-ray crystallography,[117] the significant difference in the binding strength of the two compounds in favor of benzimidazole **147** (R′=H) can be ascribed to the NH-His377(CO) H-bond that is not possible with the nonhidrogen-bond-donor benzothiazole **142**. It is to be noted, that beside binding in the catalytic center, the presence of **147** (R′=H) at the new allosteric site as well as at the benzimidazole site, discovered by that study, was also observed.[117]

Table 7 Inhibitory potency (K_i [μM]) of *N*-β-D-glucopyranosyl purines against RMGPb.[34]

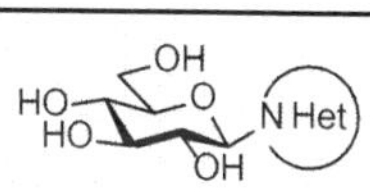

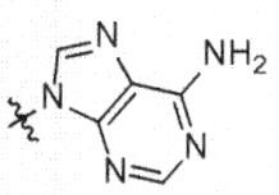	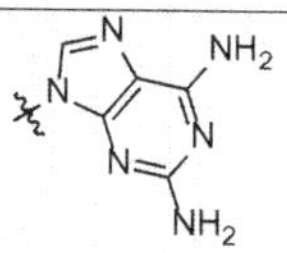	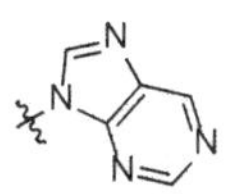	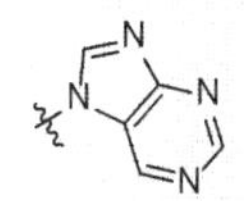
123 K_i = 315 μM *Catalytic site:* + *Inhibitor site:* +	**127** 13% inh. at 2 mM *Catalytic site:* – *Inhibitor site:* p+	**128** K_i = 1870 μM *Catalytic site:* + *Inhibitor site:* +	**129** K_i = 1870 μM *Catalytic site:* + *Inhibitor site:* +
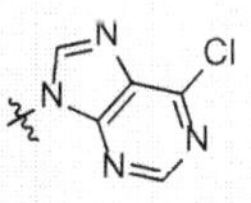	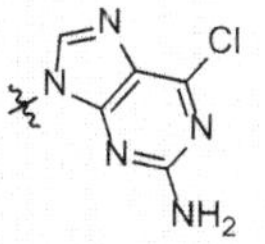	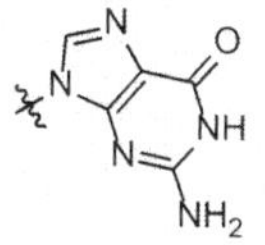	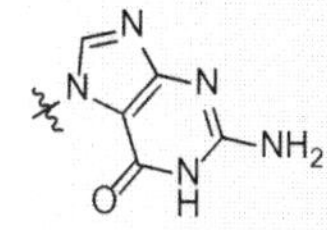
131 K_i = 760 μM *Catalytic site:* p+ *Inhibitor site:* +	**132** K_i = 850 μM *Catalytic site:* p+ *Inhibitor site:* +	**135** 13% inh. at 1 mM *Catalytic site:* – *Inhibitor site:* p+	**136** K_i = 170 μM *Catalytic site:* + *Inhibitor site:* +

Density in binding sites of RMGPb: Yes (+), partial (p+), no (−).

Table 8 Inhibitory potency of annulated *C*-(β-D-glucopyranosyl) azoles against RMGPb.

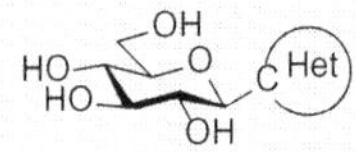

	Het	R′	Inh.	Ref.		Het	X	Inh.	Ref.
140	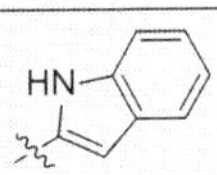	–	625[a]	72	**146**	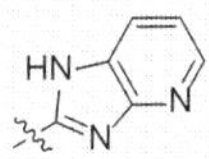	–	21.1[b]	61
142	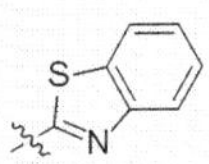	–	229[b] 76[b]	64,118	**156**	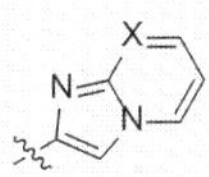	CH	28%[c]	61
					157		N	25%[c]	
147	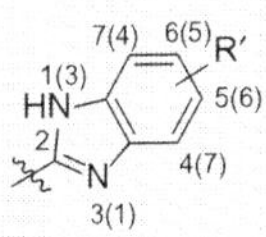	H	8.6[b] 11[b]	64,118	**158**	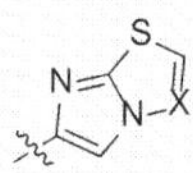	CH	15%[c]	61
		5(6)-F	55[b]	63	**159**		N	10%[c]	
		5(6)-Cl	9.7[b]		**160**		S	NI[d]	
		5(6)-Br	7.5[b]		**161**	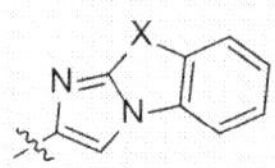	NEt	NI[d]	61
		5(6)-NO_2	179[b]		**162**		NBn	NI[d]	
		5(6)-Me	12[b]						
		4(7)-Me	2.8[b]						
		5,6-Me_2	152[b]						
148	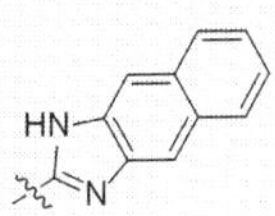	–	2.1[b]	63	**163**	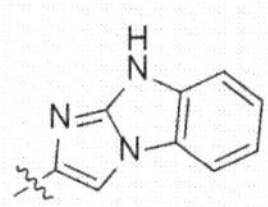	–	NI[d]	61

[a] IC_{50} [μM].
[b] K_i [μM].
[c] Inhibition (%) at 625 μM.
[d] NI: no inhibition at 625 μM.

Based on the efficacy of **147** (R′=H), the inhibitory potency of further substituted analogues **147** (substituents listed in Table 8) was also investigated.[63] The 5-chloro and 5-bromo derivatives of **147** proved to be equipotent with the unsubstituted heterocycle (R′=H), while the 5-fluoro derivative displayed weaker inhibitory activity. Introduction of a nitro group in the same position of the benzimidazole ring led to significant weakening in the inhibition. Among the monoalkylated derivatives, the 4-methylbenzimidazole was more potent than its 5-methyl counterpart, and showed ~ threefold better effect than the parent benzimidazole **147** (R′=H). 5,6-Disubstitution of the benzimidazole ring by methyl groups resulted in a weak inhibitor, while enlargement of the heterocycle by an additional benzene ring provided the most potent compound **148** of the series.[63]

Very recently, further annulated imidazoles were synthesized and tested as inhibitors of GP in order to extend the structure-activity relationships of this class of fused heterocycles.[61] The 1*H*-imidazo[4,5-*b*]pyridine **146** displayed similar potency as the benzimidazole **147** (R′=H), indicating that the presence of a N atom in the benzene ring had no significant effect on the binding strength. The constitutional isomers of compounds **146** and **147** (R′=H), such as **156** and **157**, respectively, showed negligible inhibitory activity. The lack of the efficiency of the latter compounds may probably be due to the loss of the H-bond donor NH group. Imidazoles fused by five-membered heteroaromatic rings, such as compounds **158** and **159**, also showed very weak effects, while the 5-5 bicyclic derivatives condensed with a benzene ring (**160–163**), even if having H-bond donor capacity (e.g. **163**), proved to be inactive in the investigated concentration range (max. 625 μM).

3.2 *Six-membered* N- *and* C-(β-D-*glucopyranosyl) heterocycles*

Evaluation of six-membered *N*-glucopyranosyl heterocycles as inhibitors of glycogen phosphorylase also revealed several potent molecules (Table 9). Among pyrimidine derivatives, uracyl[34, 98] **164** and most of its 5-substituted variants[34, 98, 112] displayed inhibitory potency in the low micromolar range. The 5-halogenouracils **169-171**, but

Table 9 Inhibition of RMGPb (K_i [μM]) by six-membered *N*-(β-D-glucopyranosyl) heterocycles.

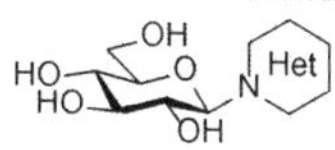

	Het	R′			Het	
164	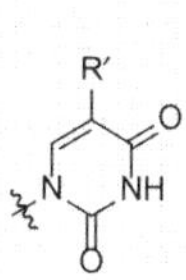	-H	6.1[34] 12.4[98]			
165		$-CH_3$	6.6[34]			
166		$-CF_3$	17.0[98]	**173**	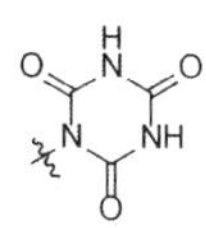	1260[34]
168		-F	5.5[34] 7.9[112]			
169		-Cl	1.0[98]			
170		-Br	3.3[98]			
171		-I	1.9[98]			
		-C≡CH	4.7[101]	**167**	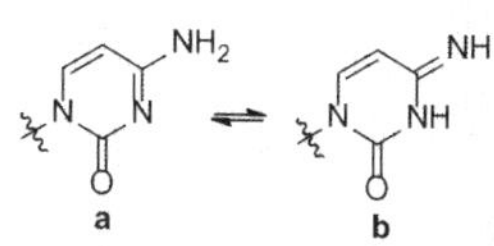	7.7[20,34]
186		$-C{\equiv}C-C_3H_7$	303[101]			
		$-C{\equiv}C-C_5H_{11}$	3204[101]			
172	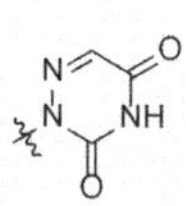	–	96[34]	**185**	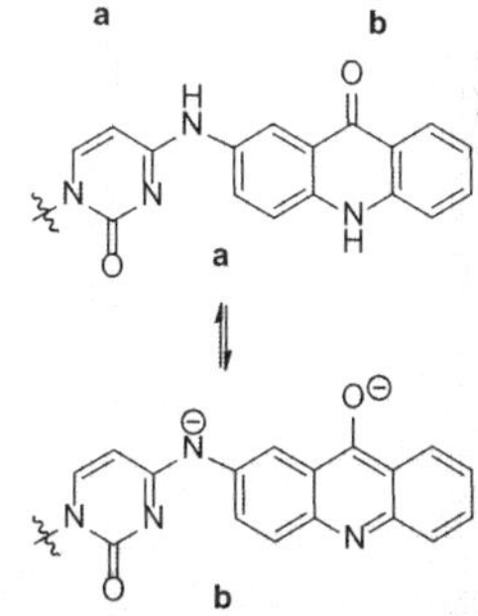	0.071[100]

Het

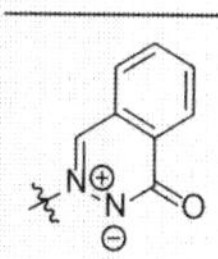	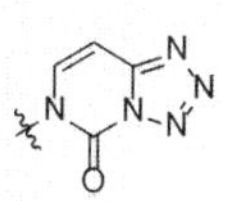	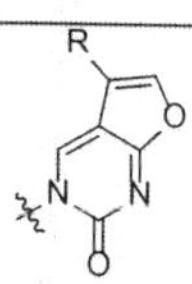	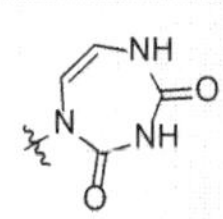
175 54.6[34]	**182** 197[34]	**187** ($R = C_3H_7$) 32.4 ($R = C_5H_{11}$) 33.4[101]	**184** 76.2[34]

not the fluoro derivative **168**, proved to be more effective than the parent compound **164**.[34, 98, 112] The improved binding affinity of **169–171** as compared to that of **164** was the result of the σ-hole-induced electrostatic interactions of the halogen atoms with the neighboring protein residue Asp339, as it was proven by a combined protein crystallographic and computational chemistry analysis.[98] Studies with 5-alkynyluracils[101] showed that the inhibitory potency of 5-ethynyl derivative **186** (R = –C≡CH) was comparable to that of **164**, while its analogues having long alkyl chains **186** (R = –C≡C–C_3H_7, –C≡C–C_5H_{11}) exhibited much weaker effects.

The decreased inhibitory potency of 6-azauracil[34] **172** as compared to uracil **164** indicated that the replacement of the C-6 carbon of the pyrimidine ring by a N-atom was disadvantageous for the inhibition. Additionally, isocyanuric acid[34] **173** as another triazine analogue was also tested had, however, no notable effect.

N-Glucosylcytosine[34] **167** was as good an inhibitor as uracil **164**. Based on molecular modeling studies and quantum chemical calculations, the binding of **167** to the catalytic site of RMGPb may probably take place via the energetically more favorable imine tautomer **167b**.[20] Very recently, N^4-substitution of cytosine with acridin-9-one was carried out resulting in compound **185** {N^1-(β-D-glucopyranosyl)-N^4-[2-acridin-9(10*H*)-onyl]cytosine; GLAC}, one of the most efficient glucose-derived inhibitors of GP known to date. The X-ray crystallographic analysis of the GLAC-RMGPb complex indicated that the bulky N-4 substituent occupied the β-channel of the enzyme.[100] The absorption spectra of GLAC along with the protein crystallographic and computational studies also revealed that GLAC could exist in a bis-anionic form (**185b**) within the catalytic site, thereby allowing participation of the aglycon part in an extended H-bonding network with specific amino acid residues of the β-cavity.[100]

Among annulated six-membered heterocycles (e.g. **175**,[34] **182**,[34] and **187**[101]), the alkylfurano[2,3-*d*]pyrimidinones[101] **187** proved to be the most effective ones. A seven-membered triazepane derivative **184** was also tested to show moderate inhibition against RMGPb.[34]

The inhibitory effect of the 2-*C*-β-D-glucopyranosyl pyrimidines[71] **189** and the tetrahydropyrimidin-2-one derivative[102] **192** was also examined; however, these compounds proved to be inactive against the RMGPb enzyme.

3.3 N- and C-Glycopyranosyl heterocycles with modified sugar units

Several modifications in the glucose unit of the most efficient *N*- and *C*-(β-D-glucopyranosyl) azoles were also carried out to investigate the variability of the sugar part (Table 10).

Changing the sugar moiety from D-glucose to D-xylose resulted in very weak or noninhibitory compounds (see the comparable pairs **5** and **195**, **62** and **196**, **107** and **197**, and **147** and **198**, respectively).[103, 105, 106] Although truncation of the glucose core by removal of the CH_2OH group was undoubtedly detrimental, the modest inhibitory effects of the imidazole **196** and 1,2,4-triazole **197** derivatives suggested that the impaired binding affinity of the glycon could partially be compensated by an effective heterocylic C-1 substituent. The 1,2,3-triazole containing 5-thioxylose **207** and the

Table 10 Inhibition of RMGPb (K_i [μM]) by glycopyranosyl azoles having a sugar part other than glucose.

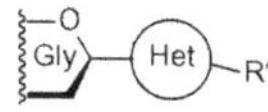

Gly	R′	Het: 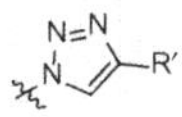		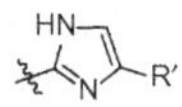		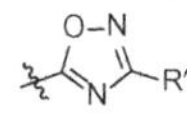		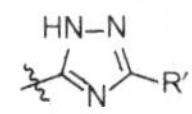		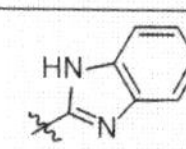	
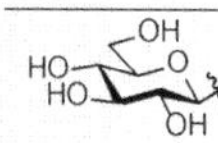	Phenyl	5	151[41]	62	0.28[73]	69	64[82]	107	7[65,90]	147	8.6[118]
	2-Naphthyl		16[41]		0.031[73]		11.6[83]		0.41[65,90]		
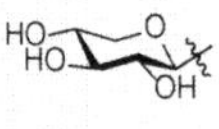	Phenyl	195	NI[a103]	196	800[105]	–	–	197	NI[a106]	198	NI[a106]
	2-Naphthyl		–		100[105]		–		491[106]		
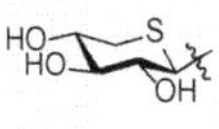	Phenyl	207	15%[103]	–	–	–	–	–	–	–	–
	2-Naphthyl		18%[103]		–		–		–		
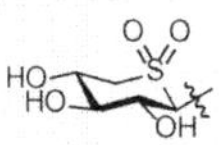	Phenyl	211	14%[103]	–	–	–	–	–	–	–	–
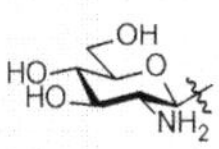	Phenyl	199	NI[a104]	200	2.67[104]	–	–	201	35.2[104]	202	107[104]
	2-Naphthyl		354[104]		0.19[104]		–		4.8[104]		
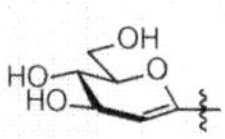	Phenyl	–	–	–	–	219	NI[a107]	–	–	–	–
	2-Naphthyl		–		–		NI[a107]		–		

[a] NI: no inhibition at 625 μM.

sulfone **211** derivatives showed very weak inhibitory potencies.[103] Isosteric replacement of the 2'-OH group of the glucopyranosyl derivatives by an NH_2 group resulted in β-D-glucosaminyl heterocycles with varied efficacies.[104] While 1,2,3-triazoles **199** as well as the benzimidazole **202** displayed no or weak inhibition, the 1,2,4-triazole and imidazole derivatives **201** and **200**, respectively, proved to be low micromolar inhibitors.[104] The inhibitor constants of the corresponding pairs of the glucose and glucosamine-based molecules (**5** vs. **199**, **62** vs. **200**, **107** vs. **201**, and **147** vs. **202**) showed that the substitution of the 2′-OH by an NH_2 group decreased the inhibitory potency by a factor of ~5–40. X-Ray crystallography revealed very similar binding modes of *C*-(β-D-glucosaminyl) heterocycles **200** and **201** as compared to those of their glucopyranosyl counterparts **62** and **107**, respectively. Due to the lack of any structural differences of the RMGPb-inhibitor complexes, the decreased inhibitory potencies of the glucosamine derivatives **200**, **201** could be explained by the less potent H-bond donor capacity of the NH_2 group as compared to that of the OH group.[104]

With the aim to mimic the structure of the transition state of the reaction catalyzed by GP that resembles a glycosylium ion,[125] glycal-derived oxadiazoles (e.g., **219**) were tested and proved, however, inactive.[107]

For the inhibition of GP some *N*-(β-D-glycopyranosyl) pyrimidines having altered glucose unit were also tested (Table 11). A comparison of the inhibitory potency of compounds **223–225** to that of the parent glucose derivatives **164**, **168**, and **167**, respectively, showed that isosteric replacement of the 3′-OH group of

Table 11 Inhibition of RMGPb (K_i [μM]) by *N*-glycopyranosyl pyrimidines having a sugar part other than glucose.

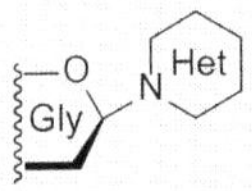

	Het							
Gly	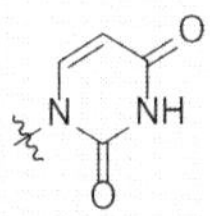		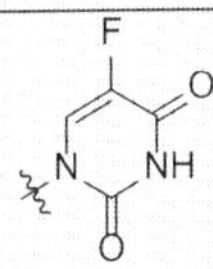		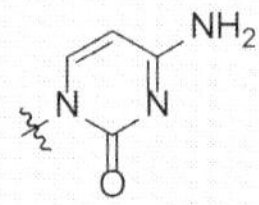		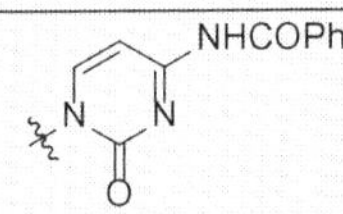	
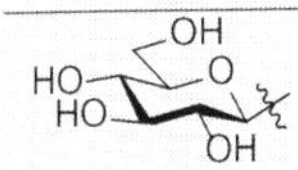	**164**	6.1[34] 12.4[98]	**168**	5.5[34] 7.9[112]	**167**	7.7[34]	–	–
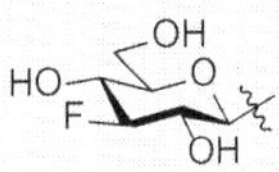	**223**	3460[111]	**224**	3670[111]	**225**	4010[111]	**226**	46[111]
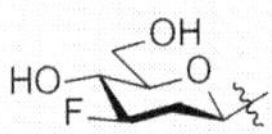	–	–	–	–	–	–	**227**	6550[111]
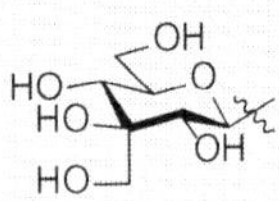	–	–	**230**	27.1[112]	–	–	–	–

the latter compounds by fluorine led to significant decrease of the inhibition.[111] Introduction of a hydrophobic aromatic substituent into the aglycon part resulted in improved binding (compare compounds **225** and **226**), thereby rendering **226** to be a micromolar inhibitor of RMGPb.[111] Removal of the 2′-OH group of the sugar unit proved detrimental for the inhibition affording a weak millimolar inhibitor **227**.[111] Attachment of an axially oriented CH_2OH substituent into the C-3 position of the glucose moiety of **168** provided a slightly weaker inhibitor **230**.[112] As evidenced by X-ray crystallography, the C_{ax}-3-CH_2OH group of **230** formed an additional direct H-bond with the PLP cofactor of the enzyme, which was absent in the **168**-RMGPb complex. Based on QM/MM-PBSA calculations, the failure to improve the effect despite the extra molecular interaction of **230** can be explained by the desolvation energy cost of the binding of **230** exceeding the enthalpy gains derived from the CH_2OH-PLP contact.[112]

Finally, it is also to be noted, that besides inhibition against RMGPb, the inhibitory potency of the most effective *C*-glycopyranosyl heterocyclic derivatives (e.g., **62**, **107**, **200**, and **201**) was also evaluated against human liver GPa to reveal similar effects of the compounds against the pharmacologically relevant isoenzyme.[72, 88, 91, 104, 119]

Additionally, in case of some *N*-glucopyranosyl pyrimidines (**169** and **186**) ex vivo experiments were also carried out to show no cytotoxicity and moderate GP inhibitory activity in HepG2 hepatocarcinoma cells.[98, 101]

4 Conclusion

The inhibition of glycogen phosphorylase (GP) is a validated therapeutic approach in the quest for new treatments of type 2 diabetes and also has an inherent potential to be utilized in combating other diseases associated with GP activity. Among the glucose-based catalytic site inhibitors of GP, glycopyranosyl heterocycles have received great attention in recent years.

For GP inhibition, a large set of *C*-glycopyranosyl azoles, a small library of *C*-glucopyranosyl pyrimidines, and several *N*-glycopyranosyl heterocyles with five- and six-membered rings were synthesized. The *C*-glycosyl compounds were mainly obtained by cyclocondensation reactions of 2,6-anhydroaldonic acid type precursors, while the *N*-glycosides were mostly achieved by Vorbrüggen type condensations of glycosyl acetates with *N*-silylated heterocyles.

The structure-activity relationships of these compound classes revealed that the nature and the constitution of the heterocycles as well as the substitution pattern of the heteroaromatic rings had strong bearings on the inhibitory efficacies. The most potent derivatives of the whole series are the N^1-(β-D-glucopyranosyl)-N^4-[2-acridin-9(10*H*)-onyl]cytosine and 2-(β-D-glucopyranosyl)-4(5)-(2-naphthyl)imidazole representing, with their low nanomolar inhibitor constants (K_i=71 and 31 nM, respectively), the best glucose-based inhibitors of GP.

Alterations of the glucose part generally result in loss or decrease of the inhibitory potential. The compounds still having low inhibitor constants (e.g., N^4-benzoyl-1-(3-deoxy-3-fluoro-β-D-glucopyranosyl)cytosine with K_i=46 μM and 2-(2-amino-2-deoxy-β-D-glucopyranosyl)-4(5)-(2-naphthyl)imidazole with K_i=0.19 μM) indicated, however, the positive impact of a well-chosen aglycon part.

Several ex vivo, in vivo, and ADMET studies revealed the potential of glucose-based GP inhibitors in drug development, and very recently, such studies have been extended to *C*-glucopyranosyl heterocycles, too.[126]

Acknowledgments

This work was supported by the National Research, Development and Innovation Office of Hungary (Grant FK 125067), and by the EU cofinanced by the European Regional Development Fund under the project GINOP-2.3.2-15-2016-00008.

I wish to thank Prof. Dr. László Somsák for his helpful advice and critical comments during compilation of this book chapter.

References

1. IDF Diabetes Atlas, 5th ed. In International Diabetes Federation [Online] Brussels, Belgium. http://www.diabetesatlas.org, 2011 (Accessed 22 May 2016).
2. IDF Diabetes Atlas, 8th ed. In International Diabetes Federation [Online] Brussels, Belgium. http://www.diabetesatlas.org, 2017 (Accessed 27 June 2018).

3. Israili, Z. H. Advances in the Treatment of Type 2 Diabetes Mellitus. *Am. J. Ther.* **2011**, *18*, 117–152.
4. Bailey, C. J. Renal Glucose Reabsorption Inhibitors to Treat Diabetes. *Trends Pharmacol. Sci.* **2011**, *32*, 63–71.
5. Fonseca, V. A. New Developments in Diabetes Management: Medications of the 21st Century. *Clin. Ther.* **2014**, *36*, 477–484.
6. Wagman, A. S.; Nuss, J. M. Current Therapies and Emerging Targets for the Treatment of Diabetes. *Curr. Pharm. Des.* **2001**, *7*, 417–450.
7. Morral, N. Novel Targets and Therapeutic Strategies for Type 2 Diabetes. *Trends Endocrinol. Metab.* **2003**, *14*, 169–175.
8. Henke, B. R. Inhibition of Glycogen Phosphorylase as a Strategy for the Treatment of Type 2 Diabetes. *RSC Drug Discov. Ser.* **2012**, *27*, 324–365.
9. Tracey, W.; Treadway, J.; Magee, W.; McPherson, R.; Levy, C.; Wilder, D.; Li, Y.; Yue, C.; Zavadoski, W.; Gibbs, E.; Smith, A.; Flynn, D.; Knight, D. A Novel Glycogen Phosphorylase Inhibitor, CP-368296, Reduces Myocardial Ischemic Injury. *Diabetes* **2003**, *52*, A135.
10. Tracey, W. R.; Treadway, J. L.; Magee, W. P.; Sutt, J. C.; McPherson, R. K.; Levy, C. B.; Wilder, D. E.; Yu, L. J.; Chen, Y.; Shanker, R. M.; Mutchler, A. K.; Smith, A. H.; Flynn, D. M.; Knight, D. R. Cardioprotective Effects of Ingliforib, a Novel Glycogen Phosphorylase Inhibitor. *Am. J. Physiol. Heart Circ. Physiol.* **2004**, *286*, H1177–H1184.
11. Xu, L.; Sun, H. Pharmacological Manipulation of Brain Glycogenolysis as a Therapeutic Approach to Cerebral Ischemia. *Mini Rev. Med. Chem.* **2010**, *10*, 1188–1193.
12. Guan, T.; Qian, Y. S.; Tang, X. Z.; Huang, M. H.; Huang, L. F.; Li, Y. M.; Sun, H. B. Maslinic Acid, a Natural Inhibitor of Glycogen Phosphorylase, Reduces Cerebral Ischemic Injury in Hyperglycemic Rats by GLT-1 Up-Regulation. *J. Neurosci. Res.* **2011**, *89*, 1829–1839.
13. Schnier, J. B.; Nishi, K.; Monks, A.; Gorin, F. A.; Bradbury, E. M. Inhibition of Glycogen Phosphorylase (GP) by CP-91,149 Induces Growth Inhibition Correlating With Brain GP Expression. *Biochem. Biophys. Res. Commun.* **2003**, *309*, 126–134.
14. Geschwind, J.-F.; Georgiades, C. S.; Ko, Y. H.; Pedersen, P. L. Recently Elucidated Energy Catabolism Pathways Provide Opportunities for Novel Treatments in Hepatocellular Carcinoma. *Expert Rev. Anticancer Ther.* **2004**, *4*, 449–457.
15. Boros, L. G.; Go, V. L. W.; Lee, W.-N.P. Glycogen Phosphorylase Inhibitor CP-320626 Inhibits Pancreatic Cancer Cell Proliferation via Inhibiting Pentose Cycle Metabolism and Glycolysis. *Pancreas* **2003**, *27*, 368–420.
16. Favaro, E.; Bensaad, K.; Chong, M. G.; Tennant, D. A.; Ferguson, D. J. P.; Snell, C.; Steers, G.; Turley, H.; Li, J.-L.; Günther, U. L.; Buffa, F. M.; McIntyre, A.; Harris, A. L. Glucose Utilization via Glycogen Phosphorylase Sustains Proliferation and Prevents Premature Senescence in Cancer Cells. *Cell Metab.* **2012**, *16*, 751–764.
17. Zois, C. E.; Harris, A. L. Glycogen Metabolism Has a Key Role in the Cancer Microenvironment and Provides New Targets for Cancer Therapy. *J. Mol. Med.* **2016**, *94*, 137–154.
18. Zois, C. E.; Favaro, E.; Harris, A. L. Glycogen Metabolism in Cancer. *Biochem. Pharmacol.* **2014**, *92*, 3–11.
19. Oikonomakos, N. G. Glycogen Phosphorylase as a Molecular Target for Type 2 Diabetes Therapy. *Curr. Protein Pept. Sci.* **2002**, *3*, 561–586.
20. Somsák, L.; Czifrák, K.; Tóth, M.; Bokor, É.; Chrysina, E. D.; Alexacou, K. M.; Hayes, J. M.; Tiraidis, C.; Lazoura, E.; Leonidas, D. D.; Zographos, S. E.; Oikonomakos, N. G. New Inhibitors of Glycogen Phosphorylase as Potential Antidiabetic Agents. *Curr. Med. Chem.* **2008**, *15*, 2933–2983.

21. Oikonomakos, N. G.; Somsák, L. Recent Advances in Glycogen Phosphorylase Inhibitor Design. *Curr. Opin. Investig. Drugs* **2008**, *9*, 379–395.
22. Gaboriaud-Kolar, N.; Skaltsounis, A.-L. Glycogen Phosphorylase Inhibitors: A Patent Review (2008-2012). *Expert Opin. Ther. Pat.* **2013**, *23*, 1017–1032.
23. Hayes, J. M.; Kantsadi, A. L.; Leonidas, D. D. Natural Products and Their Derivatives as Inhibitors of Glycogen Phosphorylase: Potential Treatment for Type 2 Diabetes. *Phytochem. Rev.* **2014**, *13*, 471–498.
24. Donnier-Maréchal, M.; Vidal, S. Glycogen Phosphorylase Inhibitors: A Patent Review (2013-2015). *Expert Opin. Ther. Pat.* **2016**, *26*, 199–212.
25. Stravodimos, G. A.; Chetter, B. A.; Kyriakis, E.; Kantsadi, A. L.; Chatzileontiadou, D. S. M.; Skamnaki, V. T.; Kato, A.; Hayes, J. M.; Leonidas, D. D. Phytogenic Polyphenols as Glycogen Phosphorylase Inhibitors: The Potential of Triterpenes and Flavonoids for Glycaemic Control in Type 2 Diabetes. *Curr. Med. Chem.* **2017**, *24*, 384–403.
26. Somsák, L. Glucose Derived Inhibitors of Glycogen Phosphorylase. *C. R. Chim.* **2011**, *14*, 211–223.
27. Praly, J.-P.; Vidal, S. Inhibition of Glycogen Phosphorylase in the Context of Type 2 Diabetes, With Focus on Recent Inhibitors Bound at the Active Site. *Mini-Rev. Med. Chem.* **2010**, *10*, 1102–1126.
28. Bokor, É.; Kun, S.; Goyard, D.; Tóth, M.; Praly, J.-P.; Vidal, S.; Somsák, L. *C*-Glycopyranosyl Arenes and Hetarenes: Synthetic Methods and Bioactivity Focused on Antidiabetic Potential. *Chem. Rev.* **2017**, *117*, 1687–1764.
29. Docsa, T.; Czifrák, K.; Hüse, C.; Somsák, L.; Gergely, P. The Effect of Glucopyranosylidene-Spiro-Thiohydantoin on the Glycogen Metabolism in Liver Tissues of Streptozotocin-Induced and Obese Diabetic Rats. *Mol. Med. Rep.* **2011**, *4*, 477–481.
30. Docsa, T.; Marics, B.; Németh, J.; Hüse, C.; Somsák, L.; Gergely, P.; Peitl, B. Insulin Sensitivity Is Modified by a Glycogen Phosphorylase Inhibitor: Glucopyranosylidene-Spiro-Thiohydantoin in Streptozotocin-Induced Diabetic Rats. *Curr. Top. Med. Chem.* **2015**, *15*, 2390–2394.
31. Goyard, D.; Kónya, B.; Chajistamatiou, A. S.; Chrysina, E. D.; Leroy, J.; Balzarin, S.; Tournier, M.; Tousch, D.; Petit, P.; Duret, C.; Maurel, P.; Somsák, L.; Docsa, T.; Gergely, P.; Praly, J.-P.; Azay-Milhau, J.; Vidal, S. Glucose-Derived Spiro-Isoxazolines Are Anti-Hyperglycemic Agents Against Type 2 Diabetes Through Glycogen Phosphorylase Inhibition. *Eur. J. Med. Chem.* **2016**, *108*, 444–454.
32. Nagy, L.; Docsa, T.; Brunyánszki, A.; Szántó, M.; Hegedűs, C.; Márton, J.; Kónya, B.; Virág, L.; Somsák, L.; Gergely, P.; Bai, P. Glycogen Phosphorylase Inhibitor *N*-(3,5-Dimethyl-Benzoyl)-*N'*-(β-D-Glucopyranosyl)urea Improves Glucose Tolerance Under Normoglycemic and Diabetic Conditions Through Rearranging Hepatic Metabolism. *PLoS One* **2013**, *8*, e69420.
33. Nagy, L.; Márton, J.; Vida, A.; Kis, G.; Bokor, É.; Kun, S.; Gönczi, M.; Docsa, T.; Tóth, A.; Antal, M.; Gergely, P.; Csóka, B.; Pacher, P.; Somsák, L.; Bai, P. Glycogen Phosphorylase Inhibition Improves β-Cell Function. *Br. J. Pharmacol.* **2018**, *175*, 301–319.
34. Gimisis, T. Synthesis of *N*-Glucopyranosidic Derivatives as Potential Inhibitors that Bind at the Catalytic Site of Glycogen Phosphorylase. *Mini-Rev. Med. Chem.* **2010**, *10*, 1127–1138.
35. Hay, M. P.; Lee, H. H.; Wilson, W. R.; Roberts, P. B.; Denny, W. A. Hypoxia-Selective Antitumor Agents. 10. Bis(nitroimidazoles) and Related Bis(nitroheterocycles)—Development of Derivatives With Higher Rates of Metabolic-Activation Under Hypoxia and Improved Aqueous Solubility. *J. Med. Chem.* **1995**, *38*, 1928–1941.

36. Patt, M.; Sorger, D.; Scheunemann, M.; Stocklin, G. Adduct of 2-[^{18}F]FDG and 2-Nitroimidazole as a Putative Radiotracer for the Detection of Hypoxia With PET: Synthesis, In Vitro- and In Vivo-Characterisation. *Appl. Radiat. Isot.* **2002**, *57*, 705–712.
37. Schweifer, A.; Hammerschmidt, F. Preparation of Nucleosides Derived From 2-Nitroimidazole and D-Arabinose, D-Ribose, and D-Galactose by the Vorbrüggen Method and Their Conversion to Potential Precursors for Tracers to Image Hypoxia. *J. Org. Chem.* **2011**, *76*, 8159–8167.
38. Kun, S.; Bokor, É.; Sipos, Á.; Docsa, T.; Somsák, L. Synthesis of New *C*- and *N*-β-D-Glucopyranosyl Derivatives of Imidazole, 1,2,3-Triazole and Tetrazole, and Their Evaluation as Inhibitors of Glycogen Phosphorylase. *Molecules* **2018**, *23*, 666.
39. He, X.-P.; Zeng, Y.-L.; Zang, Y.; Li, J.; Field, R. A.; Chen, G.-R. Carbohydrate CuAAC Click Chemistry for Therapy and Diagnosis. *Carbohydr. Res.* **2016**, *429*, 1–22.
40. Tiwari, V. K.; Mishra, B. B.; Mishra, K. B.; Mishra, N.; Singh, A. S.; Chen, X. Cu-Catalyzed Click Reaction in Carbohydrate Chemistry. *Chem. Rev.* **2016**, *116*, 3086–3240.
41. Chrysina, E. D.; Bokor, É.; Alexacou, K.-M.; Charavgi, M.-D.; Oikonomakos, G. N.; Zographos, S. E.; Leonidas, D. D.; Oikonomakos, N. G.; Somsák, L. Amide-1,2,3-Triazole Bioisosterism: The Glycogen Phosphorylase Case. *Tetrahedron-Asymmetry* **2009**, *20*, 733–740.
42. Bokor, É.; Docsa, T.; Gergely, P.; Somsák, L. Synthesis of 1-(D-Glucopyranosyl)-1,2,3-Triazoles and Their Evaluation as Glycogen Phosphorylase Inhibitors. *Bioorg. Med. Chem.* **2010**, *18*, 1171–1180.
43. Bokor, É.; Koppány, C.; Gonda, Z.; Novák, Z.; Somsák, L. Evaluation of bis-Triphenylphosphano-Copper(I)-Butyrate ($C_3H_7COOCu(PPh_3)_2$) as Catalyst for the Synthesis of 1-(D-Glycopyranosyl)-4-Substituted-1,2,3-Triazoles. *Carbohydr. Res.* **2012**, *351*, 42–48.
44. Goyard, D.; Chajistamatiou, A. S.; Sotiropoulou, A. I.; Chrysina, E. D.; Praly, J.-P.; Vidal, S. Efficient Atropodiastereoselective Access to 5,5'-bis-1,2,3-Triazoles: Studies on 1-Glucosylated 5-Halogeno 1,2,3-Triazoles and Their 5-Substituted Derivatives as Glycogen Phosphorylase Inhibitors. *Chem. Eur. J.* **2014**, *20*, 5423–5432.
45. Goyard, D.; Praly, J.-P.; Vidal, S. Synthesis of 5-Halogenated 1,2,3-Triazoles Under Stoichiometric Cu(I)-Mediated Azide-Alkyne Cycloaddition (CuAAC or 'Click Chemistry'). *Carbohydr. Res.* **2012**, *362*, 79–83.
46. Kun, S.; Nagy, G. Z.; Tóth, M.; Czecze, L.; Nguyen van Nhien, A.; Docsa, T.; Gergely, P.; Charavgi, M.-D.; Skourti, P. V.; Chrysina, E. D.; Patonay, T.; Somsák, L. Synthesis of Variously Coupled Conjugates of D-Glucose, 1,3,4-Oxadiazole, and 1,2,3-Triazole for Inhibition of Glycogen Phosphorylase. *Carbohydr. Res.* **2011**, *346*, 1427–1438.
47. Goyard, D.; Docsa, T.; Gergely, P.; Praly, J.-P.; Vidal, S. Synthesis of 4-Amidomethyl-1-Glucosyl-1,2,3-Triazoles and Evaluation as Glycogen Phosphorylase Inhibitors. *Carbohydr. Res.* **2015**, *402*, 245–251.
48. Cheng, K.; Liu, J.; Sun, H.; Bokor, É.; Czifrák, K.; Kónya, B.; Tóth, M.; Docsa, T.; Gergely, P.; Somsák, L. Tethered Derivatives of D-Glucose and Pentacyclic Triterpenes for Homo/Heterobivalent Inhibition of Glycogen Phosphorylase. *New J. Chem.* **2010**, *34*, 1450–1464.
49. Ellingboe, J. W.; Collini, M. D.; Quagliato, D.; Chen, J.; Antane, M.; Schmid, J.; Hartupee, D.; White, V.; Park, C. H.; Tanikella, T.; Bagli, J. F. Metabolites of the Angiotensin II Antagonist Tasosartan: The Importance of a Second Acidic Group. *J. Med. Chem.* **1998**, *41*, 4251–4260.

50. Couri, M. R.; Luduvico, I.; Santos, L.; Alves, R.; Prado, M. A.; Gil, R. F. Microwave-Assisted Efficient Preparation of Novel Carbohydrate Tetrazole Derivatives. *Carbohydr. Res.* **2007**, *342*, 1096–1100.
51. Rashad, A. E.; El-Sayed, W. A.; Mohamed, A. M.; Ali, M. M. Synthesis of New Quinoline Derivatives as Inhibitors of Human Tumor Cells Growth. *Arch. Pharm. (Weinheim, Ger.)* **2010**, *343*, 440–448.
52. Neochoritis, C. G.; Zhang, J.; Doemling, A. Leuckart-Wallach Approach to Sugar Isocyanides and Its IMCRs. *Synthesis* **2015**, *47*, 2407–2413.
53. Rivera, D. G.; Perez-Labrada, K.; Lambert, L.; Doerner, S.; Westermann, B.; Wessjohann, L. A. Carbohydrate-Steroid Conjugation by Ugi Reaction: One-Pot Synthesis of Triple Sugar/Pseudo-Peptide/Spirostane Hybrids. *Carbohydr. Res.* **2012**, *359*, 102–110.
54. El Remaily, M.A.E.A.A.A.; Mohamed, S. K. Eco-Friendly Synthesis of Guanidinyltetrazole Compounds and 5-Substituted 1*H*-Tetrazoles in Water Under Microwave Irradiation. *Tetrahedron* **2014**, *70*, 270–275.
55. Somsák, L.; Nagy, V. A New, Scalable Preparation of a Glucopyranosylidene-Spiro-Thiohydantoin: One of the Best Inhibitors of Glycogen Phosphorylases. *Tetrahedron-Asymmetry* **2000**, *11*, 1719–1727. Corrigendum 2247.
56. Benltifa, M.; Vidal, S.; Fenet, B.; Msaddek, M.; Goekjian, P. G.; Praly, J.-P.; Brunyánszki, A.; Docsa, T.; Gergely, P. In the Search of Glycogen Phosphorylase Inhibitors: 5-Substituted 3-*C*-Glucopyranosyl-1,2,4-Oxadiazoles From β-D-Glucopyranosyl Cyanides Upon Cyclization of *O*-Acyl-amidoxime Intermediates. *Eur. J. Org. Chem.* **2006**, 4242–4256.
57. Cecioni, S.; Argintaru, O.-A.; Docsa, T.; Gergely, P.; Praly, J.-P.; Vidal, S. Probing Multivalency for the Inhibition of an Enzyme: Glycogen Phosphorylase as a Case Study. *New J. Chem.* **2009**, *33*, 148–156.
58. Bokor, É.; Fekete, A.; Varga, G.; Szőcs, B.; Czifrák, K.; Komáromi, I.; Somsák, L. *C*-(β-D-Glucopyranosyl)formamidrazones, Formic Acid Hydrazides and Their Transformations Into 3-(β-D-Glucopyranosyl)-5-Substituted-1,2,4-Triazoles: A Synthetic and Computational Study. *Tetrahedron* **2013**, *69*, 10391–10404.
59. Misra, A. K.; Bokor, É.; Kun, S.; Bolyog-Nagy, E.; Kathó, Á.; Joó, F.; Somsák, L. Chemoselective Hydration of Glycosyl Cyanides to *C*-Glycosyl Formamides Using Ruthenium Complexes in Aqueous Media. *Tetrahedron Lett.* **2015**, *56*, 5995–5998.
60. Czifrák, K.; Szilágyi, P.; Somsák, L. Anomeric α-Azido Acid (2-Azido-2-Deoxy-Hept-2-Ulopyranosonic Acid) Derivatives En Route to Peptides Incorporating Sugar Amino Acids. *Tetrahedron-Asymmetry* **2005**, *16*, 127–141.
61. Szennyes, E.; Bokor, É.; Docsa, T.; Sipos, Á.; Somsák, L. Synthesis of *C*-β-D-Glucopyranosyl Derivatives of Some Fused Azoles for the Inhibition of Glycogen Phosphorylase. *Carbohydr. Res.* **2019**, *472*, 33–41.
62. Ősz, E.; Czifrák, K.; Deim, T.; Szilágyi, L.; Bényei, A.; Somsák, L. Preparation of 3,5-bis-(β-D-Glycopyranosyl)-1,2,4-Thiadiazoles From *C*-(β-D-Glycopyranosyl)thioformamides. *Tetrahedron* **2001**, *57*, 5429–5434.
63. Bokor, É.; Szilágyi, E.; Docsa, T.; Gergely, P.; Somsák, L. Synthesis of Substituted 2-(β-D-Glucopyranosyl)-Benzimidazoles and Their Evaluation as Inhibitors of Glycogen Phosphorylase. *Carbohydr. Res.* **2013**, *381*, 179–186.
64. Hadady, Z.; Tóth, M.; Somsák, L. C-(β-D-Glucopyranosyl) Heterocycles as Potential Glycogen Phosphorylase Inhibitors. *ARKIVOC* **2004**, *2004* (vii), 140–149.
65. Bokor, É.; Docsa, T.; Gergely, P.; Somsák, L. *C*-Glucopyranosyl-1,2,4-Triazoles as New Potent Inhibitors of Glycogen Phosphorylase. *ACS Med. Chem. Lett.* **2013**, *4*, 612–615.

66. Bokor, É.; Széles, Z.; Docsa, T.; Gergely, P.; Somsák, L. *C*-Glucopyranosyl-1,2,4-Triazol-5-Ones: Synthesis and Inhibition of Glycogen Phosphorylase. *Carbohydr. Res.* **2016**, *429*, 128–134.
67. Garcia-Lopez, M.-T.; De Las Heras, F. G.; San Félix, A. Cyanosugars IV. Synthesis of α-D-Glucopyranosyl and α-D-Galactopyranosyl Cyanides and Related 1,2-cis *C*-Glycosides. *J. Carbohydr. Chem.* **1987**, *6*, 273–279.
68. Singh, G.; Vankayalapati, H. Efficient Stereocontrolled Synthesis of *C*-Glycosides Using Glycosyl Donors Substituted by Propane 1,3-Diyl Phosphate as the Leaving Group. *Tetrahedron-Asymmetry* **2001**, *12*, 1727–1735.
69. Sipos, S.; Jablonkai, I. Preparation of 1-C-Glycosyl Aldehydes by Reductive Hydrolysis. *Carbohydr. Res.* **2011**, *346*, 1503–1510.
70. Szennyes, E.; Bokor, É.; Batta, G.; Docsa, T.; Gergely, P.; Somsák, L. Improved Preparation of 4(5)-Aryl-2-(β-D-Glucopyranosyl)-Imidazoles, the Most Efficient Glucose Analogue Inhibitors of Glycogen Phosphorylase. *RSC Adv.* **2016**, *6*, 94787–94794.
71. Szennyes, E.; Bokor, É.; Langer, P.; Gyémánt, G.; Docsa, T.; Sipos, Á.; Somsák, L. The First General Synthesis of 2-*C*-(β-D-Glycopyranosyl)pyrimidines and Their Evaluation as Inhibitors of Some Glycoenzymes. *New J. Chem.* **2018**, *42*, 17439–17446.
72. Kantsadi, A. L.; Bokor, É.; Kun, S.; Stravodimos, G. A.; Chatzileontiadou, D. S. M.; Leonidas, D. D.; Juhász-Tóth, É.; Szakács, A.; Batta, G.; Docsa, T.; Gergely, P.; Somsák, L. Synthetic, Enzyme Kinetic, and Protein Crystallographic Studies of *C*-β-D-Glucopyranosyl Pyrroles and Imidazoles Reveal and Explain Low Nanomolar Inhibition of Human Liver Glycogen Phosphorylase. *Eur. J. Med. Chem.* **2016**, *123*, 737–745.
73. Bokor, É.; Kun, S.; Docsa, T.; Gergely, P.; Somsák, L. 4(5)-Aryl-2-*C*-Glucopyranosyl-Imidazoles as New Nanomolar Glucose Analog Inhibitors of Glycogen Phosphorylase. *ACS Med. Chem. Lett.* **2015**, *6*, 1215–1219.
74. Baker, K. W. J.; Gibb, A.; March, A. R.; Paton, R. M. Generation and Cycloaddition Reactions of Pyranose-1-Carbonitrile Oxides. *Tetrahedron Lett.* **2001**, *42*, 4065–4068.
75. Dondoni, A.; Giovannini, P. P. Formyl *C*-Glycosides as Precursors to Glycosyl Nitrile Oxides and Nitrones. *Synthesis* **2002**, 1701–1706.
76. Dondoni, A.; Giovannini, P. P.; Massi, A. Assembling Heterocycle-Tethered *C*-Glycosyl and α-Amino Acid Residues via 1,3-Dipolar Cycloaddition Reactions. *Org. Lett.* **2004**, *6*, 2929–2932.
77. Kovács, L.; Herczegh, P.; Batta, G.; Farkas, I. Thiazole *C*-Nucleosides III. Synthesis of Pyranose Analogues of Tiazofurin. *Tetrahedron* **1991**, *47*, 5539–5548.
78. Dondoni, A.; Marra, A. Thiazole-Mediated Synthetic Methodology. *Chem. Rev.* **2004**, *104*, 2557–2599.
79. Kang, S. Y.; Song, K. S.; Lee, J.; Lee, S. H.; Lee, J. Synthesis of Pyridazine and Thiazole Analogs as SGLT2 Inhibitors. *Bioorg. Med. Chem.* **2010**, *18*, 6069–6079.
80. Granier, T.; Vasella, A. Synthesis and Evaluation as Glycosidase Inhibitors of 1*H*-Imidazol-2-yl *C*-Glycopyranosides. *Helv. Chim. Acta* **1995**, *78*, 1738–1746.
81. Donnier-Maréchal, M.; Goyard, D.; Folliard, V.; Docsa, T.; Gergely, P.; Praly, J.-P.; Vidal, S. 3-Glucosylated 5-Amino-1,2,4-Oxadiazoles: Synthesis and Evaluation as Glycogen Phosphorylase Inhibitors. *Beilstein J. Org. Chem.* **2015**, *11*, 499–503.
82. Benltifa, M.; Vidal, S.; Gueyrard, D.; Goekjian, P. G.; Msaddek, M.; Praly, J.-P. 1,3-Dipolar Cycloaddition Reactions on Carbohydrate-Based Templates: Synthesis of Spiro-Isoxazolines and 1,2,4-Oxadiazoles as Glycogen Phosphorylase Inhibitors. *Tetrahedron Lett.* **2006**, *47*, 6143–6147.
83. Tóth, M.; Kun, S.; Bokor, É.; Benltifa, M.; Tallec, G.; Vidal, S.; Docsa, T.; Gergely, P.; Somsák, L.; Praly, J.-P. Synthesis and Structure-Activity Relationships of *C*-Glycosylated

Oxadiazoles as Inhibitors of Glycogen Phosphorylase. *Bioorg. Med. Chem.* **2009**, *17*, 4773–4785.

84. Szőcs, B.; Tóth, M.; Docsa, T.; Gergely, P.; Somsák, L. Synthesis of 2-(β-D-Glucopyranosyl)-5-(Substituted-Amino)-1,3,4-Oxa- and -Thiadiazoles for Inhibition of Glycogen Phosphorylase. *Carbohydr. Res.* **2013**, *381*, 187–195.
85. Al-Masoudi, N.; Hassan, N. A.; Al-Soud, Y. A.; Schmidt, P.; Gaafar, A.; Weng, M.; Marino, S.; Schoch, A.; Amer, A.; Jochims, J. C. Syntheses of *C*- and *N*-Nucleosides From 1-Aza-2-Azoniaallene and 1,3-Diaza-2-Azoniaallene Salts. *J. Chem. Soc. Perkin Trans.* **1998**, *1*, 947–953.
86. Al-Masoudi, N. A.; Al-Soud, Y. A.; Ali, I. A. I. Synthesis of 1,2,4-Triazole *C*-Nucleosides From Hydrazonyl Chlorides and Nitriles. *Nucleosides Nucleotides Nucleic Acids* **2007**, *26*, 37–43.
87. Szabó, E. K.; Páhi, A.; Somsák, L. *C*-Glycosyl 1,2,4-Triazoles: Synthesis of the 3-β-D-Glucopyranosyl-1,5-Disubstituted and 5-β-D-Glucopyranosyl-1,3-Disubstituted Variants. *Tetrahedron* **2017**, *73*, 3810–3822.
88. Kun, S.; Begum, J.; Kyriakis, E.; Stamati, E. C. V.; Barkas, T. A.; Szennyes, E.; Bokor, É.; Szabó, E. K.; Stravodimos, G. A.; Sipos, Á.; Docsa, T.; Gergely, P.; Moffatt, C.; Patraskaki, M. S.; Kokolaki, M. S.; Gkerdi, A.; Skamnaki, V. T.; Leonidas, D. D.; Somsák, L.; Hayes, J. M. A Multidisciplinary Study of 3-(β-D-Glucopyranosyl)-5-Substituted-1,2,4-Triazole Derivatives as Glycogen Phosphorylase Inhibitors: Computation, Synthesis, Crystallography and Kinetics Reveal New Potent Inhibitors. *Eur. J. Med. Chem.* **2018**, *147*, 266–278.
89. Szőcs, B.; Bokor, É.; Szabó, K. E.; Kiss-Szikszai, A.; Tóth, M.; Somsák, L. Synthesis of 5-Aryl-3-*C*-Glycosyl- and Unsymmetrical 3,5-Diaryl-1,2,4-Triazoles From Alkylidene-Amidrazones. *RSC Adv.* **2015**, *5*, 43620–43629.
90. Kun, S.; Bokor, É.; Varga, G.; Szőcs, B.; Páhi, A.; Czifrák, K.; Tóth, M.; Juhász, L.; Docsa, T.; Gergely, P.; Somsák, L. New Synthesis of 3-(β-D-Glucopyranosyl)-5-Substituted-1,2,4-Triazoles, Nanomolar Inhibitors of Glycogen Phosphorylase. *Eur. J. Med. Chem.* **2014**, *76*, 567–579.
91. Kantsadi, A. L.; Stravodimos, G. A.; Kyriakis, E.; Chatzileontiadou, D. S.; Solovou, T. G.; Kun, S.; Bokor, É.; Somsák, L.; Leonidas, D. D. van der Waals Interactions Govern *C*-β-D-Glucopyranosyl Triazoles' nM Inhibitory Potency in Human Liver Glycogen Phosphorylase. *J. Struct. Biol.* **2017**, *199*, 57–67.
92. Tóth, M.; Kövér, K. E.; Bényei, A.; Somsák, L. *C*-Glycosylmethylene Carbenes: Synthesis of Anhydro-Aldose Tosylhydrazones as Precursors; Generation and a New Synthetic Route to Exo-Glycals. *Org. Biomol. Chem.* **2003**, *1*, 4039–4046.
93. Tóth, M.; Somsák, L.; Goyard, D. Preparation of 2,6-Anhydro-Aldose-Tosylhydrazones. In *Carbohydrate Chemistry: Proven Synthetic Methods*, Kováč, P., Ed.; CRC Press: Boca Raton, 2012; vol. 1, pp 355–365.
94. Nishikawa, T.; Koide, Y.; Kanakubo, A.; Yoshimura, H.; Isobe, M. Synthesis of β-Analogues of *C*-Mannosyltryptophan, a Novel *C*-Glycosylamino Acid Found in Proteins. *Org. Biomol. Chem.* **2006**, *4*, 1268–1277.
95. Smellie, I. A. S.; Fromm, A.; Fabbiani, F.; Oswald, I. D. H.; White, F. J.; Paton, R. M. Synthesis of 2-Pyranosyl Benzothiazoles, Benzimidazoles and Benzoxazoles via Nucleophilic Addition Reactions of Pyranosyl Nitrile Oxides. *Tetrahedron* **2010**, *66*, 7155–7160.
96. Dondoni, A.; Marra, A.; Mizuno, M.; Giovannini, P. P. Linear Total Synthetic Routes to β-D-*C*-(1,6)-Linked Oligoglucoses and Oligogalactoses up to Pentaoses by Iterative Wittig Olefination Assembly. *J. Org. Chem.* **2002**, *67*, 4186–4199.

97. Vojtech, M.; Petrusova, M.; Slavikova, E.; Bekesova, S.; Petrus, L. One-Pot Synthesis of 2-*C*-Glycosylated Benzimidazoles From the Corresponding Methanal Dimethyl Acetals. *Carbohydr. Res.* **2007**, *342*, 119–123.
98. Kantsadi, A. L.; Hayes, J. M.; Manta, S.; Skamnaki, V. T.; Kiritsis, C.; Psarra, A. M. G.; Koutsogiannis, Z.; Dimopoulou, A.; Theofanous, S.; Nikoleousakos, N.; Zoumpoulakis, P.; Kontou, M.; Papadopoulos, G.; Zographos, S. E.; Komiotis, D.; Leonidas, D. D. The σ-Hole Phenomenon of Halogen Atoms Forms the Structural Basis of the Strong Inhibitory Potency of C5 Halogen Substituted Glucopyranosyl Nucleosides Towards Glycogen Phosphorylase b. *ChemMedChem* **2012**, *7*, 722–732.
99. Mamais, M.; Kouloumoundra, V.; Smyrli, E.; Grammatopoulos, P.; Chrysina, E. D.; Gimisis, T. Synthesis of N^4-Aryl-β-D-Glucopyranosylcytosines: A Methodology Study. *Tetrahedron Lett.* **2015**, *56*, 5549–5552.
100. Mamais, M.; Esposti, A. D.; Kouloumoundra, V.; Gustavsson, T.; Monti, F.; Venturini, A.; Chrysina, E. D.; Markovitsi, D.; Gimisis, T. A New Potent Inhibitor of Glycogen Phosphorylase Reveals the Basicity of the Catalytic Site. *Chem. Eur. J.* **2017**, *23*, 8800–8805.
101. Kantsadi, A. L.; Manta, S.; Psarra, A. M. G.; Dimopoulou, A.; Kiritsis, C.; Parmenopoulou, V.; Skamnaki, V. T.; Zoumpoulakis, P.; Zographos, S. E.; Leonidas, D. D.; Komiotis, D. The Binding of C5-Alkynyl and Alkylfurano[2,3-*d*]pyrimidine Glucopyranonucleosides to Glycogen Phosphorylase b: Synthesis, Biochemical and Biological Assessment. *Eur. J. Med. Chem.* **2012**, *54*, 740–749.
102. Feuillastre, S.; Chajistamatiou, A. S.; Potamitis, C.; Zervou, M.; Zoumpoulakis, P.; Chrysina, E. D.; Praly, J.-P.; Vidal, S. *C*-Glucosylated Malonitrile as a Key Intermediate Towards Carbohydrate-Based Glycogen Phosphorylase Inhibitors. *Bioorg. Med. Chem.* **2012**, *20*, 5592–5599.
103. Goyard, D.; Baron, M.; Skourti, P. V.; Chajistamatiou, A. S.; Docsa, T.; Gergely, P.; Chrysina, E. D.; Praly, J.-P.; Vidal, S. Synthesis of 1,2,3-Triazoles From Xylosyl and 5-Thioxylosyl Azides: Evaluation of the Xylose Scaffold for the Design of Potential Glycogen Phosphorylase Inhibitors. *Carbohydr. Res.* **2012**, *364*, 28–40.
104. Bokor, É.; Kyriakis, E.; Solovou, T. G. A.; Koppány, C.; Kantsadi, A. L.; Szabó, K. E.; Szakács, A.; Stravodimos, G. A.; Docsa, T.; Skamnaki, V. T.; Zographos, S. E.; Gergely, P.; Leonidas, D. D.; Somsák, L. Nanomolar Inhibitors of Glycogen Phosphorylase Based on β-D-Glucosaminyl Heterocycles: A Combined Synthetic, Enzyme Kinetic and Protein Crystallography Study. *J. Med. Chem.* **2017**, *60*, 9251–9262.
105. Somsák, L.; Bokor, É.; Vágvölgyiné Tóth, M.; Juhász, L.; Czifrák, K.; Kónya, B.; Kun, S.; Páhi, A.; Szőcs, B.; Varga, G.; Gergely, P.; Docsa, T.; Kóder, L.; Nagy, K. Preparation of Imidazolyl and Triazolyl Glycosides as Glycogen Phosphorylase Inhibitors and Antitumor Agents. WO2013061105A2, 2013.
106. Somsák, L.; Bokor, É.; Czibere, B.; Czifrák, K.; Koppány, C.; Kulcsár, L.; Kun, S.; Szilágyi, E.; Tóth, M.; Docsa, T.; Gergely, P. Synthesis of *C*-Xylopyranosyl- and Xylopyranosylidene-Spiro-Heterocycles for Inhibition of Glycogen Phosphorylase. *Carbohydr. Res.* **2014**, *399*, 38–48.
107. Bokor, É.; Szennyes, E.; Csupász, T.; Tóth, N.; Docsa, T.; Gergely, P.; Somsák, L. *C*-(2-Deoxy-D-*Arabino*-hex-1-Enopyranosyl)-Oxadiazoles: Synthesis of Possible Isomers and Their Evaluation as Glycogen Phosphorylase Inhibitors. *Carbohydr. Res.* **2015**, *412*, 71–79.
108. Foster, A. B.; Hems, R.; Webber, J. M. Fluorinated Carbohydrates: Part I. 3-Deoxy-3-Fluoro-D-Glucose. *Carbohydr. Chem.* **1967**, *5*, 292–301.

109. Manta, S.; Agelis, G.; Botic, T.; Cencic, A.; Komiotis, D. Fluoro-Ketopyranosyl Nucleosides: Synthesis and Biological Evaluation of 3-Fluoro-2-Keto-β-D-Glucopyranosyl Derivatives of N^4-Benzoyl Cytosine. *Bioorg. Med. Chem.* **2007**, *15*, 980–987.
110. Manta, S.; Agelis, G.; Botic, T.; Cencic, A.; Komiotis, D. Unsaturated Fluoro-Ketopyranosyl Nucleosides: Synthesis and Biological Evaluation of 3-Fluoro-4-Keto-β-D-Glucopyranosyl Derivatives of N^4-Benzoyl Cytosine and N^6-Benzoyl Adenine. *Eur. J. Med. Chem.* **2008**, *43*, 420–428.
111. Tsirkone, V. G.; Tsoukala, E.; Lamprakis, C.; Manta, S.; Hayes, J. M.; Skamnaki, V. T.; Drakou, C.; Zographos, S. E.; Komiotis, D.; Leonidas, D. D. 1-(3-Deoxy-3-Fluoro-β-D-Glucopyranosyl)pyrimidine Derivatives as Inhibitors of Glycogen Phosphorylase b: Kinetic, Crystallographic and Modelling Studies. *Bioorg. Med. Chem.* **2010**, *18*, 3413–3425.
112. Manta, S.; Xipnitou, A.; Kiritsis, C.; Kantsadi, A. L.; Hayes, J. M.; Skamnaki, V. T.; Lamprakis, C.; Kontou, M.; Zoumpoulakis, P.; Zographos, S. E.; Leonidas, D. D.; Komiotis, D. 3'-Axial CH_2OH Substitution on Glucopyranose Does Not Increase Glycogen Phosphorylase Inhibitory Potency. QM/MM-PBSA Calculations Suggest Why. *Chem. Biol. Drug Des.* **2012**, *79*, 663–673.
113. Chrysina, E. D. The Prototype of Glycogen Phosphorylase. *Mini-Rev. Med. Chem.* **2010**, *10*, 1093–1101.
114. Watson, K. A.; Mitchell, E. P.; Johnson, L. N.; Cruciani, G.; Son, J. C.; Bichard, C. J. F.; Fleet, G. W. J.; Oikonomakos, N. G.; Kontou, M.; Zographos, S. E. Glucose Analogue Inhibitors of Glycogen Phosphorylase: From Crystallographic Analysis to Drug Prediction using GRID Force-Field and GOLPE Variable Selection. *Acta Cryst* **1995**, *D51*, 458–472.
115. Somsák, L.; Kovács, L.; Tóth, M.; Ősz, E.; Szilágyi, L.; Györgydeák, Z.; Dinya, Z.; Docsa, T.; Tóth, B.; Gergely, P. Synthesis of and a Comparative Study on the Inhibition of Muscle and Liver Glycogen Phosphorylases by Epimeric Pairs of D-Gluco- and D-Xylopyranosylidene-spiro-(thio)hydantoins and N-(D-Glucopyranosyl) Amides. *J. Med. Chem.* **2001**, *44*, 2843–2848.
116. Györgydeák, Z.; Hadady, Z.; Felföldi, N.; Krakomperger, A.; Nagy, V.; Tóth, M.; Brunyánszky, A.; Docsa, T.; Gergely, P.; Somsák, L. Synthesis of N-(β-D-Glucopyranosyl)- and N-(2-Acetamido-2-Deoxy-β-D-Glucopyranosyl)amides as Inhibitors of Glycogen Phosphorylase. *Bioorg. Med. Chem.* **2004**, *12*, 4861–4870.
117. Chrysina, E. D.; Kosmopolou, M. N.; Tiraidis, C.; Kardarakis, R.; Bischler, N.; Leonidas, D. D.; Hadady, Z.; Somsák, L.; Docsa, T.; Gergely, P.; Oikonomakos, N. G. Kinetic and Crystallographic Studies on 2-(β-D-Glucopyranosyl)-5-Methyl-1,3,4-Oxadiazole, -Benzothiazole, and -Benzimidazole, Inhibitors of Muscle Glycogen Phosphorylase *b*. Evidence for a New Binding Site. *Protein Sci.* **2005**, *14*, 873–888.
118. Anagnostou, E.; Kosmopoulou, M. N.; Chrysina, E. D.; Leonidas, D. D.; Hadjiloi, T.; Tiraidis, C.; Zographos, S. E.; Györgydeák, Z.; Somsák, L.; Docsa, T.; Gergely, P.; Kolisis, F. N.; Oikonomakos, N. G. Crystallographic Studies on Two Bioisosteric Analogues, N-Acetyl-β-D-Glucopyranosylamine and N-Trifluoroacetyl-β-D-Glucopyranosylamine, Potent Inhibitors of Muscle Glycogen Phosphorylase. *Bioorg. Med. Chem.* **2006**, *14*, 181–189.
119. Kyriakis, E.; Solovou, T. G. A.; Kun, S.; Czifrák, K.; Szőcs, B.; Juhász, L.; Bokor, É.; Stravodimos, G. A.; Kantsadi, A. L.; Chatzileontiadou, D. S. M.; Skamnaki, V. T.; Somsák, L.; Leonidas, D. D. Probing the β-Pocket of the Active Site of Human Liver Glycogen Phosphorylase With 3-(C-β-D-Glucopyranosyl)-5-(4-Substituted-Phenyl)-1,2,4-Triazole Inhibitors. *Bioorg. Chem.* **2018**, *77*, 485–493.

120. Oikonomakos, N. G.; Kosmopolou, M.; Zographos, S. E.; Leonidas, D. D.; Somsák, L.; Nagy, V.; Praly, J.-P.; Docsa, T.; Tóth, B.; Gergely, P. Binding of *N*-Acetyl-*N'*-β-D-Glucopyranosyl Urea and *N*-Benzoyl-*N'*-β-D-Glucopyranosyl Urea to Glycogen Phosphorylase *b*: Kinetic and Crystallographic Studies. *Eur. J. Biochem.* **2002**, *269*, 1684–1696.
121. Tóth, M.; Szőcs, B.; Kaszás, T.; Docsa, T.; Gergely, P.; Somsák, L. Synthesis of 2-(β-D-Glucopyranosylamino)-5-Substituted-1,3,4-Oxadiazoles for Inhibition of Glycogen Phosphorylase. *Carbohydr. Res.* **2013**, *381*, 196–204.
122. Kónya, B.; Docsa, T.; Gergely, P.; Somsák, L. Synthesis of Heterocyclic *N*-(β-D-Glucopyranosyl)carboxamides for Inhibition of Glycogen Phosphorylase. *Carbohydr. Res.* **2012**, *351*, 56–63.
123. Polyák, M.; Varga, G.; Szilágyi, B.; Juhász, L.; Docsa, T.; Gergely, P.; Begum, J.; Hayes, J. M.; Somsák, L. Synthesis, Enzyme Kinetics and Computational Evaluation of *N*-β-D-Glucopyranosyl Oxadiazolecarboxamides as Glycogen Phosphorylase Inhibitors. *Bioorg. Med. Chem.* **2013**, *21*, 5738–5747.
124. Begum, J.; Varga, G.; Docsa, T.; Gergely, P.; Hayes, J. M.; Juhász, L.; Somsák, L. Computationally Motivated Synthesis and Enzyme Kinetic Evaluation of *N*-(β-D-Glucopyranosyl)-1,2,4-Triazolecarboxamides as Glycogen Phosphorylase Inhibitors. *Med. Chem. Commun.* **2015**, *6*, 80–89.
125. Johnson, L. N.; Acharya, K. R.; Jordan, M. D.; McLaughlin, P. J. Refined Crystal Structure of the Phosphorylase-Heptulose 2-Phosphate-Oligosaccharide-AMP Complex. *J. Mol. Biol.* **1990**, *211*, 645–661.
126. Barr, D.; Szennyes, E.; Bokor, É.; Z.H. Al-Oanzi; Moffatt, C.; Kun, S.; Docsa, T.; Sipos, Á.; Davies, M. P.; Mathomes, R. T.; Snape, T. J.; Agius, L.; Somsák, L.; Hayes, J. M. Identification of *C*-β-D-Glucopyranosyl Azole-Type Inhibitors of Glycogen Phosphorylase That Reduce Glycogenolysis in Hepatocytes: In Silico Design, Synthesis, In Vitro Kinetics, and Ex-Vivo Studies. *ACS Chem. Biol.* **2019**, *14*, 1460–1470.

Recent developments in synthetic methods for sugar phosphate analogs

8

Andreia Fortuna[a,b], *Nuno M. Xavier*[a,b]
[a]Centro de Química e Bioquímica, Faculdade de Ciências, Universidade de Lisboa, Lisboa, Portugal, [b]Centro de Química Estrutural, Faculdade de Ciências, Universidade de Lisboa, Lisboa, Portugal

Chapter outline

1 Introduction

Sugar phosphates play relevant roles in various biological events. Metabolic pathways such as glycolysis and the pentose phosphate pathway use sugar phosphates as intermediates or products. Glucose 6-phosphate is metabolized in both pathways, namely to fructose 6-phosphate, intermediate in glycolysis, and to ribose 5-phosphate, the product of the pentose phosphate pathway, via various sugar phosphate intermediates. Ribose 5-phosphate is the precursor in the biosynthesis of ribonucleotides and 2-deoxyribonucleotides, which are the building blocks for nucleic acids and also serve as energy sources for cell processes [adenosine triphosphates (ATPs) and guanosine triphosphates (GTP)] and are involved in signaling pathways (ATP and GTP). ATP is also a substrate for kinases enabling the phosphorylation of biomolecules. Glycosyl phosphates (sugar 1-phosphates) are intermediates in the biosynthesis of nucleoside diphosphate sugars, which are substrates of glycosyltransferases acting as glycosyl donors for the biosynthesis of glycoconjugates and glycans. Bacteria use sugar phosphates as precursors in the biosynthesis of important components of their cell wall. In particular, the first cytoplasmic step of peptidoglycan synthesis, which is the biosynthesis of UDP-GlcNAc (uridine diphosphate *N*-acetylglucosamine), has

Recent Trends in Carbohydrate Chemistry. https://doi.org/10.1016/B978-0-12-817467-8.00008-6
© 2020 Elsevier Inc. All rights reserved.

fructose 6-phosphate as precursor.[1] In the biosynthesis of lipopolysaccharide (LPS), an essential structure of the outer membrane of Gram-negative bacteria, D-ribulose 5-phosphate, which is then converted into arabinose 5-phosphate, and sedoheptulose 7-phosphate are the biosynthetic precursors of 3-deoxy-D-*manno*-oct-2-ulosonate (KDO),[2] and heptoses,[3, 4] respectively, that constitute the LPS inner core region. In the case of mycobacteria, the arabinose 1-phosphate derivative arabinofuranosyl-1-monophosphoryldecaprenol is the key biosynthetic intermediate in the assembly of the arabinan portions of arabinogalactan and lipoarabinomannan, major components of the mycobacterial cell wall.[5] Glycosyl phosphate units are also found in phosphoglycans, which are structures present in the cell wall or in capsular polysaccharides of bacteria and yeasts, as well as in the cell surface glycocalix and secreted glycoproteins of protozoan parasites, namely of the genus *Leishmania*.[6, 7]

Due to biological relevance of sugar phosphates and that of biomolecules containing sugar phosphate moieties, namely nucleotides, the synthesis of sugar phosphates and of analogs/mimetics of these compounds has attracted considerable attention in carbohydrate and nucleotide chemistry. In particular, the access to sugar phosphate analogs comprising relatively stable phosphorous or phosphoryl-containing surrogate moieties for a phosphate group, which may mimic its size, polarity, charge distribution, or ability for interactions with protein targets, has been a focus of intensive research. These types of molecules have been especially developed to interfere with biological pathways in which sugar phosphates or molecules comprising sugar phosphate moieties are involved, within the context of a potential therapeutic application. In the case of the synthesis of nucleotide mimetics, the installation of surrogate groups for a phosphate system has also been exploited leading to various molecules of significant biological profile.[8]

Methodologies for the synthesis of glycosyl phosphates were covered in excellent reviews, devoted to biologically relevant anomeric aldose and aldulosonic acid phosphates[9] and to phosphoglycan-containing phosphodiester-linked disaccharides.[6, 7, 10] Methods for the mono-, di-, or triphosphorylation of sugar moieties at a nonanomeric position were also reviewed in the context of nucleotide synthesis.[11] Concerning sugar phosphate mimetics, early synthetic strategies for *C*-glycosyl methylphosphonates were surveyed.[12]

In this chapter, emphasis is given to methodologies for the synthesis of sugar phosphate analogs, namely sugar phosphonates, including imino sugar-based molecules, thio-modified phosphonates, sugar boranophosphates, and sugar derivatives comprising endocyclic P=O groups, covering mostly methods reported in the last 20 years.

2 Sugar phosphonates

Sugar phosphonates, in which the phosphono group is linked by the phosphorous atom to a sugar carbon atom or the phosphonate moiety is linked to the sugar through a C–C bond, involving the phosphonate carbon atom, have been exploited as potential mimetics of sugar phosphates. In the case of the replacement of an oxygen phosphate atom that is linked to the sugar by a carbon atom, the size or the geometry of the phosphorous-containing moiety is not significantly changed; therefore, the phosphonate analogs may be accommodated in the active sites of proteins that recognize the parent phosphates. Moreover, in these analogs, the phosphoester moiety cannot be hydrolytically cleaved off from the sugar moiety, contrarily to the labile phosphate group. Herein, synthetic methods toward molecules that comprise a phosphono group P-linked or linked through one or two carbon atoms to the anomeric and to a nonanomeric position of a sugar moiety are focused.

2.1 Anomeric sugar phosphonates

In this section, sugar derivatives in which the anomeric carbon is linked to a phosphono phosphorous atom (glycosylphosphonates) and those phosphonate derivatives in which the phosphonate moiety is C-C linked to the anomeric position, are covered.

Early syntheses of glycosylphosphonates include the reaction of per-*O*-acetylated glycals with dimethyl hydrogen phosphonate catalyzed by boron trifluoride, which occurred with allylic rearrangement leading to (2-enopyranosyl)phosphonates,[13] and the treatment of per-*O*-benzylated glycosyl acetates with trialkyl phosphites in the presence of trimethylsilyl triflate, which affords 1,2-*cis* glycosylphosphonates.[14] For the synthesis of (2-azidoglycosyl)phosphonates, per-*O*-acetylated and per-*O*-benzylated 2-azidoglycosyl trichloroacetimidates were shown to be the suitable glycosyl donors for analogous phosphonoylation reaction with trimethyl phosphite.[15]

Another efficient method for glycosylphosphonates is based on Michael-type addition of dimethyl phosphite to 2-nitroglycals (**1**) in the presence of potassium *tert*-butoxide (Scheme 1).[16] A mixture of anomers **2** and **3** (α/β ratio, 1:1.2) was obtained after 5 min at 0°C. However, after 2h at the same temperature, only the β-glycosyl phosphonate **3** was obtained in 60% yield via anomerization of **2** to the more thermodynamically stable β-anomer under the basic conditions of the reaction. Similar findings were observed starting from a 2-nitro-L-rhamnal derivative. Using a 2-nitroglucal, a mixture of 2-nitro α-mannosyl and β-glucosyl phosphonate were initially obtained, evolving toward the β-glucosyl phosphonate with a prolonged reaction time. Compounds **2** and **3** could be converted into the corresponding (2-acetamidogalactosyl)phosphonates by the reduction of the nitro group (platinized Raney nickel) and subsequent acetylation.

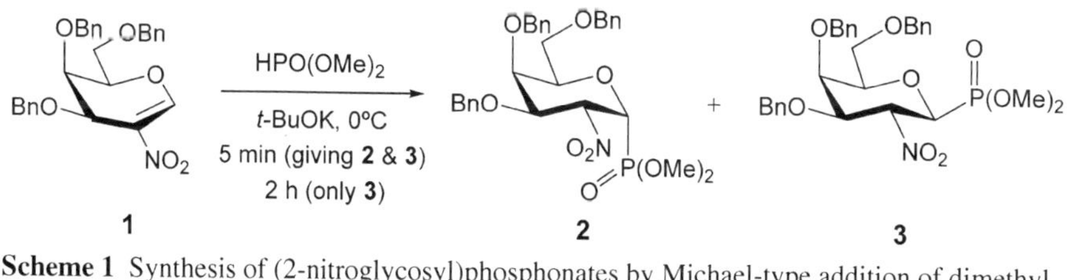

Scheme 1 Synthesis of (2-nitroglycosyl)phosphonates by Michael-type addition of dimethyl phosphite to a 2-nitroglycal.

The synthesis of *C*-glycosyl analogs (glycosylalkyl analogs) of glycosylphosphates has attracted increasing attention due to ability of the phosphonate group to act as a bioisostere of the phosphate moiety[17] and due to the metabolic stability of this moiety and that of C-C linked glycosyl derivatives.

A diversity of methodologies using various types of activated sugar precursors have been described for compounds in which the phosphonate carbon atom is linked to the anomeric position, e.g. in (glycosylmethyl)phosphonates (*C*-glycosyl methylphosphonates). A general early synthetic strategy for these types of molecules consists of a multistep procedure involving the formation of a glycosylmethyl halide and its subsequent conversion into a phosphonate via Michaelis-Arbuzov reaction with a trialkyl phosphite.[12]

This method was used in the context of obtaining a phosphonate analog of uridine diphosphate (UDP)-D-apiose.[18] Thus, *O*-benzyl protected (α-D-apiosyl)methyl bromide (**4**), prepared in nine steps from 1,2-*O*-isopropylidenexylofuranose, was reacted with trimethyl phosphite under reflux affording the corresponding methylphosphonate derivative (**5**) in good yield (72%, Scheme 2). This procedure was also successfully applied for tri-*O*-benzylated pyranosylmethyl iodides of D-*xylo* and D-*gluco* configuration, furnishing the desired (glycosylmethyl)phosphonates in good moderate to good yields (56%–77%).[18, 19]

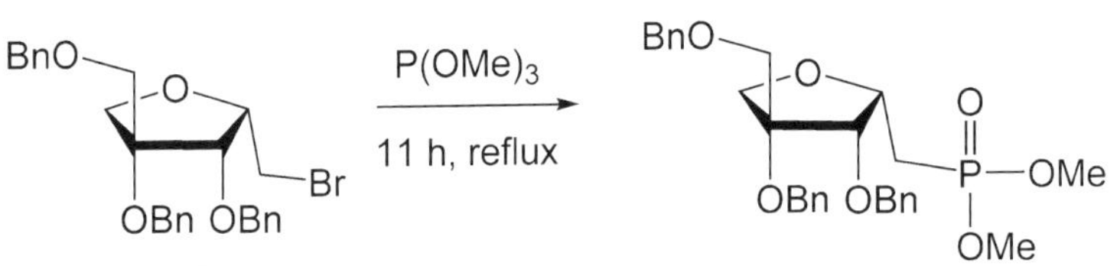

4 5

Scheme 2 Synthesis of a dimethyl (glycosylmethyl)phosphonate derivative by Michaelis-Arbuzov reaction.

Anomeric aldehydes can be useful precursors for (glycosylmethyl)phosphonates. In an early report, a tri-*O*-benzylated 2-deoxy-2-formyl fructofuranose derivative was converted into the corresponding *C*-(fructos-2-yl)methylphosphonate by Abramov reaction with diphenyl phosphite followed by deoxygenation of the intermediate α-hydroxyphosphonate.[20]

A related method was reported for the synthesis of a C1-phosphonomethyl derivative of *N*-acetylglucosamine (GlcNAc, Scheme 3) as a key step in a route toward a C-1-phosphonate analog of UDP-GlcNAc.[21] Hence, the 1-*C*-formyl GlcNAc derivative **6**, obtained from GlcNAc via anomeric C-allylation, double-bond migration and ozonolysis, was treated with the anion of dimethyl phosphite to give the α-hydroxyphosphonate derivative **7** in high diasteromeric excess (90%) and moderate yield (53%). Further deoxygenation via conversion into the methyl oxalyl ester followed by radical reduction provided the target phosphonate (**8**) in good yield (66%).

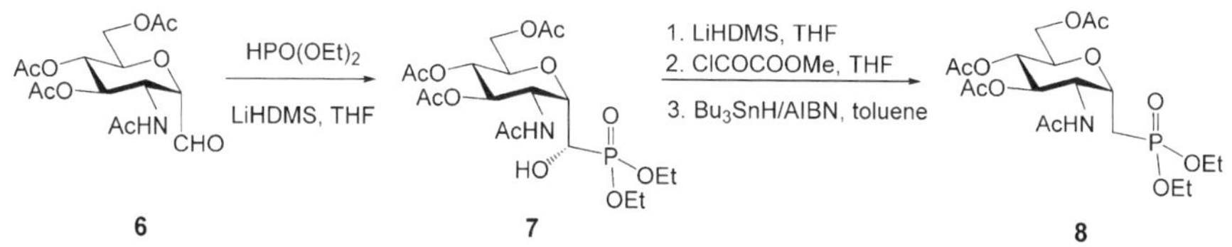

Scheme 3 Synthesis of a C-1-phosphonate analog of GlcNAc from an anomeric aldehyde.

Another practical synthetic method for *C*-glycosyl phosphonates is based on free-radical glycosylation reactions involving glycosyl radicals and vinylphosphonates.[22, 23] Starting from peracetylated glycosyl bromides (**9**), the corresponding radicals (**10**) were generated by using light or azobisisobutyronitrile (AIBN) in dry ether in the presence of tributyltin hydride or tris(trimethylsilyl)silane (TTMS) or by irradiation in the presence of TTMS as a radical source (Scheme 4). Their further addition to vinylphosphonates afforded (glycosylethyl)phosphonates **11** in moderate to good yields (17%–80%) and high stereoselectivity to α-anomers (in most cases, with α/β ratios of 98:2). This procedure was applied to aldopentoses (D-*xylo*), aldohexoses (D-*gluco*, D-*galacto* and D-*manno*), deoxy sugars (L-*fuco*), and to a disaccharide [Glc(α1→4)Glc], methyl glucuronate, and *N*-trifluoroacetylglucosamine derivatives. The method also worked reasonably well for perbenzoylated xylofuranose and different vinylphosphonates (R^2=H, COOMe, R^3=Et, Me), conducting to the desired phosphonates in moderate yields (27%–30%).

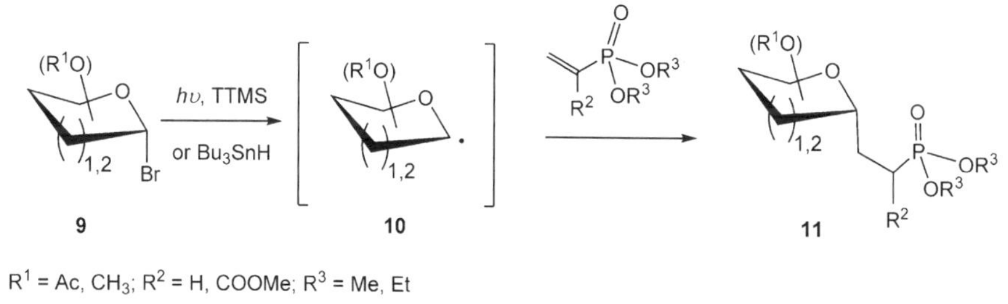

Scheme 4 Diastereoselective synthesis of (glycosylethyl)phosphonates by coupling between glycosyl radicals and dialkyl vinylphosphonates.

Some synthetic approaches toward (glycosylmethyl)phosphonates using aldonolactones have been reported. One method was based on the treatment of per-*O*-benzyl glucuronolactone **12** with lithiated dialkyl or dibenzyl methylphosphonates to afford [(1-hydroxy-β-D-glycosyl)methyl]phosphonates (Scheme 5).[24] With dibenzyl- and dimethyl lithiomethylphosphonates, formed in situ from dibenzyl and dimethyl methylphosphonates in the presence of butyl lithium, phosphonate derivatives **13** were obtained in very good yields (83%–86%).

α-Fluorination in the phosphonate moiety may lead to phosphonates that are better mimics of the phosphate group due to the increase in the acidity of the system as well as it provides hydrogen bonding-accepting ability. The α,α-difluoromethylene phosphonate system is a more effective "isopolar" and "isoelectronic" phosphate mimic, with acidity that is close to the parent phosphate.[25] Therefore, the synthesis of α-fluoromethylphosphonate analogs of sugar phosphates has been particularly explored.

A α,α-difluorinated analog of compound **13**, the [difluoro(1-hydroxyglucosyl) methyl]phosphonate **14**, could be obtained through the treatment of aldonolactone **12** with diethyl (difluoromethyl)phosphonate in presence of lithium diisopropylamide (LDA) at low temperature (− 78°C) in moderate yield.

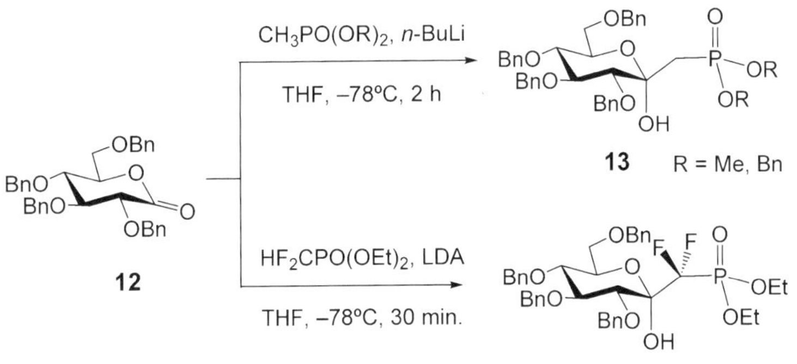

Scheme 5 Synthesis of [(1-hydroxy-β-D-glycosyl)methyl]phosphonates from per-*O*-benzyl gluconolactone.

The phosphonate derivative **13** can be strategically functionalized to different phosphonate analogs (Scheme 6).[24] Thus, treating **13** with diethylaminosulfur trifluoride (DAST) afforded the anomeric fluorinated derivative (**16**) in good yield (59%). The elimination product was also obtained (**15**) in moderate yield (38%) and could be converted into the corresponding monofluorinated derivatives by addition of an electrophilic fluorinating reagent (Selectfluor) in the presence of water as a nucleophile, thus creating chirality at the phosphonate α-carbon (C-1). [(*S*)-fluoro(1-hydroxyglucosyl)methyl]phosphonate **17** and its (*R*)-disastereoisomer **18** were obtained in yields of 18% and 54%, respectively. Moreover, treatment of the exoglycal **15** with *m*-chloroperoxybenzoic acid (*m*-CPBA) in dichloromethane and subsequent aqueous workup lead to the dihydroxyphosphonate derivative **19**. Deprotected derivatives of **13** and **17**, obtained by catalytic hydrogenation (Pd/C) followed by neutralization with NH_4OH, revealed to be substrates of a thymidylyltransferase, being completely enzymatically converted into analogs of deoxythimidine diphosphate (dTDP) sugars, as potential inhibitors of glycosyltransferases.

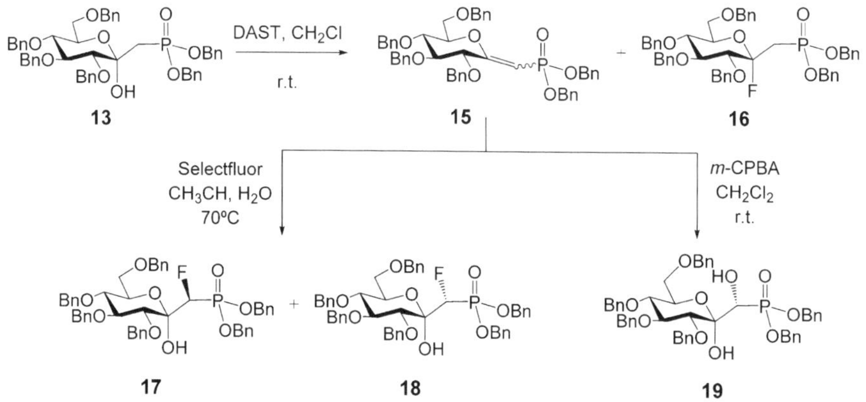

Scheme 6 Derivatization of a [(1-hydroxyglucosyl)methyl] phosphonate into a [(1-fluoroglucosyl)methyl]phosphonate, [(*S*)-/(*R*)-fluoro(1-hydroxyglucosyl)methyl] phosphonates and a [(*R*)-hydroxy(1-hydroxyglucosyl)methyl]phosphonate derivative.

Another methodology for [difluoro(glycosyl)methyl]phosphonates using an aldonolactone precursor involved its conversion into a *gem*-difluoro *exo*-glycal intermediate and subsequent phosphonyl radical addition (Scheme 7).[26] Thus, Wittig-type reaction of protected galactono-1,4-lactone (**20**) with the difluorinated ylide generated from dibromodifluoromethane and tris(dimethylamino)phosphane led to difluoroolefin **21**. Further treatment of **21** with diethyl phosphite and *tert*-butylperoxypivalate afforded the desired phoshonate **22** in moderate yield (56%).

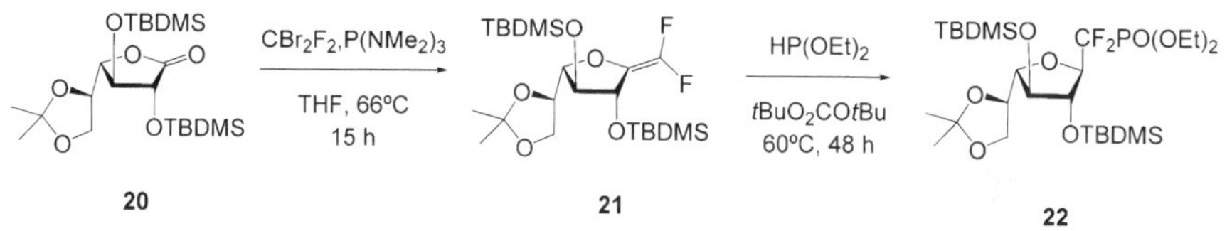

Scheme 7 Synthesis of a [difluoro(glycosyl)methyl]phosphonate from an aldonolactone *via* phosphonyl radical addition to an anomeric difluoroolefin intermediate.

The conjugate addition of lithiated (difluoromethyl)phosphonates [$LiCF_2PO(OR)_2$] to 2-nitroglycals (**23,** Scheme 8) is an efficient and general approach to obtain [difluoro(glycosyl)methyl]phosphonates **24**.[27] The method was shown to be extensible to 2-nitroglycals derived from D-glucose and D-galactose having different *O*-protecting group (**23**, R^1 = benzyl, methyl, silyl), excluding acetyl groups due to their lability toward the basic lithiated phosphonate, and to glycals obtained from L-rhamnose and L-arabinose, leading to the target compounds in moderate to excellent yields (up to 98%). Similar to the previously mentioned addition of dimethyl phosphite, a high diastereoselectivity toward 1,2-diequatorial *C*-(2-nitroglycosyl) derivatives (β-anomers) is achieved via anomerization of α-anomers through ring opening/closure. However, the introduction of rigidity in the glycal, by 4,6-*O*-di-*tert*-butylsilylene protection, creating a *trans*-fused bicyclic system, improves α-selectivity.

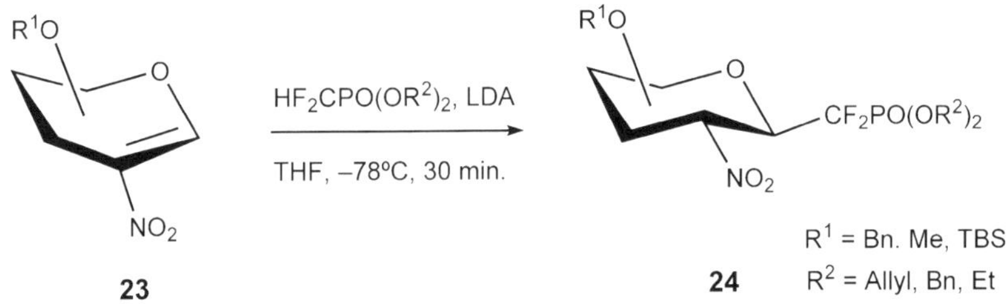

Scheme 8 Stereoselective synthesis of [difluoro(nitroglycosyl)methyl]phosphonate by conjugate addition of lithiated (difluoromethyl)phosphonates to 2-nitroglycals.

(Glycosylalkyl)phosphate mimics based on an imino sugar scaffold and comprising a difluoromethylenephosphonate moiety have been proposed as probes or inhibitors for several biological processes involving glycosylphosphates and nucleotide sugars. A method for their synthesis employs glycosylamines as key precursors, which can be obtained by condensation of protected aldoses with amines. Hence, treatment of *N*-benzyl D-arabinosyl-, D-ribosyl-, and L-xylosylamines (**25,** Scheme 9) with lithiated diethyl difluoromethylphosphonate [$LiCF_2PO(OEt)_2$] furnished a mixture of diastereomers **26** and **27**, with selectivity for the *threo* product (*threo*/*erythro* ratios from 70:30 to 85:15).[28] Mesylation of these compounds enabled an intramolecular nucleophilic displacement by attack of the amino group to form the desired [difluoro(iminoglycosyl)methyl]phosphonates (**28**, **29**) in good to excellent yields (60%–98%).

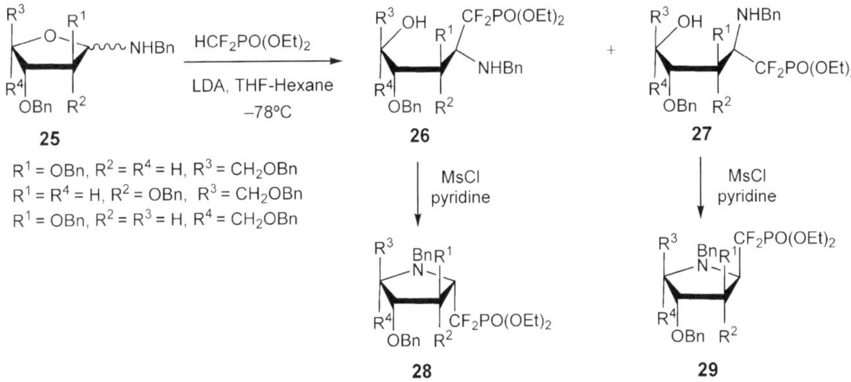

Scheme 9 Synthesis of diethyl [difluoro(iminoglycosyl)methyl]phosphonates from furanosylamines.

Another route to obtain this type of compounds consists of the diastereoselective addition of (difluoromethyl)phosphonate-based organometallic reagents to *N*-glycosyl-*S*-*tert*-butyl sulfinamides (Scheme 10).[29] These *N*-glycosyl precursors (**31**) could be obtained in good to excellent yields by *N*-glycosylation of *tert*-butanesulfinamide with *O*-protected glycofuranose or pyranose reducing sugars (**30**), namely tri-*O*-benzylated D-*arabino*/D-*xylo*-furanoses and D-xylopyranose as well as 5-*O*-silylated

2,3-*O*-isopropylidene-D-ribofuranose, using titanium ethoxide or cesium carbonate as promoter. Treatment of glycosyl sulfinamides **31** with lithiated (difluoromethyl)phosphonate or with the Grignard analog phosphono(difluoromethyl)magnesium bromide [$BrMgCF_2PO(OEt)_2$], generated by reacting diethyl bromodifluoromethylphosphonate with isopropylmagnesium chloride, afforded compounds **32** in yields ranging from 44% to 88% and modest to high 1*R* *vs* 1*S* diastereoselectivities (6:4 to 10:0).

The desired iminoglycosyl derivatives (**33**) were then accessed by mesylation followed by treatment with base, enabling intramolecular cyclization, and subsequent removal of the sulfinyl group under mild acidic conditions. In alternative, in order to access the *xylo*-configured derivative **33** the sulfinyl group was removed first and further cyclization occurred from the free amino group. The target compounds were obtained in low to good yields (10%–68%). The method allows a tuning of the pseudo-anomeric configuration of the [difluoro(iminoglycosyl)methyl]phosphonate based on the configuration at the sulfinyl sulfur atom in the *N*- glycosylsulfinamine precursor.

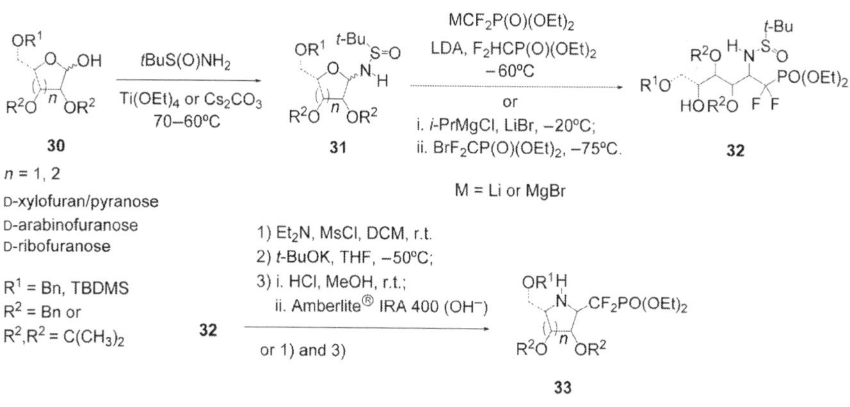

Scheme 10 Synthesis of [difluoro(iminoglycosyl)methyl]phosphonates by addition of $LiCF_2P(O)(OEt)_2$ and $BrMgCF_2P(O)(OEt)_2$ to *N*-*tert*-butanesulfinylglycosylamines.

An efficient synthesis of a nojirimycin-based glycosylphosphonate was reported (Scheme 11).[30] The key precursor was the L-sorbose-derived imine **34**, whose synthesis was accomplished in seven steps from 2,3;4,6-di-*O*-isopropylidene-α-L-sorbofuranose, involving a protection-deprotection strategy to achieve a partially deprotected derivative comprising a free hydroxy group at C6, further oxidation to the corresponding aldehyde and condensation with benzylamine. The imine **34** was then subjected to addition by phosphorous-containing nucleophiles to install a phosphono group P-linked to C-6. In one path, diethyl phosphite was reacted with trimethylsilyl chloride to give the corresponding trimethylsilyloxy P(III) derivative, which underwent addition to the imine functionality of **34** to give the α-amino phosphonate **35**, as a single diastereoisomer (6*S*) and in excellent yield (91%). On the other hand, treating **34** with diethyl phosphite in the presence of zirconium(IV) chloride led to the epimeric 6*R* α-amino phosphonate **36** in good yield (76%) and with a diastereomeric excess (de) of 85%. Compound **35** was then subjected to one pot deprotection of the benzyl ether and acetal groups and intramolecular reductive amination using hydrogen over Pd/C catalyst in aqueous trifluoroacetic acid, leading to compound **37** in 70% yield with a de of 80%. However, the conversion of epimeric **36** into deoxynojirimycin-α-1-phosphonate *via* a similar deprotection strategy proved to be ineffective since it furnished a mixture of products.

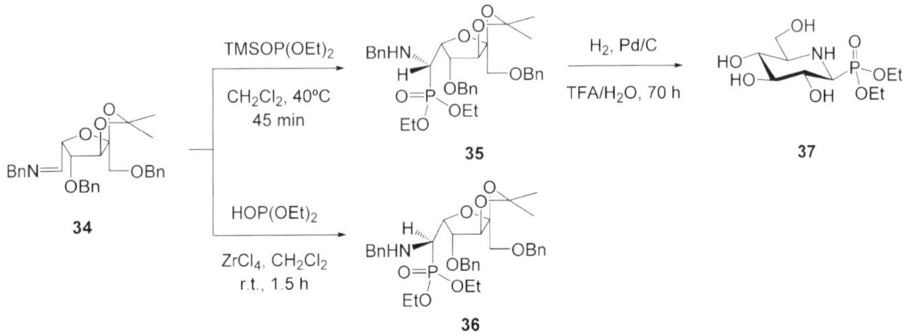

Scheme 11 Synthesis of a nojirimycin-based (iminoglycosyl)phosphonate from a L-sorbose-derived imine.

2.2 Nonanomeric sugar phosphonates

The synthesis of sugar phosphonates in which a phosphonate moiety is connected to the sugar unit at a nonanomeric position has been explored since these compounds can be precursors of nucleoside phosphonate derivatives as compounds of therapeutic potential mimicking nucleotide monophosphates. Thus, several synthetic methods have been described for their preparation and one of the most versatile ones is the Michaelis-Arbuzov reaction between halosugars and a trialkyl phosphite, as previously mentioned. This method presents some drawbacks such as the need for high temperatures, a long reaction time, which can be problematic to sensitive compounds and promote side reactions, and the removal of excess trialkylphosphite reagent. One strategy explored in order to overcome these problems is the use of microwave (MW) irradiation (Scheme 12). Nevertheless, when this procedure was applied to 6-iodo-D-*allo*- (**38**) and D-glucofuranose derivatives, the conditions that allowed the reaction to proceed were even more drastic than those used for the conventional Michaelis-Arbuzov reaction itself, with the desired sugar phosphonate (such as **39**) being formed only when a irradiation power of 300 W and temperatures above 180°C were employed.[31] In the case of a sensitive substrate such as 5′-deoxy-5′-iodo-2′,3′-*O*-isopropylidene uridine, thermal degradation occurred in these conditions. Despite this, the use of MW comparatively to conventional heating allowed a significant decrease in the amount of triethyl phosphite used in some cases (25 to 5 equivalents), as well as a decrease in the reaction time to 30 or 90 min instead of a few days. The method was shown to be compatible with the presence of various *O*-substituents such as isopropylidene, benzyl, and benzoyl groups, except acetyl groups, and proceeded well in substrates comprising vicinal iodine and hydroxy groups.[32] However, this method led to a slight decrease in the yield of the obtained phosphonate derivatives (from 60%–77% to 50%–70% using MW).

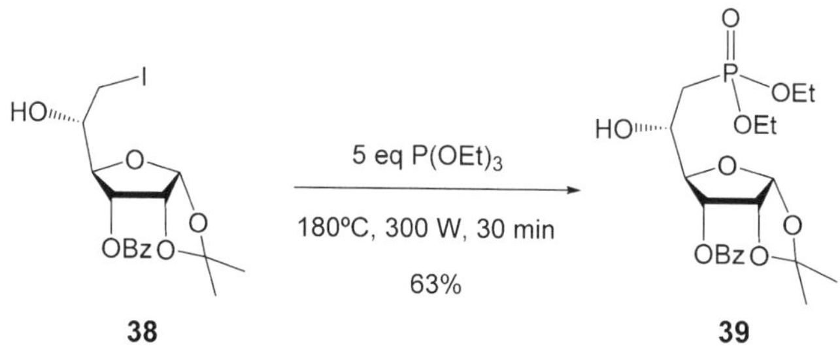

Scheme 12 Synthesis of a sugar phosphonate *via* Michaelis-Arbozuv reaction between a iodosugar and triethyl phosphite using microwave irradiation.

Protected alduronic acids can be functionalized at the carboxylic group for the introduction of a phosphonate moiety. In an early report, a 2,3-*O*-isopropylidene riburonic acid derivative (**40**) was converted into a uronic ester of *N*-hydroxy-2-thiopyridone (**41**), *via* reaction of *N*-hydroxy-2-thiopyridone in the presence of *N*-methylmorpholine[33] (Scheme 13). Photolytic addition of this intermediate to diethyl vinylphosphonate furnished the α-(piridinylthio)phosphonate **42** as a diastereomeric mixture in 65% yield, the reduction of which with tributyltin hydride gave the (ribohexofuranosid-6-yl)phosphonate **43**.

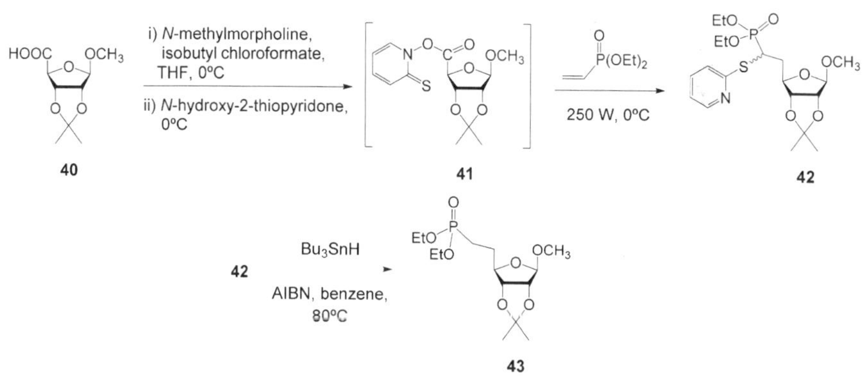

Scheme 13 Synthesis of a sugar phosphonate derivative from riburonic acid via photolysis of its 2-thioxopyridin-1-yl ester intermediate in the presence of vinyl phosphonate.

Conversion of uronic acid derivatives to the corresponding esters is also a suitable strategy for providing a useful precursor for further installation of a phosphonate moiety. Hence, the methyl glucuronate derivative **44**, which can be prepared by reaction of methyl 2,3,4-tri-*O*-benzyl-D-glucosiduronic acid with diazomethane, was converted into the sugar phosphonate **45** by reaction with (dimethylphosphonomethyl)lithium (Scheme 14) in an excellent yield (91%).[34]

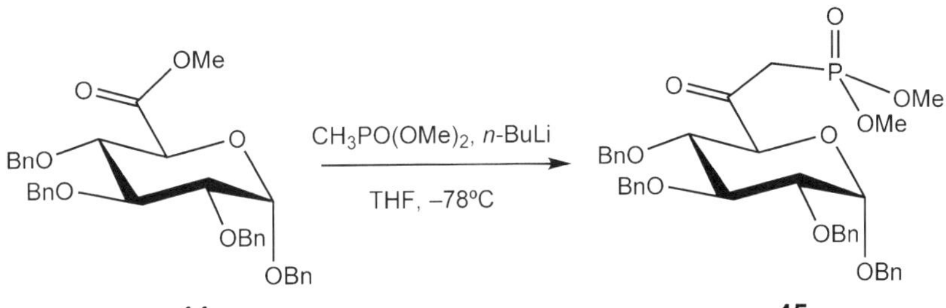

Scheme 14 Synthesis of a sugar phosphonate derivative by reaction of 2,3,4-tri-*O*-benzyl-D-glucosiduronic acid ester with dimethyl methylphosphonate.

Sugar triflates can be converted into phosphonates via nucleophilic displacement. Thus, treatment of the triflate β-D-glucopyranosides **46** with (diethylphosphonomethyl)lithium at low temperature (–78°C, Scheme 15), led to the corresponding diethyl (2-glucosylethyl)phosphonates **47** in good yields (72%–74%).[35, 36] On the other hand, by using [diethylphosphono(difluoromethyl)] lithium, the fluorinated derivatives **48** were obtained in moderate to good yields (50%–83%).

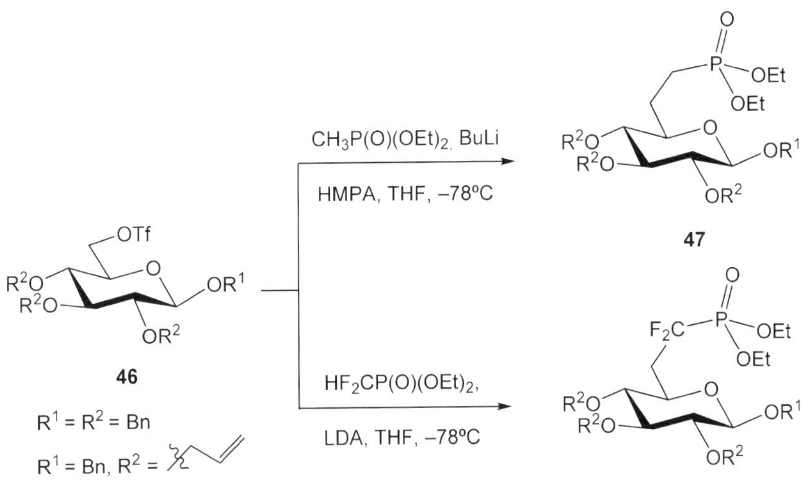

Scheme 15 Synthesis of (2-glucosylethyl)phosphonates and their fluorinated analogs from sugar triflates.

These procedures also worked for *O*-protected triflate derivatives of D-ribofuranose, D-glucofuranose, and α-D-mannopyranose with isopropylidene, allyl, or methyl groups, affording the corresponding diethyl (glycosylethyl)phosphonates in good yields (70%–75%) and the α,α-difluorophosphonate analogs in moderate yields (47%–52%).[35]

Similar to anomeric aldehydes (Scheme **3**), sugar derivatives comprising a formyl group at a nonanomeric position can be converted into α-hydroxyphosphonates through the addition of an anion of a dialkyl phosphite, as demonstrated by reacting the benzylated methyl 6-deoxy-β-D-*gluco*-heptodialdo-1,5-pyranoside **49** with diethyl phosphite in the presence of lithium bis(trimethylsilyl)amide at − 78°C (Scheme 16). A diastereoisomeric mixture of α-hydroxyphosphonate derivatives (**50**) was obtained in excellent yield (93%), with slight preference for the 7(*R*)-derivative (ratio *R*/*S* of 1.4:1).[36] Further treatment of the 7(*R*)/7(*S*) diastereoisomeric mixture (**50**) with (diethylamino)sulfur trifluoride provided stereoselectively the 7(*S*) D-*gluco*-heptopyranoside-derived α-fluorophosphonate (**51**, ratio *S*/*R* of 10:1) in modest yield (32%). This diastereoisomer could be epimerized at C-7 via deprotonation with LDA followed by quenching with acetic acid, leading to an equimolar mixture of 7(*R*)/7(*S*) diastereoisomers (**51**).

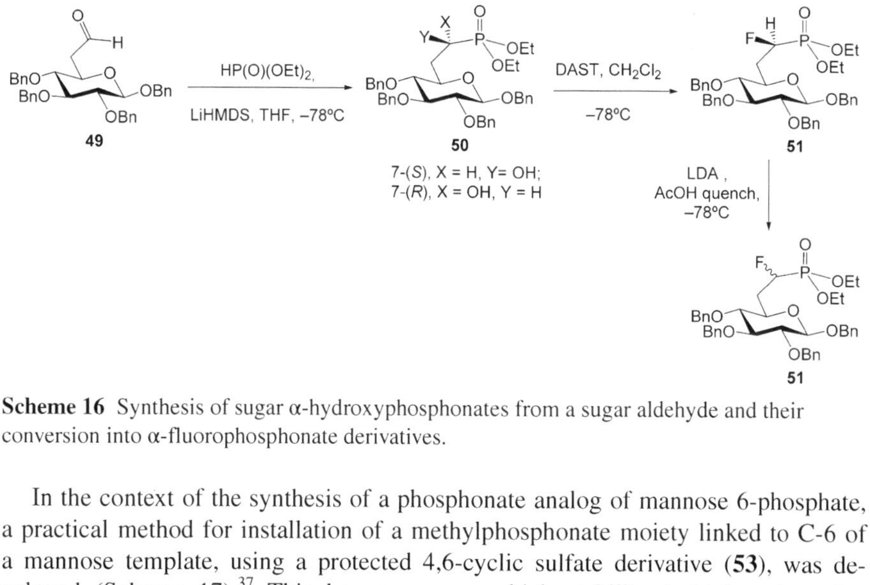

Scheme 16 Synthesis of sugar α-hydroxyphosphonates from a sugar aldehyde and their conversion into α-fluorophosphonate derivatives.

In the context of the synthesis of a phosphonate analog of mannose 6-phosphate, a practical method for installation of a methylphosphonate moiety linked to C-6 of a mannose template, using a protected 4,6-cyclic sulfate derivative (**53**), was developed (Scheme 17).[37] This key precursor, which stability is higher than that of a mannose 6-triflate derivative, could be prepared by the treatment of methyl 2,3-*O*-isopropylidene-α-D-mannopyranoside with thionyl chloride followed by oxidation of the intermediate cyclic sulfite ($NaIO_4/RuCl_3.H_2O$).[38] The cyclic sulfate **53** underwent nucleophilic replacement at C-6 with lithiated dialkyl methylphosphonates to afford the corresponding phosphonate derivatives **54**. The reaction occurred better when using lithiated diisopropyl methylphosphonate (R = *i*Pr) than with lithiated diethyl and dimethyl methylphosphonates (R = Et, Me), due to the higher stability of this reagent. The mannopyranose-derived phosphonates **54** were subsequently subjected to sugar deprotection, using Amberlyst-15 (H^+) resin in MeOH/THF, and to further *O*-dealkylation of the phosphonate moiety.

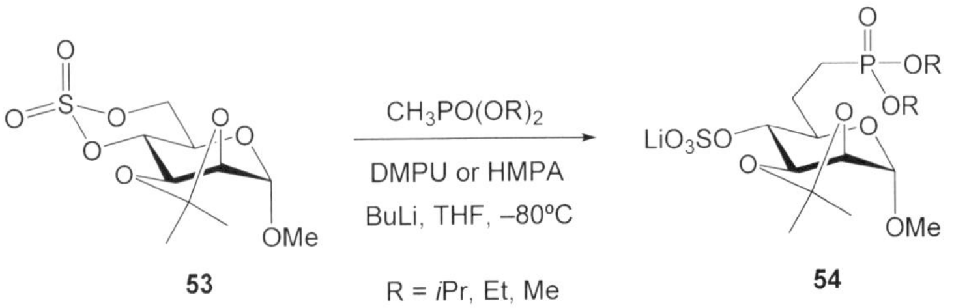

Scheme 17 Synthesis of a protected phosphonate derivative of methyl mannopyranoside via a 4,6-cyclic sulfate precursor.

Sugar phosphonates can also be obtained from sugar *gem*-difluoroalkenes via α,α-difluoro-*H*-phosphinate intermediates.[39] The radical addition of sodium hypophosphite to geminal difluoroalkenes, including furanose derivatives, in the presence of *tert*-butyl peroxypivalate (TBPP) as a radical initiator led to α,α-difluoro-*H*-phosphinates in yields ranging from 75% to quantitative. In the case of *O*-protected C-3-*exo*-difluoromethylene furanose derivative **55**, the method conduced quantitatively to α,α-difluoro-*H*-phosphinate **56** (Scheme 18). Its further treatment with trimethylsilyl chloride (TMSCl) in pyridine in the presence of oxygen led to difluorophosphonate **57** (X=O) in 86% yield. Alternatively, the exposure of the reaction medium to elemental sulfur furnished the difluorophosphonothioate **57 (X = S)** in 50% yield.

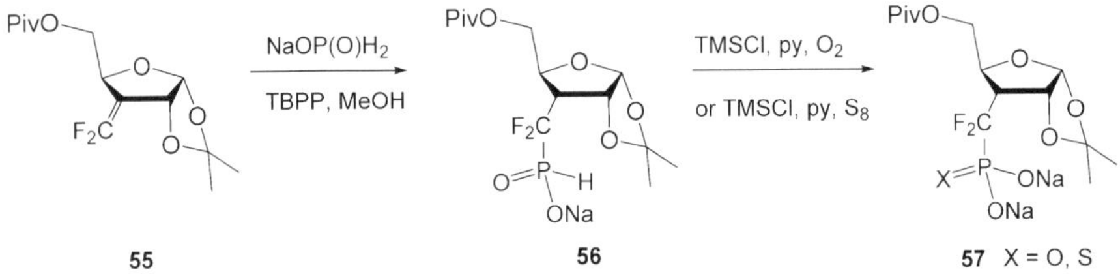

Scheme 18 Transformation of a ribofuranose-C-3-derived α,α-difluoro-*H*-phosphinate into a corresponding difluorophosphonate (X=O) and difluorophosphonothioate (X=S).

3 Glycosyl boranophosphates

Glycosyl boranophosphates are *P*-modified glycosyl phosphate analogs in which the P=O moiety is replaced by a P→BH_3 group. The boranophosphate moiety was initially developed as chemically and enzymatically stable surrogate of the intersaccharidic phosphodiester bridge in oligonucleotides.[40] In addition, boranophosphate diesters have been explored in the synthesis of dinucleoside *H*-phosphonate analogs due to their ability to be quantitatively converted into the corresponding *H*-phosphonate diesters, demonstrating that the boranophosphate moiety can be used as a stable masked analog of the relatively labile *H*-phosphonate diester as well as a protecting group throughout the synthesis.[41, 42] The isoelectronic relationship of the boranophosphate moiety with a phosphate group has motivated the development of methodologies for their installation at the anomeric position of a carbohydrate derivative, leading to rather stable glycosyl phosphate analogs as well as to useful precursors for other types of phosphoryl-containing glycosyl derivatives.

One method for their synthesis involves the boranophosphorylation of reducing sugars (**58**) with triethylammonium dimethyl boranophosphate and bis(2-oxo-3-oxazolidinyl)phosphonic chloride (Bop-Cl) as condensing agent, in the presence of 3-nitro-1,2,4-triazole (NT) as catalyst and diisopropylethylamine (i-Pr_2NEt) (Scheme 19).[43] This method was applied to various *O*-protected reducing pyranoses such as D-glucose, D-mannose, D-galactose, D-glucosamine, and glucuronic acid derivatives leading to the desired glycosyl boranophosphate triesters **59** as mixtures of α and β-anomers in good to excellent yields (56%–91%).

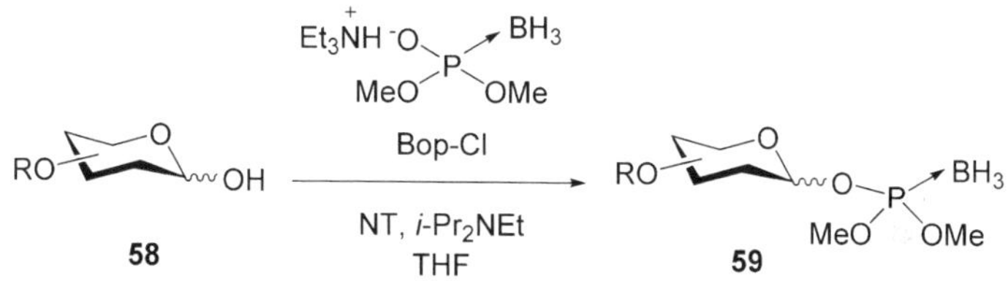

58

R = Bn, Ac

59

Scheme 19 Boranophosphorylation of reducing sugars with triethylammonium dimethyl boranophosphate leading to glycosyl boranophosphates.

Glycosyl halides can be easily converted into the corresponding glycosyl boranophosphates by reaction with dimethyl boranophosphate monopotassium salt (Scheme 20), which can be obtained by hydrolysis of trimethyl boranophosphates with potassium hydroxide.[44] Thus, reaction of **60** with the mentioned potassium boranophosphate salt afforded the desired glycosyl boranophosphate derivative (**61**) in good yield (71%). The method also worked well for peracetylated α-D-glucopyranosyl bromide, furnishing the corresponding β-D-glucopyranosyl boranophosphate derivative in good yield (82%).

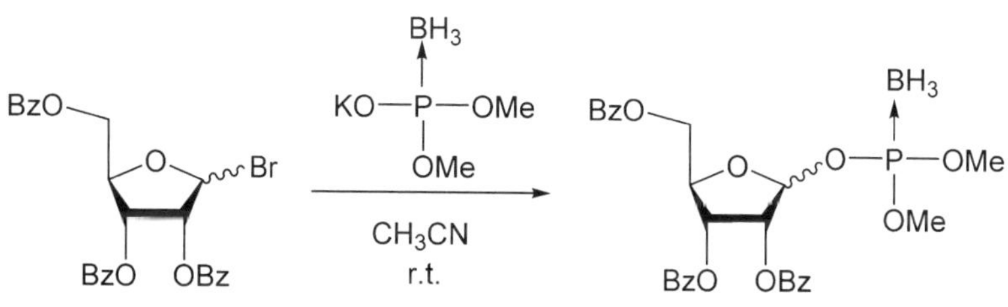

60

61

Scheme 20 Synthesis of a glycosyl boranophosphate derivative by reaction of a ribofuranosyl bromide with boranophosphate monopotassium salt.

Another efficient method for the synthesis of glycosyl boranophosphates is based on the reaction of glycosyl phosphites with a borane (Scheme 21). Thus, treatment of per-*O*-benzyl-α-D-glycosyl iodides (**62**) with dimethyl *H*-phosphonate in the presence of 1,8-bis(diethylamino)naphthalene (DMAN) afforded the glycosyl phosphite intermediates **63** with high stereoselectivity for α-anomers (α/β ratios from 75:25 to 93:7).[45] Further treatment of phosphites **63** with borane tetrahydrofuran complex (BF_3·THF) afforded glycosyl boraphosphates (**64**) in good to excellent overall yields (from 62% to 93%, from **62**).

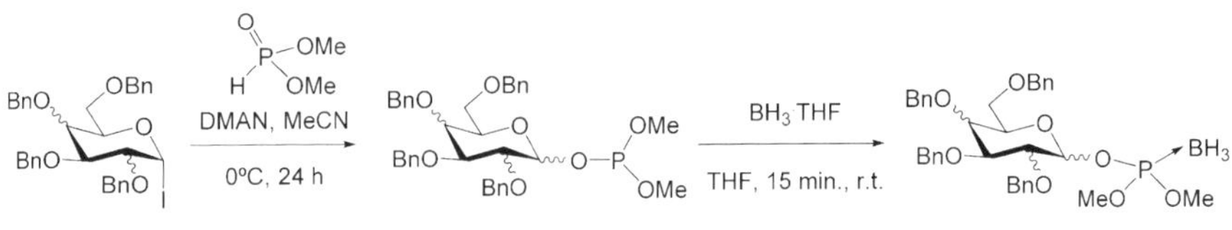

62　　**63**　　**64**

D-*gluco*, D-*manno*, D-*galacto*

Scheme 21 Synthesis of glycosyl phosphites from glycosyl iodides *via* conversion into glycosyl phoshites and further treatment with BH_3.THF.

Glycosyl boranophosphates are versatile precursors for a variety of interesting and potentially bioactive compounds as shown in Scheme 22. Their ability to act as glycosyl donors for the glycosylation of alcohols in the presence of a trityl cation (Tr^+)-containing activator, namely trityl *N*,*N*-bis(trifluoromethanesulfonic)imidate, or by using *O*-tritylated alcohols as glycosyl acceptors was demonstrated.[46] The latter method was shown to work well for perbenzylated/benzoylated glycosyl boranophosphonates of D-*gluco*, D-*manno*, and D-*galacto* configurations. For instance, treatment of tetra-*O*-benzyl/benzoyl glycosyl boranophosphotriesters (**65**) with methyl tri-*O*-benzyl-6-*O*-tritylglucoside in the presence of triflic acid afforded the corresponding disaccharides (**66**) in good yields (56%–73%). The use of *O*-tritylcholesterol as a glycosyl acceptor led to cholesteryl D-glycopyranosides (**67**) in good yields (71%–83%).

As mentioned before, glycosyl boranophosphates can be easily converted to *H*-phosphonate diesters, namely upon addition of a Tr^+ cation. These compounds are versatile precursors for glycosyl phosphates as well as to a number of *P*-substituted phosphate analogs such as phosphoramidates, phosphorothioates, alkylphosphonates, phosphoroselenoates, and phosphorofluoridates.[47] Thus, treatment of **65** (R = Ac) with 1,4-diazabicyclo[2.2.2]-octane (DABCO) enabled the demethylation of the boranophosphate triester moiety and the resulting diester was reacted with methyl trityl ether (TrOMe)/dichloroacetic acid (DCA), which generated Tr^+ cation in situ and allowed obtaining quantitatively the desired glycosyl *H*-phosphonate diester **68** as a mixture of two *P*-diastereomers.[43]

Another application of glycosyl boranophosphates is in the synthesis of glycosyl phosphate triesters that have been extensively used as highly reactive glycosyl donors.[48, 49] For example, per-*O*-acetyl and per-*O*-benzyl glucosyl boranophosphate triesters **65** were converted in the corresponding glycosyl phosphates (**69**) by oxidation with *meta*-chloroperoxybenzoic acid (*m*-CPBA) in excellent yield (91%), with retention of the anomeric configuration.[43]

Additionally, these compounds can be transesterified with alcohols leading to analogs of glycosyl phosphate-containing biomolecules, as demonstrated by the synthesis of citronellyl glucosyl boranophosphate (**70**), a mimetic of a glycosyl phosphate donor in the eukaryotic *N*-glycosylation pathway.[43] This compound was obtained from the per-*O*-acetylglycosyl boranophosphate triester **65** via demethylation and subsequent condensation with (*S*)-*β*-citronellol, which occurred almost quantitatively. It was then subjected to full deprotection, via demethylation of the boranophosphate moiety and *O*-deacetylation of the sugar ring.

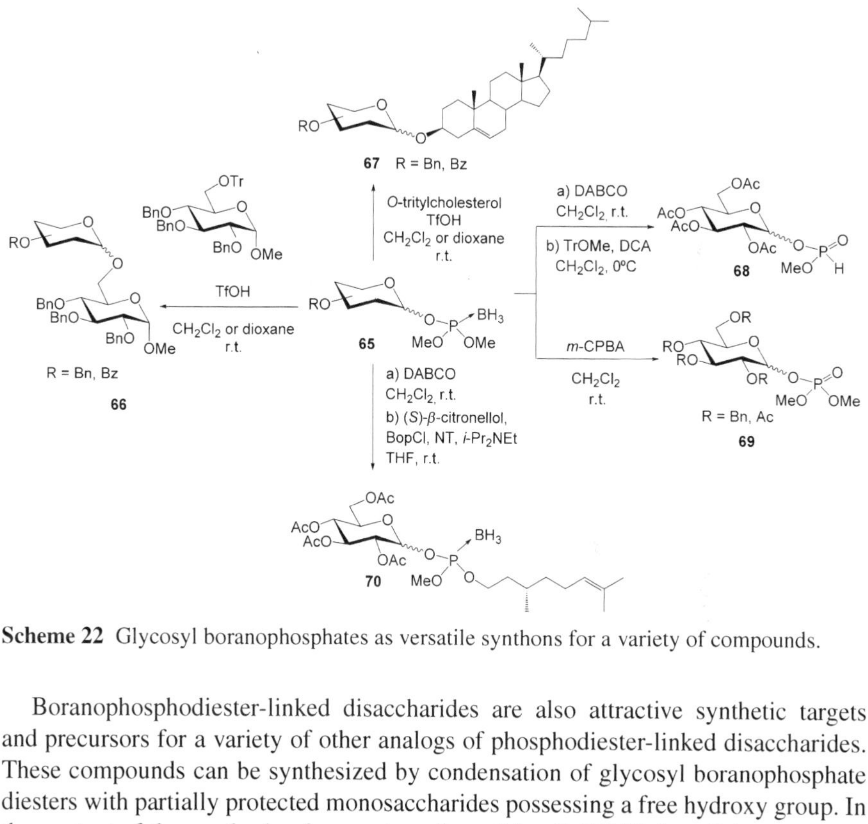

Scheme 22 Glycosyl boranophosphates as versatile synthons for a variety of compounds.

Boranophosphodiester-linked disaccharides are also attractive synthetic targets and precursors for a variety of other analogs of phosphodiester-linked disaccharides. These compounds can be synthesized by condensation of glycosyl boranophosphate diesters with partially protected monosaccharides possessing a free hydroxy group. In the context of the synthesis of an analog of α-D-Man (1 PO_3H-6)-D-Gal, a repeating unit of *Leishmania* surface lipophosphoglycans, the per-*O*-benzyl mannosyl iodide **71** was stereoselectively converted into boranophosphate diester **72** by reaction with potassium dimethyl boranophosphate and subsequent demethylation of the resulting glycosyl boranophosphonate triester with thiophenol and triethylamine (Scheme 23).[50] Condensation of **72** with partially benzoylated methyl galactoside by using 3-nitro-1,2,4-triazol-1-yl-tris(pyrrolidin-1-yl)phosphonium hexafluorophosphate (PyNTP) and further demethylation afforded the boranophosphodiester-linked disaccharide (**73**) in a very good yield (88%).

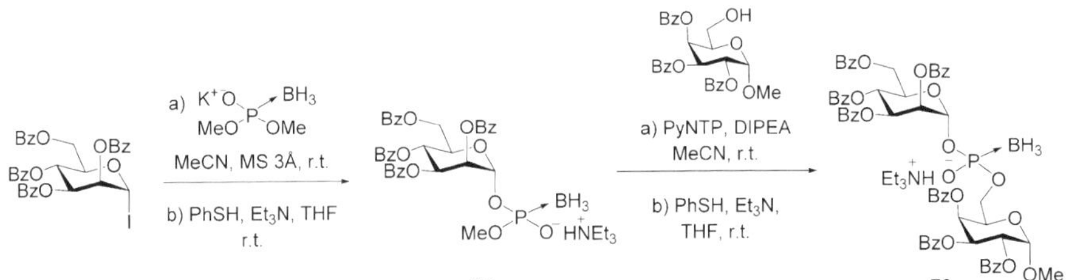

Scheme 23 Synthesis of a glycosyl boranophosphate diester from a glycosyl iodide and its further conversion into a boranophosphodiester-linked disaccharide.

The boranophosphate moiety of pseudo disaccharides such as **73** allows further conversion into various P-modified analogs, and for that efficient one-pot methods were developed[50, 51] (Scheme 24). Hence, treatment of **73** with 4,4′-dimethoxytrityl ($DMTr^+$) cation furnished the corresponding *H*-phosphonate diester in 97% yield.[50] A phosphotriester derivative (**74**) could be obtained from **73** through reaction with

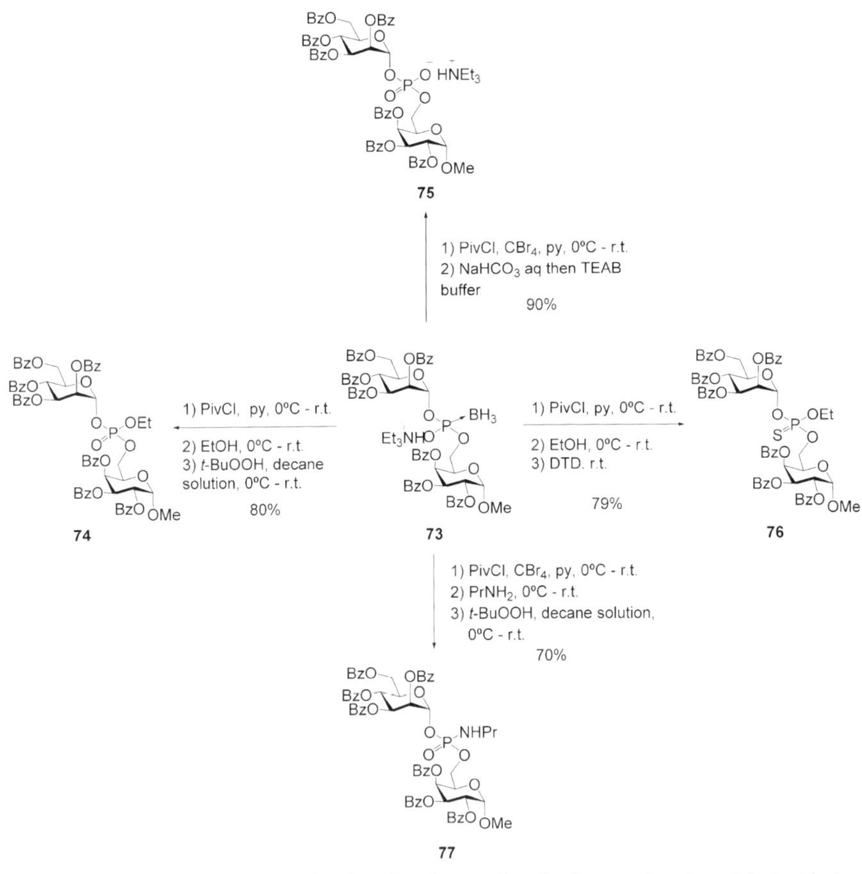

Scheme 24 One-spot syntheses of a phosphotriester, phosphodiester, phosphorothioate triester, and phosphoramidate-linked disaccharides from a boranophosphodiester-linked disaccharide.

pivaloyl chloride (PivCl) in pyridine, which generates an acyl phosphite intermediate, followed by treatment with ethanol to form a phosphite triester, and subsequent oxidation with *tert*-butyl hydroperoxide (*t*BuOOH). Treatment of **73** with PivCl and carbon tetrabromide in pyridine and further addition of a saturated sodium carbonate aqueous solution furnished the glycosyl phosphodiester **75**. Moreover, a phosphorothioate triester analog (**76**) could be obtained from **73** via a phosphite triester intermediate as for the synthesis of **74**, which underwent sulfurization with dimethylthiuram disulfide (DTD). In alternative, addition of propylamine to the acyl phosphite intermediate formed upon treatment of **73** with PivCl, to form a phosphoramidite, followed by oxidation with *t*BuOOH, furnished phosphoramidate **77**.

4 Glycosyl thiophosphates and thiophosphonates

Sugar thiophosphates (or phosphorothioates) constitute another group of isosteric and rather stable analogs of sugar phosphates, in which a bridging or nonbridging phosphate oxygen atom is replaced by sulfur. These compounds have been focused as potential inhibitors of carbohydrate-acting enzymes[52] and as constituents of oligonucleotide analogs of therapeutic interest since the internucleotide phosphorothioate moiety is relatively resistant to nucleases.[53–55]

Methodologies for glycosyl thiophosphates, in which a thiophosphate moiety is *S*-or *O*-linked to the anomeric position of a sugar moiety, use reducing monosaccharide derivatives or glycosylthiols as precursors. The synthesis of *O*-glycosyl thiophosphates by reaction of tetra-*O*-benzyl-D-gluco/mannopyranose and tetra-*O*-acetyl-D-glucopyranose (**78**) with dimethyl chlorothiophosphate (R^2=Me) in the presence of *n*-butyllithium at low temperature (−78°C) yields the corresponding *O*-glycosyl thiophosphates (**79**, X=O) in 51% to 86% yield and stereoselectivity to α-anomers (Scheme 25).[56] These compounds demonstrated their ability to serve as glycosyl donors for the glycosylation of partially deprotected monosaccharide derivatives and a steroid. In the case of the reactions between tetra-*O*-acetyl- and tetra-*O*-benzyl-D-glucopyranose and diethyl chlorothiophosphate (R^2=Et), the expected *O*-glycosyl thiophosphates (**79**) were obtained in good yields (57%–67%) with major formation of the β-anomer in the case of the *O*-acetylated derivative due to the neighboring group effect, while the stereoselectivity was reversed using the *O*-benzylated glycosyl donor, being driven by the anomeric effect.[57]

Treatment of tetra-*O*-benzyl-D-glucopyranose with diphenyl chlorothiophosphate (R^2=Ph) under similar conditions led initially to the *O*-glycosyl thiophosphate (**79,** X=O, Y=S) in modest yield (16%). Performing the reaction in the presence of 4-dimethylaminopyridine (DMAP) led exclusively to the α-anomer in 24% yield. However, in both cases, a complete isomerization of the *O*-glucopyranosyl thiophosphate to the *S*-linked derivative (X=S, Y=O), with an anomeric α/β ratio of 4:1, occurred.

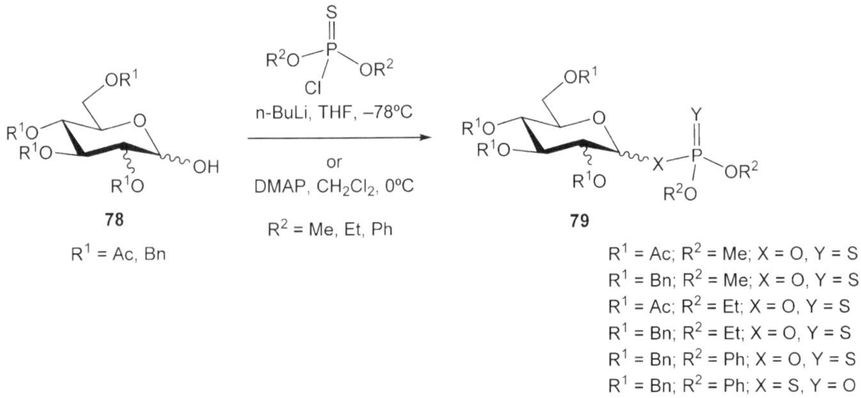

Scheme 25 Synthesis of protected α/β-D-glucopyranosyl thiophosphates by reaction of tetra-*O*-acetyl/benzyl protected monosaccharides with dialkyl chlorothiophosphate or diphenyl chlorothiophosphate.

Another method for glycosyl thiophosphates employs the Mitsunobu reaction between a monothiophosphoric acid diester and a reducing sugar (Scheme 26).[52] Treating tetra-*O*-acetyl mannopyranose (**80**) with *O*-allyl *O*-oleyl *O*-hydrogen thiophosphate in the presence of triphenylphosphane and diisopropyl azodicarboxylate (DIAD) in THF gave *S*-β- (**81**) and *O*-α/β-mannopyranosyl (**82**, α/β ratio, 2:1) thiophosphates in 35% and 25% yields, respectively, as equimolar mixtures of P-diastereomers (Scheme 28). Applying the Mitsunobu conditions to *O*-acetylated and *O*-benzylated manopyranoses and using a di-*O*-ethyl dithiophosphate, allowed obtaining *S*-mannopyranosyl dithiophosphates in yields ranging from 76% to 97%.

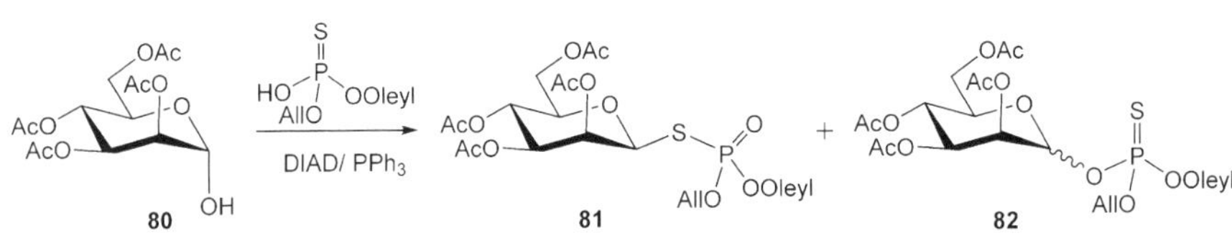

Scheme 26 Synthesis of protected *S*- and *O*-mannopyranosyl thiophosphates by Mitsunobu coupling between tetra-*O*-acetylmannopyranose and a thiophosphoric acid diester.

S-Glycosyl thiophosphates can be obtained through phosphitylation of glycosylthiols followed by oxidation. Thus, reaction of tetra-*O*-acetyl-1-thio-β-D-mannopyranose with 2-cyanoethyl oleyl *N*,*N*-diisopropylphosphoramidite, followed by treatment with *t*BuOOH, afforded the corresponding *S*-mannosyl mono- and dithiophosphates **84** (R^1 = Oleyl) and **85** in 26% and 20% yields (Scheme 27).[52] The use of dolichyl phosphoramidite led only to the monothiophosphate **84** (R^1 = Dolichyl) in moderate yield (45%).

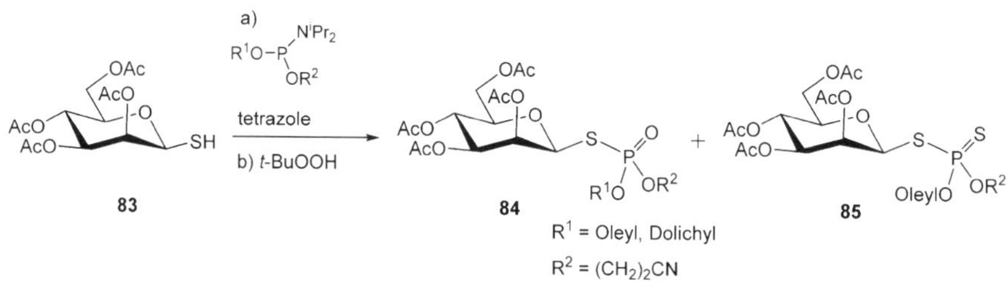

Scheme 27 Synthesis of *S*-mannopyranosyl thiophosphates by phosphitylation of tetra-*O*-acetyl-1-thio-β-D-mannopyranose followed by oxidation.

Glycosyl thiocyanates are also useful precursors for the synthesis of *S*-glycosyl thiophosphates and thiophosphonates, through Michaelis-Arbuzov-type reaction.[58] The reaction proceeds via nucleophilic attack by a phosphite phosphorous center on the thiocyanate sulfur atom, followed by cyanide-induced dealkylation of the resulting trialkoxy(glycosylthio)phosphonium salt, leading therefore to an *O,O*-dialkyl *S*-glycosyl thiophosphate and an alkyl cyanide. The method was successfully applied to perbenzoylated glycosyl thiocyanates (**86**, Scheme 28) such as β-D-glucopyranosyl, β-D-galactopyranosyl, and β-D-mannopyranosyl derivatives, leading to the target *S*-glycosyl thiophosphates (**87**, $R^1=R^2=$OEt) in moderate yields (24%–51%) when using triethylphosphite at high temperature (80–120°C). When an activated phosphorus(III) reagent such as diethyl trimethylsilyl phosphite ($R^1=R^2=$OEt, $R^3=$OTMS) was used, the reactions could be carried out with satisfactory conversion at significantly lower temperatures (room temperature or 50°C) and the desired compounds (**87**, $R^1=R^2=$OEt) were obtained in good yields (68%–82%). The scope of the methodology was also extended to the synthesis of *S*-glycosyl thiophosphonates and thiophosphinates. By using dimethyl phenylphosphonite ($R^1=R^2=$OMe, $R^3=$Ph), *S*-glycosyl thiophosphonates were obtained as a mixture of P-diastereoisomers (**87**, $R^1=$OMe, $R^2=$Ph) in moderate yields (41%–53%). The yields of these types of compounds were increased, when using

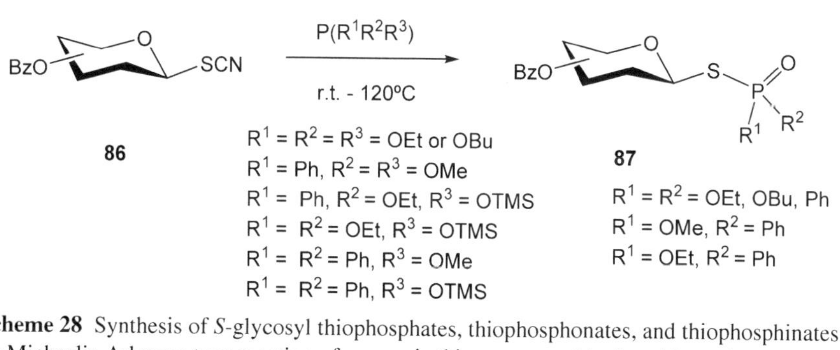

Scheme 28 Synthesis of *S*-glycosyl thiophosphates, thiophosphonates, and thiophosphinates by Michaelis-Arbuzov-type reaction of anomeric thiocyanates with phosphites, phosphonites, and phosphinites, respectively.

a silylated phenylphosphonite derivative ($R^1 = Ph$, $R^2 = OEt$, $R^3 = OTMS$), reaching 94% in the case of a *gluco*-configured derivative. When methyl diphenylphosphinite ($R^1 = R^2 = Ph$, $R^3 = OMe$) was used, the corresponding *S*-glycosyl thiophosphinates (**87**, $R^1 = R^2 = Ph$) were obtained, although in modest yields (20%–25%), which could be improved (74%–82%) by reacting the glycosyl thiocyanates with the corresponding *O*-silylated phosphinite derivative ($R^1 = R^2 = Ph$, $R^3 = OTMS$) at room temperature.

(Glycosylmethyl)thiophosphonates as noncleavable isosteric analogs of glycosylphosphates containing a sulfur atom, can be obtained from (glycosylmethyl) thiophosphates through reaction with the Lawesson's reagent, as shown in Scheme 29. Hence, the replacement of the phosphoryl oxygen atom of per-*O*-benzylated (α-D-glucosylmethyl)phosphonate (**88**) by a sulfur atom, leading to the corresponding thiophosphonate (**89**) occurred by means of Lawesson's reagent in good yield (72%).[57]

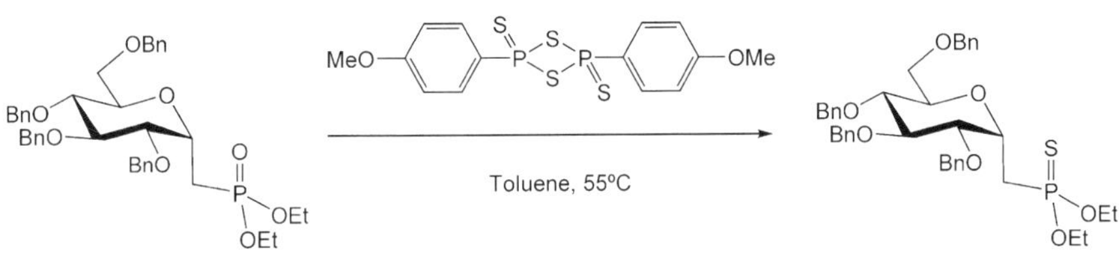

88 **89**

Scheme 29 Synthesis of a (glucosylmethyl)thiophosphonate by treatment of a (glycosylmethyl)phosphonate with the Lawesson's reagent.

5 Glycophostones

In this section, monosaccharide derivatives comprising phosphonate and phosphonamidate moieties embedded in the sugar ring are mentioned.

An interesting group of compounds named as glycophostones comprises a pentavalent phosphorus atom replacing the sugar anomeric carbon. These compounds are glycosyl versions of cyclic phosphonates and innovative anomerically modified glycomimetics. Glycophostones can be synthesized from *O*-benzylated glycals via oxidative cleavage followed by Abramov reaction in order to introduce a dialkylphosphonyl moiety (Scheme 30).[59, 60] In one example, the protected L-fucal (**90**) was subjected to ozonolysis and the obtained dicarbonyl intermediate underwent reaction at the former C1 formyl group with trimethyl phosphite in acetic acid leading to L-*fuco* and L-*talo*-configured dimethyl phosphonates (**91–92**) in a high overall yield (85%) and a 3:2 ratio, respectively.[60] Treatment of the L-*fuco*-configured derivative **91** with sodium methoxide led to P_R and P_S L-*fuco*-configured cyclic methyl phosphonates **93** and **94** in a ratio of 3:2, respectively, and an excellent yield (95%). A similar approach also allowed obtaining a cyclic phosphonate analog of *N*-acetyl neuraminic acid.

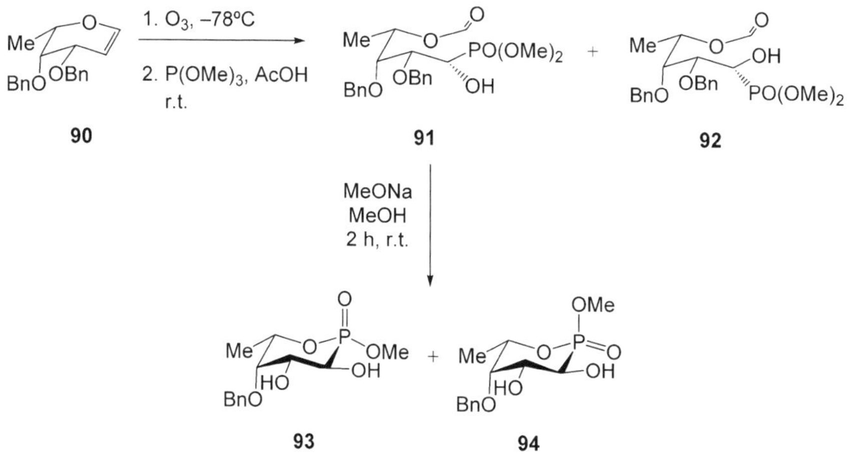

Scheme 30 Synthesis of glycophostones by oxidative cleavage of an *O*-benzylated glycal followed by Abromov reaction with trimethyl phosphite and cyclization.

In order to obtain 2-*N*-acetylglycosamine-based phostones, another methodology based on the phosphitylation of an intermediate glyco-*O*-benzyloxime was reported.[60] Hence, treatment of 2,3,5-tri-*O*-benzyl-D-arabinose (**95**) with trimethyl phosphite yielded quantitatively the oxime **96**. Further addition of trimethyl phosphite in acetic acid and spontaneous cyclization furnished a mixture of P_R and P_S D-*gluco* and D-*manno*-configured phosphonate ester diastereoisomers (**97**) in good yield (75%). *N*-Acetyl-D-gluco/D-mannosamine phostones were subsequently obtained via a sequence involving acetylation, methyl phosphonate ester hydrolysis, and benzyl group removal (Scheme 31).

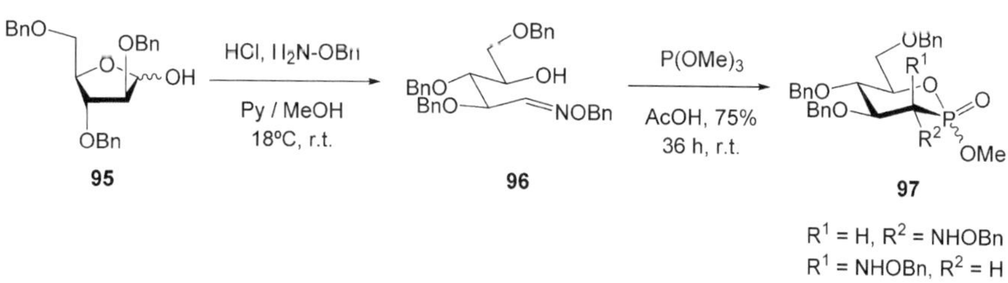

Scheme 31 Synthesis of 2-benzyloxyamino glycophostones by addition of trimethyl phosphite to a D-*arabino*-configured *O*-benzyloxime.

Glycophostone sodium salts, which can be obtained when the *O*-dealkylation of the phosphonate moiety is carried out with sodium iodide,[59] can be converted into phosphonamidate analogs, as demonstrated for the glucose-based tri-*O*-benzyl protected sodium phosphonate **98** (Scheme 32).[61] Treatment of **98** with trimethylsilyl triflate and trimethylamine and subsequent reaction with oxalyl chloride and catalytic DMF provided the phosphoryl chloride **99**. This intermediate was not isolated and was directly reacted with methylamine to afford the protected *N*-methyl phosphonamidate analog of D-glucose (**100**) in an overall yield of 30% from **98**. The corresponding full hydroxylated *N*-methyl phosphonamide derivative showed significant glycosidase inhibitory ability.

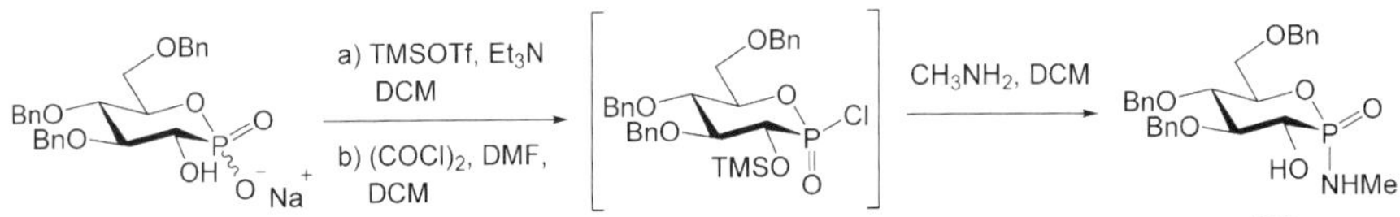

98 99 100

Scheme 32 Synthesis of a phosphonamidate analog of glucose from a tri-*O*-benzyl glucose-based phostone sodium salt.

An efficient and stereoselective method to obtain glycophostones starting from nonsaccharidic allyl-containing reagents was developed (Scheme 33).[62] The strategy starts with the addition of the alkoxides of commercially available allylic alcohols (**102**) to diphenyl allylphosphonate (**101**), furnishing the corresponding dienes (**103**, **106**) which undergo ring-closing metathesis (RCM) to give *P*-chiral allylphostones (**104, 107**). Thus, the reaction of but-3-en-2-ol (**102**, R = Me) with allylphosphonate **101** in the presence of BuLi in hexamethylphosphoramide (HMPA)/THF solution at low temperature (−78°C), afforded the diene phosphonate **103** in good yield, via displacement of the phenoxy group. The further RCM reaction catalyzed by the first-generation Grubbs catalyst provided the phostone **104**. The methodology worked also well using other allylic alcohols such as (2*R*)-1-benzyloxy-but-3-en-2-ol (**102**, R = CH_2OBn) or its triphenylmethyl protected analog (**102**, R = CH_2OTr). Dihydroxylation of **104, 107** with OsO_4/*N*-methylmorpholine-*N*-oxide (NMO) in the presence of citric acid provided the corresponding glycophostones (**105**, **108**) in very good yields and excellent selectivity. These compounds can be functionalized into vinylphostones for further stereoselective reactions, namely dihydroxylation to afford a variety of differently configured glycophostones.

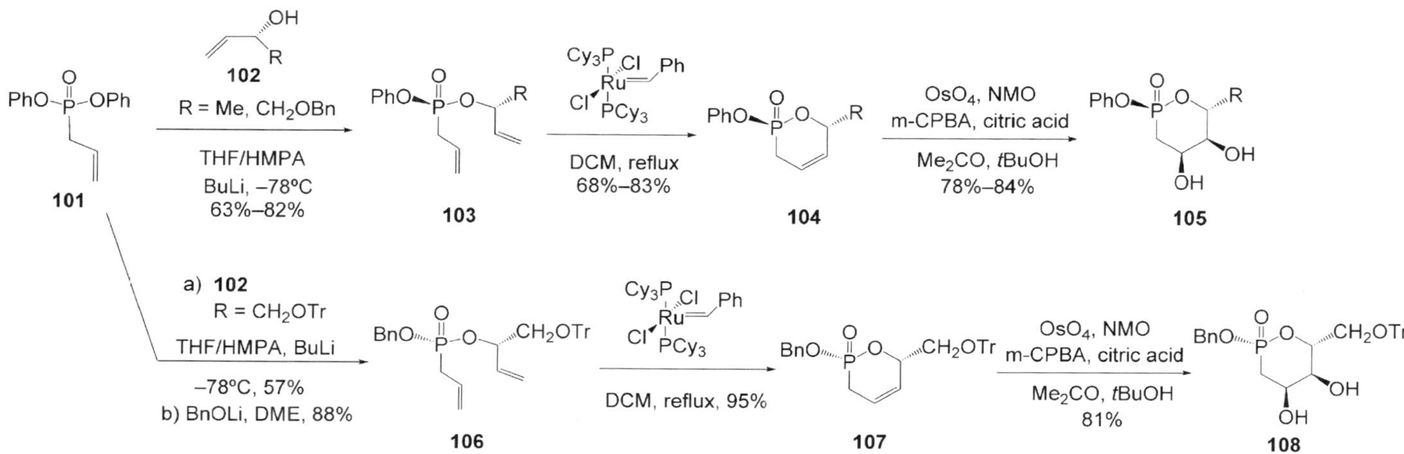

Scheme 33 Synthesis of glycophostones by addition of an allylic alkoxide to an allylphosphonate followed by RCM and subsequent dihydroxylation.

6 Conclusion

The importance of sugar phosphates as essential molecules in metabolism and as precursors or intermediates in the biosynthesis of other important biomolecules has prompted the development of analogs of these compounds comprising rather stable-related phosphorous-containing moieties.

Convenient synthetic routes toward sugar phosphonates, in which the phosphono group is C- or P-linked to the sugar moiety, as well as for sulfur-containing analogs, have been devised. Most of the reported strategies are straightforward, using sugar electrophiles such as sugar halides, glycals, sugar aldehydes, and phosphorous nucleophilic reagents, such as phosphites or activated phosphonates. Sugar-derived (α,α-difluoromethylene)phosphonates are particularly attractive synthetic targets not only due to their stability, since the phosphonate moiety is C-linked to the sugar unit but also due to the suitability of the difluorophosphonate system for the mimicry of a phosphate moiety, as an isoelectronic and isopolar surrogate of similar acidity. Glycosyl boranophosphates constitute also an interesting group of sugar phosphate analogs which are versatile synthons for a variety of P-modified compounds. Glycometic structures embodying analogous moieties of phosphate groups have also attracted significant interest, namely imino sugar derivatives or glycophostones. Efficient and practical protocols have been reported for their synthesis, which is quite challenging since it generally requires sugar ring closure and stereoselective control in key steps.

The biological potential of sugar phosphate analogs, already revealed by various groups of compounds, makes evident that the design of new molecules of these types and the exploitation of effective and selective synthetic methods for their access are relevant topics in carbohydrate chemistry and will keep encouraging research on this field.

References

1. Barreteau, H.; Kovac, A.; Boniface, A.; Sova, M.; Gobec, S.; Blanot, D. Cytoplasmic Steps of Peptidoglycan Biosynthesis. *FEMS Microbiol. Rev.* **2008**, *32*, 168–207.
2. Cipolla, L.; Polissi, A.; Airoldi, C.; Galliani, P.; Sperandeo, P.; Nicotra, F. The Kdo Biosynthetic Pathway Toward OM Biogenesis as Target in Antibacterial Drug Design and Development. *Curr. Drug Discov. Technol.* **2009**, *6*, 19–33.
3. Zamyatina, A.; Gronow, S.; Puchberger, M.; Graziani, A.; Hofinger, A.; Kosma, P. Efficient Chemical Synthesis of Both Anomers of ADP L-Glycero- and D-Glycero-D-Manno-Heptopyranose. *Carbohydr. Res.* **2003**, *338*, 2571–2589.
4. Kneidinger, B.; Marolda, C.; Graninger, M.; Zamyatina, A.; McArthur, F.; Kosma, P.; Valvano, M. A.; Messner, P. Biosynthesis Pathway of ADP-L-*glycero*-D-*manno*-Heptose in *Escherichia coli*. *J. Bacteriol.* **2002**, *184*, 363–369.
5. Birch, H. L.; Alderwick, L. J.; Bhatt, A.; Rittmann, D.; Krumbach, K.; Singh, A.; Bai, Y.; Lowary, T. L.; Eggeling, L.; Besra, G. S. Biosynthesis of Mycobacterial Arabinogalactan: Identification of a Novel a(1,3) Arabinofuranosyltransferase. *Mol. Microbiol.* **2008**, *69*, 1191–1206.

6. Nikolaev, A. V.; Botvinko, I. V.; Ross, A. J. Natural Phosphoglycans Containing Glycosyl Phosphate Units: Structural Diversity and Chemical Synthesis. *Carbohydr. Res.* **2007**, *342*, 297–344.
7. Nikolaev, A. V.; Al-Maharik, N.; Sizova, O. V. Synthetic Glycoconjugates Based on Leishmania lipophosphoglycan Structures as Potential Anti-Leishmaniasis Vaccines. In *Specialist Periodical Reports: Carbohydrate Chemistry*, Rauter, A. P., Lindhorst, T., Eds.; Royal Society of Chemistry: London, UK, **2010**; vol. 36, pp 101–126.
8. Xavier, N. M. Recent Advances on Nucleotide Analogs and Mimetics: Synthesis and Biological Properties. In *Elsevier Reference Module in Chemistry, Molecular Sciences and Chemical Engineering*, Reedijk, J., Ed.; Elsevier: Waltham, MA, USA, **2017**, pp 1–15. https://doi.org/10.1016/B978-0-12-409547-2.12655-1.
9. Zamyatina, A.; Kosma, P. Synthesis of Anomeric Phosphates of Aldoses and 2-Ulosonic Acids. In *Specialist Periodical Reports: Carbohydrate Chemistry*, Rauter, A. P., Lindhorst, T., Eds.; Royal Society of Chemistry: London, UK, **2009**; vol. 35, pp 71–98.
10. Oka, N.; Sato, K.; Wada, T. Recent Progress in the Synthesis of Glycosyl Phosphate Derivatives. *Trends Glycosci. Glycotechnol.* **2012**, *24*, 152–168.
11. Roy, B.; Depaix, A.; Périgaud, C.; Peyrottes, S. Recent Trends in Nucleotide Synthesis. *Chem. Rev.* **2016**, *116*, 7854–7897.
12. Nicotra, F. Synthesis of Glycosyl Phosphate Mimics. In *Carbohydrate Mimics: Concepts and Method*, Chapleur, Y., Ed.; Wiley-VCH Verlag GmbH: Weinheim, Germany, **1998**, pp 67–85.
13. Paulsen, H.; Thiem, J. Synthese von Glycopyranosylphosphonaten durch Dialkylphosphit-Addition an Glycale unter Allylverschiebung. *Chem. Ber.* **1973**, *106*, 3850–3876.
14. Meuwly, R.; Vasella, A. Synthesis of 1,2-*cis*-Configurated Glycosylphosphonates. *Helv. Chim. Acta* **1986**, *69*, 25–34.
15. Briner, K.; Vasella, A. Glycosylphosphonates of 2-Amino-2-Deoxy-Aldoses. Synthesis of a Phosphonate Analogue of Lipid X. *Helv. Chim. Acta* **1987**, *70*, 1341–1356.
16. Pachamuthu, K.; Figueroa-Perez, I.; Ali, I. A.; Schmidt, R. R. Synthesis of Glycosyl Phosphonates by Michael-Type Addition to 2-Nitroglycals. *Eur. J. Org. Chem.* **2004**, *19*, 3959–3961.
17. Elliott, T. S.; Slowey, A.; Ye, Y.; Conway, S. J. The Use of Phosphate Bioisosteres In Medicinal Chemistry and Chemical Biology. *Med. Chem. Commun.* **2012**, *3*, 735–751.
18. Choi, S.; Mansoorabadi, S. O.; Liu, Y.; Chien, T.; Kiu, H. Analysis of UDP-D-apiose/UDP-D-Xylose Synthase-Catalysed Conversion of UDP-D-Apiose Phosphonate to UDP-D-Xylose Phosphonate: Implications for a Retroaldol-Aldol Mechanism. *J. Am. Chem. Soc.* **2012**, *134*, 13946–13949.
19. Casero, F.; Cipolla, L.; Lay, L.; Nicotra, F.; Panza, L.; Russo, G. Stereoselective Synthesis of the Isosteric Phosphono Analogues of *N*-Acetyl-α-D-Glucosamine 1-Phosphate and *N*-Acetyl-α-D-Mannosamine 1-Phosphate. *J. Org. Chem.* **1996**, *61*, 3428–3432.
20. Meuwly, R.; Vasella, A. Synthesis of Isosteric Mono-Phosphonate Analogues of β- and α-D-Fructose 2,6-Bisphosphate. *Helv. Chim. Acta* **1986**, *69*, 751–760.
21. Hajduch, J.; Nam, G.; Kim, E. J.; Fröhlich, R.; Hanover, J. A.; Kirk, K. L. A Convenient Synthesis of the C-1-Phosphonate Analogue of UDP-GlcNAc and Its Evaluation as an Inhibitor of O-Linked GlcNAc Transferase (OGT). *Carbohydr. Res.* **2008**, *343*, 189–195.
22. Junker, H.-D.; Fessner, W.-D. Diastereoselective Synthesis of C-Glycosylphosphonates via Free-Radical Glycosylation. *Tetrahedron Lett.* **1998**, *39*, 269–272.
23. Junker, H.-D.; Phung, N.; Fessner, W.-D. Diastereoselective Free-Radical Synthesis of cx-Substituted C-Glycosyl Phosphonates, and Their Use as Building Blocks in the HWE-Reaction. *Tetrahedron Lett.* **1999**, *40*, 7063–7066.

24. Forget, S. M.; Bhattasali, D.; Hart, V. C.; Cameron, T. S.; Syvitski, R. T.; Jakeman, D. L. Synthesis and Enzymatic Evaluation of Ketose Phosphonates: The Interplay Between Mutarotation, Monofluorination and Acidity. *Chem. Sci.* **2012**, *3*, 1866–1878.
25. Blackburn, G. M.; Kent, D. E.; Kolkmann, F. The Synthesis and Metal Binding Characteristics of Novel, Isopolar Phosphonate Analogues of Nucleotides. *J. Chem. Soc. Perkin Trans.* **1984**, *1*, 1119–1125.
26. Kovensky, J.; McNeil, M.; Sinay, P. D-Galactofuranosylphosphonates. First Synthesis of UDP-C-D-Galactofuranose. *J. Org. Chem.* **1999**, *64*, 6202–6205.
27. Bouwman, S.; Orru, R. V. A.; Ruijter, E. Stereoselective Synthesis of Fluorinated Aminoglycosyl Phosphonates. *Org. Biomol. Chem.* **2015**, *13*, 1317–1321.
28. Behr, J.-B.; Evina, C. M.; Phung, N.; Guillerm, G. Synthesis of (Difluoromethyl) Phosphonate Azasugars Designed as Inhibitors for Glycosyl Transferases. *J. Chem. Soc. Perkin Trans.* **1997**, *11*, 1597–1600.
29. Cocaud, C.; Nicolas, C.; Poisson, T.; Pannecoucke, X.; Legault, C. Y.; Martin, O. R. Tunable Approach for the Stereoselective Synthesis of 1-*C*-Diethylphosphono(Difluoromethyl) Iminosugars as Glycosyl Phosphate Mimics. *J. Org. Chem.* **2017**, *82*, 2753–2763.
30. Godin, G.; Compain, P.; Masson, G.; Martin, O. R. A General Strategy for the Practical Synthesis of Nojirimycin C-Glycosides and Analogues. Extension to the First Reported Example of an Iminosugar 1-Phosphonate. *J. Org. Chem.* **2002**, *67*, 6960–6970.
31. Peyrottes, S.; Gallier, F.; Béjaud, J.; Périgaud, C. Use of Microwave Irradiation for Sugar and Nucleoside Phosphonates Synthesis. *Tetrahedron Lett.* **2006**, *47*, 7719–7772.
32. Gallier, F.; Peyrottes, S.; Périgaud, C. Ex-Chiral-Pool Synthesis of β-Hydroxyphosphonate Nucleoside Analogues. *Eur. J. Org. Chem.* **2007**, (6), 925–933.
33. Barton, D. H. R.; Géro, S. D.; Quiclet-Sire, B.; Samadi, M. New Synthesis of Sugar, Nucleoside and α-Amino Acid Phosphonates. *Tetrahedron* **1992**, *48*, 1627–1636.
34. Jarosz, S.; Mach, M. Phosphonate Versus Phosphorane Method in the Synthesis of Higher Carbon Sugars. Preparation of D-*erythro*-L-*manno*-D-Glucododecitol. *J. Chem. Soc. Perkin Trans.* **1998**, *1*, 3943–3948.
35. Berkowitz, D.; Bhuniya, D.; Peris, G. Facile Installation of the Phosphonate and (α,α-Difluoromethyl)Phosphonate Functionalities Equipped With Benzyl Protection. *Tetrahedron Lett.* **1999**, *40*, 1869–1872.
36. Berkowitz, D. B.; Bose, M.; Pfannenstiel, T. J.; Doukov, T. α-Fluorinated Phosphonates as Substrate Mimics for Glucose 6-Phosphate Dehydrogenase: The CHF Stereochemistry Matters. *J. Org. Chem.* **2000**, *65*, 4498–4508.
37. Khankin, N. A.; Montero, J.-L. Synthesis of Mannose-6-Phosphate Analogs: Large-Scale Preparation of Isosteric Mannose-6-Phosphonate via Cyclic Sulfate Precursor. *Tetrahedron Lett.* **2002**, *43*, 4017–4020.
38. Van der Klein, P. A. M.; Van Boom, J. H. Application of Cyclic Sulfates in the Synthesis of 6-deoxy-D-manno-Heptopyranose Derivatives. *Carbohydr. Res.* **1992**, *224*, 193–200.
39. Gautier, A.; Garipova, G.; Salcedo, C.; Balieu, S.; Pierre, S. R. α,α-Difluoro-*H*-Phosphinates: Useful Intermediates for a Variety of Phosphate Isosteres. *Angew. Chem. Int. Ed.* **2004**, *43*, 5963–5967.
40. Li, P.; Sergueeva, Z. A.; Dobrikov, M.; Shaw, B. R. Nucleoside and Oligonucleoside Boranophosphates: Chemistry and Properties. *Chem. Rev.* **2007**, *107*, 4746–4796.
41. Shimizu, M.; Tamura, K.; Wada, T.; Saigo, K. BH_3 as a Protecting Group for Phosphonic Acid: A Novel Method for the Synthesis of Dinucleoside *H*-Phosphonate. *Tetrahedron Lett.* **2004**, *45*, 371–374.
42. Kawanaka, T.; Shimizu, M.; Wada, T. Synthesis of Dinucleoside Phosphates and Their Analogues by the Boranophosphotriester Method Using Azido-Based Protecting Groups. *Org. Lett.* **2008**, *10*, 1557–1560.

43. Matsumura, F.; Oka, N.; Wada, T. Synthesis of Glycosyl Boranophosphates and Their Applications as Precursors of Glycosyl Phosphate Analogues. *Org. Lett.* **2008**, (8), 1557–1560.
44. Imamoto, T.; Nagato, E.; Wada, Y.; Masuda, H.; Yamaguchi, K.; Uchimaru, T. New Boranophosphorylation Reagents, Dimethyl Boranophosphate Monopotassium Salt and Tetramethyl Boranopyrophosphate. *J. Am. Chem. Soc.* **1997**, *119*, 9925–9926.
45. Matsumura, F.; Oka, N.; Wada, T. Stereoselective Synthesis of α-Glycosyl Phosphites and Phosphoramidites via O-Selective Glycosylation of H-Phosphonate Derivatives. *Org. Lett.* **2008**, *10*, 5297–5300.
46. Tatsumi, S.; Matsumura, F.; Oka, N.; Wada, T. Glycosylation of Alcohols Using Glycosyl Boranophosphates as Glycosyl Donors. *Tetrahedron Lett.* **2013**, *54*, 3731–3734.
47. Stawinksi, J.; Kraszewski, A. How to Get Most Out of Two Phosphorus Chemistries. Studies on *H*-Phosphonates. *Acc. Chem. Res.* **2002**, *35*, 952–960.
48. Plante, O. J.; Andrade, R. B.; Seeberger, P. H. Synthesis and Use of Glycosyl Phosphates as Glycosyl Donors. *Org. Lett.* **1999**, *1*, 211–214.
49. Seeberger, P. H. Automated Carbohydrate Synthesis to Drive Chemical Glycomics. *Chem. Commun.* **2003**, (10), 1115–1121.
50. Sato, K.; Oka, N.; Fujita, S.; Matsumura, F.; Wada, T. 1,2-Trans-Selective Synthesis of Glycosyl Boranophosphates and Their Utility as Building Blocks for the Synthesis of Phosphodiester-Linked Disaccharides. *J. Org. Chem.* **2010**, *75*, 2147–2156.
51. Sato, K.; Wada, T. One-Pot Conversion Reactions of Glycosyl Boranophosphates Into Glycosyl Phosphate Derivatives *via* Acyl Phosphite Intermediates. *Org. Biomol. Chem.* **2016**, *14*, 11092–11095.
52. Kulesza, A.; Frank, C. G.; Aebi, M.; Vasella, A. Synthesis of Stable Dolichylphosphomannose Analogs. *Helv. Chim. Acta* **2004**, *87*, 3106–3118.
53. Wyrzykiewicz, T. K.; Ravikumar, V. T. Efficiency of Sulfurization in the Synthesis of Oligodeoxyribonucleotide Phosphorothioates Utilizing Various Sulfurizing Reagents. *Bioorg. Med. Chem. Lett.* **1994**, *4*, 1519–1522.
54. Iannitti, T.; Morales-Medina, J. C.; Palmieri, B. Phosphorothioate Oligonucleotides: Effectiveness and Toxicity. *Curr. Drug Targets* **2014**, *15*, 663–673.
55. Eckstein, F. Phosphorothioates, Essential Components of Therapeutic Oligonucleotides. *Nucleic Acid Ther.* **2014**, *24*, 374–387.
56. Zhang, G.; Yu, B.; Deng, S.; Huia, Y. Preparation of Glycosyl Dimethylthiophosphates and Their Application as Glycosyl Donors. *J. Carbohydr. Chem.* **1998**, *17*, 547–556.
57. Lorange, M. W.; Beaton, S. A.; Lines, K. L.; Jakeman, D. L. Thiophosphate and Thiophosphonate Analogues of Glucose-1-Phosphate: Synthesis and Enzymatic Activity With a Thymidylyltransferase. *Carbohydr. Res.* **2013**, *379*, 43–50.
58. Piekutowska, M.; Pakulski, Z. Synthesis of S-Glycosyl Thiophosphates, Thiophosphonates and Thiophosphinates by the Michaelis–Arbuzov Rearrangement of Anomeric Thiocyanates. *Carbohydr. Res.* **2008**, *343*, 785–792.
59. Darrow, J. W.; Drueckhammer, D. G. Cyclic Phosphonate Analogs of Hexopyranoses. *J. Org. Chem.* **1994**, *59*, 2976–2985.
60. Hanessian, S.; Rogel, O. Synthesis of Glycophostones: Cyclic Phosphonate Analogues of Biologically Relevant Sugars. *J. Org. Chem.* **2000**, *65*, 2667–2674.
61. Darrow, J. W.; Drueckhammer, D. G. A Cyclic Phosphonamidate Analogue of Glucose as a Selective Inhibitor of Inverting Glycosidases. *Bioorg. Med. Chem.* **1996**, (8), 1341–1348.
62. Stoianova, D. S.; Whitehead, A.; Hanson, P. R. *P*-Heterocyclic Building Blocks: Application to the Stereoselective Synthesis of *P*-Sugars. *J. Org. Chem.* **2005**, *70*, 5880–5889.

Part Two

Structure-function relationships in polysaccharides

Synthetic polysaccharides

9

Yang Yu[a,b], *Martina Delbianco*[a]
[a]Department of Biomolecular Systems, Max Planck Institute of Colloids and Interfaces, Potsdam, Germany, [b]Department of Chemistry and Biochemistry, Freie Universität Berlin, Berlin, Germany

Chapter outline

1 Introduction

Polysaccharides are the most abundant polymeric organic materials in nature, serving important functions including structural and energy storage. Carbohydrate-protein interactions (CPIs) and carbohydrate-carbohydrate interactions (CCIs) regulate a variety of biochemical processes, such as cell differentiation, proliferation and adhesion, inflammation, and immune responses.[1] Despite polysaccharides abundance, for this class of biomaterials, a deep understanding of the structure and structure-function relationship is still missing. In particular, this lack of data is striking if compared to the knowledge existing on polynucleotides (DNA and RNA) and polypeptides, which are well understood at the molecular level. Automated sequencing and synthesis, combined with gene expression technologies, permitted the rapid development of genomics and proteomics. Such techniques are not yet available for carbohydrates, preventing easy access to well-defined polysaccharides. The intrinsic nature of polysaccharides, which exist as polydisperse materials, often with high degree of polymerization and modification, makes glycoscience incredibly challenging. In addition, dozens of monosaccharide units can be combined to form a large variety of complex branched structures, with different regio- and stereochemistry. Isolation from natural sources is time consuming and often inapplicable. In many cases, chemical synthesis remains the only option for the production of pure samples but requires tremendous

Recent Trends in Carbohydrate Chemistry. https://doi.org/10.1016/B978-0-12-817467-8.00009-8
© 2020 Elsevier Inc. All rights reserved.

efforts to access small quantities of "short" polysaccharides. Enzymatic synthesis, polymerization reactions, as well as new synthetic methodologies are currently being developed to provide new strategies for the quick production of synthetic polysaccharides. Such synthetic materials, with defined compositions and modifications, will fuel the development of glycoscience, with potential applications in material science, biology, and nanotechnology. Four main advantages could be mentioned:

- Synthetic polysaccharides have a defined monosaccharide composition that can confirm the composition of natural polysaccharides.
- Synthetic polysaccharides are fundamental probes for three-dimensional (3D) structural analysis helping to draw a correlation between polysaccharide structure and function.
- Synthetic polysaccharides are the perfect substrate to undoubtedly characterize enzymes activity and antibodies specificity.
- Chemical modifications can be inserted and unnatural structures can be created toward tailor-made materials with tuned properties.

1.1 Challenges in the chemical synthesis of polysaccharides

In a polysaccharide, the monosaccharide units are connected through a glycosidic bond. The glycosylation reaction that permits the linkage between a carbohydrate "acceptor" and a carbohydrate "donor" is particularly delicate (Fig. 1).[2] Several possible attachment points are available on the glycosyl acceptor (and donor) and regioisomers can be obtained. A plethora of orthogonal protecting groups (PGs) has been developed

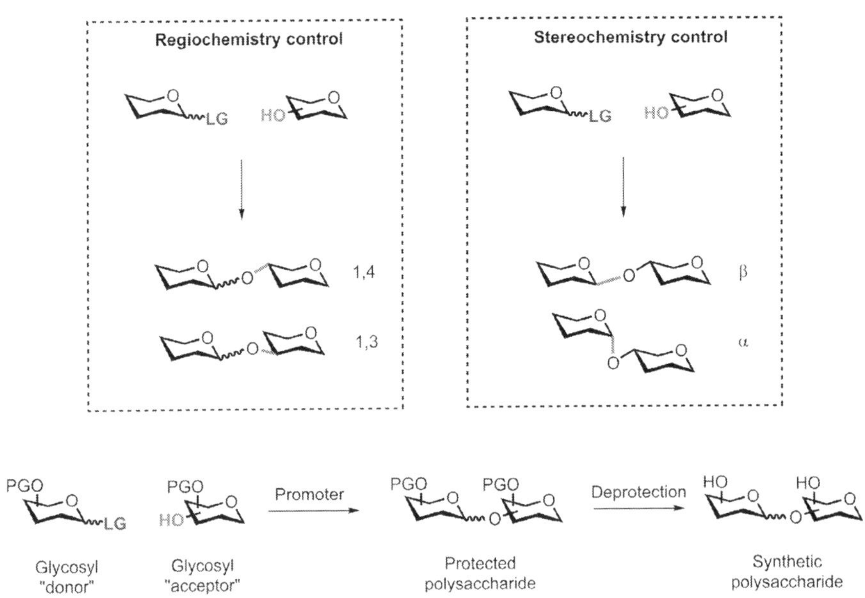

Fig. 1 Regio- and stereochemistry control during a glycosylation reaction and a general polysaccharide synthesis.

to selectively mask particular positions, leaving only the hydroxide in the desired position(s) available for glycosylation.[3, 4] Still, the preparation of the protected monosaccharides can be time-consuming, often requiring several steps to obtain the desired PG pattern. In addition, upon completion of synthesis of the polysaccharides, the PGs have to be removed to obtain the desired polysaccharides. The deprotection step(s) often proves to be challenging due to the harsh conditions required, poor solubility of the products and intermediates, and formation of aggregates decreasing the rate of reactions.[5] An additional challenge is the stereocontrol of glycosylation. In contrast to phosphate diesters in polynucleotides or amide linkages in polypeptides, the glycosidic linkage introduces a stereogenic center in the polysaccharide chain. Considerable efforts have been paid to mechanistically understand the glycosylation reaction in order to predict the stereochemical outcome.[6, 7] Nevertheless, even though much progress has been made since the Fischer glycosylation, this step remains a black box and a universal rule to guide the glycosylation process is still missing. To complicate the assembly even further, loss of stereospecificity during long polymerizations is sometimes observed. In the last decades, new leaving groups, promoters, and reaction conditions have been reported to make glycosylation more efficient.[8, 9] With these improvements, synthetic chemists were able to access a large number of natural and unnatural polysaccharides, even though generally with limited molecular weight ($M_W < 10{,}000$ g/mol). Here, the recent achievements in the field of polysaccharide synthesis are described. Polymerization reactions, as well as total synthesis, will be discussed.

1.2 Alternatives to the chemical synthesis of polysaccharides

This book chapter focuses on the chemical synthesis of polysaccharides to form structures based on monosaccharide units connected through *O*-glycosidic bonds. Other classes of synthetic polysaccharides or *pseudo*-polysaccharides exist, but they will not be described extensively (Fig. 2). A large variety of *pseudo*-polysaccharides derived from noncarbohydrate sources,[10, 11] as well as structures composed of monosaccharides connected through unnatural linkages[12–14] (e.g., amine, amide, C-C)

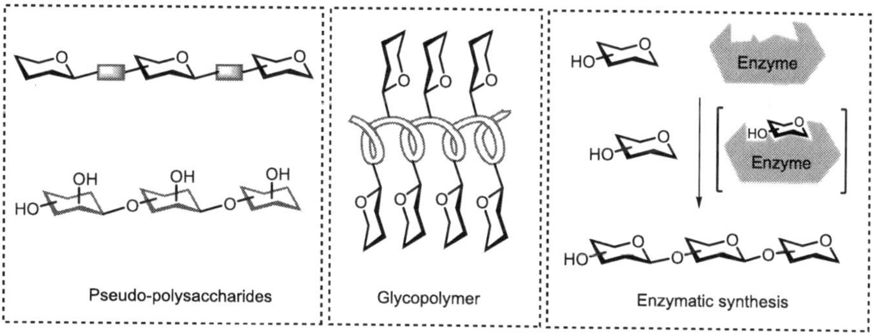

Fig. 2 Schematic representation of alternative strategies to the chemical synthesis of polysaccharides.

have been developed to increase the stability of the polymer. Glycopolymers, derived from the glycosylation of polymeric backbones, such as polyamides, polyurethanes, or natural polysaccharides, represents another important class of glyco-materials.[15, 16]

The most important alternative to the chemical synthesis of polysaccharides is enzymatic synthesis. Mimicking the natural biosynthesis, polysaccharides can be assembled using natural or modified enzymes as a catalyst for the glycosylation reaction.[17] This strategy allows for the use of mild aqueous conditions, permits to avoid the use of PGs, and provides perfect regio- and stereo-control.[18] A large variety of enzymes has been employed to produce well-defined polysaccharides. The main drawback of this technique is that chemical modifications and unnatural structures are not always compatible with the enzymatic synthesis, which very often tolerates only limited substrates variations.[19] In addition, the in vitro production of functional enzymes is sometimes troublesome.[20]

2 Polymerization reactions in polysaccharide synthesis

Most polysaccharides present a repetitive structure. The repeating unit can be a monosaccharide, a disaccharide or a more complex oligomer. Several polymerization techniques were studied to synthesize polysaccharides. Polycondensation, as well as ring-opening polymerization, was employed. With these strategies, long polymers, based on more than 500 monosaccharide units, were obtained. The current research focuses on the optimization of the polymerization conditions to obtain even longer structures (high degree of polymerization, DP) with low polydispersity index (PDI).

2.1 Condensation polymerization

A condensation reaction permits to connect two molecules, with loss of a small molecule (mostly a water molecule). When a bifunctional monomer is utilized, a linear polymer is obtained (Fig. 3B). The presence of multiple linkage points on the same monomer allows for the formation of a branched structure (Fig. 3A). The reactive anomeric position and the hydroxy groups make monosaccharides suitable substrates for condensation polymerization. With this technique, high DPs are achieved only with extremely efficient condensation reactions, but, generally, only short oligosaccharides are obtained. More reactive monomers, as well as new operating conditions (e.g., solid-phase and microwaved-assisted polymerization), were developed to improve the polymerization efficiency.

2.1.1 Polycondensation of unprotected sugars

Since the early time of carbohydrate chemistry, it is known that glucose in the presence of concentrated HCl can form a mixture of oligomers, mainly composed of disaccharides. In 1950, the first polycondensation of a glucose monosaccharide **1** to form a highly branched structure was reported (Scheme 1).[21] A fairly high DP (42) was obtained by continuously removing the water from the reaction. Replacing HCl with a

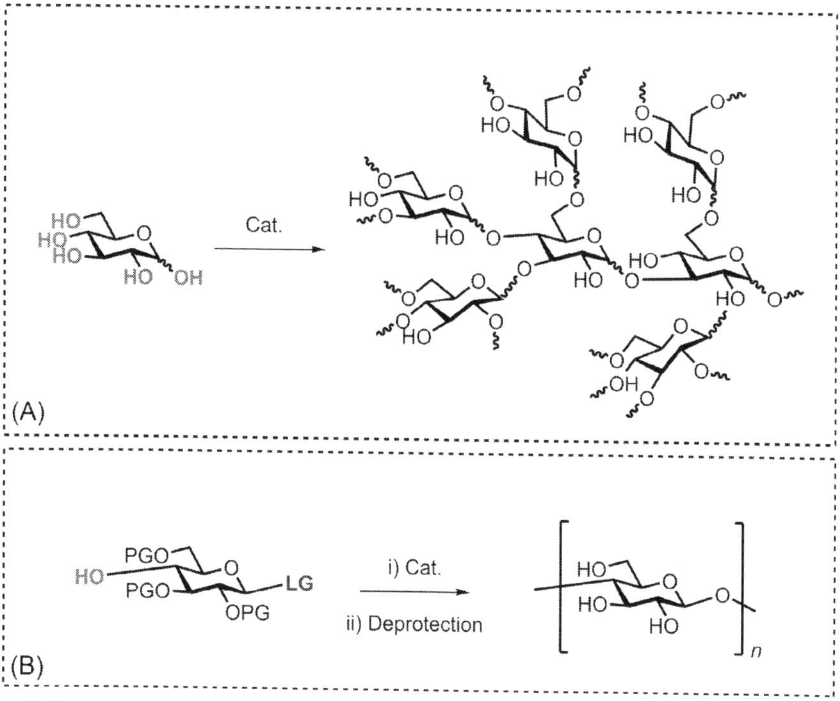

Fig. 3 Schematic representation of condensation polymerization. (A) Condensation of unprotected sugars—branched polysaccharides. (B) Condensation of protected sugars—linear polysaccharides.

catalytic amount of phosphorus acid permitted to avoid the partial decomposition observed with the HCl method.[22] Nevertheless, these early-stage strategies suffered from long reaction times and required high temperatures. In addition, the unprotected sugar used as starting material gave rise to polydisperse samples, with different regio- and stereochemistry (Scheme 1).

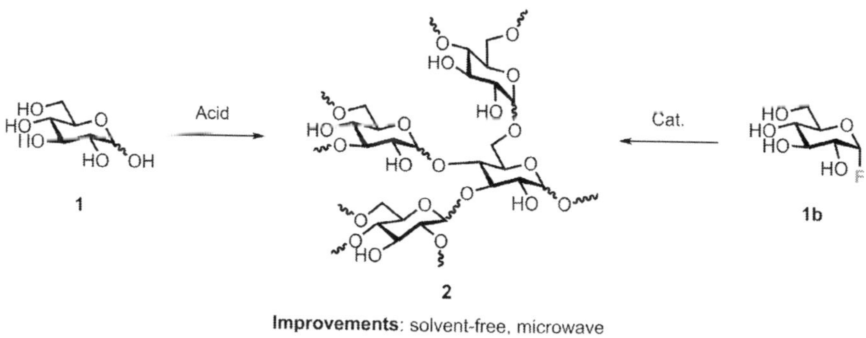

Scheme 1 Condensation polymerization starting from an unmodified saccharide or from glycosyl fluoride.

Solvent-free synthesis was investigated to reduce undesirable side reactions by lowering the reaction temperature. Moreover, the structure of the polysaccharides could be influenced by the molecular arrangement in the monomer crystal. Highly branched polysaccharides were obtained in the presence of H_3PO_4 at 100°C under a N_2 flow.[23] With the development of the microwave-assisted technology,[24] it was possible to obtain better results in a much faster timeframe (up to 2 min reaction).[25] Microwave irradiation is particularly suitable for solvent-free polymerization and found application in the synthesis of polyglucosides[26] and polymannosides.[27]

The use of glycosyl fluorides as donors permitted better control of the glycosylation process (Scheme 1). The base-promoted (NaH or LiH) glycosylation of the unprotected glycopyranosyl fluoride **1b** was used to synthesize polymers with up to 25 sugar units.[28] The different reactivity for each hydroxy groups permitted to create a numerical model to predict the level of branching in the obtained molecule. The better reactivity of the fluoride donor, compared to the unmodified saccharide, was also exploited in a powder-to-powder condensation catalyzed by $Yb(OTf)_3$ to obtain highly branched structures with an average MW of 8900 g/mol.[29]

2.1.2 Polycondensation of protected sugars

To obtained linear polymers, the use of protected building blocks (BBs) is essential. To act as donor and acceptor during the polymerization, the glycosyl unit has to be equipped with a leaving group in the anomeric position and carry an unprotected hydroxy group. Under these conditions, linear polysaccharides with narrow polydispersity can be achieved, even though issues arise from PG migrations and poor stereoselectivity of some reactions.

Bromides, imidates as well as thioethers, are used as leaving groups and standard PGs (generally esters) are used to mask the hydroxy groups. In the first attempt, the 2,3,4-tri-*O*-acetyl-α-D-glucopyranosyl bromide **3** was reacted in the presence of Ag_2O, but only short oligomers were isolated (tetrasaccharides, Scheme 2). In a similar way, most reports of polycondensation of glycosyl bromides suffer from low DP.[30] Microwave irradiation gave enough boost to the reaction permitting to isolate β-16-polyglucosides with MW of 4.76 kDa.[31]

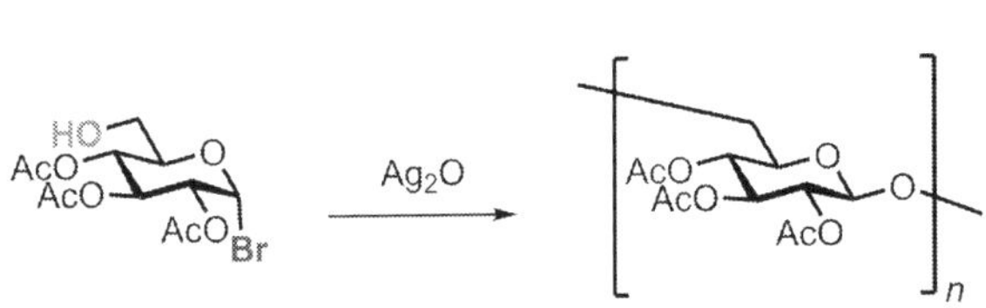

3 **4**

Scheme 2 The first attempt of glucopyranosyl bromide polymerization mediated with Ag_2O.

To increase the complexity of the polysaccharides, the polymerization can be conducted starting from dimers or more complex oligomers. The 2,3,6,2′,3′,4′-hexa-*O*-acetyl-α-maltosyl bromide **5** was used to synthesize a glucan with alternated α-1,4 and β-1,6 linkages (Scheme 3A).[32] A trichloroacetimidoyl functionalized dimer of glucosamine **7** allowed for the synthesis of chitosan analogs, even though with low DP (4.5 based on the dimer) (Scheme 3B).[33] A cellobiose disaccharide glycal was polymerized *via* 1,2-iodonium ion formation, permitting the formation of linear β-1,4 glucosides longer than 24 units.[34] Amylose-like structures **11** were obtained *via* polymerization of partially benzylated phenyl 1-thio-β-maltooctaoside **10** derived from γ-cyclodextrin **9** (Scheme 4). The oligosaccharide "monomer" used for the polycondensation permitted to overcome the low DP generally observed with polycondensation and MWs higher than 10,000 g/mol were detected.[35] Deprotection with Na in liquid ammonia afforded the desired amylose-like structure, as confirmed by NMR spectroscopy. One of the most impressive applications of polycondensation is the synthesis of the K30 bacterial capsular polysaccharides **13** (Scheme 5).[36] This approach required the synthesis of partially protected tetrasaccharide repeating units (**12**) that were employed in the *N*-iodosuccinimide/TMSOTf promoted polymerization. Products with up to seven repeating units were isolated.

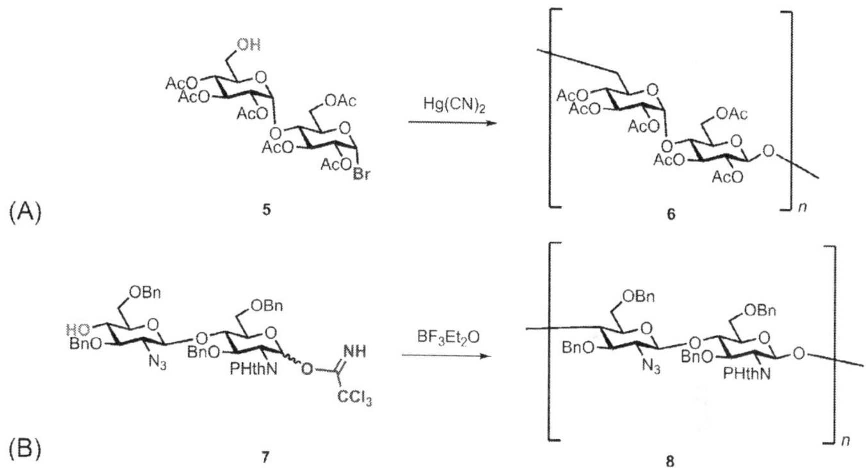

Scheme 3 Synthesis of linear polyglucosides with alternated α-1,4 and β-1,6 linkages (A) and synthesis of chitosan analogs (B).

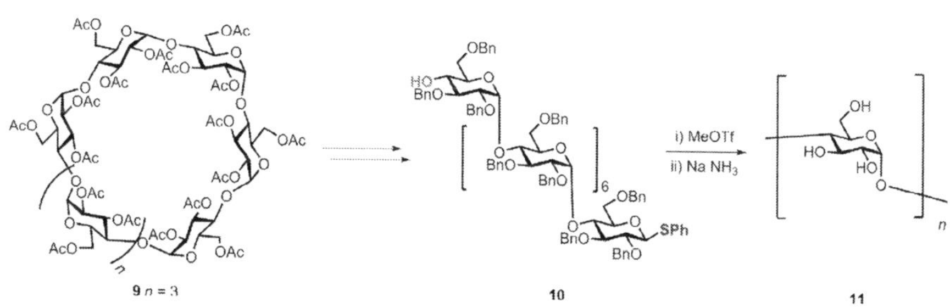

Scheme 4 Synthesis of amylose derivatives *via* polycondensation of benzylated phenyl 1-thio-β-maltooctaoside **10** derived from γ-cyclodextrin **9**.

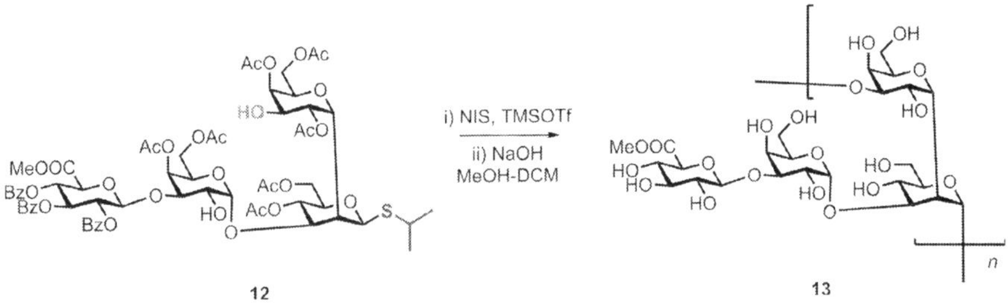

Scheme 5 Synthesis of the K30-2 bacterial capsular polysaccharide **13**.

2.1.3 Polycondensation of trityl ethers cyanoethylidene sugar derivatives

To achieve better control on the stereochemistry of glycosylation, the use of bicyclic sugars, able to generate a dioxalenium intermediate, was studied. In particular, tritylated sugar cyanoethylidene derivatives (CEDs), developed by Kochetkov, permitted the synthesis of several 1,2-*trans* homo- and heteropolysaccharides.[37, 38] The starting material is equipped with a cyanoethylidene orthoester and a trityl ether (**14**). Addition of catalytic tritylium perchlorate (Tr^+ClO4^-) initiate the condensation (**15**), promoting the formation of the dioxalenium intermediate **16**. At this point, the oxygen atom of the *O*-trityl group (of another monomer) attacks the anomeric carbon, forming the 1,2-*trans* glycoside and regenerating the tritylium cation (Fig. 4).[39] The polymerization is terminated with the addition of methanol.

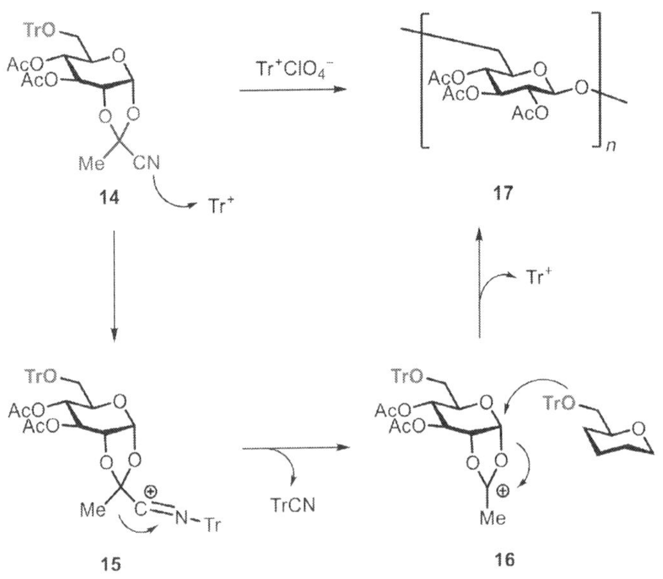

Fig. 4 Suggested mechanism of trityl-cyanoethylidene condensation.

Several CED BBs based on glucose, mannose, galactose, and rhamnose were prepared by treatment of the glycosyl bromide analog with KCN in acetonitrile (Fig. 5).[37] The position of the trityl group (primary vs secondary alcohol) does not majorly affect the reactivity, permitting the formation of 1,6- and well as 1,3- and 1,4-linkages. However, a reduced stereoselectivity is observed during the formation of particular 1,3- and 1,4-linkages.[39] Similarly, using disaccharide CEDs (Fig. 5) it was possible to synthesize mixed-linkage structures (i.e., alternated α-1,4 and β-1,6 polyglucoside)[40] as well as heteropolymers based on alternated glucose and rhamnose units.[41] Relatively low DPs, ranging from 10 to 40, were obtained. Strictly anhydrous reaction conditions and high-vacuum techniques permitted to increase the, otherwise poor (10%), condensation yields.[37]

Starting from complex tri- and tetrasaccharides (Fig. 6), it was possible to assemble bacterial polysaccharides related to *Salmonella newington,*[42] *Shigella flexneri,*[43] and *Streptococcus pneumonia* type 14.[44] Deprotection afforded the desired bacterial polysaccharides with DP around 10 (based on the tri- or tetrasaccharide unit).

Even though this method was applied to several syntheses of polysaccharides, the low DP and condensation yield remain a significant issue. In addition, loss of stereospecificity during the polymerization was observed. The complexity of the monomer preparation, that requires highly toxic potassium cyanide, poses an additional challenge.[45]

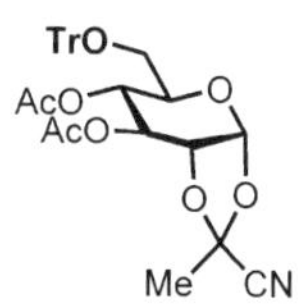

18

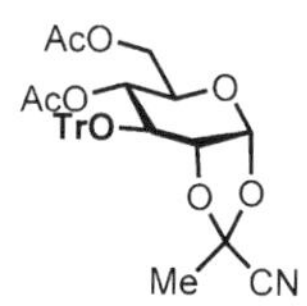

19

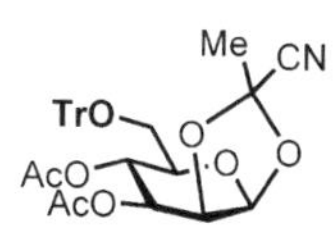

20

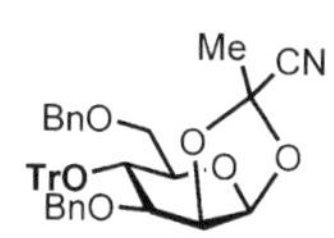

21

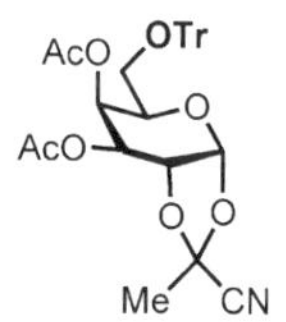

22

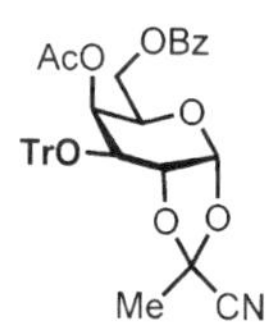

23

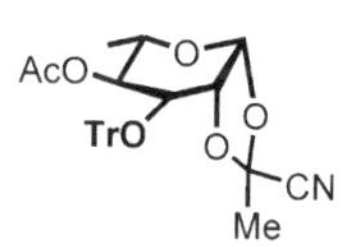

24

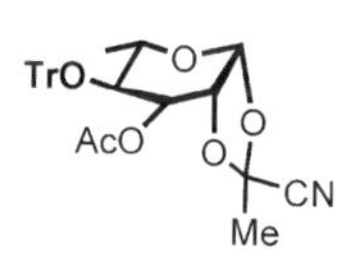

25

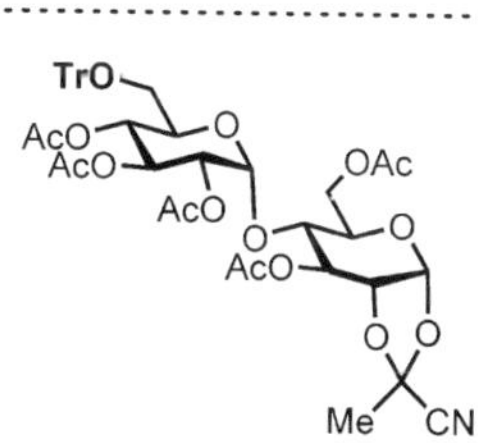

26

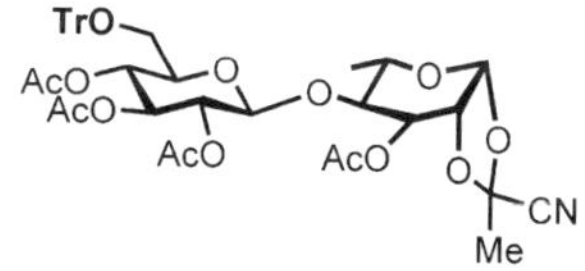

27

Fig. 5 Examples of CED building blocks based on monomers and dimers.

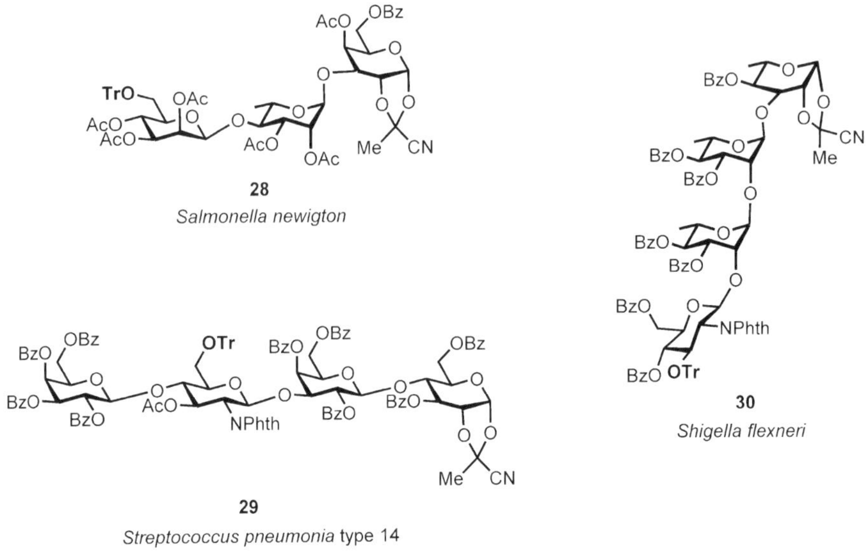

Fig. 6 CED building blocks used for the synthesis of bacterial polysaccharides.

2.1.4 *Condensation of sugar oxazolines*

Oxazoline derivatives provide extremely high stereoselectivity during glycosylation promoted by camphorsulfonic acid (CSA). These oxazoline derivatives are particularly useful for the synthesis of aminopolysaccharides (Scheme 6). Chitin derivatives (β-1,4-polyacetylglucosamines, **32**) were obtained in moderate yields but suffer from incomplete debenzylation (hydrogenolysis).[46] Slightly better results in terms of yield and deprotection efficiency were observed for the β-1,6-polyacetylglucosamine analog **34**.[47]

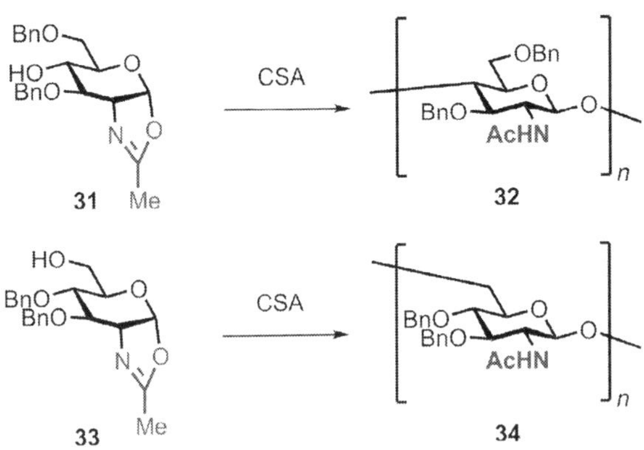

Scheme 6 Synthesis of aminopolysaccharides from sugar oxazolines.

2.2 Ring-opening polymerization

ROP is by far the most widely used method to access polymers with high MW and high stereoregularity. Cyclic monomers, upon activation, undergo ring- opening, promoted by the nucleophilic attack of a second monomer (Scheme 7). In this way, a new reactive center is created that serves for chain elongation. The relief of the steric strain of the cyclic monomers is the driving force of the polymerization. Moisture and impurities need to be strictly avoided. For this reason, ROP is often performed under high vacuum, making this process sometimes inconvenient.

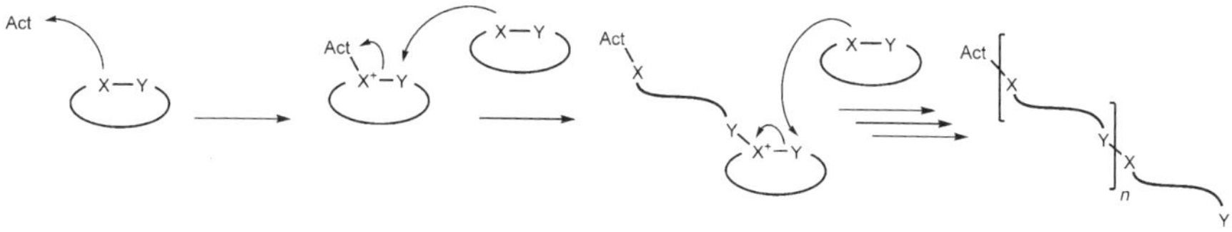

Scheme 7 General scheme of ring-opening polymerization (*Act*, activator).

2.2.1 ROP of anhydrosugars

The most common BBs for ROP are anhydrosugars. These materials can be obtained from vacuum pyrolysis of polysaccharides. Several procedures involving microwave technology are now available to make this process more efficient.[48, 49]

1,6-Anhydrosugars

Ruckel and Schuerch showed the potential of 1,6-anhydrosugar derivatives performing the first stereospecific synthesis of dextran (α-1,6-glucan) (Scheme 8).[50, 51] A benzylated BB **35** and PF_5 catalysis in anhydrous DCM at low temperature (−78°C) permitted to obtain a DP of above 200. Deprotection with Na in liquid ammonia afforded the desired polysaccharide.

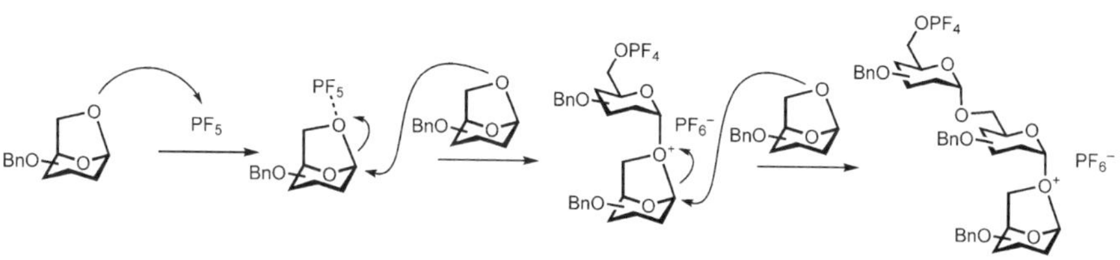

35 36 37 38

Scheme 8 First stereospecific synthesis of dextran with ROP.

A similar strategy was then applied to several 1,6-anhydrosugar BBs (Table 1). Systematic studies revealed significant differences in reactivity depending on the sugar monomer (Man > Glc > Gal > All),[53, 55, 63] the Lewis acid ($BF_3{\cdot}Et_2O$, $SbCl_5$, and PF_5),[50] and the PG pattern (benzyl vs allyl). The steric strain release, due to the reduction of the 1,3-diaxial interaction, and the conformational change of the oxonium ion are responsible for the differences in BB reactivity.[64] Depending on the configuration of the benzylated BB, a different amount of Lewis acid is required since complexation with the Bn groups

Table 1 1,6-Anhydrosugars and their use for ROP.

Monomer	Activator (mol%)	Max DP	Conversion (%)
38[52]	PF_5 (0.80)	1800	77
39[53]	PF_5 (0.94)	–	91
40[53]	PF_5 (0.80)	495	42
41[54]	PF_5 (10)	7	66
42[55]	PF_5 (4)	178	34
43[56]	PF_5 (10)	–	44
44[57]	PF_5 (10)	110	75
45[58]	PF_5 (1.9)	222	86
46[59]	PF_5 (8.6)	44	41
47[60]	$PF_5C_6H_5COF$ (5)	131	80
48[61]	PF_5 (5)	–	78
49[60]	$PF_5C_6H_5COF$ (5)	4	11
50[62]	PF_5 (5)	55	38

can compete with the polymerization initiation. Electron-donating PGs (OBn, OMe, and OAll) make ROP very efficient and high MWs can be obtained (up to 770,000 g/mol).[31] The use of the allyl PG (**43**) was particularly useful, permitting to perform the final deprotection under milder conditions, avoiding the need for high vacuum.[56] β-Linkages can also be obtained using a participating PG (benzoyl ester) in position 2 (**41**) through a double inversion mechanism.[54] Unnatural structures, bearing substituents in position 3 (F, OMe, OAc) were also synthesized to study how chemical modifications can affect the biological function (**44**–**46**).[57–59] This bottom-up approach permits to avoid the complications arising from the poorly regio-selective polysaccharide chemical modification. Azido-derivatives (**47**–**49**) were also employed to obtain amino-polysaccharides upon deprotection.[60,61] Much lower MWs were observed in particular with azides in position 2 (**49**), probably due to the implication of the azido group in termination reactions.[60] Protected amines (**50**) provided an interesting alternative, however, in most cases polymers with low DP were obtained.[62] A careful PG strategy was employed to afford synthetic polysaccharides that can be selectively deprotected in specific positions. The liberated hydroxy group can be selectively glycosylated as exemplified in the synthesis of branched dextran derivatives **53** (Scheme 9).[65]

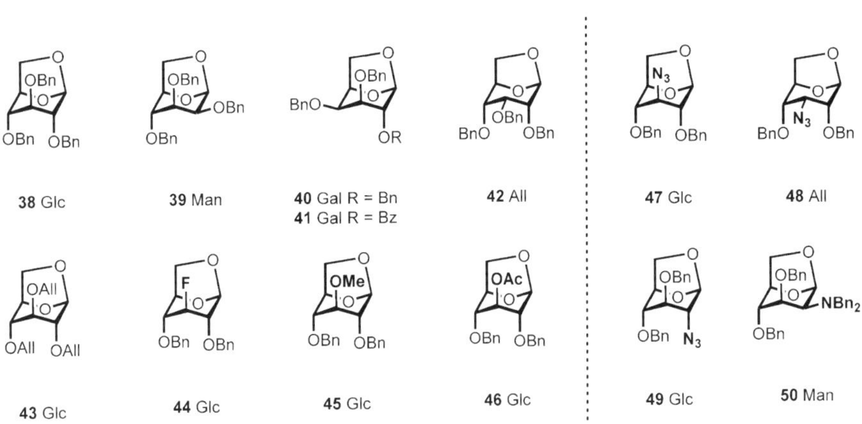

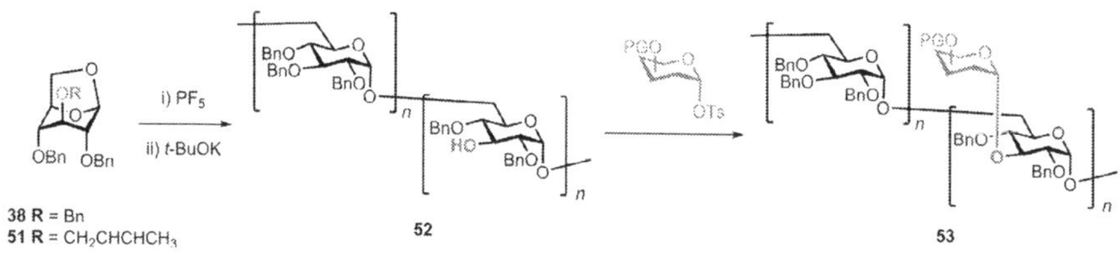

Scheme 9 Synthesis of branched dextran derivatives based on ROP.

1,4-, 1,3-, and 1,2-Anhydrosugars

The use of other anhydrosugars is less common than the widely explored 1,6-analogs (Fig. 8).

Some effort has been paid to synthesize cellulose-like polysaccharides starting from 1,4-anhydrosugars, however, always resulting in a mixture of furanosidic and pyranosidic repeating units with mixed anomeric configuration (Fig. 7).[66] Selective 1,4-scissions is necessary to obtain the cellulose-like polymer but a competing 1,5-scission promotes the formation of the unwanted furanosidic analog. The high basicity of the 1,5-anhydro ring oxygen favors the complexation of the Lewis acid in this position and the subsequent 1,5-scission. This pathway can be selectively activated at low temperature (−40°C) and can be used for the synthesis of stereoregular α-D-polyribofuranan **57** with extremely high DP (1740) (Scheme 10).[67] The use of a different Lewis acid ($SbCl_5$) and a lock conformation (**58**) permitted to suppress the 1,5-scission and yielded cellulose-like β-D-polyribopyranan **61** (Scheme 10).[68]

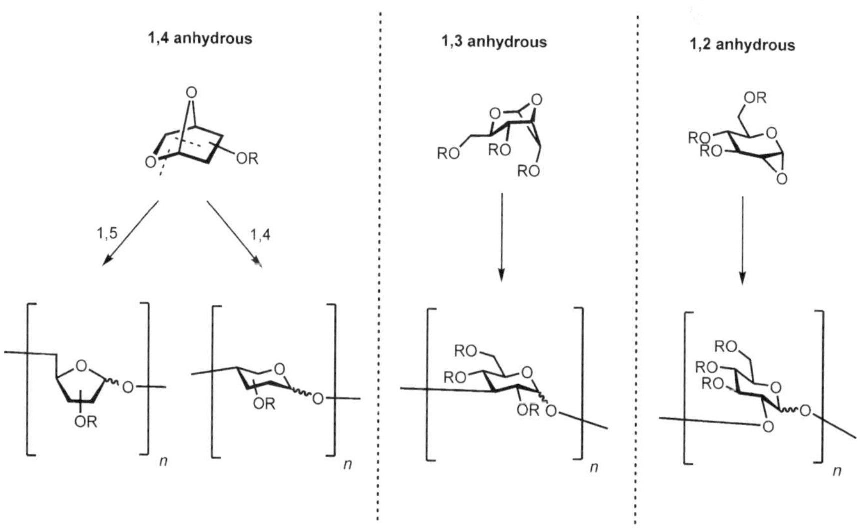

Fig. 7 General scheme of the ROP of different anhydrosugars.

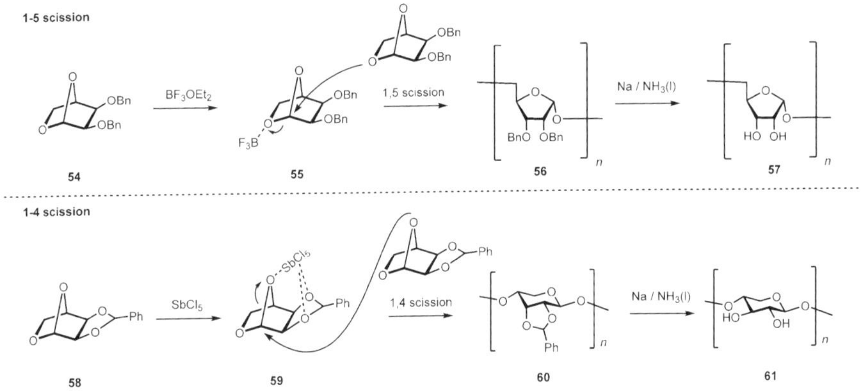

Scheme 10 ROP of stereoregular polyribopyranosides based on the selective 1,5- (*top*) and 1,4- (*bottom*) scission of 1,4-anhydroribopyranose.

α-D-Xylofuranan **65**,[69] α-D-lyxofuranan **66**,[70] and α-L-arabinofuranan **67**[71] were obtained following similar conditions. Upon sulfation, these structures were found to have potent anti-HIV and high blood anticoagulant activities (Table 2).[72]

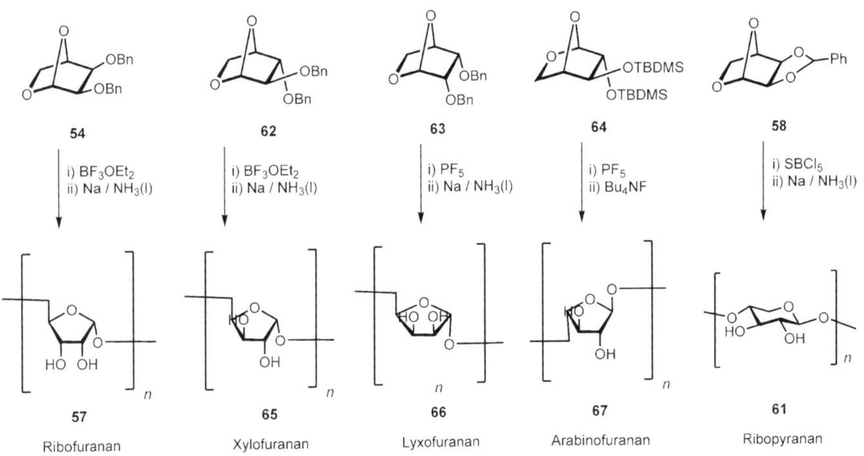

Very few examples of ROP starting from 1,3-anhydrosugars exist and most of them suffer from low stereoselectivity (Fig. 7). Standard Lewis acid catalysis (e.g., PF_5 and $SbCl_5$) gives only short oligosaccharides with no stereoregularity.[73] The use of triflic anhydride (Tf_2O) or silver triflate (AgOTf) permitted to achieve stereoregular 1,3-α-D-polymannan[74] and 1,3-β-D-polyglucan[75] with high MWs. Stabilization of the oxonium ion intermediate through a strong interaction with the catalyst counter-ion is believed to promote the polymerization reaction.[74] Similar problems were observed with 1,2-anhydrosugars. Low temperatures permitted to obtained 1,2-polyglucan, even though with poor stereoselectivity.[76]

Table 2 1,4-Anhydrosugars and their use for ROP.

Monomer	Activator (mol%)	Max DP	Conversion (%)
54	BF_3OEt_2 (2)	1740	88
62	BF_3OEt_2 (5)	477	83
63	PF_5 (5)	–	95
64	PF_5 (2)	–	90
58	$SbCl_5$ (1)	213	91

Anhydrosugar dimers

ROP is particularly useful for the synthesis of branched polysaccharides. Anhydrosugars equipped with a sugar pendant arm allow for the introduction of branches with complete control of the regio- and stereochemistry. Dimers are particularly suitable for this method, whereas longer chains suffer from low reactivity and challenging starting material preparation. Nevertheless, even though the use of dimers for ROP is quite common, it remains very sensitive to reaction conditions (e.g., temperature and catalyst loading).[77] Dimers are often copolymerized with the monomer analog to decrease issues arising from steric hindrance. Control over the branching concentration is obtained by varying the starting materials ratio (Fig. 8).[78] Anhydromaltose **68**,[77]

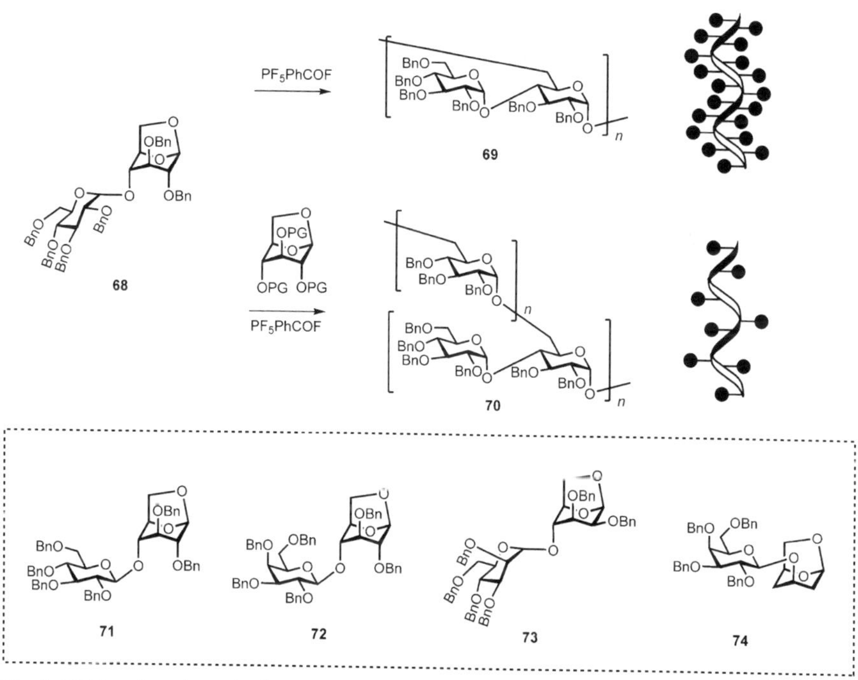

Fig. 8 ROP to form branched polysaccharides and examples of anhydrodimers used for ROP.

anhydrocellobiose **71**,[79] anhydrolactose **73**,[79] and anhydromannobiose **74**[80] are some of the several BBs used for the ROP to obtain branched polysaccharides (Fig. 8). Upon deprotection and sulfation, these polymers show interesting anti-HIV and anticoagulant activities that can be correlated to the branching level (higher activities are observed for less branched polymers).[79] Another strategy to decrease the steric hindrance and favor the polymerization used deoxyanhydrosugars as starting materials.[81] Highly branched polysaccharides were obtained since copolymerization with a monomeric unit is not necessary.

High polymerizability was observed also for some 1,4-anhydrosugars. Dimer **75**[82] and trimer **76**[83] were employed to obtain stereoregular branched α-D-ribopyranans (Scheme 11).

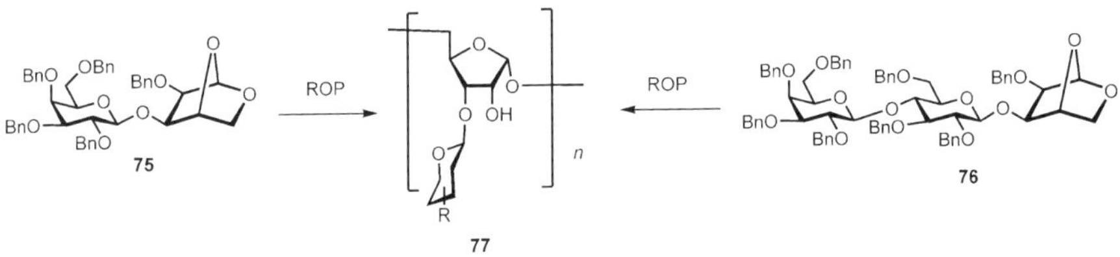

Scheme 11 Examples of 1,4-anhydrosugars used for ROP.

Unprotected anhydrosugars

Similarly to polycondensation, several unprotected anhydrosugars (1,6-anydropyranoses[84–86] and 1,6-anydrofuranoses[87]) were used for the formation of hyperbranched polysaccharides (Fig. 9). The polymerizations were performed at high temperature in solid or solution phase. Despite the success of this technique for the synthesis of linear polymers, in this case, high polarizability was always associated with a broad polydispersion index (PDI).[84] Such broad PDI complicated the analysis, resulting in poor characterization of the obtained polymers. Mixtures of linkages (α and β) as well as pyranosyl and furanosyl repeating units were detected.[86]

A significant improvement to gain better control on the final polymer composition was obtained performing ROP in metal-organic frameworks (Scheme 12).[88] The size of the one-dimensional nanochannels of the material (10.7-Å pore diameter) favors the orientation of the monomers, suppressing the formation of highly branched structures. A quasi-linear polyglucoside **84** was obtained.

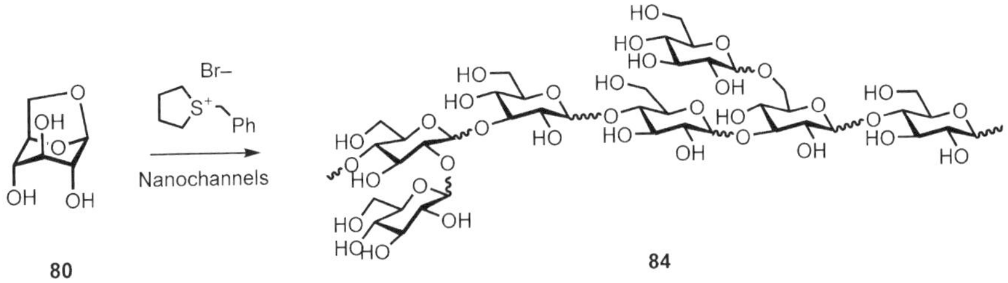

Scheme 12 Synthesis of a quasi-linear polyglucoside in nanochannels.

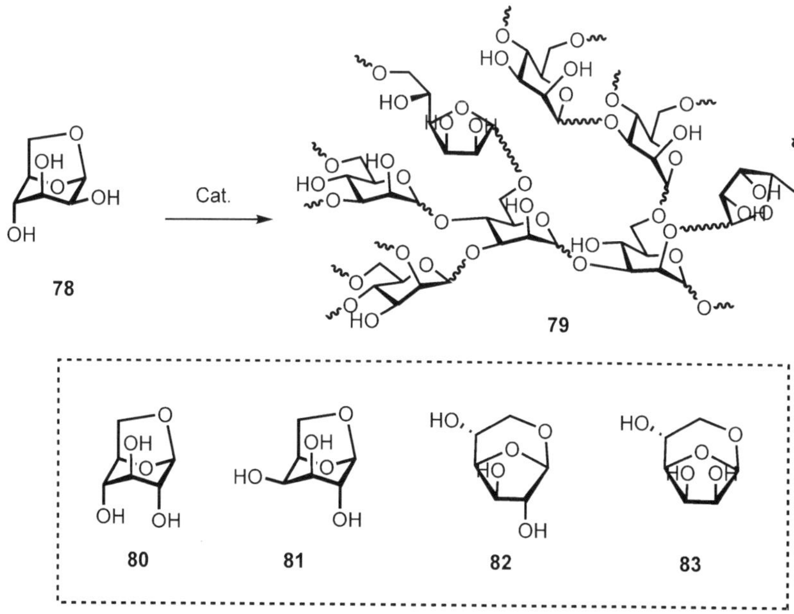

Fig. 9 ROP of 1,6-anhydromannopyranose and examples of unprotected anhydrosugars used for ROP.

2.2.2 ROP of tricyclic orthoesters

In spite of the large amount of work done in order to synthesize cellulose-like polymers using ROP, the regiospecific 1,4-scission of 1,4-anhydro-D-glucopyranose was not successful. In 1990, Nakatsubo and coworkers suggested substituting the 1,4-ether bond with the more reactive tricyclic orthoester 3,6-di-*O*-benzyl-α-D-glucopyranose 1,2,4-orthopivalate **85** to gain regiospecificity. With BB **85**, the first cationic synthesis of cellulose (DP = 20) was possible (Scheme 13).[89] Initiation was performed with Ph_3CBF_4 (5% mol concentration) and the polymerization proceeded in 2h at 20°C. Deprotection was obtained with sodium methoxide (NaOMe), followed by hydrogenation with $Pd(OH)_2$ on carbon. Acetylation permitted to confirm the regio- and stereoselectivity of the process by NMR analysis. Identical results for the synthetic and the authentic cellulose triacetate (CTA) were observed. Subsequent deacetylation allowed for X-ray diffraction analysis, confirming that the synthetic cellulose exists in the cellulose-II crystal structure.

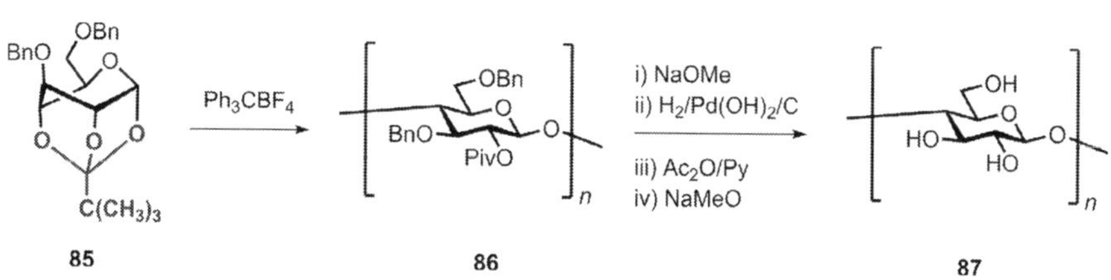

85 86 87

Scheme 13 First cationic synthesis of cellulose.

Different substituents were tested in order to better understand the reaction mechanism (**88–97**). Pivaloyl esters were systematically introduced in replacement of the benzyl ethers (Fig. 10).[90] The ROP outcomes suggested that the benzyl group at the 3-*O* position is indispensable for yielding stereoregular (1,4)-β-D-glucopyranan derivatives. Subsequent results showed that the easily removable allyl group ($PdCl_2$ in $MeOH/CHCl_3$ in 4 h) can also afford stereoregular polysaccharides.[91] Variations of the orthopivalate were not tolerated, as regioirregularities were obtained with compounds **95–97**.[92]

With these results in hands, a mechanism for the reaction was suggested (Scheme 14).[90] Initiation takes place with the formation of the oxonium ion **104** by the complexation of a triphenylcarbenium ion with the oxygen in C4 position. The trialkyloxonium ion mechanism suggested for the ROP of 1,4-anhydrosugars seems unlikely, due to the steric hindrance of the axial 3-*O*-benzyl group. On the other hand, the oxonium ion intermediate could be immediately converted to a dioxalenium ion **105** by the intramolecular backside attack of a lone-pair orbital on the oxygen at C1 or C2. The steric repulsion between trityl and the *tert*-butyl groups favors this pathway. Rapid interconversion gives the more stable dioxalenium ion with a half-chair conformation **106** that can undergo intermolecular reaction with the next monomer (forming **107**). The final compound is obtained with a sequence of chair interconversion followed by nucleophilic attack of the next monomer at the reducing end.

The flexibility of this synthetic approach allowed for the synthesis of unnatural cellulose derivatives (BBs **98–101**). Methyl and ethyl groups were introduced in position 6[93] and in positions 2 and 6,[94] showing differences in solubility and thermoresponsive behavior. Similarly, a 6-deoxy cellulose analog was synthesized.[95] Starting from $^{13}C_6$-glucose, it was also possible to produce a fully ^{13}C-labeled polysaccharide

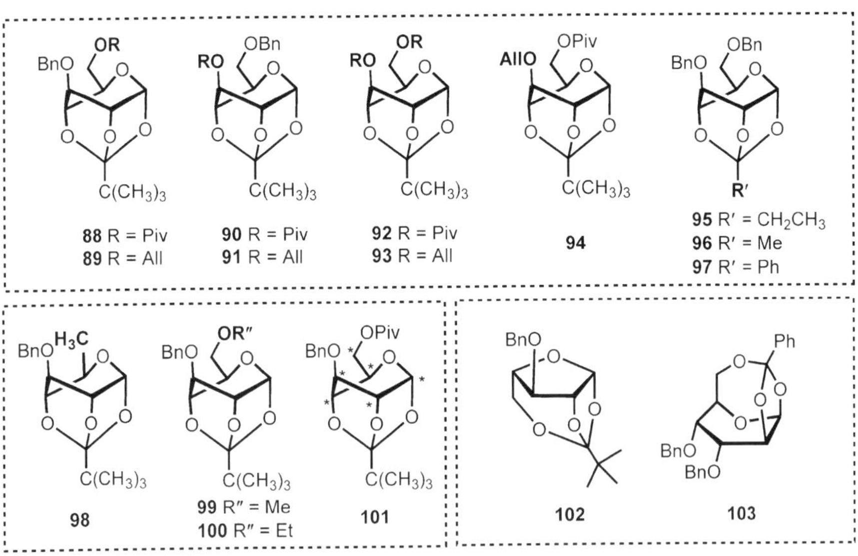

Fig. 10 Modifications of tricyclic orthoesters.

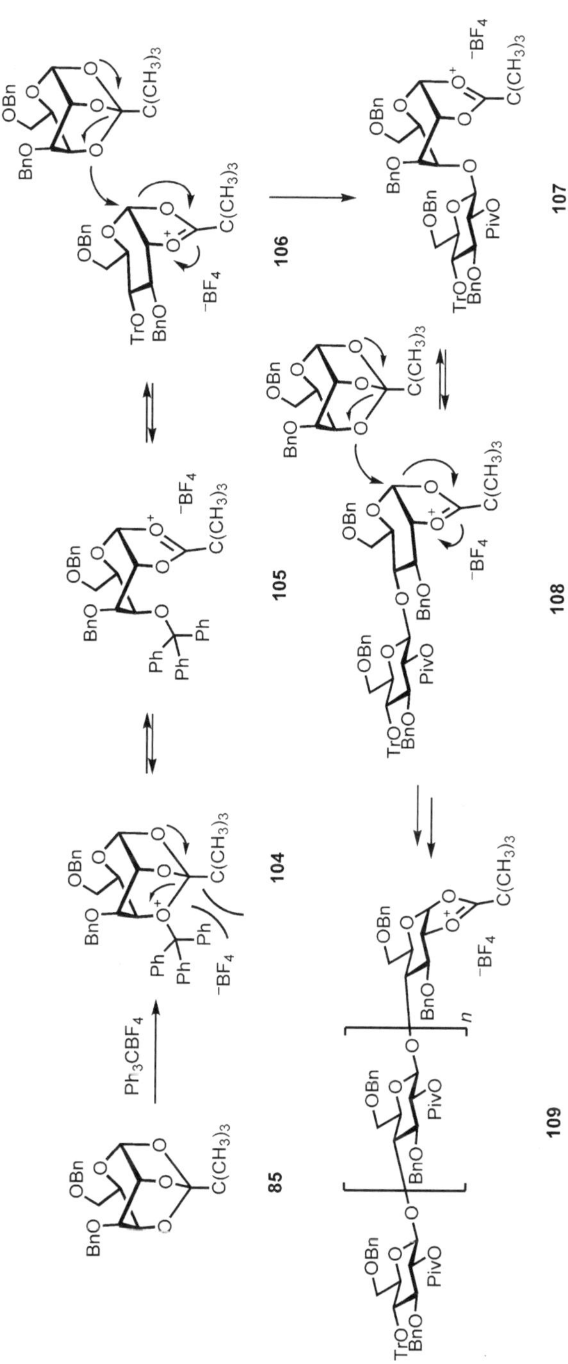

Scheme 14 Suggested mechanism for the ROP of the tricyclic orthoester.

with potential use for NMR analysis.[96] The orthoester approach was then expanded to synthesize 1,5-α-L-arabinofuranan[97] and 1,6-α-D-mannan[98–100] using BBs **102** and **103**, respectively (Fig. 10).

2.2.3 ROP of cyclodextrins (CDs)

The idea of synthesizing monodisperse polysaccharides has always been very attractive but not very applicable when small molecules are polymerized. Separation of polymers with similar MWs becomes highly challenging and often not applicable. The use of a "bigger monomer" was suggested in order to achieve good fractionation of the polymerization products to obtain a monodisperse sample. CD are macrocycles composed of glucose units (6, 7, or 8) connected through α-1,4-linkages and really attractive macromolecular monomers since they can provide a large and regular MW distribution of the polymer product.[35] The ROP of α-, β-, and γ-*O*-permethylcyclodextrins (**111–113**) catalyzed by $Et_3O^+PF_6^-$ was firstly reported in 2001 (Scheme 15).[101] Amylose-like methylated polysaccharides were obtained in good yield with the following reactivity order $\alpha < \beta < \gamma$. This surprising trend is opposite to the ring strain order, suggesting that a bigger cavity (γ-CD) makes the glycosidic bond more accessible, favoring the ROP propagation.[102] NMR studies suggested an increasing amount of β-linkages with increasing monomer conversion. This result was associated with chain transfer reactions during the polymerization and could be partially suppressed when a different activation system was used (HI-I_2).[103] Unfortunately, this method proved to be very sensitive to chemical modification as peracetylated and perbenzylated CDs could not be polymerized under these conditions, limiting the applicability of this approach.[102]

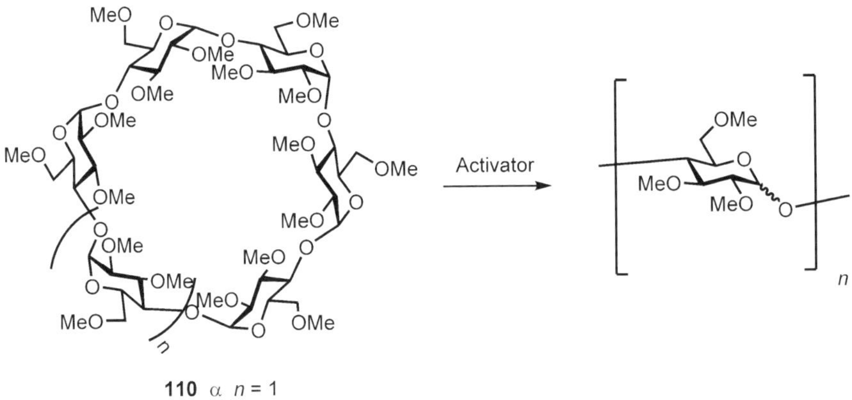

110 α *n* = 1
111 β *n* = 2
112 γ *n* = 3

113

Scheme 15 General scheme of ROP of cyclodextrins.

3 Well-defined structures: Total synthesis

With polymerization reactions, long polysaccharides are accessible in a short time. Due to the intrinsic nature of this approach, the samples are always isolated with a certain size range, which in some cases can prevent a detailed structure-function analysis. Moreover, the synthesis of non-repetitive targets with a defined composition of different BBs in a specific order is not applicable to polymerization. Total synthesis remains the only alternative to obtained well-defined structures with a defined monomer composition. Many elegant works have been reported in the last decades based on solution or solid-phase synthesis that allowed for the access to several oligosaccharides. Depending on the target molecules, different strategies were employed to minimize manual work (Fig. 11). Nevertheless, the synthesis of polysaccharides longer than 10 monosaccharide units is still considerably challenging, often requiring huge chemical efforts and synthetic steps.

3.1 Solution-phase synthesis

Depending on the target compound, the retro-synthetic approach could be extremely different (Fig. 11). Repetitive structures are always synthesized by an iterative assembly of the same BB. The synthesis of non-repetitive structures is generally more complicated and only limited examples are reported.

3.1.1 Repetitive polysaccharides

Most polysaccharides are based on repetitive structures with a repeating unit consisting of one or multiple (linear or branched) glycosidic residues. To synthesize

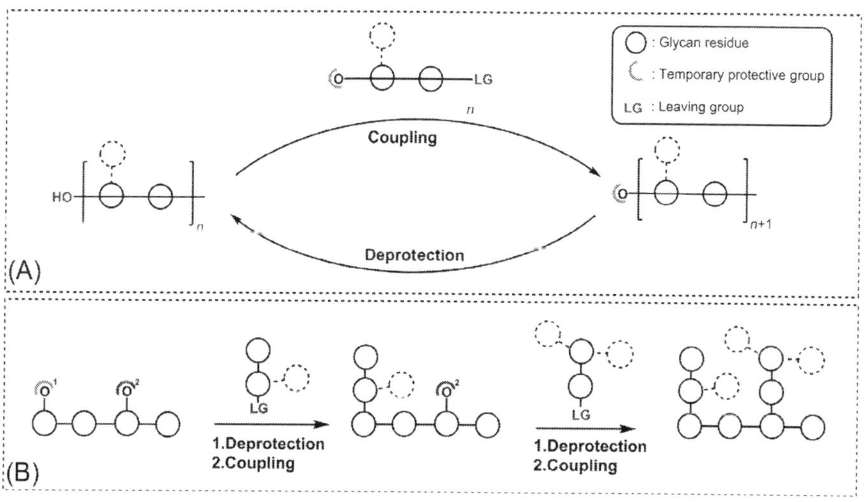

Fig. 11 General scheme of well-defined polysaccharide synthesis. (A) Repetitive structure. (B) Non-repetitive structure.

repetitive polysaccharides, a BB representing the repeating unit needs to be obtained and used iteratively (Fig. 11A). A careful PG strategy is always employed to assure the desired regio- and stereochemistry. Following this approach, the synthetic *Shigella* vaccine candidate **115** containing 16 monosaccharide residues was obtained through the "blockwise" assembly of a tetrasaccharide imidate donor **114** (Scheme 16).[104] An increased amount of BB donor was necessary to ensure completion of glycosylation as the oligosaccharide became longer. By introducing lipophilic, acyl-type PGs to facilitate the purification process, a tetracosasaccharide was then obtained. In 2016, a synthetic carbohydrate conjugate vaccine candidate against *Shigella flexneri* serotype 2a was synthesized. The carbohydrate part of this vaccine candidate consists of a polysaccharide based on three repeating pentasaccharide units and was synthesized following a [5 + 5 + 5] strategy.[105]

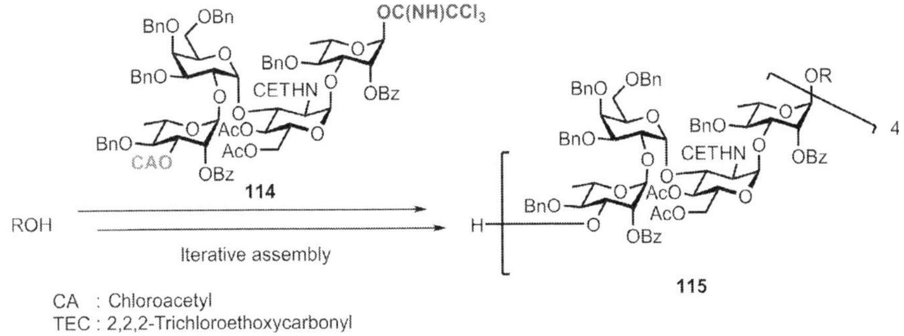

Scheme 16 16-mer obtained through the iterative assembly of a tetrasaccharide imidate donor.

Another remarkable example of solution-phase synthesis of polysaccharides permitted to obtain a series of heparin-backbone-like polysaccharides (as long as 40-mer, **117**) by iterative coupling of a tetrasaccharide BB **116** (Scheme 17).[106] The work also features the deprotection and sulfation on a 20-mer, which represents the longest purely synthetic heparin-like polysaccharide.

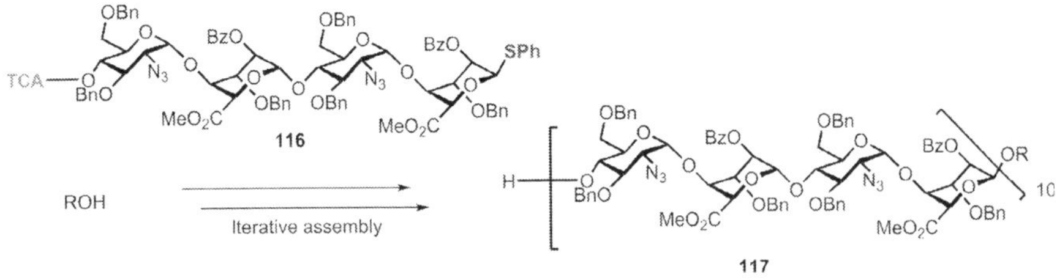

Scheme 17 Heparin backbone obtained through the iterative assembly of a tetrasaccharide thioglycoside donor.

To obtain the tetrasaccharide BBs described above, a considerable synthetic effort is required. The final PG pattern, required for the iterative assembly, is the result of several manipulation steps. Moreover, multiple glycosylation reactions are required.

Some natural products containing glycosidic bonds can be chemically transformed into synthetic BBs. This strategy can minimize the glycosylation steps in the whole synthesis and was nicely applied to the *semi*-synthesis of 6-*O*-methyl-D-glucose-containing polysaccharides **119** (as long as 20-mer) (Scheme 18).[107, 108] BBs **118** were prepared from α-, β-, and γ-cyclodextrins.

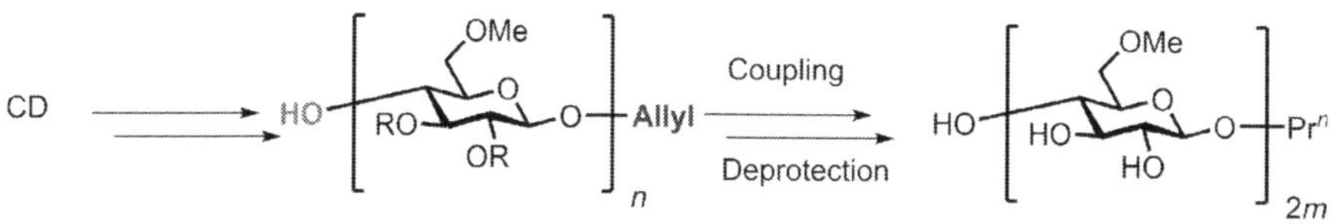

118, *n* = 6,7,8. R = Bn or Bz

119, sMGPs, *m* = 6–10

Scheme 18 Synthesis of 6-*O*-methyl-D-glucose-containing polysaccharides from modified CDs.

When the repeating unit consists of a branched structure, the *retro*-synthesis becomes more complex. As a common strategy, a linear multi-ol acceptor is firstly synthesized and the ensuing multi-glycosylation inserts all the branches in one single step. In 1993, a pentacosasaccharide poly-*N*-acetyl-lactosamine **122** was synthesized (Scheme 19).[109] The key step was the coupling between a trisaccharyl fluoride donor **120** and a decasaccharide pentaol **121** promoted by $Cp_2Hf(OTf)_2$.

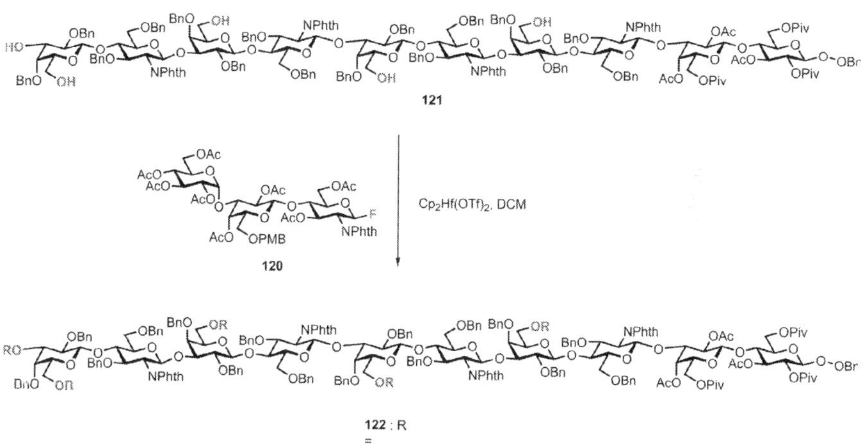

Scheme 19 Key step of synthesis of pentacosasaccharide poly-*N*-acetyl-lactosamine.

Similarly, a lipomannan component of the cell-wall complex of *Mycobacterium tuberculosis* was synthesized (Scheme 20).[110] The pentaol backbone **123** was obtained by iterative coupling of the 6-*O*-triphenylmethyl *n*-pentenylorthoester donor **124**. The 2-OH is released upon the glycosylation. To extend the linear backbone, deprotection of the triphenylmethyl group releases an additional OH in position 6.

Notably, with the use of $Yb(OTf)_3$, the glycosylation majorly takes place on the primary OH, permitting to regioselectively obtain the linear 1,6 backbone. The linear pentaol **123** was then treated with ten equivalents of trichloroacetimidate donor **125** to afford the *pseudo*-dodecasaccharide **126**.

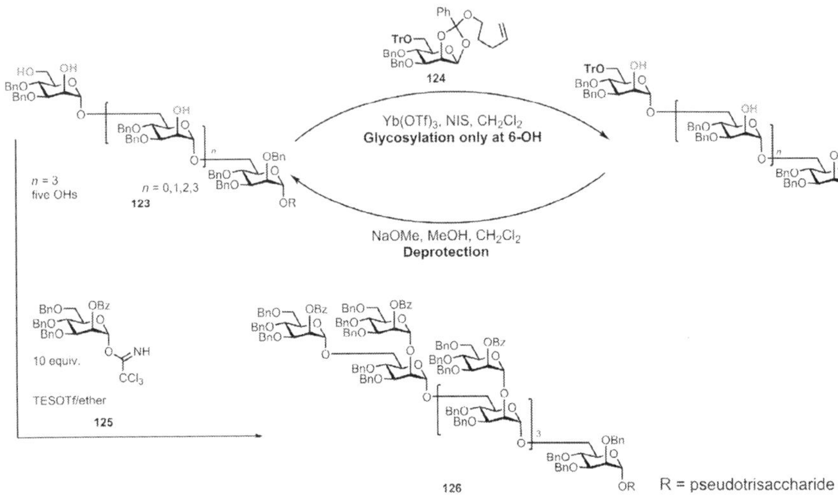

Scheme 20 Key steps of the synthesis of *Mycobacterium tuberculosis* lipomannan.

An interesting approach to obtain a series of polysaccharides with a well-defined branching pattern was reported in 2018.[111] Arabinoxylan oligosaccharides with different length and substitution pattern were converted into their fluoride analogs and subjected to polymerization mediated by glycosynthase to obtain tailor-made arabinoxylans (Scheme 21).

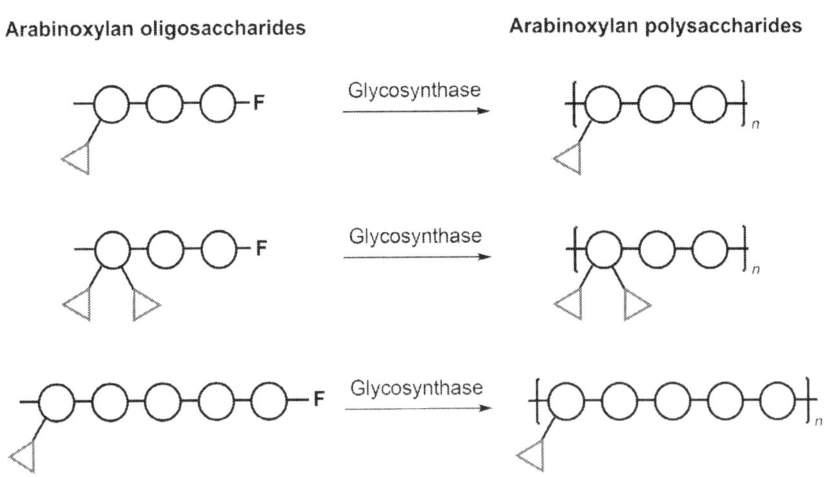

Scheme 21 Tailor-made arabinoxylans by polymerization.

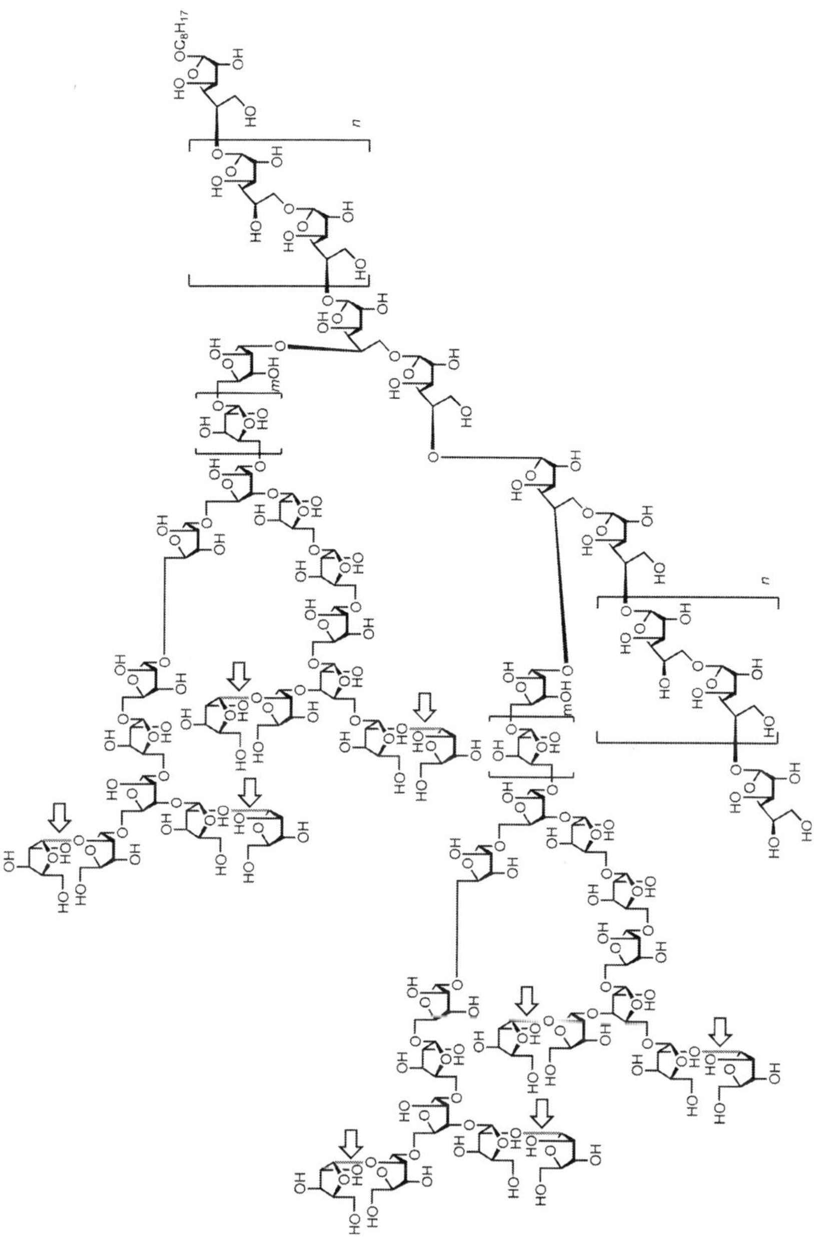

127

Fig. 12 Structure of mycobacterial arabinogalactan with highlighted β-arabinofuran linkages (*arrow*).

3.1.2 Non-repetitive polysaccharides

The synthetic strategy for non-repetitive polysaccharides highly depends on the structures of the target molecule (Fig. 11B). Till now, the effort to synthesize non-repetitive polysaccharides, larger than 10 monosaccharide units, mostly focuses on very few target molecules.

Synthesis of mycobacterial arabinogalactan

The mycobacterial arabinogalactan plays a crucial role in the infectivity, pathogenicity, and drug resistance of *Mycobacterium tuberculosis*.[112] Its structure consists of one or two highly branched arabinan chains and a linear galactan (Fig. 12). Due to its size and structural complexity, it represents one of the most challenging synthetic targets due to the following reasons:

- The synthesis of long repetitive chains calls for a high amount of skillful work.
- Laborious protective group manipulation is necessary for the highly branched arabinan fragment.
- The attachment of the β-arabinofuranose residues requires challenging stereochemical considerations.

The synthesis of mycobacterial arabinogalactan was firstly tackled targeting a short and simplified structure to overcome the synthetic difficulty.

The total synthesis of the arabinan part was accomplished in 2007 *via* a [7+5+5+5] coupling approach (Fig. 13A).[113] The challenging β-arabinofuranose residue (*cis*-linkage) was introduced with the 3,5-*O*-di-*tert*-butylsilanediyl-protected thioglycoside (Scheme 22) stereoselectively. The arabinan motif was then resynthesized following the stereoselective introduction of the β-arabinofuranose residue using a 3,5-*O*-tetraisopropyldisiloxanylidene equipped glyosidic donor (Scheme 22).[114] All the fragments were coupled in a linear sequence. The acetonide at the reducing terminal of each fragment was transformed to the corresponding thioglycoside *via* a three-step reaction for further conjugation (Scheme 23).

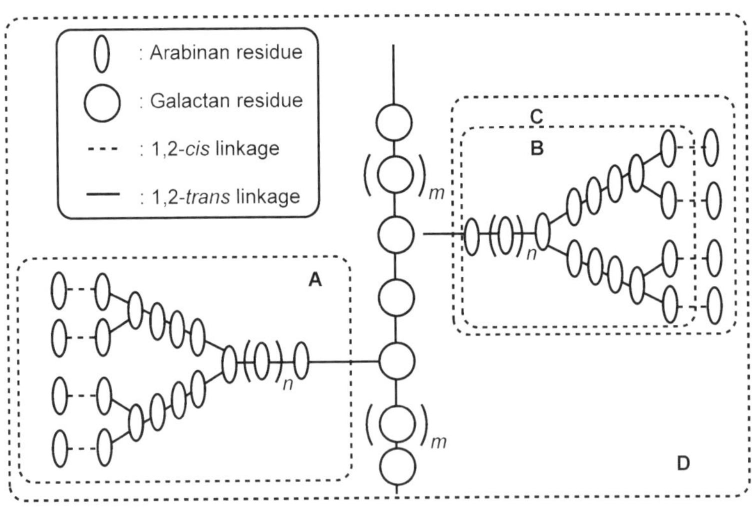

Fig. 13 General and simplified scheme of mycobacterial arabinogalactan.

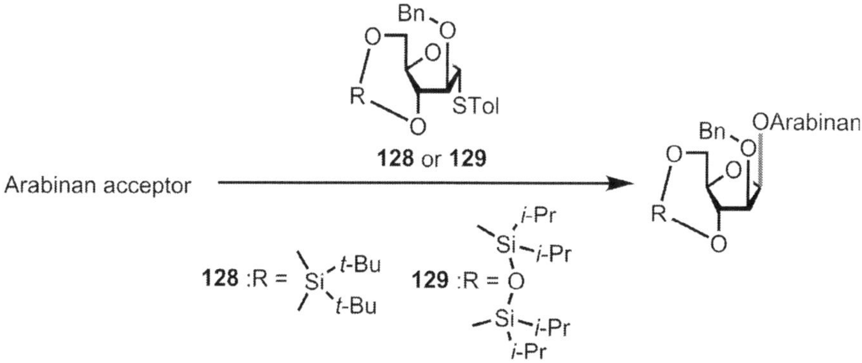

Scheme 22 Protective strategy to obtain selective β-arabinofuranosylation.

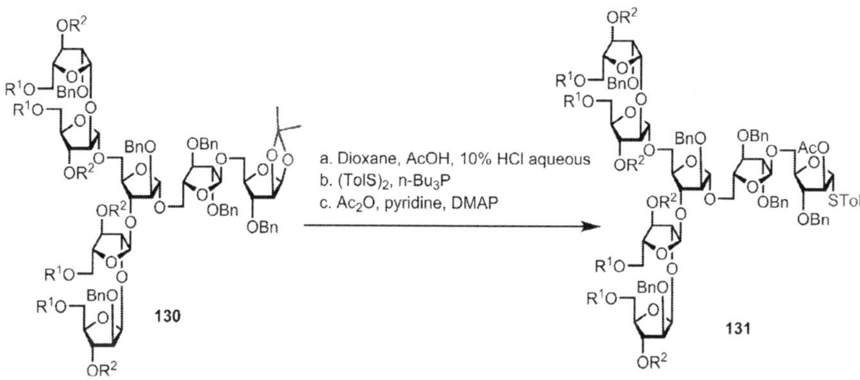

Scheme 23 Example of the transformation of the terminal acetonide moiety into the corresponding thioglycoside for further conjugation.

In 2017, a 47-step synthesis of the arabinogalactan fragment **B** (Fig. 13) was reported in 0.09% overall yield,[115] followed by the synthesis of the polysaccharide analog capped with the terminal β-arabinofuranose units **C** (Fig. 13).[116]

In 2017, an impressive total synthesis of mycobacterial arabinogalactan with 92 monosaccharide units was reported (Fig. 12, $m=13$, $n=9$, Fig. 13D), resulting in the biggest well-defined polysaccharide ever chemically synthesized.[117] The key glycosylation step was a [31 + 31 + 30] coupling, promoted by benzenesulfinyl morpholine/triflic anhydride. A preactivation-based glycosylation strategy was adopted during the assembly to decrease the need for PG manipulation.

Synthesis of mycobacterial lipoarabinomannan-related structures

Arabinomannan is another component of the cell wall of *Mycobacterium tuberculosis,* which is considered responsible for the drug resistance capability.[118, 119] The structure of arabinomannan consists of an arabinan and a lipomannan domain (Fig. 14). Both domains are highly branched. The nonreducing end of the arabinan domain is sometimes capped with a mannan residue. In the lipomannan domain, a phosphatidyl inositol mannoside (PIM) is attached at the reducing end of the mannan residues.

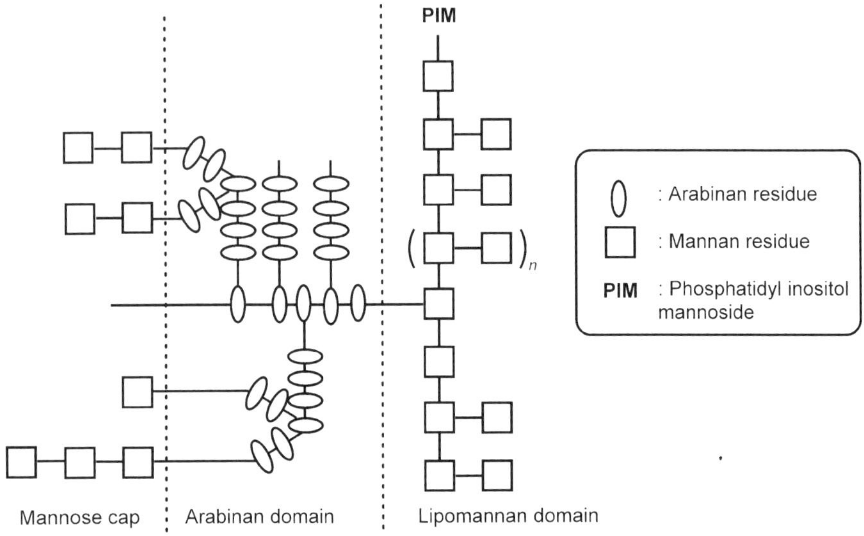

Fig. 14 Representation of the structure of mycobacterial arabinogalactan.

In 2006, a 28-mer polysaccharide core of mycobacterial lipoarabinomannan **132** was synthesized in a protected form (Fig. 15).[120] The deprotected arabinomannan 12-mer **133** was also synthesized following a [6+6] coupling between a hexamannoside and an arabinan catalyzed by TBSOTf.[121] Recently, the synthesis of a heneicosasaccharyl arabinomannan **134** was reported with an overall yield of 0.016%.[122] The structure contained both the branched arabinan domain and the mannan caps. The 1,2-*cis* glycosylation to install the arabinofuranose residue was achieved following [Au]/[Ag]-catalytic conditions.

3.2 Automated solid-phase synthesis

Automated polysaccharide synthesis was developed to reduce the manual work required by classical solution-phase synthesis and to improve the process efficiency. Several approaches have been developed, based on solution or solid-phase approaches.[123] The latter is based on the iterative addition of protected monosaccharides (or disaccharides) to a growing oligosaccharide chain bound to a solid support (Fig. 16).[124] A cleavable linker, containing a free hydroxy group, is bound to the resin and used as first anchor point. A sugar BB, equipped with a reactive leaving group, is chemically activated and coupled to the hydroxy group of the linker. After glycosylation, the excess BB and reactants are removed by filtration. The BB is equipped with a temporary protective group [generally levulinyl group (Lev) or fluorenylmethyloxycarbonyl group (Fmoc)] that permits to release the desired hydroxy group to be used in the next glycosylation step. The iterative glycosylation and deprotection cycles can be programmed and controlled by a computer. Upon completion of the assembly, the linker is cleaved form the solid support, liberating the desired fully protected polysaccharides from the resin. Removal of the PGs (generally through methanolysis of the ester-type PGs followed by hydrogenation of the ether PGs) affords the target polysaccharide.

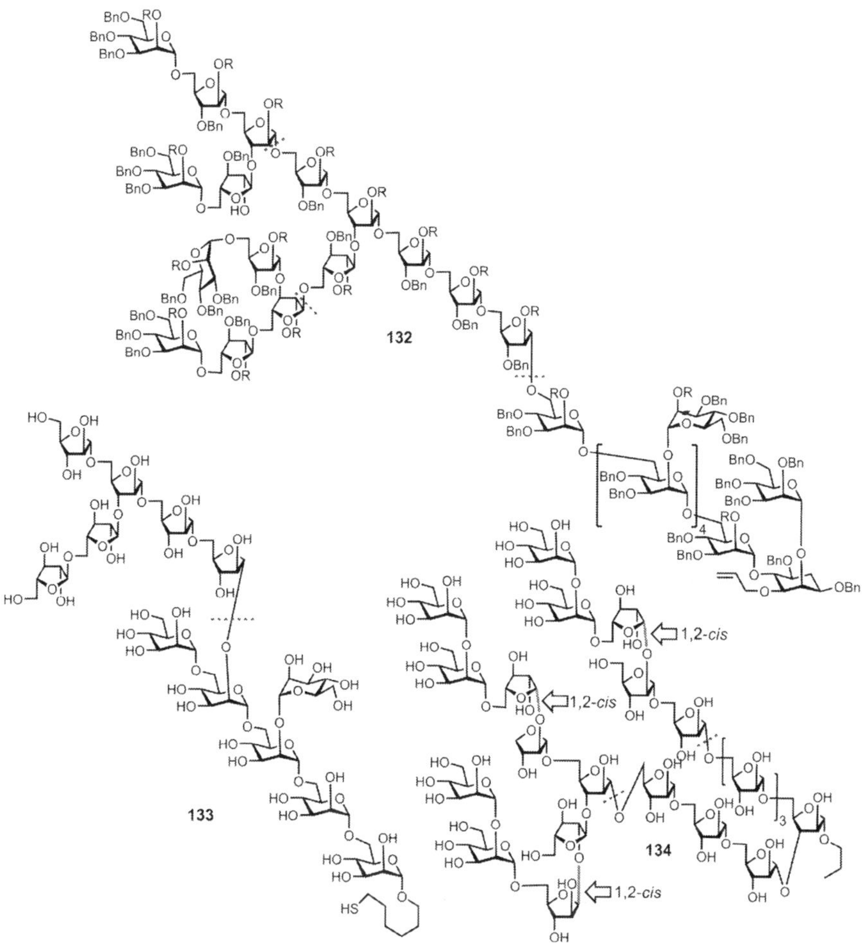

Fig. 15 Reported mycobacterial lipoarabinomannan-related structures with *retro*-synthetic cleavage approach.

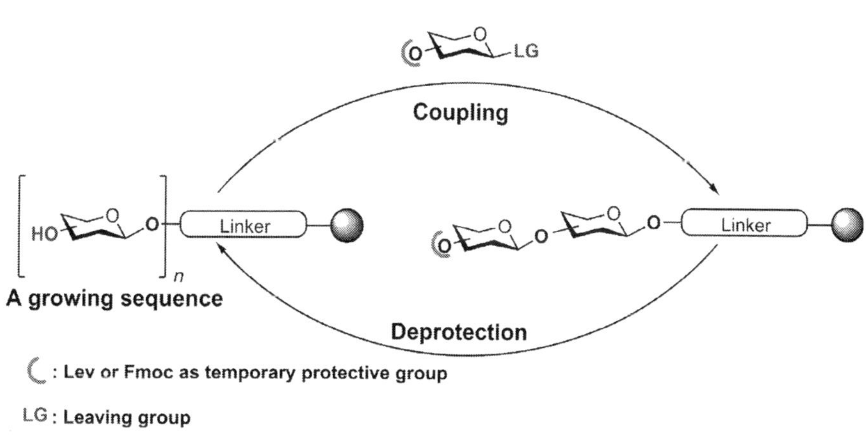

Fig. 16 General scheme of automated solid-phase polysaccharide synthesis.

Following this procedure, the 12-mer β-mannuronic acid alginate **136** was synthesized (Scheme 24).[125] This synthetic route employs Lev as temporary group and trifluoroacetimidate as leaving group. The 12-mer was obtained in a 40% yield, indicating a 90% average coupling efficiency. Similarly, the synthesis of a pentadecasaccharide fragment of hyaluronic acid **139** was reported, with 92% average coupling yield.[126] The use of the disaccharide BB **138** permitted to obtain the desired alternated polysaccharides structure. The entire process of eight cycles of automated coupling and deprotection took only 28 hours.

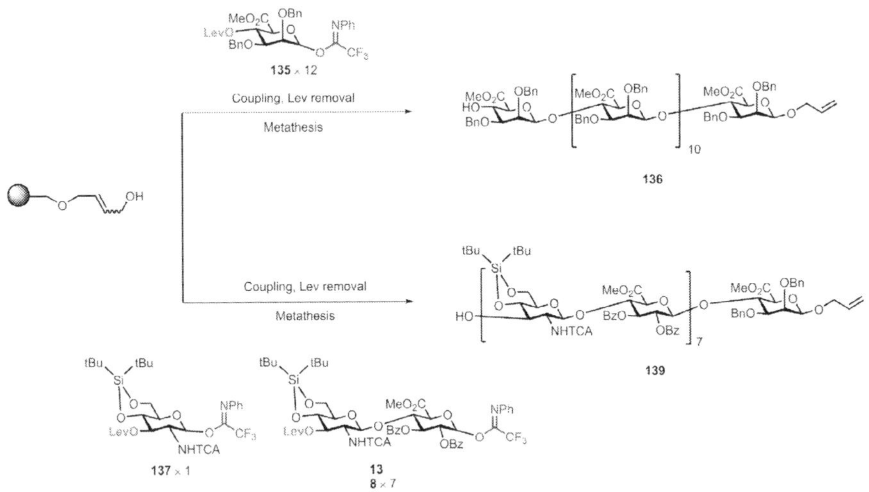

Scheme 24 Synthesis of β-mannuronic acid alginate **136** and hyaluronic acid **139**.

In 2013, a 30-mer polymannoside (**141**) was synthesized by automated glycan assembly using a monosaccharide phosphate BB **140** (Scheme 25).[127] A "catch-release" strategy was introduced to isolate the desired compound from impurities during the purification process.

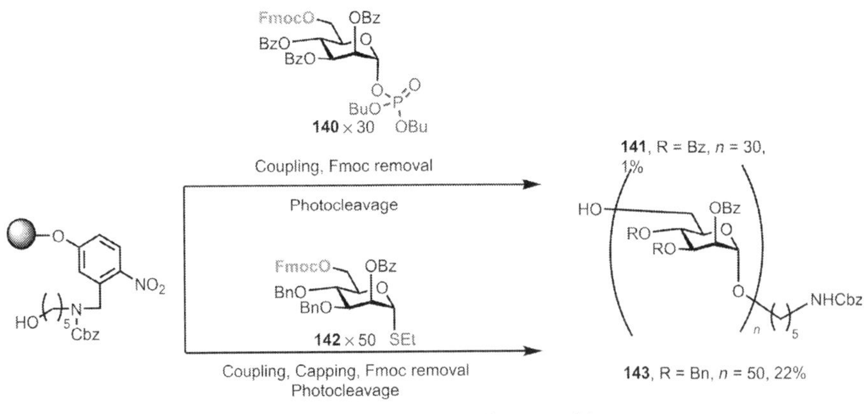

Scheme 25 Automated glycan assembly of linear polymannosides.

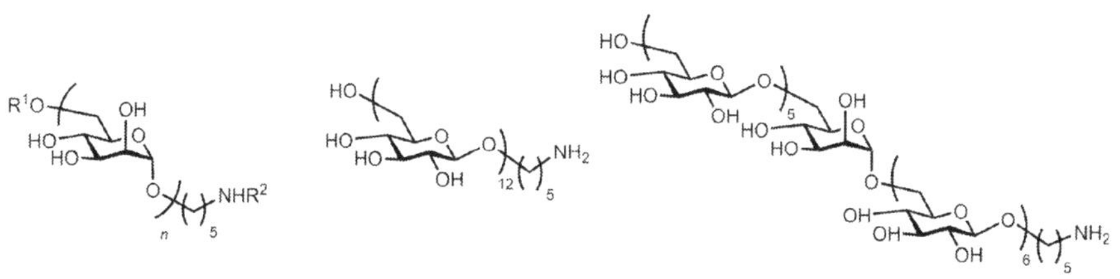

144 $n = 20$, $R^1 = H$, $R^2 = H$
147 $n = 6$, $R^1 = CH_2COOMe$, $R^2 = H$
148 $n = 6$, $R^1 = CH_2COOH$, $R^2 = Cbz$

145

146

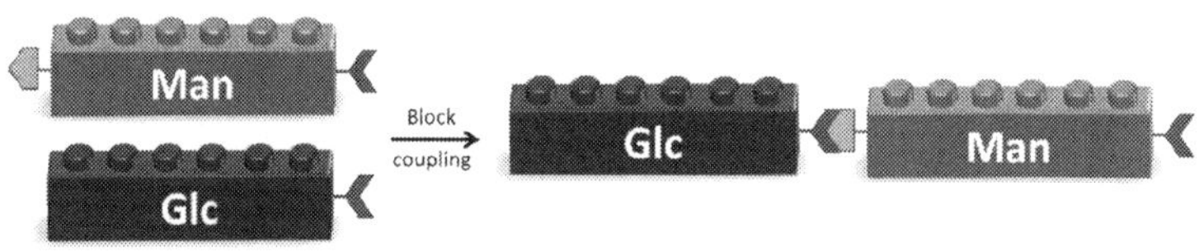

Fig. 17 Synthetic homo- and hetero-polysaccharides for structural study and schematic representation of block coupling.

To push the limit of automated glycan assembly, the 50-mer polymannoside **143** was targeted. A capping procedure was introduced after each glycosylation step, preventing the deletion sequences from growing and permitting to obtain the desired compounds in high purity and 22% yield.[128]

The same approach was extended to the synthesis of polysaccharides based on different repeating units (Fig. 17).[5] A series of polysaccharides (homo- and hetero-polymers) based on mannose, glucose, and glucosamine was synthesized and used for structural studies. The iterative addition of monomeric BBs permits the introduction of modifications (e.g., a different BB or a labeled unit) in specific positions of the chain. Single site modifications drastically affected the geometry and properties of the final compound, suggesting the potential of automated solid-phase synthesis for a detailed structure-activity study of polysaccharides. Moreover, the insertion of a terminal BB bearing a carboxylic acid introduced a linkage point that was exploited for block coupling of different oligosaccharides fragments.

4 Conclusions

The last century has seen a tremendous development of synthetic methodologies to provide artificial polysaccharides, either by polymerization or total synthesis. Nevertheless, chemists are still facing striking challenges stemming from laborious PGs manipulation, stereocontrol of glycosylation, and post-polymerization processes. Microwave-assisted polycondensation of unprotected sugars is an attractive way to access highly branched polysaccharides. PGs manipulation is always necessary to obtain linear polysaccharides, requiring extensive manual work, during both the preparation of the BBs and the polysaccharide deprotection. To achieve high MW and high stereoregularity, ROP remains the most powerful option. Several anhydrosugar analogs allow for the control of the linkage pattern and polymerization reactivity. Nevertheless, strict reaction conditions are sometimes required which may hinder the ROP potential.

Total synthesis of polysaccharides remains the only option to obtain structurally well-defined compounds, but is still extremely challenging. Multiple steps of glycosylations and complex PGs manipulations are required. Nevertheless, for particular target compounds, a careful *retro*-synthetic analysis could considerably minimize the total volume of manual work, and impressive results were obtained. Automated glycan assembly provides another powerful alternative to access well-defined polysaccharides, while minimizing the manual work.

References

1. Varki, A.; Cummings, R.; Esko, J.; Freeze, H.; Stanley, P.; Hart, G.; Seeberger, P. H. *Essentials of Glycobiology*, 3rd ed.; Cold Spring Harbor Laboratory Press: Cold Spring Harbor, N.Y., 2017.
2. Mydock, L. K.; Demchenko, A. V. Mechanism of Chemical O-Glycosylation: From Early Studies to Recent Discoveries. *Org. Biomol. Chem.* **2010**, *8*, 497–510.
3. Wang, C.-C.; Lee, J.-C.; Luo, S.-Y.; Kulkarni, S. S.; Huang, Y.-W.; Lee, C.-C.; Chang, K.-L.; Hung, S.-C. Regioselective One-Pot Protection of Carbohydrates. *Nature* **2007**, *446*, 896.
4. Guo, J.; Ye, X.-S. Protecting Groups in Carbohydrate Chemistry: Influence on Stereoselectivity of Glycosylations. *Molecules* **2010**, *15*, 7235.
5. Delbianco, M.; Kononov, A.; Poveda, A.; Yu, Y.; Diercks, T.; Jiménez-Barbero, J.; Seeberger, P. H. Well-Defined Oligo- and Polysaccharides as Ideal Probes for Structural Studies. *J. Am. Chem. Soc.* **2018**, *140*, 5421–5426.
6. Nigudkar, S. S.; Demchenko, A. V. Stereocontrolled 1,2-cis Glycosylation as the Driving Force of Progress in Synthetic Carbohydrate Chemistry. *Chem. Sci.* **2015**, *6*, 2687–2704.
7. Adero, P. O.; Amarasekara, H.; Wen, P.; Bohé, L.; Crich, D. The Experimental Evidence in Support of Glycosylation Mechanisms at the SN1–SN2 Interface. *Chem. Rev.* **2018**, *118*, 8242–8284.
8. Crich, D. Mechanism of a Chemical Glycosylation Reaction. *Acc. Chem. Res.* **2010**, *43*, 1144–1153.
9. Nielsen, M. M.; Pedersen, C. M. Catalytic Glycosylations in Oligosaccharide Synthesis. *Chem. Rev.* **2018**, *118*, 8285–8358.
10. Komada, H.; Okada, M.; Sumitomo, H. Syntheses and Properties of Polysaccharide-Type Polymers, 2. Poly(cis-3,4-Dihydroxytetrahydropyran-6,2-Dihyloxymethylene). *Makromol. Chem.* **1978**, *179*, 2859–2866.
11. Xiao, R.; Grinstaff, M. W. Chemical Synthesis of Polysaccharides and Polysaccharide Mimetics. *Prog. Polym. Sci.* **2017**, *74*, 78–116.
12. Xiao, R.; Dane, E. L.; Zeng, J.; McKnight, C. J.; Grinstaff, M. W. Synthesis of Altrose Poly-Amido-Saccharides With β-N-(1→2)-d-Amide Linkages: A Right-Handed Helical Conformation Engineered in at the Monomer Level. *J. Am. Chem. Soc.* **2017**, *139*, 14217–14223.
13. Dane, E. L.; Grinstaff, M. W. Poly-Amido-Saccharides: Synthesis via Anionic Polymerization of a β-Lactam Sugar Monomer. *J. Am. Chem. Soc.* **2012**, *134*, 16255–16264.

14. Mikami, K.; Lonnecker, A. T.; Gustafson, T. P.; Zinnel, N. F.; Pai, P.-J.; Russell, D. H.; Wooley, K. L. Polycarbonates Derived From Glucose via an Organocatalytic Approach. *J. Am. Chem. Soc.* **2013**, *135*, 6826–6829.
15. Delbianco, M.; Bharate, P.; Varela-Aramburu, S.; Seeberger, P. H. Carbohydrates in Supramolecular Chemistry. *Chem. Rev.* **2016**, *116*, 1693–1752.
16. Oh, Y. I.; Sheng, G. J.; Chang, S.-K.; Hsieh-Wilson, L. C. Tailored Glycopolymers as Anticoagulant Heparin Mimetics. *Angew. Chem. Int. Ed.* **2013**, *52*, 11796–11799.
17. Kadokawa, J.-i. Precision Polysaccharide Synthesis Catalyzed by Enzymes. *Chem. Rev.* **2011**, *111*, 4308–4345.
18. Kobayashi, S. Challenge of Synthetic Cellulose. *J. Polym. Sci. A Polym. Chem.* **2005**, *43*, 693–710.
19. Wen, L.; Edmunds, G.; Gibbons, C.; Zhang, J.; Gadi, M. R.; Zhu, H.; Fang, J.; Liu, X.; Kong, Y.; Wang, P. G. Toward Automated Enzymatic Synthesis of Oligosaccharides. *Chem. Rev.* **2018**, *118*, 8151–8187.
20. Pfrengle, F. Synthetic Plant Glycans. *Curr. Opin. Chem. Biol.* **2017**, *40*, 145–151.
21. Pacsu, E.; Mora, P. T. Polycondensation of D-Glucose and Other Simple Sugars in Presence of Acids. *J. Am. Chem. Soc.* **1950**, *72*, 1045.
22. Mora, P. T.; Wood, J. W. Synthetic Polysaccharides. I. Polycondensation of Glucose1. *J. Am. Chem. Soc.* **1958**, *80*, 685–692.
23. Kanazawa, A.; Okumura, S.; Suzuki, M. Powder-to-Powder Polycondensation of Natural Saccharides. Facile Preparation of Highly Branched Polysaccharides. *Org. Biomol. Chem.* **2005**, *3*, 1746–1750.
24. Gawande, M. B.; Shelke, S. N.; Zboril, R.; Varma, R. S. Microwave-Assisted Chemistry: Synthetic Applications for Rapid Assembly of Nanomaterials and Organics. *Acc. Chem. Res.* **2014**, *47*, 1338–1348.
25. Wang, H.; Shi, Y.; Le, G. Rapid Microwave-Assisted Synthesis of Polydextrose and Identification of Structure and Function. *Carbohydr. Polym.* **2014**, *113*, 225–230.
26. Li, X.-M.; Le, G.-W.; Cheng, J.-X.; Wang, F.; Shi, Y.-H. Optimization of Microwave-Assisted Solid-Phase Oligosaccharides Synthesis Reaction and Analysis of Components and Structure of Synthetic Product. *Carbohydr. Polym.* **2006**, *64*, 408–418.
27. Wang, H.; Cheng, X.; Shi, Y.; Le, G. Preparation and Structural Characterization of Poly-Mannose Synthesized by Phosphoric Acid Catalyzation Under Microwave Irradiation. *Carbohydr. Polym.* **2015**, *121*, 355–361.
28. Steinmann, A.; Thimm, J.; Matwiejuk, M.; Thiem, J. Formation of Homooligosaccharides Using Base-Promoted Glycosylation of Unprotected Glycosyl Fluorides. *Macromolecules* **2010**, *43*, 3606–3612.
29. Kanazawa, A.; Namiki, S.; Suzuki, M. Thermal Polycondensation of Sugar Fluoride to Form Highly Branched Polysaccharide. *J. Polym. Sci. A Polym. Chem.* **2007**, *45*, 3851–3860.
30. Haq, S.; Whelan, W. J. 873. The Chemical Synthesis of Polysaccharides. Part I. Synthesis of Gentiodextrins. *J. Chem. Soc. (Resumed)* **1956**, 4543–4549.
31. Li, L.; Franckowiak, E. A.; Xu, Y.; McClain, E.; Du, W. Efficient Synthesis of β-(1,6)-Linked Oligosaccharides Through Microwave-Assisted Glycosylation. *J. Polym. Sci. Part A: Polym. Chem.* **2013**, *51*, 3693–3699.
32. Kochetkov, N. K.; Klimov, E. M.; Ovchinnikov, M. V. Synthesis of Polysaccharides. 9. Polycondensation of 2,3,6,2′,3′,4′-hexa-0-Acetyl-α-Maltosyl bromide. *Russ. Chem. Bull.* **1977**, *26*, 1726–1730.
33. Kawada, T.; Yoneda, Y.; Shimizu, K.; Itoh, C. A novel Polycondensation Method for the Synthesis of a Two-Faced β-1,4-Glucan. *Monatsh. Chem.* **2009**, *140*, 1257–1260.

34. Ogawa, Y.; Hinou, H.; Matsuoka, K.; Terunuma, D.; Kuzuhara, H. Synthetic Conversion of Cellobiose Into the Glycal-Type Monomers and Their Polymerization. *Tetrahedron Lett.* **1998**, *39*, 5789–5792.
35. Nishiki, M.; Ousaka, Y.; Nishi, N.; Tokura, S.; Sakairi, N. Chemical Synthesis of an Amylose-Like Polysaccharide by Polymerization of Partially Benzylated Phenyl 1-thio-β-Maltooctaoside Derived From Γ-Cyclodextrin. *Carbohydr. Polym.* **1999**, *39*, 1–6.
36. Kong, L.; Almond, A.; Bayley, H.; Davis, B. G. Chemical Polyglycosylation and Nanolitre Detection Enables Single-Molecule Recapitulation of Bacterial Sugar Export. *Nat. Chem.* **2016**, *8*, 461.
37. Kochetkov, N. K. Tetrahedron Report Number 218: Synthesis of Polysaccharides With a Regular Structure. *Tetrahedron* **1987**, *43*, 2389–2436.
38. Bochkov, A. F.; Obruchnikov, I. V.; Kalinevich, V. M.; Kochetkov, N. K. Synthesis of β-1→6-d-Glucan; A New Reaction of Polycondensation. *Tetrahedron Lett.* **1975**, *16*, 3403–3406.
39. Kitov, P. I.; Tsvetkov, Y. E.; Backinowsky, L. V.; Kochetkov, N. K. Reactivity of 1,2-O-Cyanoalkylidene Sugar Derivatives in Trityl-Cyanoalkylidene Condensation. *Russ. Chem. Bull.* **1993**, *42*, 1423–1428.
40. Kochetkov, N. K.; Obruchnikov, I. V. Stereospecific Block-Polycondensation of Oligosaccharides. Synthesis of a Regular Glucan With Alternating α-1-4 and β-1-6-Linkages. *Tetrahedron Lett.* **1977**, *18*, 57–60.
41. Kochetkov, N.; Klimov, E. Synthesis of Polysaccharides. Glucorhamnan. *Tetrahedron Lett.* **1981**, 337–340.
42. Kochetkov, N. K.; Betaneli, V. I.; Ovchinnikov, M. V.; Backinowsky, L. V. Synthesis of the O-Antigenic Polysaccharide of Salmonella newington and of Its Analogue Differing in Configuration at the Only Glycosidic Centre. *Tetrahedron* **1981**, *37*, 149–156.
43. Kochetkov, N. K.; Byramova, N.É.; Tsvetkov, Y. E.; Backinowsky, L. V. Synthesis of the O-Specific Polysaccharide of Shigellaflexneri. *Tetrahedron* **1985**, *41*, 3363–3375.
44. Kochetkov, N. K.; Nifant'ev, N. E.; Backinowsky, L. V. Synthesis of the Capsular Polysaccharide of Streptococcus Pneumoniae Type 14. *Tetrahedron* **1987**, *43*, 3109–3121.
45. Kong, F. Recent Studies on Reaction Pathways and Applications of Sugar Orthoesters in Synthesis of Oligosaccharides. *Carbohydr. Res.* **2007**, *342*, 345–373.
46. Kadokawa, J.-i.; Watanabe, Y.; Karasu, M.; Tagaya, H.; Chiba, K. Synthesis of a Dibenzylchitin-Type Polysaccharide by Acid-Catalyzed Polymerization. *Macromol. Rapid Commun.* **1996**, *17*, 367–372.
47. Kadokawa, J.-i.; Kasai, S.; Watanabe, Y.; Karasu, M.; Tagaya, H.; Chiba, K. Synthesis of Natural- and Non-Natural-Type Aminopolysaccharides: 2-Acetamido-2-deoxy-β-d-glucopyranan Derivatives by Acid-Catalyzed Polymerization of 2-Methyl(3,6- and 3,4-di-O-benzyl-1,2-dideoxy-α-d-Glucopyrano)-[2,1-d]-2-Oxazolines Involving Stereoregular Glycosylation. *Macromolecules* **1997**, *30*, 8212–8217.
48. Yoshida, T. Synthesis of Polysaccharides Having Specific Biological Activities. *Prog. Polym. Sci.* **2001**, *26*, 379–441.
49. Ko, Y.-C.; Tsai, C.-F.; Wang, C.-C.; Dhurandhare, V. M.; Hu, P.-L.; Su, T.-Y.; Lico, L. S.; Zulueta, M. M. L.; Hung, S.-C. Microwave-Assisted One-Pot Synthesis of 1,6-Anhydrosugars and Orthogonally Protected Thioglycosides. *J. Am. Chem. Soc.* **2014**, *136*, 14425–14431.
50. Ruckel, E. R.; Schuerch, C. Preparation of High Polymers From 1,6-Anhydro-2,3,4-tri-O-Substituted β-D-Glucopyranose. *J. Organomet. Chem.* **1966**, *31*, 2233–2239.

51. Ruckel, E. R.; Schuerch, C. Chemical Synthesis of a Stereoregular Linear Polysaccharide. *J. Am. Chem. Soc.* **1966**, *88*, 2605–2606.
52. Uryu, T.; Tachikawa, H.; Ohaku, K. I.; Terui, K.; Matsuzaki, K. Synthesis of 2,3,4-tri-O-benzyl-[1→6]-α-D-Glucopyranan and [1→6]-α-D-Glucopyranan With High Molecular Weight by Polymerization of 1,6-anhydro-2,3,4-tri-O-benzyl-β-D-Glucopyranose. *Makromol. Chem.* **1977**, *178*, 1929–1940.
53. Lin, J.W.-P.; Schuerch, C. Synthesis and Properties of Stereoregular 2,3,4-tri-o-acetyl-[1→6]-α-D-gluco-, -manno-, and -Galactopyranans. *J. Polym. Sci. A-1 Polym. Chem.* **1972**, *10*, 2045–2060.
54. Ichikawa, H.; Kobayashi, K.; Sumitomo, H.; Schuerch, C. Synthesis of a β-(1→6)-Linked Polysaccharide via Ring-Opening Polymerization With Neighboring-Group Participation. *Carbohydr. Res.* **1988**, *179*, 315–320.
55. Uryu, T.; Sakamoto, Y.; Hatanaka, K.; Matsuzaki, K. Chemical Synthesis of a New Polysaccharide. Ring-Opening Polymerization of 1,6-anhydro-2,3,4-tri-O-benzyl-.beta.-D-Allopyranose and Preparation of Stereoregular (1→6)-α-D-Allopyranan. *Macromolecules* **1984**, *17*, 1307–1312.
56. Kakuchi, T.; Kusuno, A.; Miura, M.; Kaga, H. Cationic ring-Opening Polymerization of 1,6-anhydro-2,3,4-tri-O-allyl-β-D-Glucopyranose as a Convenient Synthesis of Dextran. *Macromol. Rapid Commun.* **2000**, *21*, 1003–1006.
57. Kobayashi, K.; Kondo, T. Synthesis of a Regiospecifically Fluorinated Polysaccharide 3-Deoxy-3-fluoro-(1→6)-α-d-Glucopyranan via Ring-Opening Polymerization. *Macromolecules* **1997**, *30*, 6531–6535.
58. Kobayashi, K.; Sumitomo, H. Regioselectively Modified Stereoregular Polysaccharides. 2. Synthesis of 3-O-methyl-(1→6)-α-D-Glucopyranan. *Macromolecules* **1981**, *14*, 250–253.
59. Kobayashi, K.; Sumitomo, H.; Yasui, A. Regioselectively Modified Stereoregular Polysaccharide. 1. Polymerization of 1,6-Anhydro-3-O-acetyl-2,4-di-O-benzyl-β-D-Glucopyranose and Synthesis of 2,4-Di-O-benzyl-(1→6)-α-D-Glucopyranan. *Macromolecules* **1979**, *12*, 1019–1023.
60. Uryu, T.; Hatanaka, K.; Matsuzaki, K.; Kuzuhara, H. Chemical Synthesis of Amino-Group-Containing (1→6)-α-D-Glucan Derivatives by Ring-Opening Polymerization of 1,6-Anhydroazido Sugars. *Macromolecules* **1983**, *16*, 853–858.
61. Hattori, K.; Yoshida, T.; Uryu, T. Ring-Opening Polymerization and Copolymerization of a New Benzylated 1,6-anhydro-3-azido-3-deoxy-β-D-Allopyranose and Synthesis of Amino-Polysaccharides With 1,6-α-Allopyranosidic Structure. *Macromol. Chem. Phys.* **1997**, *198*, 29–39.
62. Hattori, K.; Yoshida, T. Synthesis of a New 2-Amino-Glycan, Poly-(1→6)-α-D-Mannosamine, by Ring-Opening Polymerization of 1,6-Anhydro-Mannosamine Derivatives. *J. Polym. Sci. A Polym. Chem.* **2012**, *50*, 4524–4531.
63. Hattori, K.; Yoshida, T. Ring-Opening Polymerization of a Benzylated 1,6-Anhydro-β-d-talopyranose and Synthesis of a New Polysaccharide, (1→6)-α-d-Talopyranan. *Macromolecules* **2009**, *42*, 6044–6049.
64. Kobayashi, K.; Schuerch, C. Copolymerization of Anhydroglucose and Anhydromannose Derivatives: Structure, Reactivity, and Conformational Analyses. *J. Polym. Sci. Polym. Chem. Ed.* **1977**, *15*, 913–926.
65. Ito, H.; Schuerch, C. Synthesis of .alpha.-(1 .fwdarw. 3)-Branched Dextrans by Copolymerization and .alpha.-D-Glucosidation. *J. Am. Chem. Soc.* **1979**, *101*, 5797–5806.
66. Kops, J.; Schuerch, C. The Polymerization of 1,4-Anhydro Sugar Derivatives. *J. Polym. Sci., C Polym. Symp.* **1965**, *11*, 119–138.

67. Uryu, T.; Yamanouchi, J.; Kato, T.; Higuchi, S.; Matsuzaki, K. Selective Ring-Opening Polymerization of di-O-Methylated and di-O-Benzylated 1,4-anhydro-.alpha.-D-Ribopyranoses and Structure Proof of Synthetic Cellulose-Type Polysaccharide (1 .fwdarw. 4)-.beta.-D-Ribopyranan and (1 .fwdarw. 5)-.alpha.-D-Ribofuranan. *J. Am. Chem. Soc.* **1983**, *105*, 6865–6871.
68. Uryu, T.; Kitano, K.; Ito, K.; Yamanouchi, J.; Matsuzaki, K. Selective Ring-Opening Polymerization of 1,4-anhydro-α-D-Ribopyranose Derivatives and Synthesis of Stereoregular (1→4)-β-D-Ribopyranan. *Macromolecules* **1981**, *14*, 1–9.
69. Uryu, T.; Yamanouchi, J.; Hayashi, S.; Tamaki, H.; Matsuzaki, K. Selective Ring-Opening Polymerization of 1,4-anhydro-2,3-di-O-benzyl-α-D-Xylopyranose and Synthesis of Stereoregular (1→5)-.alpha.-D-Xylofuranan. *Macromolecules* **1983**, *16*, 320–326.
70. Hagino, A.; Yoshida, S.; Shinpuku, T.; Matsuzaki, K.; Uryu, T. Selective Ring-Opening Polymerization of 1,4-anhydro-α-D-Lyxopyranose Derivatives and Synthesis of Stereoregular (1→5)-α-D-Lyxofuranan. *Macromolecules* **1986**, *19*, 1–7.
71. Yoshida, T.; Kida, M.; Uryu, T. Selective Ring-Opening Polymerization of Di-O-t-Butyldimethylsilylated and Di-O-p-Bromobenzylated 1,4-Anhydro-α-L-Arabinopyranoses and Structural Analysis of Free Arabinans. *Polym. J. (Tokyo, Jpn.)* **1987**, *19*, 923.
72. Yoshida, T.; Kang, B. W.; Hattori, K.; Mimura, T.; Kaneko, Y.; Nakashima, H.; Premanathan, M.; Aragaki, R.; Yamamoto, N.; Uryu, T. Anti-HIV Activity of Sulfonated Arabinofuranan and Xylofuranan. *Carbohydr. Polym.* **2001**, *44*, 141–150.
73. Ito, H.; Schuerch, C. Polymerization of 1,3-anhydro-2,4,6-tri-O-benzyl-.beta.-D-Glucopyranose, a Substituted 2,6-Dioxabicyclo[3.1.1]heptane. *Macromolecules* **1981**, *14*, 246–249.
74. Kong, F.; Schuerch, C. Synthesis of (1→3)-α-D-Mannopyranan by Stereoregular Cationic Polymerization of Substituted 2,6-Dioxabicyclo[3.1.1]heptanes. *Macromolecules* **1984**, *17*, 983–989.
75. Good, F. J.; Schuerch, C. Synthesis of (1 .fwdarw. 3)-.alpha.-D-Glucopyranan by Stereoregular Cationic Polymerization of Substituted 2,6-Dioxabicyclo[3.1.1]heptanes: 1,3-anhydrotri-(p-Substituted-benzyl)-.beta.-D-Glucopyranoses. *Macromolecules* **1985**, *18*, 595–599.
76. Sharkey, P. F.; Eby, R.; Schuerch, C. Chemical Synthesis of a (1→2)-d-Glucopyranan. *Carbohydr. Res.* **1981**, *96*, 223–229.
77. Veruovic, B.; Schuerch, C. The Preparation of Comb-Shaped Polysaccharides by Polymerization of a Maltose Derivative. *Carbohydr. Res.* **1970**, *14*, 199–206.
78. Lindenberger, W. H.; Schuerch, C. Copolymerization of 1,6-Anhydroglucose and 1,6-Anhydromaltose Derivatives. *J. Polym. Sci. Polym. Chem. Ed.* **1973**, *11*, 1225–1235.
79. Masura, V.; Schuerch, C. Polymerization of a Cellobiose Derivative to Comb-Shaped Oligosaccharides. *Carbohydr. Res.* **1970**, *15*, 65–72.
80. Kobayashi, K.; Nomura, K.; Okada, M. Chemical Synthesis of a Comb-Shaped, Branched Stereoregular Polysaccharide, 4-O-α-d-Mannopyranosyl-(1→6)-α-d-Mannopyranan. *Carbohydr. Res.* **1993**, *242*, 161–166.
81. Ichikawa, H.; Kobayashi, K.; Sumitomo, H. Synthesis of a Comb-Shaped Branched Polysaccharide via Ring-Opening Polymerization of a Reactive Anhydro Disaccharide Derivative. *Macromolecules* **1990**, *23*, 1884–1886.
82. Yoshida, T.; Endo, K. Synthesis of Branched Ribo-Polysaccharides With Defined Structures by Ring-Opening Polymerization of 1,4-Anhydro Ribo-Disaccharide Monomer. *Macromol. Biosci.* **2001**, *1*, 298–304.
83. Yoshida, D.; Han, S.; Narita, K.-I.; Hattori, K.; Yoshida, T. Synthesis of Oligosaccharide-Branched Ribofuranan by Ring-Opening Polymerization of a New Anhydroribo Trisaccharide Monomer. *Macromol. Biosci.* **2009**, *9*, 687–693.

84. Abe, H.; Prins, W. Addition Polymerization of Anhydrosugar Derivatives. II. The Molecular Weight Distribution of Polyanhydroglucose Addition Polymers. *Makromol. Chem.* **1960**, *42*, 216–229.
85. Bhattacharya, A.; Schuerch, C. Polymerization of Anhydro Sugar Derivatives. III. 1,6-Anhydro-β-D-Galactopyranose and Its 2-O-Methyl Ether. *J. Organomet. Chem.* **1961**, *26*, 3101–3104.
86. Satoh, T.; Imai, T.; Ishihara, H.; Maeda, T.; Kitajyo, Y.; Narumi, A.; Kaga, H.; Kaneko, N.; Kakuchi, T. Synthesis of Hyperbranched Polysaccharide by Thermally Induced Cationic Polymerization of 1,6-Anhydro-β-d-mannopyranose. *Macromolecules* **2003**, *36*, 6364–6370.
87. Hoai, N. T.; Sasaki, A.; Sasaki, M.; Kaga, H.; Kakuchi, T.; Satoh, T. Synthesis, Characterization, and Lectin Recognition of Hyperbranched Polysaccharide Obtained From 1,6-Anhydro-d-hexofuranose. *Biomacromolecules* **2011**, *12*, 1891–1899.
88. Kobayashi, Y.; Horie, Y.; Honjo, K.; Uemura, T.; Kitagawa, S. The Controlled Synthesis of Polyglucose in One-Dimensional Coordination Nanochannels. *Chem. Commun.* **2016**, *52*, 5156–5159.
89. Nakatsubo, F.; Kamitakahara, H.; Hori, M. Cationic Ring-Opening Polymerization of 3,6-Di-O-benzyl-α-d-glucose 1,2,4-Orthopivalate and the First Chemical Synthesis of Cellulose. *J. Am. Chem. Soc.* **1996**, *118*, 1677–1681.
90. Kamitakahara, H.; Hori, M.; Nakatsubo, F. Substituent Effect on Ring-Opening Polymerization of Regioselectively Acylated α-d-Glucopyranose 1,2,4-Orthopivalate Derivatives. *Macromolecules* **1996**, *29*, 6126–6131.
91. Karakawa, M.; Kamitakahara, H.; Takano, T.; Nakatsubo, F. The Utility of a 3-O-Allyl Group as a Protective Group for Ring-Opening Polymerization of α-d-Glucopyranose 1,2,4-Orthopivalate Derivatives. *Biomacromolecules* **2002**, *3*, 538–546.
92. Hori, M.; Kamitakahara, H.; Nakatsubo, F. Substituent Effects of the Orthoester Group on Ring-Opening Polymerization of α-d-Glucopyranose 1,2,4-Orthoester Derivatives. *Macromolecules* **1997**, *30*, 2891–2896.
93. Kamitakahara, H.; Funakoshi, T.; Nakai, S.; Takano, T.; Nakatsubo, F. Syntheses of 6-O-ethyl/methyl-Celluloses via Ring-Opening Copolymerization of 3-O-Benzyl-6-O-Ethyl/Methyl-α-d-Glucopyranose 1,2,4-Orthopivalates and Their Structure–Property Relationships. *Cellulose* **2009**, *16*, 1179.
94. Kamitakahara, H.; Funakoshi, T.; Takano, T.; Nakatsubo, F. Syntheses of 2,6-O-Alkyl Celluloses: Influence of Methyl and Ethyl Groups Regioselectively Introduced at O-2 and O-6 Positions on Their Solubility. *Cellulose* **2009**, *16*, 1167.
95. Hori, M.; Nakatsubo, F. Substituent Effects of the C6-Position on Ring-Opening Polymerization of Glucose Ortho Esters: Synthesis of Stereoregular 6-Deoxy-(1→4)-β-d-Glucopyranan. *Macromolecules* **2001**, *34*, 2476–2481.
96. Adelwöhrer, C.; Takano, T.; Nakatsubo, F.; Rosenau, T. Synthesis of 13C-Perlabeled Cellulose With More Than 99% Isotopic Enrichment by a Cationic Ring-Opening Polymerization Approach. *Biomacromolecules* **2009**, *10*, 2817–2822.
97. Hori, M.; Nakatsubo, F. Ring-Opening Polymerization of 3-O-Benzyl-β-l-Arabinofuranose 1,2,5-Orthopivalate and Synthesis of Stereoregular (1→5)-α-l-Arabinofuranan. *Macromolecules* **2000**, *33*, 1148–1151.
98. Yongyat, C.; Ruchirawat, S.; Boonyarattanakalin, S. Polymerization of Mannosyl Tricyclic Orthoesters for the Synthesis of α(1–6) Mannopyranan—The Backbone of Lipomannan. *Bioorg. Med. Chem.* **2010**, *18*, 3726–3734.
99. Wattanasiri, C.; Paha, J.; Ponpuak, M.; Ruchirawat, S.; Boonyarattanakalin, S. Synthesis of Synthetic Mannan Backbone Polysaccharides Found on the Surface of Mycobacterium tuberculosis as a Vaccine Adjuvant and Their Immunological Properties. *Carbohydr. Polym.* **2017**, *175*, 746–755.

100. Leelayuwapan, H.; Kangwanrangsan, N.; Chawengkirttikul, R.; Ponpuak, M.; Charlermroj, R.; Boonyarattanakalin, K.; Ruchirawat, S.; Boonyarattanakalin, S. Synthesis and Immunological Studies of the Lipomannan Backbone Glycans Found on the Surface of Mycobacterium tuberculosis. *J. Organomet. Chem.* **2017**, *82*, 7190–7199.
101. Suzuki, M.; Numata, O.; Shimazaki, T. Cyclodextrin as a Macrocyclic Monomer: Cationic Ring-Opening Polymerization of O-Permethylcyclodextrins. *Macromol. Rapid Commun.* **2001**, *22*, 1354–1357.
102. Suzuki, M.; Numata, O.; Shimazaki, T. Novel Cationic Ring-Opening Polymerization of Cyclodextrin: A Uniform Macrocyclic Monomer With Unique Character. *Macromol. Symp.* **2004**, *215*, 255–266.
103. Suzuki, M.; Shimazaki, T. Preparation of a Unique Glucan With Large Intervals in Molecular Weight Distribution. Controlled Ring-Opening Polymerization of O-Permethylcyclodextrin. *Org. Biomol. Chem.* **2003**, *1*, 604–608.
104. Pozsgay, V. Synthetic Shigella Vaccines: A Carbohydrate–Protein Conjugate With Totally Synthetic Hexadecasaccharide Haptens. *Angew. Chem. Int. Ed.* **1998**, *37*, 138–142.
105. van der Put, R. M. F.; Kim, T. H.; Guerreiro, C.; Thouron, F.; Hoogerhout, P.; Sansonetti, P. J.; Westdijk, J.; Stork, M.; Phalipon, A.; Mulard, L. A. A Synthetic Carbohydrate Conjugate Vaccine Candidate Against Shigellosis: Improved Bioconjugation and Impact of Alum on Immunogenicity. *Bioconjug. Chem.* **2016**, *27*, 883–892.106.
106. Hansen, S. U.; Miller, G. J.; Cliff, M. J.; Jayson, G. C.; Gardiner, J. M. Making the Longest Sugars: A Chemical Synthesis of Heparin-Related [4]n Oligosaccharides From 16-mer to 40-mer. *Chem. Sci.* **2015**, *6*, 6158–6164.
107. Meppen, M.; Wang, Y.; Cheon, H.-S.; Kishi, Y. Synthetic 6-O-Methylglucose-Containing Polysaccharides (sMGPs): Design and Synthesis. *J. Organomet. Chem.* **2007**, *72*, 1941–1950.
108. Cheon, H.-S.; Lian, Y.; Kishi, Y. Practical and Scalable Synthesis of α-(1→4)-Linked Polysaccharides Composed of 6-O-Methyl-d-Glucose. *Org. Lett.* **2007**, *9*, 3327–3329.
109. Matsuzaki, Y.; Ito, Y.; Nakahara, Y.; Ogawa, T. Synthesis of Branched Poly-N-Acetyl-Lactosamine Type Pentaantennary Pentacosasaccharide: Glycan Part of a Glycosyl Ceramide From Rabbit Erythrocyte Membrane. *Tetrahedron Lett.* **1993**, *34*, 1061–1064.
110. Jayaprakash, K. N.; Lu, J.; Fraser-Reid, B. Synthesis of a Lipomannan Component of the Cell-Wall Complex of Mycobacterium tuberculosis Is Based on Paulsen's Concept of Donor/Acceptor "Match". *Angew. Chem. Int. Ed.* **2005**, *44*, 5894–5898.
111. Senf, D.; Ruprecht, C.; Kishani, S.; Matic, A.; Toriz, G.; Gatenholm, P.; Wågberg, L.; Pfrengle, F. Tailormade Polysaccharides With Defined Branching Patterns: Enzymatic Polymerization of Arabinoxylan Oligosaccharides. *Angew. Chem. Int. Ed.* **2018**, *57*, 11987–11992.
112. Brennan, P. J. Structure, Function, and Biogenesis of the Cell Wall of Mycobacterium tuberculosis. *Tuberculosis* **2003**, *83*, 91–97.
113. Joe, M.; Bai, Y.; Nacario, R. C.; Lowary, T. L. Synthesis of the Docosanasaccharide Arabinan Domain of Mycobacterial Arabinogalactan and a Proposed Octadecasaccharide Biosynthetic Precursor. *J. Am. Chem. Soc.* **2007**, *129*, 9885–9901.
114. Ishiwata, A.; Ito, Y. Synthesis of Docosasaccharide Arabinan Motif of Mycobacterial Cell Wall. *J. Am. Chem. Soc.* **2011**, *133*, 2275–2291.
115. Thadke, S. A.; Mishra, B.; Islam, M.; Pasari, S.; Manmode, S.; Rao, B. V.; Neralkar, M.; Shinde, G. P.; Walke, G.; Hotha, S. [Au]/[Ag]-Catalysed Expedient Synthesis of Branched Heneicosafuranosyl Arabinogalactan Motif of Mycobacterium tuberculosis Cell Wall. *Nat. Commun.* **2017**, *8*, 14019.

116. Pasari, S.; Manmode, S.; Walke, G.; Hotha, S. A Versatile Synthesis of Pentacosafuranoside Subunit Reminiscent of Mycobacterial Arabinogalactan Employing One Strategic Glycosidation Protocol. *Chem. Eur. J.* **2018**, *24*, 1128–1139.
117. Wu, Y.; Xiong, D.-C.; Chen, S.-C.; Wang, Y.-S.; Ye, X.-S. Total Synthesis of Mycobacterial Arabinogalactan Containing 92 Monosaccharide Units. *Nat. Commun.* **2017**, *8*, 14851.
118. Besra, G. S.; Khoo, K.-H.; McNeil, M. R.; Dell, A.; Morris, H. R.; Brennan, P. J. A New Interpretation of the Structure of the Mycolyl-Arabinogalactan Complex of Mycobacterium tuberculosis as Revealed Through Characterization of Oligoglycosylalditol Fragments by Fast-Atom Bombardment Mass Spectrometry and 1H Nuclear Magnetic Resonance Spectroscopy. *Biochemistry* **1995**, *34*, 4257–4266.
119. Mishra, A. K.; Driessen, N. N.; Appelmelk, B. J.; Besra, G. S. Lipoarabinomannan and Related Glycoconjugates: Structure, Biogenesis and Role in Mycobacterium tuberculosis Physiology and Host–Pathogen Interaction. *FEMS Microbiol. Rev.* **2011**, *35*, 1126–1157.
120. Fraser-Reid, B.; Lu, J.; Jayaprakash, K. N.; López, J. C. Synthesis of a 28-mer Oligosaccharide Core of Mycobacterial lipoarabinomannan (LAM) Requires Only Two n-Pentenyl Orthoester Progenitors. *Tetrahedron Asymmetry* **2006**, *17*, 2449–2463.
121. Hölemann, A.; Stocker, B. L.; Seeberger, P. H. Synthesis of a Core Arabinomannan Oligosaccharide of Mycobacterium tuberculosis. *J. Organomet. Chem.* **2006**, *71*, 8071–8088.
122. Islam, M.; Shinde, G. P.; Hotha, S. Expedient Synthesis of the Heneicosasaccharyl Mannose Capped Arabinomannan of the Mycobacterium tuberculosis Cellular Envelope by Glycosyl Carbonate Donors. *Chem. Sci.* **2017**, *8*, 2033–2038.
123. Pardo-Vargas, A.; Delbianco, M.; Seeberger, P. H. Automated Glycan Assembly as an Enabling Technology. *Curr. Opin. Chem. Biol.* **2018**, *46*, 48–55.
124. Seeberger, P. H. The Logic of Automated Glycan Assembly. *Acc. Chem. Res.* **2015**, *48*, 1450–1463.
125. Walvoort, M. T. C.; van den Elst, H.; Plante, O. J.; Kröck, L.; Seeberger, P. H.; Overkleeft, H. S.; van der Marel, G. A.; Codée, J. D. C. Automated Solid-Phase Synthesis of β-Mannuronic Acid Alginates. *Angew. Chem. Int. Ed.* **2012**, *51*, 4393–4396.
126. Walvoort, M. T. C.; Volbeda, A. G.; Reintjens, N. R. M.; van den Elst, H.; Plante, O. J.; Overkleeft, H. S.; van der Marel, G. A.; Codée, J. D. C. Automated Solid-Phase Synthesis of Hyaluronan Oligosaccharides. *Org. Lett.* **2012**, *14*, 3776–3779.
127. Calin, O.; Eller, S.; Seeberger, P. H. Automated Polysaccharide Synthesis: Assembly of a 30mer Mannoside. *Angew. Chem. Int. Ed.* **2013**, *52*, 5862–5865.
128. Yu, Y.; Kononov, A.; Delbianco, M.; Seeberger, P. H. A Capping Step During Automated Glycan Assembly Enables Access to Complex Glycans in High Yield. *Chem. Eur. J.* **2018**, *24*, 6075–6078.

Linear and cyclic amyloses: Beyond natural

10

Shinichi Kitamura, Shiho Suzuki
Center for Research and Development of Bioresources, Organization for Research Promotion, Osaka Prefecture University, Osaka, Japan

Chapter outline

1 Introduction

Amylose, a polysaccharide first isolated from starch,[1] has been widely studied for its interesting solution properties, particularly its ability to form complexes with iodine and some organic reagents. In addition to natural amyloses, which have branching chains, completely linear amylose (LA) can be synthesized *in* vitro using enzymatic reactions.

Unlike LA, which has a relatively long history,[2–5] cyclic amylose (CA) (i.e., large-ring cyclodextrins or cycloamylose, referred to as CA) was discovered recently as a result of the action of recombinant potato D-enzyme (4-α-glucanotransferase, EC 2.4.1.25) on synthetic linear amylase.[6] The degree of polymerization (DP) of CA isolated from reaction mixtures has been found to range from 17 to several hundred. However, information regarding the conformation and molecular characteristics of CA is still limited.[7–9]

Recent Trends in Carbohydrate Chemistry. https://doi.org/10.1016/B978-0-12-817467-8.00010-4
© 2020 Elsevier Inc. All rights reserved.

This chapter outlines the progress of recent studies of synthetic linear and cyclic amyloses. In the first section, the basic principles of the preparation of synthetic linear and cyclic amyloses are described. In the second section, the dilute solution properties of these synthetic amyloses in terms of their primary structures and conformation are discussed. In the third section, we deal with their characteristic behavior, such as the formation of complexes of these amyloses. In the final section, potential industrial uses of these materials, particularly in relation to their use in the biotechnology and chemical industry and as a renewable resource of increasing significance, are presented.

In this chapter, we refer to our recent studies as well as to recently published results by others in this field. The reader should note that we have not attempted a comprehensive coverage of all the available literature on this subject.

2 Preparation of linear and cyclic amyloses

2.1 Linear amylose and linear-amylose-containing polymers

Synthetic LA can be produced using either glucan phosphorylase[10–13] or the coupling of sucrose phosphorylase and glucan phosphorylase.[14, 15] The chain length can be controlled by changing the reaction conditions. The resulting amyloses are strictly linear molecules and have a narrow molecular-weight distribution. Scheme 1 shows the coupling reaction of sucrose phosphorylase (sph) [EC.2.4.1.7] and (1→4)-α-D-glucan phosphorylase (gph) [EC.2.4.1.1]. In the first step, D-fructose and glucose 1-phosphate (G-1-P) are synthesized from sucrose and inorganic phosphate, and then G-1-P is used for amylose chain elongation of the primer by glucan phosphorylase. Phosphate is reused in this system.

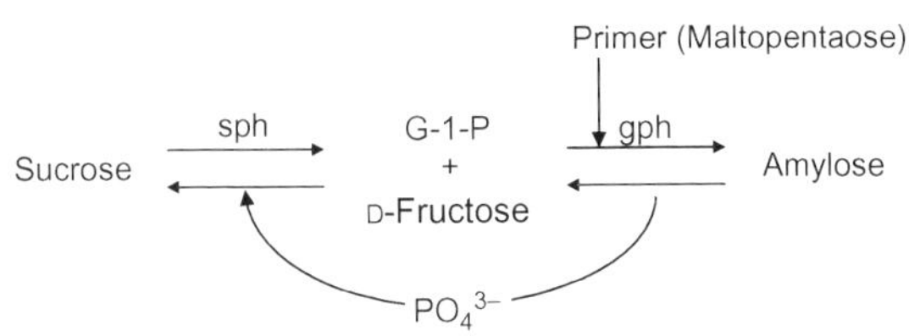

Scheme 1 The sucrose phosphorylase (sph)-(1→4)-α-D-glucan phosphorylase (gph) coupled enzyme system.

Fig. 1 shows the DP distribution of two amylose samples synthesized by the coupling reaction using maltopentaose as a primer. Fig. 1A shows that the DP for a short LA sample was well simulated by the Poisson distribution function. This is the expected result, as all the nonreducing ends of amylose elongate at an equal rate, similarly to in the radical polymerization of synthetic polymers.[16–18] Although the distribution of the DP for long LA molecules is much broader than the Poisson distribution, it is much narrower than that of natural amylose. Recently, two of our collaborators, the Ezaki Glico Co., Ltd. and Sanwa

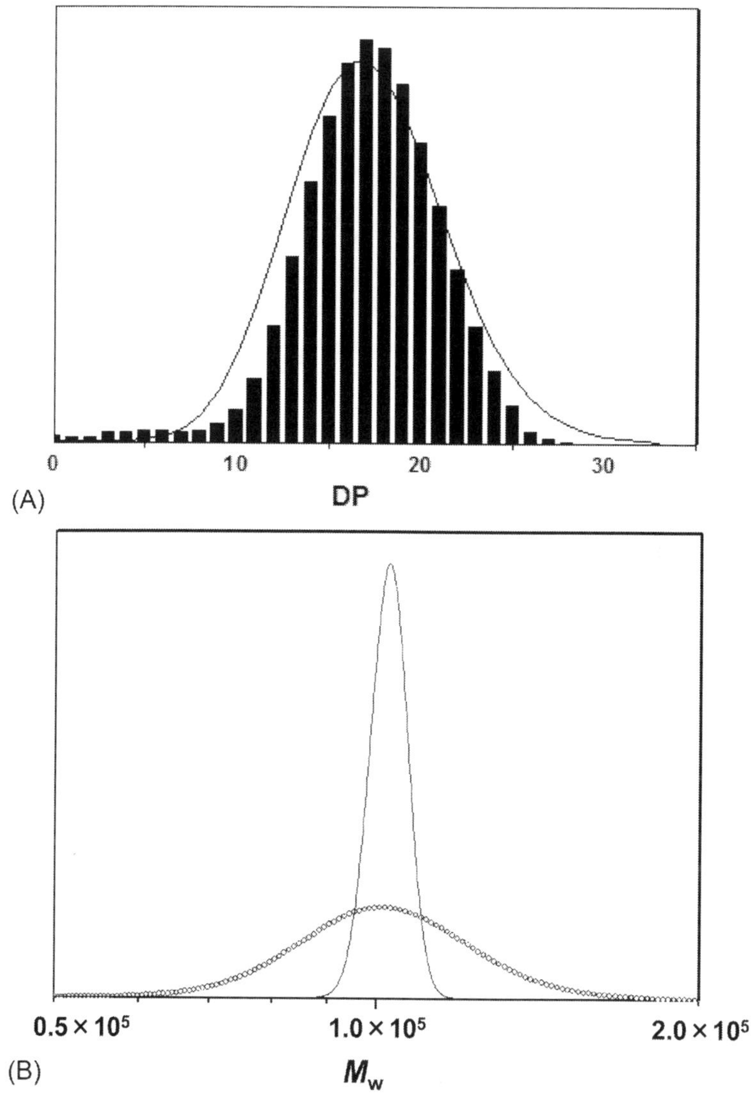

Fig. 1 DP or molecular-weight distribution for short (A) and long (B) synthetic amylose samples synthesized from the coupling reaction of sucrose phosphorylase from *Streptococcus mutans* and glucan phosphorylase from rice. Theoretical Poisson distributions (solid lines) for the two samples are also shown in the figures.

Starch Co., Ltd., have successfully mass-produced synthetic amylose from sucrose.[17] Thus, synthetic amyloses can be purchased from PS-biotech Inc. and Glico Nutrition Co., Ltd.

The chain elongation of amylose using glucan phosphorylases has been used for the preparation of amphiphilic A-B block copolymers, namely, ω-methoxypoly(ethylene oxide)-amylose copolymers (MPEO-amylose).[19] MPEO-amylose self-assembled in chloroform showed unique binding behavior with methyl orange (MO). The complexation of MPEO-amylose with MO was significantly enhanced in the amylose domain of the copolymer in chloroform. Thus, MPEO-block-amylose can be considered a new amphiphilic polymer with a molecular recognition site.[20] In another study, a polymer which amphiphilicity could be controlled using a chain-elongation reaction triggered by the addition of glucan phosphorylase was developed. By exploiting this property, the micelle-to-vesicle transition of mixed lipid/primer systems can be controlled.[21] Using this system, the membrane protein bacteriorhodopsin was transferred from a micelle to a liposome. Kadokawa reviewed amylose-grafting heteropolysaccharides and amylose-grafting synthetic polymers have been synthesized by the combined method of the α-glucan phosphorylase-catalyzed polymerization with the chemical reaction.[22]

Chain elongation using phosphorylase was also used in the preparation of silica gel-amylose. Maltopentaose was first oxidized to form potassium gluconate at the reducing terminal.[23] After the enzymatic elongation was performed using this primer, the amylose end was lactonized and immobilized on 3-aminopropyl-silanized silica gel through amide bond formation. The amylose-conjugated silica gels thus obtained were treated with 3,5-dimethylphenyl isocyanate to convert the hydroxy groups of amyloses to the corresponding carbamate residues. The materials prepared by this method showed chiral recognition. It should be noted that Pfannemüller[24] and coworker first showed that chain elongation using glucan phosphorylase could produce many types of polymers that form complexes with guest molecules.[23]

2.2 *Cyclic amylose*

CAs with DP values of up to several hundreds have been obtained from the reaction of recombinant potato D-enzyme (EC 2.3.1.25) and LA. Recent improvements in the production of CA using recombinant D-enzyme[25] and amylomaltase from *Thermus aquaticus* have permitted the production of larger amounts of CA. It is well known that final reaction products using CGTase on amylose consist of α-, β-, and γ-cyclodextrins, namely CA6, CA7 and CA8, respectively. Recently, Terada et al.[26] reported that these small CAs are not the major products by the initial action of CGTase on amylose; instead, the initial products appear to be large CAs with DPs in excess of 60.

Fig. 2 shows the DP distribution of CAs produced from LA using D-enzyme and amylomaltase. In contrast to CGTase, these two enzymes cannot produce small CAs such as α-, β-, and γ-cyclodextrins. The minimum DPs of the CAs produced by D-enzyme and amylomaltase are 17 and 22, respectively. The term CA*n* is used to represent a general CA, where *n* represents the DP. The mechanisms by which the minimum DP and the DP distribution are determined in these enzymatic reactions have been studied, and the DP has been discussed in terms of the possible conformation of CA and the binding of CA with enzymes. Shimada et al.[27] showed that the distribution of CA could be simulated by a double helical model rather than a single circularized model. Imamura et al. determined the 3D-structure of D-enzyme with CA26.[28] This structure is of interest in terms of the resulting distribution of CA. We constructed a model of CA26 and D-enzyme in which each end of CA26 was bound to the D-enzyme, while the other CA residues were considered to be flexible. In order for the two ends to connect, the minimum DP is 17 (unpublished result). Fig. 3 shows

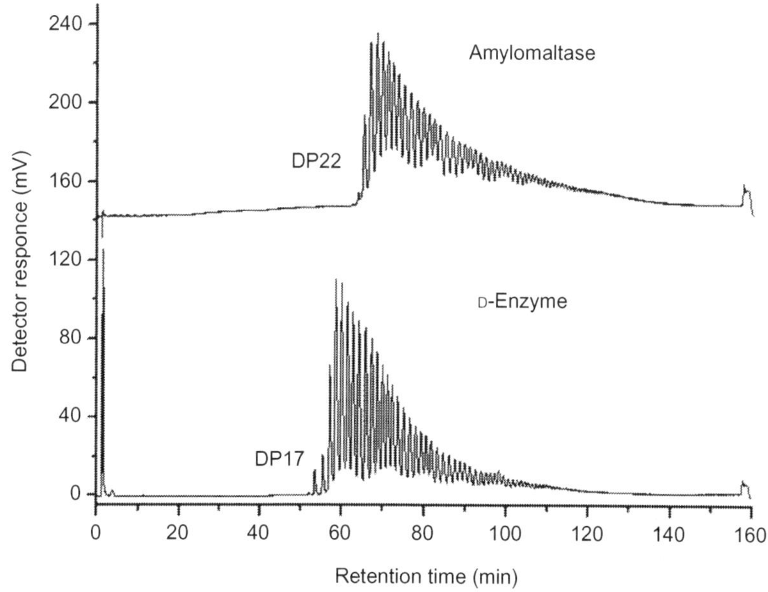

Fig. 2 DP distribution of cyclic amylose synthesized from linear amylose using amylomaltase and D-enzyme.

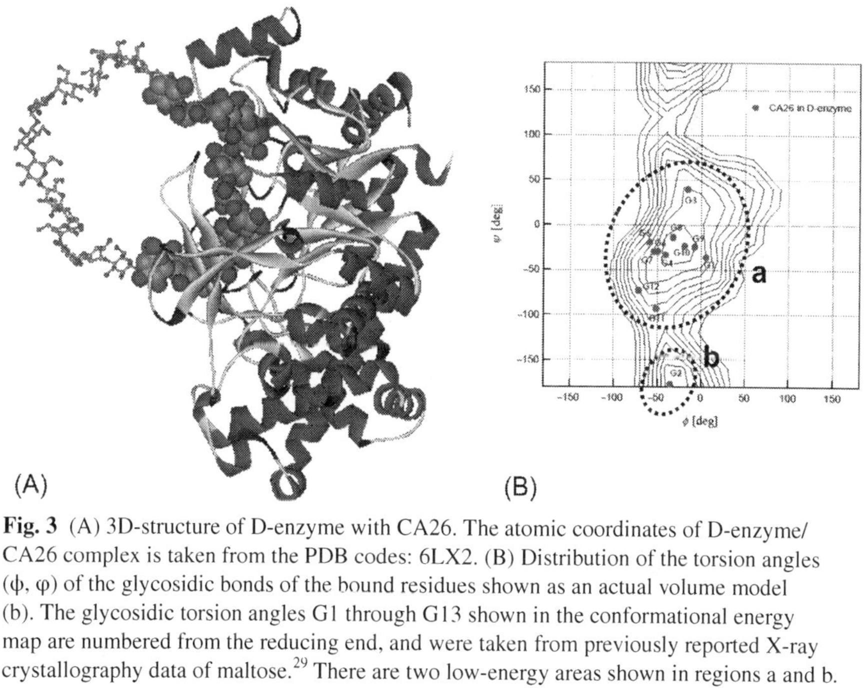

Fig. 3 (A) 3D-structure of D-enzyme with CA26. The atomic coordinates of D-enzyme/CA26 complex is taken from the PDB codes: 6LX2. (B) Distribution of the torsion angles (ϕ, φ) of the glycosidic bonds of the bound residues shown as an actual volume model (b). The glycosidic torsion angles G1 through G13 shown in the conformational energy map are numbered from the reducing end, and were taken from previously reported X-ray crystallography data of maltose.[29] There are two low-energy areas shown in regions a and b.

the distribution of the glycosidic torsion angles ϕ and φ of the CA chains bound to the D-enzyme. One torsion angle is located in the local minimum of region B, while the other 7 are located in main minimum region A. Most recently, Roth et al.[30] determined the crystal structure of a complex of *T. aquaticus* amylomaltase and CA34, revealing that 15 glucose residues interact with the enzyme. They discussed a molecular ruler mechanism to determine the minimum ring length of the CA products based on the structure of the complex.

3 Conformation and dilute solution properties

3.1 Linear amylose

Studies of the basic conformation and solution properties of amylose molecules require well-characterized samples. The use of natural amylose fractions for hydrodynamic studies has often been criticized due to the possibility of contamination with amylopectin. Furthermore, natural amyloses, which are carefully isolated from starches, are usually slightly branched, although the degree of branching and specific structure depend on the source and the method of isolation.[31]

Fig. 4 is a plot of the *z*-average mean-square radius of gyration $\langle S^2 \rangle_z$ versus the molecular weight (M_w) of amyloses in dimethyl sulfoxide (DMSO). The data obtained by our group for synthetic amylose is represented by open circles.[32] The straight-line fitting of the double-log plot of these data points has a slope of 1.2, which is the exponent expected for long flexible chains in good solvents. The other symbols in the figure represent $\langle S^2 \rangle_z$ data reported by other groups. Surprisingly, all

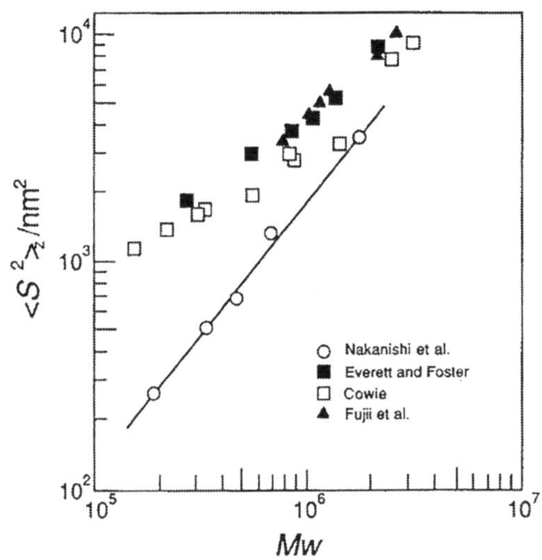

Fig. 4 Molecular-weight dependence of $\langle S^2 \rangle_z$ for amylose in DMSO. Data for synthetic (○) and natural (■, □, and ▲) amyloses were reported by the indicated groups.[32–35]

the data from other authors are considerably higher than ours, and the discrepancy is greater at lower M_w. Actually, Fujii et al.[33] observed pronounced downward curvatures at low scattering angle for a sample with a M_w of 7.8×10^5 in Zimm plots. These curvatures may produce large $\langle S^2 \rangle_z$ values. Nakanishi et al.[32] also showed that the characteristic ratio at infinite molecular weight is 5 in DMSO; this value was similar to those obtained for aqueous amylose. Thus, it was concluded that without excluded-volume effects, high-molecular-weight amylose exhibits a similar random coil conformation in water and in DMSO. However, it should be noted that hydrodynamic measurements can only provide information about the overall conformation of the molecule.

We have shown that the experimental $\langle S^2 \rangle_z$ for synthetic amylose in DMSO can be simulated using model amylose chains with an excluded-volume chain sequence. The model chains are generated by a Monte Carlo method using a conformational energy map of the dimer segment, that is, α-maltose.[36] The excluded-volume effects were introduced by placing a hard-core sphere of radius, r, at the center of mass of each glucose unit. Fig. 5 shows a perspective drawing of a typical instantaneous conformation of a LA chain with excluded-volume effects ($r = 0.1$ nm). The normalized distribution function of the end-to-end distance R for a given DP has also been found to shift to larger R on the introduction of the excluded-volume effect. Without the excluded-volume effect, the distribution is approximated well by a Gaussian function for DP larger than 200.[29]

It is well established that LA chains behave as random coils with sequences that have significant pseudo-helical character in aqueous solutions.[4, 37–40] This model is compatible with the concept that thermal segmental motion is responsible for the imperfect helical structure, and promotes the random-coil character of the LA chain in solution.[41]

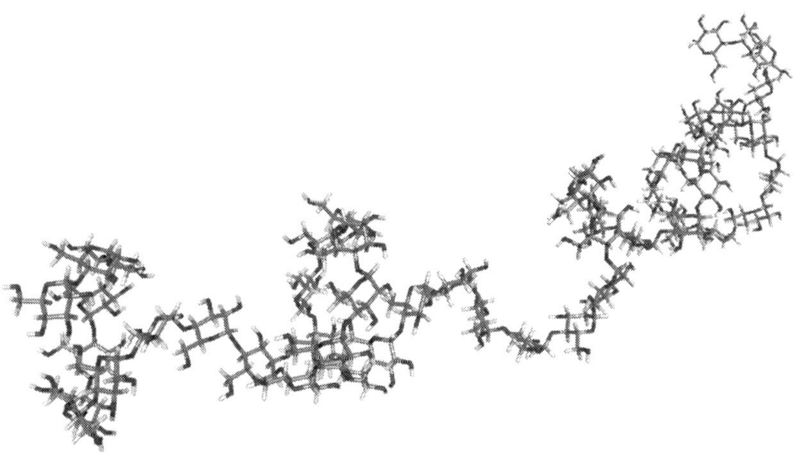

Fig. 5 A 3D-structure of a representative amylosic chain (DP = 50) with excluded volume in chain sequences. The chain was generated using a previously reported Monte Carlo method.[29]

3.2 Cyclic amylose

Seven CA samples with weight-average molecular weights ranging from 5×10^3 to 1.8×10^4 were prepared and studied by sedimentation equilibrium in DMSO at 25°C and by dynamic light scattering in 0.5 N aqueous sodium hydroxide (NaOH) at 25°C, which is a good solvent for LA.[42] The measured translational diffusion coefficients *D* agreed fairly well with Monte Carlo results for cyclic glucan chains with excluded volume effects.[43] It was concluded that the translational diffusion behavior of CA in 0.5 N aqueous NaOH was consistent with the conformational characteristics derived from the conformational energy of maltose and dilute-solution data for LA.[42]

Computer modeling of CA suggested two plausible conformations: a circularized single-helical structure, and a double-helical structure with fold-backs at each end.[27] Experimental tests of these predictions were carried out using small-angle X-ray scattering (SAXS) on CA21.[44] Calculations of the dependence of the scattering function, $P(q)$, on the scattering vector q, were conducted, starting from several CA21 conformers proposed on the basis of modeling. The results clearly showed that, for CA21, a closed three-turn single-helical structure (structure b in Fig. 6)

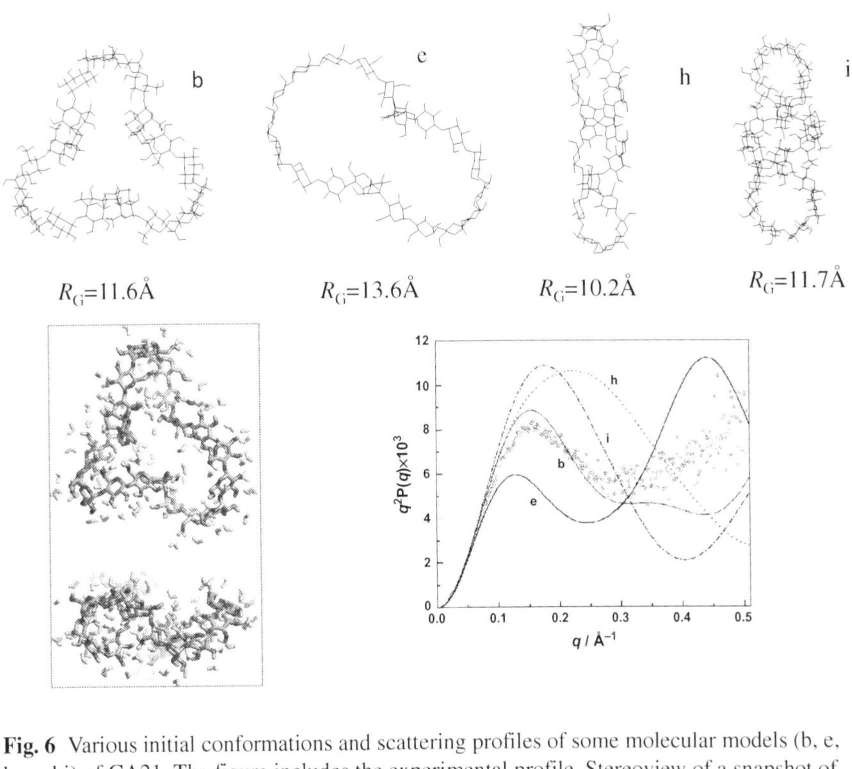

Fig. 6 Various initial conformations and scattering profiles of some molecular models (b, e, h, and i) of CA21. The figure includes the experimental profile. Stereoview of a snapshot of CA21 with hydrogen-bonded water molecules. This view was obtained by a 180-ps molecular dynamic simulation in water at 300 K. The right and left views are separated by 36 along the horizontal direction.[44]

conformed better to the experimental scattering data than the double-helical forms (h and i) did. The calculated scattering functions were based on all-atom models, which take into account the cross section of the molecule, as well as its contour. Molecular dynamic simulations in water showed that overall, the conformation of the three-turn single-helical form was maintained at 300 K, although glucose unit replacements were observed. The best simulation results were achieved when the computed scattering function was averaged over an ensemble of conformations that were generated during a molecular dynamics simulation. The radius of gyration R_G fluctuated within 0.7 Å of its mean, in agreement with the SAXS experimental results.[44] For one model, it was estimated that 96 water molecules were interacting with the CA21 molecule, judging from hydrogen bonding distances and angles (Fig. 6).

The amount of unfrozen water associated with linear and CA and oligomers including cyclodextrins were experimentally determined by differential scanning calorimetry.[45]

4 Formation of complexes with guest molecules

Linear amyloses are well known for their ability to incorporate a variety of organic compounds into their cavities. They can form inclusion complexes with many types of materials, ranging from iodine to aliphatic and aromatic acids, esters, and many other aromatic compounds; the aliphatic chain and aromatic rings lie within the helical structure of amylose in such complexes.

Saenger and collaborators[46] studied the crystal structure of CA26, which is shown in Fig. 7B. This structure is constructed from two turns of sixfold (6_1) helices connected by two glucose molecules, as shown in the figure. The crystal structures are clearly different from the solution structures; these two structures can interconvert.

In this section, we discuss our studies of the folding that occurs during the inclusion of guest molecules in aqueous solution. As discussed in the previous section, if the DP is sufficient to allow a CA molecule to behave as a flexible chain, its local conformation may be similar to that of a long LA chain.

We would like to emphasize here that isothermal titration calorimetry, in which the affinity constant and binding enthalpy are directly and simultaneously determined from data obtained at a constant temperature, has considerable advantages over other methods, in which thermodynamic quantities are determined from the van't Hoff equation by assuming an invariant heat capacity over the temperature range employed.[47] The titration calorimetry described in this section was performed using an MCS titration microcalorimeter (Micro Cal., Inc., MA) at a constant temperature of 25°C.

CAs with DPs from 25 to 28 were titrated with sodium dodecyl sulfate (SDS). Most of these CAs exhibited titration curves consistent with 1:1 complex formation with SDS (see CA25 in Fig. 8A), and had binding constants (K_1) on the order of 10^3 M^{-1} (see Table 1). However, CA26 showed different binding behavior (Fig. 8B), and could be analyzed using a model assuming 1:2 complex formation with identical interacting sites, as shown in Fig. 9. The binding constants K_1 and K_2 ($K_1 < K_2$), which

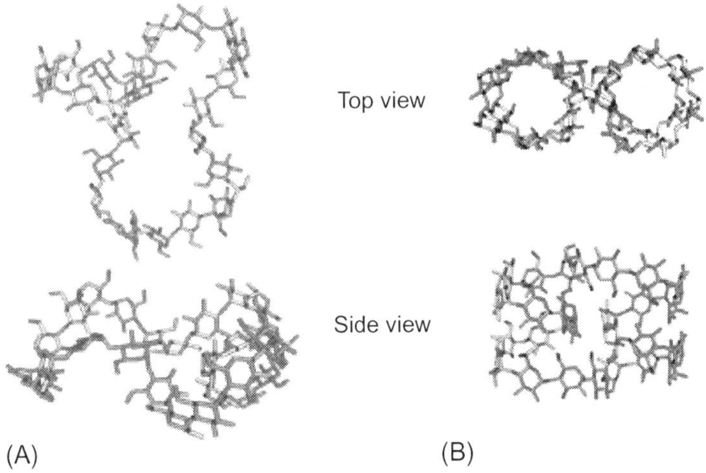

Fig. 7 Top and side views of the solution structure (A)[7] and crystal structure (B) of CA26.[42]

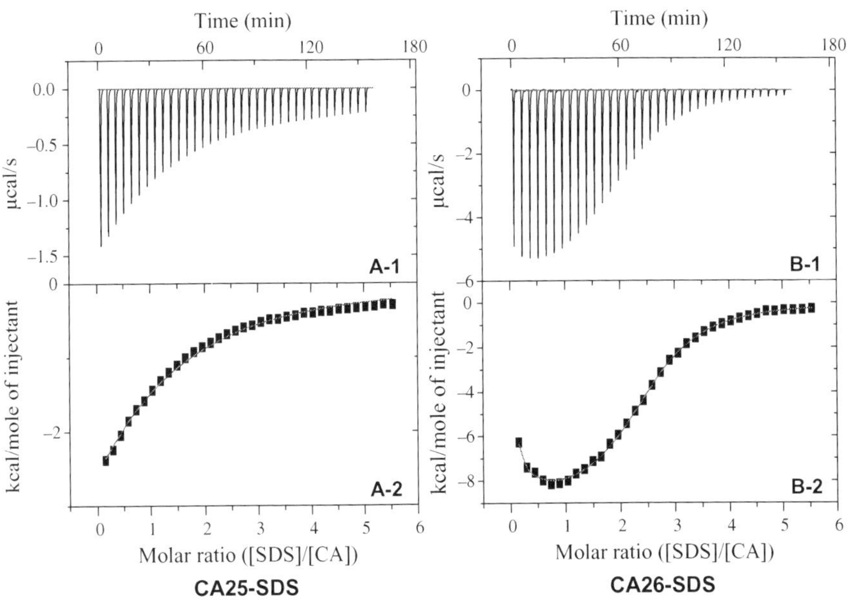

Fig. 8 (A-1) and (A-2): Experimental data for titration of 7 μL aliquots of 2.5 mM SDS aqueous solutions into 100 μM aqueous solutions of CD25 and CD26 at 25°C. 35 injections of the SDS solution were made at 4.5 min intervals. (B-1) and (B-2): Data points obtained from the integration of the injection peaks, and the titration curves (solid lines) obtained by fitting the data points.[8, 48]

Table 1 Thermodynamic data for the complex formation of CAs with SDS.

DP	Guest molecule	K_1	K_2	ΔH_1	ΔH_2	ΔG_1	ΔG_2	$T\Delta S_1$	$T\Delta S_2$
		($\times 10^3 M^{-1}$)		(kcal/mol)		(kcal/mol)		(kcal/mol)	
25	SDS	7.17	–	−7.98	–	−5.27	–	−2.71	–
26	2 SDS	13.2	134	−7.03	−14.0	−5.63	−7.01	−1.40	−6.99
27	SDS	8.29	–	−7.12	–	−5.35	–	−1.76	–
28	SDS	2.58	–	−14.5	–	−4.66	–	−10.7	–

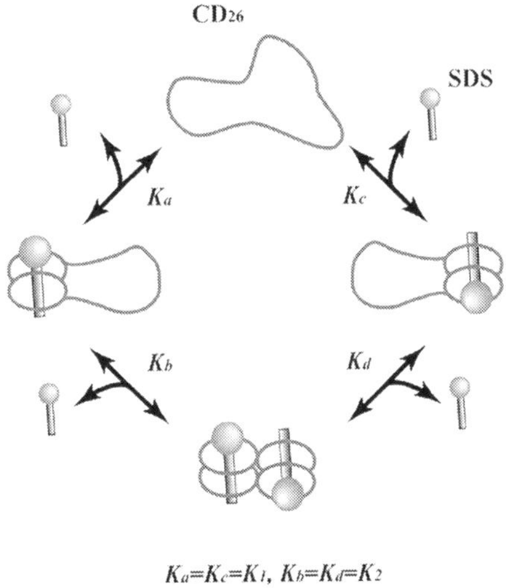

Fig. 9 A model assuming 1:2 complex formation with identical interacting sites. For the two identical sites, the binding constants $K_a = K_c = K_1$ and $K_b = K_d = K_2$, and the binding enthalpies ΔH_1 and ΔH_2 are defined relative to the progress of saturation.

were defined relative to the progress of saturation, and the corresponding enthalpies of binding were determined. These thermodynamic data indicated that the binding process was enthalpically driven, and that the binding of a second SDS molecule to CA26 was rather strong ($K_2 = 1.34 \times 10^5\ M^{-1}$). The full thermodynamic data for the various CA-SDS complexes are given in Table 1.

Since the binding constant of the second SDS molecule to CA26 was large, we attempted to simulate the binding process using molecular dynamics (Fig. 10). We assumed that one SDS molecule was already incorporated in the cavity of CA26, and

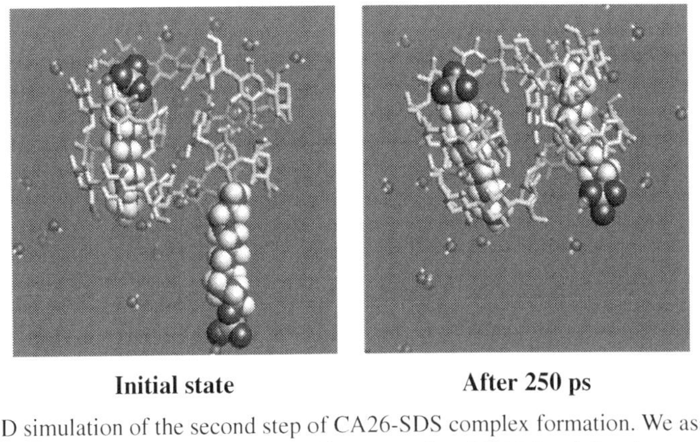

Fig. 10 MD simulation of the second step of CA26-SDS complex formation. We assumed that one SDS molecule was already incorporated in one cavity of CA26 and that the second SDS was located near the "entrance" to the second cavity. Approximately 6000 water molecules were located in the surroundings.

that the second SDS was located near the "entrance" of the cavity. Approximately 6000 water molecules were located in the surroundings. Bound water molecules in a cavity are also shown.[8, 48]

Molecular modeling also suggested that the longer hydrocarbon chains of CA38, CA50, and CA62 can interact with surfactants specifically. The possible folding structures for CA38, CA50, and CA62, determined using the models, exhibited longer channels than CA26 (Fig. 11).

In order to determine the effect of the length of the hydrocarbon chains of surfactants on their interactions with CA, we extended our study of CA complex formation to a series of surfactants, namely, 3-(*N,N*-dimethylmyristylammonio)propanesulfonate (SB3-14), 3-(*N,N*-dimethylpalmitylammonio)propanesulfonate (SB3-16), and hexadecyltrimethylammonium bromide (CTAB). The above surfactants were added to an aqueous solution of a mixture of CAs (CA22 to CA120) to form precipitates.[8, 48] The DPs of the precipitated CAs were determined by high-performance anion exchange chromatography with a pulsed amperometric detector (Fig. 12). The results indicated that SB3-14 and SB3-16 interacted strongly with CA50 and CA62, respectively, to form the precipitated complexes. The powder X-ray diffraction patterns of the precipitates indicated that these complexes consisted of crystalline structures containing 6_1 helixes. The thermal stability of the complexes was determined using differential scanning calorimetry. SB3-14 dissociated from the complexes at temperatures of 95°C, which is higher than that of LA (53°C) (Fig. 13). These results clearly showed that the molecular complex of CA with surfactants is more stable than that of LA.

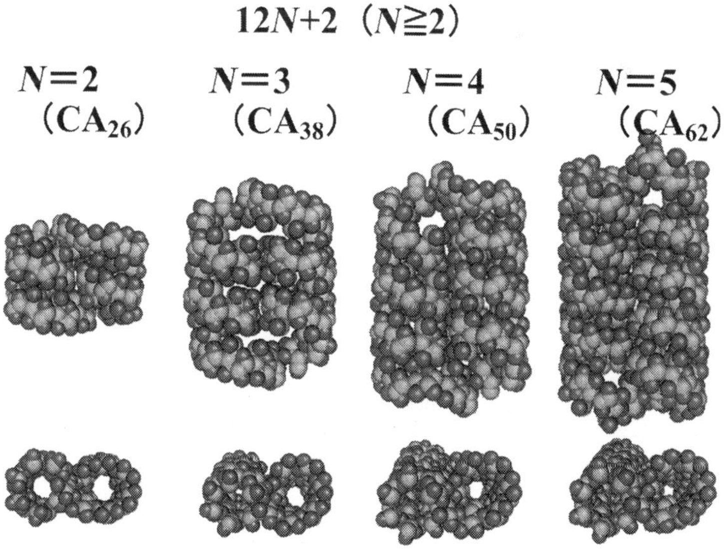

Fig. 11 Molecular models with two cavities for CA38, CA50, and CA62 based on the 3D structure of CA26 determined by X-ray crystallography.

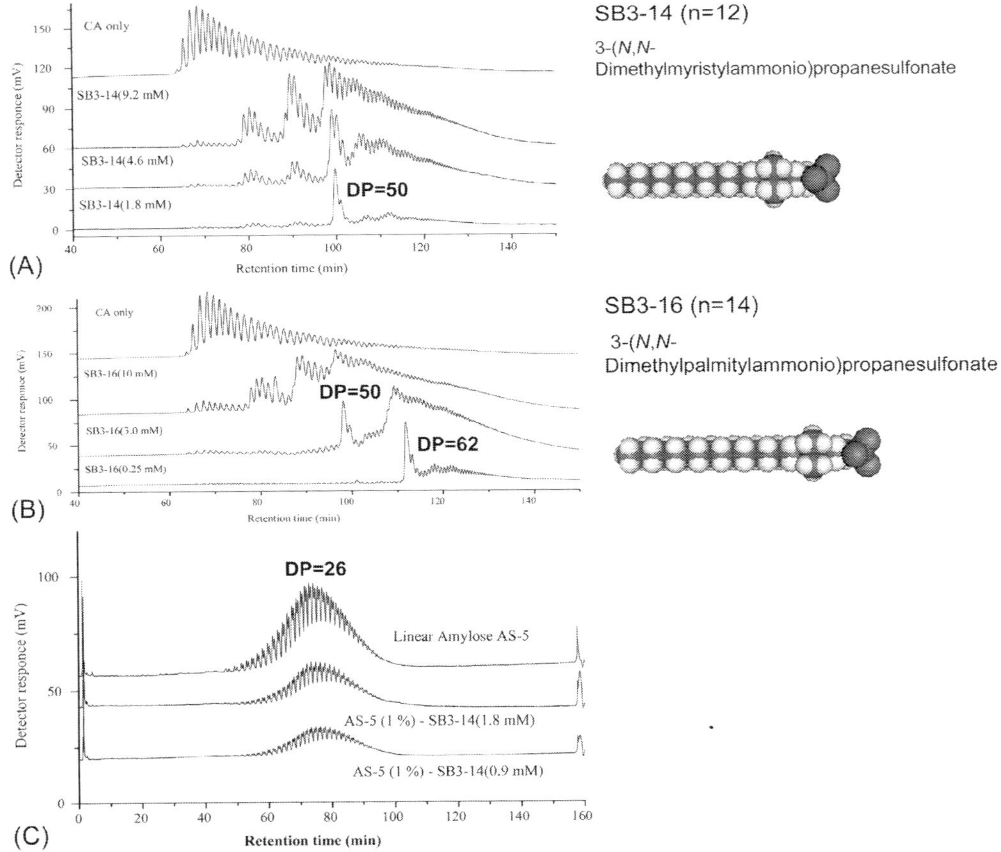

Fig. 12 High-performance anion exchange chromatograms with pulsed amperometric detection of the precipitants from CA mixtures (A, B) and an LA mixture (C) after the addition of surfactants at the concentrations indicated in the figures.

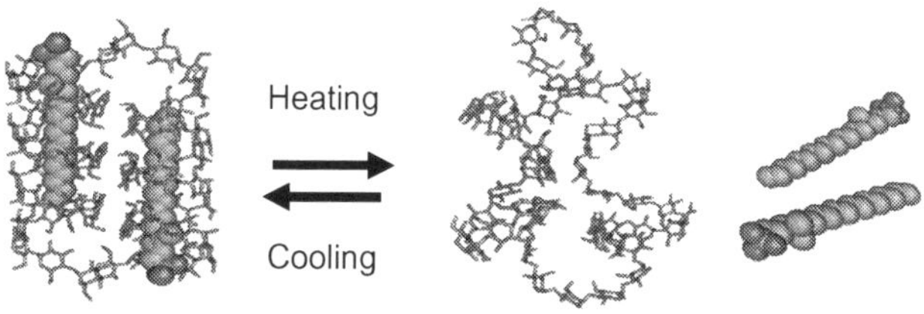

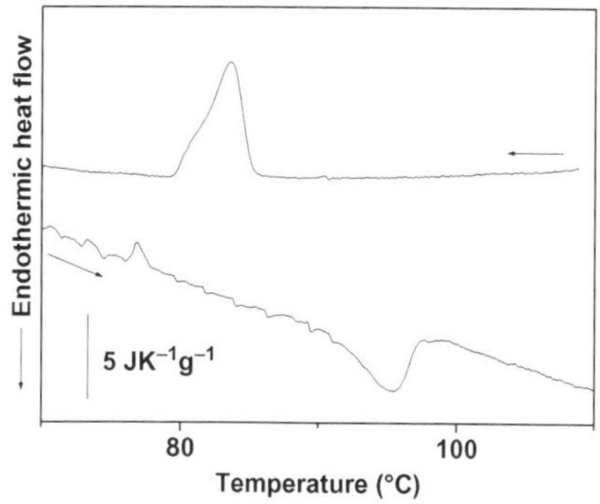

	Melting		Complex formation	
	T_p (°C)	ΔH (J/g)	T_p (°C)	ΔH (J/g)
CA	95	28.5	84	−29.9
LA	53	12.5	48	−11.9

Fig. 13 DSC curves of the melting (dissociation) and reformation of complexes (precipitants) from 5% CA mixtures after the addition of 1.8 mM SB3-14.

5 Stability of amylose in aqueous solution

While aqueous solutions of LA with DPs in the range 12–200 are very unstable in aqueous solution, shorter and longer LA molecules are much more soluble.[3, 37, 49] Not only maltooligosaccharides, but also linear synthetic amyloses with DP > 4000 are soluble in water and can remain stable for long periods of time. For example, a 1% aqueous solution of LA (DP 5200) was stable for more than 1 month at room temperature. Fig. 14 provides a schematic representation of the phase diagram for LA in aqueous solution as a function of DP and concentration. In contrast to LA, CA is stable

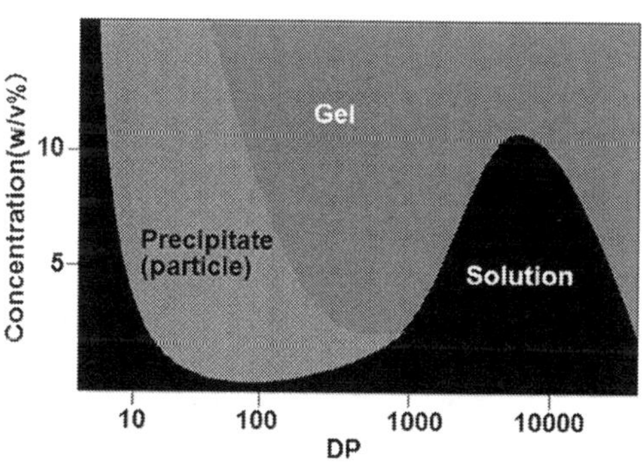

Fig. 14 Schematic representation of the phase diagram of linear amylose and water as a function of the DP and concentration.

in water even in the DP range of 20–200. Thus, the double helix formation of LA in aqueous solution was expected to be related to its aqueous solubility.

5.1 Self-assembly and double helix formation

Linear amylose in aqueous solution has a tendency toward molecular association, which includes the self-assembly of amylose molecules through the formation of double helices. Dissociation followed by reassociation leads to the formation of double helical structure.

Putaux et al. recently succeeded in preparing a single crystal of low-molecular-weight LA.[50] Fig. 15 shows the double helix architecture in the crystal structure. The morphological analysis of faceted single crystals combined with electron diffraction data confirmed that the double helices were oriented along the long dimension of the crystal and packed into lamellae with a parallelogram cross section defined by the *a* and *b* directions of the monoclinic unit cell of A-type[51] of amylose crystal. The lamellae are stacked along the *c*-axis, which is oriented parallel but opposite to the growth direction of the crystal.

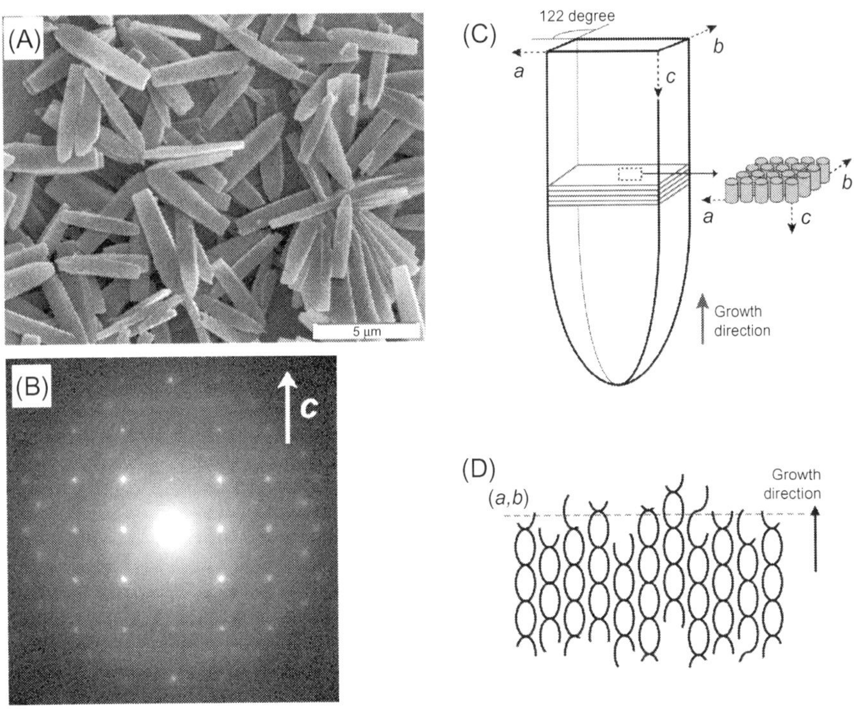

Fig. 15 (A) SEM image of crystals prepared from low-molecular-weight LA (number-average of DP = 20.9). (B) Example of the electron diffraction pattern of the crystals. (C) Schematic representation of the orientation of the double helices in the long A-type[51] of amylose single crystals. The gray cylinders correspond to amylose double helices. (D) Longitudinal view of a crystalline lamella showing the parallel packing of the double helices.[50]

In addition to A-type, B-type of amylose crystal, which has a more open structure containing significantly more water between helices, has been known.[52, 53]

Two enzymatically synthesized amyloses with different molecular weights, A-22 ($M_w = 2.2 \times 10^4$) and A-850 ($M_w = 8.5 \times 10^5$), were used to prepare amylose nanofibers on mica from an aqueous solution. The aggregation of amylose was controlled by varying the solvent, amylose concentration, and drying temperature. Fig. 16 shows an Atomic force microscopy (AFM) image of the amylose A-850 nanofibers formed on mica at a drying temperature of 70°C. The nanofibers exhibited a uniform height of 8–10 Å, and consisted of amylose double helices in a two-dimensional array. Renaturing A-22 dissolved in DMSO by dilution in water and drying at 60°C produced an ordered dot pattern consisting of submicron-sized (100–200 nm) amylose aggregates 7 nm in height that were arrayed at 1 μm intervals (Fig. 17).

Amylose spherules with uniformity in size can be produced from LA dissolved in DMSO or alkali solution by dilution in water.[54, 55] These spherules showed B-type crystallinity.

5.2 *Amylose gels*

At high concentrations, LA solutions form a gel, which then undergoes aging. For chains of DP > 250, gelation occurs at concentrations of greater than 2.0%. The mechanism of amylose gelation may be more complicated than that of precipitation from aqueous dilute solution because of extensive nucleation and the formation of very small particles. Due to their close proximity, the amylose particles tend to aggregate and form a three-dimensional network throughout the solution.

The mechanism of the gelation and retrogradation of amylose has been extensively studied on a molecular basis using both small and wide angle X-ray diffraction, NMR, differential scanning calorimetry, Raman spectroscopy, fluorescence probe techniques, and electron microscopy.[10, 56–62] NMR and X-ray diffraction experiments have demonstrated that amyloses precipitated from dilute aqueous solution contain only B-type aggregated double helices, and that amylose gels contain rigid double-helical "junction zones" interconnected by more mobile amorphous single-chain segments.

Yuguchi et al.[63] recently monitored the gelation and structure formation of amylose in an alkaline solution via in situ neutralization using time-resolved SAXS. Sharp increases in the SAXS profile in the lower angle region were observed after gelation. The results suggested that the aggregation of amylose chains led to a gelation point along with crystal growth. The aggregates appeared to function as junction zones, and the structure of the aggregates depended on the molecular weight of amylose.

6 Progress toward the industrial application of LA and CA

Amylose is one of the two components of starch; however, natural amylose has long been considered difficult to process and use in industrial applications because its branched structure and average molecular weight vary depending on the plant species from which it is extracted.[5, 31] Synthetic linear and cyclic amyloses are now commercially available,

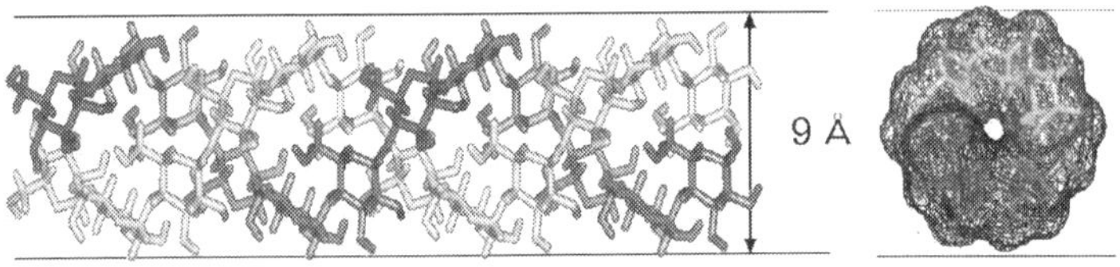

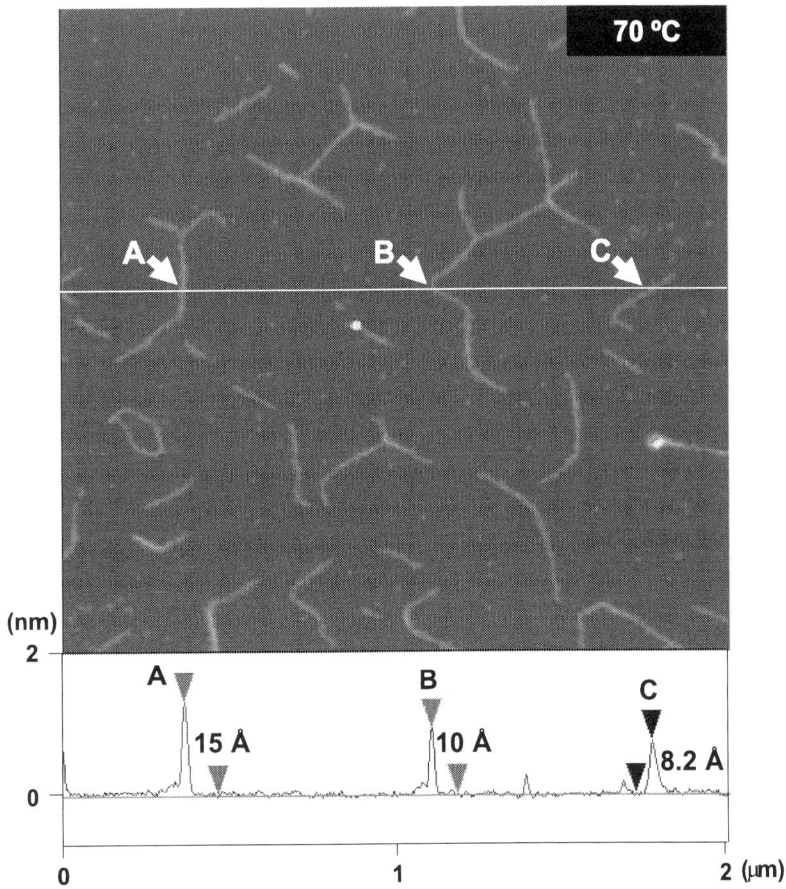

Fig. 16 Illustration of an amylose double helix generated with molecular modeling (*top*). AFM image and height data at position A, B and C of the amylose A-850 nano-fibers formed on mica when deposited and dried at 70°C (*bottom*).

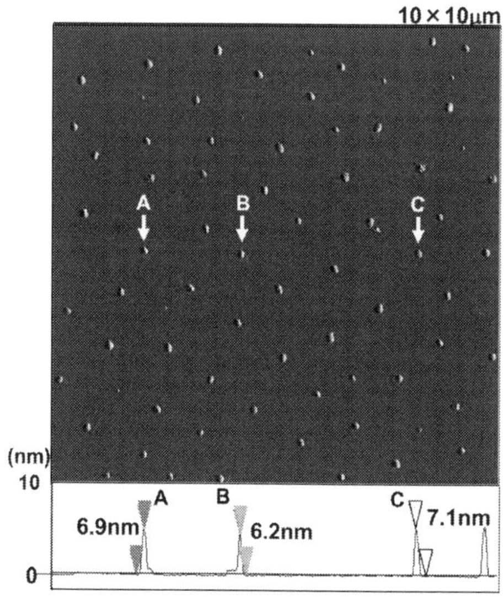

Fig. 17 AFM image of an amylose dot pattern (phase mode) and height information of amylose particles A, B and C on mica prepared by renaturing A-22 dissolved in DMSO by dilution in water and drying at 60°C.

and many applications are now in development. Synthetic amylose can be formed into films, gels, or fibers with a diameter of one nanometer, and are expected to find use in various industrial products, such as polarized films for liquid crystal displays (LCDs) and seamless capsules. In this section, some examples of applications that take advantage of the characteristic properties of synthetic amylose are presented.

6.1 Amylose films

As described previously, the characteristic aggregation behaviors of amylose, such as gelation, precipitation, and film formation, are strongly dependent on its molecular weight. At molecular weights greater than 1.0×10^5, liner amylose forms films. The films prepared by casting aqueous solutions of liner amyloses with DP > 4000 are tough, flexible, and transparent, and have high gas barrier properties. This subsection describes the functional properties of amylose films, including mechanical and gas barrier properties, as well as the applications of iodine-doped amylose films as polarized films and amylose-chitosan hybrid films as antibacterial films.

6.1.1 Optical polarization properties of amylose-iodine films

The formation of amylose-iodine complexes has been known since the early 1800s,[64] and the crystal structure of an amylose-iodine complex was reported in the mid-1900s.[65] Iodine is used as a dichroic dye in polyvinyl alcohol (PVA) complexes for polarizer films in the LCD industry. This context describes optical properties of a stretched amylose film containing iodine.

Fig. 18A shows the absorption spectra of natural and synthetic amylose films. In the spectrum of natural amylose, the absorbance was greater than 0.2 at most wavelengths, and was considerably higher between 200 and 300 nm. Due to their low transmittance, which is mainly caused by retrogradation, native amylose films have not been used for optical applications. In contrast, in the spectrum of synthetic amylose, the optical density (OD) is quite low in most of the visible light range. That is, synthetic amylose has high transmittance, which, combined with its favorable mechanical properties, indicates its potential for use in optical films.

Fig. 18B shows the typical polarizing properties of an iodine-doped amylose film and a PVA polarized film.[66] Crossing the two films (an iodine-amylose film with draw ratio

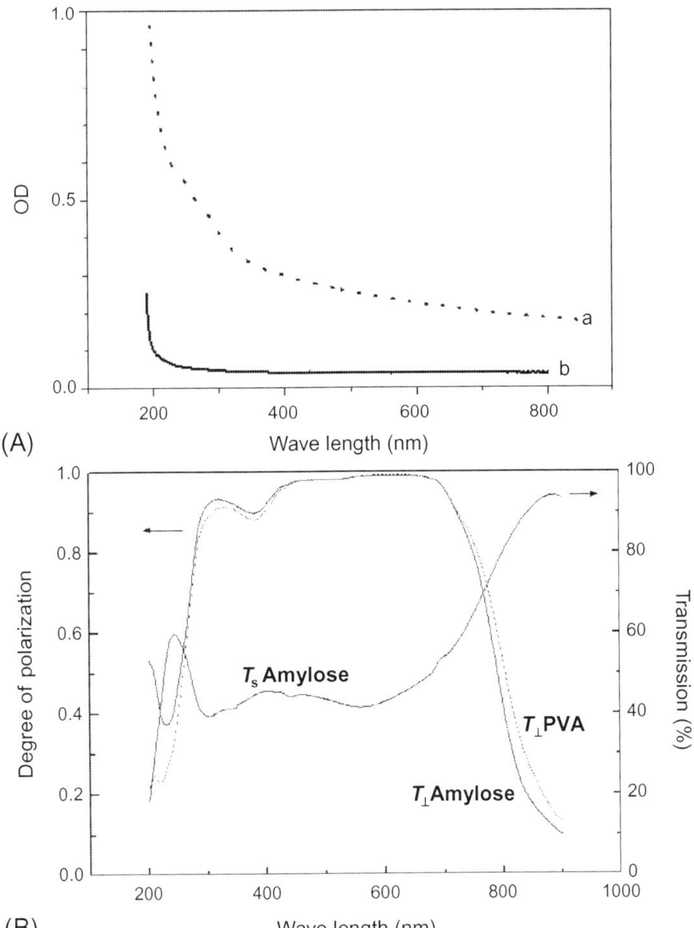

Fig. 18 (A) Absorbance spectra of native (a) and synthetic (b) amyloses. (B) Optical properties of amylose-iodine film and PVA-iodine film. Transmission of two crossed polarizers for amylose-iodine film ($T_{\perp\,\mathrm{Amylose}}$) and PVA-iodine film ($T_{\perp\,\mathrm{PVA}}$). Both of the two films had threefold draw ratio. Transmission of a single amylose-iodine film drawn 3.5 times ($T_{\mathrm{s\,Amylose}}$).

threefold and commercial polarizing film) at 90 degree to one another resulted in the high absorption of transmitted light in the range from 300 to 700 nm. Amylose-iodine film has a sufficient polarization even at a draw ratio of about 2.5-fold, as compared to the PVA polarizing film that can exhibit sufficient polarization effects in a lower draw ratio.

Most recently a novel approach for preparing low-density graphic films was developed using iodine-doped LA.[67]

6.1.2 Amylose-chitosan blend film

Cast films prepared from natural amylose, a major component of starch, are known to be excellent compared to those of starch or amylopectin. Synthetic amylose films are tough and flexible.[68] The addition of a small amount of chitosan to amylose films has been shown to increase their elongation remarkably, that is, the hybrid films are stronger than the films of the individual components alone.[69, 70] To impart amylose films with increased strength and additional functionalities associated with chitosan, such as enhanced gas permeability and antibacterial activity, we prepared a hybrid chitosan-amylose film. Synthetic amylose with a M_w of 1.7×10^6 and chitosan with a viscosity-averaged M_w of 1.4×10^6 and a deacetylation degree of 44.1% were used. The addition of a small amount of chitosan (< 10%) improved the mechanical parameters of amylose film; in particular, the elastic modulus and elongation of the blend films were greater than those of amylose or chitosan films.[68, 69]

Linear amylose having high molecular weight has also been reported to act as a strong barrier to several gasses. Its gas barrier properties are similar to those of PVA film, which shows the lowest gas permeability among synthetic polymer films. The addition of a small amount of chitosan significantly increased the permeability of the gases including nitrogen, oxygen, carbon dioxide, and ethylene.

Fig. 19A shows the antibacterial activity of the hybrid films against *E. coli*. Surprisingly, the chitosan film did not show any antibacterial activity. In contrast, all the hybrid films had strong antibacterial properties, even those with very low chitosan concentrations. These results indicated that the antibacterial properties of chitosan were restored when it was blended with amylose. The antibacterial activity of chitosan has been attributed to the free amino groups on the chitosan molecules. Therefore, the existence of free amino groups on the surfaces of the films was studied using zeta potential measurements and X-ray photoelectron spectroscopy (XPS).[70] The zeta potential measurements revealed that the surface of the hybrid film had a highly positive potential, while the surface of the chitosan film was almost uncharged, indicating that the surfaces of the films with strong antibacterial activity were positively charged. XPS measurements showed that the hybrid film had many amino groups on the surface layer. Especially on the thin layer close to the surface, the hybrid film had nearly as many amino groups as the chitosan film.

A model was developed for the chitosan molecules in the chitosan and hybrid films based on the results of the zeta-potential measurements, XPS measurements, and crystal structure of a hydrated crystal of chitosan. In this model, all the amino groups of chitosan molecule are oriented parallel, rather than perpendicular, to the film surface in a chitosan-only film (Fig. 19B-1). This accounts for the loss of antibacterial activity of the chitosan film surface. In contrast, the formation of this sheet-like chitosan

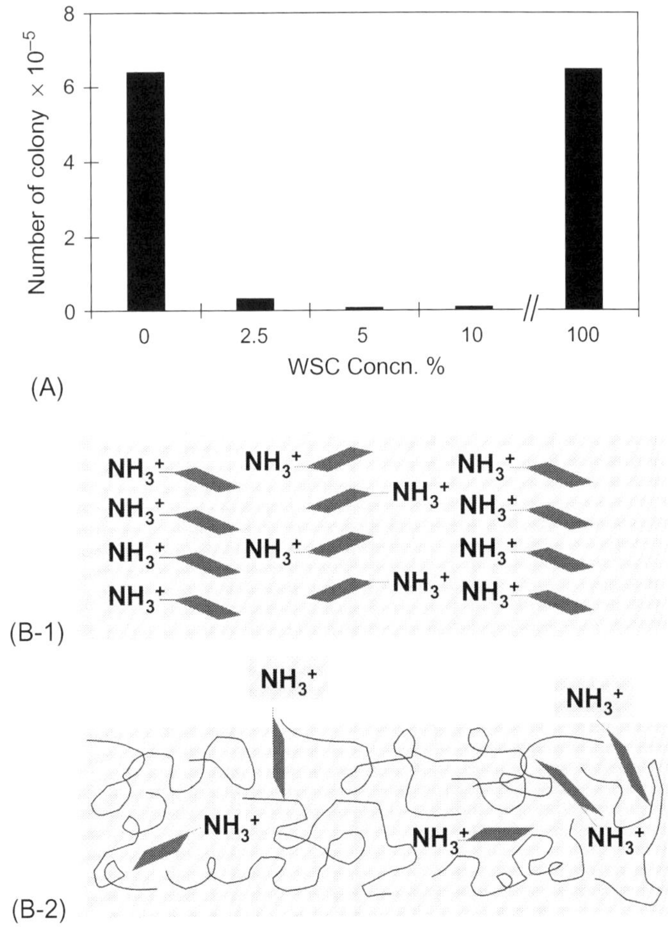

Fig. 19 (A) Test of the antibacterial activity of amylose-chitosan films against *E. coli*. (B) Models of the chitosan molecules in the films: (B-1) chitosan-only film, and (B-2) chitosan-amylose hybrid film. The chitosan and amylose molecules are illustrated as parallelograms with an amino group and curved lines, respectively.

structure was disturbed by amylose molecules in the hybrid films, exposing the amino groups at the film surface (Fig. 19B-2).

Hybrid amylose nanogels and microgels containing anionic and cationic polysaccharides were also prepared.[71] The ionic polysaccharides increased the surface charge of the amylose gels, which enhanced the stability of the nanogels in water by preventing aggregation of amylose molecules.

6.2 CA as an artificial chaperone for protein refolding

Protein refolding is one of the limiting steps during the recovery of active proteins from inclusion bodies. As shown in Fig. 20, in refolding processes, the inclusion

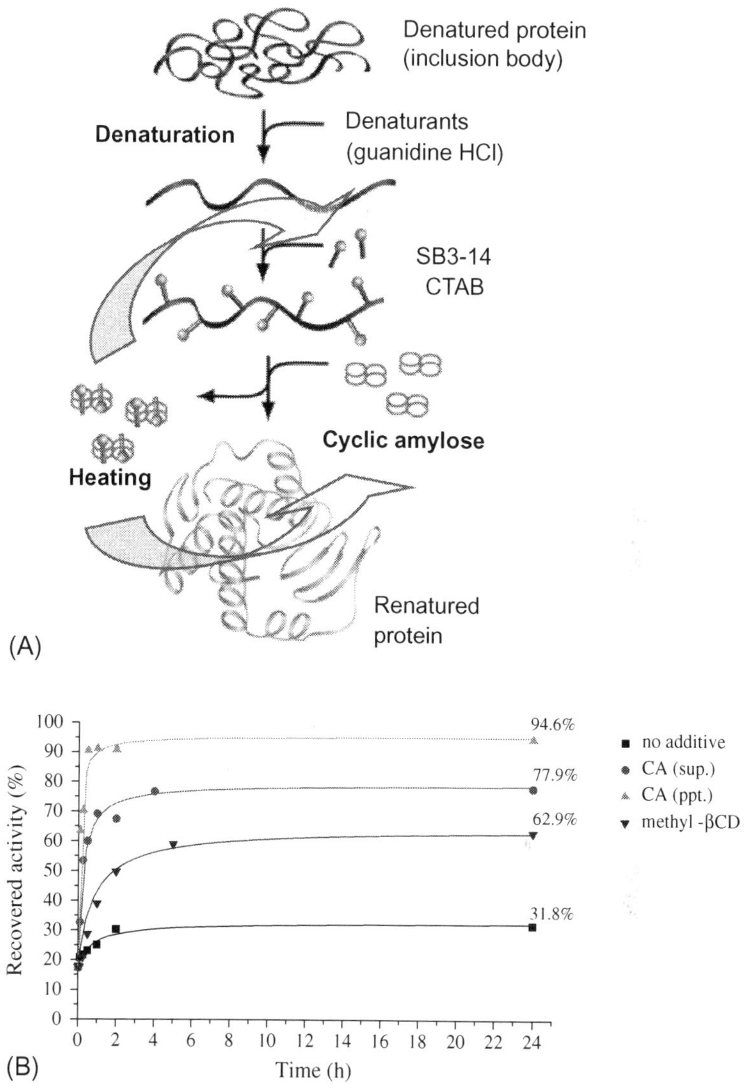

Fig. 20 (A) Schematic representation of the protein refolding system using cyclic amylose. (B) Time-dependent recovery of the lysozyme activity in cyclic amylose-assisted refolding. Lysozyme denatured by guanidine hydrochloride and dithiothreitol was diluted with 1.4 mM SB3-14/200 mM tris-acetate (pH 8.1) and incubated for 1 h at room temperature. In the next step, 1.1 mM (final concentration) of ●: CA (sup.); ▲: CA (ppt.); ▼: methyl-β-cyclodextrin; or ■: no additive was added. Small aliquots were withdrawn at the times indicated and assayed for enzymatic activity. All values are the percentage relative to the initial catalytic rate of the native lysozyme.

bodies are first solubilized in denaturing concentrations, using urea or guanidine hydrochloride as a denaturant. Then, the selected denaturing agent (SB3-14 or CTAB) is removed to start the procedure in what is considered the key step of the operation. Machida showed that high-molecular-weight CA is one of the most effective artificial chaperones to improve renaturation.[72] The CA plays role to take off these surfactants from denatured protein.

This method was further improved in several respects by utilizing high-molecular-weight CA fractionated precipitated complexes of CA and the surfactant SB3-14. The CA was divided into CA (ppt.), that is, the CA obtained from precipitation, and CA (sup.), that is, the CA obtained from the supernatant. The refolding of chemically denatured lysozyme was investigated using CA (ppt.), CA (sup.), and methyl-β-cyclodextrin (methyl-βCD). The recovery rate and the recovery speed followed the order CA (ppt.) > CA (sup.) > methyl-βCD. Moreover, the inclusion properties of CA effectively allowed the SB3-14 used for protein refolding to be excluded from the final products by collecting the precipitated CA-SB3-14 complexes. The collected complexes could be separated into CA and SB3-14 and reused.

6.3 Amylose derivatives

Studies of the solution properties and conformations of derivatives of linear and cyclic amyloses in a series of organic solvents have provided very interesting results in terms of the flexibility of their helical nature and their chiral recognition properties. Terao et al. systematically investigated the properties of dilute solutions of carbamate derivatives of LA and CA and the local helical conformations in various solvents using light and SAXS.[73–87] Their results have been summarized in a chapter referencing a series of papers related to this subject.[88]

Derivatized amyloses, such as 3,5-dimethylphenyl carbamate, show much better enantioselectivity and chromatographic properties compared to the LA. Chiral stationary phases based on amylose derivatives has been commercialized and experimentally applied for both analytical and preparative enantioseparations.[89]

Acknowledgments

The author's own studies described in this review were carried out with many collaborators, whom we greatly appreciate. The authors thank Professor B. Christensen of the Norwegian University of Science and Technology for their encouragement, comments, and suggestions.

References

1. Schoch, T. J. The Fractionation of Starch. In *Advances in Carbohydrate Chemistry*, Pigman, W. W., Wolfrom, M. L., Eds; Academic Press, **1945**; Vol. 1, pp 247–277.
2. Banks, W.; Greenwood, C. T. Starch and Its Components. In *Starch—Stärke*, Edinburgh University Press: Edinburgh (England), **1975**; Vol. 27, pp 325.

3. Kitamura, S. *Starch Polymers, Natural and Synthetic*; CRC Press: New York, NY, USA, **1996**.
4. Kitamura, S.; Kuge, T. Conformation and Physical Properties of Amylose in Aqueous Solutions. *Food Hydrocoll.* **1989**, *3*(4), 313–326.
5. Young, A. H. Chapter VIII: Fractionation of Starch. In *Starch: Chemistry and Technology*, 2nd ed.; Whistler, R. L., Bemiller, J. N., Paschall, E. F., Eds; Academic Press: San Diego, **1984**, pp 249–283.
6. Takaha, T.; Yanase, M.; Takata, H.; Okada, S.; Smith, S. M. Potato D-Enzyme Catalyzes the Cyclization of Amylose to Produce Cycloamylose, a Novel Cyclic Glucan. *J. Biol. Chem.* **1996**, *271*(6), 2902–2908.
7. Kitamura, S. Cyclic Oligosaccharides and Polysaccharides. In *Cyclic Polymers*, Springer: Dordrecht, **2000**; pp 125–160.
8. Kitamura, S.; Terada, Y. Large-Ring Cyclodextrins. *Kobunshi* **2005**, *54*(11), 813–816.
9. Larsen, K. L. Large Cyclodextrins. *J. Incl. Phenom. Macrocycl. Chem.* **2002**, *43*(1), 1–13.
10. Gidley, M. J.; Bulpin, P. V. Aggregation of Amylose in Aqueous Systems: The Effect of Chain Length on Phase Behavior and Aggregation Kinetics. *Macromolecules* **1989**, *22*(1), 341–346.
11. Kitamura, S.; Yunokawa, H.; Mitsuie, S.; Kuge, T. Study on Polysaccharide by the Fluorescence Method. 2. Micro-Brownian Motion and Conformational Change of Amylose in Aqueous-Solution. *Polym. J.* **1982**, *14*(2), 93–99.
12. Pfannemuller, B. Einfluβ der kompetitiven Substrathemmung auf die Kinetik der Phosphorylase aus Kartoffeln. *Starch* **1968**, *20*(11), 351–362.
13. Whelan, W. J.; Bailey, J. M. The Action Pattern of Potato Phosphorylase. *Biochem. J.* **1954**, *58*(4), 560–569.
14. Waldmann, H.; Gygax, D.; Bednarski, M. D.; Randall Shangraw, W.; Whitesides, G. M. The Enzymic Utilization of Sucrose in the Synthesis of Amylose and Derivatives of Amylose, Using Phosphorylases. *Carbohydr. Res.* **1986**, *157*, c4–c7.
15. Yanase, M.; Takaha, T.; Kuriki, T. α-Glucan Phosphorylase and Its Use in Carbohydrate Engineering. *J. Sci. Food Agric.* **2006**, *86*(11), 1631–1635.
16. Flory, P. J. Molecular Size Distribution in Ethylene Oxide Polymers. *J. Am. Chem. Soc.* **1940**, *62*(6), 1561–1565.
17. Kitamura, S.; Suzuki, S.; Ogawa, K. Enzymatically Synthesized Amyloses. In *New Development of Bio-Based Materials*, Kimura, Y., Ohara, H., Eds; CMC Publishing Co., Ltd.: Tokyo, **2007**; pp 45–51.
18. Niemann, C.; Saenger, W.; Pfannemüller, B.; Eigner, W. D.; Huber, A. Phosphorolytic Synthesis of Low-Molecular-Weight Amyloses with Modified Terminal Groups. In *Biotechnology of Amylodextrin Oligosaccharides*, American Chemical Society, **1991**; Vol. 458, pp 189–204.
19. Akiyoshi, K.; Kohara, M.; Ito, K.; Kitamura, S.; Sunamoto, J. Enzymatic Synthesis and Characterization of Amphiphilic Block Copolymers of Poly(Ethylene Oxide) and Amylose. *Macromol. Rapid Commun.* **1999**, *20*(3), 112–115.
20. Akiyoshi, K.; Maruichi, N.; Kohara, M.; Kitamura, S. Amphiphilic Block Copolymer With a Molecular Recognition Site: Induction of a Novel Binding Characteristic of Amylose by Self-Assembly of Poly(Ethylene Oxide)-Block-Amylose in Chloroform. *Biomacromolecules* **2002**, *3*(2), 280–283.
21. Morimoto, N.; Ogino, N.; Narita, T.; Kitamura, S.; Akiyoshi, K. Enzyme-Responsive Molecular Assembly System With Amylose-Primer Surfactants. *J. Am. Chem. Soc.* **2007**, *129*(3), 458–459.
22. Kadokawa, J.-i. Precision Polysaccharide Synthesis Catalyzed by Enzymes. *Chem. Rev.* **2011**, *111*(7), 4308–4345.

23. Enomoto, N.; Furukawa, S.; Ogasawara, Y.; Akano, H.; Kawamura, Y.; Yashima, E.; Okamoto, Y. Preparation of Silica Gel-Bonded Amylose Through Enzyme-Catalyzed Polymerization and Chiral Recognition Ability of Its Phenylcarbamate Derivative in HPLC. *Anal. Chem.* **1996**, *68*(17), 2798–2804.
24. Andresz, H.; Richter, G. C.; Pfannemuller, B. Chemische Synthese Verzweigter Polysaccharide, 5. Kopplung von oligosacchariden und amylose an verschiedene trager durch hydrazonbindung. *Makromol. Chem.* **1978**, *179*(2), 301–312.
25. Terada, Y.; Fujii, K.; Takaha, T.; Okada, S. Thermus Aquaticus ATCC 33923 Amylomaltase Gene Cloning and Expression and Enzyme Characterization: Production of Cycloamylose. *Appl. Environ. Microbiol.* **1999**, *65*(3), 910–915.
26. Terada, Y.; Yanase, M.; Takata, H.; Takaha, T.; Okada, S. Cyclodextrins Are Not the Major Cyclic α-1, 4-Glucans Produced by the Initial Action of Cyclodextrin Glucanotransferase on Amylose. *J. Biol. Chem.* **1997**, *272*(25), 15729–15733.
27. Shimada, J.; Handa, S.; Kaneko, H.; Takada, T. Conformation of Novel Cycloamylose: Topological Aspects and Simulations. *Macromolecules* **1996**, *29*(20), 6408–6421.
28. Imamura, K. Study on Catalytic Function of Potato D-Enzyme Based on Protein Structure. Doctoral Thesis, Osaka Prefecture University, **2005**.
29. Kitamura, S.; Okamoto, T.; Nakata, Y.; Hayashi, T.; Kuge, T. A Monte-Carlo Study of the Excluded-Volume Effect on the Amylosic Chain Conformation. *Biopolymers* **1987**, *26*(4), 537–548.
30. Roth, C.; Weizenmann, N.; Bexten, N.; Saenger, W.; Zimmermann, W.; Maier, T.; Sträter, N. Amylose Recognition and Ring-Size Determination of Amylomaltase. *Sci. Adv.* **2017**, *3*(1), e1601386.
31. Hizukuri, S.; Takeda, Y.; Yasuda, M.; Suzuki, A. Multi-Branched Nature of Amylose and the Action of Debranching Enzymes. *Carbohydr. Res.* **1981**, *94*(2), 205–213.
32. Nakanishi, Y.; Norisuye, T.; Teramoto, A.; Kitamura, S. Conformation of Amylose in Dimethyl-Sulfoxide. *Macromolecules* **1993**, *26*(16), 4220–4225.
33. Fujii, M.; Honda, K.; Fujita, H. Dilute Solution of Amylose in Dimethylsuldoxide. *Biopolymers* **1973**, *12*(5), 1177–1195.
34. Cowie, J. M. G. Studies on Amylose and Its Derivatives. Part IV. The Viscosity Behaviour of Amylose in Mixed Solvents. *Makromol. Chem.* **1963**, *59*(1), 189–200.
35. Everett, W. W.; Foster, J. F. The Subfractionation of Amylose and Characterization of the Subfractions by Light Scattering1, 2. *J. Am. Chem. Soc.* **1959**, *81*(13), 3459–3464.
36. Nakata, Y.; Kitamura, S.; Takeo, K.; Norisuye, T. Conformation of Amylose and Excluded-Volume Effects on Its Chain Dimensions. *Polym. J.* **1994**, *26*(9), 1085–1089.
37. Kitamura, S.; Kobayashi, K.; Tanahashi, H.; Ozaki, T.; Kuge, T. Dilute Solution Properties of Starch Related Polysaccharides. Part 1. On the Mark-Houwink-Sakurada Equation for Amylose in Aqueous Solvents. *J. Jpn. Soc. Starch Sci.* **1989**, *36*(4), 257–264.
38. Norisuye, T. Conformation and Properties of Amylose in Dilute Solution. *Food Hydrocoll.* **1996**, *10*(1), 109–115.
39. Shimada, J.; Kaneko, H.; Takada, T.; Kitamura, S.; Kajiwara, K. Conformation of Amylose in Aqueous Solution: Small-Angle X-Ray Scattering Measurements and Simulations. *J. Phys. Chem. B* **2000**, *104*(9), 2136–2147.
40. Goebel, K.; Brant, D. The Configuration of Amylose and Its Derivatives in Aqueous Solution. Experimental Results. *Macromolecules* **1970**, *3*, 634–643.
41. Kitamura, S.; Tanahashi, H.; Kuge, T. Study of Polysaccharides by the Fluorescence Method .3. Chain-Length Dependence of the Micro-Brownian Motion of Amylose in an Aqueous-Solution. *Biopolymers* **1984**, *23*(6), 1043–1056.

42. Nakata, Y.; Amitani, K.; Norisuye, T.; Kitamura, S. Translational Diffusion Coefficient of Cycloamylose in Aqueous Sodium Hydroxide. *Biopolymers* **2003**, *69*(4), 508–516.
43. Nakata, Y.; Norisuye, T.; Kitamura, S. Monte Carlo Study of Cycloamylose: Chain Conformation, Radius of Gyration, and Diffusion Coefficient. *Biopolymers* **2002**, *64*(2), 72–79.
44. Kitamura, S.; Isuda, H.; Shimada, J.; Takada, T.; Takaha, T.; Okada, S.; Mimura, M.; Kajiwara, K. Conformation of Cyclomaltooligosaccharide ("cycloamylose") of dp21 in Aqueous Solution. *Carbohydr. Res.* **1997**, *304*(3-4), 303–314.
45. Suzuki, S.; Kitamura, S. Unfrozen Water in Amylosic Molecules Is Dependent on the Molecular Structures—A Differential Scanning Calorimetric Study. *Food Hydrocoll.* **2008**, *22*(5), 862–867.
46. Gessler, K.; Usón, I.; Takaha, T.; Krauss, N.; Smith, S. M.; Okada, S.; Sheldrick, G. M.; Saenger, W. V-Amylose at Atomic Resolution: X-Ray Structure of a Cycloamylose With 26 Glucose Residues (Cyclomaltohexaicosaose). *Proc. Natl. Acad. Sci.* **1999**, *96*(8), 4246–4251.
47. Kitamura, S.; Nakatani, K.; Takaha, T.; Okada, S. Complex Formation of Large-Ring Cyclodextrins With Iodine in Aqueous Solution as Revealed by Isothermal Titration Calorimetry. *Macromol. Rapid Commun.* **1999**, *20*(12), 612–615.
48. Kitamura, S. Molecular Inclusion of Large-Ring Cyclodextrins. *J. Appl. Glycosci.* **2003**, *50*(2), 321–325.
49. Burchard, V. W. Das viskositätsverhalten von amylose in verschiedenen lösungsmitteln. *Makromol. Chem.* **1963**, *64*(1), 110–125.
50. Montesanti, N.; Veronese, G.; Buleon, A.; Escalier, P. C.; Kitamura, S.; Putaux, J. L. A-Type Crystals from Dilute Solutions of Short Amylose Chains. *Biomacromolecules* **2010**, *11*(11), 3049–3058.
51. Imberty, A.; Chanzy, H.; Pérez, S.; Bulèon, A.; Tran, V. The Double-Helical Nature of the Crystalline Part of A-Starch. *J. Mol. Biol.* **1988**, *201*(2), 365–378.
52. Gidley, M. J.; Bulpin, P. V. Crystallisation of Malto-Oligosaccharides as Models of the Crystalline Forms of Starch: Minimum Chain-Length Requirement for the Formation of Double Helices. *Carbohydr. Res.* **1987**, *161*(2), 291–300.
53. Kainuma, K.; French, D. Nägeli Amylodextrin and Its Relationship to Starch Granule Structure. I. Preparation and Properties of Amylodextrins From Various Starch Types. *Biopolymers* **1971**, *10*(9), 1673–1680.
54. Kitamura, S.; Hakozaki, K.; Kuge, T. Effects of Molecular Weight on the Retrogradation of Amylose. In *Food Hydrocolloids*, Springer, 1994; pp 179–182.
55. Creek, J. A.; Ziegler, G. R.; Runt, J. Amylose Crystallization From Concentrated Aqueous Solution. *Biomacromolecules* **2006**, *7*(3), 761–770.
56. Müller, J. J.; Gernat, C.; Schulz, W.; Müller, E.-C.; Vorwerg, W.; Damaschun, G. Computer Simulations of X-Ray Scattering Curves: Gelation and Crystallization Process in Amylose Solutions. *Biopolymers* **1995**, *35*(3), 271–288.
57. I'Anson, K. J.; Miles, M. J.; Morris, V. J.; Ring, S. G.; Nave, C. A Study of Amylose Gelation Using a Synchrotron X-Ray Source. *Carbohydr. Polym.* **1988**, *8*(1), 45–53.
58. Bulkin, B. J.; Kwak, Y.; Dea, I. C. M. Retrogradation Kinetics of Waxy-Corn and Potato Starches; A Rapid, Raman-Spectroscopic Study. *Carbohydr. Res.* **1987**, *160*, 95–112.
59. Miles, M. J.; Morris, V. J.; Ring, S. G. Gelation of Amylose. *Carbohydr. Res.* **1985**, *135*(2), 257–269.
60. Miles, M. J.; Morris, V. J.; Orford, P. D.; Ring, S. G. The Roles of Amylose and Amylopectin in the Gelation and Retrogradation of Starch. *Carbohydr. Res.* **1985**, *135*(2), 271–281.

61. Nakazawa, F.; Noguchi, S.; Takahashi, J.; Takada, M. Gelatinization and Retrogradation of Rice Starch Studied by Differential Scanning Calorimetry. *Agric. Biol. Chem.* **1984**, *48*(1), 201–203.
62. Leloup, V. M.; Colonna, P.; Ring, S. G.; Roberts, K.; Wells, B. Microstructure of Amylose Gels. *Carbohydr. Polym.* **1992**, *18*(3), 189–197.
63. Yamamoto, K.; Suzuki, S.; Kitamura, S.; Yuguchi, Y. Gelation and Structural Formation of Amylose by In Situ Neutralization as Observed by Small-Angle X-ray Scattering. *Gels* **2018**, *4*(3), 57.
64. Colin, J. J.; de Claubry, H. G. *Ann. Chim.* **1814**, *1*(90), 87–100.
65. Rundle, R. E. Configuration of Starch in the Starch-Iodine Complex. V. Fourier Projections From X-Ray Diagrams. *J. Am. Chem. Soc.* **1947**, *69*, 1769–1772.
66. Kitamura, S.; Terada, Y.; Terada, A.; Sunako, M.; Hosoya, K.; Suzuki, S.; Junichi, T.; Takaha, T. A Polarizing Film and a Production Method Thereof JP4599576B2, **2004**.
67. Yan, B.; Matsushita, S.; Suzuki, S.; Kitamura, S.; Kaiho, T.; Akagi, K. Low-Density Graphitic Films Prepared From Iodine-Doped Enzymatically Synthesized Amylose Films as Carbonization Precursors. *Carbohydr. Polym.* **2018**, *196*, 332–338.
68. Suzuki, S. Physical and Functional Properties of Novel Polysaccharide Films. Osaka Prefecture University Doctoral Thesis, **2008**.
69. Suzuki, S.; Shimahashi, K.; Takahara, J.; Sunako, M.; Takaha, T.; Ogawa, K.; Kitamura, S. Effect of Addition of Water-Soluble Chitin on Amylose Film. *Biomacromolecules* **2005**, *6*(6), 3238–3242.
70. Suzuki, S.; Ying, B.; Yamane, H.; Tachi, H.; Shimahashi, K.; Ogawa, K.; Kitamura, S. Surface Structure of Chitosan and Hybrid Chitosan-Amylose Films—Restoration of the Antibacterial Properties of Chitosan in the Amylose Film. *Carbohydr. Res.* **2007**, *342*(16), 2490–2493.
71. Suzuki, S.; Nishioka, J.; Kitamura, S. Characterization of Amylose Nanogels and Microgels Containing Ionic Polysaccharides. *J. Appl. Glycosci.* **2017**, *64*, 21–25.
72. Machida, S.; Ogawa, S.; Xiaohua, S.; Takaha, T.; Fujii, K.; Hayashi, K. Cycloamylose as an Efficient Artificial Chaperone for Protein Refolding. *FEBS Lett.* **2000**, *486*(2), 131–135.
73. Ryoki, A.; Yokobatake, H.; Hasegawa, H.; Takenaka, A.; Ida, D.; Kitamura, S.; Terao, K. Topology-Dependent Chain Stiffness and Local Helical Structure of Cyclic Amylose Tris (3, 5-Dimethylphenylcarbamate) in Solution. *Macromolecules* **2017**, *50*(10), 4000–4006.
74. Ryoki, A.; Kim, D.; Kitamura, S.; Terao, K. Linear and Cyclic Amylose Derivatives Having Brush Like Side Groups in Solution: Amylose tris (n-Octadecylcarbamate)s. *Polymer* **2018**, *137*, 13–21.
75. Jiang, X.; Kitamura, S.; Sato, T.; Terao, K. Chain Dimensions and Stiffness of Cellulosic and Amylosic Chains in an Ionic Liquid: Cellulose, Amylose, and an Amylose Carbamate in BmimCl. *Macromolecules* **2017**, *50*(10), 3979–3984.
76. Terao, K.; Shigeuchi, K.; Oyamada, K.; Kitamura, S.; Sato, T. Solution Properties of a Cyclic Chain Having Tunable Chain Stiffness: Cyclic Amylose Tris(n-Butylcarbamate) in Theta and Good Solvents. *Macromolecules* **2013**, *46*(13), 5355–5362.
77. Oyamada, K.; Terao, K.; Suwa, M.; Kitamura, S.; Sato, T. Lyotropic Liquid Crystallinity of Armylose Tris(alkylcarbamates): Cholesteric and Smectic Phase Formation in Different Solvents. *Macromolecules* **2013**, *46*(11), 4589–4595.
78. Asano, N.; Kitamura, S.; Terao, K. Local Conformation and Intermolecular Interaction of Rigid Ring Polymers Are Not Always the Same as the Linear Analogue: Cyclic Amylose Tris(phenylcarbamate) in Theta Solvents. *J. Phys. Chem. B* **2013**, *117*(32), 9576–9583.

79. Tsuda, M.; Terao, K.; Kitamura, S.; Sato, T. Solvent-Dependent Conformation of a Regioselective Amylose Carbamate: Amylose-2-acetyl-3,6-bis(Phenylcarbamate). *Biopolymers* **2012**, *97*(12), 1010–1017.
80. Terao, K.; Maeda, F.; Oyamada, K.; Ochiai, T.; Kitamura, S.; Sato, T. Side-Chain-Dependent Helical Conformation of Amylose Alkylcarbamates: Amylose Tris(ethylcarbamate) and Amylose Tris(n-Hexylcarbamate). *J. Phys. Chem. B* **2012**, *116*(42), 12714–12720.
81. Terao, K.; Asano, N.; Kitamura, S.; Sato, T. Rigid Cyclic Polymer in Solution: Cycloamylose Tris(phenylcarbamate) in 1,4-Dioxane and 2-Ethoxyethanol. *ACS Macro Lett.* **2012**, *1*(11), 1291–1294.
82. Arakawa, S.; Terao, K.; Kitamura, S.; Sato, T. Conformational Change of an Amylose Derivative in Chiral Solvents: Amylose tris(n-Butylcarbamate) in Ethyl Lactates. *Polym. Chem.* **2012**, *3*(2), 472–478.
83. Tsuda, M.; Terao, K.; Nakamura, Y.; Kita, Y.; Kitamura, S.; Sato, T. Solution Properties of Amylose Tris(3,5-dimethylphenylcarbamate) and Amylose Tris(phenylcarbamate): Side Group and Solvent Dependent Chain Stiffness in Methyl Acetate, 2-Butanone, and 4-Methyl-2-Pentanone. *Macromolecules* **2010**, *43*(13), 5779–5784.
84. Terao, K.; Murashima, M.; Sano, Y.; Arakawa, S.; Kitamura, S.; Norisuye, T. Conformational, Dimensional, and Hydrodynamic Properties of Amylose Tris(n-butylcarbamate) in Tetrahydrofuran, Methanol, and Their Mixtures. *Macromolecules* **2010**, *43*(2), 1061–1068.
85. Sano, Y.; Terao, K.; Arakawa, S.; Ohtoh, M.; Kitamura, S.; Norisuye, T. Solution Properties of Amylose tris(n-butylcarbamate). Helical and Global Conformation in Alcohols. *Polymer* **2010**, *51*(18), 4243–4248.
86. Terao, K.; Fujii, T.; Tsuda, M.; Kitamura, S.; Norisuye, T. Solution Properties of Amylose Tris(Phenylcarbamate): Local Conformation and Chain Stiffness in 1,4-Dioxane and 2-Ethoxyethanol. *Polym. J.* **2009**, *41*(3), 201–207.
87. Fujii, T.; Terao, K.; Tsuda, M.; Kitamura, S.; Norisuye, T. Solvent-Dependent Conformation of Amylose Tris(phenylcarbamate) as Deduced from Scattering and Viscosity Data. *Biopolymers* **2009**, *91*(9), 729–736.
88. Terao, K.; Sato, T. Conformational Properties of Polysaccharide Derivatives. In *Bioinspired Materials Science and Engineering*, **2018**; pp 167–183.
89. Shen, J.; Okamoto, Y. Efficient Separation of Enantiomers Using Stereoregular Chiral Polymers. *Chem. Rev.* **2016**, *116*(3), 1094–1138.

Modification of xanthan in the ordered and disordered states

11

Marianne Øksnes Dalheim[a], *Bjørn E. Christensen*[a], *Sébastien Comesse*[b], *Frédéric Renou*[c]
[a]Department of Biotechnology and Food Science, NTNU-Norwegian University of Science and Technology, Trondheim, Norway, [b]Laboratoire URCOM, EA 3221, INC3M-CNRS-FR 3038, Université Le Havre Normandie, Le Havre, France, [c]Institut des Molécules et Matériaux du Mans, UMR CNRS 6283, Le Mans University, Le Mans, France

Chapter outline

1 Introduction

Xanthan, or xanthan gum, is an extracellular polysaccharide produced by strains of the bacterium *Xanthomonas campestris*. It was discovered in the early 1960s as a result of an USDA screening program for new industrial gums[1] and was soon industrialized.[2] Xanthan was approved as a food additive in 1968. It is widely used not only in foods but also relevant as a viscosifier in areas like increased oil recovery (IOR).[3] The basis for most applications is the unusual solution properties: aqueous xanthan has a very high intrinsic viscosity at low shear rates (Newtonian regime), typically in the range 4–6000 mL/g, and show pronounced shear thinning at high shear rates. Despite being a polyelectrolyte, the intrinsic viscosity of xanthan is—contrary to, for example, alginate or carboxymethyl cellulose (CMC)—essentially independent of the ionic strength (at moderate temperatures). Moreover, xanthan is, compared to most other

Recent Trends in Carbohydrate Chemistry. https://doi.org/10.1016/B978-0-12-817467-8.00011-6
© 2020 Elsevier Inc. All rights reserved.

polysaccharides, quite stable against chemical and enzymatic degradation, but is at the same time considered biodegradable. These—and other—properties are linked to the formation of a double-stranded, ordered structure. Xanthan shows a characteristic pH and ionic strength-dependent conformational transition at elevated temperatures, whereupon many of the unique solution and stability properties may converge toward a more CMC-like behavior.

2 Chemical structure

The full chemical structure of xanthan was published in 1975.[4] It consists of a branched pentasaccharide repeating unit (Fig. 1A). The main chain consists of 1,4-linked β-D-glucose (glucopyranoside), being hence identical to that of cellulose. Every second glucose residue is substituted at *O*-3 by the trisaccharide β-D-Man*p*-(1→4)-β-D-GlcA*p*-(1→2)-α-D-Man*p*-(1→3). Moreover, the terminal β-D-mannose may contain an acetal-linked pyruvate (4,6-linked) and the inner α-D-mannose may contain an *O*-acetyl at position 6. The degree of pyruvate and acetate substitution depends to some extent on the bacterial strain and the culturing conditions. It was recently shown that xanthan consists of six types of repeating units with identical sugar composition, but different pyruvate and acetate substitution, including a previously undescribed *O*-acetyl group on the terminal β-D-mannose (Fig. 1B).[5]

Some interesting but little explored variations of the oligosaccharide repeating units have indeed been described.[6] By selectively inactivating genes coding for transferases responsible for assembling the lipid-linked oligosaccharide repeating units

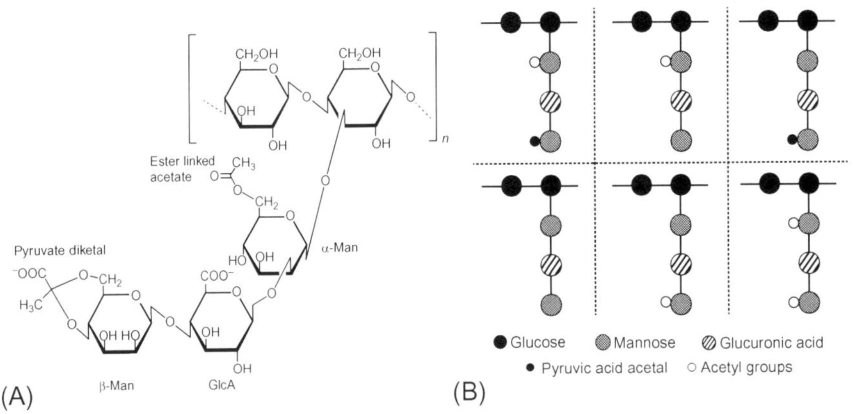

Fig. 1 (A) Pentasaccharide repeating unit of xanthan (Haworth formula). The amount of pyruvate and acetate may vary, depending on the bacterial strain and culturing conditions. (B) Repeating units identified in xanthan following enzymatic degradation.
Reproduced from Kool, M. M.; Gruppen, H.; Sworn, G.; Schols, H. A., Comparison of Xanthans by the Relative Abundance of Its Six Constituent Repeating Units. Carbohydr. Polym. **2013**, *98* (1), 914–921 (with permission).

(prior to the polymerization step) xanthans having truncated side chains could be obtained, although at low yields. However, the physical properties of these variants have so far not been further investigated, except for an initial comparison of solution viscosities.[7]

3 Conformation, order-disorder transition, and polyelectrolyte properties

Xanthan has a relatively high linear charge density due to glucuronic acid and pyruvate in the side chains. Fully pyruvated xanthan has two negative charges per repeating units. The charge density thus corresponds to that of CMC (DS=1.0) or alginate. The latter displays a high ionic strength dependence of the intrinsic viscosity (Fig. 2B), typical for semiflexible chains (the intrinsic persistence length of alginate is around 12 nm[8]), whereas the rigidity of xanthan (persistence length of about 120 nm[9]) prevents chain compaction when charges are screened at high ionic strength. The intrinsic viscosity is therefore essentially independent of the ionic strength (Fig. 2A). This also applies to the radius of gyration.

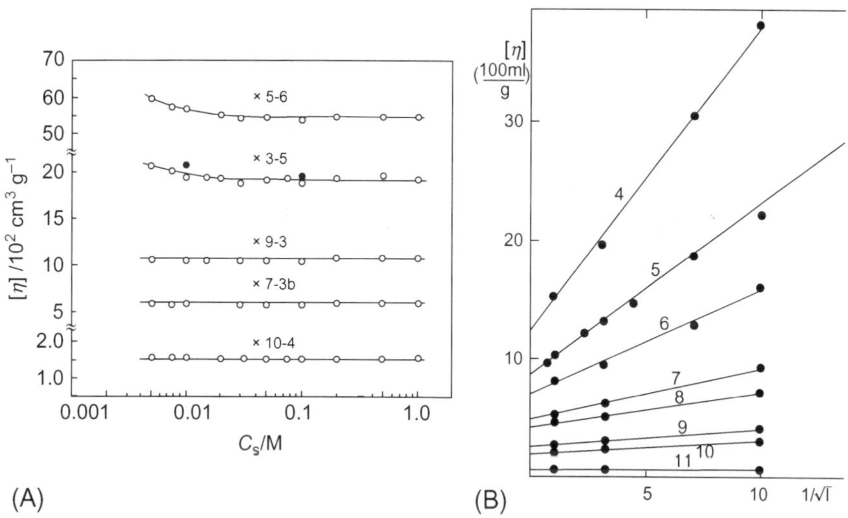

Fig. 2 (A) Ionic strength dependence of the intrinsic viscosity at 25°C of five xanthan fractions with different molecular weights. (B) Ionic strength dependence of the intrinsic viscosity at 20°C of alginates of different molecular weights. Note X-axis is $1/\sqrt{I}$.
Part (A) reproduced from Sho, T.; Sato, T.; Norisuye, T., Viscosity Behavior and Persistence Length of Sodium Xanthan in Aqueous Sodium-Chloride. Biophys. Chem. **1986**, *25* (3), 307–313 (with permission) and (B) from Smidsrød, O.; Haug, A., Estimation of the Relative Stiffness of the Molecular Chain in Polyelectrolytes From Measurements of Viscosity of Different Ionic Strengths. Biopolymers **1971**, *10*, 1213–1227 (with permission).

Heating dilute solutions of xanthan produces a characteristic cooperative transition resembling protein denaturation. The transition can thus be detected by, for example, polarimetry (optical rotation),[10–12] ellipsometry (circular dichroism (CD)),[10–12] differential scanning calorimetry (DSC),[12, 13] NMR (nuclear magnetic resonance), or rheological measurements. The midpoint of the thermal transition (T_m) increases with increasing ionic strength. It decreases with increasing content of pyruvate (higher charge density) but increases with increasing content of acetate. Lowering pH, that is, neutralizing the negative charges, also leads to an increase in T_m.[13] The ionic strength and pH dependencies are illustrated by the optical rotation study shown in Fig. 3. The molecular phenomena underlying the transition markers detected by the methods described above remain elusive but are probably related to conformational changes in the side chains rather than the main chain. Some support for this hypothesis comes from the observation that gradual removal of the terminal β-mannose by partial acid hydrolysis resulted in a corresponding (linear) decrease in specific optical rotation, the main CD band (204 nm), and the transition enthalpy, whereas the transition temperature was essentially unchanged.[13]

The nature of the ordered ($T<T_m$) and disordered ($T>T_m$) conformation has been heavily debated in the literature for decades. Some investigators have observed essentially no change in the molecular weight when heating above T_m, which was interpreted as an intramolecular single helix-to-single coil transition.[12] Other investigators did, however, observe a halving in M_w upon heating above T_m.[14] Further, the 50% reduction in M_w and change from rod-like to coil-like behavior upon dissolution in cadoxen (cellulose solvent) were interpreted as a duplex-to-single coil transition.[9]

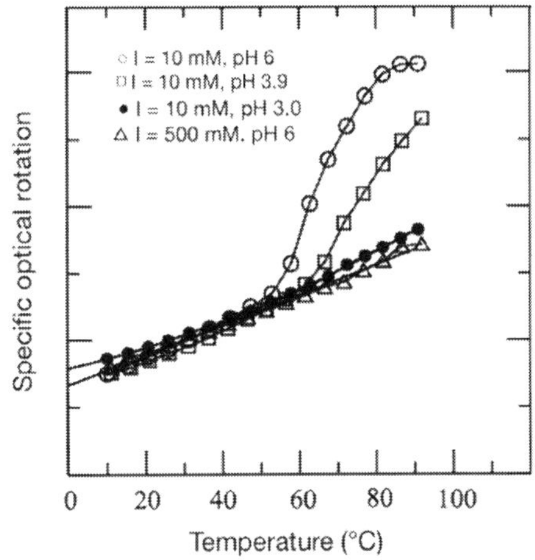

Fig. 3 Temperature and pH dependences of the optical rotation of xanthan. Reproduced from Christensen, B. E.; Smidsrød, O., Hydrolysis of Xanthan in Dilute Acid—Effects on Chemical-Composition, Conformation, and Intrinsic-Viscosity. Carbohydr. Res. **1991**, *214* (1), 55–69 (with permission).

In any case there has long been a general consensus that the ordered state has a rod-like character with persistence length approximately one order of magnitude larger than water-soluble cellulose derivatives such as CMC.

The double-stranded nature of the ordered state was further supported by electron microscopy (Fig. 4A). Two separate studies[15, 16] showed local chain dissociation (loops) in addition to characteristic stiff chains. Recently, an AFM study also revealed a detailed helical topology of double-stranded xanthan (Fig. 4B).[17] Support for the double-stranded behavior also comes from the complex and strongly concentration-dependent geometries formed upon renaturation.[18, 19] Interestingly, a double helix model was suggested as early as 1977.[20] In a recent review[21] Morris suggested that conformational ordering could in fact be a two-step process, which may explain many of the apparently non-reconciling results and interpretations in earlier literature. He noted that conformational ordering monitored by molecular weight measurements did not coincide with ordering monitored chiroptically (Fig. 5). The two-step model is, however, at variance with the "all-or-none" behavior observed for denaturation in cadoxen-water mixtures.[23]

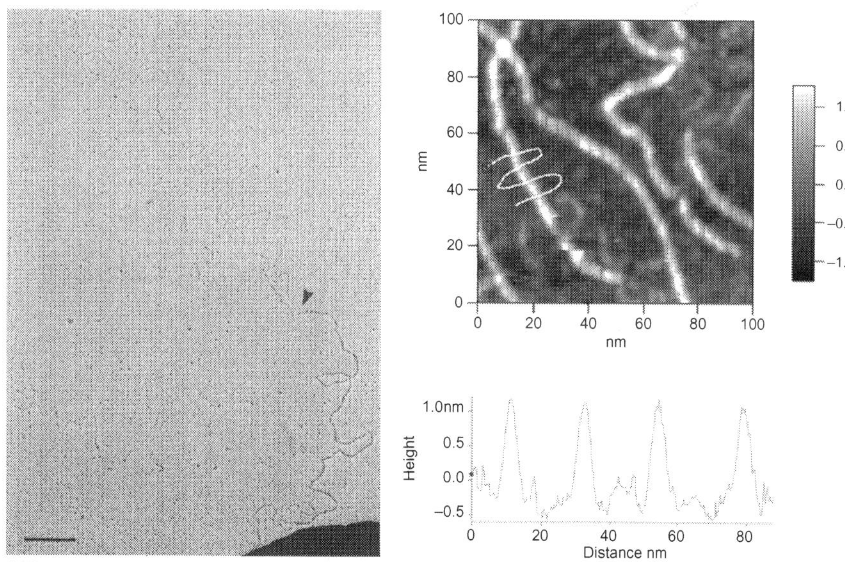

Fig. 4 (A) Electron micrograph of xanthan in 50% glycerol, 2 mM ammonium acetate, and pH 7. Note that the chain divides into two thinner strands (arrow). Bar: 200 nm. (B) AFM showing the helical pitch (*top*) and height profiles of double-stranded xanthan.

Part (A) reproduced from Stokke, B. T.; Elgsæter, A.; Smidsrød, O., Electron-Microscopic Study of Single-Stranded and Double-Stranded Xanthan. Int. J. Biol. Macromol. **1986**, *8* (4), 217–225 (with permission) and (B) from Moffat, J.; Morris, V. J.; Al-Assaf, S.; Gunning, A. P., Visualisation of Xanthan Conformation by Atomic Force Microscopy. Carbohydr. Polym. **2016**, *148*, 380–389 (with permission).

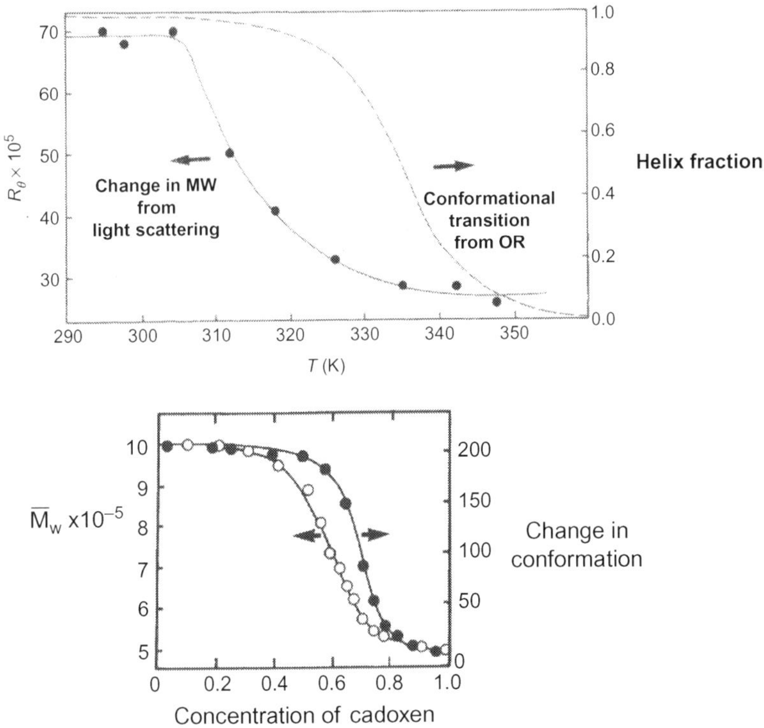

Fig. 5 Compilation of data by Morris[21] based on results published by Norton et al.[12] (*top*) and Kitagawa et al. (*bottom*).[22]
Reproduced from Morris, E. R., Ordered Conformation of Xanthan in Solutions and "Weak Gels": Single Helix, Double Helix—or Both? Food Hydrocoll. **2019**, *86*, 18–25 (with permission).

Denaturation-renaturation cycles may be reversible (although hysteresis is commonly observed) in terms of ORD (optical rotatory dispersion) and CD, but usually not in terms of rheology. The physical structure and solution properties of renatured xanthan depend particularly on factors such as molecular weight, concentration, ionic strength, and thermal history.[18, 19] Complete dissociation into single strands may be obtained by heating salt-free dilute solutions (below 1 mg/mL) above 80°C, but cooling tended to produce hairpin-type duplexes. At higher concentrations (10 mg/mL) the data fitted a model where duplexes only partially unwind from the chain termini. In the renaturation process, more complex aggregates are formed (Fig. 6). At even higher concentrations a thermal denaturation-renaturation cycle can be applied to form xanthan hydrogels.[24] Although the first cycle has large rheological consequences subsequent cycles seem generally to be rheologically reversible.

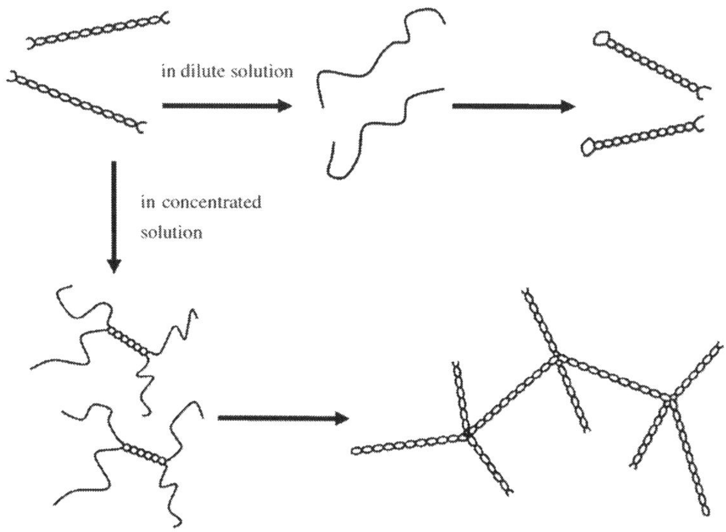

Fig. 6 Schematic diagram of the denaturation and renaturation of the xanthan double helix. Reproduced from Matsuda, Y.; Biyajima, Y.; Sato, T., Thermal Denaturation, Renaturation, and Aggregation of a Double-Helical Polysaccharide Xanthan in Aqueous Solution. Polym. J. **2009**, *41* (7), 526–532 (with permission).

4 Stability and degradation: Role of conformational states

4.1 Acid hydrolysis

The partial acid hydrolysis of xanthan indeed strongly depends on conformational state, which is again dependent on temperature, ionic strength, and pH as partly described above.[25, 26] The conformationally stabilizing effect of lowering pH leads to the counterintuitive observation that, at low ionic strength (10 mM, 80°C), the degradation (decrease in M_w) is faster at pH 4 (disordered) than at pH 2 (ordered), whereas the "normal" pH relationship is restored at high ionic strength (500 mM, 80°C).[25]

Relatively mild acid hydrolysis rapidly removes the terminal pyruvate before other sugars are affected.[27] Next, the terminal β-D-mannose hydrolyzes relatively fast, leading to its gradual release while retaining high molecular weight. Upon more extensive hydrolysis α-D-mannose also hydrolyzes, leading to complete loss of the side chain. Even in this case xanthan may retain high molecular weight[25] and display the characteristic order-disorder transition.[13]

The cellulose-type backbone appears very resistant to hydrolysis as the decrease in M_w corresponds to a rate of hydrolysis orders of magnitude below the side chain mannoses. However, reducing end assays reveal "hidden" chain breaks in the duplexes.[28] Moreover, single-stranded fragments appear after extensive hydrolysis,[26] and heating

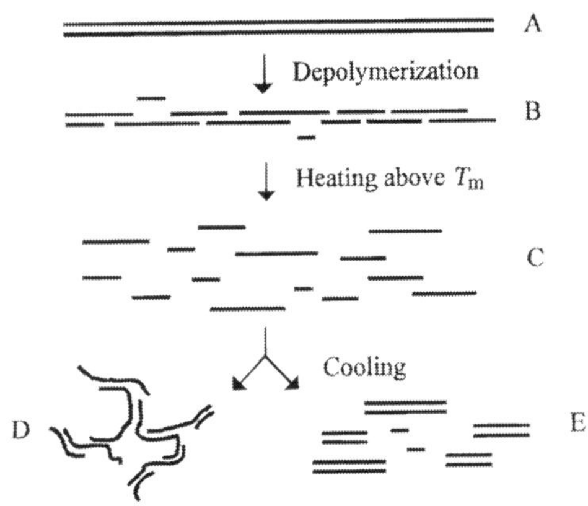

Fig. 7 Schematic representation of the depolymerization of double-stranded xanthan (A–B). Random depolymerization will induce a metastable structure (B) consisting of overlapping chain fragments and a fraction of short fragments ($DP<DP_c$) unable to remain incorporated in a duplex. A subsequent heat treatment well above T_m results in partial or complete dissociation of the duplexes. Complete dissociation results in dispersed single strands of various lengths (C). The cooling step may either induce an aggregated state (D) or perfectly matched duplexes (E), largely depending on the polymer concentration.
Reproduced from Christensen, B. E.; Myhr, M. H.; Smidsrød, O., Degradation of Double-Stranded Xanthan by Hydrogen Peroxide in the Presence of Ferrous Ions: Comparison to Acid Hydrolysis. Carbohydr. Res. **1996**, *280* (1), 85–99 (with permission).

of partially hydrolyzed xanthan above T_m (dilute solutions) followed by cooling revealed their metastable structures and rearrangements toward perfectly matched duplexes of lower M_w (Fig. 7).[28]

5 Rheological properties

The rheological properties of xanthan were soon after its discovery the subject of numerous studies, both academic and industrial. Indeed, xanthan has a strong viscosifying ability at very low concentrations and is strongly shear thinning[29] (Fig. 8A). In the presence of salt, the viscosity of water increases by, typically three orders of magnitude for a xanthan concentration of about 1 g/L. The concentration dependence of the viscosity of xanthan solutions allows determination of the different concentration regimes of xanthan characterized by critical concentrations, C* and C**, where C* corresponds to the transition from diluted to non-entangled semi-diluted regime and C** the transition from non-entangled to entangled semi-diluted regime (Fig. 8B).[30] It may be noted that Takada et al.[31] found a molecular weight dependence in an analysis analogous to that presented in Fig. 8B, which was accounted for by a "fuzzy cylinder model" rather than a flexible chain model.

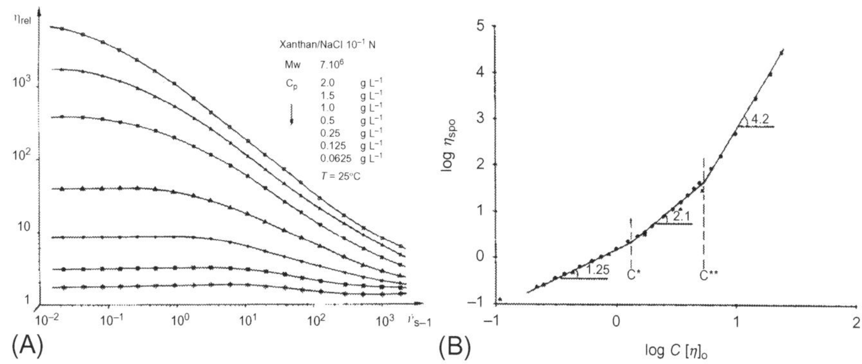

Fig. 8 (A) Flow curves of xanthan at different concentrations in the presence of 0.1 M NaCl. (B) Determination of C* and C** for xanthan in the ordered conformation.
Part (A) reproduced from Milas, M.; Rinaudo, M.; Tinland, B., The Viscosity Dependence on Concentration, Molecular Weight and Shear Rate of Xanthan Solutions. Polym. Bull. **1985**, *14*, 157–164 (with permission) and (B) from Cuvelier, G.; Launay, B., Concentration Regimes in Xanthan Gum Solutions Deduced From Flow and Viscoelastic Properties. Carbohydr. Polym. **1986**, *6*, 321–333 (with permission).

The rheological properties, including viscosity, C*, and C** of a xanthan solution depend strongly on the conformation. Typical values of C* and C** are presented in Table 1 and one can observe that both C* and C** are small for disordered conformation. In the absence of salt, the negative charges of the side chains are not screened. Consequently, disordered chains become stretched due to electrostatic repulsion, and their hydrodynamic volumes increase.

Xanthan solutions have very interesting viscoelastic properties, even at quite low concentrations. At low temperature (below the order to disorder transition temperature) they display gel-like behavior characterized by a storage modulus (G′) higher than the loss modulus (G″), both being weakly dependent on the frequency. At high temperatures and/or very low concentration (when xanthan is in the disordered conformation) they flow as Maxwell fluids. Nevertheless, several authors[35–37] have demonstrated that the viscoelastic behavior at high and low temperatures can be superimposed *via* the construction of frequency-temperature master curves (Fig. 9A and B). It has also been demonstrated that frequency-temperature-concentration master curves can be obtained at different salt concentrations (Fig. 10A).[35] This demonstrates that the viscoelastic properties of aqueous solutions of xanthan are independent of temperature,

Table 1 Average values of the critical concentrations for xanthan in ordered and disordered conformations.[30, 32–34]

Xanthan	C* (g/L)	C** (g/L)
Ordered conformation	0.2	1
Disordered conformation	0.07	0.4

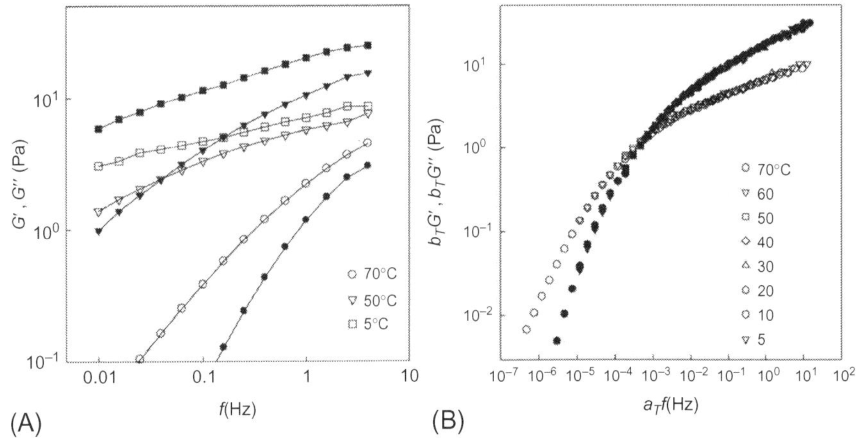

Fig. 9 (A) Frequency dependence of G′ (filled symbols) and G″ (open symbols) of a xanthan solution at C=10 g/L without salt at T=5°C, 20°C, and 70°C. (B) Frequency-temperature master curve of a xanthan solution at C=10 g/L without salt. The reference temperature is 20°C.
Reproduced from Choppe, E.; Puaud, F.; Nicolai, T.; Benyahia, L., Rheology of Xanthan as a Function of Temperature, Concentration and Ionic Strength. Carbohydr. Polym. **2010**, *82*, 1228–1235 (with permission).

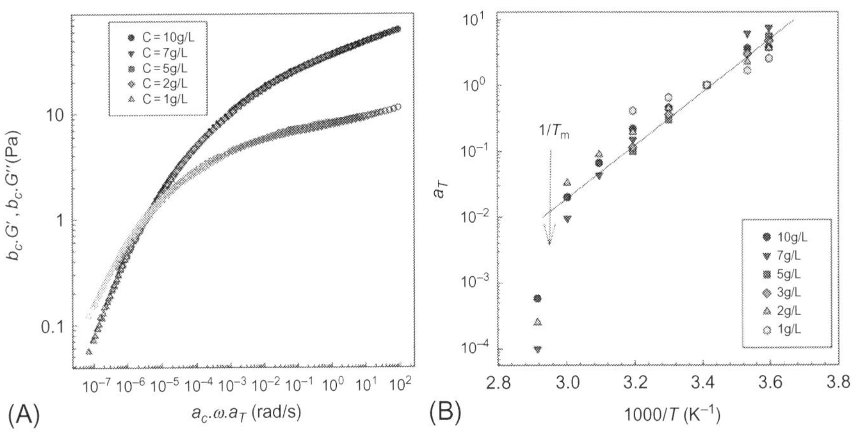

Fig. 10 (A) Frequency-temperature-concentration master curve of G′ (filled symbols) and G″ (open symbols) for a xanthan solution with 0.01 M NaCl. The reference temperature and concentration are 20°C and 10 g/L, respectively. (B) Arrhenius plots of the frequency shift factors used to obtain the frequency-temperature master curves for different xanthan concentrations. The solid line represents Arrhenius law fit for temperatures below T_m.
Reproduced from Renou, F.; Petibon, O.; Malhiac, C.; Grisel, M., Effect of Xanthan Structure on Its Interaction With Locust Bean Gum: Toward Prediction of Rheological Properties. Food Hydrocoll. **2013**, *32*, 331–340 (with permission).

concentration, and ionic strength both in its ordered and disordered conformations. Hence, the conformational transition does not have a significant effect on the chain dynamics, showing that the relaxation process of xanthan chains is the same for both conformations. However, we should stress that the conformational transition strongly accelerates the relaxation. It was shown that the relaxation process follows Arrhenius behavior when xanthan is in its ordered conformation (for temperature below the transitional temperature T_m) with an activation energy around 70–80 kJ/mol.[35, 37] This is not the case when xanthan is disordered (Fig. 10B).

6 Chemical modification of xanthan

The physicochemical properties of xanthan have been extensively studied over the years (described above). A more recent trend in the field is to do various chemical modifications in order to improve the solubility of xanthan and/or to confer new properties such as amphiphilicity by grafting hydrophobic moieties. Nevertheless, these modifications remain scarce compared to other bio-sourced polysaccharides and are closely related to the existing protocols employed: either a random derivatization approach, mostly directed toward the alcohol functions, or a more selective one targeting the carboxyl groups in the xanthan side chains. In the following sections, most of the chemical modifications published to date regarding these two approaches are presented.

6.1 Chemical modification targeting both the alcohol and the acid function of xanthan

Qiu et al. have reported chemical modification of xanthan gum targeting both the alcohol and the acid function.[38] The authors have synthetized aminated xanthan ($XG\text{-}NH_2$) in two steps starting from xanthan (Scheme 1). The first step led to the multi-chlorinated xanthan (XG-Cl) by the action of thionyl chloride ($SOCl_2$) in DMF at 80°C and the second step was the addition of ethylenediamine (EDA) onto XG-Cl in water at 80°C. Formation of the desired $XG\text{-}NH_2$ derivative was confirmed by FT-IR spectra and solid-state ^{13}C NMR. This approach is not very chemoselective since both the alcohol and carboxyl functions are potentially modified during the process.

6.2 Chemical modification targeting the alcohol functions of xanthan

This second approach, targeting only the alcohol functions, is more chemoselective than the previous one. Nevertheless, considering the number of hydroxy groups in the xanthan backbone, this tactic remains poorly regioselective.

In order to form hydrogels from xanthan derivatives, Hamcerencu et al.[39] employed, for the first time, esterification reactions targeting the alcohol functions of xanthan (probably onto the more reactive secondary alcohols at C6) for the introduction of

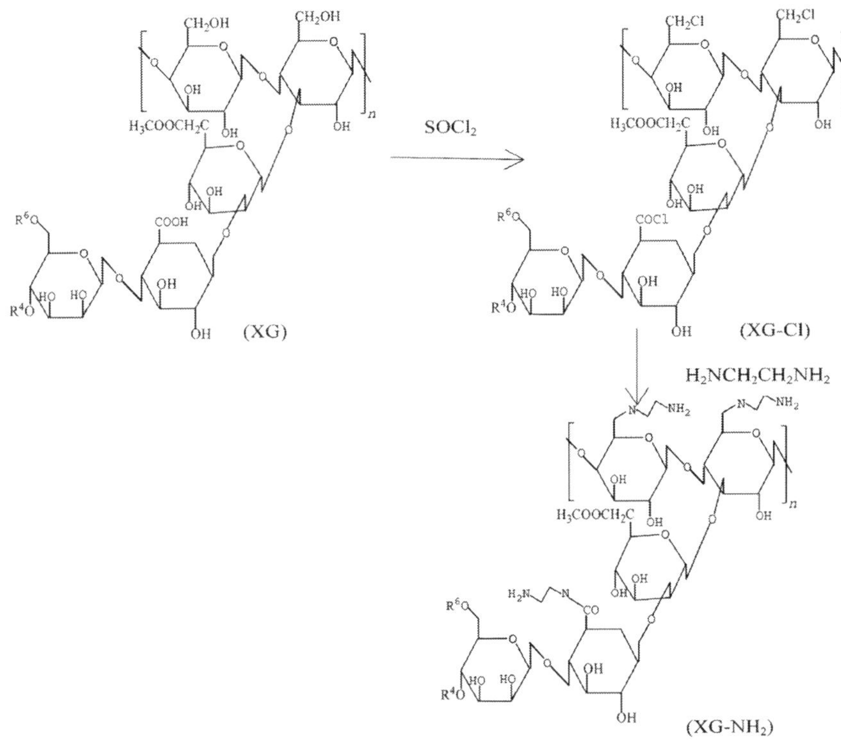

Scheme 1 Chemical synthetic scheme of $XG\text{-}NH_2$.
Reproduced from Qiu, H. Y.; Yan, J. H.; Lan, G. H.; Liu, Y. Q.; Song, X. Q.; Peng, W. X.; Cui, Y. Y., Removal of Cu2+ From Wastewater by Modified Xanthan Gum (XG) With Ethylenediamine (EDA). RSC Adv. **2016**, *6* (86), 83226–83233 (with permission).

an activated carbon-carbon double bond. For that purpose, the authors have explored three different pathways using (a) acrylic acid, (b) acryloyl chloride, and (c) maleic anhydride (Scheme 2).

Acrylate xanthan (A-Xan) could be obtained by employing either acrylic acid or acryloyl chloride. In the first strategy, xanthan was dissolved in water (homogeneous reaction condition) and acrylic acid (11 mole of reactant per mole of xanthan repeating unit) and *N'*-[3-(dimethylamino)propyl]-*N*-ethylcarbodiimide hydrochloride (1.1 mole/mole of acrylic acid) [Scheme 2 pathway (A)] were added. After 624 h at various temperatures, in order to find the optimal conditions, the polymers were precipitated with acetone and further purified and analyzed (Table 2). Due to the low degree of substitution (DS=0.003), an alternative pathway for making A-Xan was considered, starting this time with the more reactive acryloyl chloride [Scheme 2 pathway (B)]. Due to the competitive hydrolysis of this reactant in water, the synthesis was performed in an organic medium in which xanthan is insoluble (called heterogeneous reaction condition by the authors). Xanthan was suspended in DMF and stirred for 12 h at room temperature. Triethylamine (1.1 mol/mol acryloyl chloride) was then added before

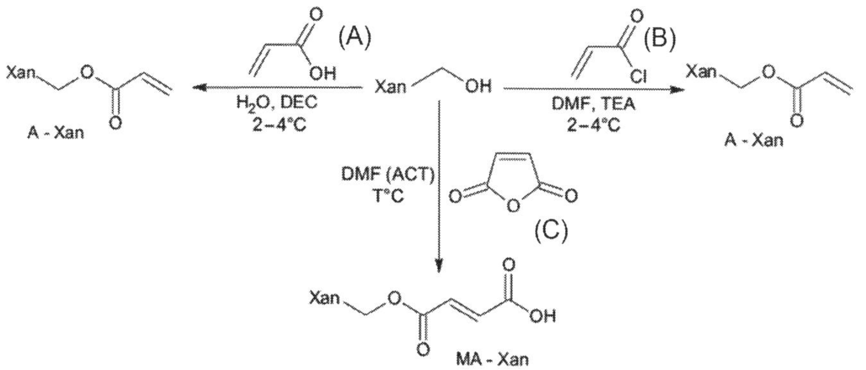

Scheme 2 Strategies for the modification of xanthan: (A) acrylic acid, (B) acryloyl chloride, and (C) maleic anhydride.
Reproduced from Hamcerencu, M.; Desbrieres, J.; Popa, M.; Khoukh, A.; Riess, G., New Unsaturated Derivatives of Xanthan Gum: Synthesis and Characterization. Polymer **2007**, *48* (7), 1921–1929 (with permission).

acryloyl chloride (11 mole of reactant per mole of xanthan repeating unit, dissolved in DMF) was added dropwise. After stirring at 2–4°C for 2, 6, and 24 h, acetone was added and the resulting precipitate was separated by filtration, purified, and analyzed (Table 2). However, this second approach did not lead to higher DS.

The authors then turned their attention to the third pathway, namely the maleic anhydride approach (MA-Xan), which was carried out (heterogeneously) in acetone or DMF to prevent hydrolysis side reactions [Scheme 2, pathway (C)]. In both cases, xanthan was reacted with maleic anhydride using the same ratio as above, while employing various temperatures and reaction times (Table 3). This strategy proved to be much more efficient, especially in DMF, where DS up to 0.108 could be obtained.

More recently, Wang et al.[40] have modified xanthan gum by cross-linking with poly(maleic anhydride/octadec-1-ene) (PMAO) in DMSO at 70°C (Scheme 3).

Table 2 Influence of reaction conditions (time, temperature, and chemical nature of esterification agent, respectively) on DS for the synthesis of A-Xan.

Xan				
Esterification agent	**Reaction time (h)**	**Reaction temperature (°C)**	**DS[a] (%)**	**Double bond per repeating unit**
AA	24	2–4	0.3	0.03
		70	0.3	0.03
AC	6	2–4	0.1	0.01
	24		0.4	0.04

[a] As determined by ^{1}H NMR spectroscopy (in D_2O, T=85°C, D_1=30 s, c_{pol}=around 3 g L^{-1}).
Reproduced from Hamcerencu, M.; Desbrieres, J.; Popa, M.; Khoukh, A.; Riess, G., New Unsaturated Derivatives of Xanthan Gum: Synthesis and Characterization. Polymer **2007**, *48* (7), 1921–1929 (with permission).

Table 3 Effect of reaction parameters (time, temperature, nature of dispersion phase and MA/Xan molar ratio) on DS for the synthesis of maleate substituted xanthan MA-Xan.

MA/Xan (mol/mol)	Reaction time (h)	Reaction temperature (°C)	DS[a] (%)	Double bond per repeating unit
DMF as dispersion medium				
11/1	6	20	1.7	0.2
	24		3.6	0.4
	24	50	4.5	0.5
	6	60	2.9	0.33
	24		9.7	1
	6	70	8	0.9
	24		10.8	1.18
ACT as dispersion medium				
1/1	24	50	0.4	0.045
11/1	24		1.7	0.19

[a] As determined by ^{1}H NMR spectroscopy (in D_2O, T=85°C, D_1=30 s, c_{pol}=around 3 g L^{-1}).
Reproduced from Hamcerencu, M.; Desbrieres, J.; Popa, M.; Khoukh, A.; Riess, G., New Unsaturated Derivatives of Xanthan Gum: Synthesis and Characterization. Polymer **2007**, *48* (7), 1921–1929 (with permission).

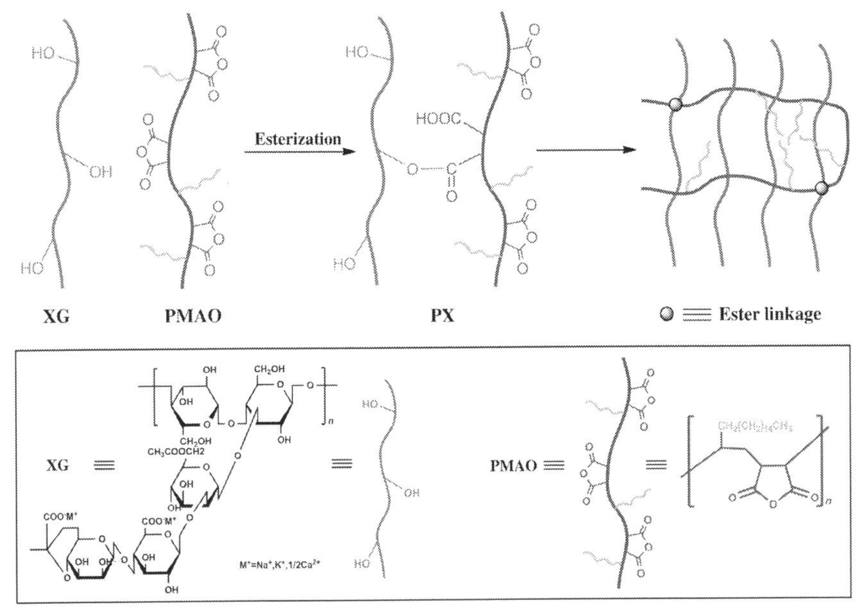

Scheme 3 Esterification of xanthan with PMAO.
Reproduced from Wang, X. J.; Xin, H. P.; Zhu, Y. J.; Chen, W. J.; Tang, E. G.; Zhang, J.; Tan, Y. B., Synthesis and Characterization of Modified Xanthan Gum Using Poly(Maleic Anhydride/1-Octadecene). Colloid Polym. Sci. **2016**, *294* (8), 1333–1341 (with permission).

This esterification pathway is related to that of Hamcerencu et al.[39] (compare Scheme 2 pathway c and Scheme 3), however, the use of maleic anhydride derivative resulted in multiple ester linkages to hydroxy groups in the xanthan backbones, and hence cross-linking of the system. The formation of the cross-linked copolymer was confirmed by FT-IR, static light scattering, and ^{1}H NMR.

Interestingly, it is also possible to form cross-linked starch-xanthan gum hydrogel systems by the use of sodium trimetaphosphate (STMP) as a cross-linker, as demonstrated by Shalviri et al.[41] For that purpose, various amounts of xanthan were gradually added to an aqueous solution of starch and then placed in boiling water for an hour under vigorous stirring. Then different quantities of STMP and 20% (based on the weight of dry starch) sodium sulfate were added at 55°C. The resulting starch-xanthan gum was insoluble in water and in polar organic solvents such as DMSO. Cross-linking was further confirmed by FT-IR spectra and solid ^{31}P NMR. Their combined results indicated that the reaction resulted in a complex mixture of cross-linked xanthan and starch, phosphorylated starch, phosphorylated starch as well as cross-linked starch and possibly cross-linked xanthan.

An esterification approach based on the one previously described by Hamcerencu[39] was used by Skender et al.[42] to obtain hydrophobically modified xanthan for use as a surfactant for stabilizing suspensions of carbon nanotubes. Xanthan was here modified with phthalic anhydride (XPA) and diphenylmaleic anhydride (XDPMA) [compare Scheme 2, pathway (C) and Scheme 4, pathways (A) and (B)]. In the same paper, they have also showed the synthesis of xanthan modified with epichlorhydrin-phenol (XEPH) by a two-step sequence in water: (i) ring opening of 2-(chloromethyl)oxirane in acidic conditions in the presence of $Zn(BF_4)_2$ and (ii) a Williamson reaction

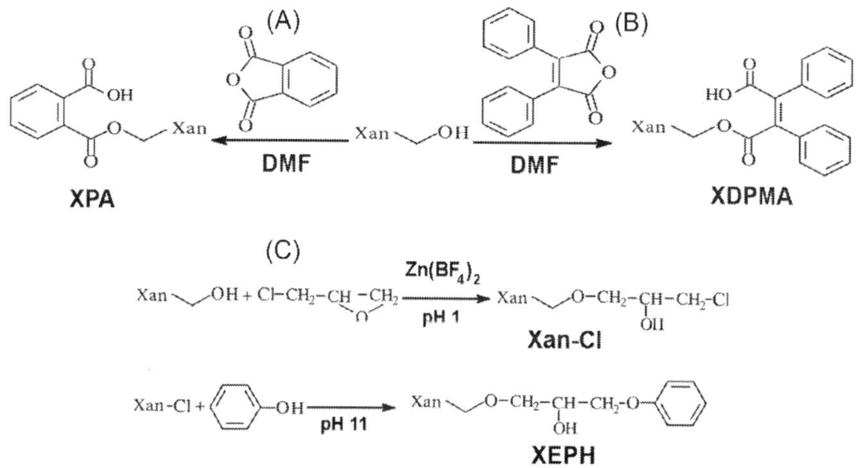

Scheme 4 Procedures for the modification of xanthan: (A, B) heterogeneous media; (C) homogenous media.
Reproduced from Skender, A.; Hadj-Ziane-Zafour, A.; Flahaut, E., Chemical Functionalization of Xanthan Gum for the Dispersion of Double-Walled Carbon Nanotubes in Water. Carbon **2013**, *62*, 149–156 (by permission).

between the resulting Xan-Cl and the phenol in basic conditions [Scheme 4 pathway (C)]. The authors have added 11 mole of each reactant per mole of xanthan repeating unit considering that there is 11 hydroxyl functions per repeating unit.

Successful chemical modification was confirmed by ^1H NMR spectra recorded in D_2O (Fig. 11). Indeed, characteristic peaks of aromatic protons were observed in all cases between 7 and 8.6 ppm albeit with a low intensity, suggesting a low efficiency of the grafting process. DS was therefore not determined.

Carboxymethylation of xanthan was first reported by Ahuja et al.[43] and later by Bakshi et al.,[44] and Yahoum et al.[45] In all the cases, a Williamson reaction between the alcohol functions of xanthan and chloroacetic acid in the presence of a base was employed. In the case of Ahuja et al.,[43] an NaOH solution (45%, w/w) was used and the media was heated up to 70°C. The same base was selected by Yahoum et al.,[45] but here water was replaced by ethanol (heterogeneous conditions) and the reaction was performed at 50°C. Bakshi et al.,[44] on the other hand, carried out the reaction in DMF between 50°C and 65°C with sodium hydride as the base leading to the desired

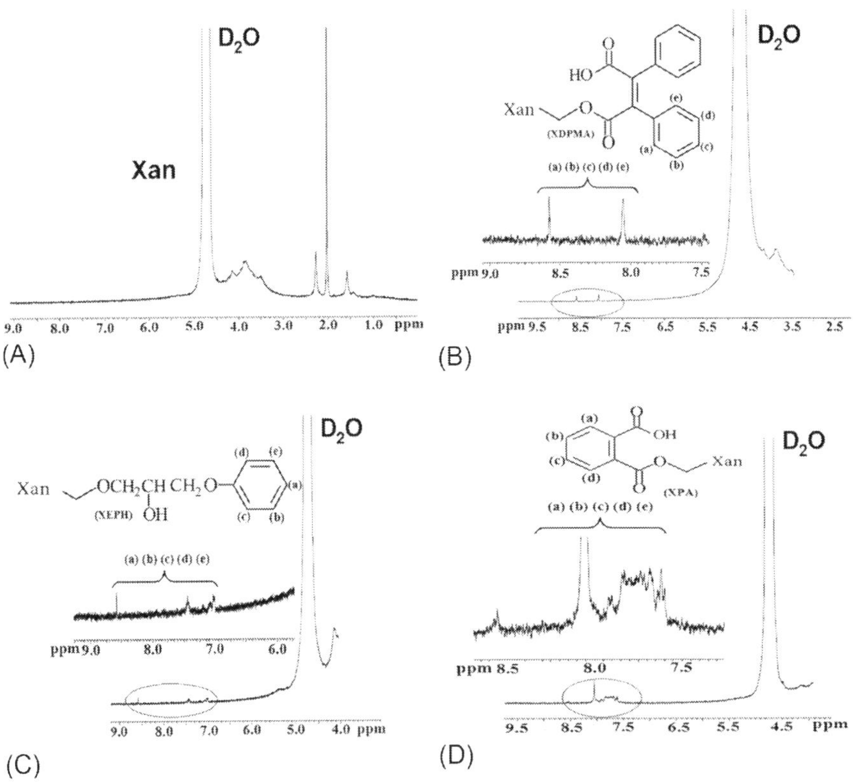

Fig. 11 ^1H NMR spectra in D_2O of Xan (A), XDPMA (B), XEPH (C), and XPA (D). Reproduced from Skender, A.; Hadj-Ziane-Zafour, A.; Flahaut, E., Chemical Functionalization of Xanthan Gum for the Dispersion of Double-Walled Carbon Nanotubes in Water. Carbon **2013**, *62*, 149–156 (with permission).

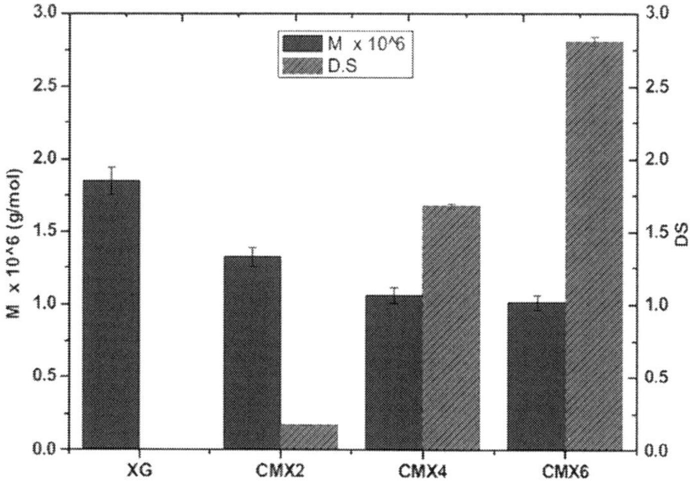

Fig. 12 Variations in the molecular weight and the degree of substitution of carboxymethyl xanthan derivatives (CMX2, CMX4, and CMX6).
Reproduced from Yahoum, M. M.; Moulai-Mostefa, N.; Le Cerf, D., Synthesis, Physicochemical, Structural and Rheological Characterizations of Carboxymethyl Xanthan Derivatives. Carbohydr. Polym. **2016**, *154*, 267–275 (with permission).

modified xanthan with a low DS. In all these cases de-*O*-acetylation is expected due to the alkaline conditions, and probably also alkaline β-elimination of the terminal mannose. Yahoum et al.[45] observed that the higher the DS, the lower the molecular weight, demonstrating that the conditions used for carboxymethylation were too harsh and thus led to the cleavage (hydrolysis) of covalent bonds in the polymer (Fig. 12).

The alcohol functions of xanthan were also targeted by Su et al.[46] to chemically modify xanthan with formaldehyde directly in the fermentation broth at low pH. Even though the authors have confirmed that the xanthan was modified by forming acetals, they did neither discuss the efficacy of the process, nor provide any macromolecular parameters.

Free radical polymerization of ethyl acetate (EA) onto xanthan has been reported by Pandey et al.[47] The reaction was initiated by potassium peroxydisulfate (KPS) in water, and was reported to lead to the desired product XG-g-PEA (Scheme 5). In order to find the best grafting conditions, various parameters were investigated; the concentration of monomer (EA), the KPS and xanthan gum concentrations, the reaction time, and the temperature. The authors have demonstrated that such copolymerization was efficient at 45°C allowing high percentages of ethyl acrylate conversion (%C) and grafting efficiency (%E) even when a low ethyl acrylate/gum ratio was used. This study was followed up by another article from the same team a few years later,[48] demonstrating that the same reaction conditions could be used to graft methyl methacrylate onto xanthan.

Xanthan-g-poly(acrylamide) was first reported by Kumar et al.[49] where they have compared microwave-assisted and ceric ammonium nitrate-induced graft copolymerization

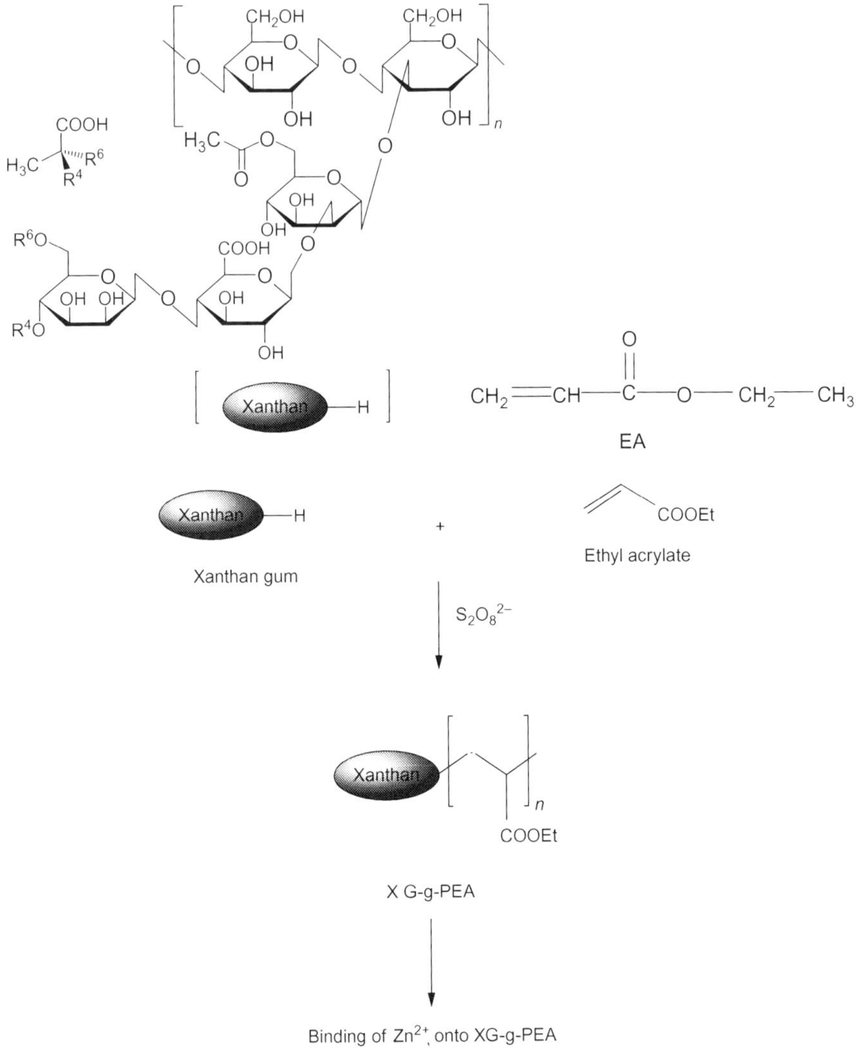

Scheme 5 Schematic diagram of the graft copolymerization.
Reproduced from Pandey, S.; Mishra, S. B., Graft Copolymerization of Ethylacrylate Onto Xanthan Gum, Using Potassium Peroxydisulfate as an Initiator. Int. J. Biol. Macromol. **2011**, *49* (4), 527–535 (with permission).

in water. They have observed that microwave-assisted graft copolymerization was more efficient than the one using ceric ammonium nitrate, in terms of degree of grafting (%G), grafting efficiency (%GE), and percentage of conversion (%C) as summarized in Table 4.

Few years later, xanthan-g-poly(acrylamide) was synthetized by Ghorai et al.[50] using a free radical polymerization technique employing KPS as presented above (Scheme 5). The obtained copolymer was then used to form modified xanthan gum/silica hybrid nanocomposites as adsorbents.

Table 4 Grafting parameters from various batches of microwave-assisted and ceric-induced grafting.

Microwave		Grafting (%G)	Grafting efficiency (%GE)	Conversion (%C)	Add on (%A)	Homopolymer (%H)
Power	Time (s)					
40	40	12.76	4.49	39.70	11.32	95.50
40	100	120.30	42.35	77.57	54.60	57.64
60	40	110.33	38.84	74.06	52.45	61.15
60	100	129.86	45.72	80.93	56.49	54.27
80	40	126.36	44.49	79.70	55.82	55.50
80	100	149.63	52.68	87.89	59.94	47.31
100	40	87.15	30.68	65.89	46.56	69.31
100	100	190.28	67.00	102.21	65.55	33.00
Ceric-induced grafting		62.87	2233	57.35	38.60	77.86

Reproduced from Kumar, A.; Singh, K.; Ahuja, M., Xanthan-g-Poly(Acrylamide): Microwave-Assisted Synthsis, Characterization and In Vitro Release Behavior. Carbohydr. Polym. **2009**, *76*, 261–267 (with permission).

The synthesis of carboxymethyl xanthan grafted with poly(acrylamide) (CMXG-g-PAAm) has been published by Badwaik et al.,[51] copolymerizing this time carboxymethyl xanthan (CMXG, for the carboxymethylation of xanthan see above) and acrylamide. The initiator was ammonium persulfate under a nitrogen atmosphere. The optimum reaction conditions were investigated by varying the concentrations of CMXG, monomer, and initiator as well as the temperature and reaction time.

6.3 Chemical modification targeting the carboxylate functions of xanthan

More selective functionalization targeted to the carboxylic acid functions of xanthan was first described by Yalpani et al.,[52] presenting two different strategies for amidation. The first approach was *via* carbodiimide-mediated coupling of an amine label in the presence of 1-ethyl-3-[3-(dimethylamino)propyl]carbodiimide (EDAC) to form the corresponding product **20** (Fig. 13). In this efficient process, 43% of the carboxyl groups were modified. The second approach was *via* a direct condensation between xanthan gum and primary amines in the presence of a small amount of water to form **21** and **22** (Fig. 13). For the latter, that is, the grafting of *n*-octadecylamine (**22**), the authors could not determine DS, but the C/N ratio corresponded approximately to a fully substituted product (determined by combustion analysis). However, electron spin resonance (ESR) spectroscopy revealed that, despite the purification procedure, a small amount of adsorbed (unreacted) amine remained.

The coupling of L-cysteine to xanthan in an aqueous solution in the presence of 1-ethyl-3-[3-(dimethylamino)propyl]carbodiimide hydrochloride (EDAC) and *N*-hydroxysuccinimide (NHS) was recently reported by Laffleur and Michalek[53] (Scheme 6).

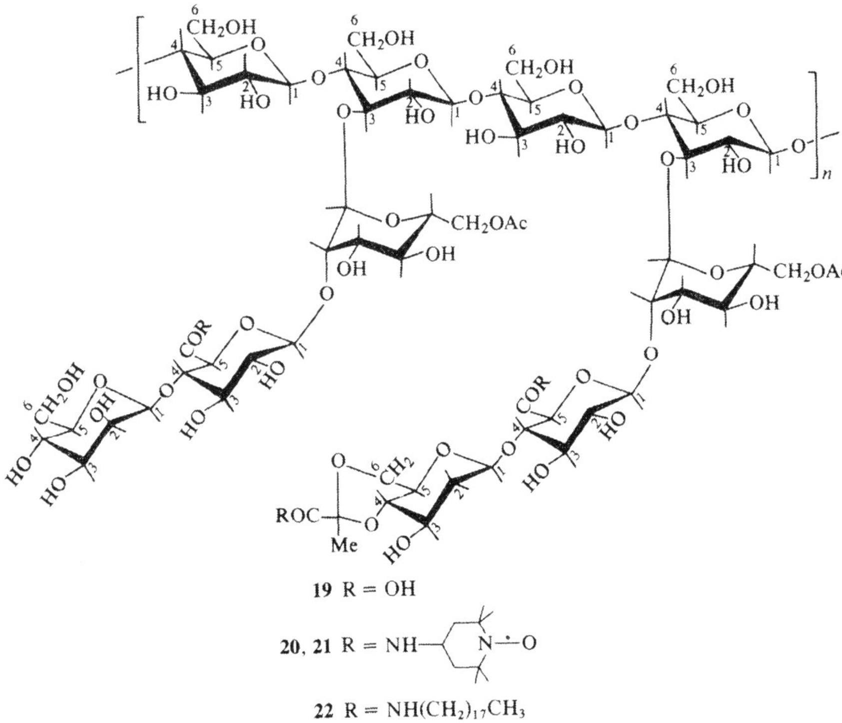

Fig. 13 Chemical modification of carboxylic acids groups in xanthan.
Reproduced from Yalpani, M.; Hall, L. D., Some Chemical and Analytical Aspects of Polysaccharide Modifications. 1. Nitroxide Spin-Labeling Studies of Alginic Acid, Cellulose, and Xanthan Gum. Can. J. Chem. **1981**, *59* (21), 3105–3119 (with permission).

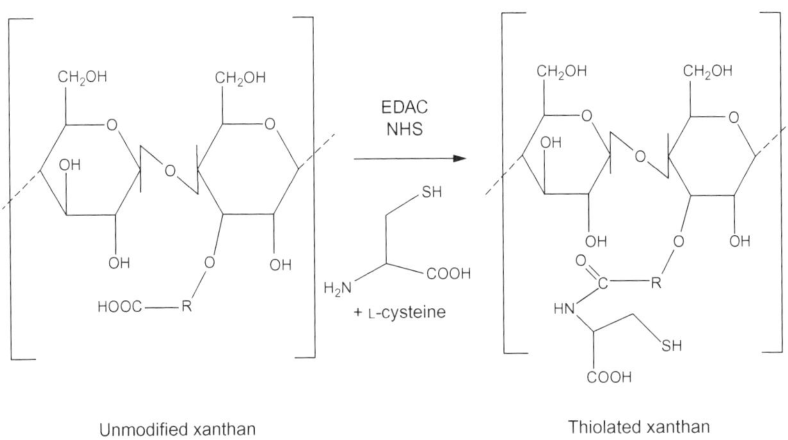

Scheme 6 Thiolation of xanthan.
Reproduced from Laffleur, F.; Michalek, M., Modified Xanthan Gum for Bucal Delivery—A Promising Approach in Treating Sialorrhea. Int. J. Biol. Macromol. **2017**, *102*, 1250–1256 (with permission).

Most of the studies on chemical modifications of xanthan do not consider the fact that xanthan can adopt two highly distinct conformations. Indeed, as summarized in this overview, authors have changed many parameters in order to modify xanthan gum efficiently, such as solvent, temperature, additives, etc. It is therefore difficult, if not impossible, to be sure of the conformation adopted during (and after) the modification. This is a key point since the adopted conformation may have a strong impact on the resulting polymer behavior.

This issue was recently taken into account by Mendes et al.[54] while working on preparation of phospholipid-xanthan microcapsules. The coupling reaction between phospholipid 1,2-dioleoyl-*sn*-glycerophosphoethanolamine (DOPE) and the carboxylate groups of xanthan was performed at room temperature in the presence of NHS and ethyl(dimethylaminopropyl)carbodiimide (EDC) in water (Scheme 7). The reaction was terminated after one night by the addition of an excess of cold ethanol, resulting in the precipitation of the modified xanthan. The authors have determined a DS of about 1.16±0.024 (considering the two carboxylic groups in the xanthan repeating unit). They have also demonstrated that the microcapsules based on xanthan modified in the ordered conformation had an improved stability compared to the native one.

Renou and coworkers were interested in gaining a more in-depth understanding of the impact of the xanthan conformation during chemical modification and its repercussions on its physical and chemical properties. For that purpose, they developed two pathways involving the same coupling partner, *n*-octylamine, but different grafting

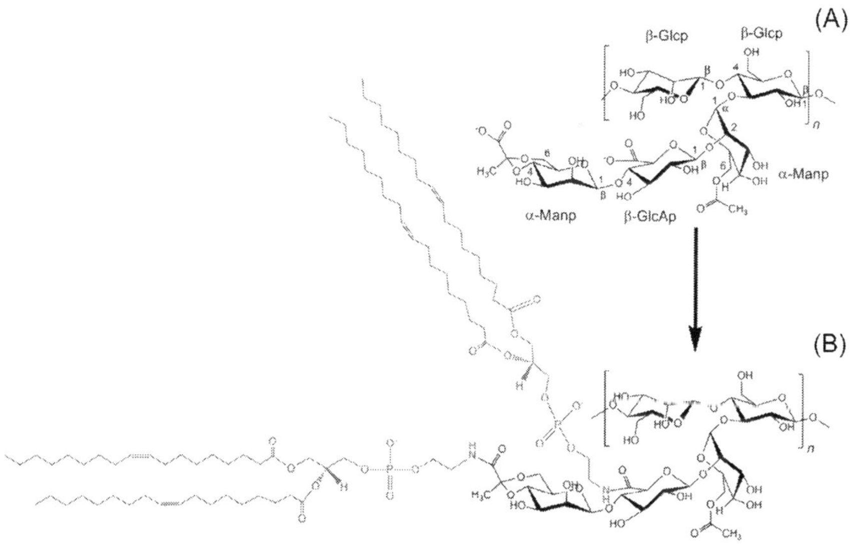

Scheme 7 Chemical structure of the xanthan repeating unit (A) and xanthan–DOPE (B), with grafted DOPE molecules conjugated to the carboxylic groups of xanthan.
Reproduced from Mendes, A. C.; Baran, E. T.; Reis, R. L.; Azevedo, H., Fabrication of Phospholipid-Xanthan Microcapsules by Combining Microfluidics With Self-Assembly. Acta Biomater. **2013**, *9*, 6675–6685 (with permission).

conditions implicating either the ordered conformation or the disordered one. A key point in this study was to compare and control the regioselectivity of the grafting process, that is, pyruvic vs glucuronic acid functionalization, and to correlate it to the physical and chemical properties of modified xanthans. In the first paper,[55] xanthan was hydrophobically modified in the ordered conformation (Scheme 8, ordered pathway). Here, *n*-octylamine was grafted onto the carboxylate groups of xanthan in an aqueous solution at room temperature in the presence of 1-ethyl-3-[3-(dimethylamino) propyl]carbodiimide hydrochloride (EDAC) as the coupling agent and NHS as a co-reactant. The authors have demonstrated via monitoring by optical rotation that xanthan was in the ordered conformation during the amidation.

In the second paper[56] xanthan was hydrophobically modified in its disordered conformation (Scheme 8, disordered pathway). For that purpose, the reaction with *n*-octylamine was carried out in DMSO (xanthan was first acidified to insure its solubility in this solvent) in the presence of *O*-(benzotriazol-1-yl)-*N,N,N′,N′*-tetramethyluronium hexafluorophosphate (HBTU), and triethylamine (Et_3N). The authors have confirmed by optical rotation that xanthan was in its disordered conformation during the reaction.

It is worth noting that a wide range of hydrophobically modified xanthan derivatives could be synthesized by simply tuning the stoichiometry of the coupling reagents (Fig. 14B). Moreover, this process proved to be highly efficient even with low loading of coupling agents, particularly for the disordered pathway in DMSO. The observed

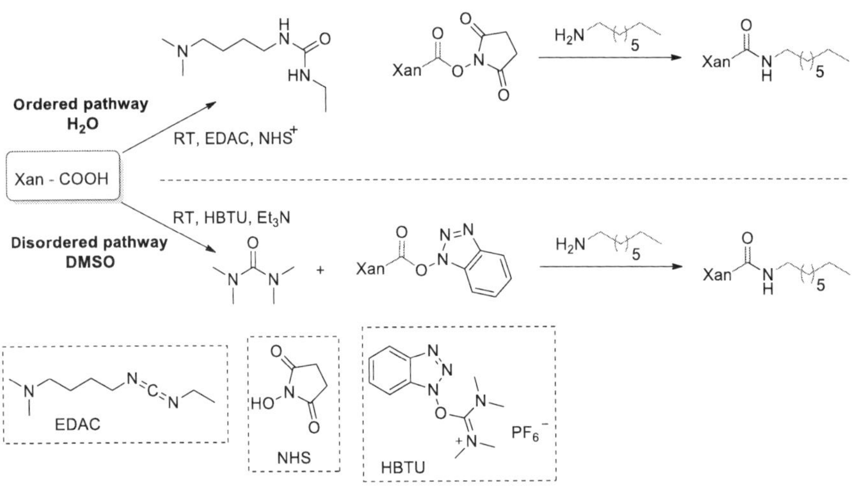

Scheme 8 Compared syntheses of hydrophobically modified xanthan gum with *n*-octylamine, modified in its ordered and disordered conformations, respectively.
Adapted from Roy, A.; Comesse, S.; Grisel, M.; Hucher, N.; Souguir, Z.; Renou, F., Hydrophobically Modified Xanthan: An Amphiphilic But Not Associative Polymer. Biomacromolecules **2014**, *15* (4), 1160–1170 and Fantou, C.; Roy, A. N.; De, E.; Comesse, S.; Grisel, M.; Renou, F., Chemical Modification of Xanthan in the Ordered and Disordered States: An Open Route for Tuning the Physico-Chemical Properties. Carbohydr. Polym. **2017**, *178*, 115–122 (with permission).

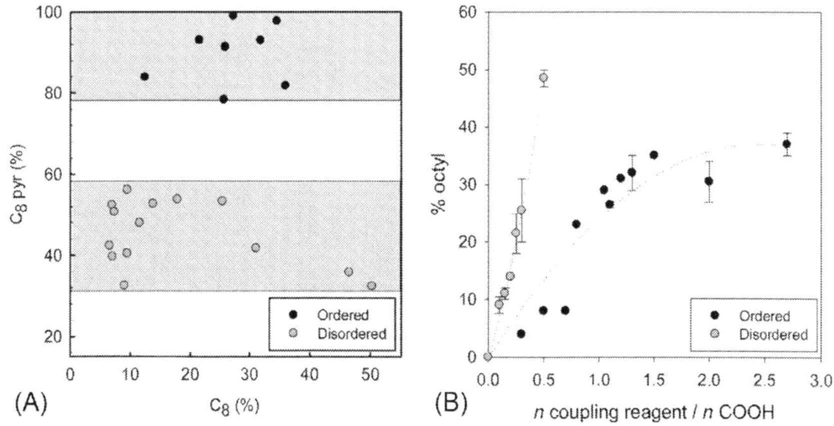

Fig. 14 (A) Evolution of the relative pyruvic acid grafting density as a function of the overall grafting density. (B) Evolution of the overall grafting density as a function of the coupling reagent quantity.
Reproduced from Fantou, C.; Roy, A. N.; De, E.; Comesse, S.; Grisel, M.; Renou, F., Chemical Modification of Xanthan in the Ordered and Disordered States: An Open Route for Tuning the Physico-Chemical Properties. Carbohydr. Polym. **2017**, *178*, 115–122 (with permission).

maximum grafting density of 40%, in the case of the ordered conformation, was ascribed to the availability of the pyruvic acids, which is further described below.

The key point of these two highly efficient chemical modifications is that there is a dramatic change in the grafting pattern depending on the conformation adopted during modification (Fig. 14A). Indeed, while the pyruvic acid becomes selectively modified in the ordered form of xanthan, these selectivity changes are in favor of the glucuronic acid residue when xanthan is modified in its disordered conformation, indicating that the two grafting sites (pyruvic and glucuronic acids) do not have the same accessibility in the two conformations. This can be tentatively explained by the fact that the glucuronic acid functions are thought to stabilize the ordered conformation by hydrogen bonds, and therefore are less accessible when the chemical modification is performed in the ordered conformation.

To conclude, it has been clearly demonstrated that the conditions used, and therefore the conformation of xanthan, have a major impact on the chemical modification. In consequence, very different physicochemical properties are observed. These aspects will be fully discussed in the following section.

7 Physicochemical properties of modified xanthan

Despite its useful rheological and thermal properties, xanthan has been the subject of several studies where the aim is to improve existing or confer new properties in order to widen its area of applications, such as controlled release and drug delivery.[44, 49, 53]

To achieve this, xanthan modification may consist of varying the amount of acetate and/or pyruvate substituents,[27, 57–60] removing subunits of the side chain,[46] or grafting molecules on the hydroxy groups[41, 50, 61] or carboxylic acid groups.[52, 55, 56, 62] Some examples of grafting reactions have been presented in the previous sections.

One of the goals of xanthan modification has been to modify, and in some cases tune, its flow properties. For instance, some authors were able to decrease the viscosity of xanthan solutions (i)[42, 43, 45, 63] while others chemically modified xanthan in order to improve its flow properties (ii).[40] For the former purpose, Yahoum et al.[45] obtained carboxymethyl derivatives of xanthan with different grafting densities. They further demonstrated that carboxymethylated xanthan solutions are less viscous than native xanthan at the same concentration. The decrease in viscosity is more pronounced when the grafting density increases (Fig. 15). On the other hand, carboxymethyl xanthans present a less pronounced shear-thinning behavior as demonstrated by the n parameter of the Cross model which evolves from 0.767 for native xanthan to 0.845 for the most grafted derivative. The authors have attributes this observation to the greater hydrophilicity (higher charge density) of carboxymethyl derivatives compared to unmodified xanthan. This increase in hydrophilicity could also explain the higher dissolution rate observed by the authors. The same tendency was also observed by Ahuja et al.[43] for carboxymethyl xanthan.

Yahoum et al.[45] also showed that the viscoelastic properties were altered by carboxymethylation. G′, G″, and their frequency dependence were modified. As illustrated in Fig. 16, carboxymethyl derivatives present an elasticity down to 1.5 decade lower than native xanthan. In addition, the frequency dependence is more pronounced after modification confirming the greater flowability of carboxylated xanthan.

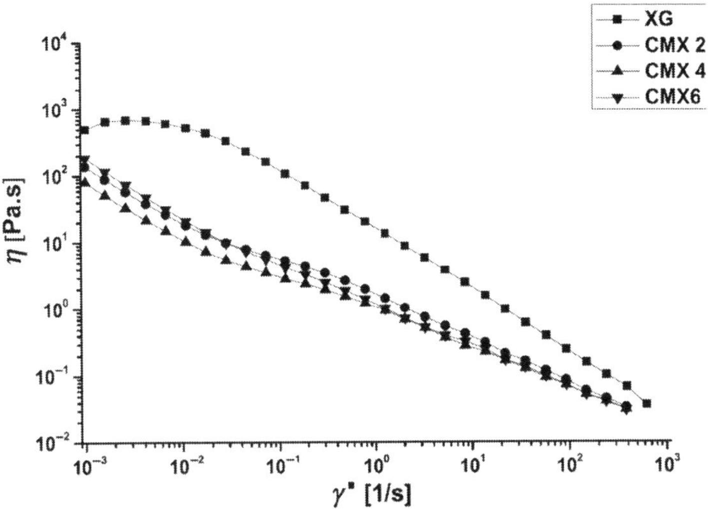

Fig. 15 Flows curves of native and carboxymethyl xanthan derivatives at 1% (w/w) in water. Reproduced from Yahoum, M. M.; Moulai-Mostefa, N.; Le Cerf, D., Synthesis, Physicochemical, Structural and Rheological Characterizations of Carboxymethyl Xanthan Derivatives. Carbohydr. Polym. **2016**, *154*, 267–275.

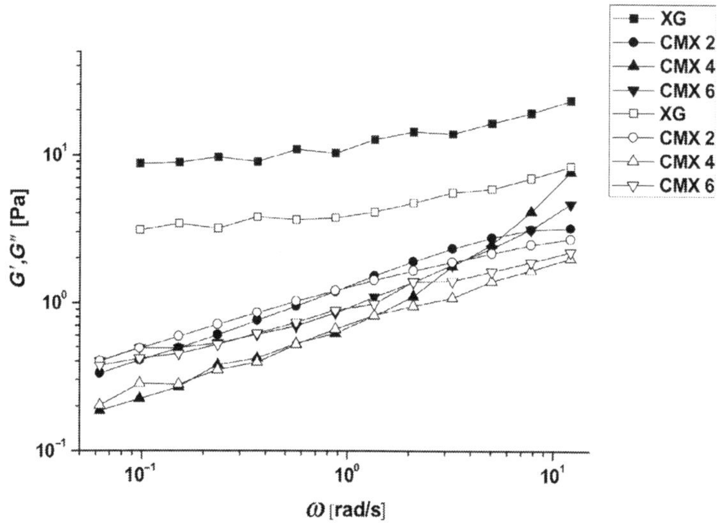

Fig. 16 Frequency dependence of G′ (filled symbols) and G″ (unfilled symbols) for native and carboxymethyl xanthan at 1% (w/w) in water.
Reproduced from Yahoum, M. M.; Moulai-Mostefa, N.; Le Cerf, D., Synthesis, Physicochemical, Structural and Rheological Characterizations of Carboxymethyl Xanthan Derivatives. Carbohydr. Polym. **2016**, *154*, 267–275 (with permission).

Another strategy to alter the physical and chemical properties of xanthan has been to introduce various hydrophobic groups. Similar to the above-mentioned works, Skender et al.[42] demonstrated that grafting of hydrophobic moieties to the hydroxy groups (see Scheme 4) alter the viscosity of xanthan solutions (Fig. 17). It was observed that both the apparent viscosity and its shear rate dependency are affected by hydrophobization. The results here clearly show that hydrophobic modification of xanthan also leads to an alteration of its flow properties. However, the observed effects could not be directly linked to hydrophobicity. On one hand, grafting of diphenylmaleic anhydride (DMPA) and phthalic anhydride (PA) had almost the same impact on the flow properties of xanthan while DPMA is much more hydrophobic than PA. On the other hand, epichlorhydrin-phenol (EPH) has a hydrophobicity similar to PA but its impact on the viscosity was much more pronounced.

In total, these studies illustrate that chemical modification of xanthan leads to an alteration of the flow properties of xanthan derivatives in solution regardless of whether the grafted moieties are hydrophilic or hydrophobic. Besides, it is not clear whether the molecular weight (or conformation) is always affected by modification, which adds complexity to the studies.

Xanthan has also been modified in order to form hydrogels for different purposes such as controlled drug delivery systems. Two different approaches have been investigated, direct cross-linking of xanthan chains and cross-linking using another polymer. An example of direct cross-linking was presented by Jampala et al.,[61] who were

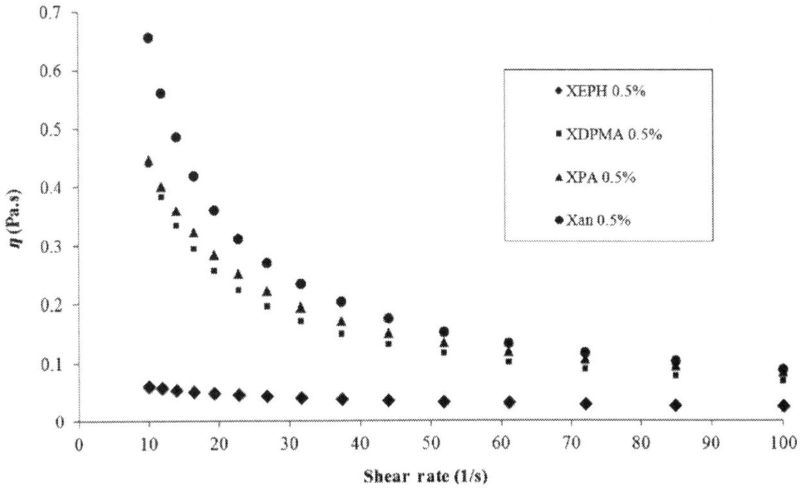

Fig. 17 Flow curves of native (Xan) and hydrophobically modified xanthan (XDPMA, XPA, and XEPH) at pH 7.
Reproduced from Skender, A.; Hadj-Ziane-Zafour, A.; Flahaut, E., Chemical Functionalization of Xanthan Gum for the Dispersion of Double-Walled Carbon Nanotubes in Water. Carbon **2013**, *62*, 149–156 (with permission).

able to form stable gels by first functionalizing xanthan with primary amines using cold-plasma treatment, followed by cross-linking of the functionalized xanthan using glutaraldehyde (see Scheme 9).

The authors have demonstrated that both unmodified xanthan and plasma-treated xanthan cross-linked with glutaraldehyde behaved as soft gels at 2 wt% within the investigated frequency range. However, the dynamic moduli (both G′ and G″) are much higher (half a decade) for plasma-treated xanthan over the entire range of frequency, except at very low frequency (~10^{-3} Hz), as illustrated in Fig. 18. This suggests a strengthening of the network due to cross-linking of the plasma-modified xanthan. The authors have pointed out that the difference in elasticity (G′) between the unmodified and cross-linked xanthan decreases with increasing xanthan concentration. This was attributed to the fact that the number of junction points and entanglements increase at high concentrations, which masks the effect of cross-linking. On the other hand, the authors have demonstrated that the elasticity was improved and the frequency dependence decreased with increasing time of plasma treatment. For instance, the gel strength of 1 wt% XG-15 (15 min of plasma treatment) was comparable to that of 4 wt% XG-5 (5 min of plasma treatment). This suggests that cross-linking is enhanced by increasing the duration of the plasma treatment.

Xanthan hydrogels have also been obtained *via* a more conventional pathway. Bejenariu et al.[62] used adipic acid dihydrazide as the cross-linking agent, and the obtained hydrogels present behavior similar to those obtained by Jampala et al.[61]

The other strategy for obtaining xanthan hydrogels is to chemically graft and cross-link xanthan with another polymer, notably in order to take benefit of specific

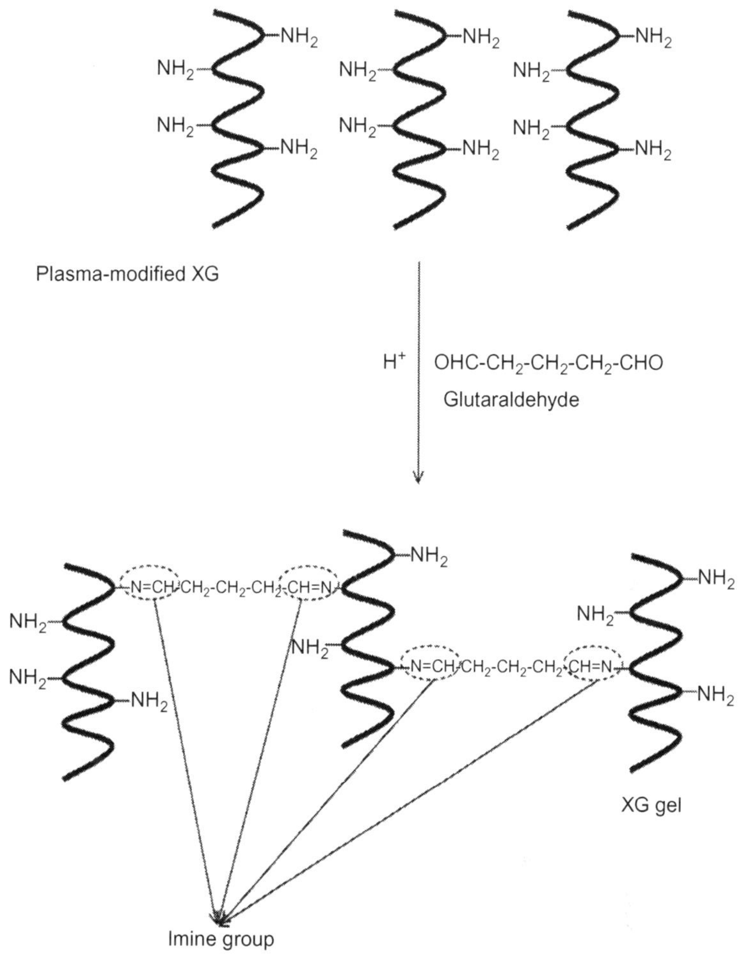

Scheme 9 Schematic illustration of the cross-linking of plasma-treated xanthan with glutaraldehyde.
Reproduced from Jampala, S. N.; Manolache, S.; Gunasekaran, S.; Denes, F. S., Plasma-Enhanced Modification of Xanthan Gum and Its Effect on Rheological Properties. J. Agric. Food Chem. **2005**, *53*, 3618–3625 (with permission).

characteristics of each polymer such as biocompatibility and biodegradability of xanthan and, for instance, high resistance to enzymes for the other polymer. One example of this approach is the work of Wang et al.[40] who esterified xanthan with PMAO. As one can see in Scheme 3, the obtained polymer has both the structure of a hydrogel network through cross-linking with PMAO (ester linkage), and an amphiphilic character due to the presence of hydrophobic octadecene groups in the PMAO chain. Different syntheses were done by varying the amount of PMAO and the authors have stated that all samples were only slightly cross-linked since they remained easy to dissolve in water below 5 g/L. In aqueous solution, the authors observed an increase

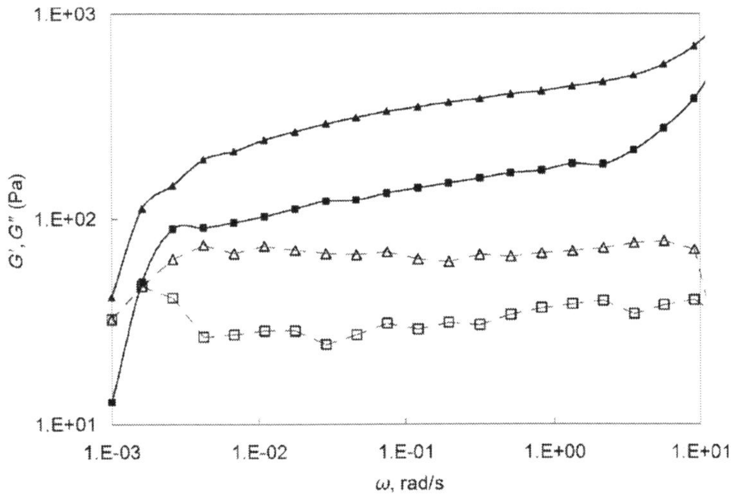

Fig. 18 Frequency dependence of the elastic (G′, filled symbols) and viscous (G″, open symbols) moduli of 2 wt% unmodified XG (squares) and cross-linked XG-5 (5 min of plasma treatment) (triangles) at T=50°C.
Reproduced from Jampala, S. N.; Manolache, S.; Gunasekaran, S.; Denes, F. S., Plasma-Enhanced Modification of Xanthan Gum and Its Effect on Rheological Properties. J. Agric. Food Chem. **2005**, *53*, 3618–3625 (with permission).

in molecular mass and a more compact conformation for the modified xanthan compared to unmodified xanthan, which is attributed to both hydrophobic aggregation and cross-linking of the modified xanthan.

Concerning the flow properties, it was demonstrated that PMAO-grafted xanthans possess a strong shear-thinning behavior similar to unmodified xanthan. However, both the plateau viscosity (Newtonian viscosity) and the flow consistency increased with increasing grafting density, while the flow index was not affected by the modification.[40] The authors have concluded that this enhanced resistance to shear is a consequence of both the formation of a micro-cross-linked structure and hydrophobic association. In addition, a decrease in viscosity with temperature was observed at about 40°C for unmodified xanthan and about 50°C for modified ones, due to the conformational change (order-to-disorder transition). The authors have concluded that the thermal stability in solution was enhanced by the hydrophobic association of the alkyl chains.

Viscoelasticity was also measured for different derivatives and compared with unmodified xanthan under the same conditions (Fig. 19). All modified xanthans present higher G′ and G″ than the unmodified xanthan due to the combined effect of physical cross-links via hydrophobic association and chemical cross-links via ester linkages. The increased elasticity becomes more pronounced as the grafting density increases. In addition, the overlapping frequency decreases with increase in the content of PMAO, which demonstrates that the relaxation of the network slows down with increase in PMAO content.

All in all, there are relatively few studies dealing with chemical modification of xanthan in the literature. Some of them render xanthan more hydrophilic but

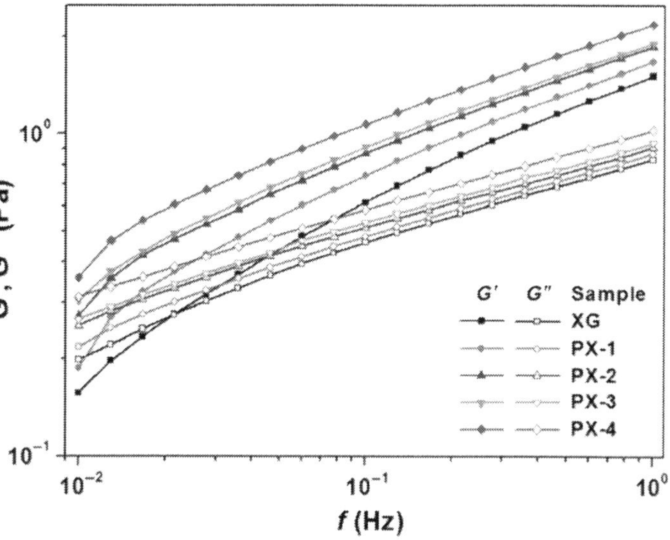

Fig. 19 Frequency dependence of G′ and G″ for unmodified xanthan (XG) and modified xanthans (PX) within the linear regime.
Reproduced from Wang, X. J.; Xin, H. P.; Zhu, Y. J.; Chen, W. J.; Tang, E. G.; Zhang, J.; Tan, Y. B., Synthesis and Characterization of Modified Xanthan Gum Using Poly(Maleic Anhydride/1-Octadecene). Colloid Polym. Sci. **2016**, *294* (8), 1333–1341 (with permission).

the majority were performed to form hydrogels or introduce hydrophobic groups. However, in most cases, the grafting density was not quantified. In addition, none of the above-mentioned authors paid attention to the conformation adopted by xanthan neither during nor after the modification, even though it has been demonstrated that depending on its conformation, xanthan presents very different physicochemical properties. To the best of our knowledge, Mendes et al.[54] were the first authors who investigated the chain organization after xanthan modification. In this study, the authors have grafted DOPE onto the carboxylic functions of xanthan in its ordered conformation (Scheme 7) and measured CD spectra before and after grafting. They have demonstrated that the ordered conformation to some extent is preserved after grafting. However, only one grafting density was considered in this study, and the rheological properties were not characterized. Hence, based on this observation, it is neither possible to draw any trend nor conclusion. Based on this, Renou and coworkers[55, 56] were interested in the impact of xanthan conformation during and after modification on its structural and rheological properties. Two different modification protocols to hydrophobically modify xanthan on its COOH groups with alkyl moieties (*n*-octylamine, C8) were developed; one protocol for the modification in the ordered conformation[55] and the other in the disordered conformation[56] (Scheme 8). The grafting density can be tuned depending on the stoichiometry of the reactant for each protocol; hence it was possible to determine the impact of grafting density on both conformation and rheological properties of hydrophobically modified xanthans for both protocols.

Xanthan exhibits two specific CD bands in the ordered conformation: one positive band at 210 nm, corresponding to the carboxylates, and one negative band at 225 nm attributed to the acetyl groups of the xanthan side chain. More specifically, the degree of ordering is characterized by the intensity of the band at 210 nm, the higher the intensity the more ordered the xanthan.[10, 14] In their study, Fantou et al.[56] obtained CD spectra of native xanthan (precursor) and different derivatives characterized by their grafting density (%C8) for both derivatives modified in the ordered and disordered conformations. Fig. 20 presents the evolution of ellipticity at 210 nm as a function of the grafting density for modified xanthans (ordered and disordered conformations). On one hand, it was demonstrated that for xanthan modified in the ordered conformation, the presence of hydrophobic moieties reinforces the chain ordering. On the other hand, a progressive disordering with increasing grafting density is observed for xanthan modified in the disordered conformation. Thus, this clearly shows that the two modification protocols have opposite impact on chain organization and rigidity in aqueous solutions.

It is well established that rheological properties of unmodified xanthans are strongly correlated to the conformation of the chains in solution. Since the degree of ordering of xanthan is affected by both the conformation of the xanthan during the modification and the grafting density, the authors have studied the viscoelastic properties of hydrophobized xanthans, modified in both ordered and disordered conformations.

For xanthan modified in the ordered conformation, Roy et al.[55] measured the frequency dependence of G′ and G″ for unmodified and modified xanthans having different grafting densities at various temperatures. Fig. 21A shows the

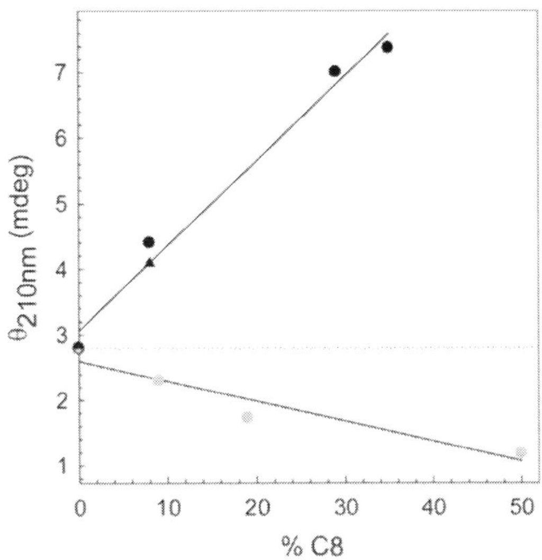

Fig. 20 Evolution of ellipticity at 210 nm as a function of grafting density for xanthan modified in ordered (*black* symbols) and disordered conformations (*gray* symbols). Reproduced from Fantou, C.; Roy, A. N.; De, E.; Comesse, S.; Grisel, M.; Renou, F., Chemical Modification of Xanthan in the Ordered and Disordered States: An Open Route for Tuning the Physico-Chemical Properties. Carbohydr. Polym. **2017**, *178*, 115–122 (with permission).

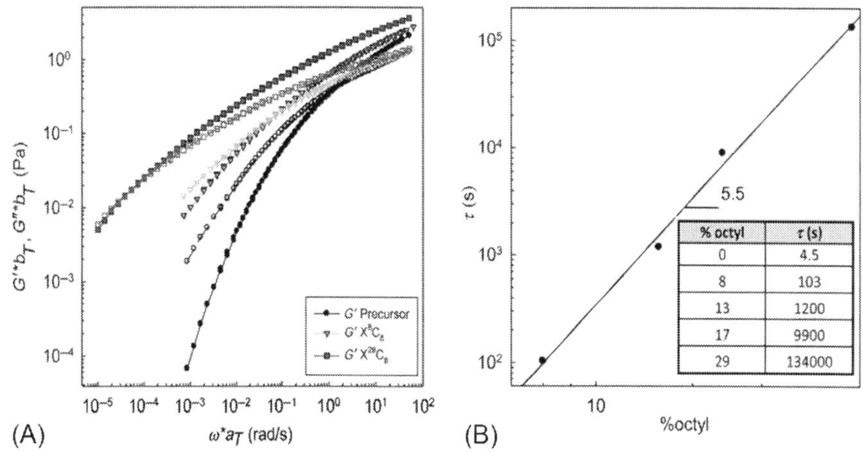

Fig. 21 (A) Frequency-temperature master curves of G′ (filled symbols) and G″ (open symbols) for precursor and xanthan modified in the ordered conformation X^8C_8, $X^{29}C_8$ at 2 g/L (T_{ref}=20°C). (B) Evolution of the relaxation time as a function of the grafting density for xanthans modified in the ordered conformation and precursor.
Reproduced from Roy, A.; Comesse, S.; Grisel, M.; Hucher, N.; Souguir, Z.; Renou, F., Hydrophobically Modified Xanthan: An Amphiphilic But Not Associative Polymer. Biomacromolecules **2014**, *15* (4), 1160–1170 (with permission).

frequency-temperature master curves of G′ and G″ for the precursor and modified xanthans with grafting densities of 8% and 29% (called X^8C_8 and $X^{29}C_8$, respectively). Xanthans modified in the ordered conformation exhibit behavior very similar to that of the precursor, except that the relaxation is strongly shifted to lower frequencies reflecting a pronounced decrease in the chain dynamics. The dynamics of the system has been characterized by its relaxation time $\tau = 1/\omega_r$ (ω_r corresponds to the frequency where G′ and G″ cross) which is plotted as a function of the grafting density in Fig. 21B. The relaxation time shows a strong power law dependency with the grafting density which proves that the latter strongly affect the relaxation of xanthan chains. Here the authors have concluded that xanthan modified in the ordered conformation behave similar to unmodified xanthan with a relaxation time that increases with increasing grafting density due to sticky relaxation, which is attributed to the increase in xanthan backbone rigidity.

On the other side, xanthan modified in the disordered conformation exhibits completely different viscoelastic properties. Fig. 22A shows the frequency dependence of both G′ and G″ at different temperatures for a 2-g/L solution of modified xanthan in the disordered conformation with a grafting density of 9% (called X^9C_{8dis}). At low temperatures (up to 40°C) X^9C_{8dis} exhibits viscoelastic liquid features similar to the precursor. Above 50°C, a dramatic change in both moduli is observed with G′ much higher than G″, and both moduli being frequency independent which is typical of a chemical gel-like behavior. Thus, the construction of a frequency-temperature master curve becomes impossible. The abrupt transition from a viscoelastic liquid to a gel is shown in Fig. 22B, with a transition temperature between 40°C and 50°C.

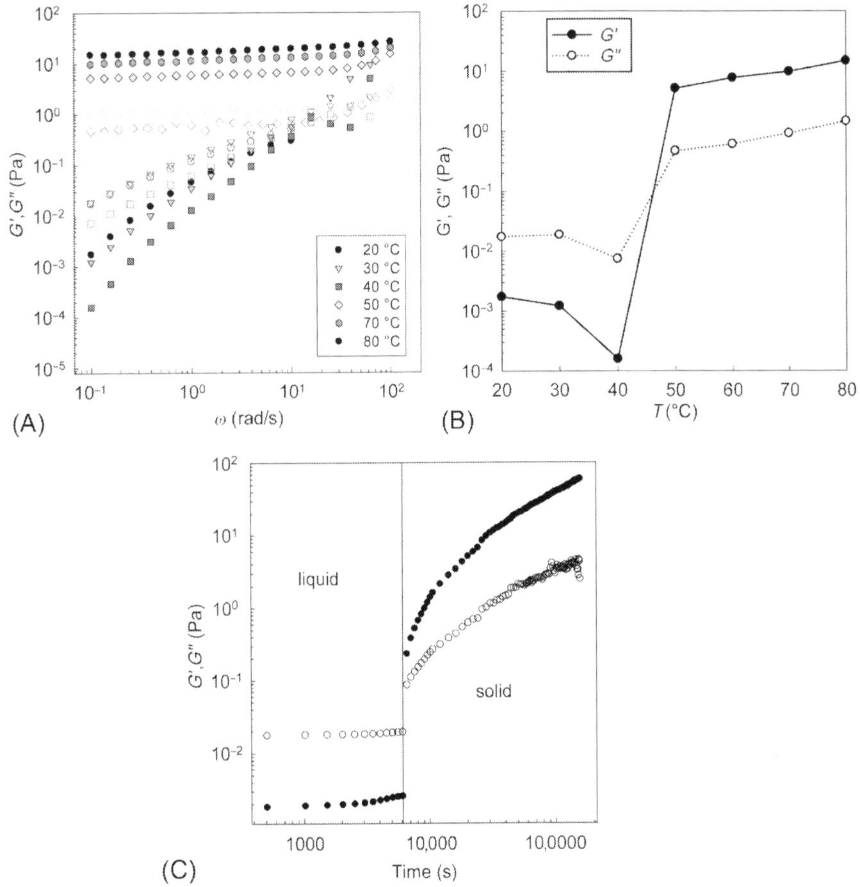

Fig. 22 (A) Frequency dependence of G′ (filled symbols) and G″ (open symbols) at different temperatures for X^9C_{8dis} at 2 g/L. (B) Temperature dependence of G′ and G″ at ω=0.1 rad/s. (C) Time dependence of G′ and G″ at 20°C and ω=0.1 rad/s for X^9C_{8dis}.
Reproduced from Fantou, C.; Roy, A. N.; De, E.; Comesse, S.; Grisel, M.; Renou, F., Chemical Modification of Xanthan in the Ordered and Disordered States: An Open Route for Tuning the Physico-Chemical Properties. Carbohydr. Polym. **2017**, *178*, 115–122 (with permission).

The authors have demonstrated that this phenomenon also occurs with time (Fig. 22C). This sudden transition from liquid to solid behavior with both temperature and time was attributed to a percolation mechanism of modified xanthan chains due to the creation of hydrophobic microdomains, which is attributed to the increased flexibility of the xanthan backbone.

To conclude on this part, it has been shown that the conformation adopted by xanthan during its modification has not only an effect on the regioselectivity of the reaction (between the pyruvic and glucuronic acids, Fig. 14A), but also a major impact on both the conformation of xanthan and the rheological properties in solution (Fig. 23).

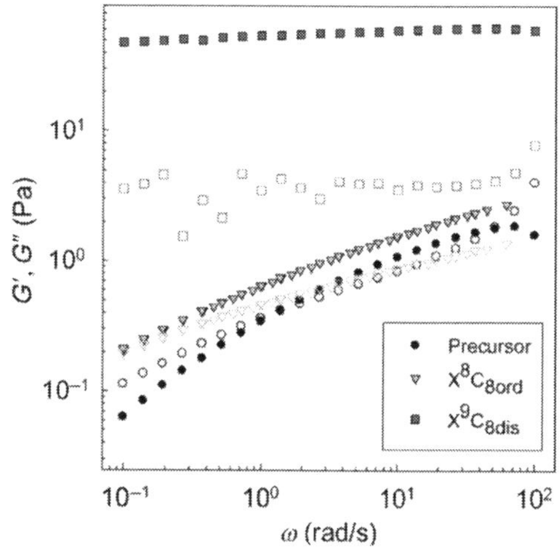

Fig. 23 Comparison of the frequency dependence of precursor and modified xanthans under ordered (X^8C_{8ord}) and disordered (X^9C_{8dis}) conformation at 2 g/L. Filled symbols: G′, open symbols: G″.
Reproduced from Fantou, C.; Roy, A. N.; De, E.; Comesse, S.; Grisel, M.; Renou, F., Chemical Modification of Xanthan in the Ordered and Disordered States: An Open Route for Tuning the Physico-Chemical Properties. Carbohydr. Polym. **2017**, *178*, 115–122 (with permission).

Indeed, modification in the ordered conformation allows a more selective grafting on the pyruvic acids, while there appears to be no preference between pyruvic acid and glucuronic acid when the modification is carried out in the disordered conformation. As a consequence, the helical structure of xanthan is reinforced (higher T_m) when it is modified in the ordered conformation, which results in a behavior similar to the precursor, except that the relaxation of the chains is up to 30,000 times slower than the precursor for the most grafted xanthan. This atypical behavior is attributed to the higher rigidity of the xanthan helices, which do not allow self-association of the hydrophobic grafts. Thus, the modified xanthan in the ordered form is amphiphilic, but not associative. On the other hand, the xanthan modified in the disordered conformation is able to partially reorder in solution, but becomes less and less ordered, and therefore more flexible, when the grafting density increases. This flexibility favors the hydrophobic interactions that dominate the organization of xanthan in solution thus leading to a gel-like behavior. The impact of the conformation of xanthan during its modification can be summarized as illustrated in Scheme 10.

8 Conclusions

Xanthan differs from many other polysaccharides in that it appears to be produced mostly in an ordered, double-stranded state with high molecular weight and very high chain stiffness, and hence, have very interesting rheological properties. A temperature,

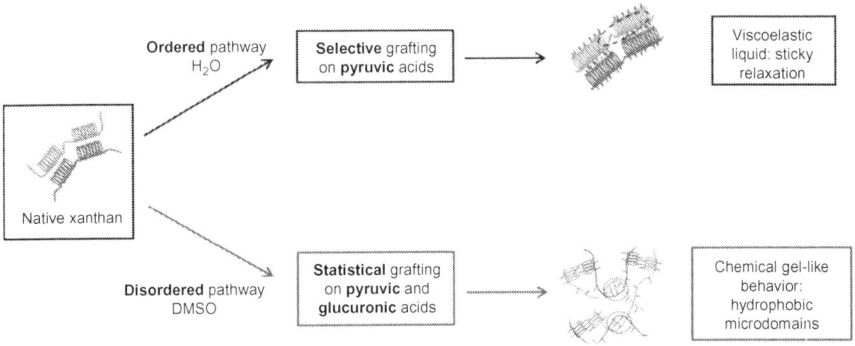

Scheme 10 Schematic representation of the impact of the conformation during modification on xanthan organization and viscoelasticity in solution.
Adapted from Fantou, C.; Roy, A. N.; De, E.; Comesse, S.; Grisel, M.; Renou, F., Chemical Modification of Xanthan in the Ordered and Disordered States: An Open Route for Tuning the Physico-Chemical Properties. Carbohydr. Polym. **2017**, *178*, 115–122 (with permission).

ionic strength, and pH-dependent cooperative order-disorder transition, which is readily detected by, for example, chiroptical methods, NMR, or DSC, is a hallmark of xanthan. The transition seems to be directly associated with the side chains, but induces changes also in the overall macromolecular behavior, in dilute, semi-dilute, and concentrated solutions. This includes partial or complete strand separation in the disordered state, and a range of reassociation modes upon ordering.

A range of chemical modifications, including both removal of substituents (pyruvate, acetate, and partial removal of side chain sugars) and conventional chemical derivatization, have been presented in the literature. In many cases, focus has been on the chemistry of the modification, including regioselectivity and DS, but sometimes without considering depolymerization or relation to the conformational state, making interpretation of rheological studies is challenging. However, in recent studies modification of the carboxyl groups (amide formation with alkylamines) has been carried out in both conformational states. Results clearly demonstrate that the selectivity of the reaction (pyruvate vs. glucuronate) is different in the two states. The resulting hydrophobically modified xanthans display very different viscoelastic properties, particularly the relaxation behavior, which is partly attributed to the difference in chain stiffness of the two states. In the ordered and macromolecularly rigid state the modified xanthans display amphiphilic, but not associative, properties. By contrast, the higher chain flexibility of the disordered state favors hydrophobic interactions leading to a gel-like behavior.

Modified xanthans, despite being relatively little explored, offer a potential to produce a wide range of rheologically interesting materials, either alone or in mixture with other macromolecules. One key area is biodegradability, which generally depends on both the conformation and the type, as well as on distribution and DS. Clearly, other macromolecular properties, as well as possibly new applications in a wide range of areas including food science, pharmacy, and biomedicine, and even oil recovery, are excellent subjects of future research on xanthan.

References

1. Sloneker, J. H.; Jeanes, A. Exocellular Bacterial Polysaccharide From Xanthomonas campestris Nrrl B-1459 .1. Constitution. *Can. J. Chem.* **1962**, *40*(11), 2066–2071.
2. Roscher, N. M.; Sandford, P. A. Allene R. Jeanes 1906–1995. In *Advances in Carbohydrate Chemistry and Biochemistry*; Horton, D., Ed.; vol. 53; Academic Press, 1998; pp 6–10.
3. Taylor, K. C.; Nasr-El-Din, H. A. Water-Soluble Hydrophobically Associating Polymers for Improved Oil Recovery: A Literature Review. *J. Pet. Sci. Eng.* **1998**, *19*(3-4), 265–280.
4. Jansson, P. E.; Kenne, L.; Lindberg, B. Structure of Extracellular Polysaccharide From Xanthomonas-Campestris. *Carbohydr. Res.* **1975**, *45*, 275–282.
5. Kool, M. M.; Gruppen, H.; Sworn, G.; Schols, H. A. Comparison of Xanthans by the Relative Abundance of Its Six Constituent Repeating Units. *Carbohydr. Polym.* **2013**, *98*(1), 914–921.
6. Vanderslice, R. W.; Doherty, D. H.; Capage, M. A.; Betlach, M. R.; Hassler, R. A.; Henderson, N. M.; Ryangraniero, J.; Tecklenburg, M. Genetic-Engineering of Polysaccharide Structure in Xanthomonas campestris. In *Biomedical and Biotechnological Advances in Industrial Polysaccharides*; 1989; pp 145–156.
7. Hassler, R. A.; Doherty, D. H. Genetic-Engineering of Polysaccharide Structure—Production of Variants of Xanthan Gum in Xanthomonas-Campestris. *Biotechnol. Prog.* **1990**, *6*(3), 182–187.
8. Vold, I. M. N.; Kristiansen, K. A.; Christensen, B. E. A Study of the Chain Stiffness and Extension of Alginates, In Vitro Epimerized Alginates, and Periodate-Oxidized Alginates Using Size-Exclusion Chromatography Combined With Light Scattering and Viscosity Detectors. *Biomacromolecules* **2006**, *7*, 2136–2146.
9. Sato, T.; Norisuye, T.; Fujita, H. Double-Stranded Helix of Xanthan—Dimensional and Hydrodynamic Properties in 0.1-M Aqueous Sodium-Chloride. *Macromolecules* **1984**, *17*(12), 2696–2700.
10. Morris, E. R.; Rees, D. A.; Young, G.; Walkinshaw, M. D.; Darke, A. Order-Disorder Transition for a Bacterial Polysaccharide in Solution—Role for Polysaccharide Conformation in Recognition between Xanthomonas Pathogen and Its Plant Host. *J. Mol. Biol.* **1977**, *110*(1), 1–16.
11. Milas, M.; Rinaudo, M. Conformational Investigation on the Bacterial Polysaccharide Xanthan. *Carbohydr. Res.* **1979**, *76*, 189–196.
12. Norton, I. T.; Goodall, D. M.; Frangou, S. A.; Morris, E. R.; Rees, D. A. Mechanism and Dynamics of Conformational Ordering in Xanthan Polysaccharide. *J. Mol. Biol.* **1984**, *175*(3), 371–394.
13. Christensen, B. E.; Knudsen, K. D.; Smidsrød, O.; Kitamura, S.; Takeo, K. Temperature-Induced Conformational Transition in Xanthans With Partially Hydrolyzed Side-Chains. *Biopolymers* **1993**, *33*(1), 151–161.
14. Capron, I.; Brigand, G.; Muller, G. About the Native and Renatured Conformation of Xanthan Exopolysaccharide. *Polymer* **1997**, *38*(21), 5289–5295.
15. Stokke, B. T.; Elgsæter, A.; Smidsrød, O. Electron-Microscopic Study of Single-Stranded and Double-Stranded Xanthan. *Int. J. Biol. Macromol.* **1986**, *8*(4), 217–225.
16. Kitamura, S.; Takeo, K.; Kuge, T.; Stokke, B. T. Thermally Induced Conformational Transition of Double-Stranded Xanthan in Aqueous Salt-Solutions. *Biopolymers* **1991**, *31*(11), 1243–1255.
17. Moffat, J.; Morris, V. J.; Al-Assaf, S.; Gunning, A. P. Visualisation of Xanthan Conformation by Atomic Force Microscopy. *Carbohydr. Polym.* **2016**, *148*, 380–389.

18. Matsuda, Y.; Sugiura, F.; Okumura, K.; Tasaka, S. Renaturation Behavior of Xanthan With High Molar Mass and Wide Molar Mass Distribution. *Polym. J.* **2016**, *48*(5), 653–658.
19. Matsuda, Y.; Biyajima, Y.; Sato, T. Thermal Denaturation, Renaturation, and Aggregation of a Double-Helical Polysaccharide Xanthan in Aqueous Solution. *Polym. J.* **2009**, *41*(7), 526–532.
20. Holzwarth, G.; Prestridge, E. B. Multistranded Helix in Xanthan Polysaccharide. *Science* **1977**, *197*(4305), 757–759.
21. Morris, E. R. Ordered Conformation of Xanthan in Solutions and "Weak Gels": Single Helix, Double Helix—or Both? *Food Hydrocoll.* **2019**, *86*, 18–25.
22. Kitagawa, H.; Sato, T.; Norisuye, T.; Fujita, H. Optical-Rotation Behavior of Xanthan in Mixtures of Water and Cadoxen. *Carbohydr. Polym.* **1985**, *5*(6), 407–422.
23. Sato, T.; Norisuye, T.; Fujita, H. Double-Stranded Helix of Xanthan—Dissociation Behavior in Mixtures of Water and Cadoxen. *Polym. J.* **1985**, *17*(5), 729–735.
24. Oviatt, H. W.; Brant, D. A. Viscoelastic Behavior of Thermally Treated Aqueous Xanthan Solutions in the Semidilute Concentration Regime. *Macromolecules* **1994**, *27*(9), 2402–2408.
25. Christensen, B. E.; Smidsrød, O. Hydrolysis of Xanthan in Dilute Acid—Effects on Chemical-Composition, Conformation, and Intrinsic-Viscosity. *Carbohydr. Res.* **1991**, *214*(1), 55–69.
26. Christensen, B. E.; Smidsrød, O.; Elgsæter, A.; Stokke, B. T. Depolymerization of Double-Stranded Xanthan by Acid-Hydrolysis—Characterization of Partially Degraded Double Strands and Single-Stranded Oligomers Released From the Ordered Structures. *Macromolecules* **1993**, *26*(22), 6111–6120.
27. Holzwarth, G.; Ogletree, J. Pyruvate-Free Xanthan. *Carbohydr. Res.* **1979**, *76*, 277–280.
28. Christensen, B. E.; Myhr, M. H.; Smidsrød, O. Degradation of Double-Stranded Xanthan by Hydrogen Peroxide in the Presence of Ferrous Ions: Comparison to Acid Hydrolysis. *Carbohydr. Res.* **1996**, *280*(1), 85–99.
29. Milas, M.; Rinaudo, M.; Tinland, B. The Viscosity Dependence on Concentration, Molecular Weight and Shear Rate of Xanthan Solutions. *Polym. Bull.* **1985**, *14*, 157–164.
30. Cuvelier, G.; Launay, B. Concentration Regimes in Xanthan Gum Solutions Deduced From Flow and Viscoelastic Properties. *Carbohydr. Polym.* **1986**, *6*, 321–333.
31. Takada, Y.; Sato, T.; Teramoto, A. Dynamics of Stiff-Chain Polymers in Isotropic Solution .2. Viscosity of Aqueous-Solutions of Xanthan, a Rigid Double-Helical Polysaccharide. *Macromolecules* **1991**, *24*(23), 6215–6219.
32. Wyatt, N. B.; Liberatore, M. W. Rheological and Viscosity Scaling of the Polyelectrolyte Xanthan Gum. *J. Appl. Polym. Sci.* **2009**, *114*, 4076–4084.
33. Milas, M.; Rinaudo, M.; Knipper, M.; Schuppiser, J. C. Flow and Viscoelasticproperties of Xanthan Gum Solutions. *Macromolecules* **1990**, *23*, 2507–2511.
34. Southwick, J.; Brant, D. A. Quasielastic Light-Scattering Studies of Xanthan in Solution; ACS Symposium Series; 1981.
35. Choppe, E.; Puaud, F.; Nicolai, T.; Benyahia, L. Rheology of Xanthan as a Function of Temperature, Concentration and Ionic Strength. *Carbohydr. Polym.* **2010**, *82*, 1228–1235.
36. Pelletier, E.; Viebke, C.; Meadows, J.; Williams, P. A. A Rheological Study of the Order-Disorder Conformation Transition of Xanthan Gum. *Biopolymers* **2001**, *59*, 339–346.
37. Renou, F.; Petibon, O.; Malhiac, C.; Grisel, M. Effect of Xanthan Structure on Its Interaction With Locust Bean Gum: Toward Prediction of Rheological Properties. *Food Hydrocoll.* **2013**, *32*, 331–340.
38. Qiu, H. Y.; Yan, J. H.; Lan, G. H.; Liu, Y. Q.; Song, X. Q.; Peng, W. X.; Cui, Y. Y. Removal of Cu2+ From Wastewater by Modified Xanthan Gum (XG) With Ethylenediamine (EDA). *RSC Adv.* **2016**, *6*(86), 83226–83233.

39. Hamcerencu, M.; Desbrieres, J.; Popa, M.; Khoukh, A.; Riess, G. New Unsaturated Derivatives of Xanthan Gum: Synthesis and Characterization. *Polymer* **2007**, *48*(7), 1921–1929.
40. Wang, X. J.; Xin, H. P.; Zhu, Y. J.; Chen, W. J.; Tang, E. G.; Zhang, J.; Tan, Y. B. Synthesis and Characterization of Modified Xanthan Gum Using Poly(Maleic Anhydride/1-Octadecene). *Colloid Polym. Sci.* **2016**, *294*(8), 1333–1341.
41. Shalviri, A.; Liu, Q.; Abdekhodaie, M. J.; Wu, X. Y. Novel Modified Starch-Xanthan Gum Hydrogels For Controlled Drug Delivery: Synthesis and Characterization. *Carbohydr. Polym.* **2010**, *79*(4), 898–907.
42. Skender, A.; Hadj-Ziane-Zafour, A.; Flahaut, E. Chemical Functionalization of Xanthan Gum for the Dispersion of Double-Walled Carbon Nanotubes in Water. *Carbon* **2013**, *62*, 149–156.
43. Ahuja, M.; Kumar, A.; Singh, K. Synthesis, Characterization and In Vitro Release Behavior of Carboxymethyl Xanthan. *Int. J. Biol. Macromol.* **2012**, *51*(5), 1086–1090.
44. Bakshi, P.; Sadhukhan, S.; Maiti, S. Design of Modified Xanthan Mini-Matrices for Monitoring Oral Discharge of Highly Soluble Soluplus (R)-Glibenclamide Dispersion. *Mater. Sci. Eng. C Mater. Biol. Appl.* **2015**, *54*, 169–175.
45. Yahoum, M. M.; Moulai-Mostefa, N.; Le Cerf, D. Synthesis, Physicochemical, Structural and Rheological Characterizations of Carboxymethyl Xanthan Derivatives. *Carbohydr. Polym.* **2016**, *154*, 267–275.
46. Su, L.; Ji, W. K.; Lan, W. Z.; Dong, X. Q. Chemical Modification of Xanthan Gum to Increase Dissolution Rate. *Carbohydr. Polym.* **2003**, *53*(4), 497–499.
47. Pandey, S.; Mishra, S. B. Graft Copolymerization of Ethylacrylate Onto Xanthan Gum, Using Potassium Peroxydisulfate as an Initiator. *Int. J. Biol. Macromol.* **2011**, *49*(4), 527–535.
48. Pandey, S.; Mishra, S. B. Chromatographic Resolution of Racemic Alpha-Amino Acids: Chiral Stationary Phase Derived From Modified Xanthan Gum. *Carbohydr. Polym.* **2013**, *92*(2), 2201–2205.
49. Kumar, A.; Singh, K.; Ahuja, M. Xanthan-g-Poly(Acrylamide): Microwave-Assisted Synthsis, Characterization and In Vitro Release Behavior. *Carbohydr. Polym.* **2009**, *76*, 261–267.
50. Ghorai, S.; Sarkar, A. K.; Panda, A. B.; Pal, S. Effective Removal of Congo Red Dye From Aqueous Solution Using Modified Xanthan Gum/Silica Hybrid Nanocomposite as Adsorbent. *Bioresour. Technol.* **2013**, *144*, 485–491.
51. Badwaik, H. R.; Sakure, K.; Nakhate, K. T.; Dhongde, H.; Kashayap, P.; Tripathi, D. K. Microwave Assisted Eco-Friendly Synthesis, Characterization and In Vitro Release Behavior of Carboxymethyl Xanthan Gum. *Curr. Microwav Chem.* **2016**, *3*(3), 203–211.
52. Yalpani, M.; Hall, L. D. Some Chemical and Analytical Aspects of Polysaccharide Modifications. 1. Nitroxide Spin-Labeling Studies of Alginic Acid, Cellulose, and Xanthan Gum. *Can. J. Chem.* **1981**, *59*(21), 3105–3119.
53. Laffleur, F.; Michalek, M. Modified Xanthan Gum for Bucal Delivery—A Promising Approach in Treating Sialorrhea. *Int. J. Biol. Macromol.* **2017**, *102*, 1250–1256.
54. Mendes, A. C.; Baran, E. T.; Reis, R. L.; Azevedo, H. Fabrication of Phospholipid-Xanthan Microcapsules by Combining Microfluidics With Self-Assembly. *Acta Biomater.* **2013**, *9*, 6675–6685.
55. Roy, A.; Comesse, S.; Grisel, M.; Hucher, N.; Souguir, Z.; Renou, F. Hydrophobically Modified Xanthan: An Amphiphilic But Not Associative Polymer. *Biomacromolecules* **2014**, *15*(4), 1160–1170.
56. Fantou, C.; Roy, A. N.; De, E.; Comesse, S.; Grisel, M.; Renou, F. Chemical Modification of Xanthan in the Ordered and Disordered States: An Open Route for Tuning the Physico-Chemical Properties. *Carbohydr. Polym.* **2017**, *178*, 115–122.

57. Bradshaw, I. J.; Nisbet, B. A.; Kerr, M. H.; Sutherland, I. W. Modified Xanthan: Its Preparation and Viscosity. *Carbohydr. Polym.* **1983**, *3*, 23–38.
58. Smith, I. H.; Symes, K. C.; Lawson, C. J.; Morris, E. R. Influence of the Pyruvate Content of Xanthan on Macromolecular Association in Solution. *Int. J. Biol. Macromol.* **1981**, *3*, 129–134.
59. Tako, M.; Nakamura, S. Rheological Properties of Deacetylated Xanthan in Aqueous Media. *Agric. Biol. Chem.* **1984**, *48*(12), 2987–2993.
60. Tako, M.; Nakamura, S. Rheological Properties of Depyruvated Xanthan in Aqueous Solution. *Agric. Biol. Chem.* **1988**, *52*(6), 1585–1586.
61. Jampala, S. N.; Manolache, S.; Gunasekaran, S.; Denes, F. S. Plasma-Enhanced Modification of Xanthan Gum and Its Effect on Rheological Properties. *J. Agric. Food Chem.* **2005**, *53*, 3618–3625.
62. Bejenariu, A.; Popa, M.; Le Cerf, D.; Picton, L. Stiffness Xanthan Hydrogels: Synthesis, Swelling Characteristics and Controlled Release Properties. *Polym. Bull.* **2008**, *61*, 631–641.
63. Badwaik, H. R.; Sakure, K.; Alexander, A.; Ajazuddin; Dhongade, H.; Tripathi, D. K. Synthesis and Characterisation of Poly(Acryalamide) Grafted Carboxymethyl Xanthan Gum Copolymer. *Int. J. Biol. Macromol.* **2016**, *85*, 361–369.

Further reading

64. Sho, T.; Sato, T.; Norisuye, T. Viscosity Behavior and Persistence Length of Sodium Xanthan in Aqueous Sodium-Chloride. *Biophys. Chem.* **1986**, *25*(3), 307–313.
65. Smidsrød, O.; Haug, A. Estimation of the Relative Stiffness of the Molecular Chain in Polyelectrolytes From Measurements of Viscosity of Different Ionic Strengths. *Biopolymers* **1971**, *10*, 1213–1227.

Derivatized polysaccharides on silica and hybridized with silica in chromatography and separation—A mini review

12

Hubert Hettegger[a], *Wolfgang Lindner*[b], *Thomas Rosenau*[a]
[a]Department of Chemistry, Institute for Chemistry of Renewable Resources, University of Natural Resources and Life Sciences, Vienna (BOKU), Austria, [b]Department of Analytical Chemistry, University of Vienna, Vienna, Austria

Chapter outline

1 Introduction

Polysaccharides (PS) and polysaccharide-modified porous silica materials have already been gaining immense attention in separation and chromatography technologies for a long time. PS, such as cellulose and starch derivatives, are both major components of ordinary paper and filter devices used in chemistry or biotechnology-related processes, as well as for coffee filters, etc. Also many membrane filters as well as cigarette filters are composed of cellulose acetate representing a technical PS derivative.[1] In addition to conventional household use, there are many scientifically highly specialized applications. The separation of organic compounds on a stationary phase by means of a mobile transport medium has been invented and elaborated by the Russian chemist Mikhail Tsvet around the turn to the 20th century. He also coined the term "chromatography," literally "color writing"; his last name, coincidentally, meaning "color" in Russian.[2] The invention of paper chromatography by Martin and Synge

Recent Trends in Carbohydrate Chemistry. https://doi.org/10.1016/B978-0-12-817467-8.00012-8
© 2020 Elsevier Inc. All rights reserved.

in 1943 (Nobel Laureates in 1952) was a major step in the field of chromatography.[3] Subsequently, columns filled with water-saturated porous silica gel have been used for the chromatographic separation of amino acid mixtures from wool hydrolysates.[4] The first separation of enantiomeric amino acids using cellulose as a chiral resolving agent was reported by Kotake et al. in 1951.[5] Later on PS-modified porous silica materials have been used successfully as chiral stationary phases (CSPs) for the resolution of enantiomers.[6]

The separation of enantiomeric compounds represents a challenge not only in analytical chemistry, but also in organic synthesis on all scales due to identical physical and chemical properties of enantiomers in an achiral environment. Since the first resolution of racemic tartrate into the respective enantiomers by Louis Pasteur 170 years ago,[7] a large variety of analytical and preparative methods for the stereoselective resolution of diverse chiral compounds has been developed. Early approaches were based on the formation of stable diastereomers by coupling of the respective enantiomers of a racemic or nonracemic mixture with an auxiliary chiral reagent in high enantiomeric excess (*ee*), so that conventional achiral separation techniques could subsequently be used in the so-called "indirect" resolution of enantiomers (Fig. 1).[8] However, for preparative applications, the cleavage of the resolved diastereomeric derivatives is necessary in order to obtain the neat enantiomers of interest.

Nowadays the "direct" separation of enantiomers via a "chiral environment" strategy is achieved by the formation of intermediate diastereomeric species with so-called chiral selectors.[9] Phenomenologically speaking, the formation and dissociation of the diastereomeric molecular adducts are thus driven by thermodynamic parameters, in contrast to, for example, the kinetic resolution as in the case of enantioselective enzymatic transformation. Modern analytical methods applied to the direct separation

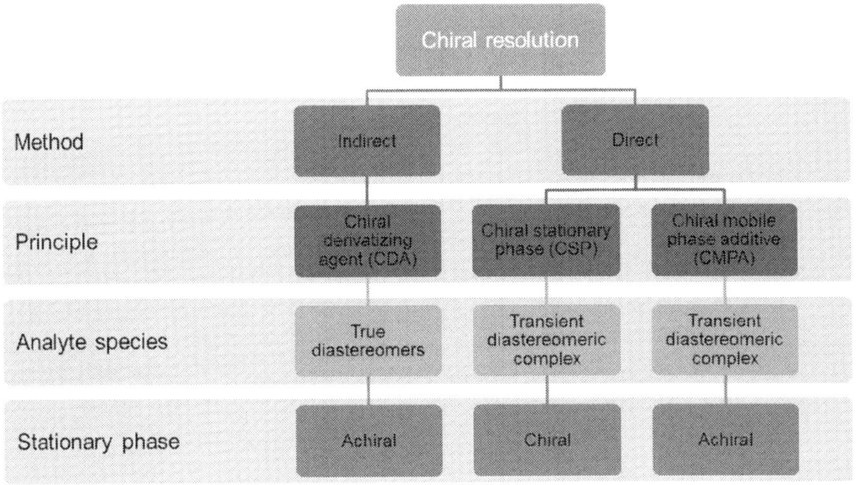

Fig. 1 Direct and indirect approaches in chiral resolution.

of enantiomers are stereoselective high-performance liquid chromatography (HPLC), gas chromatography (GC), capillary electrophoresis (CE), and supercritical fluid chromatography (SFC). The continuous simulated moving-bed chromatography (SMB) methodology gained high popularity for the preparative resolution of enantiomers, particularly in the pharmaceutical industry. The SMB is applied both in the LC (liquid chromatography) and in the SFC mode.

Currently, "chiral" HPLC is the most popular enantiomer separation method,[10] in which enantioselectivity is governed by the interaction of a chiral stationary phase with the solute and the mobile phase. Hitherto hundreds of different silica-based CSPs comprising chiral selectors adsorbed or chemically immobilized on the silica surface have been developed, of which more than 100 have been commercialized, although over the years many of them have again disappeared from the market. The respective chiral selectors are either low- or high-molecular-weight compounds and the respective CSPs can be roughly differentiated into π-donor-acceptor (Pirkle)-type phases (ion-exchanger-type phases),[11] ligand-exchange-type phases,[12] chiral cavity-type phases (encompassing, e.g., cyclodextrins or chiral crown ethers as chiral selectors, but also chiral imprints fall into this category),[13] chiral macrocyclic antibiotic-type phases,[14] protein-, polysaccharide-, and synthetic helical polymer-type phases.[15]

This chapter focuses only on PS-type CSPs. In the early 1980s, systematic studies of differently derivatized PS-type CSPs have been undertaken by Okamoto[16]; several of those phases became commercially available early, thus stimulating the field with long-lasting effects. Along that line substantial contributions by Chankvetadze[17] and Francotte[18] should be mentioned in particular. The first successfully applied PS-type CSP was reported by Hesse and Hagel in 1973 using solid cellulose triacetate particles in columns for the direct separation of enantiomers by LC.[19] As there are already excellent review articles and book chapters available focusing on the history, synthesis, chiral recognition mechanism as well as applications of polysaccharide-based CSPs for enantioseparation including future prospects[6, 10, 15, 20] we aim in this chapter only on very recent developments, including applications.

As can be drawn from the literature most of the developed PS-derived CSPs are carbamate- or ester-type derivatives of cellulose and amylose, respectively, in which the hydroxy groups are derivatized with diverse phenyl isocyanates or acid chlorides. It turned out that 3,5-dimethylphenyl, 3,5-dichlorophenyl, and positional isomers of the chloro-methyl-phenyl carbamate derivatives, which contain an electron-donating methyl- and an electron withdrawing chloro-substituent at the respective phenyl group, lead to CSPs with exceptionally high and broadly applicable chiral recognition abilities[21]; more details are provided in the following.

Nowadays the application of polysaccharide-type CSPs for the direct chiral resolution allows for the separation of >90% of the tested chiral analytes and racemic drugs.[22] Most often the amylose derivatives exhibit higher chiral recognition ability compared to the respective cellulose derivatives, presumably due to a different order of the helical structure of the main chain of the polysaccharide backbone, although contrary observations have been made as well, depending on the chemical structure of the particular analytes.[23] Besides, the same derivatives of cellulose or amylose occasionally lead to a reversal of the elution order for the same pair of enantiomeric analytes, which in turn is a result of differences in the molecular structure (α- and β-glycosidic bond) of amylose and cellulose, respectively.[24]

Today PS-type CSPs are commercially available both in physically coated (first generation) and chemically immobilized (second generation) forms. Chemical immobilization in this case means covalent linkage of the respective PS derivative to the silica particle. One major disadvantage of coated-type CSPs is their incompatibility with many organic solvents that could potentially serve as respective mobile phases. Thus the composition of the mobile phase is often limited to a certain range of solvents. The use of "strong" solvents such as acetone, dimethyl sulfoxide (DMSO), *N,N*-dimethylacetamide (DMAc), *N,N*-dimethylformamide (DMF), ethyl acetate (EtOAc), tetrahydrofuran (THF), or chlorinated solvents, such as chloroform or dichloromethane, may swell or in the worst case even dissolve the respective PS derivatives coated onto the silica support resulting in strong column bleeding and loss of chiral recognition and separation performance.[25]

This limitation was *inter alia* overcome by an innovation of E. Francotte et al. through cross-linking the coated PS chains, thus making the PS-type chiral selector insoluble and quasi-"immobilized" onto the silica surface.[26] To a certain extent this "immobilization" technique of polysaccharide derivatives may result in somewhat altered chiral recognition ability due to a slight change in the spatial structure and rigidity of the chiral polymer. Thus only a low degree of cross-linking is preferred, which is well controllable. For details on other chemical immobilization methods the reader is referred to Shen et al.[10]

The structures of selected immobilized polysaccharide-type CSPs, which, for example, have been commercialized by DAICEL Corporation under the trademark CHIRALPAK, are shown in Fig. 2. Similar CSPs principally comprising the same types of chiral selectors have been commercialized by several other companies as well.

This mini review focuses on summarizing only recent trends and developments in the field of polysaccharide-type CSPs representing PS-modified porous silica materials for HPLC and SFC applications with an emphasis on the developments in the last 2–3 years. In particular the developments in synthesis, material modification, analytical applications, and related findings are presented.

In a separate sub-chapter also polysaccharide-based other related materials, such as PS-silica hybrid and composite materials for oil/water separation and heavy-metal adsorption, are discussed briefly. It is important to point out that these materials are structurally completely different from the surface-modified porous silica materials. Detailed review articles of such sol-gel derived biopolymer-silica hybrid and composite materials, for example, for water remediation, air purification, metallic implants, protein separation, catalysis, antibacterial materials, enzyme immobilization, sensors, biomedical applications, and drug delivery, etc., have recently been published by Singh et al.,[27] Salama,[28] as well as by Pandey and Mishra,[29] thus related topics will not be discussed here in detail.

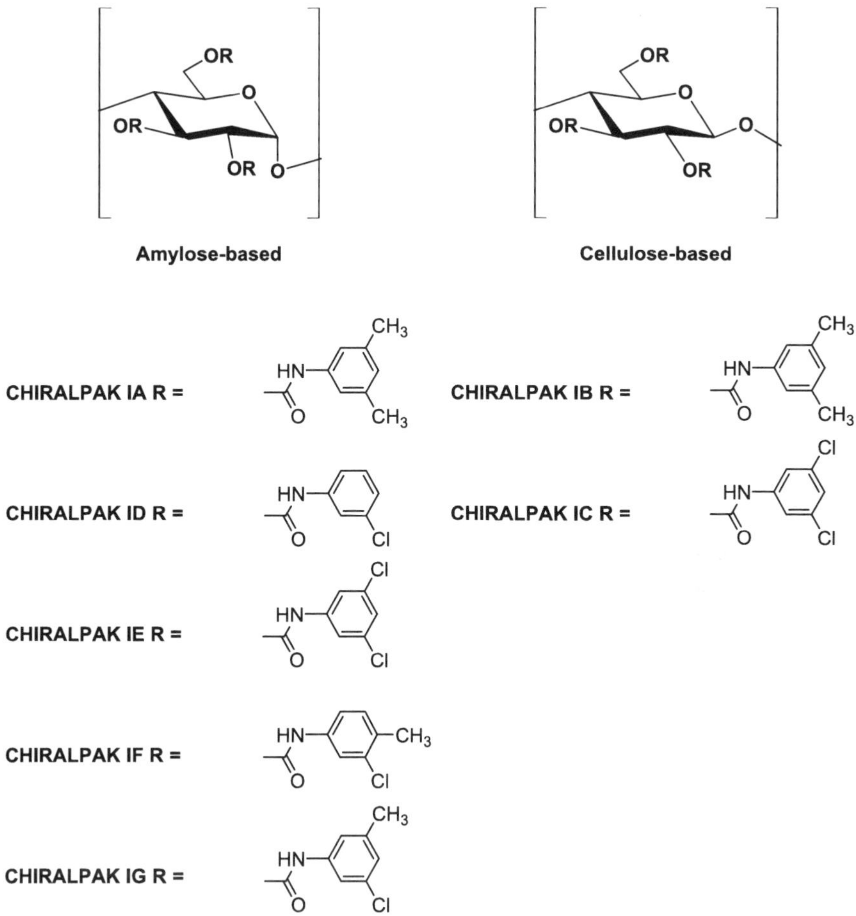

Fig. 2 Chemical structures of the PS-type selectors of CHIRALPAK CSPs.

2 Porous silica materials surface-modified with PS derivatives

2.1 *Coated PS-type CSPs*

A very recent and often discussed "hot" topic in the field of polysaccharide-type CSPs are materials on the basis of superficially porous silica particles (SPP), which generally have a more uniform particle-size distribution and shorter diffusion paths than the fully porous silica particle (FPP) counterparts currently most often used in LC. Initial studies and first reports by the Chankvetadze group in 2012 (see Lomsadze et al.[30]) have pointed out that these materials are characterized by higher enantioselectivity (α) at comparable selector loading, combined with limited dependence of the plate height (H) on the flow rate of the mobile phase, as well as higher resolution (R_s) and plate numbers (N) per unit time. Chiral compounds are thus separated with still high resolution within a comparably short analysis time (high-speed separation). Diverging conclusions in later studies and skepticism led to further experiments and a deeper understanding of the separation mechanism.[31] It is important to consider the higher surface density of the chiral selector on the CSP. The use of denser core-shell particles resulted in significantly different characteristics of the respective columns compared to FPP particles. In earlier studies, core-shell particles were coated with amounts of chiral selectors comparable to porous silica particles, which—due to the lower surface area—quickly led to overloading of the particles and changed particle surface morphology. Kharaishvili et al. have demonstrated that with only 2 wt% of chiral selector loading high-speed separations resulted in high column efficiency (N > 100,000 plates per meter) paired with adequate resolutions (particle morphology: only 2.8 and 3.6 μm nominal particle diameter, 30 and 20 nm nominal pore size, respectively; chiral selectors: coated cellulose and amylose tris(3,5-dimethylphenylcarbamate).[31] It was pointed out that the selector loading amount is a very essential aspect for the preparation of PS-type superficially porous silica materials for chiral separations. However, an appropriately high detector data acquisition rate also needs to be taken into account due to the resulting high speed of separation (analysis time 15–25 s, in which the analyte only spends about 5 s on column).

In the case of high-molecular-weight chiral selectors, such as carbamoylated cellulose and amylose derivatives, the use of wide-pore silica is preferable. The effect of pore size on the chromatographic performance in the case of SPPs coated with chiral selectors was studied by Bezhitashvili et al.[32] Observations as discussed above were further confirmed. A selector loading of 5% in the case of 600 Å nominal pore size led to agglomeration of the particles, thus significantly increasing the particle size distribution. However, lower selector loadings could also lead to agglomeration when materials with comparably lower nominal pore size were used. Under optimized conditions high separation performance using wide-pore core-shell particles in combination with PS-type chiral selectors was demonstrated by very small reduced plate heights of h = 1.42–1.70 for the first eluting enantiomers of *trans*-stilbene oxide, flavanone, and benzoin as the respective analytes (see Fig. 3).

Recently, the first core-shell material comprising covalently immobilized cellulose 3,5-dichlorophenylcarbamate as a chiral selector was reported by the Chankvetadze group.[33] In a proprietary technique 3.6 μm SPPs (50 nm pore size) were modified with the PS derivative (2 wt%). The applicability with respect to chiral recognition was evaluated on the basis of different chiral sulfoxides (including etozoline) and *trans*-stilbene oxide in pure methanol (MeOH) and acetonitrile (ACN) as the respective mobile phases with analysis times of less than 30 s. Due to immobilization of the chiral selector the use of such polar eluents was enabled. In the case of the chiral sulfoxides lower enantioselectivity was observed in methanol as the mobile phase in comparison to ACN, indicating hydrogen-bonding-type interactions in chiral recognition.

The influence of the molecular weight of the cellulose starting material used for the preparation of respective tris(3,5-dimethylphenylcarbamate) chiral selectors was studied by Okada et al.[34] In this study fully porous wide-pore silica gel (7 μm mean

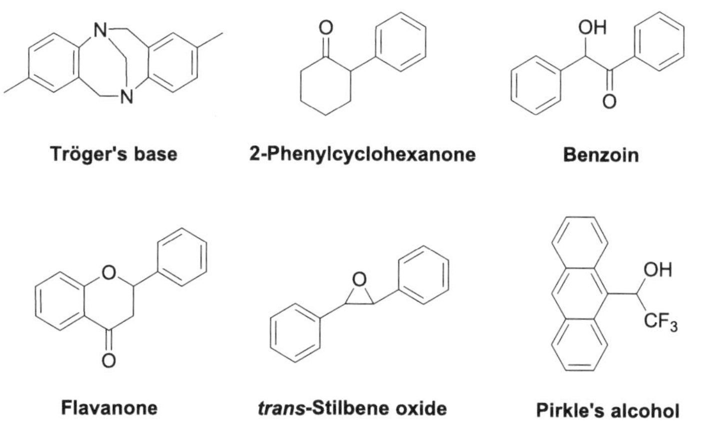

Fig. 3 Chemical structures of representative racemates for the evaluation of chiral recognition ability.
Redrawn and adapted from Shen, J.; Ikai, T.; Okamoto, Y., Synthesis and Application of Immobilized Polysaccharide-Based Chiral Stationary Phases for Enantioseparation by High-Performance Liquid Chromatography. J. Chromatogr. A **2014**, *1363*, 51–61 with permission from Elsevier.

particle size, 100 nm mean pore diameter) was used. Cellulose oligomers with a varying degree of polymerization (DP=7–52) were prepared by acid hydrolysis starting from microcrystalline cellulose (MCC) according to a protocol developed by Isogai and Usuda.[35] After derivatization with 3,5-dimethylphenyl isocyanate the respective tris(3,5-dimethylphenylcarbamate) derivatives of the cellooligomers were coated onto silica. Due to the rather high solubility of the shorter carbamoylated oligomer derivatives in "traditional" normal phase solvent mixtures (e.g., hexane/2-propanol 9:1 v/v) a reduced isopropanol content of maximum 2% was used for the evaluation of the chiral separation potential. The chemical structures of some tested racemates representing typical compounds for the evaluation of chiral recognition ability in the case of PS-type CSPs are shown in Fig. 3.

On the other hand, the high solubility allowed for detailed studies of the molecular interaction and chiral separation mechanism by ^{1}H NMR (nuclear magnetic resonance) spectroscopy in deuterated chloroform ($CDCl_3$). It was demonstrated that the chiral recognition in case of oligomer derivatives with DP=7 and 11 was much lower in comparison to oligomers with DP=18 or 52, so the authors concluded that a minimum DP of 18 seemed to be sufficient for the establishment of chiral recognition. As was additionally shown by circular dichroism (CD) spectroscopic studies at about DP=26 the formation of a regular helical structure of the oligomeric cellulose tris(3,5-dimethylphenylcarbamate) derivative may be possible in solution. However, it has to be kept in mind that the respective PS derivatives are coated onto the silica particles and thus the spatial arrangement does not necessarily correspond to that in solution, which can possibly lead to differences in the heterogeneous chiral selector-analyte interaction from the homogeneous one in solution.

Besides the DP of the oligo- or polysaccharide starting material on the one hand and the silica morphology on the other hand, the chiral recognition ability is also influenced by the respective substituents at positions 2, 3, and 6 of the glucose subunit of the polymeric backbone as demonstrated by Shen et al. for carbamoylated amylose.[36] In this study, a variety of different phenylcarbamoylated amylose derivatives was synthesized. Selective protection of positions 2 and 6 was carried out in a homogeneous system by the reaction with bulky thexyldimethylchlorosilane in DMAc/LiCl and imidazole according to a protocol developed by Koschella, Heinze, and Klemm yielding respective di-*O*-dimethylthexylsilyl ethers.[37] The hydroxy groups in position 3 were subsequently carbamoylated with phenyl, 4-chlorophenyl, or 3,5-dichlorophenyl isocyanate in pyridine. After deprotection applying tetra-*n*-butylammonium fluoride in THF, the hydroxy groups at 2,6-positions were derivatized with different isocyanates: *m*-tolyl, *p*-tolyl, 3,5-dimethylphenyl, phenyl, 3-chlorophenyl, 4-chlorophenyl, and 3,5-dichlorophenyl isocyanates, respectively, yielding a total of 18 different amylose carbamate-derived chiral selectors. The PS derivatives were coated onto aminopropyl silanized macroporous silica gel and the influence of the regioselective derivatization was studied on the basis of 10 racemic compounds and compared to a commercially available CHIRALPAK AD column (amylose tris(3,5-dimethylphenylcarbamate) coated onto silica). Although extensively studied, clear trends and conclusions could not be derived with respect to the influence of regioselective derivatization; each CSP in this complex study possessed characteristic chiral discrimination power, which was influenced by the chemical nature as well as respective position and number of substituents.

In a similar approach, regioselectively substituted cellulose carbamates have also been synthesized by Shen et al. and evaluated with respect to chiral resolution properties.[38] After protection of the hydroxy group in position 6 of the respective glucose units with trityl chloride yielding 6-*O*-triphenylmethyl cellulose, positions 2 and 3 were carbamoylated with four different isocyanates including aryl and alkyl derivatives (3,5-dimethylphenyl, 3,5-dichlorophenyl, 4-chlorophenyl, and cyclohexyl isocyanate, respectively). Cleavage of the trityl group and subsequent carbamoylation with the above-mentioned isocyanates yielded a total of nine different cellulose derivatives, which were subsequently coated onto aminopropyl silanized silica gel. The influence of regioselective derivatization in terms of chiral separation was evaluated by HPLC with basically the same set of compounds as used in an earlier study[36] and compared to a commercially available CHIRALCEL OD column (cellulose tris(3,5-dimethylphenylcarbamate) coated onto silica). It was demonstrated that the introduction of two different carbamate moieties resulted in comparably higher chiral recognition ability, in some cases even higher than the commercially available counterparts. The authors also concluded that chiral selectors based on cellulose 2,3-(3,5-dimethylphenylcarbamate) derivatives possessed better chiral recognition than the respective other derivatives.

2.2 Immobilized PS-type CSPs

Two interesting immobilization approaches based *inter alia* on concepts described in now expired patents[26, 39] were recently published by Francotte et al.[40] In a first approach, 4-methylbenzoylcellulose (chiral selector) equipped with photoreactive

2,3-dimethylmaleimide moieties was cross-linked after coating onto silica gel by UV irradiation using a high-pressure mercury immersion lamp. During the cross-linking process, a [2+2] cycloaddition reaction of two dimethylmaleimide residues takes place, forming a covalent linkage between the cellulose chains via a cyclobutane ring (see Fig. 4).[40a] The resulting cross-linked chiral selector was described to be insoluble in almost all organic solvents. Thus, it can be used with an extended range of mobile phases.

It was shown that a spacer containing an aromatic moiety (see, e.g., respective aryl residues $X=CH_2C_6H_5$ or C_6H_5 in Fig. 4) positively influences chiral recognition compared to a sole aliphatic residue ($X=CH_2$). As expected, selectivities (α) and column efficiency (N) were consistently higher in case of the nonimmobilized reference materials.

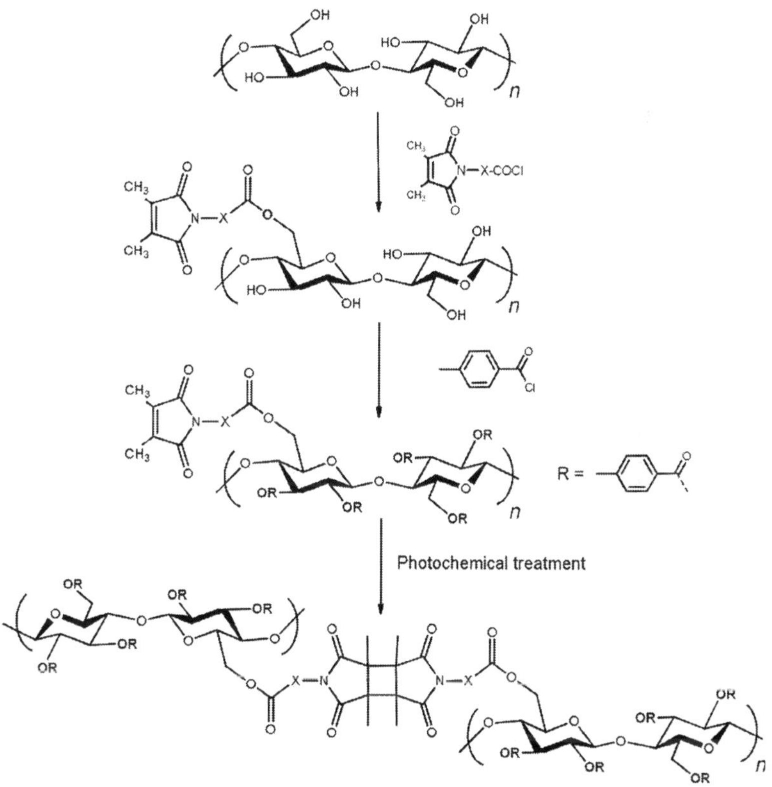

Fig. 4 Synthesis and immobilization of a photoreactive cellulose-derived chiral selector via cross-linking in a [2+2] cycloaddition reaction forming a cyclobutane ring. X = alkyl or aryl residue.

Reprinted from Francotte, E.; Zhang, T., Preparation and Evaluation of Immobilized 4-Methylbenzoylcellulose Stationary Phases for Enantioselective Separations. J. Chromatogr. A **2016**, *1467*, 214–220 with permission from Elsevier.

The tedious synthesis and immobilization process of the CSPs (in total six steps) led to further studies in this field.[40b] In a second approach, aminopropyl silanized silica was coated with cellulose tris(4-methylbenzoate) and exposed to UV light without the addition of any photopolymerizable functional group, radical initiator, or photosensitizer. In model studies, when *p*-methylbenzoyl cellobiose was used as a model compound, no oligomeric species was detected, suggesting that a polymeric species was formed during immobilization.

However, the actual immobilization mechanism was not clarified further and it could also not be elucidated whether immobilization is due to cross-linking of cellulose chains or covalent linkage to the silica gel carrier. It was also shown that the typical negative influence of immobilization on selectivity was less pronounced, presumably due to the absence of any further functional substituents on the cellulose derivative. The influences of suspending medium and irradiation time on immobilization effectiveness were investigated. Compared to pure ACN or aqueous suspension media containing a high organic content the use of 25% MeOH in water gave slightly higher rates of immobilization (87%).

2.3 Analytical applications

The high and outstandingly broad chiral resolving power of PS-type CSPs renders them ideal not only for preparative but also for analytical applications, whenever enantioselective chromatography is needed. Thus commercially available CSPs are typically used for both analytical method development[41] and preparative scale applications.[42] The use of immobilized stationary phases (=PS-type selector covalently linked to silica particles) allows for a virtually unlimited variation in mobile phase mixtures due to high robustness, by which target sample solubility issues in typical hexane-isopropanol mixtures can be overcome. Also a potentially reduced chiral recognition ability in polar eluents can be counterbalanced by the possibility of changing or adjusting the eluent composition. Thus, adjustments of the selectivity and chiral recognition mechanism are rather easily possible compared to coated-type CSPs. An overview of exemplary analytical applications of PS-type CSPs including comparative studies, method development, evaluation of commercially available CSPs, chiral recognition studies, etc., is illustrated in Table 1. This selection on the one hand aims at giving an overview of recent applications and analytical methods (HPLC, SFC, CEC, etc.), but on the other hand it is meant to demonstrate the high number of chemically diverse compounds, which can be resolved by carbamoylated PS-type CSPs.

By using 175 chiral acidic, basic, and neutral compounds the effect of coating versus immobilization was studied by Beridze et al. for cellulose 3,5-dichlorophenylcarbamate and amylose 3,5-dimethylphenylcarbamate as the chiral selectors.[25] The authors concluded that in the case of the cellulose carbamate selector immobilization yielded higher success rates; the opposite was true in the case of the amylose carbamate selector. Matarashvili et al. have shown that the use of aqueous-organic mobile phases does not necessarily result in a typical reversed-phase chromatographic behavior (especially in aqueous ACN at low water content <20% v/v).[55] Even a reversal of elution order was observed for enantiomers depending on the water content of the

Table 1 Compounds tested applying different chromatographic techniques using PS-type CSPs.

Tested compounds	Method	Aim and topic of the study	Reference
10 cathinone and amphetamine derivatives	CEC, HPLC, SFC	Comparison of CEC, SFC, and HLPC modes	Albals et al.[43]
6 flavanone derivatives	HPLC	Separation behavior on experimental columns	Fanali et al.[44]
8 β-agonists	HPLC	Separation behavior and elution order	Gumustas et al.[45]
6 neutral and cationic hetero[6]helicenes	HPLC	Enantiomeric separation	Labrador et al.[46]
6 β-carboline derivatives	HPLC	Comparison of PS- and strong cation exchanger-based CSPs	Lajkó et al.[47]
1 melatoninergic ligand	SFC	Design of experiment	Landagaray et al.[48]
2 flufiprole derivatives (pesticides)	UPLC	LC-MS/MS method development	Li and Zhang et al.[41]
4 3-carboxamido-5-aryl isoxazole derivatives	SFC	Enantioseparation and thermodynamic parameters	Zehani et al.[49]
4 proton-pump inhibitors	HPLC	Method development	Zhao et al.[50]
6 antihistamines	HPLC	Enantiomeric separation	Zhou et al.[51]
8 flavanone derivatives	nano-LC, CEC, pCEC	Comparative study	D´Orazio et al.[52]
5 chiral sulfoxides	nano-LC, CEC	Optimization for high-speed separation	D´Orazio et al.[53]
52 biologically active compounds (profens, flavanones, thiazide diuretics, calcium channel blockers, phenothiazines, amphetamines, benzofuran derivatives, β-blockers)	SFC	Evaluation of enantioselective potential of ACQUITY UPC^2 Trefoil CEL1 and ACQUITY UPC^2 Trefoil AMY1	Kucerova et al.[54]
26 weak chiral acids (including profens)	HPLC	Study of nonreversed phase behavior and elution order	Matarashvili et al.[55]

Continued

Table 1 Continued

Tested compounds	Method	Aim and topic of the study	Reference
14 terphenyl isomers and PAHs	HPLC	Achiral and chiral recognition behavior	Shibata et al.[56]
4 chiral sulfoxides	HPLC	Study of chiral recognition behavior	Carradori et al.[57]
28 pharmaceuticals (α- and β-blockers, antifungal and anti-inflammatory drugs, sedative hypnotics, anticancer drugs, anti-histaminics, antiarrhythmic drugs, adrenaline, flavonoids, etc.)	HPLC	Evaluation of CHIRALPAK IG	Ghanem and Wang[58]
24 chiral sulfoxides	SFC	Study of chiral recognition behavior	West et al.[59]
6 β-hydroxy-1,2,3-triazole derivatives	HPLC, SFC	Enantiomeric separation	Alvarenga et al.[60]
6 indole derivatives	HPLC, SFC	Comparison of PS- and zwitterionic *Cinchona*-based CSPs	Bajtai et al.[61]
175 chiral acidic, basic, and neutral compounds	HPLC	Coated vs. covalently immobilized CSPs	Beridze et al.[25]
Metconazole	HPLC	Method development	He et al.[62]
171 achiral probes and 97 racemates	HPLC, SFC	Comparative study	Khater et al.[63]
13 phenylthiohydantoin-amino acids	SFC	Thermodynamic enantioseparation behavior	Khater et al.[64]
5 chiral sulfoxides	HPLC	Study of chiral recognition behavior	Khatiashvili et al.[65]
8 anticholinergic drugs	HPLC	Enantiomeric separation	Li et al.[66]
4 phenothiazine derivatives (including levomepromazine)	HPLC	Method development	Niedermeier et al.[67]
14 chiral sulfoxides	HPLC	Study of chiral recognition behavior	Shedania et al.[68]

CEC, capillary electrochromatography; *PAH*, polycyclic aromatic hydrocarbon; *pCEC*, pressure-assisted CEC; *SFC*, supercritical fluid chromatography.

eluent. Similar observations were reported by Shedania et al.[68] All these interesting phenomena should be studied further for a better understanding of the chiral recognition mechanism, which includes the conformational flexibility of the helical structure of the chiral polymer. The influence of halogen substituents on enantioseparation and molecular recognition of halogenated chiral compounds, for example, has been reviewed by Peluso et al.[69]

An evaluation of the CHIRALPAK IG CSP recently introduced by DAICEL Corporation (see Fig. 2) was presented by Ghanem and Wang.[58] The stationary phase consisting of an immobilized amylose tris(3-chloro-5-methylphenylcarbamate) selector was evaluated by means of 28 different pharmaceuticals. It was demonstrated that the immobilized CSP could be operated with traditionally "prohibited" eluents, such as dichloromethane, THF, or methyl *tert*-butyl ether (MTBE). CHIRALPAK IG was superior to CHIRALPAK IA, which uses an immobilized amylose tris(3,5-dimethylphenylcarbamate) selector. The latest developments in this field have shown that selectors having both an electron-donating methyl- and an electron-withdrawing chloro-substituent at the respective phenylcarbamate group are superior in terms of the breadth of chiral resolutions compared to the dichlorophenyl or dimethylphenyl derivatives and thus may become prevalent in future, which is also true for immobilized CSPs.

Besides separation of chiral molecules, also the separation of achiral compounds by PS-type CSPs has been reported by Shibata et al.[56] A set of hard-to-separate aromatic positional isomers (terphenyl, triphenylene, and PAHs) was used to evaluate the potential of polysaccharide-based chiral phases with respect to their achiral recognition behavior. The experiments have demonstrated the outstanding potential in isomer recognition and thus the authors suggested that chiral phases can be used as powerful tools even for difficult achiral separation issues. The same holds true for the resolution of diastereomers.

3 Polysaccharide-silica hybrid and composite materials

In the following, a selection of three exemplary applications of PS-silica hybrid materials is briefly discussed. Besides the use of surface-modified porous silica materials for chromatography, efforts in R&D are also made toward true PS-silica hybrid materials as chiral packing materials and other separation applications, both on analytical and preparative scales, including oil-water separations and heavy metal adsorption. However, these materials are structurally different from the above-mentioned CSPs, since they represent true hybrid/composite materials. This selection does not claim to be complete. It rather gives a small insight into other potential applications and research directions as well as material preparation methods.

3.1 Chiral packing materials

In addition to the modification of ready-made fully porous silica particles by coating and/or immobilization of PS-type chiral selectors Weng et al. published a method based on the methodology proposed by Ikai et al.[70] for the de novo preparation of porous organosilica hybrid particles, which were obtained by a sol-gel method.[71]

A mechanically stable PS-silica hybrid chiral packing material with mesoporous structure (narrow pore size of 4 nm) and high surface area was prepared by carefully controlling the synthesis parameters such as pH (key factor), amounts of reagents, time, and temperature. The Brunauer-Emmett-Teller (BET) surface area was determined to be $522\,m^2\,g^{-1}$ (S_{BET}). The organic content of the optimum material was approximately 50 wt% (TGA analysis). Chiral selector loading was thus at least 2–3 times higher than for the conventional CSP materials for HPLC.[72] The organosilica hybrid particles were prepared in three steps: (1) homogeneous carbamoylation of cellulose with 3,5-dimethylphenyl isocyanate and 3-(triethoxysilyl)propyl isocyanate in DMAc/LiCl/pyridine; (2) cross-linking of the cellulose derivative with tetraethoxysilane and catalytic trimethylsilyl chloride in THF/heptan-1-ol/H_2O until completion of the sol-gel reaction; (3) aging of the sol by pouring it into hot aqueous SDS (sodium dodecyl sulfate) solution. The obtained solid hybrid spheres were subsequently slurry packed and evaluated with respect to chiral separation. Due to the high loading of chiral selector and excellent solvent durability even in chlorinated "strong" eluents, such as chloroform, preparative chiral separations have been proposed. In comparison to simply coated CSPs, which are used with, for example, hexane/isopropanol (9:1 v/v) mixtures as a mobile phase, chloroform-containing mobile phases allow for much higher amounts of injected sample due to significantly increased solubility. Chiral resolution was tested with racemic 2,2,2-trifluoro-1-(9-anthryl)ethanol and the β-blocker pindolol. A selectivity of $\alpha = 1.6$ was achieved for 2,2,2-trifluoro-1-(9-anthryl)ethanol, which was however lower compared to conventional CSP materials as described by Ikai et al. ($\alpha = 1.9$).[70] Furthermore, the use of a broad range of solvents and mixtures as mobile phases in case of immobilized selectors additionally offers the opportunity of easy modulation of the retention time. Thus, a shortening of analysis time is possible for strongly retained compounds.

3.2 Oil-water separation

Taking into account the huge amounts of oily wastewater and frequent oil spill accidents worldwide (2 million tons of refined oil products per year),[73] it is evident that research efforts are needed toward the development of hydrophobic oil-absorbent materials as well as filter membranes for oil-water separation. Wang et al. have developed a nanocomposite material by modifying a bacterial cellulose (BC) organogel using methylene diphenyl diisocyanate followed by subsequent silanization using methyltrimethoxysilane as a precursor in acidic media.[74] The modification of BC with the aromatic diisocyanate in the first stage allowed for a more homogeneous and uniform distribution of the condensed silica particles due to pre-hydrophobization of the initially hydrophilic cellulose fibers. After freeze-drying the obtained hydrophobic (water contact angle = 134 degrees), highly porous and flexible composite material was used for oil-water separation in a sorbent batch mode, in which the sponge-like material was simply dipped into an oil-water system. Recyclability was proven after squeezing and washing. Even after five cycles about 93% of the initial amounts were still adsorbed. In a chemically more complex modification procedure involving block copolymerization of 3-(trimethoxysilyl)propyl acrylate-*block*-myrcene in a RAFT (reversible addition-fragmentation chain transfer) approach a cellulose membrane with superhydrophobic (water contact angle > 160 degrees) and nonfouling character was obtained.[75] The efficient adsorption of oil in both sorbent batch and continuous filtration mode has been demonstrated.

An opposite mode of action was chosen by Du et al. for the separation of oily emulsions.[76] A chitosan-TiO_2 composite layer on a cellulose acetate membrane yielded a superoleophobic membrane material which was used for emulsion separation under a high flux in continuous mode (>6000 $L\,m^{-2}h^{-1}$ for a hexadecane-in-water emulsion; separation efficiency of 97%). Both cellulose acetate and chitosan were chosen in this study due to their beneficial film-forming properties, easy modification, and biodegradability. Another approach in which nylon fabrics as a support material were coated with a hydrophilic cellulose-starch-silica composite for oil-water separation was recently reported by Zhang et al.[77] Due to the use of nylon fabric as a carrier material and consequently high mechanical strength a membrane flux of about 32000 $L\,m^{-2}h^{-1}bar^{-1}$ was reached. The separation efficiency for different synthetic and natural oils was >97% after 100 cycles. The outstandingly high flux and facile preparation as well as the high separation efficiency and reusability of the superhydrophilic/underwater superoleophobic material were the reasons for rendering them promising materials for larger-scale oil-water separation.

Superhydrophobic (water contact angle up to 168 degrees) cellulose nanofiber (CNF)-assembled highly porous aerogels with high roughness were prepared by suspending CNFs and SiO_2 nanoparticles followed by cross-linking with methyltrimethoxysilane and subsequent freeze-drying.[78] CNFs were used as rigid nanoscale building blocks for the fabrication of the highly porous aerogel material, which was used for gravity-driven separation of water-in-oil emulsions ($1910 \pm 60\,L\,m^{-2}h^{-1}$ without external pressure). Using a cellulose aerogel as a reference it was shown that the composite material exhibited superior separation properties. In the case of the reference still a milky white filtrate was obtained due to the lower separation efficiency of the cellulose fibers.

Surfactant-stabilized oil-water emulsions with a droplet size in the nanoscale representing a bigger challenge compared to phase-separated oil-water mixtures were separated using hydrophobic cellulose acetate-SiO_2/TiO_2 hybrid microsphere composite aerogel films prepared by a phase-inversion method of dissolved cellulose acetate and *co*-condensation of hydrolyzed 3-aminopropyltrimethoxysilane and tetrabutyl titanate.[79] The achievable flux was up to 667 $L\,m^{-2}h^{-1}$ with a separation efficiency for the nano-droplets up to 99.99 wt% due to the hierarchical micro-/nanostructure of the composite film.

3.3 Heavy metal adsorption

Besides the pollution of water with oil, the emission of toxic heavy metals, such as Cu^{2+}, Ni^{2+}, Pb^{2+}, Cd^{2+}, and others, by industry and intensive farming represents another major environmental problem. Due to the low mechanical stability of, for example, cellulose- or chitosan-derived adsorbent materials, their application on a commercial scale has not been successful. Improving the mechanical stability by the preparation of biopolymer-silica composite materials would thus lead to higher stability. Further functionalization with negatively charged groups (e.g., phosphonic, sulfonic, or carboxylic acid groups) will eventually yield stable ion-exchange-type materials for the removal of heavy metal ions. One example for the recovery of heavy metal ions by polysaccharide-silica hybrid materials was recently published by Srivastava et al.[80] In a two-step modification protocol starting from cellulose triacetate first a polysaccharide-silica hybrid composite was prepared using (3-aminopropyl) triethoxysilane as a siloxane precursor. Subsequent phosphorylation was carried out at room temperature using a mixture of formaldehyde and phosphorous acid. Different parameters were investigated with respect to heavy metal adsorption: the effect of time and temperature of adsorption, the influence of pH, adsorbate concentration and adsorbent dose, kinetic models, thermodynamic parameters, and adsorption isotherms. The adsorbents showed excellent adsorption capacity for different bivalent metal ions; Ni^{2+} was shown to be adsorbed with comparatively high capacity.

4 Conclusions and outlook

In the field of surface-modified porous silica materials the development of immobilized polysaccharide-type CSPs was a major step forward in "chiral" chromatography. Chemically covalently linked chiral selectors broaden the range of applicable eluents significantly, nowadays allowing the use of "strong" eluents such as acetone, DMSO, DMAc, DMF, ethyl acetate, THF, or chlorinated solvents. Latest developments in this field are concerned with the use of surface-modified superficially porous silica CSPs, in which only 2 wt% of chiral selector loading allow for high-speed separations (analysis times 15–25 s) at still high resolution. It is important to point out that the recording speed for the detection data is critical in this case. The successful separation of hard-to-separate achiral isomeric compounds by chiral phases offers additional benefits and analytical opportunities. Two-dimensional LC applications including chiral × chiral as well as chiral × achiral correlation in LC as *inter alia* reported by Woiwode et al. will further broaden analytical applications and give valuable and analytically interesting (stereochemical) information in future.[81]

Surface-modified porous silica materials are also used on a preparative scale, and PS-silica hybrid as well as composite materials offer great possibilities and potential on a technical (industrial) scale. Materials, for example, for oil-water separation and heavy metal adsorption combining the advantageous properties of silica and polysaccharides are just two of a variety of potential options.

References

1. Li, B.; Cui, X.; Zhao, L.; Wang, L.; Xie, G.; Deng, N. Pressure and Gas Flow Distribution Inside the Filter of a Non-Filter Ventilated Lit Cigarette During Puffing. *Beitr. Tabakforsch. Int.* **2017**, *27*, 113–124.
2. Zwet, M. Adsorptionsanalyse und chromatographische Methode. Anwendung auf die Chemie des Chlorophylls. *Ber. Dtsch. Botan. Ges.* **1906**, *24*, 384–393.

3. Haslam, E. Vegetable tannins—Lessons of a Phytochemical Lifetime. *Phytochemistry* **2007**, *68*(22), 2713–2721.
4. Martin, A. J. P.; Synge, R. L. M. A New Form of Chromatogram Employing Two Liquid Phases. I. A Theory of Chromatography. II. Application to the Microdetermination of the Higher Monoamino Acids in Proteins. *Biochem. J.* **1941**, *35*, 1358–1368.
5. Kotake, M.; Sakan, T.; Nakamura, N.; Senoh, S. Resolution Into Optical Isomers of Some Amino Acids by Paper Chromatography. *J. Am. Chem. Soc.* **1951**, *73*, 2973–2974.
6. Chankvetadze, B. Recent Developments on Polysaccharide-Based Chiral Stationary Phases for Liquid-Phase Separation of Enantiomers. *J. Chromatogr. A* **2012**, *1269*, 26–51.
7. Gal, J. When Did Louis Pasteur Present His Memoir on the Discovery of Molecular Chirality to the *Académie des Sciences*? Analysis of a Discrepancy. *Chirality* **2008**, *20*, 1072–1084.
8. Slegel, P.; Vereczkey-Donath, G.; Ladanyi, L.; Toth-Lauritz, M. Enantiomeric Separation of Chiral Carboxylic Acids, as Their Diastereomeric Carboxamides, by Thin-Layer Chromatography. *J. Pharm. Biomed. Anal.* **1987**, *5*, 665–673.
9. Davankov, V. A. Analytical Chiral Separation Methods. *Pure Appl. Chem.* **1997**, *69*, 1469–1474.
10. Shen, J.; Ikai, T.; Okamoto, Y. Synthesis and Application of Immobilized Polysaccharide-Based Chiral Stationary Phases for Enantioseparation by High-Performance Liquid Chromatography. *J. Chromatogr. A* **2014**, *1363*, 51–61.
11. Laemmerhofer, M.; Lindner, W. Liquid Chromatographic Enantiomer Separation and Chiral Recognition by *Cinchona* Alkaloid-Derived Enantioselective Separation Materials. *Adv. Chromatogr. (Boca Raton, FL)* **2008**, *46*, 1–107.
12. Ravikumar, M.; Satish Varma, M.; Satyanarayana Raju, T.; Suchitra, P.; Yadagiri Swamy, P. Enantiomeric Separation of Moxifloxacin and Its (*R*,*R*)-Isomer by Ligand-Exchange Chiral Chromatography. *Chromatographia* **2009**, *69*, 85–89.
13. Beesley, T. E.; Lee, J. T. Updates in the Technology and Application of Chiral Stationary Phases. *LC-GC Eur.* **2003**, *16*, 33–36.
14. Armstrong, D. W.; Tang, Y.; Chen, S.; Zhou, Y.; Bagwill, C.; Chen, J.-R. Macrocyclic Antibiotics as a New Class of Chiral Selectors for Liquid Chromatography. *Anal. Chem.* **1994**, *66*(9), 1473–1484.
15. Okamoto, Y.; Kaida, Y. Resolution by High-Performance Liquid Chromatography Using Polysaccharide Carbamates and Benzoates as Chiral Stationary Phases. *J. Chromatogr. A* **1994**, *666*, 403–419.
16. Okamoto, Y.; Kawashima, M.; Hatada, K. Chromatographic Resolution. 7. Useful Chiral Packing Materials for High-Performance Liquid Chromatographic Resolution of Enantiomers: Phenylcarbamates of Polysaccharides Coated on Silica Gel. *J. Am. Chem. Soc.* **1984**, *106*, 5357–5359.
17. Chankvetadze, B.; Chankvetadze, L.; Sidamonidze, S.; Yashima, E.; Okamoto, Y. High-Performance Liquid Chromatography Enantioseparation of Chiral Pharmaceuticals Using Tris(Chloromethylphenylcarbamate)s of Cellulose. *J. Pharm. Biomed. Anal.* **1996**, *14*, 1295–1303.
18. Francotte, E.; Wolf, R. M.; Lohmann, D.; Mueller, R. Chromatographic Resolution of Racemates on Chiral Stationary Phases. I. Influence of the Supramolecular Structure of Cellulose Triacetate. *J. Chromatogr.* **1985**, *347*, 25–37.
19. Hesse, G.; Hagel, R. Complete Separation of a Racemic Mixture by Elution Chromatography on Cellulose Triacetate. *Chromatographia* **1973**, *6*, 277–280.

20. (a) Chen, X.; Yamamoto, C.; Okamoto, Y. Polysaccharide Derivatives as Useful Chiral Stationary Phases in High-Performance Liquid Chromatography. *Pure Appl. Chem.* **2007**, *79* (9), 1561–1573; (b) Ikai, T.; Yamamoto, C.; Kamigaito, M.; Okamoto, Y. Immobilized-Type Chiral Packing Materials for HPLC Based on Polysaccharide Derivatives. *J. Chromatogr. B Anal. Technol. Biomed. Life Sci.* **2008**, *875*(1), 2–11; (c) Zhang, T. F. Pilar, Analytical and Preparative Potential of Immobilized Polysaccharide-Derived Chiral Stationary Phases. In *Chiral Separation Techniques*, Subramanian, G., Ed.; WILEY-VCH Verlag GmbH & Co. KGaA: Weinheim, Germany, 2007, pp 99–132; (d) Okamoto, Y. Precision Synthesis, Structure and Function of Helical Polymers. *Proc. Jpn. Acad. Ser. B* **2015**, *91*, 246–261; (e) Chankvetadze, B. Enantioseparations by High-Performance Liquid Chromatography Using Polysaccharide-Based Chiral Stationary Phases: An Overview. *Methods Mol. Biol. (N.Y., U.S.A.)* **2013**, *970*, 81–111; (f) Shen, J.; Okamoto, Y. Efficient Separation of Enantiomers Using Stereoregular Chiral Polymers. *Chem. Rev.* **2016**, *116*(3), 1094–1138; (g) Ding, P.; Chang, B.; Qing, G.; Sun, T. New Approach For Chiral Separation: From Polysaccharide-Based Materials to Chirality-Responsive Polymers. *SCIENCE CHINA Chem.* **2014**, *57*(11), 1492–1506; (h) Ikai, T.; Yamamoto, C.; Kamigaito, M.; Okamoto, Y. Immobilized Polysaccharide-Based Chiral Stationary Phases for HPLC. *Polym. J.* **2006**, *38*(2), 91–108; (i) Okamoto, Y. Chiral Polymers for Resolution of Enantiomers. *J. Polym. Sci. A Polym. Chem.* **2009**, *47*(7), 1731–1739; (j) Okamoto, Y.; Ikai, T.; Shen, J. Controlled Immobilization of Polysaccharide Derivatives for Efficient Chiral Separation. *Israel J. Chem.* **2011**, *51*(10), 1096–1106.
21. Chankvetadze, B.; Yashima, E.; Okamoto, Y. Dimethyl-, Dichloro- and Chloromethyl-phenylcarbamates of Amylose as Chiral Stationary Phases for High-Performance Liquid Chromatography. *J. Chromatogr. A* **1995**, *694*, 101–109.
22. Ikai, T.; Okamoto, Y. Preparation and Chiral Recognition of Polysaccharide-Based Selectors. In *Chiral Recognition in Separation Methods: Mechanisms and Applications*, Berthod, A., Ed.; Springer: Berlin, Heidelberg, **2010**; pp 33–52.
23. (a) Chankvetadze, B.; Kartozia, I.; Yamamoto, C.; Okamoto, Y. Comparative Enantioseparation of Selected Chiral Drugs on Four Different Polysaccharide-Type Chiral Stationary Phases Using Polar Organic Mobile Phases. *J. Pharm. Biomed. Anal.* **2002**, *27*, 467–478; (b) Witte, D. K.; Bruggeman, F. K.; Franke, J. P.; Copinga, S.; Jansen, J. M.; De Zeeuw, R. A. Comparison Between Cellulose and Amylose Tris(3,5-dimethylphenylcarbamate) Chiral Stationary Phases for Enantiomeric Separation of 17 Amidotetralins. *Chirality* **1993**, *5*, 545–553.
24. Wang, T.; Chen, Y. W.; Vailaya, A. Enantiomeric Separation of Some Pharmaceutical Intermediates and Reversal of Elution Orders by High-Performance Liquid Chromatography Using Cellulose and Amylose Tris(3,5-Dimethylphenylcarbamate) Derivatives as Stationary Phases. *J. Chromatogr. A* **2000**, *902*(2), 345–355.
25. Beridze, N.; Tsutskiridze, E.; Takaishvili, N.; Farkas, T.; Chankvetadze, B. Comparative Enantiomer-Resolving Ability of Coated and Covalently Immobilized Versions of Two Polysaccharide-Based Chiral Selectors in High-Performance Liquid Chromatography. *Chromatographia* **2018**, *81*, 611–621.
26. Francotte, E.; Zhang, T. Photochemically Crosslinked Polysaccharide Derivatives Having No Photopolymerizable Functional Groups. WO9704011A1, 1997.
27. Singh, V.; Srivastava, P.; Singh, A.; Singh, D.; Malviya, T. Polysaccharide-Silica Hybrids: Design and Applications. *Polym. Rev.* **2016**, *56*(1), 113–136.
28. Salama, A. Polysaccharides/Silica Hybrid Materials: New Perspectives for Sustainable Raw Materials. *J. Carbohydr. Chem.* **2016**, *35*(3), 131–149.
29. Pandey, S.; Mishra, S. B. Sol-Gel Derived Organic-Inorganic Hybrid Materials: Synthesis, Characterizations and Applications. *J. Sol-Gel Sci. Technol.* **2011**, *59*, 73–94.

30. Lomsadze, K.; Jibuti, G.; Farkas, T.; Chankvetadze, B. Comparative High-Performance Liquid Chromatography Enantioseparations on Polysaccharide Based Chiral Stationary Phases Prepared by Coating Totally Porous and Core-Shell Silica Particles. *J. Chromatogr. A* **2012**, *1234*, 50–55.
31. Kharaishvili, Q.; Jibuti, G.; Farkas, T.; Chankvetadze, B. Further Proof to the Utility of Polysaccharide-Based Chiral Selectors in Combination With Superficially Porous Silica Particles as Effective Chiral Stationary Phases for Separation of Enantiomers in High-Performance Liquid Chromatography. *J. Chromatogr. A* **2016**, *1467*, 163–168.
32. Bezhitashvili, L.; Bardavelidze, A.; Ordjonikidze, T.; Chankvetadze, L.; Chity, M.; Farkas, T.; Chankvetadze, B. Effect of Pore-Size Optimization on Performance of Polysaccharide-Based Superficially Porous Chiral Stationary Phases for Separation of Enantiomers in High-Performance Liquid Chromatography. *J. Chromatogr. A* **2017**, *1482*, 32–38.
33. Bezhitashvili, L.; Bardavelidze, A.; Mskhiladze, A.; Gumustas, M.; Ozkan, S. A.; Volonterio, A.; Farkas, T.; Chankvetadze, B. Application of Cellulose 3,5-Dichlorophenylcarbamate Covalently Immobilized on Superficially Porous Silica for the Separation of Enantiomers in High-Performance Liquid Chromatography. *J. Chromatogr. A* **2018**, *1571*, 132–139.
34. Okada, Y.; Yamamoto, C.; Kamigaito, M.; Gao, Y.; Shen, J.; Okamoto, Y. Enantioseparation Using Cellulose Tris(3,5-Dimethylphenylcarbamate) as Chiral Stationary Phase for HPLC: Influence of Molecular Weight of Cellulose. *Molecules* **2016**, *21*, 1484/1–1484/10.
35. Isogai, A.; Usuda, M. Preparation of Low-Molecular Weight Celluloses Using Phosphoric Acid. *Mokuzai Gakkaishi* **1991**, *37*, 339–344.
36. Shen, J.; Li, G.; Yang, Z.; Okamoto, Y. Synthesis and Chiral Recognition of Amylose Derivatives Bearing Regioselective Phenylcarbamate Substituents at 2,6- and 3-Positions for High-Performance Liquid Chromatography. *J. Chromatogr. A* **2016**, *1467*, 199–205.
37. Koschella, A.; Heinze, T.; Klemm, D. First Synthesis of 3-O-functionalized Cellulose Ethers via 2,6-di-*O*-Protected Silyl Cellulose. *Macromol. Biosci.* **2001**, *1*, 49–54.
38. Shen, J.; Wang, F.; Bi, W.; Liu, B.; Liu, S.; Okamoto, Y. Synthesis of Cellulose Carbamates Bearing Regioselective Substituents at 2,3- and 6-Positions for Efficient Chromatographic Enantioseparation. *J. Chromatogr. A* **2018**, *1572*, 54–61.
39. Francotte, E. Photochemically Cross-Linked Polysaccharide Derivatives as Supports for the Chromatographic Separation of Enantiomers. WO9627615A1, 1996.
40. (a) Francotte, E.; Zhang, T. Preparation and Evaluation of Immobilized 4-Methylbenzoylcellulose Stationary Phases for Enantioselective Separations. *J. Chromatogr. A* **2016**, *1467*, 214–220; (b) Francotte, E.; Huynh, D.; Zhang, T. Photochemically Immobilized 4-Methylbenzoyl Cellulose as a Powerful Chiral Stationary Phase for Enantioselective Chromatography. *Molecules* **2016**, *21*, 1740/1–1740/12.
41. Li, J.; Zhang, Y.; Cheng, Y.; Yuan, S.; Liu, L.; Shao, H.; Li, H.; Li, N.; Zhao, P.; Guo, Y. Simultaneous Enantioselective Determination of Phenylpyrazole Insecticide Flufiprole and Its Chiral Metabolite in Paddy Field Ecosystem by Ultra-High Performance Liquid Chromatography/Tandem Mass Spectrometry. *J. Pharm. Biomed. Anal.* **2016**, *121*, 261–270.
42. Nelson, T. D.; Welch, C. J.; Rosen, J. D.; Smitrovich, J. H.; Huffman, M. A.; McNamara, J. M.; Mathre, D. J. Effective Use of Preparative Chiral HPLC in a Preclinical Drug Synthesis. *Chirality* **2004**, *16*(9), 609–613.
43. Albals, D.; Heyden, Y. V.; Schmid, M. G.; Chankvetadze, B.; Mangelings, D. Chiral Separations of Cathinone and Amphetamine-Derivatives: Comparative Study Between Capillary Electrochromatography, Supercritical Fluid Chromatography and Three Liquid Chromatographic Modes. *J. Pharm. Biomed. Anal.* **2016**, *121*, 232–243.

44. Fanali, C.; Fanali, S.; Chankvetadze, B. HPLC Separation of Enantiomers of Some Flavanone Derivatives Using Polysaccharide-Based Chiral Selectors Covalently Immobilized on Silica. *Chromatographia* **2016**, *79*, 119–124.
45. Gumustas, M.; Ozkan, S. A.; Chankvetadze, B. Separation and Elution Order of the Enantiomers of Some β-Agonists Using Polysaccharide-Based Chiral Columns and Normal Phase Eluents by High-Performance Liquid Chromatography. *J. Chromatogr. A* **2016**, *1467*, 297–305.
46. Labrador, G. M.; Bosson, J.; Breitbach, Z. S.; Lim, Y.; Francotte, E. R.; Sabia, R.; Villani, C.; Armstrong, D. W.; Lacour, J. High-Performance Liquid Chromatographic Resolution of Neutral and Cationic Hetero[6]Helicenes. *Chirality* **2016**, *28*, 282–289.
47. Lajko, G.; Grecso, N.; Megyesi, R.; Forro, E.; Fulop, F.; Wolrab, D.; Lindner, W.; Peter, A.; Ilisz, I. Enantioseparation of β-Carboline Derivatives on Polysaccharide- and Strong Cation Exchanger-Based Chiral Stationary Phases. A Comparative Study. *J. Chromatogr. A* **2016**, *1467*, 188–198.
48. Landagaray, E.; Yous, S.; Vaccher, C.; Lipka, E. Design of Experiments for Enantiomeric Separation in Supercritical Fluid Chromatography. *J. Pharm. Biomed. Anal.* **2016**, *120*, 297–305.
49. Zehani, Y.; Lemaire, L.; Ghinet, A.; Millet, R.; Chavatte, P.; Vaccher, C.; Lipka, E. Exploring Chiral Separation Of 3-Carboxamido-5-Aryl Isoxazole Derivatives by Supercritical Fluid Chromatography on Amylose and Cellulose Tris Dimethyl- and Chloromethyl Phenylcarbamate Polysaccharide Based Stationary Phases. *J. Chromatogr. A* **2016**, *1467*, 473–481.
50. Zhao, P.; Deng, M.; Huang, P.; Yu, J.; Guo, X.; Zhao, L. Solid-Phase Extraction Combined With Dispersive Liquid-Liquid Microextraction and Chiral Liquid Chromatography-Tandem Mass Spectrometry for the Simultaneous Enantioselective Determination of Representative Proton-Pump Inhibitors in Water Samples. *Anal. Bioanal. Chem.* **2016**, *408*, 6381–6392.
51. Zhou, J.; Luo, P.; Chen, S.; Meng, L.; Sun, C.; Du, Q.; Sun, F. Enantioseparation of Six Antihistamines With Immobilized Cellulose Chiral Stationary Phase by HPLC. *J. Chromatogr. Sci.* **2016**, *54*, 531–535.
52. D'Orazio, G.; Fanali, C.; Karchkhadze, M.; Chankvetadze, B.; Fanali, S. Enantiomeric Separation of Some Chiral Analytes Using Amylose 3,5-Dimethylphenylcarbamate Covalently Immobilized on Silica by Nano-Liquid Chromatography and Capillary Electrochromatography. *J. Chromatogr. A* **2017**, *1520*, 127–134.
53. D'Orazio, G.; Kakava, R.; Volonterio, A.; Fanali, S.; Chankvetadze, B. An Attempt for Fast Separation of Enantiomers in Nano-Liquid Chromatography and Capillary Electrochromatography. *Electrophoresis* **2017**, *38*, 1932–1938.
54. Kucerova, G.; Kalikova, K.; Tesarova, E. Enantioselective Potential of Polysaccharide-Based Chiral Stationary Phases in Supercritical Fluid Chromatography. *Chirality* **2017**, *29*, 239–246.
55. Matarashvili, I.; Ghughunishvili, D.; Chankvetadze, L.; Takaishvili, N.; Khatiashvili, T.; Tsintsadze, M.; Farkas, T.; Chankvetadze, B. Separation of Enantiomers of Chiral Weak Acids With Polysaccharide-Based Chiral Columns and Aqueous-Organic Mobile Phases in High-Performance Liquid Chromatography: Typical Reversed-Phase Behavior? *J. Chromatogr. A* **2017**, *1483*, 86–92.
56. Shibata, T.; Shinkura, S.; Ohnishi, A.; Ueda, K. Achiral Molecular Recognition of Aromatic Position Isomers by Polysaccharide-Based CSPs in Relation to Chiral Recognition. *Molecules* **2017**, *22*, 38/1–38/11.

57. Carradori, S.; Secci, D.; Faggi, C.; Cirilli, R. A Chromatographic Study on the Exceptional Chiral Recognition Of 2-(Benzylsulfinyl)Benzamide by an Immobilized-Type Chiral Stationary Phase Based on Cellulose Tris(3,5-Dichlorophenylcarbamate). *J. Chromatogr. A* **2018**, *1531*, 151–156.
58. Ghanem, A.; Wang, C. Enantioselective Separation of Racemates Using CHIRALPAK IG Amylose-Based Chiral Stationary Phase Under Normal Standard, Non-Standard and Reversed Phase High Performance Liquid Chromatography. *J. Chromatogr. A* **2018**, *1532*, 89–97.
59. West, C.; Konjaria, M.-L.; Shashviashvili, N.; Lemasson, E.; Bonnet, P.; Kakava, R.; Volonterio, A.; Chankvetadze, B. Enantioseparation of Novel Chiral Sulfoxides on Chlorinated Polysaccharide Stationary Phases in Supercritical Fluid Chromatography. *J. Chromatogr. A* **2017**, *1499*, 174–182.
60. Alvarenga, N.; Porto, A. L. M.; Barreiro, J. C. Enantioselective Separation of (±)-β-Hydroxy-1,2,3-Triazoles by Supercritical Fluid Chromatography and High-Performance Liquid Chromatography. *Chirality* **2018**, *30*, 890–899.
61. Bajtai, A.; Lajko, G.; Szatmari, I.; Fulop, F.; Lindner, W.; Ilisz, I.; Peter, A. Dedicated Comparisons of Diverse Polysaccharide- and Zwitterionic Cinchona Alkaloid-Based Chiral Stationary Phases Probed With Basic and Ampholytic Indole Analogs in Liquid and Subcritical Fluid Chromatography Mode. *J. Chromatogr. A* **2018**, *1563*, 180–190.
62. He, R.; Fan, J.; Tan, Q.; Lai, Y.; Chen, X.; Wang, T.; Jiang, Y.; Zhang, Y.; Zhang, W. Enantioselective Determination of Metconazole in Multi Matrices by High-Performance Liquid Chromatography. *Talanta* **2018**, *178*, 980–986.
63. Khater, S.; Lozac'h, M.-A.; Adam, I.; Francotte, E.; West, C. Comparison of Liquid and Supercritical Fluid Chromatography Mobile Phases for Enantioselective Separations on Polysaccharide Stationary Phases. *J. Chromatogr. A* **2016**, *1467*, 463–472.
64. Khater, S.; Canault, B.; Azzimani, T.; Bonnet, P.; West, C. Thermodynamic Enantioseparation Behavior of Phenylthiohydantoin-Amino Acid Derivatives in Supercritical Fluid Chromatography on Polysaccharide Chiral Stationary Phases. *J. Sep. Sci.* **2018**, *41*, 1450–1459.
65. Khatiashvili, T.; Kakava, R.; Matarashvili, I.; Tabani, H.; Fanali, C.; Volonterio, A.; Farkas, T.; Chankvetadze, B. Separation of Enantiomers of Selected Chiral Sulfoxides With Cellulose Tris(4-Chloro-3-Methylphenylcarbamate)-Based Chiral Columns in High-Performance Liquid Chromatography With Very High Separation Factor. *J. Chromatogr. A* **2018**, *1545*, 59–66.
66. Li, M.; Zhang, B.; Yu, J.; Wang, J.; Guo, X. Enantiomeric Separation and Simulation Study of Eight Anticholinergic Drugs on an Immobilized Polysaccharide-Based Chiral Stationary Phase by HPLC. *New J. Chem.* **2018**, *42*, 11724–11731.
67. Niedermeier, S.; Matarashvili, I.; Chankvetadze, B.; Scriba, G. K. E. Simultaneous Determination of Dextromepromazine and Related Substances 2-Methoxyphenothiazine and Levomepromazine Sulfoxide in Levomepromazine on a Cellulose Tris(4-Methylbenzoate) Chiral Column. *J. Pharm. Biomed. Anal.* **2018**, *158*, 294–299.
68. Shedania, Z.; Kakava, R.; Volonterio, A.; Farkas, T.; Chankvetadze, B. Separation of Enantiomers of Chiral Sulfoxides in High-Performance Liquid Chromatography With Cellulose-Based Chiral Selectors Using Methanol and Methanol-Water Mixtures as Mobile Phases. *J. Chromatogr. A* **2018**, *1557*, 62–74.
69. Peluso, P.; Mamane, V.; Cossu, S. Liquid Chromatography Enantioseparations of Halogenated Compounds on Polysaccharide-Based Chiral Stationary Phases: Role of Halogen Substituents in Molecular Recognition. *Chirality* **2015**, *27*, 667–684.

70. Ikai, T.; Yamamoto, C.; Kamigaito, M.; Okamoto, Y. Organic-Inorganic Hybrid Materials for Efficient Enantioseparation Using Cellulose 3,5-Dimethylphenylcarbamate and Tetraethyl Orthosilicate. *Chem. Asian J.* **2008**, *3*(8–9), 1494–1499.
71. Weng, X.; Bao, Z.; Zhang, Z.; Su, B.; Xing, H.; Yang, Q.; Yang, Y.; Ren, Q. Preparation of Porous Cellulose 3,5-Dimethylphenylcarbamate Hybrid Organosilica Particles for Chromatographic Applications. *J. Mater. Chem. B* **2015**, *3*, 620–628.
72. Ong, T.-T.; Wang, R.-Q.; Muderawan, I. W.; Ng, S.-C. Synthesis and Application of Mono-6-(3-Methylimidazolium)-6-Deoxyperphenylcarbamoyl-β-Cyclodextrin Chloride as Chiral Stationary Phases for High-Performance Liquid Chromatography and Supercritical Fluid Chromatography. *J. Chromatogr. A* **2008**, *1182*, 136–140.
73. Hussain, Z.; Mohammed, S. A.; Ibraheem, H.; Khan, N.; Abdullah, B. M.; Yousif, E. Photo-Fenton Reaction: How to Remediation Waste-Water Using (Fe^{+2}/UV). *Res. J. Pharm., Biol. Chem. Sci.* **2016**, *7*, 1072–1078.
74. Wang, Q.; Asoh, T.-A.; Uyama, H. Facile Fabrication of Flexible Bacterial Cellulose/Silica Composite Aerogel for Oil/Water Separation. *Bull. Chem. Soc. Jpn.* **2018**, *91*(7), 1138–1140.
75. Kollarigowda, R. H.; Abraham, S.; Montemagno, C. D. Antifouling Cellulose Hybrid Biomembrane for Effective Oil/Water Separation. *ACS Appl. Mater. Interfaces* **2017**, *9*, 29812–29819.
76. Du, Y.; Li, Y.; Wu, T. A Superhydrophilic and Underwater Superoleophobic Chitosan-TiO_2 Composite Membrane for Fast Oil-in-Water Emulsion Separation. *RSC Adv.* **2017**, *7*, 41838–41846.
77. Zhang, X.; Wang, C.; Liu, X.; Wang, J.; Zhang, C.; Wen, Y. A Durable and High-Flux Composite Coating Nylon Membrane for Oil-Water Separation. *J. Clean. Prod.* **2018**, *193*, 702–708.
78. Zhou, S.; You, T.; Zhang, X.; Xu, F. Superhydrophobic Cellulose Nanofiber-Assembled Aerogels for Highly Efficient Water-in-Oil Emulsions Separation. *ACS Appl. Nano Mater.* **2018**, *1*(5), 2095–2103.
79. Yang, X.; Ma, J.; Ling, J.; Li, N.; Wang, D.; Yue, F.; Xu, S. Cellulose Acetate-Based SiO_2/TiO_2 Hybrid Microsphere Composite Aerogel Films for Water-in-Oil Emulsion Separation. *Appl. Surf. Sci.* **2018**, *435*, 609–616.
80. Srivastava, N.; Thakur, A. K.; Shahi, V. K. Phosphorylated Cellulose Triacetate-Silica Composite Adsorbent for Recovery of Heavy Metal Ion. *Carbohydr. Polym.* **2016**, *136*, 1315–1322.
81. Woiwode, U.; Reischl, R. J.; Buckenmaier, S.; Lindner, W.; Lämmerhofer, M. Imaging Peptide and Protein Chirality via Amino Acid Analysis by Chiral × Chiral Two-Dimensional Correlation Liquid Chromatography. *Anal. Chem.* **2018**, *90*(13), 7963–7971.

Index

Note: Page numbers followed by *f* indicate figures and *t* indicate tables.

U

V

W

X

Z

Made in United States
Troutdale, OR
10/09/2025

40066990R00275